DISCRETE MATHEMATICS

For Engineering Students

by

Dr. ANJANA GUPTA
(Associate Professor)
Department of Mathematics
Delhi Technological University, Delhi

BHASKAR KAPOOR
(Assistant Professor)
Department of Information Technology
Maharaja Agrasen Institute of Technology, GGSIPU
New Delhi

S.K. KATARIA & SONS®
Publisher of Engineering Books
4885/109, Prakash Mahal, Dr. Subhash Bhargav Lane,
Opposite Delhi Medical Association, Daryaganj, New Delhi–110002 (INDIA)
Phone: +91-11-43551243 Mobile: +91-9871775858, 7678425209
e-mail: katariabook@gmail.com, katariabooks@yahoo.com
Website: www.skkatariaandsons.com
(e-books, subscriptions and bulk buying also available on our website)

Published by:

S.K. Kataria & Sons®

4885/109, Prakash Mahal, Dr. Subhash Bhargav Lane,
Opposite Delhi Medical Association, Daryaganj, New Delhi-110002 (INDIA)
Phone: +91-11-43551243 Mobile: +91-9871775858, 7678425209
e-mail: katariabook@gmail.com, katariabooks@yahoo.com
Website: www.skkatariaandsons.com

Send all correspondence to : **M/s S.K. Kataria & Sons,** New Delhi
bmail: katariabook@gmail.com

First Edition : August, 2002; Fourth Edition: June, 2012; Reprint: 2025

Laser Typesetting :
PLUS COMPUTERS, Shahdara, Delhi

Printed at :
Repro India Limited, Delhi

Preface to the Fourth Edition

The revised edition of the book has been written to serve diverse set of students keeping in view their requirements as per the syllabi at the graduate level of some of the universities & to make the book more illustrative & self contained.

The sections have been re-arranged & several new exercises & solved examples have been incorporated to reinforce & solidify the understanding of some of the chapters. New chapters on finite state machine has been introduced. All the different concepts have been elucidated with care to provide conceptual clarity to the readers.

Dr. Anjana Gupta
Author

Bhaskar Kapoor
Co-Author

June, 2012

Preface to the First Edition

It gives me a great pleasure to offer discrete mathematics to engineering students to teach them certain aspects of noncontinuous mathematics in set theory, logic, combinatorics, relations, lattices, Boolean algebra and graph theory as to help them develop the skill of mathematical reasoning, deduction, proof and recursive thinking.

It is possible to make excellent and effective use of computers without involving oneself in mathematical considerations. But any understanding of how computers operate, in terms of either their hardware or software, inevitably involves mathematics, even if one ignores the fact that many of the applications of computing lie in facilitating numerical and combinational calculations.

Text assumes that student has gone through the first year course in engineering mathematics as a prerequisite. It is intended to be a textbook for a course in "Foundation to computer Science" in the third semester for B.Tech programme, of Guru Gobind Singh Indraprashtha University, Delhi.

I emphasize that the chapters have been written in a manner so that order can be changed without difficulty and loss of continuity. I hope this book will serve its purpose and prove beneficial to its readers.

Anjana Gupta
Author

August, 2002

Acknowledgement

We express our sincere gratitude to **Dr. Nand Kishore Garg,** Chairman Maharaja Agrasen Institute of Technology for his constant inspiration and encouragement. I am thankful to **Prof. M.L. Goyal**, Director, Maharaja Agrasen Institute of Technology, for his continued guidance and inspiration.

I have the benefit of valuable advice and suggestions from the faculty teaching in the colleges of GGSIPU.

I am also thankful to all the bright students' who with their inquisitive questions, have helped me to elaborate on the topics written.

I will be failing in my duties if I do not acknowledge the boundless support and envouragement of my family members. I express my sincere thanks to my husband **Nitin Gupta,** who offered every possible help and was a driving force behind this effort. My daughter **Srishti** and son **Kshitij** deserve a special thanks for their patience shown in my busy hours. I am also thankful to my mother in law **Mrs. Pramila Gupta.** The book would have not been in the present form, without the blessings of my loving mother, **Smt. Pushpa Gupta** and my loving father, **Shri Puran Chand Gupta**.

Finally, we wish to thank the staff of Kataria Publications especially **Mr. Sanjeev Kataria**, for their unfailing co-operation they executed during the course of writing this book.

Dr. Anjana Gupta
Author

June, 2012

Contents

CHAPTER 1

SET THEORY

This Chapter Covers :

1.1. INTRODUCTION

This chapter gives the foundation of discrete mathematics. Much of discrete mathematics is devoted to the study of discrete structures, which are used to represent discrete objects. All discrete structures are built up from sets, which are collection of objects, used extensively in counting ; relations, which are set of ordered pairs that represent relationships between objects ; graphs, which are set of vertices & connected edges.

In this chapter we cover some of the basic set languages and notations that will be used throughout the text. *Venn diagrams* will be introduced in order to give clear picture of set theory.

1.2. SYMBOLS AND NOTATIONS

Following are the special symbols to denote the sets which will be used very often in the text.

N Set of natural numbers
Z Set of integers
Q Set of rational numbers
R Set of real numbers
C Set of complex numbers

1.3. SET

A set is a collection of well defined objects. The objects that constitute a set are called elements or members of the set.

Usually set is denoted by capital letters $A, B, C,$ .... and its elements are denoted by lower case letters $a, b, c,$ .... For example, if A is any set having elements x_1, x_2, x_3 then we write

$$A = \{x_1, x_2, x_3\}$$

If an element x is in the set A or belongs to the set A then it is denoted by $x \in A$ otherwise $x \notin$ A.

For example : $x_1 \in A$ but $a \notin A$

Usually sets are denoted by two ways : One way is to define a set by listing the elements and the second way is to state those properties which characterise the elements in the set.

EXAMPLE 1.1. If X is the set having four elements say 2, 4, 6, 8 then x can be given by;

$$X = \{x \mid x \text{ is an even positive integer less than } 10\}$$

or

$$X = \{2, 4, 6, 8\}$$

Here first way is by defining the property and the second way is by listing the elements.

Subsets : Let X and Y be two sets. If every element of the set X is also the element in Y then X is called subset of Y and is denoted by $X \subseteq Y$, and it means either X is in Y or X equals to Y. If X is not equal to Y then X is called proper subset or trivial of Y and is denoted by $X \subset Y$.

EXAMPLE 1.2. X = set of natural numbers = $\{x \mid x \in N\}$

Y = set of integers = $\{y \mid y \in Z\}$

then $X \subset Y$

EXAMPLE 1.3. $X = \{1, 2, 3, 5, 7, 9\}$

$Y = 1, 2, 3, 5, 7, 9, 8\}$

then $X \subset Y$

Equal sets : Two sets A and B are said to be equal if they have same elements & is denoted by $A = B$.

For example $A = \{1, 2, 3\}$, $B = \{x \in N \mid 1 \leq x \leq 3\}$

then $A = B$.

EXAMPLE 1.4. $X = \{x : x^2 - 3x + 2 = 0\}$

$Y = \{1, 2\}$

$$Z = \left\{1, 2, \frac{10}{5}, \frac{30}{15}, 2, 1, 1\right\}$$

It is notable in the example (1.4) that all the above three sets are equal, therefore, the set does not depend on the way in which its elements are displayed.

Do not use $\{\phi\}$ because this means the set containing the null set ϕ. The null set is a subset of every set.

$$\phi \subseteq A \quad \text{means} \quad x \in \phi \rightarrow x \in A$$

This is true by default.

Power Set :

Power set of A is the set of all subsets of A, called $P(A)$. So if we are given

$$B = \{a, b, c, d\}$$

$$P(B) = \begin{cases} \phi, \{a\}, \{b\}, \{c\}, \{d\}, \{a,b\}, \{a,c\}, \{a,d\}, \{b,c\}, \{b,d\}, \\ \{c,d\}, \{a,b,c\}, \{a,b,d\}, \{b,c,d\}, \{a,b,c,d\} \end{cases}$$

There are 2^n subsets in a power set, where n represents the number of sets in the original set. Cardinality of a set in the number of members in a set, if that number is a non-negative integer.

The cardinality of B :

$$|B| = 4$$

Note: $$|P(s)| = 2^{|s|}$$

Proper subset—a subset which is not equal to the original set.

$$A \subset B$$

1.4. MULTI-SETS AND MULTIPLICITY

If a set contains exactly m distinct elements where m denotes some non-negative integers, then the set is said to be a **finite set** otherwise the set is said to be an **infinite set**.

For example, the set {1, 2, 5, 7, 9} is a finite set whereas N, the set of natural number is infinite set.

As we have discussed above that we can define a set by different ways. Now, we define a **multiset** which is used to define a collection of objects which are not necessarily distinct.

For example {1, 1, 2, 2, 2}, $\{a, b, c\}$, { }, {1, 1} are all multisets.

Multiplicity of a Set : Number of times an element appearing in a set is called multiplicity of that element. As in the above first example {1, 1, 2, 2, 2} multiplicity of 1 is 2 & multiplicity of 2 is 3.

Definition : The **sum of the multiplicities** of the elements in the sets P and Q is denoted by multiplicity in the set $P + Q$.

EXAMPLE 1.5. Let $P = \{a, a, b, b, c\}$

$$Q = \{a, b, c\}$$
$$P + Q = \{a, a, a, b, b, b, c, c\}$$

Here multiplicity of a in the set P is 2 and in the set Q is 1, therefore multiplicity of a in $P + Q$ is 3.

$P + Q$ is called **sum of two multisets.**

Definition : Difference of the multisets P and Q is denoted by $P - Q$ and the multiplicity of an element in $P - Q$ is equal to the multiplicity in the set P of that element minus the multiplicity in the set Q of that element, if the difference is positive and if the difference is negative then the multiplicity is considered to be 0 as the number of times an element appearing in the set should not be negative.

EXAMPLE 1.6. Let P and Q be two multiset defined as follow:

$$P = \{a, a, b, b, c, c\}$$
$$Q = \{a, b, c, c, c\}$$
$$P - Q = \{a, b\}$$

Here multiplicity of a & b in $P - Q$ is 1 but multiplicity of c in $P - Q$ is 0.

Theorem 1.1 : Let A and B are two sets then $A = B$ if and only if $A \subseteq B$ and $B \subseteq A$.

That is two sets are equal if an only it they are subset of each other.

Theorem 1.2 : Every set A is a subset of the universal set U.

[*Hint :* Since all the elements of A belong to U]

Theorem 1.3 : Every set is a subset of itself.

Theorem 1.4 : If $A \subseteq B$ and $B \subseteq C$ then $A \subseteq C$.

1.5. VENN DIAGRAM

A venn diagram is a pictorial representation of sets in which sets are represented by enclosed areas in the plane. In Venn diagram a set is generally represented by a region inside a geometrical configuration (figure) for example a circle, a rectangle. Venn diagram ensures the questions like : Why certain theorems of set theory should hold? What could be the relations between the set operations? Venn diagram are certainly suggesting about the truth of the theorem. Shading are generally used to represent the set operations which follow the interpretations made to extract the result.

The universal set U is represented by the interior of a rectangle and the other sets are represented by disks lying within the rectangle.

For example, $A \subset U$ is given by figure 1.1.

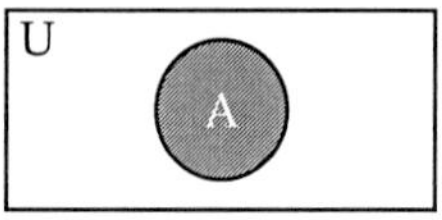

Fig. 1.1.

1.6. SET OPERATIONS

Union of Two Sets

Let A and B be two sets then the union of two sets A and B is denoted by $A \cup B$ is the set of elements which are in A or in B or in both, *i.e.*,

$$A \cup B = \{x : x \in A \text{ or } x \in B\}$$

By use of Venn diagram, the union of two sets is given by the shaded portion in figure 1.2.

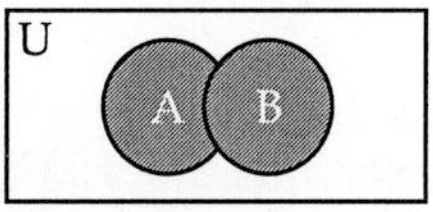

Fig. 1.2.

EXAMPLE 1.7. $A = \{1, 2, 3, 4, 5, 6\}$

$B = \{5, 7, 4, 3, 9\}$

$A \cup B = \{1, 2, 3, 4, 5, 6, 7, 9\}$

Intersection of Two Sets

For two sets A and B the intersection gives a set having the elements in A as well as in B and is denoted by $A \cap B$.

i.e., $A \cap B = \{x : x \in A \text{ and } x \in B\}$

By Venn diagram $A \cap B$ is given by the following shaded portion in figure 1.3.

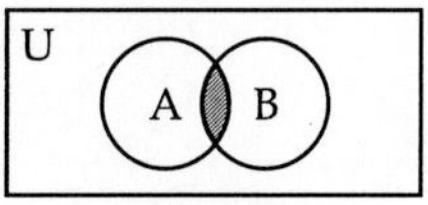

Fig. 1.3.

EXAMPLE 1.8. Consider the set of example 1.7. then $A \cap B = \{3, 4, 5\}$.

Disjoint Set : If for any two sets A and B, $A \cap B = \phi$ *i.e.*, no element is common then A and B are called disjoint sets & by Venn diagram.

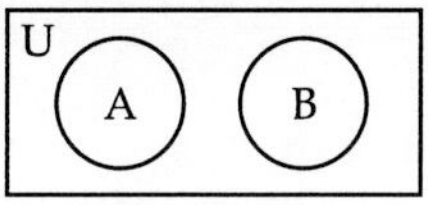

Fig. 1.4.

EXAMPLE 1.9. $A = \{1, 2, 3, 4, 5\}$

$B = \{10, 11, 12\}$

$A \cap B = \phi$

Complement of a Set : The complement of a set A is the set of elements which belongs to U but not to A. It is denoted by A^c or A' or $\overline{A}$.

i.e., $A^c = \{x : x \in U, \ x \notin A\}$

By Venn diagram complement of the set is given by the shaded portion in figure 1.5.

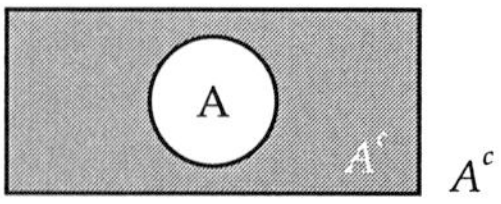

Fig. 1.5.

Relative Complement

The relative complement of a set B with respect to a set A or simply, the difference of A and B is denoted by A/B or $A \sim B$ is the set of elements which belong to A but not B.

i.e., $A \sim B = \{x : x \in A \text{ and } x \notin B\}$

By Venn diagram $A \sim B$ is given by the shaded portion in figure 1.6.

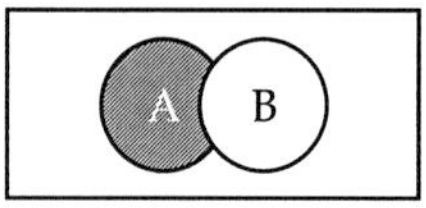

Fig. 1.6.

EXAMPLE 1.10. If $A = \{1, 2, 3, 4\}$

$B = \{3, 4, 5, 6, 7\}$

$A \sim B = \{1, 2\}$

Symmetric Difference

The symmetric difference of the sets A and B is denoted by $A \oplus B$ consists of those elements which belongs to A or B but not both *i.e.*,

$$A \oplus B = (A \cup B) \sim (A \cap B)$$
$$= (A \sim B) \cup (B \sim A)$$

By Venn Diagram the symmetric difference of two sets is given by the shaded portion, in figure 1.7.

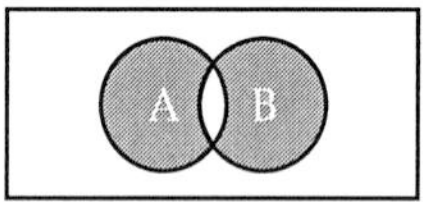

Fig. 1.7.

EXAMPLE 1.11. Let $A = \{1, 2, 3, 4, 5, 6\}$

$B = \{4, 5, 6, 7, 8, 9\}$

$A \sim B = \{1, 2, 3\}$

$B \sim A = \{7, 8, 9\}$

$A \oplus B = \{1, 2, 3, 7, 8, 9\}$

Fundamental Products :

Consider n distinct sets A_1, A_2, ... A_n. A fundamental product of these sets is the set of the form.

$$A_1^* \cap A_2^* \cap A_3^* \cap \ldots\ldots \cap A_n^*$$

where A_i^* is either A_i or complement of A_i.

Properties :

1. There are 2^n such fundamental products for n sets. (which are same as number of rows in the truth table which you will do in the next chapter)

2. Any two such fundamental products are disjoint.
3. Universal set U is the union of all the fundamental products.

EXAMPLE 1.12. Consider any three sets A, B and C then there are 2^3 fundamental products which are given as follows :

Let us take number 1 for the sets A, B & C and 0 for their complements. Now consider the table 1.

	A	B	C
$P_1 = A \cap B \cap C$	1	1	1
$P_2 = A \cap B \cap C'$	1	1	0
$P_3 = A \cap B' \cap C$	1	0	1
$P_4 = A \cap B' \cap C'$	1	0	0
$P_5 = A' \cap B \cap C$	0	1	1
$P_6 = A' \cap B' \cap C$	0	0	1
$P_7 = A' \cap B \cap C'$	0	1	0
$P_8 = A' \cap B' \cap C'$	0	0	0

In the given table there are $8 = 2^3$ row & each row gives you the corresponding fundamental products.

1.7. ALGEBRA OF THE SET

Following are the properties which give the algebra of sets for the given sets A, B and C.

1. $A \cup A = A$ — Idempotent laws
   $A \cap A = A$

2. $(A \cup B) \cup C = A \cup (B \cup C)$ — Associative laws
   $(A \cap B) \cap C = A \cap (B \cap C)$

3. $A \cup B = B \cup A$ — Commutative laws
   $A \cap B = B \cap A$

4. $(A \cup (B \cap C) = (A \cup B) \cap (A \cup C)$ — Distributive laws
   $(A \cap (B \cup C) = (A \cap B) \cup (A \cap C)$

5. $A \cup \phi = A$ — Identity laws
   $A \cap \phi = \phi$
   $A \cup U = U$
   $A \cap U = A$

where ϕ is the empty set and U is the universal set.

6. $A \cup A^c = U$ — Complement laws
   $A \cap A^c = \phi$
   $(A^c)^c = A$
   $U^c = \phi$
   $\phi^c = U$

7. $(A \cup B)^c = A^c \cap B^c$ De Morgan's laws
$(A \cap B)^c = A^c \cup B^c$

Students can verify the results with the help of Venn diagrams.

Duality

Suppose E is an equation of set algebra. The dual E^* of E is the equation obtained by replacing union by intersection and universal set by the empty set & *vice-versa*.

EXAMPLE 1.13. Dual of the given equation

$$(U \cup B) \cap (U \cup A) = U$$

is

$$(\phi \cap B) \cup (\phi \cap A) = \phi$$

1.8. DIMENSION OF A SET

The total number of elements in a set A are called dimension of a set or cardinality of a set and is denoted by $n(A)$, card (A), $\#A$, $|A|$.

Fundamental Counting Principle

Theorem 1.5 : If A and B are finite sets then $A \cup B$ and $A \cap B$ are also finite and

$$n(A \cup B) = n(A) + n(B) - n(A \cap B)$$

COR (1.5.1) : If A and B are finite disjoint sets then

$$n(A \cup B) = n(A) + n(B)$$

Theorem 1.6 : If A, B and C are finite sets then

$$n(A \cup B \cup C) = n(A) + n(B) + n(C) - n(A \cap B) - n(A \cap C) - n(B \cap C) + n(A \cap B \cap C)$$

Power sets : For a given set S, the collection of all subsets of S is called the power set of S and is denoted by Power (S).

If the set S is finite then so is power(S). The number of elements in Power(S) of S having n elements is 2^n.

EXAMPLE 1.14. If $S = \{1, 2, 3\}$ then

$$\text{Power } (S) = \{\phi, \{1\}, \{2\}, \{3\}, \{1, 2\}, \{1, 3\}, \{2, 3\}, \{1, 2, 3\}\}$$

$$n\,(\text{Power } (S)) = 8 = 2^3.$$

1.9. PARTITION OF A SET S

Let S be any set such that

$$S = \bigcup_{i=1}^{n} A_i$$

that is A_i's are subsets of S.

A partition of a set S is a collection $\{A_i\}$ of nonempty subsets of S such that:

(*i*) Each a in S belongs to one of the A_i

(*ii*) The sets of $\{A_i\}$ are mutually disjoint *i.e.*, if

$$A_i \neq A_j \quad \text{for } i \neq j \text{ then}$$

$$A_i \cap A_j = \phi$$

The subsets in a partition are called cells.

EXAMPLE 1.15. Consider the following collection of subsets of S = {1, 2, 3, ..., 8, 9}

(*i*) $A_1 = \{\{1, 3, 5\}, \{2, 6\}, \{4, 8, 9\}\}$

(*ii*) $A_2 = \{\{1, 3, 5\}, \{2, 4, 6, 8\}, \{5, 7, 9\}\}$

(*iii*) $A_3 = \{\{1, 3, 5\}, \{2, 4, 6, 8\}, \{7, 9\}\}$

Here A_1 is not a partition of the set S as 7 is missing *i.e.*, violating the first condition.

A_2 is not partition of the set S as {1, 3, 5} & {5, 7, 9} are not disjoint.

A_3 is a partition of the set S.

1.10. CARTESIAN PRODUCT

Given two sets A and B, the cartesian product $A \times B$ is defined by

$$A \times B = \{(a, b) \mid a \in A, b \in B\}$$

i.e., $A \times B$ is a set of ordered pairs

e.g., $A = \{1, 2, 3\}, B = \{a, b, c\}$

$$A \times B = \{1, a) ; (1, b) ; (1, c); (2, a) ; (2, b); (2, c); (3, a) ; (3, b) ; (3, c)\}$$

It is to be noted that $A \times B$ may or may not be equal to $B \times A$.

What is $A \times A$?

The Cartesian product of A itself.

e.g., $A = \{1, 2, 3\}$

$A \times A = \{(1, 1), (1, 2) (1, 3), (2,1), (2, 2), (2, 3) (3,1), (3, 2) (3,3)\}$

Theorem 1.7 : For any two finite, non-empty sets A and B,

$|A \times B| = |A|\,|B|$ *i.e.*

dimension $A \times B$ = (dimension A) (dimension B)

Proof. Let $|A| = m$ and $|B| = n$

The elements of $A \times B$ are ordered pairs in which first element belongs to the Set A and second element belongs to set B. *e.g.*, $(x, y) \in A \times B$ where $x \in A$ and $y \in B$.

There are n ways to choose $x \in A$ for the first position in ordered pair and m ways to choose $y \in A$ for second position in the ordered pair. So by, **Multiplication principle*** there are $m \times n$ ways to form an order pair (a, b). *i.e.*

$$|A \times B| = m\,n = |A|\,|B|$$

***M-ultiplication Principle**

Suppose that two tasks T_1 and T_2 are to be performed in sequence. If T_1 can be performed in n_1 ways and T_2 can be performed in n_2 ways then the sequence $T_1\ T_2$ can be performed in $n_1 n_2$ ways.

EXAMPLE 1.16. $A \equiv \{x \mid 3x^2 - 7x - 6 = 0\}$

$B \equiv \{x \mid 6x^2 - 5x - 6 = 0\}$

Find $A \cap B$

Solution :

$$3x^2 - 7x - 6 = 0$$

$$\therefore\ 3x^2 - 9x + 2x - 6 = 0$$

$$3x(x-3) + 2(x-3) = 0$$

$$\therefore\quad (3x+2)(x-3) = 0$$

$$\therefore\quad x = -\frac{2}{3} \quad \text{or} \quad x = 3$$

$$\therefore\quad A \equiv \left\{-\frac{2}{3}, 3\right\}$$

$$6x^2 - 5x - 6 = 0$$

$$6x^2 - 9x + 4x - 6 = 0$$

$$3x(2x-3) + 2(2x-3) = 0$$

$$(3x+2)(2x-3) = 0$$

$$\therefore\quad x = -\frac{2}{3} \quad \text{or} \quad x = \frac{3}{2}$$

$$\therefore\quad B \equiv \left\{-\frac{2}{3}, \frac{3}{2}\right\}$$

$$\therefore\quad A \cap B = \left\{-\frac{2}{3}\right\}$$

EXAMPLE 1.17. If N is set of natural number and Z^+ is the set of positive integers then which of the following are true ?

(*i*) $Z^+ \subseteq N$

(*ii*) $N \cup \{\ \} = Z^+$

Solution : $N = \{1,2,3,4, ...\}$

$Z^+ = \{0, 1, 2, 3, 4, ... \}$

$\therefore$ $N \subseteq Z^+$ hence, (*i*) is not true

$\{\}$ represents null set (empty set)

$N \cup \{\} \neq Z^+$

$\therefore$ (*ii*) is also not true.

EXAMPLE 1.18. Two finite sets have x and y number of elements. The total number of subsets of the first set is four times the total number of subsets of the second set. Find the value of $x - y$.

Solution : Let A and B are two sets having dimensions x and y respectively.

The number of subsets of the two sets are 2^x and 2^y respectively.

It is given that the total number of subsets of the first set is four times the total number of subsets of the second set.

$$\Rightarrow\quad 2^x = 4\ 2^y$$

$$\Rightarrow\quad \frac{2^x}{2^y} = 4$$

$$\Rightarrow \qquad 2^{x-y} = 4$$
$$\Rightarrow \qquad 2^{x-y} = 2^2$$
$$\Rightarrow \qquad x - y = 2$$

EXAMPLE 1.19. In an examination 70% of the candidates passed in English, 65% in Mathematics, 27% failed in both the subjects and 248 passed in both the subjects. Find the total number of candidates.

Solution : Let the set E and M represent students who passed in English and Mathematics respectively.

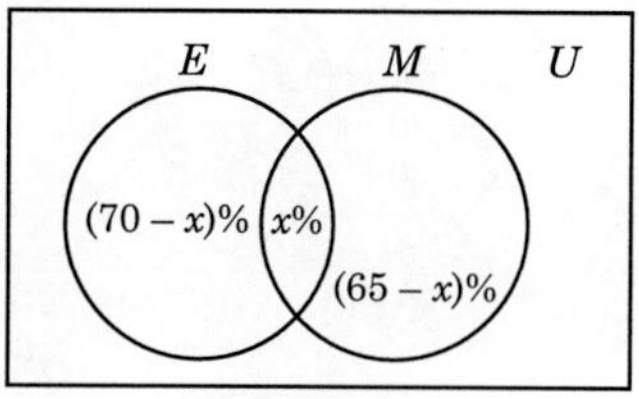

Fig. 1.8.

$$n(E \cup M) = (100 - 27)\% = 73\%$$

By fundamental counting principle

$$n(E \cup M) = n(E) + n(M) - n(E \cap M)$$
$$73\% = 70\% + 65\% - x\%$$
$$\Rightarrow \qquad x\% = 62\%$$

Now $\qquad 62\% \equiv 248 \qquad$ (given)

$$\therefore \text{ Total number of candidates} = \frac{248 \times 100}{62} = 400$$

EXAMPLE 1.20. In a survey of 100 students it was found that 50 used the college library, 40 had their own library and 30 borrowed books. Of these 20 used both the college library and their own, 15 used their own library and borrowed books and 10 used the college library and borrowed books. How many students used all the three sources of books ?

Solution : A : Students using the college library.

B : Students using their own library.

C : Students who borrow books.

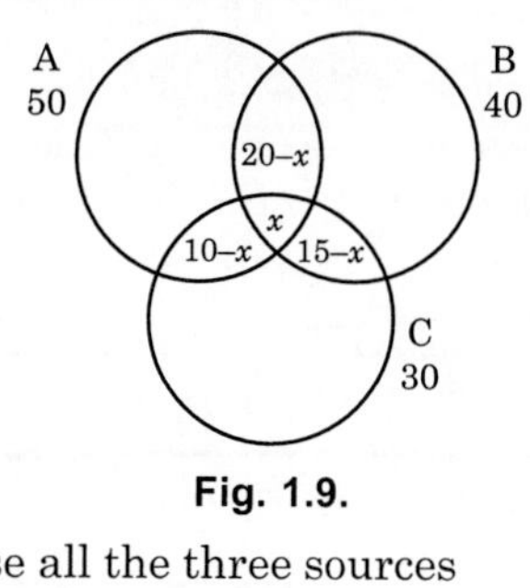

Fig. 1.9.

Suppose x students use all the three sources

By fundamental counting principle

$$n(A \cup B \cup C) = n(A) + n(B) + n(C) - n(A \cap B) - n(A \cap C) - n(B \cap C) + n(A \cap B \cap C)$$

$\therefore \quad 100 = 50 + 40 + 30 - 20 - 15 - 10 + x$

$\therefore \quad x = 25$

EXAMPLE 1.21. A firm has 40 workers working in the factory premises, 30 working in its office and 20 working in both the factory and the office. How many workers are there in the firm ? How many are working in (*i*) the factory alone (*ii*) office alone ?

Solution : Let F = Factory
O = Office

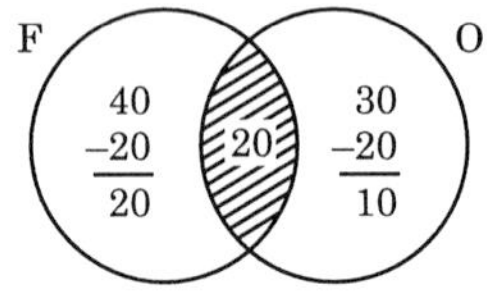

Fig. 1.10.

$n(F) = 40 \qquad n(O) = 30 \qquad n(F) \cap O) = 20$

The number of workers working in the factory only $= n(F) - n(F \cap O)$
$= 40 - 20$
$= 20.$

The number of workers working in the office only $= n(O) - n(F \cap O)$
$= 30 - 20 = 10$

Total number of workers $= 20 + 20 + 10 = 50$

EXAMPLE 1.22. In a group of 20 adults there are 8 females, 9 literate and 6 female literate. Find the number of male illiterates in the group.

Solution :

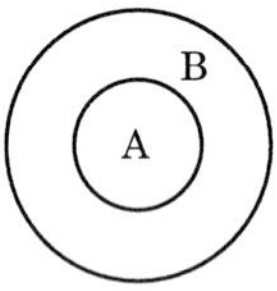

Fig. 1.11.

Number of males = 20 – 8 = 12

Number of male literate = 9 – 6 = 3

∴ Number of male illiterate = 12 – 3 = 9

EXAMPLE 1.23. Set A has 4 elements and set B has 7 elements. What can be the minimum number of elements in $A \cup B$?

Solution :

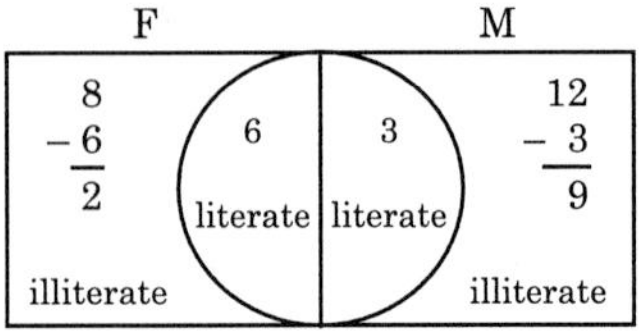

Fig. 1.12.

$A \cup B$ will have minimum number of elements if set A is a subset of B.

$\therefore n(A \cup B) = n(B) = 7$

For example (1.24) and (1.25) :

In the market survey of a particular brand of toothpaste, three attributes were tested - ability to clean teeth (T), ability to protect and strengthen gums (G), ability to prevent bad breath (B).

The following data was collected :

Number of persons interviewed	= 17,500
Number of persons who attributed T to the toothpaste	= 8,000
Number of persons who attributed B to the toothpaste	= 12, 000
Number of persons attributed B and G to the toothpaste	= 3,000
Number of persons attributing B & T to the toothpaste	= 3,000
Number of persons attributed B and G to the toothpaste	= 1,000
Number of persons attributed T and G to the toothpaste	= 2,000
Number of persons attributing T, B and G to the toothpaste	= 500

EXAMPLE 1.24. Find the number of persons attributing only B to the toothpaste.

Solution : Only (B) = (B) – (B, T) – (B, G) + (B, T, G)

Only (B) = 12000 – 3000 – 1000 + 500

= 8500

EXAMPLE 1.25. Find the number of persons attributing only T and G to the toothpaste

Solution : Only $(T, G) = (T, G) - (B, T, G)$

$= 2000 - 500$

$= 1500$

Ordered n-tuples :

Ordered pair :

(x, y) points in the cartesian plane.

In an ordered pair, order matters !

$$(1, 2) \neq (2, 1)$$

Ordered triple :

(x, y, z) "Ordered three-tuple".

If we kept going we would continue untill we reached an ordered n-tuple.

Set of ordered triples.

$$A \times B \times C$$

$$A \times B \times C = \{(x, y, z) \mid x \in A, y \in B, z \in C\}$$

Ordered n-tuple :

$$A_1 \times A_2 \times \ldots\ldots \times A_n = \{(a_1, \ldots\ldots, a_n) \mid a_i \in A_i, i \leq i \leq n, i \in Z\}$$

EXERCISE 1

1. In a class of 100, 84% of the students have taken politics and 16% of the students have taken History. How many students have taken both subjects if all the students take at least one of these subjects?

2. At a recent examination every candidate took English or Hindi ; 75.8 per cent took English and 49.4 percent took Hindi. If the total number of candidates was 2500, how many took both English and Hindi ?

3. Describe the shaded region in the following Venn diagram :

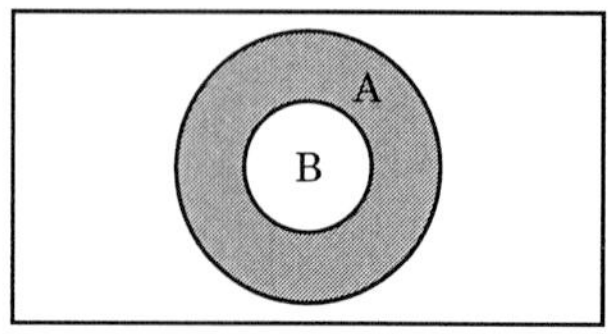

Fig. 1.13.

4. The set of all the points inside, or on the lines bordering a polygon, is referred to by the capital letters representing the vertices of the polygon, *e.g.*, AFGE means the set of all the points contained in the quadrilateral obtained by joining point A to point F to point G to point E.

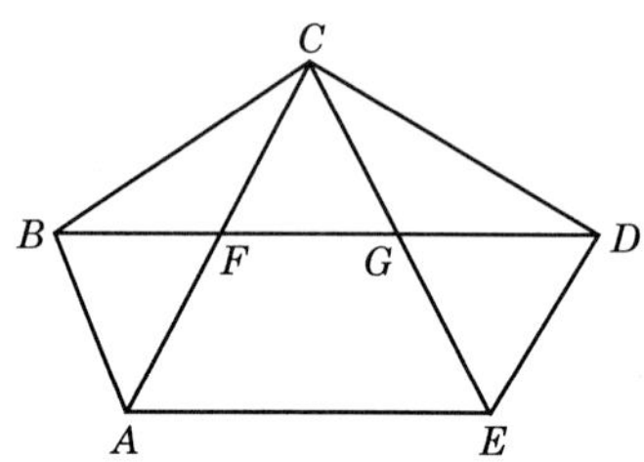

Fig. 1.14.

Using the diagram simplify

(*a*) AFBA ∪ FBCF

(*b*) AGEA ∩ CBDC

(*c*) CBGC ∩ CFDC

(*d*) CGDC ∩ GDEG

(*e*) CGFC ∩ DGED

(*f*) BFAB ∩ GDEG

(*g*) AEDBA ∩ CAEC

(*h*) ABCDEA ∪ ACEA

(*i*) ABCEA ∪ CDEC

(*j*) (BCGB ∪ GDEG) ∩ CGDC

5. Prove that for any two sets A and B

$$(A \cup B)^c = A^c \cap B^c$$
$$(A \cap B)^c = A^c \cup B^c$$

By using Venn Diagram.

6. If A and B are two sets such that neither A is a subset of B nor B is a subset of A then which of the following statement is true.

1. $A \cup B \subseteq A$ 2. $A \subseteq A \cap B$ 3. $A \subseteq A \cup B$ 4. $A \cup B \subseteq B$

7. The shaded region is represented by

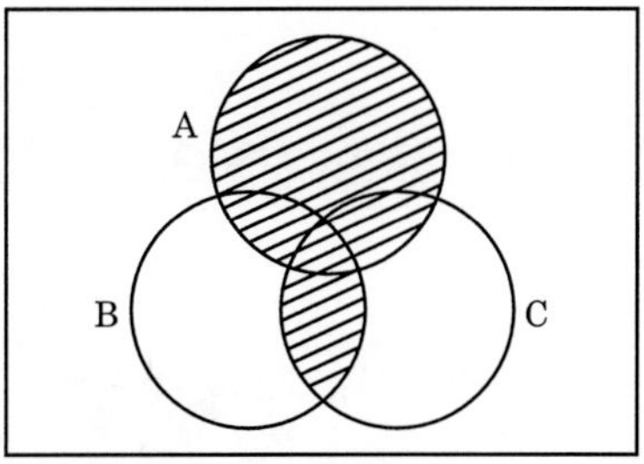

Fig. 1.15.

(*i*) A ∪ B) ∪ (A ∪ C)
(*ii*) (A ∩ B) ∪ (A ∩ C)
(*iii*) (*A* ∪ *B*) ∩ (*A* ∪ *C*)
(*iv*) None of these

8. The market research of a certain breakfast cereal firm interviewing 100 people, found that on a certain morning for breakfast 72 had cereals, 39 had an egg, 75 had toast; 32 had cereal and egg, 53 had cereal and toast, 26 had toast and egg. Of these 21 had cereal, toast and egg. How many of those interviewed had neither cereal, toast nor egg ?
   (*i*) 6 (*ii*) 2.4 (*iii*) 5 (*iv*) 8
9. A class of 30 students comprises boys who can play Cricket, Hockey and Football .3 boys play only cricket, 3 boys play Hockey and 2 play only football. 4 boys could play all three games, while 11 could play Football and Cricket, and 10 boys could play Football and Hockey.
   (*a*) How many boys played Cricket and Hockey but not Football ?
   (*b*) How many boys can play at least two games ?

1.11. FUNCTIONS

In the previous section we have done set theory, that is, what is a set and what are the operations on the set. Now we will learn how to make a link between two or more sets. The link can be between the elements of the same set. The link between the sets is nothing but the link between the corresponding elements of the sets. These types of links are called mapping or functions.

Definition : Let *A* and *B* be two non- empty sets. A function *f* from a set *A* into a set *B* is a mapping from *A* into *B* if for all $a \in A$, there exist a unique $b \in B$ such that $(a, b) \in f$ and we write this as:

$$f : A \to B$$
$$f(a) = b \quad \forall \quad a \in A, \text{b} \in B$$

The element *b* is called **image** of *a* under *f* and *a* is called **pre image** of *b* under *f*. The set *A* is called the **Domain** of the function *f*. and the set *B* is called the **co-domain** of the function *f*.

The subset of *B* which contains only those members which have pre-images in *A* is called **range** of *f*.

1.11.1. Type of Mappings

One-One, (1-1) mapping

A mapping $f : A \rightarrow B$ is called one to one mapping (function) or injective mapping (function) if for $a_1, a_2 \in A$

$$f(a_1) = f(a_2)$$
$$\Rightarrow a_1 = a_2$$

or if

$$a_1 \neq a_2$$
$$\Rightarrow f(a_1) \neq f(a_2)$$

Thus a function (mapping) is said to be injective from a set A to a set B if all the members of A are related to different members of B.

EXAMPLE 1.26. Let $f : N \rightarrow N$ where N is the set of natural numbers such that

$$f(x) = x + 2 , x \in N$$

then f is ingective as:

Let $x_1, x_2 \in N$ such that

$$f(x_1) = f(x_2)$$
$$\Rightarrow x_1 + 2 = x_2 + 2$$
$$\Rightarrow x_1 = x_2$$

1.11.2. Onto Mapping or Surjective Mapping

A mapping $f : A \rightarrow B$ is called onto or surjective if for each element b in B there exists an element a in A such

$$f(a) = b$$

that is if range of f equals B.

The example 1.26. is also surjective.

1.11.3. Bijective Mapping or one-to-one Mappping

A mapping which is both one-one and onto is called Bijective mapping.

In case of set theory, two sets are said to have the same cordinality if there exists a bijection between them

Example 1.26 is the example of Bijective function.

EXAMPLE 1.27. If Z = Set of integers

Define a mapping

$f : Z \rightarrow$ Z such that

$$f(x) = 2x$$

f is injective as

$$f(x_1) = f(x_2) , \quad x_1, x_2 \in \text{Z}$$
$$\Rightarrow \quad 2x_1 = 2x_2$$
$$\Rightarrow \quad x_1 = x_2$$

f is not surjective as:

for $1 \in Z$ we cannot find any $x \in Z$ such that

$$f(x) = 1$$

That is $1 \in Z$ has no pre-image

EXAMPLE 1.28. (Function which is surjective but not injective)

Let R = set of real numbers.

Define a mapping

$$f : R \rightarrow \{5\} \text{ such that}$$
$$f(x) = 5 \ \forall \ x \in R$$

***f* is onto**

f is onto by the mapping as every element in R maps to 5.

***f* is not one-one**

for $x_1, x_2 \in R$

Let $f(x_1) = f(x_2)$

$$\Rightarrow 5 = 5$$

$$\not\Rightarrow x_1 = x_2$$ (as we cannot say anything about

the variables x_1 and x_2)

Therefore *f* is surjective but not injective.

EXAMPLE 1.29. Let $A = \{a,b.c\}$ and $B = \{x, y, z, t\}$ and define a mapping

$$f : A \rightarrow B \quad \text{by}$$
$$f(a) = y$$
$$f(b) = z$$
$$f(c) = x$$

then *f* is injective but not surjective, as $t \in B$ has no pre-image in A..

Composition of Functions

Let $f : A \rightarrow B$ and $g : B \rightarrow C$ be two functions.

Then the composition of *f* with *g* written *gof*, is a function from A into C defined by:

$$(gof)(x) = g(f(x)) \ \forall \ x \in A.$$

Theorem 1.8 : If $f : A \rightarrow B$ and $g : B \rightarrow C$ are injective (surjective) then so is gof.

Theorem 1.9: Let $f : A \rightarrow B$ and $g : B \rightarrow C$ and $h : C \rightarrow D$

then $ho\,(gof) = (hog)\,of$

That is, associativity law holds.

Proof : We need to prove for all $x \in A$

$$(h(g\,f))\,(x) = ((h\,g)\,f)\,(x)$$

The proof can be done by the definition of composition.

LHS. $h(g\,f)\,(x) = h(g(f(x)))$

RHS. $((h\,g)f)\,(x) = (h\,g)\,(f(x))$

$$= h(g(f(x)))$$

Hence the result is proved.

Theorem 1.10 : Let $f : A \to B$ then f^{-1} (Inverse of f) exists if and only if f is one to one and onto.

Proof :

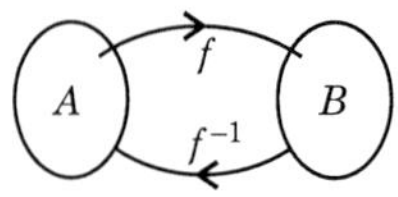

Fig. 1.16.

Theorem 1.11 : If $f : A \to B$ and $g : B \to C$ are injective functions then $go\,f : A \to C$ is an injective function.

Proof. given $\quad f(a) = f(b) \quad \Rightarrow a = b \quad \forall\ a \in A,\ b \in B.$

$\qquad g(b) = g(c) \quad \Rightarrow b = c \quad \forall\ b \in B,\ c \in C$

Let

$$\Rightarrow (gof)\ (a) = (gof)\ (b)$$
$$\Rightarrow\ g(f(a)) = g(f(b))$$

as g is injective

$$\Rightarrow f(a) = f(b)$$

as f is injective

$$\Rightarrow \quad a = b \text{ as } f \text{ is injective}$$

$\Rightarrow go\,f$ is injective.

Theorem 1.12 : If $f : A \to B$ and $g : B \to C$ are surjections, then $gof : A \to C$ is a surjection.

Proof: $f : A \to B$ is surjection $\Rightarrow \forall\ b \in B,\ b = f(a)$ for some $a \in A$

$g : B \to C$ is surjection $\Rightarrow \forall\ c \in C,\ c = g(b)$ for some $b \in B$

Now to show that gof is surjection.

Let $c \in C$ we have to show that, there exists an $a \in A$: such that $(gof)(a) = c$

Let $c \in C$, as g is surjection, therefore,

$$c = g(b)\ ,\ b \in B$$

as f is surjection

$$b = f(a), \qquad a \in A$$
$$\Rightarrow c = g(f(a))$$
$$\Rightarrow c = (gof)(a)$$

$\Rightarrow gof$ is surjection

Remark : Converse of the above result does not hold.

Identity function

The identity function, under composition, is a function from A to A denoted by I. :

$$fo\,I = I\,of = f\ \forall \quad \text{functions from } A \text{ to } A$$

i.e., $\qquad I(a) = a\ \forall\ a \in A.$

EXAMPLE 1.30. If $\quad A = \{1, 2, 3\}$

then the identity function $I : A \to A$ is denoted by

$$I(1) = 1$$

$$I(2) = 2$$
$$I(3) = 3$$

EXAMPLE 1.31. $I : R \to R$

$\therefore \quad I(x) = x \quad \forall\, x \in R.$

Inverse function

Let $f : A \to A$

If there exists a function $g : A \to A : g\,o\,f = f\,o\,g = I$ then g is called the inverse of the function f and is denoted by f^{-1}, read "f inverse". Such functions are called invertible functions.

As we know that inverse of an element is unique therefore a function is invertible only when it is bijective.

i.e., Let $f : A \to A$

such that $f(a) = b \quad a, b \in A$

then $f^{-1} : A \to A$ is given as

$: f^{-1}(b) = a \quad a, b \in A$

EXAMPLE 1.32. Let $A = \{1, 2, 3\}$

$$f : A \to A$$
$$f(1) = 2$$
$$f(2) = 3$$
$$f(3) = 1$$

then $f^{-1} : A \to A$ is

$$f^{-1}(1) = 3$$
$$f^{-1}(2) = 1$$
$$f^{-1}(3) = 2$$

It is to be noted that

$$\boldsymbol{fo\,f^{-1} = f^{-1}\,o\,f = I}$$
$$fo\,f^{-1}(1) = f(f^{-1}(1)) = f(3) = 1 = I(1)$$
$$fo\,f^{-1}(2) = f(f^{-1}(2)) = f(1) = 2 = I(2)$$
$$fo\,f^{-1}(3) = f(f^{-1}(3)) = f(2) = 3 = I(3)$$

EXAMPLE 1.33. Let the functions f and g be defined by $f(x) = 2x + 1$ and $g(x) = x^2 - 2$. Find the formula defining the composition function gof.

$$go\,f(x) = g(f(x)) = g(2x + 1)$$
$$= (2x + 1)^2 - 2$$
$$= 4x^2 + 4x - 1$$

EXAMPLE 1.34. Let $N = \{0, 1, 2, 3, ...\}$. Define functions f, g and h from set N to N by:

$$\mathrm{f}(n) = n + 1$$
$$g(n) = 2n$$
$$h(n) = \begin{cases} 0 & \text{if } n \text{ is even} \\ 1 & \text{if } n \text{ is odd} \end{cases}$$

Compute $go(fog)\,oh$.
Is the function h is inversible?
Is the function f is onto ?

Solution : $(go(fog)oh)(n)$

$$= go(fog)(h(n))$$
$$= go(fog)(x) \qquad \text{where } x = \begin{cases} 0 & \text{if } n \text{ is even} \\ 1 & \text{if } n \text{ is odd} \end{cases}$$

Now,
$$\begin{aligned} fog(x) &= f(g(x)) \\ &= f(2x) \\ &= 2x + 1 \quad \text{odd always} \\ go(fog)x &= g((fog)(x)) \\ &= g(2x + 1) \\ &= 2(2x + 1) \\ &= 4x + 2 \\ &= \begin{cases} 2 & \text{if } n \text{ is even} \\ 6 & \text{if } n \text{ is odd} \end{cases} \end{aligned}$$

The function h is not invertible as h is not one-one.
The function f is not onto as there does not exist pre-image of 0.

1.12. FLOOR AND CEILING FUNCTIONS

Let $x \in R$, then x lies between two integers called floor and ceiling of x.

$\lfloor x \rfloor$, called the floor of x, denotes the greatest integer that does not exceed x.

$\lceil x \rceil$, called the ceiling of x, denotes the least integer that is not less than x.

$$I_1 \le x \le I_2$$
$$\downarrow \qquad \downarrow$$
Floor. Ceiling.

EXAMPLE 1.35. Give the value of $\lfloor 7\cdot 5 \rfloor, \lfloor -7\cdot 5 \rfloor, \lfloor -18 \rfloor$

$$\lfloor 7\cdot 5 \rfloor = 7$$
$$\lfloor -7\cdot 5 \rfloor = -8$$
$$\lfloor -18 \rfloor = -18$$

EXAMPLE 1.36. Give the value of $\lceil 7\cdot 5 \rceil, \lceil -7\cdot 5 \rceil, \lceil -18 \rceil$

$$\lceil 7\cdot 5 \rceil = 8$$
$$\lceil -7\cdot 5 \rceil = -7$$
$$\lceil -18 \rceil = -18$$

Congruence Relation (mod Function)

K (mod M)

$a \equiv b$ (mod M) iff M divides $b - a$ that is a is the remainder

M is called the modulus and $a \equiv b$ (mod M) is read "a is congruent to b modulo M".

e.g., $0 \equiv M \,(\text{mod M})$ as M/m

and $a \pm M \equiv a(\text{mod } M)$ as $M/((a \pm M) - a)$

$\Rightarrow M| \pm M.$

If k is negative then remainder is $M - r$ where $r \equiv K\,(\text{mod } M)$, if k is positive.

EXAMPLE 1.37. Find (*a*) 25 (Mod 7)

(*b*) 25 (mod 5)

(*c*) – 35 (mod 11)

(*d*) – 3 (mod 8)

Solution :

(*a*) 25 (mod 7) = 4

(*b*) 25 (mod 5) = 0

(*c*) – 35 (mod 11) = 11 – 2 = 9

– 3 (mod 8) = 8 – 3 = 5

EXAMPLE 1.38. Let $A = Z$ and Let $B = \{0, 1\}$. Let $f : A \to B$ be defined by

$$f(a) = 0 \quad \text{if } a \text{ is even}$$
$$1 \quad \text{if } a \text{ is odd.}$$

Then f is a function, since each set $f(a)$ consists of a single element.

Every where defined function : Let f be function from A to B. Then we say that f is everywhere defined if Dom (f) = A.

EXAMPLE 1.39. Let $A = \{a_1, a_2, a_3\}$, $B = \{b_1, b_2, b_3\}$, $C = \{c_1, c_2\}$ and $D = \{d_1, d_2, d_3, d_4\}$. Consider the following functions, from A to B, A to D, B to C and D to B respectively.

(*a*) $f_1 = \{(a_1, b_2), (a_2, b_3), (a_3, b_1)\}$

(*b*) $f_2 = \{(a_1, d_2,), (a_2, d_1), (a_3, d_4)\}$

(*c*) $f_3 = \{(b_1, c_2), (b_2, c_2), (b_3, c_1)\}$

(*d*) $f_4 = \{(d_1, b_1), (d_2, b_2), (d_3, b_1)\}$

Check the 1 – 1, onto properties of all functions

Solution :

f_1	1 – 1,	onto
f_2	1 – 1,	not onto
f_3	not 1 – 1,	onto
f4	not 1 – 1,	not onto

Invertible Function

A function $f : A \to B$ is said to be invertible if its inverse relation, f^{-1} is also a function.

Theorem: Let $f : A \to B$ be a f^n

(*a*) Then f^{-1} is a function from B to A iff it is 1 – 1.

(*b*) If f^{-1} is a function then the function f^{-1} is also 1 – 1.

(*c*) f^{-1} is onto iff if f is everywhere defined.

1.13. PERMUTATION FUNCTION

In the last section we have discussed what a function is ? What are the properties of functions ? We have done one-to-one and ontoness of the function. The function is nothing but the link between the elements of the two or more sets. Now we are going to define the link between the elements of the same set.

Permutation

A **permutation** of a set A is a function from A into A that is both one-to-one and onto. In other words, a permutation of A is one – to – one or bijective function from A onto A.

EXAMPLE 1.40. Let $A = \{1, 2, 3\}$

consider the maps $p : A \to A$ such that

$$p(1) = 2$$
$$p(2) = 3$$
$$p(3) = 1$$

EXAMPLE 1.41. Let $A = \{1, 2, 3, 4, 5\}$ and let p be a permutation on A is defined as:

$$p(1) = 4$$
$$p(2) = 2$$
$$p(3) = 5$$
$$p(4) = 3$$
$$p(5) = 1$$

We now introduce a notation to represent permutations, which can be given as:

$$p = \begin{pmatrix} 1 & 2 & 3 & 4 & 5 \\ 4 & 2 & 5 & 3 & 1 \end{pmatrix}$$

Now define another function $q : A \to A$ such that

$$q(1) = 3$$
$$q(2) = 2$$
$$q(3) = 4$$
$$q(4) = 3$$
$$q(5) = 5$$

It is to be noted that this function q is not a permutation as 3 appears twice while 1 does not appear at all on the Right hand side.

Cyclic Permutation

Let $x_1, x_2, ... x_r$ be distinct elements of the set

$X = \{x_1, x_2, ... x_r\}$. Then the permutation $p : X \to X$ is defined as follows :

$$p(x_1) = x_2$$
$$p(x_2) = x_3$$
$$p(x_3) = x_4$$
$$\vdots$$

$$\vdots$$

$$p(x_{r-1}) = x_r$$

$$p(x_r) = x_1$$

The above is called a cyclic permutation of length r and is written as $(x_1, x_2, x_3, \ldots x_r)$ and is given by the diagram.

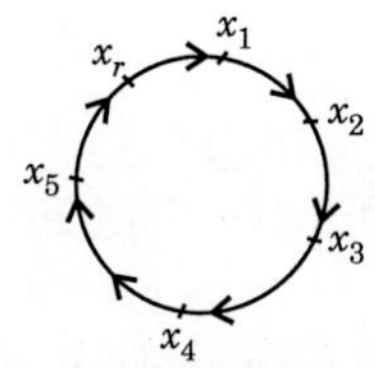

Fig. 1.17.

The cycle indicated by arrows acts to move these elements in a clockwise direction $x_1 \to x_2$, $x_2 \to x_3$, ... $x_{r-1} \to x_r$ and $x_r \to x_1$.

EXAMPLE 1.42. Let $A = \{1, 2, 3, 4, 5\}$

Compute (2 1 3 4) and (5, 4, 2)

Now we have

$$(2\ 1\ 3\ 4) = \begin{pmatrix} 1 & 2 & 3 & 4 & 5 \\ 3 & 1 & 4 & 2 & 5 \end{pmatrix}$$

and

$$(5\ 4\ 2) = \begin{pmatrix} 1 & 2 & 3 & 4 & 5 \\ 1 & 5 & 3 & 2 & 4 \end{pmatrix}$$

Disjoint Cycles

Two cycles of a set A are called disjoint if they do not have element (or) elements of A as common element.

EXAMPLE 1.43. Let $A = \{1, 2, 3, 4, 5\}$

then the cycles (1, 2, 4) and (3, 5) are disjoint. But (1, 2, 4) and (1, 3, 5) are not disjoints.

Let $p_1 = (x_1, x_2, \ldots x_n)$ and $p_2 = (y_1\, y_2 \ldots y_n)$ be two disjoint cycles of a set A, then

$$p_1 \circ p_2 = p_2 \, o \, p_1$$

that is commutative property holds in it but this is not true in case when p_1 and p_2 are not disjoint.

Theorem 1.13 : Every permutation of a finite set A is a cycle or it can be expressed as a product of disjoint cycles.

Proof : $f : A \to$ A be any permutation of A. We select any element a_1 in A and find $f(a_1)$, which we call a_2. If $a_1 = a_2$ then (a_1) is the first cycle is the factorisation of f. If not, we find $f(a_2) = a_3$ say. Continuing like this having found a_k as the image of a_{k-1}, we determine $f(a_k)$. Since f is one-one, $f(a_k)$ cannot be equal to a_2, a_3, ... a_{k-1}. If $f(a_k) = a_1$, we close the cycle and $(a_1, a_2 \ldots.. a_k)$ is the first factor of f.

If $f(a_k) \neq a_1$, we write $f(a_k) = a_{k+1}$ and continue. Since A is finite, this process must terminate after finite number of steps. After finding the first factor, we select any remaining element of A and continue as above to find second factor.

Continuing this way we exhaust the set A and find the factorization of f into disjoint cycles.

EXAMPLE 1.44. In the above example compute (1 2 4) ∘ (3 5) and (3 5) ∘ (1 2 4) and check for commutativity.

$$(1\ 2\ 4) \circ (3\ 5) = \begin{pmatrix} 1 & 2 & 3 & 4 & 5 \\ 2 & 4 & 3 & 1 & 5 \end{pmatrix} o \begin{pmatrix} 1 & 2 & 3 & 4 & 5 \\ 1 & 2 & 5 & 4 & 3 \end{pmatrix} = \begin{pmatrix} 1 & 2 & 3 & 4 & 5 \\ 2 & 4 & 5 & 1 & 3 \end{pmatrix}$$

$$(3\ 5) \circ (1\ 2\ 4) = \begin{pmatrix} 1 & 2 & 3 & 4 & 5 \\ 1 & 2 & 5 & 4 & 3 \end{pmatrix} \circ \begin{pmatrix} 1 & 2 & 3 & 4 & 5 \\ 2 & 4 & 3 & 1 & 5 \end{pmatrix} = \begin{pmatrix} 1 & 2 & 3 & 4 & 5 \\ 2 & 4 & 5 & 1 & 3 \end{pmatrix}$$

∴ (1 2 4) o (3 5) = (3 5) o (1 2 4) as both are disjoint.

Transposition

A cycle of length 2 is called transposition. In other words, a transposition is a cycle denoted as

$p(a_i, a_j)$, $p(a_i) = a_j$ and $p(a_j) = a_i$

Theorem 1.14 : Every cycle can be written as a product of transpositions.

$$(x_1\ x_2\ \ldots.\ x_r) = (x_1\ x_r) \circ (x_1\ x_{r-1}) \circ (x_1\ x_{r-2}) \circ \ldots \circ (x_1\ x_3) \circ (x_1\ x_2)$$

Proof : Now, we will use method of mathematical induction to prove the result.

Let $p(r) : (x_1, x_2 \ldots x_r) = (x_1\ x_r)\ o\ (x_1\ x_{r-1})\ o\ldots..o\ (x_1\ x_3)\ o\ (x_1\ x_2)$

Basic step

If $r = 2$.

$p(2) : (x_1\ x_2)$ already is a transposition.

Induction step

Let the result is true for $r = k$, that is, $p(k)$ is true.

i.e., $(x_1\ x_2 \ldots x_k) = (x_1\ x_k)\ (x_1\ x_{k-1})\ (x_1\ x_{k-2})\ (x_1\ x_{k-3}) \ldots (x_1\ x_3)\ (x_1\ x_2)$

Let's prove the result for $n = k + 1$

$$\begin{aligned}(x_1\ x_2 \ldots x_{k+1}) &= (x_1\ x_{k+1}) \cdot (x_1\ x_2 \ldots x_k) \\ &= (x_1\ x_{k+1}) \cdot (x_1\ x_k)\ (x_1\ x_{k-1}) \ldots (x_1\ x_2) \quad \text{[By hypothesis]}\end{aligned}$$

⇒ $p(k + 1)$ is true.

Therefore by induction the result is true.

Theorem 1.15: Every permutation of a finite set with at least two elements can be written as a product of transposition.

Proof. We have seen every permutation can be written as product of cycles and every cycle can be written as product of transposition. Therefore every permutation can be written as product of transpositions.

We now simplify the notation in example 1.40. Since in each representation, the first row is same, we omit writing it and simply represent p by means of cyclic permutation as (1 4 3 5 1) (2)

Again in representation we can omit 2 also and simply we can write (1 4 3 5 1)

The absence of 2 automatically means that 2 maps itself.

(1 4 3 5 1) and (2) are called cycles of the permutation p.

Theorem 1.16 : Every transposition $(x\ y)$ is its own inverse. That is $(xy)^{-1} = (xy)$.

Proof : By the definition of inverse b will be the inverse of a if $ab = e$(Identity)

Now, if we prove that

$$(xy)(xy)^{-1} = I$$

then nothing to prove

$$\begin{aligned}(xy)(xy)^{-1} &\\ &= (xy)\ (yx)\\ &= \begin{pmatrix} x & y \\ y & x \end{pmatrix}\begin{pmatrix} x & y \\ y & x \end{pmatrix}\\ &= \begin{pmatrix} x & y \\ x & y \end{pmatrix}\\ &= I\end{aligned}$$

⇒ the result holds.

EXAMPLE 1.45. If $A = \{1, 2, 3\}$ then obtain all permutations of A.

Solution : All permutations of A are:

$$I = \begin{pmatrix} 1 & 2 & 2 \\ 1 & 2 & 3 \end{pmatrix} = (1)\ (2)\ (3)$$

$$p_1 = \begin{pmatrix} 1 & 2 & 3 \\ 1 & 3 & 2 \end{pmatrix} = (2\ \ 3)\ (1)$$

$$p_2 = \begin{pmatrix} 1 & 2 & 3 \\ 2 & 1 & 3 \end{pmatrix} = (1\ \ 2)\ (3)$$

$$P_3 = \begin{pmatrix} 1 & 2 & 3 \\ 3 & 2 & 1 \end{pmatrix} = (1\ \ 3)\ (2)$$

$$P_4 = \begin{pmatrix} 1 & 2 & 3 \\ 2 & 3 & 1 \end{pmatrix} = (1\ \ 2\ \ 3)$$

$$p_5 = \begin{pmatrix} 1 & 2 & 3 \\ 3 & 1 & 2 \end{pmatrix} = (1\ \ 3\ \ 2)$$

EXAMPLE 1.46. In the above example compute $p_2 \circ p_4$, $p_2 \circ p_3$

$$p_2 \circ p_4 = \begin{pmatrix} 1 & 2 & 3 \\ 2 & 1 & 3 \end{pmatrix} \circ \begin{pmatrix} 1 & 2 & 3 \\ 2 & 3 & 1 \end{pmatrix} = \begin{pmatrix} 1 & 2 & 3 \\ 1 & 3 & 2 \end{pmatrix} = (2\ 3)$$

$$p_2 \circ p_3 = \begin{pmatrix} 1 & 2 & 3 \\ 2 & 1 & 3 \end{pmatrix} \circ \begin{pmatrix} 1 & 2 & 3 \\ 3 & 2 & 1 \end{pmatrix} = \begin{pmatrix} 1 & 2 & 3 \\ 3 & 1 & 2 \end{pmatrix} = (1\ \ 3\ \ 2)$$

EXAMPLE 1.47. Compute p_2^{-1}, p_4^{-1}.

Treating p_2 as a function, we obtain

$$p_2 = \{(1,2), (2, 1), (3, 3)\}$$

Solution : $$p_2^{-1} = \{(2, 1), (1, 2) (3, 3)\}$$

$$\therefore \quad p_2^{-1} = \begin{pmatrix} 1 & 2 & 3 \\ 2 & 1 & 3 \end{pmatrix} = p_2 \text{ is it self.}$$

$$p_4 = \{(1, 2), (2, 3), (3, 1)\}$$

$$\therefore \quad p_4^{-1} = \{(2, 1), (3, 2), (1, 3)\}$$

$$= \begin{pmatrix} 1 & 2 & 3 \\ 3 & 1 & 2 \end{pmatrix}$$

$$= (1\ 3\ 2) = p_5$$

Thus p_5 is the inverse of p_4.

Verification.

$$p_4 \, o \, p_5 = \begin{pmatrix} 1 & 2 & 3 \\ 2 & 3 & 1 \end{pmatrix} o \begin{pmatrix} 1 & 2 & 3 \\ 3 & 1 & 2 \end{pmatrix} = \begin{pmatrix} 1 & 2 & 3 \\ 1 & 2 & 3 \end{pmatrix} = (1)\ (2)\ (3) = I.$$

Even and Odd Permutation

A permutation is called even permutation if it can be expressed as a product of even number of transposition and if the permutation is expressible as a product of an odd number of tansposition then the permutation is called odd permutation.

EXAMPLE 1.48. Is the permutation $\begin{pmatrix} 1 & 2 & 3 & 4 & 5 & 6 & 7 \\ 3 & 4 & 1 & 5 & 6 & 7 & 2 \end{pmatrix}$ an odd permutation ?

Solution : $\begin{pmatrix} 1 & 2 & 3 & 4 & 5 & 6 & 7 \\ 3 & 4 & 1 & 5 & 6 & 7 & 2 \end{pmatrix}$

$= (1\ 3)\ (2\ 4\ 5\ 6\ 7)$

$= (1\ 3)\ (2\ 7)\ (2\ 6)\ (2\ 5)\ (2\ 4)$

= product of an odd number of transposition.

$\therefore$ the given permutation is an odd permutation.

It is clear from the above definition that:

(*a*) the product of two even permutations is even,

(*b*) the product of two odd permutations is even, and

(*c*) the product of an even and an odd permutation is odd.

EXAMPLE 1.49. Check whether the given permutations are even or odd.

(*i*) $\begin{pmatrix} a & b & c & d & e & f \\ b & a & e & f & d & c \end{pmatrix}$

(*ii*) $\begin{pmatrix} 1 & 2 & 3 & 4 & 5 & 6 \\ 2 & 3 & 1 & 6 & 5 & 4 \end{pmatrix}$

Solution :

(*i*) $\begin{pmatrix} a & b & c & d & e & f \\ b & a & e & f & d & c \end{pmatrix}$

$= (a\ b)(c\ e\ d\ f)$

$= (a\ b)(c\ f)(c\ d)(c\ e)$

The given permutation is an even permutation.

(*ii*) $\begin{pmatrix} 1 & 2 & 3 & 4 & 5 & 6 \\ 2 & 3 & 1 & 6 & 5 & 4 \end{pmatrix}$

$= (1\ 2\ 3)\ (4\ 6)\ (5)$

$= (1\ 3)(1\ 2)(4\ 6)$

The given permuatation is an odd permutation.

EXAMPLE 1.50. Express the following permutations as product of disjoint cycles and then product of transpositions.

(*i*) $(a\ b\ c)\ (b\ c\ d)\ (d\ e\ f)\ (f\ g)$

(*ii*) (1 2) (1 3) (1 4) (1 5) (1 6)

(*iii*) (2 4) (2 6) (2 8) (1 3) (1 5) (1 7)

(*iv*) (1 2 3) (4 5 7 6)

(*v*) (1 2 3 4 5) (1 2 3) (4 5)

Solution : Now applying theorem (1.13) we will get

(*i*) $(a\ c)(a\ b)(b\ d)\ (b\ c)\ (d\ f)\ (d\ e)\ (f\ g)$

Product of transposition

(*ii*) (1 2) (1 3) (1 4) (1 5) (1 6)

= (1 6 5 4 3 2)

Product of disjoint cycle.

(*iii*) (2 4) (2 6) (2 8) (1 3) (1 5) (1 7)

= (2 8 6 4) (1 7 5 3)

Product of disjoint cycle.

(*iv*) (1 3)(1 2)(4 6)(4 7)(4 5)

Product of transposition.

(*v*) (1 2 3 4 5) (1 2 3) (4 5)

= (1 5)(1 4)(1 3)(1 2)(1 3)(1 2)(4 5)

= (1 5)(1 4)(1 3)(1 2)(4 5)

Product of transposition.

EXAMPLE 1.51. Consider the following permutations.

$$p_1 = \begin{pmatrix} 1 & 2 & 3 & 4 & 5 & 6 \\ 3 & 1 & 4 & 5 & 6 & 2 \end{pmatrix}$$

$$p_2 = \begin{pmatrix} 1 & 2 & 3 & 4 & 5 & 6 \\ 2 & 4 & 1 & 3 & 6 & 5 \end{pmatrix}$$

$$p_3 = \begin{pmatrix} 1 & 2 & 3 & 4 & 5 & 6 \\ 5 & 2 & 4 & 3 & 1 & 6 \end{pmatrix}$$

Compute

(*a*) $p_1\, p_2$ (*b*) $p_1\, p_2{}^2$ (*c*) $p_1{}^2\, p_3$ (*d*) $p_2\, p_1{}^{-1}$ (*e*) $p_1\, p_2{}^{-1}\, p_3$

Solution : $p_1 = \begin{pmatrix} 1 & 2 & 3 & 4 & 5 & 6 \\ 3 & 1 & 4 & 5 & 6 & 2 \end{pmatrix}$

$$p_2 = \begin{pmatrix} 1 & 2 & 3 & 4 & 5 & 6 \\ 2 & 4 & 1 & 3 & 6 & 5 \end{pmatrix}$$

$$p_3 = \begin{pmatrix} 1 & 2 & 3 & 4 & 5 & 6 \\ 5 & 2 & 4 & 3 & 1 & 6 \end{pmatrix}$$

(*a*)
$$p_1 p_2 = \begin{pmatrix} 1 & 2 & 3 & 4 & 5 & 6 \\ 3 & 1 & 4 & 5 & 6 & 2 \end{pmatrix}\begin{pmatrix} 1 & 2 & 3 & 4 & 5 & 6 \\ 2 & 4 & 1 & 3 & 6 & 5 \end{pmatrix}$$
$$= \begin{pmatrix} 1 & 2 & 3 & 4 & 5 & 6 \\ 1 & 5 & 3 & 4 & 2 & 6 \end{pmatrix}$$
$$= (1)\,(2\;5)\,(3)\,(4\;6)$$
$$= (2\;5)(4\;6)$$

(*b*)
$$p_1 {p_2}^2 = p_1p_2p_2$$
$$= \begin{pmatrix} 1 & 2 & 3 & 4 & 5 & 6 \\ 1 & 5 & 3 & 4 & 2 & 6 \end{pmatrix} \cdot \begin{pmatrix} 1 & 2 & 3 & 4 & 5 & 6 \\ 5 & 2 & 4 & 3 & 1 & 6 \end{pmatrix} \qquad \text{(By above part)}$$
$$= \begin{pmatrix} 1 & 2 & 3 & 4 & 5 & 6 \\ 2 & 5 & 4 & 3 & 1 & 6 \end{pmatrix}$$
$$= (1\;2\;5)(3\;4)(6) = (1\;2\;5)(3\;4)$$

(*c*)
$$p_1{}^2 p_3 = \begin{pmatrix} 1 & 2 & 3 & 4 & 5 & 6 \\ 3 & 1 & 4 & 5 & 6 & 2 \end{pmatrix} \cdot \begin{pmatrix} 1 & 2 & 3 & 4 & 5 & 6 \\ 3 & 1 & 4 & 5 & 6 & 2 \end{pmatrix} \cdot \begin{pmatrix} 1 & 2 & 3 & 4 & 5 & 6 \\ 5 & 2 & 4 & 3 & 1 & 6 \end{pmatrix}$$
$$= \begin{pmatrix} 1 & 2 & 3 & 4 & 5 & 6 \\ 4 & 3 & 5 & 6 & 2 & 1 \end{pmatrix} \cdot \begin{pmatrix} 1 & 2 & 3 & 4 & 5 & 6 \\ 5 & 2 & 4 & 3 & 1 & 6 \end{pmatrix}$$
$$= \begin{pmatrix} 1 & 2 & 3 & 4 & 5 & 6 \\ 2 & 3 & 6 & 5 & 4 & 1 \end{pmatrix}$$
$$= (1\;2\;3\;6)(4\;5)$$

(*d*)
$$p_2 p_1{}^{-1} = \begin{pmatrix} 1 & 2 & 3 & 4 & 5 & 6 \\ 2 & 4 & 1 & 3 & 6 & 5 \end{pmatrix} \cdot \begin{pmatrix} 1 & 2 & 3 & 4 & 5 & 6 \\ 3 & 1 & 4 & 5 & 6 & 2 \end{pmatrix}^{-1}$$
$$= \begin{pmatrix} 1 & 2 & 3 & 4 & 5 & 6 \\ 2 & 4 & 1 & 3 & 6 & 5 \end{pmatrix}\begin{pmatrix} 3 & 1 & 4 & 5 & 6 & 2 \\ 1 & 2 & 3 & 4 & 5 & 6 \end{pmatrix}$$
$$= \begin{pmatrix} 1 & 2 & 3 & 4 & 5 & 6 \\ 4 & 5 & 2 & 1 & 3 & 6 \end{pmatrix}$$
$$= (1\;4)(2\;5\;3)(6)$$
$$= (1\;4)(2\;5\;3)$$

(e) $p_1p_2^{-1}p_3 = \begin{pmatrix}1&2&3&4&5&6\\3&1&4&5&6&2\end{pmatrix}\cdot\begin{pmatrix}1&2&3&4&5&6\\2&4&1&3&6&5\end{pmatrix}^{-1}\cdot\begin{pmatrix}1&2&3&4&5&6\\5&2&4&3&1&6\end{pmatrix}$

$$= \begin{pmatrix}1&2&3&4&5&6\\3&1&4&5&6&2\end{pmatrix}\cdot\begin{pmatrix}2&4&1&3&6&5\\1&2&3&4&5&6\end{pmatrix}\cdot\begin{pmatrix}1&2&3&4&5&6\\5&2&4&3&1&6\end{pmatrix}$$

$$= \begin{pmatrix}1&2&3&4&5&6\\4&3&5&1&2&6\end{pmatrix}\cdot\begin{pmatrix}1&2&3&4&5&6\\5&2&4&3&1&6\end{pmatrix}$$

$$= \begin{pmatrix}1&2&3&4&5&6\\2&3&1&5&4&6\end{pmatrix}$$

$= (1\ 2\ 3)(4\ 5)(6)$

$= (1\ 2\ 3)(4\ 5)$

EXAMPLE 1.52. Which of the following functions from $R \to R$ are permutations of R ?

(i) $p : R \to R$ as
$p(x) = x + 1$

(ii) $p : R \to R$.
$p(x) = x^2$

(iii) $p : R \to R$.
$p(x) = e^x$

Ans. (i) and (iii) are permutations.

1.14. CHARACTERISTIC FUNCTIONS

In the last section we defined function and some basic concept of the function. In this section we study more functions commonly used in computer science applications. We will discuss the function from the universal set $\cup$ to the set {0, 1}. A one-to-one correspondence exists between these functions and the sets. With the help of these functions, the statements about the sets and their operations can easily be represented and manipulated on computer as binary numbers.

Definition : Characterstic function : Let $\cup$ be a universal set and let X be a subset of $\cup$. The characterstic function Ψ of X is denoted by $\Psi_x : \cup \to \{0, 1\}$ defined by

$$\Psi_x = \begin{cases}1 & \text{if } x \in X\\ 0 & \text{if } x \notin X\end{cases}$$

EXAMPLE 1.53. Let $A = \{b, c, f\}$ and $\cup = \{a, b, c, \ldots, z\}$, then the characteristic function Ψ_x is given by

$\Psi_A(b) = 1$

$\Psi_A(c) = 1$

$\Psi_A(f) = 1$

and for $x \in \cup$ but $x \notin A$

$$\Psi_A(x) = 0$$

It is to be noticed that $\Psi_A(5)$ is not defined as $5 \notin \cup$.

1.14.1. Properties of Characteristic Functions

The following are some properties of characteristic functions which are useful for determining the set operations.

Let A and B are any two subsets of the universal set $\cup$. Then $\forall\, x \in \cup$

1. $\Psi_A(x) = 0 \Leftrightarrow A = \phi$
2. $\Psi_A(x) = 1 \Leftrightarrow A = \cup$
3. $\Psi_A(x) \le \Psi_B(x) \Leftrightarrow A \subseteq B$
4. $\Psi_A(x) = \Psi_B(x) \Leftrightarrow A = B$
5. $\Psi_{A \cap B}(x) = \Psi_A(x) * \Psi_B(x)$
6. $\Psi_{A \cup B}(x) = \Psi_A(x) + \Psi_B(x) - \Psi_{A \cap B}(x)$
7. $\Psi_{\sim A} = 1 - \Psi_A$

All the above properties can easily be proved using the definition of characteristics functions.

For example the property (5) can be proved as follows :

If $x \in A \cap B$

$\Leftrightarrow x \in A$ and $x \in B$

$$\Leftrightarrow \Psi_A(x) = 1 \text{ and } \Psi_B(x) = 1$$

and $\qquad \Psi_{A \cap B}(x) = 1 * 1 = 1$

If $x \notin A \cap B$ then.

$$\Psi_{A \cap B}(x) = 0 \text{ and } \Psi_A(x) = 0 \text{ or } \Psi_B(x) = 0$$

Consequently $\Psi_A(x) * \Psi_B(x) = 0$

Many set identities can be proved by using characteristic functions. Consider the following example.

EXAMPLE 1.54. Show that $A \cap (B \cup C) = (A \cap B) \cup (A \cap C)$

Solution : $\Psi_{A \cap (B \cup C)}(x) = \Psi_A(x) * \Psi_{B \cup C}(x)$ [By property 5]

$$\begin{aligned}
&= \Psi_A(x)\,(\Psi_B(x) + \Psi_C(x) - \Psi_{B \cap C}(x)) && \text{(By property 6)}\\
&= \Psi_A(x) * \Psi_B(x) + \Psi_A(x) * \Psi_C(x) - \Psi_A(x) * \Psi_{B \cap C}(x) &&\\
&= \Psi_{A \cap B}(x) + \Psi_{A \cap C}(x) - \Psi_{A \cap (B \cap C)}(x) && \text{(By property 5)}\\
&= \Psi_{A \cap B}(x) + \Psi_{A \cap C}(x) - \Psi_{(A \cap B) \cap (A \cap C)}(x) &&\\
&= \Psi_{(A \cap B) \cup (A \cap C)}(x) && \text{(By property 6)}
\end{aligned}$$

EXAMPLE 1.55. Show that $\sim \sim A = A$

Solution :

$$\begin{aligned}
\Psi_{\sim \sim A}(x) &= 1 - \Psi_{\sim A}(x) &&\\
&= 1 - [1 - \Psi_A(x)] && \text{(By property 7)}\\
&= \Psi_A(x) &&
\end{aligned}$$

EXAMPLE 1.56. Show that (*i*) $(A \cup B)' = A' \cap B'$, (*ii*) $(A \cap B)' = A' \cup B'$

Solution : Proof (*i*) $(A \cup B)' = A' \cap B'$

Case I : $\Psi_{A \cup B} = 1$

$\Rightarrow x \in (A \cup B)'$

$\Rightarrow x \notin (A \cup B)$

$\Rightarrow x \notin A$ and $x \notin B$

$\Rightarrow x \in A'$ and $x \in B'$

$\Rightarrow x \in A' \cap B'$

$\Rightarrow \Psi_{A' \cap B'} = 1$

$\Rightarrow \Psi_{(A \cap B)'} = \Psi_{A' \cap B'}$

By Property (4) $(A \cup B)' = A' \cap B'$

Case II : $\Psi_{(A \cup B)'} = 0$

$\Rightarrow x \notin (A \cup B)'$

$\Rightarrow x \in A \cup B$

$\Rightarrow x \in A$ or $x \in B$ or $x \in A \cap B$

$\Rightarrow x \notin A'$ and $x \notin B'$

$\Rightarrow x \notin A' \cap B'$

$\Rightarrow \Psi_{A' \cap B'} = 0$

$\Rightarrow \Psi_{(A \cup B)'} = \Psi_{A' \cap B'}$

By Property (4)

$$(A \cup B)' = A' \cap B'$$

Part (*ii*) is left for the reader.

The characteristic function can be used to name the subsets of a finite set as given in the following example.

EXAMPLE 1.57. Consider a set $A = \{1, 2, 3\}$.

The subsets of A are ϕ, {1}, {2}, {3}, {1,2}, {1, 3}, {2, 3}, {1, 2, 3}.

The values of the characteristic functions of these subsets are given by the following table which consists of three binary digits or binary triplets.

x	ϕ	{1}	{2}	{3}	{1,2}	{1, 3}	{2, 3}	{1, 2, 3}
1	0	1	0	0	1	1	0	1
2	0	0	1	0	1	0	1	1
3	0	0	0	1	0	1	1	1

= {(0, 0, 0), (0, 0, 1), (0, 1, 0), (1, 0, 0), (1, 1,0),

Let B = (1, 0, 1), (0, 1, 1), (1, 1, 1)}

$P(A)$ = {Q, {1}, {2}, {3}, {1, 2}, {1, 3}, {2, 3}, {1, 2, 3}}

One can define a mapping from the power set of A to the B. This mapping is one one and onto and hence describes a one-to-one correspondence between the power sets of A and the set B.

1.15. SPECIAL TYPE FUNCTIONS

Even function

A function $f : A \rightarrow B$ is called an even function if $f(-x) = f(x)$ for all $x \in A$.

e.g.,

$$f(x) = x^2 + 4$$
$$f(-x) = (-x)^2 + 1 = x^2 + 4$$
$$\therefore \quad f(x) = f(-x)$$
$$f(1) = 1^2 + 4 = 5$$
$$f(-1) = (-1)^2 + 4 = 5$$

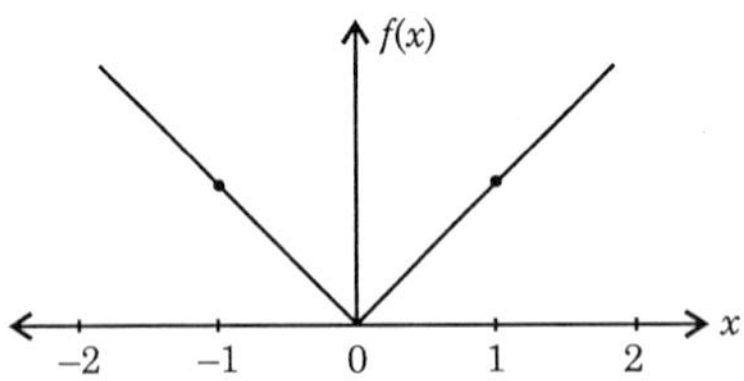

Fig. 1.18.

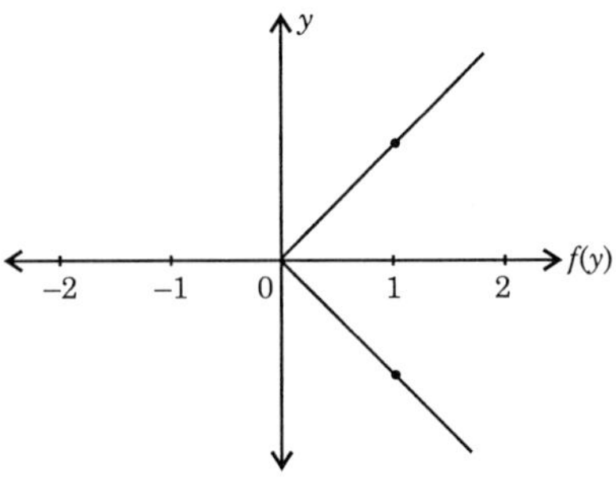

Fig. 1.19.

The graph of an even function is **symmetric** about the function axis.

Alternatively,

In fig. 1.18, for the function $y = f(x)$, if we take any positive value of x, *e.g.,* $x = a$, $y = f(a)$. For the same value of x, but in the negative axis *i.e.,* $x = -a$, we get $y = f(-a)$. In the graph, we can observe that, $f(a) = f(-a)$ *i.e.,* for both positive and negative values of x we get a positive value of y. Thus, $y = f(x)$ is an even function.

Similar argument holds true if we get a negative value of y for both negative and positive values of x, *i.e.,* $y = -|f(x)|$.

Similarly, in fig. 1.19, x is a function of y and the same arguments holds for $x = f(y)$.

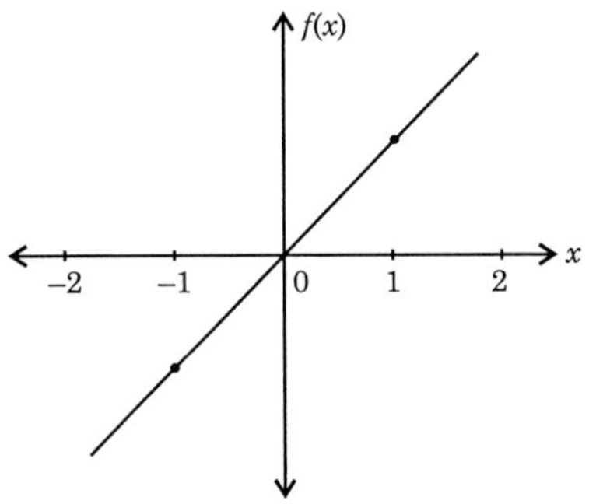

Fig. 1.20.

Odd function

A function $f: A \to B$ is called an odd function if $f(-x) = -f(x)$ for all $x \in A$.

e.g.,
$$f(x) = x^3$$
$$f(-x) = (-x)^3 = -x^3$$
$$\therefore f(x) = -f(-x)$$
$$f(1) = 1^3 = 1$$
$$f(-1) = (-1)^3 = -1$$

The graph of an odd function is a *double reflection*. first in the function axis and then in the other axis.

Alternatively,

For the function $y = f(x)$, if we take any positive value of x, for example, $x = a$, $y = f(a)$. For the same value of x but in the negarive axis, *i.e.*, $x = -a$ we get $y = f(-a)$. In the graph, we can see that $f(-a) = -f(a)$.

Note : There are functions, which are neither even nor odd.

e.g.,
$$f(x) = x^3 + x^2 + x + 5.$$

Inverse function

A one-one onto function $f: A \to B$ has an inverse function f^{-1} if for each $y \in B$, $f^{-1}(y) \in A$ and is unique.

$f^{-1}: B \to A$ is a function defined by

$f^{-1}(y) = x$ if and only if $f(x) = y$

f^{-1} is the inverse of function of f.

e.g., $f: A \to B$, $f^{-1}: B \to A$

$f(x) = 2x$, where $x \in A$

$f^{-1}(x) = \dfrac{x}{2}$, where $x \in B$

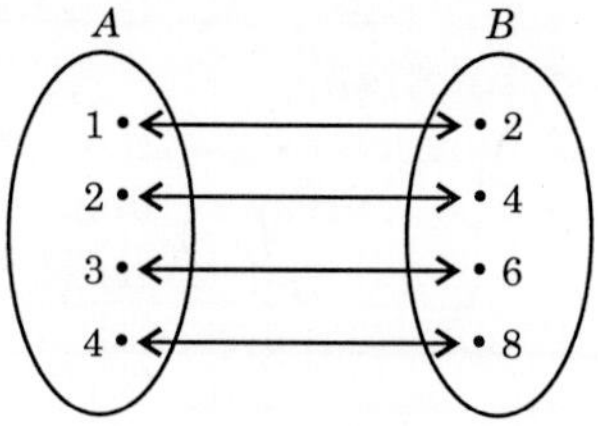

Fig. 1.21.

Note : Inverse function should not be confused by interpreting inverse of $f(x)$ as $f\left(\dfrac{1}{x}\right)$.

Constant function

A function $f: A \to B$ is a constant function if each element of domain is associated with a single element of the co-domain B *i.e.*, the range of the function is a singleton set, $f(x)$ = constant.

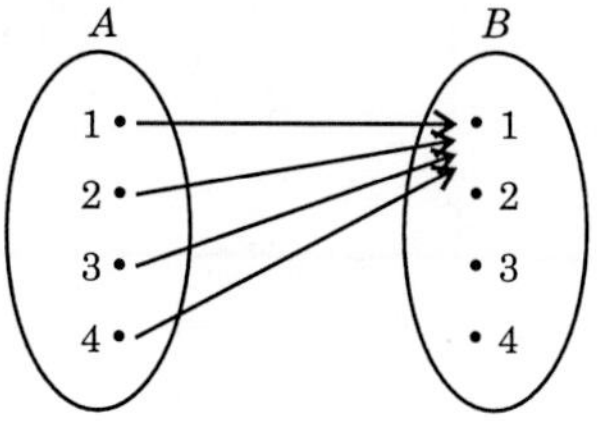

Fig. 1.22.

Graphically, a constant function will be a straight line parallel to either of the two axes. *e.g.*, $f(x) = 3$, *i.e.*, $y = 3$

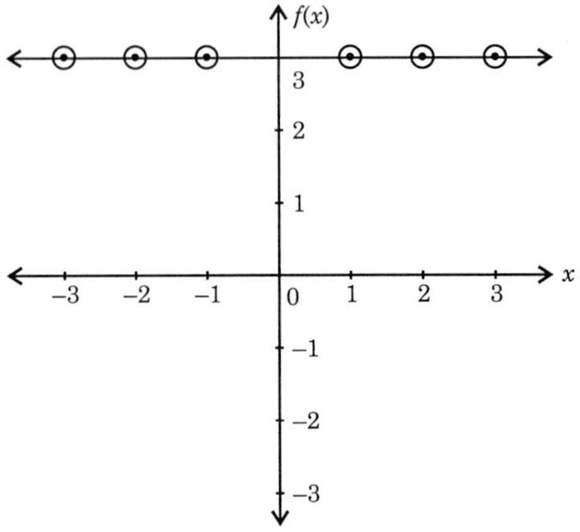

Fig. 1.23.

x	1	2	3	–1	–2	–3
$f(x)$	3	3	3	3	3	3

Step function

A function $f : A \to B$ is a step function, also called an **integer function**, if $f(x) = [x]$ *i.e.*, greatest integer less than or equal to x for all $x \in A$.

e.g., If $f(x) = [x]$, then $f([1.\ 35]) = 1$, $f([2.89]) = 2$, $f([3]) = 3$, $f([-2.29]) = -3$ etc.

Graphical representation of a step function is as shown below :

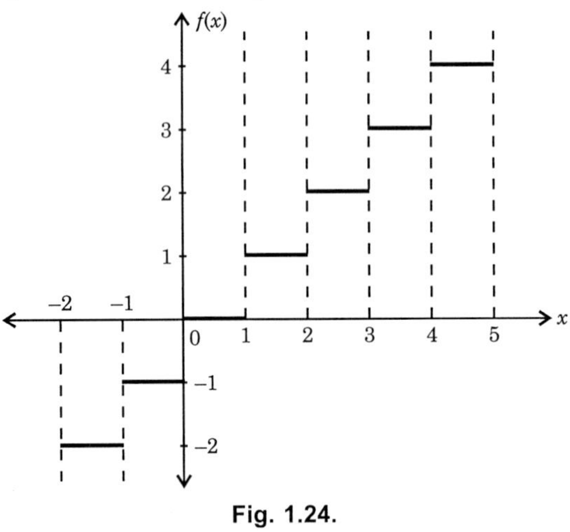

Fig. 1.24.

$$f(x) = 0, \text{ where } 0 \le x < 1$$
$$f(x) = 1, \text{ where } 1 \le x < 2$$
$$f(x) = -1, \text{ where } -1 \le x < 0$$

Modulus function

A function $f: A \rightarrow B$ is called a modulus function if $f(x) = |x|$ *i.e.*, $f(x)$ takes only the magnitude of x.

$$\begin{aligned} f(x) &= -x, \text{ if } x < 0 \\ &= x \text{ if } x \geq 0 \end{aligned}$$

The graph of a modulus function, $f(x) = |x| = -1$ is as shown below :

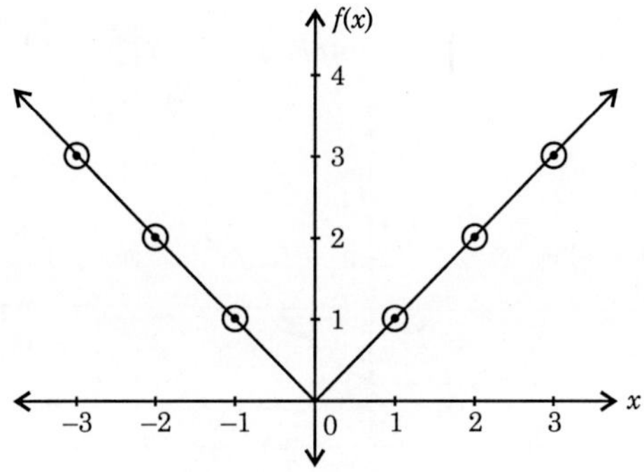

Fig. 1.25.

$$\begin{aligned} f(1) &= f(-1) = 1 \\ f(2) &= f(-2) = 2 \\ f(3) &= f(-3) = 3 \end{aligned}$$

Note : A modulus function is also an even function.

Linear function

A function $f: A \rightarrow B$ of the form $f(x) = ax + b$, where a and b are real numbers is called a linear function.

The graph of a linear function is always a straight line.

e.g., Graph of $f(x) = 4x - 2$ is as shown below:

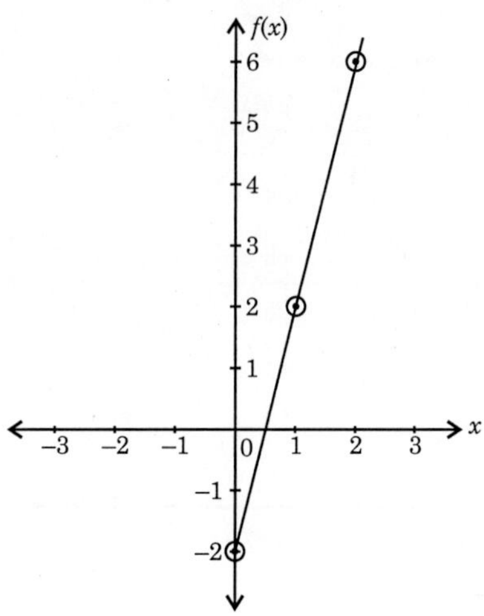

Fig. 1.26.

x	1	0	2
$f(x)$	2	–2	6

Exponential function

A function $f : A \to B$ of the form $f(x) = a^x$, where $a > 0$ and $x \in R$ is called an exponential function.

The graph of the exponential function $f(x) = 2^x$ is shown below :

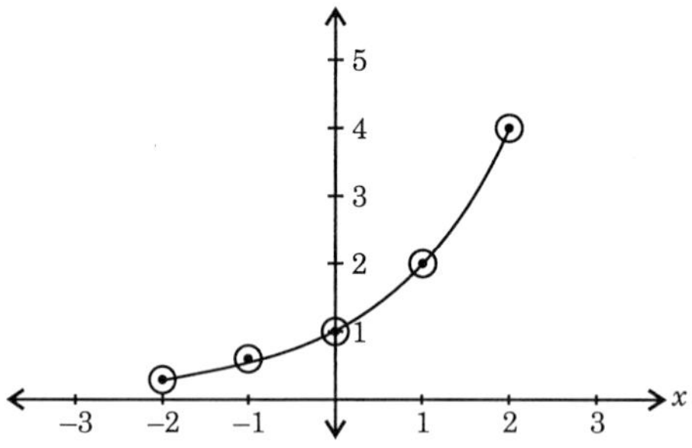

Fig. 1.27.

x	–2	–1	0	1	2	∞	$-\infty$
$f(x)$	1/4	1/2	1	2	4	∞	0

Note : The exponential functions of 2^x will never be negative for any value of x. The least value that it can have is 0 for x tends to negative infinity.

Quadratic function

A function $f : A \to B$ is called a quadratic function if it is of the form $y = ax^2 + bx + c$, where a, b and c are real numbers and $a \neq 0$, *i.e.*, quadratic function is nothing but a polynomial of degree 2.

The graph of such a function is a curve called parabola.

Graph of $f(x) = 2x^2$ is as shown below :

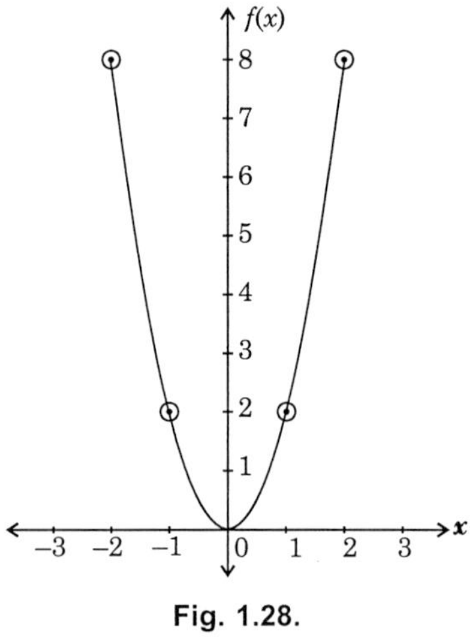

Fig. 1.28.

x	-2	1	-1	2
$f(x)$	8	2	2	8

SOME MORE EXAMPLES

EXAMPLE 1.58. Find $f(4)$ if $f(x) = x^4 - 3x^3 + 6x^2 - 10x + 16$

Solution :
$$f(x) = x^4 - 3x^3 + 6x^2 - 10x + 16$$
$$f(4) = (4)^4 - 3(4)^3 + 6(4)^2 - 10(4) + 16$$
$$= 256 - 192 + 96 - 40 + 16 = 136.$$

EXAMPLE 1.59. Find $f(-1.5)$, if $f(z) = 2^{z-2}$

Solution :
$$f(z) = 2^{z-2}$$
$$f(-1.5) = 2^{-1.5-2} = 2^{-3.5}$$
$$= 2^{-\frac{7}{2}} = \frac{1}{\sqrt{128}} = \frac{1}{8\sqrt{2}}$$

EXAMPLE 1.60. $f(t) = 2t^2 + \dfrac{2}{t^2} + \dfrac{5}{t} + 5t$. Find $f\left(\dfrac{1}{t}\right)$.

Solution :
$$f(t) = 2t^2 + \frac{2}{t^2} + \frac{5}{t} + 5t$$
$$f\left(\frac{1}{t}\right) = 2\left(\frac{1}{t}\right)^2 + \frac{2}{\left(\frac{1}{t}\right)^2} + \frac{5}{\left(\frac{1}{t}\right)} + 5\left(\frac{1}{t}\right)$$
$$= \frac{2}{t^2} + 2t^2 + 5t + \frac{5}{t} = f(t)$$

EXAMPPLE 1.61. If $f(x) = 2x$ and $g(x) = x + 1$. Find $(gof)\ (x)$.

Solution : $(gof)\ (x) = g(f(x)) = g(2x) = (2x) + 1 = 2x + 1$

EXAMPLE 1.62. If $f(x,y) = 3x^2 - 2xy - y^2 + 4$, find $f(1, -1)$.

Solution :
$$f(x, y) = 3x^2 - 2xy - y^2 + 4$$
$$f(1, -1) = 3(1)^2 - 2(1)(-1) - (-1)^2 + 4 = 3 + 2 - 1 + 4 = 8$$

EXAMPLE 1.63. If $f(x) = 3x - 5$ and $f[g(x)] = 2x$, then find $g(x)$.

Solution :
$$f(x) = 3x - 5$$
$$f[g(x)] = 3g(x) - 5 = 2x$$
$$\therefore \quad 3g(x) = 2x + 5$$
$$g(x) = \frac{2x+5}{3}$$

EXAMPLE 1.64. If $f(x) = ax^2 + bx + 2$. $f(1) = 3$ and $f(4) = 42$, find 'b'

Solution :
$$f(x) = ax^2 + bx + 2$$
$$f(1) = a(1)^2 + b(1) + 2 = 3$$
$$\therefore \quad a + b + 2 = 3$$

$$\therefore \quad a + b = 1 \qquad ...(i)$$

$$f(4) = a(4)^2 + b(4) + 2 = 42$$

$$\therefore 16a + 4b + 2 = 42$$

$$16a + 4b = 40 \qquad ...(ii)$$

Solving (*i*) and (*ii*) simultaneously, $b = -2$

EXAMPLE 1.65. Is $f(x) = 2x^2 - 4$ an even or odd function?

Solution : $f(x) = 2x^2 - 4$

$$f(-x) = 2(-x)^2 - 4 = 2x^2 - 4$$

$\therefore f(x) = f(-x)$ Hence, $f(x) = 2x^2 - 4$ is an even function.

For example 1.66 and 1.67 : $f(x) = 2x^2 + 1$ if $x \geq 1$

$$= \frac{2}{x} + 1 \text{ if } 0 < x < 1$$

$$= 3|x| \text{ if } x \geq 0$$

EXAMPLE 1.66. Which of the following is true ?

(*i*) $f\left(\frac{1}{2}\right) < f(1)$ (*ii*) $f(-2) = f(2)$ (*iii*) $f(-3) = f(2)$

Solution :

$$f\left(\frac{1}{2}\right) = \frac{2}{\frac{1}{2}} + 1 = 5$$

$$f(1) = 2(1)^2 + 1 = 3$$

$$f(2) = 2(2)^2 + 1 = 9$$

$$f(-2) = 3|-2| = 6$$

$f(-3) = 3|-3| = 9$ $\therefore f(-3) = f(2)$ is true. $\therefore$ (*iii*) is true.

EXAMPLE 1.67. Find $f[f(x)]$ if $x < 1$ land x is an integer.

Solution : $f(x) = 3|x|$ as $x < 1$

Now, $3|x| > 3$

$$\therefore \quad f[f(x)] = 2[3|x|]^2 + 1 = 2 \times 9x^2 + 1 = 18x^2 + 1$$

For examples 1.68 to 1.70 : The funcrtion of f and g are defined for natural numbers.

$$f(x) = 1 + x^2, \text{ if } x \text{ is even}$$

$$= (1 + x)^2, \text{ otherwise}$$

$$g(x) = (x - 1)^2, \text{ if } x \text{ is prime}$$

$$= x^2 - 1, \text{ otherwise}$$

EXAMPLE 1.68. Find $f[g(x)]$, if $x = 3$.

Solution : $g(x) = (3 - 1)^2, = 4$

$$f[g(x)] = f(4) = 1 + 4^2 = 17$$

EXAMPLE 1.69. For which number is $f[g(x)] = 1 + [(x - 1)^2]^2$ true.

Solution : $g(x) = (x - 1)^2$ which implies that x is prime.

$$f[g(x)] = 1 + [(x - 1)^2]^2 \; i.e.,$$

$f(x) = 1 + x^2$ is applied *i.e.*, $(x - 1)^2$ is even.

The only prime which is even, is 2 for which $(x - 1)^2 = (2 - 1)^2 = 1$ *i.e.*, odd

$\therefore f[g(x)] = 1 + [(x - 1)^2]^2$ is applicable for all prime numbers except 2.

EXAMPLE 1.70. Which of the following is not true?

(*i*) $f(2) + g(1) = f(2)$ (*ii*) $f(5) = g(7)$ (*iii*) $\dfrac{g(3)}{f(3)} = 4$

Solution :

$$g(1) = 1^2 - 1 = 0$$

$$\therefore f(2) + g(1) = f(2) \qquad \therefore (i) \text{ is true.}$$

$$f(5) = (1 + 5)^2 = 6^2$$

$$g(7) = (7 - 1)^2 = 6^2 \qquad \therefore (ii) \text{ is true.}$$

$$g(3) = (3 - 1)^2 = 2^2 = 4$$

$$f(3) = (1 + 3)^2 = 4^2 = 16$$

$$\therefore \frac{g(3)}{f(3)} = \frac{4}{16} = \frac{1}{4} \qquad \therefore (iii) \text{ is not true.}$$

For example, 1.71 and 1.72 : Determine whether the graphs given below are even or odd or otherwise with respect to each other.

EXAMPLE 1.71.

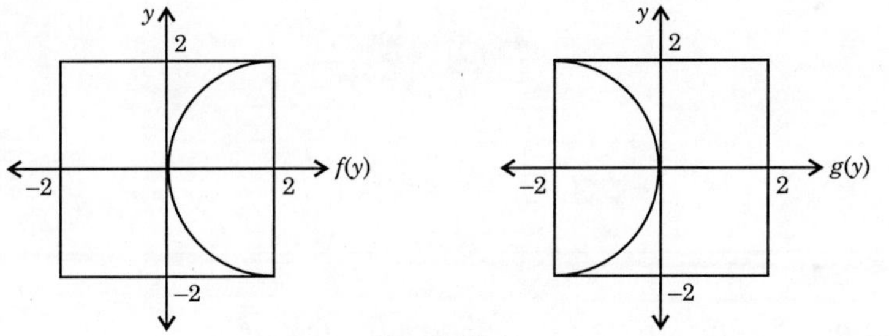

Fig. 1.29.

Solution : As the graph $g(y)$ is a double reflection of the function $f(y)$, first about the function axis $f(x)$ and then about the other axis *i.e.*, y, it represents an odd function.

Alternatively,

$f(y)$ and $g(y)$ by themselves are even function. $f(y)$ and $g(y)$ are curves of a parabola, where $g(y) = -y^2$ and $f(y) = y^2$.

$\therefore g(y) = -f(y)$ and $f(y) = -g(y)$.

Thus, $g(y)$ and $f(y)$ are odd with respect to each other.

EXAMPLE 1.72.

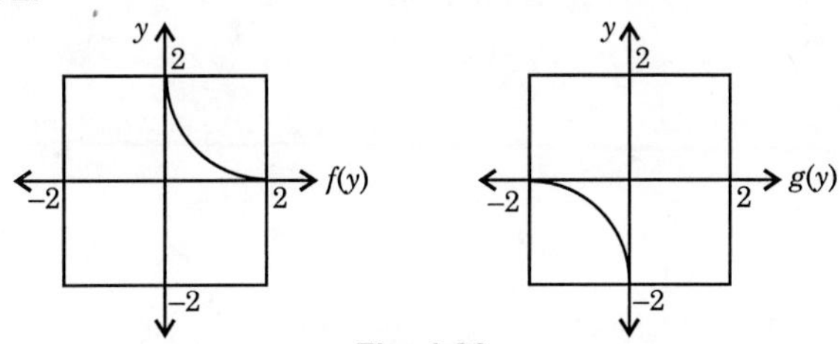

Fig. 1.30.

Solution : As the graph $g(y)$ is a double reflection of the function $f(y)$, first about the function axis $f(x)$ and then about the other axis, *i.e.*, y. it represents an odd function.

Alternatively,

For positive y, *e.g.*, $y = a$, we get $f(y)$ as positive and for $y = -a$, $g(y)$ as negative. The nature of $f(y)$ and $g(y)$ is the same but they are in the first and third quadrants, respectively. Thus, $|f(a)| = -|g(a)|$.

$\therefore$ The functions $f(y)$ and $g(y)$ are odd respect to each other.

EXAMPLE 1.73. Show that if n is an odd integer,

$$\left\lceil \frac{n^2}{4} \right\rceil = \left(\frac{n^2+3}{4} \right)$$

Solution: If n is an odd integer, $n = 2k + 1$ for some integer k

$$\frac{n^2}{4} = \frac{(2k+1)^2}{4} = \frac{4k^2+4k+1}{4} = k^2 + k + \frac{1}{4}$$

k is an integer so $k^2 + k$ is also an integer.

Therefore, ceiling $\left(\frac{n^2}{4}\right) = k^2 + k + 1$. Now

$$\frac{n^2+3}{4} = \frac{(2k+1)^2+3}{4} = k^2 + k + 1$$

So, ceiling $\left(\frac{n^2}{4}\right) = \left(\frac{n^2+3}{4}\right)$.

EXERCISE 2

Directions for questions 1 to 5 : Solve the following:

1. If $f(x) = \sqrt{x+1}$ then find the value of $f(a^2 + 2a)$
2. If $f : X \to Y$ such that $f(x) = x + 1$ where $x \in X$ and $g : Y \to Z$ such that $g(y) = y^2 + 3$ when $y \in Y$, find $(gof)(-2)$.
3. If $f(x) = 1$ If x is a rational number
   $= 0$ otherwise

   Find the value of $f(\pi)$ and $f\left(\frac{22}{7}\right)$.
4. If $f(x) = x^4 - 13x^2 - 2x - 15$ then, find $f(5)$.
5. If $f(x, y, z) = x^y y^2 + xy^z - yz^y + 15z^y$ then find $f(4, 3, 2)$.

Directions for questions 6 to 8 : Choose the correct alternative.

6. $f(x) = x + 5$

   $g(x) = 2x + 4$

   Find $f[g(x)] - g|f(x)]$

   1. $4x - 5$ 2. $4x + 23$

3. -5 4. Cannot be determined

7. $f(x) = ax^2 + bx + 2$ and $f(1) = 3$. $f(3) = -7$. Find the value of a and b
   1. $-3, -2$ 2. $-2, 3$ 3. $-3, 2$ 4. $2, 3$

8. Find x if $f(x) = 1 + x - x^2$ and $f(x + 1) = f(x + 2)$
   1. 0 2. 1 3. -1 4. 2

Directions for questions 9 and 10 : Refer to the function defined below:

$f(x) = (x - 1)(x - 2)(x + 3), -2 \le x \le 4$

$= (5 - x)$ otherwise

$g(x) = x^3 - 1, x \ge 0$

$= \dfrac{1}{x}$ when $x < 0$

9. Find the values of x for which $f(x) = 0$
   1. 3, 5 2. 1, 2, -3, 5
   3. $-1, -2, 3, -5$ 4. 1, 2, 5

10. Which of the following is / are true ?
    (*i*) If $x = 1$, $f(x) = g(x)$
    (*ii*) $f(6) = g(-1)$
    (*iii*) $\dfrac{f(0)}{g(0)} = -6$
    1. (*i*) and (*iii*) 2. (*ii*) and (*iii*)
    3. (*i*), (*ii*) and (*iii*) 4. (*i*) and (*ii*)

Directions for questions 11 to 15 :

Let 'x' and 'y' be real numbers and let $f(x, y) = |x + y|$; $F[f(x, y)] = -f(x, y)$; $G[f(x, y)] = -F[f(x, y)]$

11. $G[f(x, y)] = ?$
    1. $-|x + y|$ 2. $F(f(x, y))$ 3. $f(x, y)$ 4. None of these

12. $G[f(x, -x)] = ?$
    1. $f(x, -x)$ 2. $F[f(x, -x)]$ 3. 0 4. All of these

13. $G[f(x, x)] + f(x, x) = ?$
    1. $F[f(2x, 2x)]$ 2. $F[f(2x, -2x)]$ 3. $G[f(2x, -2x)]$ 4. $G[f(2x, 2x)]$

14. $[f(x, y)]^2 = ?$
    1. $(x + y)^2$ 2. $[F(f(x, y))]^2$ 3. $[G(f(x, y))]^2$ 4. All of these

15. $f(x, y) - G[f(x, y)] = ?$
    1. $2F[f(x, y)]$ 2. $G[f(x, y)] - F[f(x, y)]$
    3. 0 4. $F[f(x, y)] - G[f(x. y)]$

Directions for question 16 to 20 : A pair of graphs F(x) and F′ (x) is given.

If $F'(x) = F(x)$, choose the answer *a*.

If $F'(x) = F(-x)$, choose the answer *b*.

If $F'(x) = -F(-x)$. choose the answer *c*.

And if none of the above is true, choose the answer d.

16. 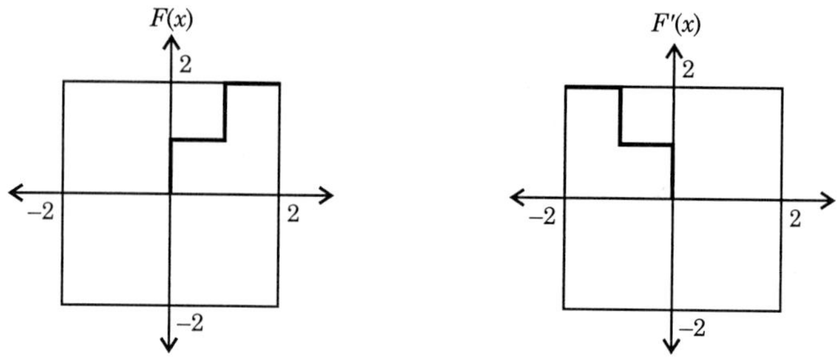

1. a 2. b 3. c 4. d

17. 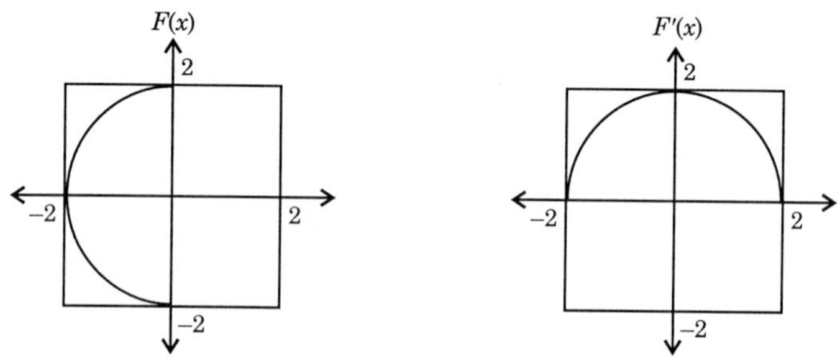

1. a 2. b 3. c 4. d

18. 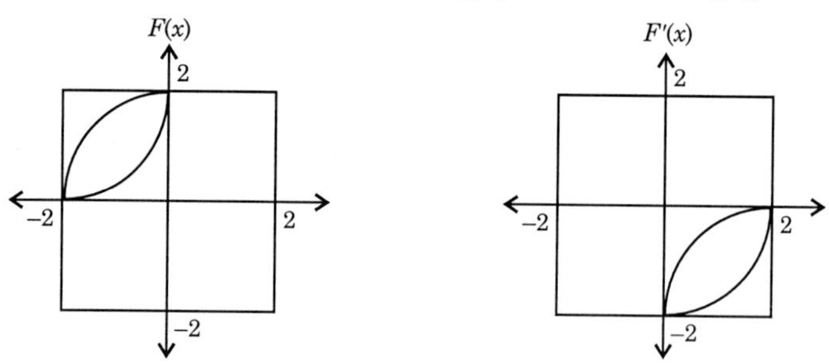

1. a 2. b 3. c 4. d

19. 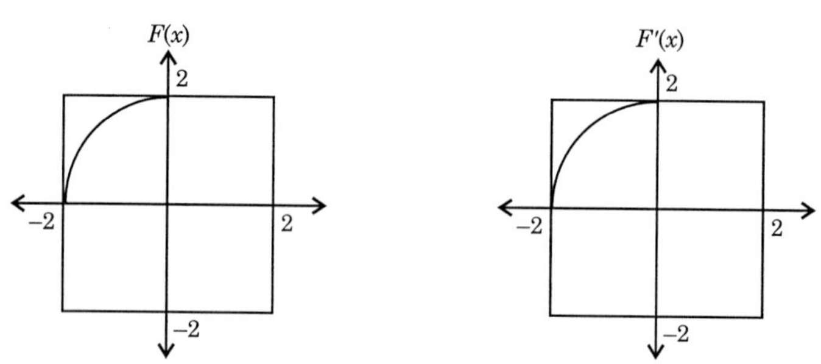

1. a 2. b 3. c 4. d

20. 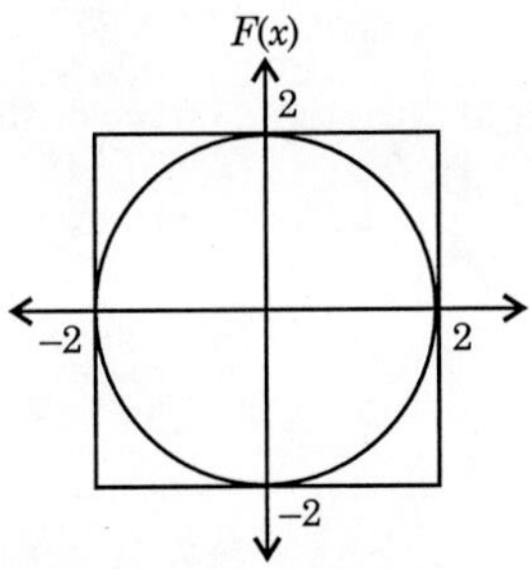

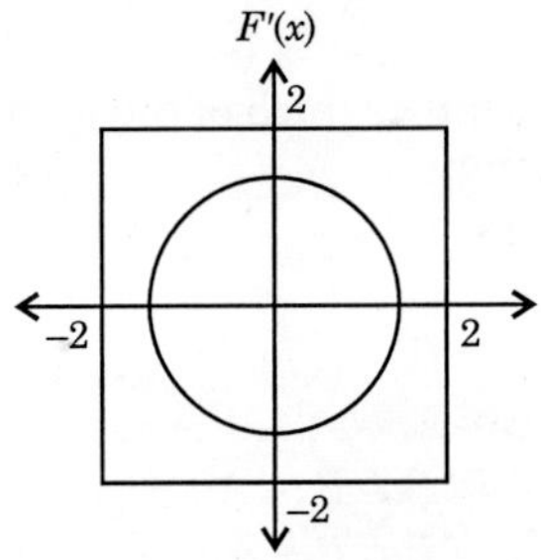

1. *a* 2. *b* 3. *c* 4. *d*

EXERCISE 3

1. Given that ∪ = all students at a university.
   A = day students.
   B = Mathematics majors
   C = graduate students
   Draw individual Venn diagram to illustrate the following sets.
   (*a*) Evening students.
   (*b*) Undergraduate mathematics majors.

2. Let the set A, B, C and ∪ be as in the problem 1.
   Let # ∪ = 16000
   # A = 9000
   # B = 300
   # C = 1000
   Also assume that the number of days students are mathematics majors is 250, 50 of which are graduate students and that the total number of day graduate students is 700. Determine the number of students who are :
   (*a*) Evening students.
   (*b*) Non-mathematics students.
   (*c*) Undergraduates (day or evening)
   (*d*) Day graduate non-mathematics majors
   (*e*) Evening graduate students.
   (*f*) Evening graduate mathematics majors.
   (*g*) Evening undergraduate non-mathematics majors.

3. In the following problem, describe the set by listing its elements.
   (*a*) $\{x \in R \mid x^2 = 3\}$
   (*b*) $\{x \in z \mid xy = 60 \text{ for some } y \in z\}$
   (*c*) $\{x \in z\}\ xy = 60 \text{ for some } y \in z\}$
   (*d*) $\{x \in z \mid x^2 - x < 115\}$

4. Let $A = \{0, 2, 3\}$
   $B = \{2, 3\}$
   $C = \{1, 5, 9\}$

$D = \{3, 2\}$
$E = \{2, 3, 2\}$

Determine which of the following statements are true. Give reasons for your decisions.

(*a*) $A = B$
(*b*) $B = C$
(*c*) $B = D$
(*d*) $B = E$
(*e*) $A \cap B = B \cap A$
(*f*) $A \cup B = B \sim A$
(*g*) $A \sim B = B \sim A$
(*h*) $A \times B = B \times A$

5. Let $\cup = \{0, 1, 2, 3, 4, 5, 6, 7, 8, 9\}$
$A = \{x \in \cup : x \text{ is a multiple of } 3\}$
$B = \{x \in \cup : x^2 - 5 \geq 0\}$

Determine

(*a*) $A \cup B$
(*b*) $A \cap B$
(*c*) B^C

6. Let $A = \{1, 2, 3, 4, 5, 6\}$
$B = \{3, 4, 5, 6, 7, 8\}$
$C = \{6, 7, 8, 9, 10, 11\}$
$D = \{2, 4, 6, 8, 10\}$

Find.

(*a*) $A \oplus B$
(*b*) $A \oplus C$
(*c*) $B \oplus D$
(*d*) $C \oplus D$
(*e*) $A \oplus D$
(*f*) $A \cap (B \oplus C)$
(*g*) $(A \cap B) \oplus (C \cap D)$
(*h*) $(A \oplus B) \cap (C \oplus D)$
(*i*) $(A \oplus B) \sim (C \oplus D)$
(*j*) $(A \cup B) \oplus (C \cup D)$
(*k*) $A \oplus B \oplus C$

7. Prove the following results.

(*a*) $A \cap (B \oplus C) = (A \cap B) \oplus (A \cap C)$
(*b*)
$$\begin{aligned} A \oplus B &= (A \sim B) \cup (B \sim A) \\ &= (B - A) \cup (A \sim B) \\ &= B \oplus \text{A} \end{aligned}$$

8. Find the power set $P(A)$ of $A = \{a, b, c, d, e\}$

9. Find the power set $P(A)$ of $A = \{\{1, 2\}, \{3\}, \{4, 5, 6\}\}$

10. Consider the class of sets $B = \{1, 3, 5\}, \{2, 4, 6\}, \{0\}$.
    List the elements of B and determine wheter each of the following statement is true or false.
    (*a*) $\phi \subseteq B$
    (*b*) $5 \in B$
    (*c*) $\{1, 3, 5\} \subseteq B$
    (*d*) $\{\{1, 3, 5\}, \{0\}\} \subseteq B$
    (*e*) $\{0, 1, 2, 3, 4, 5, 6\} \in B$
    (*f*) $\{0\} \in B$
11. Suppose that 100 of the 120 mathematices students at a college take at least one of the language English, French and German. Also suppose.
    65 study English — 20 study English and French.
    45 study French — 25 study English and German.
    42 study German — 15 study French and German.
    (*a*) Find the number of students who study all three languages.
    (*b*) Determine the number of students who study exactly one languge.
    (*c*) Determine the number of students who study exactly two languag.
    (*d*) Give the number of students who study English and French but not German.
12. Is the power set $P(\phi)$ of the empty set ϕ empty ?
13. Find the number of elements in the power set of each of the following sets :
    (*a*) $\{x : x$ is a day of the weak$\}$
    (*b*) $\{x : x$ is a positive divisor of $6\}$
    (*c*) $\{x : x$ is a season of the year$\}$
    (*d*) $x : x$ is a letter in the word "India"$\}$.
14. List all the fundamental products of the three sets A, B and C and represent all of them by using Venn diagram.
15. Identify whether each of the following sets is infinite or finite.
    (*a*) {Days in a month}
    (*b*) {Positive integers}
    (*c*) {lines through the origin}
    (*d*) {Vertices in a prallogram}
    (*e*) {Weeks in the leap year}
    (*f*) {Number of pigeons in India}
    (*g*) $\{2, 4, 6, 10, 12, 14, 16, 18, 20\}$
    (*h*) $\{x \mid x$ is a solution to $x^3 = 64\}$
    (*i*) $\{x \mid x > x + 3\}$
16. In a survey of 90 people, it was found that 30 read Hindustan Times, 40 read Times of India, and 25 reads . The Hindu. Also 15 read both Hindustan Times and times of India, 20 reads both Times of india and the Hundu and 30 reads both Hindustan Times and the Hindu and 10 reads all the three Newspapers.

(*a*) Give Venn Diagram representation of the given problem.
(*b*) Find the number of people who read no newspaper at all.
(*c*) Determine the number of people who read exactly one magazine.

17. Let $\cup = \{1, 2, 3, 4, 5, 6, 7, 8, 9\}$
A = Set of even numbers in $\cup$.
B = Set of prime number in $\cup$.
C = Set of odd numbers in $\cup$.
Find :
(1) $A \cup B$
(2) $A \cup C$
(3) $B \cup C$
(4) $A \cap B \cap C$
(5) $A \cap C$
(6) $A \oplus B$
(7) $A - C$
(8) $A \cup (B \cap C)$ and also verify De' Morgan's law.
(9) A^C, B^C, C^C
(10) $(A \cup B)^C$
(11) $(A \cap B) \sim C$
(12) $(A \sim B)^C$

18. Give the following operations by using Venn Diagram.
(1) $A \cap B^C$
(2) $A \cap (B \cup C)$
(3) $(A \cup B) \cap (A \cup C)$
(4) $A \cup B^C \cup C$
(5) $(A \cup B) \sim C$

Answers for Exercise 3

2. (*a*) 7000 (*b*) 157000 (*c*) 15000 (*d*) 650
(*e*) 300

3 (*a*) $\left\{-\sqrt{3}, \sqrt{3}\right\}$ (*b*) ϕ
(*c*) $\{\pm 1, \pm 2, \pm 3, \pm 4, \pm 5, \pm 6, \pm 10, \pm 12, \pm 15, \pm 20, \pm 30, \pm 60\}$
(*d*) ϕ.

4. (*a*) False (*b*) False (*c*) True (*d*) True
(*e*) True (*f*) True (*g*) False (*h*) False

5. (*a*) $\{3,6,9,4,5,7,8\}$ (*b*) $\{3, 6, 9\}$ (*c*) $\{0, 1, 2)$

8. $P(A)$ = List of all the subsets of A
There must be $2^5 = 32$ elements Q in $P(A)$
$P(A) = \{\phi, \{a\}, \{b\}, \{c\}, \{d\}, \{e\},$
$\{a, b\}, \{a, c\}, \{a, d\}, \{a, e\}, \{b,c\}, \{b,d\}, \{b,e\}, \{c,d\}, \{c,e\}, \{d,e\},$
$\{a, b, c\}, \{a, b, d\}, \{a, b, e\}, \{a, c, d\}, \{a, c, e\}, \{a,d, e\},$

{b,c, d}, {b,c,e} {b, d, e}, {c, d, e},
{a, b, c, d}, {a, b, c, e}, {a, c, d, e}, {a, b, d, e},
{b, c, d, e},
{a, b, c, d, e}}

9. Since A has 3 elements therefore
   $P(A)$ has $2^3 = 8$ elements.
   $P(A) = \{A, \{\{1, 2\}, \{3\}\}, \{\{1, 2\}, \{4,5,6\}\}, \{\{3\} \{4,5, 6\}\}, \{1, 2\}, \{3\}, \{4, 5, 6\}, \phi\}$

10 (*a*) True. As empty set is a subset of every set.
   (*b*) False, while 5 is an element of one of the sets which is an element of the class of sets B, it is not one of the elements of B.
   (*c*) False, as {1, 3, 5} is as element not a subset.
   (*d*) True. (*e*) False (*g*) True.

12. No, $P(\phi) = \{\{\phi\}\}$

13. (*a*) $2^7 = 128$
   (*b*) the positive divisors of 6 are 1, 2, 3, 6.
   then required number is
   $2^4 = 16$
   (*c*) $2^4 = 16$
   (*d*) $2^5 = 32$

15. (*a*) Finite. (*b*) Infinite. (*c*) Infinite. (*d*) finite.
   (*e*) finite. (*f*) Infinite. (*g*) finite. (*h*) finite.
   (*i*) finite

17. $A = \{1, 4, 6, 8\}, B = \{2, 3, 5, 7\}$
   $C = \{1, 3, 5, 7, 9\}$
   (1) $A \cup B = \{2, 3, 4, 5, 6, 7, 8\}$ (2) $A \cup C = \{1, 2, 3, 4, 5, 6, 7, 8, 9\}$
   (3) $B \cup C = \{1, 2, 3, 5, 7, 9\}$ (4) $A \cap B \cap C = \phi$
   (5) ϕ (6) {4, 5, 6, 7, 8}
   (7) {2, 4, 6, 8} (8) {2, 3, 4, 5, 6, 7, 8}
   (9) {1, 3, 5, 7, 9} (10) {1, 9}
   (11) 2 (12) {1, 2, 3, 5, 7,9}

EXERCISE 4

1. $X = \{x : 0 < x < 17, x \text{ integral}\}$, $P = \{4, 8, 12, 16)$, and $Q = \{3, 6, 9, 12, 15\}$. Draw a Venn diagram to show these sets and their inter-relationships. Mark all elements in the correct regions of the diagram. From the diagram list $P \cap Q$ and $(P \cup Q)'$.

2. D = {Adam, Beryl, Chandrakant, Desmond, Elaine, Fearon, Guy, Henrietta ldris, Jacqueta, Kathryn, Leo}, thc set of data processing staff; the set of Cobol programmers, C = {Beryl, Desmond, ldris, Kathryn, Leo}, and the set of Fortran programmers, F = {Adam, Beryl, Desmond, Guy, Jacquetta, Leo}.

Show all this information in full detail using a Venn diagram with labels. Use this to answer the following questions.

(*i*) How many people can program in Fortran ?

(*ii*) Who can program both in Cobol and in Fortran ?

(*iii*) Who can program in neither Cobol nor Fortran ?

(*iv*) How many people can program in Cobol but not in fortran ?
(The set is $C \cap F'$).

(*v*) How many people can program in Fortran but not in Cobol ?

(*vi*) List those who can program either in Cobol or in Fortran or in both.

3. $N = \{x : 12 < x < 29, x \text{ integral}\}$, $P = \{14, 21, 28\}$, $Q = \{18, 24\}$, and $R = \{16, 20, 24, 28\}$. Draw a Venn diagram to show these sets and their inter-relationships. Use the diagram to list the following.

(*i*) $P \cup Q$ (*v*) $(P \cup Q \cup R)'$

(*ii*) $P \cap Q$ (*vi*) $P \cap R$

(*iii*) $P \cap Q \cap R$ (*vii*) $Q \cap R$

(*iv*) $P \cap Q'$ (*viii*) $P \cup R'$

4. Write the elements of the following sets ... notation may be used, if necessary.

(*a*) $A = \{x : x \text{ is a positive odd integer}\}$

(*b*) $B = \{x : x \text{ is a multiple of 3 and } 0 < x < 12\}$

(*c*) $C = \{x : x^2 = 9\}$

(*d*) $D = \{x : x \text{ is a country whose first letter is } X\}$

5. Do you agree with the following statements ? Briefly explain your answer.

(*a*) $\phi \subseteq \phi$

(*b*) If A and B are two sets and the third set $C = \{x \mid x \in A \wedge x \in B)$, then $C \subset A$.

(*c*) If $A \subset B$ but A and B are disjoint, then the set $A = \phi$.

(*d*) $\{\phi\} \subseteq \{\phi\}$

(*e*) $\{a, b\} \subseteq \{a, b, c, \{a, b, c\}\}$

(*f*) $\{a, \phi\} \subseteq \{a, \{a, \phi\}\}$

(*g*) Every subset of an infinite set is infinite.

(*h*) $\{2, 3\} = \{2, 2, 3, 2\}$

(*i*) $\{2, 4, 2, 3, 4, 3\} \subset \{2, 3, 4\}$

(*j*) Every subset of a finite set is finite.

6. (*a*) If $A \subset B$ and $C = (x : x \in A \wedge x \in B)$, then $C = A$.

(*b*) If $A \subset B$ and $D = \{x : x \in A \vee x \in B)$, then $D = B$.(Connectives "$\wedge$" and "$\vee$" would mean "and" and "or", respectively).

7. (*a*) Prove that for any sets A and B, $A \cap B \subset A \subset A \cup B$

(*b*) Prove that $A \subset B$ if and only if $A \cap B = A$

(*c*) Determine $\{\phi\} \cap \{a, d, \{\phi\}\}$

(*d*) Let A, B, C be subsets of U. If $A \cap B = A \cap C$ and $\overline{A} \cap B = \overline{A} \cap C$, is $B = C$? If so, explain it.

(*e*) Let $(A \cap C) \subseteq (B \cap C)$

$(A \cap \overline{C}) \subseteq (B \cap \overline{C})$, show that $A \subseteq B$

(*f*) Given that $A \cup B = A \cup C$, is it necessary that $B = C$?
$A \cap B = A \cap C$, is it necessary that $B = C$?

8. (*a*) In the Venn diagram represent the following relations
(*i*) $X \sim Y$ (*ii*) $X^c \sim Y^c$ (*iii*) $X \cap Y^c$

(*b*) Use a Venn diagram to represent the associative property for intersection of three sets.

(*c*) Use a Venn diagram to illustrate De Morgan's laws
$(A \cap B)' = A' \cup B$; $(A \cup B)' = A' \cap B'$

(*d*) Let $A \subset B$, use a Venn diagram to show that $B' \subset A'$.

(*e*) Illustarte with Venn diagram the fact that if $A \subset B$, then $A \cap B = A$.

(*f*) Use a Venn diagram to show $(A \cup B)' \cap (A \cup C)'$

(*g*) On the Venn diagram represent $A \cap (B \cup C)$.

EXERCISE 5

1. Redefine each of the following sets, using set-builder notation.
(*a*) $\{-2, -4, -6, ...\}$
(*b*) $\{0, 3, -3, 6, -6, 9, ...\}$
(*c*) $\{2, 3, 4, 6, 8, 9, 10, 12, 14, 15, ...\}$
(*d*) $\left\{1, \frac{1}{2}, \frac{1}{3}, \frac{1}{4}, ...\right\}$
(*e*) $\{0, 1, 2, ..., 99, 100\}$

2. Let $A = \{x : x$ is rational and $0 \le x \le 2\}$, and let $B = \{x : x$ is rational and $1 \le x \le 3)$. Indicate whether each of the following is true or false.
(*a*) $0.5 \in A$ (*b*) $2^{1/2} \in B$
(*c*) $2.6 \in A$ or $2.6 \in B$ (*d*) $\exists x\{x \in A$ and $x \in B\}$

3. Determine which of the following pairs of sets are equal.
(*a*) $S = \{x : x$ is an integer divisible by both 3 and 2$\}$ and $Q = \{6, 12, 18, 24 ...\}$
(*b*) $X = \{x : x$ is real and $x^2 < x\}$ and $T = \{x : x$ is real and $0 < x < 1\}$

4. Answer true or false :
(*a*) $A' \cup B' = (A \cup B)'$ (*b*) $A' = \cup \backslash A$
(*c*) $A \cup (B \cup C) = (A \cup B) \cup C$ (*d*) $A \cup (B \cap C) = (A \cup B) \cap C$
(*e*) $A \backslash (B \backslash C) = (A \backslash B) \backslash C$

5. Prove that $A \cup A' = U$ and $A \cap A' = \phi$.

6. Let A and B be the following subsets of the real numbers : $A = \{x : 0 < x < 5\}$ and $B = \{x : 2 < x < 8\}$. Express $A \cup B$ as the union of three disjoint sets. (Use Proposition 2 of this section).

7. For each natural number k, let A_k be the subset of the real number defined by $A_K = \{x : -k \le x \le k\}$. Find each of the following:
(*a*) A_4 (*b*) $A_5 \cap A_6$ (*c*) $A_3 \cup A_7$
(*d*) $\bigcup_{k=1}^{x} A_k$ (*e*) $\bigcap_{k=1}^{\infty} A_k$

8. Let U be the set of real numbers. For each natural number n, let $A_n = \{x : -n \le x \le 1/n\}$. Find each of the following.

(a) $\left(\bigcup_{n=1}^{\infty} A_n\right)'$ (b) $\bigcup_{n=1}^{\infty} A_n'$

(c) $\left(\bigcap_{n=1}^{\infty} A_n\right)'$ (d) $\bigcap_{n=1}^{\infty} A_n'$

9. For each natural number m, let $A_m = \{x : 1/(m+2) \le x \le 1 - [1/(m+1)]\}$.

Let $B = \left\{x : x \le \frac{1}{3}\right\}$. Find each of the following:

(a) $B \cap \left(\bigcup_{m=1}^{\infty} A_m\right)$ (b) $\bigcup_{m=1}^{\infty} (B \cap A_m)$

(c) $B \cup \left(\bigcap_{m=1}^{\infty} A_m\right)$ (d) $\bigcap_{m=1}^{\infty} (B \cup A_m)$

10. List all the members of the power set of each of the following sets.
(a) $A = \{a, b, 2, 3\}$ (b) $B = \{\text{cat, dog, mouse}\}$
(c) $C = \{\{a\}, \{b\}\}$ (d) $D = \{\phi, \{\phi\}\}$

11. Let $A = \{a, b, c, d, e\}$, $B = \{a, b\}$, $C = \{B, \phi\}$, and $D = \{a, b, \{a, b\}\}$. Find $A \cap B$, $C \cap D$, $A \cap D$, $C \cap p(A)$, and $D \cap p(A)$. Indicate whether each of the following is true or false.
(a) $A \in p(A)$ (b) $C \subset p(A)$
(c) $D \subset p(A)$ (d) $B \supset D$
(e) $B \in D$ (f) $\{a, b\} \in C$

12. Prove that if $A \subset B$, then $p(A) \subset p(B)$.

13. Let $A = \{1, 2\}$, $B = \{a, b, c\}$, and $C = \{x, y\}$. List all the elements in each of the following sets.
$A \times C$, $B \times B$, $A \times B \times C$, $(A \times A) \times (C \times C)$

14. List of the some elements of each of the following sets. Express each a cartesian product or indicate why this cannot be done.
(a) $\{(x, y, z) : x$ and y are digits and z is any integer$\}$.
(b) $\{(x, y, z) : x, y,$ and z are real and $x^2 + y^2 + z^2 = 1\}$
(c) $\{(x_1, x_2, x_3, x_4,) :$ where x_k is real and $0 < x_4 < 1$ for $k = 1, 2, 3, 4\}$

15. List some of the elements in each of the following sets. Express each as a cartesian product or indicate why this cannot be done.
(a) $\{(x, y) : x$ is a positive integer and y is a negative integer$\}$
(b) $\{(x, y) : x$ is a digit and y is a letter$\}$
(c) $\{(x, y) : x \in N$ and $y = x + 1)\}$

16. For each of the following pairs of sets A and B, determine whether $A \subset B$ and, if so, determine whether the containment is proper.

(*a*) $A = \{x : x \text{ is real and } 0 \le x^2 \le 1\}$ and $B = \{x : -1 \le x \le 1)\}$
(*b*) $A = \{x : x$ is an integer multiple of 2$\}$ and $B = \{x : x$ is an integer multiple of 4$\}$
(*c*) $A = \{x : 2x$ is an integer multiple of 12$\}$ and $B = \{x : x$ is divisible by both 2 and 3$\}$

17. Prove : $B \subset A$ where $A = \{x : x^2$ is an integer multiple of 9$\}$ and $B = \{x : x$ is an integer divisible by 6$\}$.

18. Prove : $A = B$ where $A = \{x : x^2$ is divisible by 9$\}$ and $B = \{x : x$ is divisible by 3$\}$.

19. Prove each of the following set identities with an argument like that given to prove Theorem 1 in Section 2. 3.
(*a*) $A \cap B = A \backslash B'$
(*b*) $A \cup (B \backslash A) = A \cup B$
(*c*) $A \backslash (B \cup C) = (A \backslash B) \cap (A \backslash C)$
(*d*) $A \backslash (B \cap C) = (A \backslash B) \cup (A \backslash C)$

20. Let A and B be sets. The *symmetric difference* of A and B, denoted by $A \Delta B$, is defined as follows :
$$A \Delta B = (A \backslash B) \cup (B \backslash A)$$
(*a*) Let $A = \{1, 2, 3, ... 10\}$, $B = \{2, 4, 6, ... 10, 12\}$, and $C = \{3, 6, 9, 12\}$. Find $A \Delta B$, $B \Delta C$, $(A \Delta B) \Delta C$, and $A \Delta (B \Delta C)$.
(*b*) Prove that for arbitrary sets, S, T, and R, we have
$$(S \Delta T) \Delta R = S \Delta (T \Delta R)$$

21. Let $A = \{a, b, c\}$, $B = \{a, b\}$, and $C = \{a, b, \{a, b\}\}$. List all the elements in $\mathcal{P}(A)$, $\mathcal{P}(C)$, and $\mathcal{P}(A) \cap \mathcal{P}(C)$. Indicate whether each of the following statements is true or false.

(*a*) $B \subset \mathcal{P}(A)$	(*b*) $B \in \mathcal{P}(A)$	(*c*) $B \subset C$
(*d*) $B \in C$	(*e*) $\mathcal{P}(C) \subset \mathcal{P}(A)$	(*f*) $\mathcal{P}(B) \subset \mathcal{P}(C)$
(*g*) $B \in \mathcal{P}(C)$	(*h*) $A \cap \mathcal{P}(A) = \phi$	(*i*) $C \cap \mathcal{P}(C) = \phi$

22. Let $A = \{a, b\}$, $B = \{t, v\}$, and $C = \{0, 1, 2, 4\}$. List all the elements of each of the following cartesian products.

(*a*) $A \times C$	(*b*) $B \times B$
(*c*) $\mathcal{P}(A) \times C$	(*d*) $A \times B \times C$

23. Express each of the following sets as a cartesian product, or show that this is impossible.
(*a*) The rectangle and its interior in $R \times R$ with corners at (0, 1), (0, 4), (3, 1), and (3, 4).
(*b*) The set of all elements (x, y) where x is an even integer and y is an odd integer.
(*c*) The set of all elements (n, m) of $N \times N$ such that $n = m + 1$.
(*d*) The first quadrant in $R \times R$.
(*e*) The union of the first and third quadrants in $R \times R$.

24. Each of the following sets is defined inductively. Describe each, using set-builder notation.

(*a*) (1) $\frac{1}{2}$ is in S.

(2) If x is in S, then $\left(\frac{1}{2}\right) x$ is in S.

(3) Only number so formed are in S.

(*b*) (1) The numbers 3 and 5 are in Q.

(2) If x and y are in Q, then $x + y$ and $x - y$ are in Q.

(3) Only numbers so formed are in Q.

(*c*) Define the language L over the alphabet $\{a, b\}$ as follows :

(1) The word b is in L.

(2) If x is a word in L, then xbb is in L and aax is in L.

(3) Only words so formed are in L.

EXERCISE 6

1. Redefine each of the following sets, using set-builder notation.
   (*a*) $\{-2, -4, -6, \ldots\}$
   (*b*) $(0, 3, -3, 6, -6, 9, \ldots\}$
   (*c*) $\{2, 3, 4, 6, 8, 9, 10, 12, 14, 15, \ldots\}$
   (*d*) $\left\{1, \frac{1}{2}, \frac{1}{3}, \frac{1}{4}, \ldots\ldots\right\}$
   (*e*) $\{0, 1, 2, \ldots, 99, 100\}$

2. Let $A = \{x : x$ is rational and $0 \le x \le 2\}$, and let $B = \{x : x$ is rational and $1 \le x \le 3\}$. Indicate whether each of the following is true or false.
   (*a*) $0.5 \in A$ (*b*) $2^{1/2} \in B$
   (*c*) $2.6 \in A$ or $2.6 \in B$ (*d*) $\exists x\{x \in A$ and $x \in B\}$

3. Determine which of the following pairs of sets are equal.
   (*a*) $S = \{x : x$ is an integer divisible by both 3 and 2$\}$ and $Q = \{6, 12, 18, 24, \ldots\}$
   (*b*) $X = \{x : x$ is real and $x^2 < x\}$ and $T = \{x : x$ is real and $0 < x < 1\}$

4. Answer true or false :
   (*a*) $A' \cup B' = (A \cup B)'$ (*b*) $A' = \cup \backslash A$
   (*c*) $A \cup (B \cup C) = (A \cup B) \cup C$ (*d*) $A \cup (B \cap C) = (A \cup B) \cap C$
   (*e*) $A \backslash (B \backslash C) = (A \backslash B) \backslash C$

5. Prove that $A \cup A' = \cup$ and $A \cap A' = \theta$.

6. Let A and B be the following subsets of the real numbers : $A = \{x : 0 < x < 5\}$ and $B = \{x : 2 < x < 8\}$. Express $A \cup B$ as the union of three disjoint sets. (Use Proposition 2 of this section.)

7. For each natural number k, let A_k be the subset of the real numbers defined by $A_k = \{x : -k \le x \le k\}$. Find each of the following:
   (*a*) A_4 (*b*) $A_5 \cap A_6$ (*c*) $A_3 \cup A_7$
   (*d*) $\bigcup_{k=1}^{x} A_k$ (*e*) $\bigcap_{k=1}^{x} A_k$

8. Let ∪ be the set of real numbers. For each natural number $r \in R+$, let $A_r = \{x : -n \le x \le 1/n\}$. Find each of the following:

(a) $\left(\bigcup_{n=1}^{x} A_n\right)'$ (b) $\bigcup_{n=1}^{x} A_n'$

(c) $\left(\bigcap_{n=1}^{x} A_n\right)'$ (d) $\bigcap_{n=1}^{x} A_n'$

9. For each natural number m, let $A_m = \{x : 1/m + 2) \le x \le 1 - [1/(m + 1)]\}$.

Let $B = \left\{x : x \le \frac{1}{3}\right\}$. Find each of the following:

(a) $B \cap \left(\bigcup_{m=1}^{x} A_m\right)$ (b) $\bigcup_{m=1}^{x} (B \cap A_m)$

(c) $B \cup \left(\bigcup_{m=1}^{x} A_m\right)$ (d) $\bigcap_{m=1}^{x} (B \cup A_m)$

10. List all the members of the power set of each of the followings sets:
(a) $A = \{a, b, 2, 3\}$ (b) $B = \{\text{cat, dog, mouse}\}$
(c) $C = \{\{a\}, \{b\}\}$ (d) $D = \{\theta, \{\theta\}\}$

11. Let $A = \{a, b, c, d, e\}$, $B = \{a, b\}$, $C = \{B, \theta\}$, and $D = \{a, b, \{a, b\}\}$. Find $A \cap B$, $C \cap D$, $A \cap D$, $C \cap \mathcal{P}(A)$, and $D \cap \mathcal{P}(A)$. Indicate whether each of the following is true or false.
(a) $A \in \mathcal{P}(A)$ (b) $C \subset \mathcal{P}(A)$
(c) $D \subset \mathcal{P}(A)$ (d) $B \subset D$
(e) $B \in D$ (f) $\{a, b\} \in C$

12. Prove that if $A \subset B$, then $\mathcal{P}(A) \subset \mathcal{P}(B)$.

13. Let $A = \{1, 2\}$, $B = \{a, b, c\}$, and $C = \{x, y\}$. List all the elements in each of the following sets.

$$A \times C,\ B \times B,\ A \times B \times C,\ (A \times A) \times (C \times C)$$

14. List some elements of each of the following sets. Express each as a cartesian product or indicate why this cannot be done.
(a) $\{(x, y, z) : x \text{ and } y \text{ are digits and } z \text{ is any integer}\}$
(b) $\{(x, y, z) : x, y, \text{ and } z \text{ are real and } x^2 + y^2 + z^2 = 1\}$
(c) $\{(x_1, x_2, x_3, x_4) : \text{where } x_4 \text{ is real and } 0 < x_k < 1 \text{ for } k = 1, 2, 3, 4\}$

15. List some of the elements in each of the following sets. Express each as a cartesian product or indicate why this cannot be done.
(a) $\{(x, y) : x \text{ is a positive integer and } y \text{ is a negative integer}\}$
(b) $\{(x, y) : x \text{ is a digit and } y \text{ is a letter}\}$
(c) $\{(x, y) : x \in N \text{ and } y = x + 1\}$

16. For each of the following pairs of sets A and B, determine whether $A \subset B$ and, if so, determine whether the containment is proper.
(a) $A = \{x : x \text{ is real and } 0 \le x^2 \le 1\}$ and $B = \{x : -1 \le x \le 1\}$

(*b*) $A = \{x : x$ is an integer multiple of 2$\}$ and $B = \{x : x$ is an integer multiple of 4$\}$

(*c*) $A = \{x : 2x$ is an integer multiple of 12$\}$ and $B = \{x : x$ is divisible by both 2 and 3$\}$

17. Prove : $B \subset A$ where $A = \{x : x^2$ is an integer multiple of 9$\}$ and $B = \{x : x$ is an integer divisible by 6$\}$.

18. Prove : $A = B$ where $A = \{x : x^2$ is divisible by 9$\}$ and $B = \{x : x$ is divisible by 3$\}$

19. Prove each of the following set identities :

(*a*) $A \cap B = A \backslash B'$ (*b*) $A \cup (B \backslash A) = A \cup B$

(*c*) $A \backslash (B \cup C) = (A \backslash B) \cap (A \backslash C)$ (*d*) $A \backslash (B \cap C) = (A \backslash B) \cup (A \backslash C)$

20. Let A and B be sets. The symmetric difference of A and B, denoted by $A \Delta B$, is defined as follows :

$$A \Delta B = (A \backslash B) \cup (B \backslash A)$$

(*a*) Let $A = \{1, 2, 3, \ldots, 10\}$,
$B = \{2, 4, 6, \ldots, 10, 12\}$, and
$C = \{3, 6, 9, 12\}$.

Find $A \Delta B$, $B \Delta C$, $(A \Delta B) \Delta C$, and $A \Delta (B \Delta C)$.

(*b*) Prove that for arbitrary sets S, T, and R, we have
$(S \Delta T) \Delta R = S \Delta (T \Delta R)$

21. Let $A = \{a, b, c\}$, $B = \{a, b\}$ and $C = \{a, b, \{a, b\}\}$. List all the elements in $\mathcal{P}(A)$. $\mathcal{P}(C)$, and $\mathcal{P}(A) \cap \mathcal{P}(C)$. Indicate whether each of the following statements is true or false.

(*a*) $B \subset \mathcal{P}(A)$ (*b*) $B \in \mathcal{P}(A)$ (*c*) $B \subset C$

(*d*) $B \in C$ (*e*) $\mathcal{P}(C) \subset \mathcal{P}(A)$ (*f*) $\mathcal{P}(B) \subset \mathcal{P}(C)$

(*g*) $B \in \mathcal{P}(C)$ (*h*) $A \cap \mathcal{P}(A) = \theta$ (*i*) $C \cap \mathcal{P}(C) = \theta$

22. Let $A = \{a, b\}$, $B = \{s, t, v\}$, and $C = \{0, 1, 2, 4\}$. List all the elements of each of the following cartesian products

(*a*) $A \times C$ (*b*) $B \times B$

(*c*) $P(A) \times C$ (*d*) $A \times B \times C$

23. Express each of the following sets as a cartesian product, or show that this is impossible.

(*a*) The rectangle and its interior in $R \times R$ with corners at (0, 1), (0, 4), (3, 1), and (3, 4).

(*b*) The set of all elements (x, y) where x is an even integer and y is an odd integer.

(*c*) The set of all elements (n, m) of $N \times N$ such that $n = m + 1$.

(*d*) The first quadrant in $R \times R$.

(*e*) The union of the first and third quadrants in $R \times R$.

24. Each of the following sets is defined inductively. Describe each, using set-builder notation.

(*a*) (1) 1/2 is in S.

(2) If x is in S, then $(1/2)\, x$ is in S.
(3) Only numbers so formed are in S.

(*b*) (1) The numbers 3 and 5 are in Q.
(2) If x and y are in Q, then $x + y$ and $x - y$ are in Q.
(3) Only numbers so formed are in Q.

(*c*) Define the language L over the alhabet $\{a, b\}$ as follows :
(1) The word b is in L.
(2) If x is a word in L, then xbb is in L and aax is in L.
(3) Only words so formed are in L.

(*d*) Define the language S over the alphabet $E = \{a, b\}$ as follows :
(1) The empty word $\wedge$ is in S.

25. (*a*) Prove the following:
$n(A \cup B) = n(A) + n(B) - n(A \cap B)$

(*b*) Determine the number of integers between 1 and 250 that are divisible by 2, 3, 5, 7. ***(Terminal Exam., GGSIPU, 2001)***

26. (*a*) Let $N = \{0, 1, 2, 3, ...\}$ Define functins f, g and h from Set N to N by
$f(n) = n + 1$, $g(n) = 2n$ and $h(n) = \{0$ if n is even
$\{1$ if n is odd

Compute go (fog) oh. Is the functional h invertible?
Is the function f on to ?

OR

Define characterstic function of a set and use it to prove $A \cap (B \cup C) = (A \cap B) \cup (A \cap C)$

(*b*) Define the terms (*i*) disjoint permutations (*ii*) even and odd permutations. Give one example in each case.
Compute $f.^{|}\, g.\, f$ when $f^{\,|} = (235)\,(12)$, $g = (1579)$

$$\text{Let } Q = \begin{pmatrix} 1\,2\,3\,4\,5\,6\,7\,8\,9 \\ 3\,1\,2\,a\,b\,7\,8\,9\,6 \end{pmatrix}$$

If Q is an Even Permutations, Find 'a' and 'b'

(Terminal Exam., GGSIPU, 2001)

27. (*a*) What do you mean by disjoint permutations in set theory, give one example.

(*b*) Let a permutation

$$P = \begin{pmatrix} 1 & 2 & 3 & 4 & 5 & 6 & 7 & 8 \\ 2 & 5 & \text{a} & 3 & \text{b} & 1 & 8 & 4 \end{pmatrix}$$

Find the values of 'a' and 'b' so that permutation P becomes an even permutation.

(*c*) Let f, g and h be the functions defined from set A to set A as mentioned below:
$f(x) = x + 1$, $g(x) = 2(x)$, $h(x) = 0$ if x is odd and $h(x) = 1$ if x is even when $A = (0, 1, 2, 3, ...)$. Which of the above functions are one to one or onto? Also compute fogoh. ***(End-term Exam., GGSIPU, 2002)***

28. (*a*) Define the characteristic function of a set.
    Let A and B be two subsets of a universal set E.
    Prove that $\Psi_{A \cap B}(X) = \Psi_A(X) \times \Psi_B(X)$
    (*b*) Is the function $f : R \to R$ defined by $f(x) = 4x + 5$ invertible?
    If yes, then find f^1 when R is the set of real numbers.

(End-term Exam., GGSIPU, 2002)

29. Find the product of two permutations and show that it is used commutative.

$$f = \begin{pmatrix} 1 & 2 & 3 & 4 \\ 2 & 1 & 4 & 3 \end{pmatrix},\ g = \begin{pmatrix} 1 & 2 & 3 & 4 \\ 3 & 2 & 1 & 4 \end{pmatrix}$$

MULTIPLE CHOICE QUESTIONS

1. Which one of the pairs of sets comprises of disjoint sets
   (*a*) {1, 2, 3} and {2, 4, 7, 9} (*b*) {0, 1} and {1, 2}
   (*c*) {0} and {1} (*d*) None of these
2. If $A = \{3, 5\}$ then $A \times A$ is equal to:
   (*a*) {(3, 3), (3, 5), (5, 3), (5, 5)} (*b*) {(3, 3), (5, 5)}
   (*c*) {(3, 5), (3, 3), (5, 5)} (*d*) none of these
3. The number of elements in powerset of the set $\{\{\phi\}, 1, \{4, 5\}\}$ is
   (*a*) 0 (*b*) 2 (*c*) 4 (*d*) 6
4. Let P and Q be sets and P' and Q' represent the complements of the set P and Q. The set $(P - Q) \cup (Q - P) \vee (P \wedge Q)$ is equivalent to:
   (*a*) $P \cup Q'$ (*b*) $P \vee Q$ (*c*) $P \wedge Q'$ (*d*) $P' \wedge Q'$
5. Let P is a finite set of size n, the number of elements in the power set of $P \times P$ is:
   (*a*) 2^n (*b*) n^{2n} (*c*) 2^{n2} (*d*) None of these
6. If A = {red, blue, green, violet} then which of the following is a partition of A
   (*a*) [{red, blue, violet}] (*b*) [{blue}, {red, violet, green}]
   (*c*) ϕ, {red, blue} {green, violet} (*d*) None of these
7. If out of two sets, neither A is a subset of B nor B is subset of A, then which of the following statement is True?
   (*a*) $A \cup B$ (*b*) $A \leq A \cap B$ (*c*) $A \leq A \cup B$ (*d*) $A \cup B \leq B$
8. Red $A = \{x : x \text{ is rational and } 0 \leq x \leq 2\}$ and $B = \{x : x \text{ is irrational and } 1 \leq x \leq 3\}$. Which of the following is false?
   (*a*) $0.5\ \varepsilon$ A (*b*) $2^{1/2}\ \varepsilon\ B$ (*c*) $2.6\ \varepsilon\ A$ (*d*) $0.3\ \varepsilon\ B$
9. Let $A = \{\phi, \{\phi\}\}$. What is the value of $P(A)$
   (*a*) $\{\phi, \{\phi\}\}, \{\{\phi\}\}, \{\phi, \{\phi\}\}$ (*b*) $\{\phi, \{\phi\}, \{\{\phi\}\}\}$
   (*c*) $\{\{\phi\}, \{\{\phi\}\}, \{\phi, \{\phi\}\}, \phi\}$ (*d*) None of these
10. $A \cap A^C$ is equivalent to:
    (*a*) $\cup$ (*b*) A (*c*) ϕ (*d*) A^C

Answers

1. (*c*)	2. (*a*)	3. (*d*)	4. (*a*)	5. (*c*)	6. (*b*)
7. (*b*)	8. (*c*)	9. (*c*)	10. (*c*)		

❐❐❐

CHAPTER 2

PROPOSITIONAL CALCULUS AND FORMAL LOGIC

This Chapter Covers :

2.1. INTRODUCTION

Logic is the study of reasoning ; it is specifically concerned with whether reasoning is correct. It in the basis on which all the sentences are built. It has practical application to the design of computing machines, to artificial intelligence, to computer programming, to programming languages and to the other areas of computer science.

Logic is the basis on which all the sentences are built. Logic was extensively studied and developed in ancient Greece. But the mathematical theory of logic, called symbolic logic, only came into its own in the 19th century. This algebraic way of studying the arguments was developed by the English Mathematician

George Boole (1815-1864). The main objective of logic is to formulate rules which may be helpful in taking decisions whether any argument or reasoning is "valid".

2.2. SYMBOLS AND NOTATIONS

Following are the symbols and notations which we will use very often in this chapter.

N	Set of natural numbers
R	Set of real numbers
$p \vee q$	p or q (p, q being statements)
$p \oplus q$	either p or q but not both
$p \wedge q$	p and q
$\sim p$	not p
$p \to q$	p implies q p only if q p is sufficient for q
$p \leftrightarrow q$	p if and only if q p is necessary and sufficient for q p implies and is implied by q
$p \Rightarrow q$	if p is true the q is true
$p \Leftrightarrow q$	p is true if and only if q is true
$p \equiv q$	p is eqvavalent to q
T	Tautalogy
F	Contradiction
$\therefore$	therefore
$\forall$	for all
$\exists$	there exists
$\exists!$	there exists one and only one
$P(X)$	Power set of X
CNF	Conjunctive normal form
DNF	disjunctive normal form
$PCNF$	Principle conjunctive normal form
$PDNF$	Principle Disjunctive normal form

2.3. PROPOSITIONS

Consider the sentence "The Taj Mahal is red". When you read this sentence, we can immediately decide whether it is ture or false. It would not happen that some people will say it is true and other will say it is false. Everybody would have the same opinion. So this sentence is either universally true or universally false but not both. Similarly "Delhi is the capital of India" is a declarative sentence which is either true or false but not both. In mathematical logic we call such sentences as **statements** or **propositions**. Therefore "**proposition** is a declarative sentence which is either universally true or universally false but not both". It is denoted by lower case latters p, q etc. If the proposition is universally true we give it a truth value T (or 1) otherwise F (or 0).

EXAMPLE 2.1. Consider the following sentences and check whether the given sentences are proposition or not.

1. The president of India is a woman.
2. An elephant weights more than a human being.
3. If a stone is thrown in the air, it will fall to the ground.
4. Watch the movie.
5. How wonder ful !
6. What is your name ?

Here the first sentence is the declarative sentence which is universally false therefore is a proposition. Similarly, second sentence is also a proposition (think about it). The sentence 3rd is a proposition with the truth value T. But the 4th, 5th and 6th sentences are not even the declarative sentences and therefore none of them are propositions.

From the above example we conclude that the sentences that are not statements include questions, exclamations and expression of opinions are not proposition.

EXAMPLE 2.2. Check whether the following sentences are propositions or not. Give reason also.

1. The equation $x^2 + 25$ has two solutions.
2. $3 + 3 = 6$
3. $x^2 > 0, x \in R$
4. If x and y are even numbers then the sum $x + y$ is also an even number.
5. $x + y > 0$

Solution : 1. It is a proposition being a declarative sentence which is universally true.

2, 3 and 4 are propositions [same reason as in (1)], but sentence given by 5 is not a proposition as it is neither universally true nor universally false, everything depends on the value or x and y.

2.4. LOGICAL CONNECTIVES

When we are talking to someone we never use simple sentences always. We use more complicated sentences which are formed by two or more simple sentences by using the words like 'and', 'or', if ... then etc. These words or phrases or connections or links are called logical connectives as in mathematics we use +, –, *, / - There are six such connectives which are given as:

2.4.1 Disjunction

The disjunction of two original statements p and q is a new compound statement written as $p \vee q$ and is nothing but p or q. Consider a sentence "Rahul or Rachna went to see a movie". This sentence can be written as "Rahul went to see a movie or Rachana went to see a movie". So the given statement is actually made up of two simple said above statements by the connective called disjunction ($\vee$).

The truth value of the compound statements $p \vee q$ was depends on the truth values of the atomic statements therefore the truth values of the compound statement $(p \vee q)$ are given by the following truth table (2.1).

Table 2.1

p	q	$p \vee q$
T	T	T
T	F	T
F	T	T
F	F	F

You find your first number of rows in a table

$$= (\text{number of propositions})^2$$

Since we have two propositions in this example, we know 2^2 or 4,so the number of rows are 4. (Similarly three propositions would be 8; four would be 16 and so on). The second column is found by having the first column. For example, we have two T's in the first column and therefore we assign those two T's one true value & one false value (in that order). We have two F's also so we assign those two F's one true value & one false value (in that order). Now, how do we find $p \vee q$.

For the two statements p and q following cases will give the clarity of first two columns of the truth table.

1. p is true, q is true
2. p is true, q is false
3. p is false, q is true
4. p is false, q is false.

That is for a single value of p there are two possibilities of q so consider the following figure

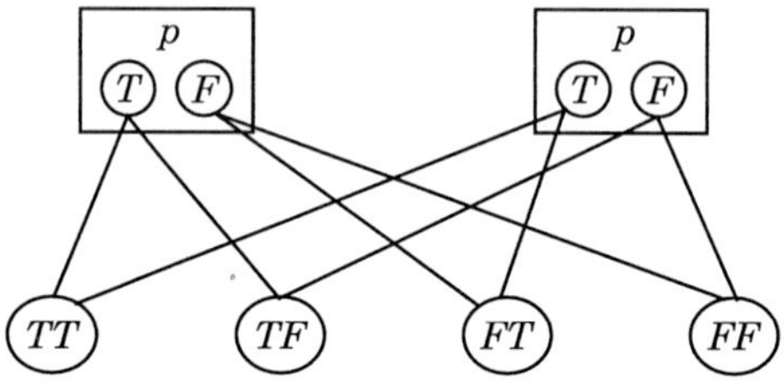

From the above truth table it is clear that even if one of the p or q is true then the resultant compound statement is true otherwise false. Disjunction has the same properties as union. It is also called **inclusive or.**

Now we define **"exclusive or"** for two propositon p & q is denoted by $p \oplus q$ is true when exactly one of the proposition is true otherwise false that is $p \oplus q$ is true when either p is true or q is true not both. Again, the truth value of the compound statement. $p \oplus q$ depends on the truth value of the atomic statements, therefore for the truth value of $p \oplus q$ consider the following truth table 2.2.

Table 2.2.

p	q	$p \oplus q$
T	T	F
T	F	T
F	T	T
F	F	F

EXAMPLE 2.3. Let p be the statement "Sonmoy is an architect" and q is "Ashok is a draftman". Then $p \vee q$ is:

"Sonmoy is an architect or Ashok is a draftman"

EXAMPLE 2.4. Let A and B be sets. Then

$$A \cup B = \{x : x \in A \text{ or } x \in B\}$$
$$= \{x : x \in A \vee B\}$$

2.4.2. Conjunction

If p and q are two propositions then their conjunction is "p and q" denoted by $p \wedge q$ gives the new compound statement which is true only when both of them are true. If at least one of the statement is false then the compound statement is false. Sometimes the symbols 'and', 'dot(·)', '&', are used to mean the conjunction. It is notable that conjunction have the same properties as intersection.

The truth values of the compound statement '$p \wedge q$' are given by the table 2.3.

Table 2.3.

p	q	$p \wedge q$
T	T	T
T	F	F
F	T	F
F	F	F

EXAMPLE. 2.5. Let p : Tajmahal is white.

q : Tajmahal is white and is in Agra.

$p \wedge q$: Tajmahal is white and is in Agra.

The truth value of $p \wedge q$ is T. (By case 1 in the table 2.3).

EXAMPLE 2.6. Consider the following statements :

1. Mumbai is in Maharashtra and $2 + 2 = 4$
2. Mumbai is in U.P. and $2 + 2 = 4$
3. Banglore is in Mysore and $2 + 2 = 4$
4. Banglore is in Gujarat and $2 + 2 = 5$

The statements (1) and (3) are true and the statements (2) and (4) are false.

2.4.3. Negation (Not)

If p is a proposition, its negation is denoted by ~p gives the statement which is true only when p is false. Negation has the same properties as compliment operation in a set. Its truth value is given by truth table 2.4.

Table 2.4.

p	$\sim p$
T	F
F	T

EXAMPLE 2.7. p : Ram is a good boy
then $\sim p$: Ram is not a good boy

EXAMPLE 2.8. p : The sun is shining.
$\sim p$: The sun is not shining.

EXAMPLE 2.9. p : $8 + 1 = 5$
$\sim p$: $8 + 1 \neq 5$

2.4.4. Conditional Operator [If... then]

If p and q are statements, the compound statement if p then q is denoted by $p \to q$, is called a **conditional statement** or **implication**. The statement p is called the **antecedent** or **hypothesis** and the statement q is called the **consequent** or **conclusion.**

The connective if ... then is denoted by the symbol $\to$.

EXAMPLE 2.10. p : I will do my work.
q : I will go to see movie.
then $p \to q$ implies If p then q, *i.e.*,

If I will do my work then I will go to see movie is the resultant compound statement.

Truth values of $p \to q$ are given by the table 2.5.

Table 2.5.

p	q	$p \to q$
T	T	T
T	F	F
F	T	T
F	F	T

By the truth table it is clear that the compound statement $p \to q$ is false only if p is true and q is false.

EXAMPLE 2.11. Let p & q be two statements, such that

p : Two triangles are congruent and

q : Two triangles are similar.

Then $p \to q$ is the statement :

If two triangles are congruent, then they are similar

or two triangles are congruent implies that they are similar

or two triangles are congruent only if they are similar

or the necessary condition for the two trianges to be similar is that they are congruent.

EXAMPLE 2.12. Assume your father gave you the following promise :

'If you receive a grade of 95 or better in the final examination then you will get a car as reward". Your father has made a promise to you (placed a condition with you).

If you fulfill his condition you expect the conclusion (getting a car) to be forth coming. Your final result has come to you. Has your father told the truth (kept the promise) or is he guilty of a falsehood?

Case I : Your final exam score was greater than 95 (Condition is true) and you receive a car (conclusion is true). The father told the truth.

Case II : Your final exam score was greater than 95 and you did not get a car. The father lied.

Case III : Your final exam score was less than 95 and you get a car. The father told the truth (perhaps your overall result was excellent).

Case IV : Your final exam score was less than 95 and you did not get a car. The father told the truth.

There are other forms in which a conditional statement may appear. For any conditional statement $p \to q$, three other related statements are as follows :

Statement : $p \to q$ If p, then q
Converse : $q \to p$ If q, then p
Inverse : $\sim p \to \sim q$ If not p, then not q
Contrapositive : $\sim q \to \sim p$ If not q, then not p

EXAMPLE 2.13.

Statement : If it is raining then I get wet.
Negation : If is not the case that if it is raining then I get wet.
Converse : If I get wet then it is raining
Inverse : It it is not raining then will not get wet.
Contrapositive : I will not get wet if it is not raining.

EXAMPLE 2.14.

Statement : If x is positive, then $x \neq 0$.
Negation : If x is positive then $x = 0$.
Converse : If $x \neq 0$ then x is positive.
Inverse : If x is negative, then $x = 0$.
Contrapositive : If $x = 0$ then x is negative.

The following is a truth table for these related statements of a conditional statement $p \to q$.

Table 2.6.

p	q	$\sim p \to \sim q$	$q \to p$	$\sim q \to \sim p$	$p \to q$
T	T	T	T	T	T
T	F	T	T	F	F
F	T	F	F	T	T
F	F	T	T	T	T

2.4.5. Biconditional Operator (if and only if)

If p and q are propositions, the biconditional statement p if and only if q is denoted by $p \leftrightarrow q$ is true when p and q both have the same truth values.

$p \leftrightarrow q$ is same as $p \to q$ and $q \leftrightarrow p$ *i.e.*,

$p \leftrightarrow q$ is the conjunction of $p \rightarrow q$ and $q \rightarrow p$

i.e., $p \leftrightarrow q$ is $(p \rightarrow q) \wedge (q \rightarrow p)$

The truth values of the statement $p \leftrightarrow q$ are given by the table 2.7.

Table 2.7.

p	q	$p \rightarrow q$	$q \rightarrow p$	$(p \rightarrow q) \wedge (q \rightarrow p)$ or $p \leftrightarrow q$
T	T	T	T	T
T	F	F	T	F
F	T	T	F	F
F	F	T	T	T

EXAMPLE 2.15. Compute the truth table of the statement $(p \rightarrow q) \leftrightarrow (\sim q \rightarrow \sim p)$

p	q	$p \rightarrow q$	$\sim q$	$\sim p$	$\sim q \rightarrow \sim p$	$(p \rightarrow q) \leftrightarrow (\sim q \rightarrow \sim p)$
T	T	T	F	F	T	T
T	F	F	T	F	F	T
F	T	T	F	T	T	T
F	F	T	T	T	T	T

An expression (compound statement) involving logical variables that is true in all the cases is called a tautology as in the above example the given expression is **tautology**. For example, $p \vee \sim p$ and $q \rightarrow (p \vee q)$.

A statement which is always false is called contradiction or an absurdity. For example, $p \wedge \sim p$ and $(p \vee q) \wedge (\sim p) \wedge (\sim q)$ are **contradictions**.

Equavalence. Let S be a set of propositions and let r and s be propositions generated by S. The two proposition r and s are equivalent if they have the same truth value. The equivalence of the propositions r & s is denoted by $r \equiv s$.

For any two statements p & q

$$p \vee q \equiv q \vee p$$
$$p \wedge q \equiv q \wedge p$$
$$p \rightarrow q \equiv \sim q \rightarrow \sim p \qquad \text{[By truth table (2.6)]}$$
$$p \rightarrow q \equiv (\sim p) \vee q$$

Precedance Rule

While dealing with operations on numbers, we have realised the need for applying the BODMAS Rule.

Similarly while calculating the truth value of compound propositions involving more than one connective, we have a similar rule which tells us which connective to apply first. This rule is named as "the rule of precedence". According to this rule the order of preference in which the connectives are applied in a formula of propositions that has no bracket is:

(*i*) $\sim$

(*ii*) $\wedge$

(*iii*) $\vee$ and $\oplus$

(*iv*) $\rightarrow$ and $\leftrightarrow$

Note that ∨ and ⊕ *i.e.*, 'inclusive or' and 'exclusive or' are both third in order of preferce. Therefore we can apply either of them first with the same end result. Similarly, for position $\leftrightarrow$ and $\rightarrow$.

In order to clear this rule consider the following example.

EXAMPLE 2.16. $p \rightarrow q \wedge \sim r \leftrightarrow r \oplus q$

Solution : To find the truth value of the following compound statement we use precedence Rule. Then we will write this statement as $(P \rightarrow (q \wedge (\sim r)) \leftrightarrow (r \oplus q)$.

Consider the truth table

p	q	r	$\sim r$	$r \oplus q$	$q \wedge (\sim r)$	$p \rightarrow (q \wedge \sim pr)$	$p \rightarrow (q \wedge \sim r) \leftrightarrow r \oplus q$
T	T	T	F	F	F	F	T
T	T	F	T	T	T	T	T
T	F	T	F	T	F	F	F
T	F	F	T	F	F	F	T
F	T	T	F	F	F	T	F
F	T	F	T	T	T	T	T
F	F	T	F	T	F	T	T
F	F	F	T	F	F	T	F

Tautologies

Consider the following composite statements and their truth tables :

(*i*) $p \longleftrightarrow p$

p	$p \leftrightarrow p$
T	T
F	T

(*ii*) $p \vee \sim p$

p	$\sim p$	$p \vee \sim p$
T	F	T
F	T	T

(*iii*) $p \rightarrow p \vee q$

p	q	$p \vee q$	$p \rightarrow p \vee q$
T	T	T	T
T	F	T	T
F	T	T	T
F	F	F	T

In the above truth tables we conclude that the statements $p \leftrightarrow p$, $p \vee \sim p$ and $p \rightarrow p \vee q$ are always true irrespective of the truth values of p and q. Such statements are called **Tautologies**. Thus **Tautology** is a composite statement which is always true irrespective of the truth value of the component statement.

A **Contradiction** is a composite statement which is always false irrespective of the truth values of the component Statements.

For example, $r : p \wedge \sim p$

p	$\sim p$	$p \wedge \sim p$
T	F	F
F	T	F

$r : (p \vee q) \wedge (\sim p) \wedge (\sim q)$

p	q	$p \vee q$	$\sim p$	$\sim q$	r
T	T	T	F	F	F
T	F	T	F	T	F
F	T	T	T	F	F
F	F	F	T	T	F

It is clear that the negation of a tautology is a contradiction and vice-versa.

EXAMPLE 2.17. Determine whether the following statements are tautologies :

1. $r : (p \wedge \sim q) \vee (q \wedge \sim p)$
2. $r : p \wedge q \rightarrow p$
3. $(p \rightarrow q) \vee (q \rightarrow p)$

Solution : 1.

p	q	$\sim p$	$\sim q$	$p \wedge \sim q$	$q \wedge \sim p$	r
T	T	F	F	F	F	F
T	F	F	T	T	F	T
F	T	T	F	F	T	T
F	F	T	T	F	F	F

The given compound statement is not a Tautology.

2.

p	q	$p \wedge q$	$p \wedge q \rightarrow p$
T	T	T	T
T	F	F	T
F	T	F	T
F	F	F	T

The given compound statement is Tautology.

3.

p	q	$p \rightarrow q$	$q \rightarrow p$	$(p \rightarrow q) \vee (q \rightarrow p)$
T	T	T	T	T
T	F	F	T	T
F	T	T	F	T
F	F	T	T	T

The given compound statement is Tautology.

2.5. THE LAWS OF LOGIC

Let T stand for tautology and F stand for contradiction and p is any propositions then the identity law is given by:

$$p \vee T = T$$
$$p \wedge T = p$$
$$p \wedge F = \text{F}$$
$$p \vee F = p$$

Verification is given by the following truth table

p	T	F	$p \vee T$	$p \wedge T$	$p \wedge F$	$p \vee F$
T	T	F	T	T	F	T
F	T	F	T	F	F	F

The 3rd, 4th, 5th and 6th columus give the result,
Similarly we can give all the basic law of logic which are given by the table (2.7)

Table 2.7.

Basic Logical Laws		
1.	$p \vee q \equiv q \vee p$ $p \wedge q \equiv q \wedge p$	Commutative law
2.	$(p \vee q) \vee r \equiv p \vee (q \vee r)$ $(p \wedge q) \wedge r \equiv p \wedge (q \wedge r)$	Associative law
3.	$p \wedge (q \vee r) \equiv (p \wedge q) \vee (p \wedge r)$ $p \vee (q \wedge r) \equiv (p \vee q) \wedge (p \vee r)$	Distributive laws.
4.	$p \vee F \equiv p, p \wedge F = F$ $p \wedge T \equiv p \text{ , } p \vee T = T$	Identity law
5.	$p \wedge {\sim}p \equiv F$ $p \vee {\sim}p \equiv T$	Negation law
6.	$p \vee p \equiv p$ $p \wedge p \equiv p$	Idempotent law
7.	$p \wedge (p \vee q) \equiv p$ $p \vee (p \wedge q) \equiv p$	Absorption law
8.	${\sim}(p \vee q) \equiv ({\sim}p) \wedge ({\sim}q)$ ${\sim}(p \wedge q) \equiv ({\sim}p) \vee ({\sim}q)$	De Morgan's law
9.	${\sim}({\sim}p) \equiv p$	Involution law (double negative)
10.	$p \rightarrow q \equiv {\sim}p \vee q$	(equivalent of conditional)
11.	$p \leftrightarrow q \equiv (p \wedge q) \vee ({\sim}p \wedge {\sim}q)$	(equivalent of Bi-conditional)

All the above results can be proved by the use of truth table.

2.6. SET: THEORY AND LOGIC THEORY

T	1	represents included
F	0	represents excluded.

Instead of calling them truth tables, we call them Membership tables. They have the exact same format as what learned in the truth table if we examine them closely.

Here are the parallels :

Set	*Logic*
$\cap$	$\wedge$
$\cup$	$\vee$
—	$\neg$
ϕ	F
U	T

EXAMPLE 2.18. Show that $A \cap (B \cup C) = (A \cap B) \cup (A \cap C)$

A	B	C	$(B \cup C)$	$A \cap (B \cup C)$	$A \cap B$	$A \cap C$	$(A \cap B) \cup (A \cap C)$
1	1	1	1	1	1	1	1
1	1	0	1	1	1	0	1
1	0	1	1	1	0	1	1
1	0	0	0	0	0	0	0
0	1	1	1	0	0	0	0
0	1	0	1	0	0	0	0
0	0	1	1	0	0	0	0
0	0	0	0	0	0	0	0

2.7. PREDICATES AND LOGICAL QUANTIFIERS

Consider the declarative sentence "He is a good student". This sentence is not a proposition as it is not clear that whether it is true of false unless we know. 'who is he'. Such sentences are the examples of **propositional functions** or **predicates**. Thus the predicate in a variable x is a sentence $p(x)$ that becomes a proposition when we give x a particular value. Let us take, "he is Ram", then the given sentence becomes "Ram is a good student" is a proposition as it will have a definite truth value either T or F.

Consider the another example $x + 2 = y$. It is again a proposition function unless we assign the values to x and y from a given set. Normally the propositional functions are denoted by $P(x)$.

EXAMPLE 2.19. Let $p(x)$ be "$x^2 - 6x + 9 = 0$". Then $p(x)$ is a propositional function on the set of integers Z. $p(x)$ is true only when x is 3.

Let $p(x)$ be a propositional function on a set X. Then the set of all elements of $a \in X$ with due property that $p(a)$ is true is called the turn set T of $p(x)$. In other words,

$$T = \{a; a \in S, p(a) \text{ is true}\}$$

$$\text{or simply } T = \{a; p(a)\}$$

EXAMPLE 2.20. Let $p(x)$ be "$x + 3 > 3$". Then the truth set of $p(x)$ on N, the set of natural numbers is N.

Most of the statements in mathematics and computer science use terms such as "for every" and "for some". For example, in mathematics we have the following theorem :

For every triangle T, the sum of the angles of T is equal to 180°.

In computer science, we have the theorem :

for some program P, the output of P is P itself.

Now we extend the logical system so that we can handle statements that includes "for every" and "for some". Symbolically "for every" is denoted by $\forall$ is called universal quantifier. and "for some" is denoted by $\exists$ is called existential quantifier.

Alternative ways to write universal quantifiers are as follows:

for every x, $p(x)$

are for all x, $p(x)$

and for any x, $p(x)$

i.e., the symbol ∀ may be read "for every", "for all", or "for any".

Therefore the statement $\forall x\, p(x)$ is called universal quantified statement.

Alternative ways to write existential quantifier are as follows :

for some x, $p(x)$

and for at least one x, $p(x)$

and there exists x such that, $p(x)$.

i.e., the symbol ∃ may be read "for some", "for at least one" or "there exists".

Thus the statement $\exists\, x\, p(x)$ is called existentially quantified statement.

We can say that the statement $(\exists\, x \in S)\, p(x)$ is equivalent to the statement that the truth set of $p(x)$ is non empty. That is $\{x;\, x \in S,\, p(x)\} \neq \phi$.

Quantifiers		
Statement	**When is it true ?**	**When is it false ?**
$\forall x\, p(x)$	$p(x)$ is true for every x	This is an x for which $P(x)$ is false.
$\exists x\, p(x)$	There is an x for which $p(x)$ is true	$p(x)$ is false for every x

EXAMPLE 2.21. The statement "for some x, $x^2 - 6x + 9 = 0$" Can be written $\exists\, x\, (x^2 - 6x + 9 = 0)$.

This is a true statement since

$$\{x : x^2 - 6x + 9 = 0\} = \{2, 3\} \neq \phi$$

EXAMPLE 2.22. The statement $(\exists\, x \in N)(3x + 9 = 6x + 2)$ is false. Since there is no integer x such that $3x + 9 = 6x + 2$, and therefore $\{x : 2x + 3 = 5x + 1\} = \phi$.

EXAMPLE 2.23. The statement "for every natural number n, $n + 3 > 1$" can be written

$$\text{“}(\forall\, n \in N)(n + 5) > 1\text{”}$$

This is a true statement since

$$\{\, n : n + 5 > 1\} = \{1, 2, 3, ...\} = N$$

EXAMPLE 2.24. The statement $(\forall n \in N)(n + 1 > 3)$ is false since

$$\{\, n : n + 1 > 5\} = \{5, 6, 7, ...\} \neq N$$

EXAMPLE 2.25. The statement

for every real number x, $x^2 \geq 0$

is a universally quantified statement. The domain of discourse is the set of real numbers. The statement is true therefore have truth value T.

EXAMPLE 2.26. To verify that the existentially quantified statement for some real number x, $\dfrac{1}{1+x^2} > 1$ is false.

Now we have to showe that the proposition $\dfrac{1}{1+x^2} > 1$ for some real number x is false is equivalent to prove $\dfrac{1}{1+x^2} \leq 1$ is true.

Now consider $\frac{1}{1+x^2} \leq 1$

Since $x^2 \geq 0$

$\Rightarrow 1 + x^2 \geq 1$

$\Rightarrow \frac{1}{1+x^2} \leq 1$

$\therefore$ the statement $\frac{1}{1+x^2} \leq 1$ is true for every real number x.

Thus the statement $\frac{1}{1+x^2} > 1$ is false for every real number x.

Therefore, the result is proved.

Theorem 2.1. Generalised, *De* Morgan Laws for Logic

If p is a propositional function, each pair of propositions in (*a*) and (*b*) has the same truth values (*i.e.,* either both are true or both are false).

(*a*) $\overline{\forall x p(x)}$, $\exists x \overline{p(x)}$

(*b*) $\overline{\exists x p(x)}$, $\forall x \overline{p(x)}$

Proof : (*a*) Suppose that the proposition $\overline{\forall x p(x)}$ is true then the proposition $\forall x P(x)$ is false.

The proposition $\forall x p(x)$ is false if $P(x)$ is false for at least one x in the domain of discourse, *i.e.*, D. But if $p(x)$ is false for at least one x in D then $\overline{p(x)}$ is true for at least one x in D.

$\therefore$ the proposition $\exists x \overline{p(x)}$ is true.

Thus if the proposition $\overline{\forall x p(x)}$ is true then $\exists x \overline{p(x)}$ is true.

Similarly if the proposition $\overline{\forall x p(x)}$ is false then the proposition $\exists x \overline{p(x)}$ is false.

Part (*b*) is left for the students.

EXAMPLE 2.27. Consider the proposition

"There is a criminal who has committed every crime".

In symbolic form

A : Set of criminals

B : Set of crimes

$(\exists\, x \in A)\ (\forall\, y \in B)$ (x has committed y)

$(\exists\, x \in A)\ (\forall y \in B)\ p(x, y)$

Now consider its negation

$\sim (\exists x \in A)\ (\forall y \in B)\ p(x, y)$

$\equiv (\forall x \in A) \sim(\forall y \in B)\ p(x, y)$

$\equiv (\forall x \in A)\ (\exists y \in B) \sim p(x, y)$

$\equiv (\forall x \in A)\ (\exists y \in B)$ (x has not committed y)

EXAMPLE 2.28. Since $\sim(p \vee q) = \sim p \wedge \sim q$, so we have

$$\sim(\forall\, x\, p(x) \vee \exists\, y\, p(y)) = \sim(\forall\, x\, p(x)) \wedge \sim (\exists\, y\, p(y))$$
$$= \exists\, x \sim p(x) \wedge \forall\, y \sim p(y)$$

2.8. NORMAL FORM

Let $A(P_1, P_2, ... P_n)$ be a statement formula where $P_1, P_2, ... P_n$ are the atomic variables, if we consider all possible assignments of the truth values to $P_1, P_2, ... P_n$ and obtain the resulting truth values of the formula A, then we get a resulting truth table for A..

If all values for A are true then A is identically true or a *tautology*.

If all values for A are false then A is identically false or a *contradiction*.

If A has the resulting truth value true for at least one combination of truth value assigned to $P_1, P_2, .. P_n$ then A is said to be *satisfiable*.

Decision Problem

The problem of determining in a finite number of steps whether a given statement formula is tautology or a contradiction or at least satisfiable is known as decision problem. The decision problem always has a solution. We use the word 'product' in place of 'conjunction'. We use the word 'sum' in place of 'disjunction'.

Elementary Product

A product (conjunction) of the variables or their negation in a formula is called Elementary product.

e.g., p, $\sim p$, $\sim p \wedge q$, $p \wedge q$, $\sim p \wedge q \wedge \sim p$, $p \wedge \sim p$, $q \wedge \sim p$ etc.

Elementary Sum

A sum (Disjunction) of the variable and their negation in a formula is called elementary sum.

e.g. p_1, $\sim p$, $\sim p \vee p$, $\sim p \vee q$, $p \vee \sim q$, $p \vee q \vee \sim p$ etc.

Factor

Any part of an elementary sum or product which is itself an elementary sum or product is called a factor of the original elementary sum or product *e.g.*,

$\sim p$, $p \wedge \sim p$ are the factors of $\sim p \wedge q \wedge p \wedge \sim p$,

Disjunctive Normal Form (DNF)

A formula which is equivalent to a given formula and which consists of a sum of elementary products is called a disjunctive normal form of the given formula. Disjunctive normal form of the given formula is not unique.

EXAMPLE 2.29. Obtain *DNF* of

(*a*) $p \wedge (p \to q)$

(*b*) $\sim(p \vee q) \leftrightarrow (p \wedge q)$

Solution : (*a*) $p \wedge (p \to q) \equiv p \wedge (\sim p \vee q)$

$$\equiv (p \wedge \sim p) \vee (p \wedge q) \quad \text{(By Distributive law)}$$

is the required DNF.

(*b*) $\sim(p \vee q) \leftrightarrow (p \wedge q)$

$$\equiv [\sim(p \vee q) \wedge (p \wedge q)] \vee [(p \vee q) \wedge \sim(p \wedge q)]$$

$$\equiv [(\sim p \wedge \sim q) \wedge (p \wedge q)] \vee [(p \vee q) \wedge (\sim p \vee \sim q)]$$

$$\equiv (\sim p \wedge \sim q \wedge p \wedge q) \vee [(p \wedge \sim p) \vee (p \wedge \sim q) \vee (q \wedge \sim p) \vee (q \wedge \sim q)]$$

By Distributive Law

$$= (\sim p \wedge \sim q \wedge p \wedge q) \vee (p \wedge \sim p) \vee (p \wedge \sim q) \vee (q \wedge \sim p) \vee (q \wedge \sim q)$$

It is clear from the above formula that DNF is identically false if every elementary product appearing in its DNF is identically false if at least one of them is true then the formula is true.

Minterms

Let p and q be two statement variables. All possible formulas which consists of conjunction of p (or its negations) and q (or its negation). None of the formula should contain both variable and its negation. These formulas are called minterms. No two minterms are equivalent.

For n variables $p_1, p_2, ...p_z$ there are 2^n such minterms. In particular for 2 varibles say p and q there are 2^2 such formulas $p \wedge q, p \wedge \sim q, \sim p \wedge q$, and $\sim p \wedge \sim q$.

$p \wedge \sim p$ is not a minterm.

One can find the minterms by using truth table. Each row in the truth table will give one minterm by using conjunction between the variables. For example, consider the table having two variables attached

x	y	**minterms**
T	T	$x \wedge y$
T	F	$x \wedge \sim y$
F	T	$\sim x \wedge y$
F	F	$\sim x \wedge \sim y$

An in case of truth table there are 2^n rows for n number of variables, these rows gives each minterms. Similarly in the case of three variables.

x	y	z	**minterms**
T	T	T	$x \wedge y \wedge z$
T	T	F	$x \wedge y \wedge \sim z$
T	F	T	$x \wedge \sim y \wedge z$
T	F	F	$x \wedge \sim y \wedge \sim z$
F	T	T	$\sim x \wedge y \wedge z$
F	T	F	$\sim x \wedge y \wedge \sim z$
F	F	T	$\sim x \wedge \sim y \wedge z$
F	F	F	$\sim x \wedge \sim y \wedge \sim z$

Principal Disjunctive Normal form (PDNF)

For a given formula, an equivalent formula consisting of disjunction of minterms only is known as principal disjunctive normal form or "Sum of product canonical form".

EXAMPLE 2.30. Write *PDNF* of the following :

(*i*) $(\sim p \vee \sim q) \rightarrow (p \leftrightarrow \sim q)$

(*ii*) $q \wedge (p \vee \sim q)$

Solution : (*i*) $\sim \quad (p \wedge q) \rightarrow (p \wedge \sim q) \vee (\sim p \wedge q)$

$$\equiv \sim\sim(p \wedge q) \vee [(p \wedge \sim q) \vee (\sim p \wedge q)]$$

$$\equiv (p \wedge q) \vee (p \wedge \sim q) \vee (\sim p \wedge q)$$

Disjunction of minterms, therefore is the required PDNF.

(*ii*) $q \wedge (p \vee \sim q)$

$\equiv (q \wedge p) \vee (q \wedge \sim q)$

$\equiv (q \wedge p) \vee F$

$\equiv (q \wedge p)$ is the required form.

Conjunctive Normal form (CNF)

A formula which is equivalent to a given formula which consists of a product of elementary sum is called conjunctive normal form of the given formula. For a given formula conjunctive normal form is not unique. For a given formula CNF is identically true if every elementary sum is identically true.

Principal Conjunctive Normal Form (PCNF)

To define the principal conjunctive normal form, we first define formula which are called maxterms.

Maxterms

For a given number of variables, the maxterms consists of disjunctions in which each variable or its negation but not both appears only once. Maxterms are the dual of minterms. One can find maxterms by using truth table as in the case of minterms by replacing the sign conjunction ($\wedge$) by the disjunction ($\vee$).

For a given formula an equivalent formula consisting of conjunction of the maxterms only is known as Principle conjunctive normal form or 'product of sums canonical form. Method, for obtaining *PCNF* is similar to the one described for *PDNF*.

EXAMPLE 2.31. Obtain the principle conjunctive normal form of:

$(\sim p \to r) \wedge (q \leftrightarrow p)$

$\equiv [\sim(\sim p) \vee r] \wedge [(q \to p) \wedge (p \to q)]$

$\equiv (p \vee r) \wedge [(\sim q \vee p) \wedge (\sim p \vee q)]$

$\equiv [(p \vee r) \vee F] \wedge [(\sim q \vee p) \vee F] \wedge [(\sim p \vee q) \vee F]$

$\equiv [(p \vee r) \vee (q \wedge \sim q)] \wedge [(\sim q \vee p) \vee (r \wedge \sim r)] \wedge [(\sim p \vee q) \vee (r \wedge \sim r)]$

$\equiv (p \vee r \vee q) \wedge (p \vee r \vee \sim q) \wedge (\sim q \vee p \vee r) \wedge (\sim q \vee p \vee \sim r) \wedge$
$(\sim p \vee q \vee r) \wedge (\sim p \vee q \vee \sim r)$

$\equiv (p \vee q \vee r) \wedge (p \vee r \vee \sim q) \wedge (p \vee \sim q \vee \sim r) \wedge (\sim p \vee q \vee r) \wedge (\sim p \vee q \vee \sim r)$

EXAMPLE 2.32. Obtain the conjunctive normal form of:

(*i*) $p \wedge (p \to q)$

(*ii*) $\sim(p \vee q) \leftrightarrow (p \wedge q)$

Solution : (*i*) $p \wedge (p \to q)$

$p \wedge (\sim p \vee q)$

is the required conjunctive normal form.

(*ii*) $\sim(p \vee q) \leftrightarrow (p \wedge q) \equiv (\sim(p \vee q) \to (p \wedge q)) \wedge ((p \wedge q) \to \sim(p \vee q))$

$\equiv ((p \vee q) \vee (p \wedge q)) \wedge (\sim(p \wedge q) \vee \sim(p \vee q))$

$\equiv (p \vee q) \wedge (p \vee q) \wedge ((\sim p \vee \sim q) \vee (\sim p \wedge \sim q)$

$= (p \vee q) \wedge (\sim p \vee \sim q) \wedge (\sim p \vee \sim q)$

$= (p \vee q) \wedge (\sim p \vee \sim q)$

EXAMPLE 2.33. Obtain the Principle Conjunctive Normal (PCNF) form of

(*i*) $\sim (p \vee q)$

(*ii*) $\sim (p \rightarrow q)$

Solution : $\sim (p \vee q) \equiv \sim p \wedge \sim q$

$$\begin{aligned} &\equiv (\sim p \vee F) \wedge (\sim q \vee F) \\ &\equiv (\sim p \vee (q \wedge \sim q)) \wedge (\sim q \vee (p \wedge \sim p)) \\ &\equiv (\sim p \vee q) \wedge (\sim p \vee \sim q) \wedge (\sim q \vee p) \wedge (\sim q \vee \sim p) \\ &\equiv (\sim p \vee q) \wedge (\sim p \vee \sim q) \wedge (\sim q \vee p) \end{aligned}$$

EXAMPLE 2.34. Put the function $f = [(x \wedge y')' \vee z'] \wedge (x' \vee z)'$ in the Conjunctive Normal Form (CNF).

Solution : $((x \wedge y')' \vee z') \wedge (x' \vee z)'$

$$\begin{aligned} &= (x' \vee y \vee z') \wedge (x \wedge z') \\ &= (x \wedge z') \wedge (x' \vee y \vee z') \\ &= (x \wedge z' \wedge x') \vee (x \wedge z' \wedge y) \vee (x \wedge z \wedge z') \end{aligned}$$

EXAMPLE 2.35. Put the function $f = x \wedge (y \vee z)$ in the conjunctive normal form.

Solution : The given function is already in conjunctive normal form.

EXAMPLE 2.36. Find the *CN* form of the function $f = [x \wedge (y' \vee z)] \vee z'$ and then find its *DN* form from it.

Solution :

$$\begin{aligned} f &= (x \wedge (y' \vee z)) \vee z' \\ &= (x \vee z') \wedge [(y' \vee z) \vee z'] \\ &= (x \vee z') \wedge (y' \vee (z \vee z')) \\ &= (x \vee z') \wedge (y' \vee T) \\ &= (x \vee z') \wedge T \\ &= x \vee z' \\ &= (x \vee z') \vee F \\ &= (x \vee z') \vee (y \wedge y') \\ &= (x \vee z' \vee y) \wedge (x \vee y' \vee z') \text{ is the } CN \text{ form} \end{aligned}$$

Now consider CN form of the function f

$$f(x, y, z) = (x \vee z' \vee y)\,(x \vee y \vee z')$$

As $$f = ((f))' = [[(x \vee y \vee z') \wedge (x \vee y' \vee z')]']'$$

$$\begin{aligned} &= [(x \vee y \vee z') \vee (x \vee y' \vee z')']' \\ &= [(x' \wedge y' \wedge z) \vee (x' \wedge y \wedge z)]' \end{aligned}$$

$$= (x' \wedge z)'$$

$$= x \vee z'$$

$$= (x \wedge T) \vee (z' \wedge T)$$

$$= [x \wedge (y \vee y')] \vee [z' \wedge (y \vee y')$$

$$= (x \wedge y) \vee (x \wedge y') \vee (z' \wedge y) \wedge (z' \wedge y')$$

is the required DN fom.

SOLVED PROBLEMS

PROBLEM 1. Which of the following are logically equivalent to $\forall x\ \exists y p(x, y)$? Explain.

1. $\exists x\ \forall y\ \overline{P(x, y)}$ 2. $\forall x\ \forall y\ \overline{P(x, y)}$

Solution: 1. $\overline{\forall x\, \exists y\, P(x, y)} \equiv \exists x\ \forall y\ \overline{P(x, y)}$

To show this more clearly : $\overline{\forall x\, \exists y\, P(x, y)}$ By De Morgan's Laws

$$\equiv \exists x\ \overline{(\exists y\ P(x, y)}$$

$$= \exists x\ \forall y\ \overline{P(x, y)}$$

Thus the given statement is logically equivalent, (*i.e.,*)

$$\forall x\ \overline{\exists y\, P(x, y)} \equiv \forall x\ \forall y\ \overline{P(x, y)}$$

But $\quad \forall x\ \forall y\ \overline{P(x, y)} \not\equiv \exists x\ \forall y\ \overline{P(x, y)}$

So the statement is not logically equivalent.

Note: Many people claimed that the two statements were logically equivalent by setting $P(x, y) = x < y$ the domain D of all integers. Doing this would get a truth value of false for each statement, but this does not necessarily mean that the two statements are logically equivalent :

Let $\quad D =$ all integers.

$$P(x, y) = x < y$$

$\overline{\forall x\, \exists y\, P(x, y)} \equiv \exists x\ \forall y\ \overline{P(x, y)}$ Means that there exists $x \in D$ such that for all $y \in D$.

$x \geq y$. This is of course is false since we can always choose $y = x + 1$. Now the second statement says $\forall x\ \overline{\exists y\, P(x, y)} \equiv \forall x\ \forall y\ \overline{P(x, y)}$ which claims that for every x, for every y, $x \geq y$, which is also false. However the statements

1. There exists an x such that for all y, $x \geq y$, and

2. for every x, for every y, $x \geq y$ are not equivalent, although they are both false.

Furthermore, we can show that (2) implies (1) but is not equal to (1).

PROBLEM 2. Give the meaning of each statement given below. Clarify each statement by re-phrasing it.

1. Economist Robert J. Samuelson : Every environmental problem is not a tragedy.
2. Headline over a Martha Stewart Column : All Lampshades can't be cleaned.
3. From Newsweek: Formal investigations are a sound practice in the right circumstances, but every circumstance is not right.

Solution:

1. *As stated:* No evnrionment problem is a tragedy.

   Intended: Not every environmental problem is a tragedy.
2. *As stated*: Every lampshade cannot be cleaned or no lampshade can be cleaned.

   Intended: Not every lampshade can be cleaned.
3. *As stated*: No circumstance is right. Formal investigation are never a sound practice.

   Intended: Not every circumstance is right, Investigations sometimes sound.

Give an argument using rules of inference to show that the conclusion follows from the hypothesis.

EXERCISE

1. Give two propositions each, the truth values of which are T and F respectively. Also give two examples of the sentences that are not propositions.
2. Give the truth value of the conjunction of the following propositions.
   (*i*) "$3 \div 5 = 2$", "Ram is in Delhi".
   (*ii*) "$5 + 7 = 12$", $x^2 \geq 0 \ \forall \ x$
   (*iii*) "$x^2 - 3x + 5 = 0$ has two solutions". "$9 \geq 7$".
3. Write down two propositions p and q for wihch $q \to p$ is true but $p \leftrightarrow q$ is false.
4. Write down the truth table of $p \to q \wedge {\sim}r \leftrightarrow r \wedge q$
5. Form the truth table of $p \wedge q \vee {\sim}r$ and $(p \wedge q) \vee ({\sim}r)$ and explain where differ.
6. How would you bracket the following formulae to correctly interpret them?
   (*i*) ${\sim}p \vee q$ (*ii*) ${\sim}p \to {\sim}q$
   (*iii*) $p \to q \leftrightarrow {\sim}p \vee q$ (*iv*) $p \oplus q \wedge \mathrm{r} \to {\sim}p \vee q \leftrightarrow p \wedge r$
7. Use the laws given in this section to show that ${\sim}({\sim}p \wedge q) \wedge (p \vee q) \equiv p$
8. Write down suitable mathematical statements that can be represented by the following symbolic propositions. Also write down their negations. Give the truth value of the given propositions.
   (*i*) $(\forall \ x) \ (\exists \ y)p$ (*ii*) $(x \geq 0) \to (\exists \ y) \ (y^2 = x)$
   (*iii*) $(\exists \ x) \ (\exists! \ y) \ (x \ y = 0)$ (*iv*) ${\sim}(\exists x) \ (\exists! \ y) \ (x + y = 0)$
9. Obtain the disjunctive normal form of the Boolean expression
   (*i*) $f(x_1, x_2, x_3) = (x_1{}' \wedge x_2) \vee (x_1 \wedge x_3)$
   (*ii*) $= (x_1 \vee x_2)' \vee (x_1{}' \wedge x_3)$

10. Obtain the CNF of the Boolean expression
    (*i*) $f(x_1, x_2, x_3) = (x_1 \wedge x_2)' \vee (x'_1 \wedge x_3)$
    (*ii*) $f(x_1, x_2, x_3) = (x_1 \wedge x_2') \vee (x_1' \wedge x_3')'$
11. Reduce the following Boolean expression to a simple form.
    (*i*) $f(x_1, x_2) = (x_1 \wedge x_2) \wedge (x_1 \wedge x_2')$
    (*ii*) $f(x_1, x_2, x_3) = (x_1 \wedge x_2) \vee (x_1 \wedge x_2' \wedge x_3) \vee (x_1 \wedge x_3)$
    (*iii*) $f(x_1, x_2, x_3) = (x_1 \wedge x_2) \vee ((x_1 \wedge x_2) \wedge x_3) \vee (x_2 \wedge x_3)$
12. Check whether the following sentences are statements or not.
    (*i*) Come in !
    (*ii*) What are you doing?
    (*iii*) 7 is a lucky number.
    (*iv*) This statement you are reading is false.
13. State the following argument in symbolic form and test its validity.
    "If the parcel is not propely addressed or is too large, then the Postoffice will not accept it. The parcel is not too large. If Ram wrote the address on the parcel, then it is properly addressed. Hence if Ram wrote the address the Postoffice will accept the parcel." ***(End-term Exam., GGSIPU, 2001)***
14. Prove the following implication :
    $P \to Q \cdot Q \to S \cdot P \vee Q \Rightarrow S \vee R$ ***(End-term Exam., GGSIPU, 2001)***
15. Determine whether the following wffs are tautology, a contradiction.
    (*i*) $(P \to r) \vee (r \to \rceil q) \wedge \rceil r$
    (*ii*) $((P \vee q \to r) \wedge p) \to (r \to p)$ ***(End-term Exam., GGSIPU, 2001)***
16. (*a*) Show that the following arguments is a fallacy :
    $P \to q, \neg p \vdash \neg q$ where symbol $\neg$ represents not (~).
    (*b*) Write the Converse, Inverse and Contrapositive of the implication :
    "If it is rose, then it is red"
    (*c*) Negate the statement:
    "Every student has at least one Course where the teacher is an Asstt. Professor." ***(Terminal Exam., GGSIPU, 2001)***
17. (*a*) Show that:

$$\neg P \wedge (\neg q \wedge r) \vee (q \wedge r) \vee (p \wedge r) \equiv r$$

    (*b*) Obtain the conjunctive Normal form of $(P \to Q) \wedge (Q \vee (P \wedge R))$ and determine whether or not it is a tautology.
    (*c*) Show that $\neg (P \wedge Q)$ follows from $\neg P \vee \neg Q$ using poof by contradiction.
    (Terminal Exam., GGSIPU, 2001)
18. Use resolution principle in testing the validity of the following argument:
    "If the musicians sing the songs, then crowds will come, if the ticket prices are not too high.
    If the musicians sing the songs, then the ticket prices are not too high. Therefore, if the musicians sing the song, then crowds will come"
    (End-term Exam., GGSIPU, 2002)

19. (*a*) Write negetion, converse, inverse and contra positive statements of the following implication:
"If it is cold, then she wears a coat"
(*b*) Prove the De Morgan's Rule in Predicate logic
$\sim(\forall\, x\, P(x)) \Leftrightarrow \exists x \sim P(x)$
Also negate the statement" Every locality in New Delhi is clean"
(*c*) Define Conjunctive Normal Form (CNF) and Disjunctive Normal Form (DNF) in prepositional logic Also convert $P \leftrightarrow Q$ into DNF.
(End-term Exam., GGSIPU, 2002)

20. (*a*) Show that $\sim(P \wedge Q)$ logically follows from $\sim P \wedge \sim Q$ without using Truth Tables.
(*b*) If $P \to (Q \to R)$ is true then prove that $Q \to (P \to R)$ is true.
(End Term Exam., GGSIPU, 2001)

ANSWERS

1. $p : x^2 < 0\ \forall\ x \in R$" , the truth value of p is F
q : Taj Mahal is in Agra", the truth value of q is T.
Example of the sentences that are not propositions.
1.Do work.
2.$x + y \geq 9\ \forall x, y \in R$,

2. (*i*) Let $p : 3 \div 5 = 2$
q : Ram is in Delhi
The truth value of p is F
The truth value of q is T
Therefore truth value of $p \wedge q$ = F
(*ii*) Let $p : 5 + 7 = 12$
$q : x^2 \geq 0\ \forall\ x$
The truth value of p is T
The truth value of q is T
Therefore truth value of $p \wedge q$ is T
(*iii*) Let $p : x^2 - 3x + 5 = 0$ has two solutions.
$q : 9 \geq 7$
The truth value p is T
The truth value q is T
The truth value $p \wedge q$ is T.

5. (*i*) $(\sim p) \vee q$
(*ii*) $(\sim p) \to (\sim q)$
(*iii*) $(p \to q) \leftrightarrow ((\sim p) \vee q))$
(*iv*) $((p \oplus q) \wedge r) \to ((\sim p) \vee q) \leftrightarrow (p \wedge r)$

6. $\sim(\sim p \wedge q) \wedge (p \vee q)$
$\equiv (p \vee \sim q) \wedge (p \vee q)$

$\equiv p \vee (\sim q \wedge q)$

$\equiv p \vee 0$

$\equiv p$

7. $(p \rightarrow (q \wedge (\sim r)) \leftrightarrow (r \wedge q)$

p	q	r	$\sim r$	$q \wedge (\sim r)$	$p \rightarrow (q \wedge \sim r)$	$r \wedge q$	$LHS \leftrightarrow RHS$
T	T	T	F	F	F	T	F
T	T	F	T	T	T	F	F
T	F	T	F	F	F	F	T
T	F	F	T	F	F	F	T
F	T	T	F	F	T	T	T
F	T	F	T	T	T	F	F
F	F	T	F	F	T	F	F
F	F	F	T	F	T	F	F

8. (*i*) True
   (*ii*) True
   (*iii*) True
   (*iv*) is equivalent to $(\forall x \in R)(\sim (\exists! y \in R)(x + y = 0))$ *i.e.*, for every x there is no unique y such that $x + y = 0$
   This is clearly false because for every x these is a unique $y(= - x)$ such that $x + y = 0$

12. All the given sentences are not propositions as:
   (*i*) Exclamation
   (*ii*) Question
   (*iii*) and (*iv*) are expressions of opinion.

DO IT BY YOURSELF

1. Using the statements R : Mark is rich H : Mark is happy.
   Write the following statements in symbolic form:
   (*a*) Mark is poor but happy (*b*) Mark is rich or unhappy
   (*c*) Mark is neither rich nor unhappy
   (*d*) Mark is poor or he is both rich and unhappy
2. Assuming that p and q are F and q and s are T, find the truth value of each proposition:
   (*a*) $p \rightarrow q$ (*b*) $\neg p \rightarrow \neg q$
   (*c*) $\neg (p \rightarrow q)$ (*d*) $(p \rightarrow q) \wedge (q \rightarrow r)$
   (*e*) $(p \rightarrow q) \rightarrow r$ (*f*) $p \rightarrow (q \rightarrow r)$
   (*g*) $(s \rightarrow (p \wedge \neg r) \wedge ((p \rightarrow (r \vee q)) \wedge s)$ (*h*) $(p \wedge \neg q) \rightarrow (q \wedge r) \rightarrow (s \vee \neg q)$
3. Rewrite the following statements without using the conditional:
   (*a*) If it is cold, he wears a hat.
   (*b*) If productivity increases, then wages rise.
4. Determine the converse, inverse and contrapositive of each statement:
   (*a*) If john is a poet, then he is poor.
   (*b*) Only if Marc studies will he pass the test.

5. Write the negation of each statement as simply as possible:
   (*a*) It she works, she will earn money.
   (*b*) He swims if and only i the water is warm.
   (*c*) If it snows, then they do not drive the car.
6. Formulate the symbolic expressions in words using
   p: Today is morning, q: It is raining, r. It is hot.
   (*a*) $p \to q$
   (*b*) $\neg q \to (r \wedge p)$
   (*c*) $\neg p \to (q \vee r)$
   (*d*) $\neg (p \vee q) \leftrightarrow r$
   (*e*) $(p \wedge (q \vee r)) \to (r \vee (q \vee p))$
7. Show that: $(p \to q) \equiv (\neg p \vee q)$, $(\neg p \leftrightarrow q) \equiv (p \wedge q) \vee (\neg p \wedge \neg q)$
8. (*a*) Write truth tables for exclusive OR, NAND and NOR.
   (*b*) Prove that $\{\uparrow\}$, $\{\downarrow\}$, $\{\neg\ \wedge\}$ are functionally complete sets.
   (i.e. define $\wedge$, $\vee$, $\neg$, $\to$, $\leftrightarrow$ using connective(s) of particular set only)
   (*c*) Express $P \uparrow Q$ in terms of $\downarrow$ only. Verify the DeMorgan law, $\neg (p \wedge q) \equiv \neg p \vee \neg q$.
9. Define the truth table for *imp*2 by

p	***q***	***p* imp 2 *q***
T	T	T
T	F	F
F	T	T
F	F	F

   (*a*) Show that $(p \text{ imp2 } q) \wedge (q \text{ imp2 } p) \neq p \leftrightarrow q$.
   (*b*) Show that above equation remains true if we alter imp2 so that if p is false and q is true, then p imp2 q is false.
10. Obtain the principal disjunctive and conjunctive normal forms of the following formulas.
   (*a*) $(\neg P \vee \neg Q) \to (P \leftrightarrow \neg Q)$
   (*b*) $Q \wedge (P \vee \neg Q)$
   (*c*) $P \vee (\neg P \to (Q \vee (\neg Q \to R)))$
   (*d*) $P \to (P \wedge (Q \to P))$
   (*e*) $(Q \to P) \wedge (\neg P \wedge Q)$
   Which of the above formulas are tautologies?
11. Test the validity of each argument:
   (*a*) If it rains, Erik will be sick.
   It did not rain.
   ∴ Erik was not sick.
   (*b*) If it rains, Erik will be sick.
   Erik was not sick.
   ∴ It did not rain.
   (*c*) If it study, then I will not fail mathematics.
   If I do not play basketball, then I will study.
   But I failed mathematics.
   Therefore I must have played basketball.
12. Show that $R \wedge (P \vee Q)$ is a valid conclusion from the premises $P \vee Q$, $Q \to R$, $P \to M$ and $\neg M$.

13. Show that $\neg (p \wedge q)$ follows from $\neg p \wedge \neg q$.
14. Show that $R \rightarrow S$ can be derived from the premises $P \rightarrow (Q \rightarrow S)$, Q and $\neg R \vee$ P.
15. Show that following premises are inconsistent:
    (a) If Jack misses many classes through illness, then he fails high school.
    (b) If Jack fails high school, then he is uneducated.
    (c) If Jack reads a lot of books, then he is not educated.
    (d) Jack misses many classes through illness and reads a lot of books.
16. Show that the following premises are inconsistent:
    (a) $P \rightarrow Q$, $P \rightarrow R$, $Q \rightarrow \neg R$, P
17. Show without using Truth table that if $P \rightarrow (Q \rightarrow R)$ is true then, $Q \rightarrow (P \rightarrow R)$ is also True.
18. Let P(x): x is a person.
    F (x, y); x is the father of y.
    M (x, y): x is the mother of y.
    Write the predicate "x" is the father of the mother of y.
19. Symbolize the expression "All the world loves a lover." Using:
    P(x): x is a person
    L(x): x is a lover.
    R(x, y): x loves y.
20. Negate the following statements:
    (a) If the teacher is absent, then some students do not complete their homework.
    (b) All the students completed their homework and the teacher is present.
    (c) Some of the students did not complete their homework or the teacher is absent.
    (d) Every student has at least one course where the lecturer is not a teaching assistant.
21. Determine the truth value of each statement. The domain of discourse is the set of real numbers. Justify your answers.
    (a) For every real number x, if $x > 1$, then $x + 1 > 1$.
    (b) For every real number x, $x^2 - 1 > 0$.
    (c) For some positive integer n, if n, is prime, then $n + 1$, $n + 2$, $n + 3$ and $n + 4$ are not prime.
    (d) For every x, for some y, $x + y = 0$.
    (e) For every x, for every y, if $x^2 < y^2$, then $x < y$.
22. Let T (x, y) be the propositional function "x is taller than y." The domain of discourse consists of three students: Garth, who is 5 feet 11 inches tall; Erin, who is 5 feet 6 inches tall; and Mary who is 6 feet tall. Write the following propositions in word and tell whether it is T or F. Also write negation of each proposition in words and symbolically.
    (a) $\forall x \, \forall y \; T(x, y)$
    (b) $\exists x \, \forall y \; T(x, y)$
    (c) $\forall x \, \exists y \; T(x, y)$
    (d) $\exists x \, \exists y \; T(x, y)$
23. Test the validity:
    Every 2nd year student likes FCS.

If a student likes FCS, then he/she is happy.
Therefore, every 2nd year student is happy.

24. Consider the statement "Given any positive integer n, there is a greater positive integer." Symbolize this statement with and without using the set of positive integers as the Universe of discourse.
25. Prove that:
$(\exists x)\ (P(x) \wedge Q\ (x)) \Rightarrow (\exists x)\ P\ (x) \wedge (\exists x)\ Q\ (x)$
26. Show that from
$(a)(\exists x)\ (F(x) \wedge S\ (x)) \rightarrow (y)\ (M\ (y) \rightarrow W(y))$
$(b)(\exists y)\ (M\ (y) \wedge \neg\ W(y))$
the conclusion (x) $(F(x) \rightarrow S\ (x)$ follows.
27. Show that
$(x)\ (p\ (x) \vee q(x)) \Rightarrow (x)p(x)\ \ (\exists\ x)\ q(x)$
28. Verify the existentially quantified statement:
For some real number ×, $\dfrac{1}{x^2+1} > 1$ is false.
29. Show that for universally and existentially quantified statements, the correct connectives that should be used are the conditional and conjunction respectively.
30. Negate each of the following statements:
   (*a*) If there is riots, then someone is killed.
   (*b*) It is daylight and all the people have arisen.

MULTIPLE CHOICE QUESTIONS

1. The number of rows in the truth table of a compound statement having n distinct primary or atomic statement is:
   (*a*) $n + 1$ (*b*) n^2
   (c) 2^n (*d*) 2^{n-1}
2. Out of these, which one is a tautology?
   (*a*) $p \wedge p$ (*b*) $\sim p \wedge \sim p$
   (c) $p \vee p$ (*d*) $\sim \{p \wedge (\sim p)\}$
3. Which of the following statement has truth value "TRUE"?
   (*a*) $4 + 5 = 8$ or $0 > 1$ (*b*) $1 > 0$ or $3 + 8 - 1 = 23$
   (c) Delhi is in Rome and Lotus temple in Delhi
   (*d*) $p \vee (p \wedge q) \equiv \sim p$
4. $(p \rightarrow q) \wedge (r \rightarrow q)$ is equivalent to:
   (*a*) $(p \vee r) \rightarrow q$ (*b*) $p \vee (q \rightarrow r)$
   (*c*) $p \vee (r \rightarrow q)$ (*d*) $p \rightarrow (q \rightarrow r)$
5. $(A \vee q) \wedge (A \rightarrow C) \wedge (B \rightarrow D)$ is equivalent to:
   (*a*) $D \vee C$ (*b*) $D \rightarrow C$
   (*c*) $D \vee C$ (*d*) None of these
6. Negation of $p \rightarrow q$ is:
   (*a*) $p \vee \sim q$ (*b*) $\sim p \vee q$

(c) $p \wedge \sim q$ (d) $\sim p \wedge q$

7. Assume the value of $p \rightarrow q$ is false. Determine the value of $(\sim p \vee \sim q) \rightarrow q$
   (a) True (b) False
   (c) Uncertain (d) None of these.
8. Determine the negation of the following statement $\forall\, x \,\exists\, y\ P(x, y)$
   (a) $\forall\, x \,\exists\, y \sim P(x, y)$ (b) $\forall\, x \,\forall\, y \sim P(x, y)$
   (c) $\exists\, x \,\exists\, y \sim P(x, y)$ (d) $\exists\, x \,\forall\, y \sim P(x, y)$
9. If $P(x) : x$ is mortal and $Q(x) : x$ is man then symbolic representation of the statement "All man are mortal" is:
   (a) $\exists x\, [Q(x) \rightarrow P(x)]$ (b) $\exists x\, [Q(x) \rightarrow P(x)]$
   (c) $\forall x\, [P(x) \rightarrow Q(x)]$ (d) $\exists x\, [P(x) \rightarrow Q(x)]$
10. Determine the validity of the conclusion from the set of premises, $p \rightarrow q$, $r \vee \sim s$, $q \rightarrow s$ and $\sim p \rightarrow r$
   (a) $p \vee q$ (b) r
   (c) $\sim q$ (d) None of these.

Answers

1. (c)	2. (d)	3. (b)	4. (a)	5. (c)	6. (c)
7. (b)	8. (d)	9. (a)	10. (b)		

❑❑❑

CHAPTER 3

METHODS OF PROOF

This Chapter Covers :

3.1. INTRODUCTION

In the previous chapter we studied about statements and their truth values. In this chapter we will discuss the way in which statements can be linked to form a logically valid argument. we will talk about what a theorem is and what constitutes a mathematical acceptable proof ?

The statement that is proved to be true is called a ***theorem.*** Special kinds of theorems are referred to as ***lemmas*** and ***corollaries***. A ***lemma*** is a theorem that is usually not too interesting in its own right but is useful in proving another theorem. A ***corollary*** is a theorem that follows quickly from another theorem. A ***conjecture*** is a statement whose truth value is unknown. When a proof of conjecture is found, the conjecture becomes a theorem. Many times conjectures are shown to be false, so they are not theorems. An argument that establishes the truth of the theorem is called a ***proof*** and ***logic*** is a tool for the analysis of proofs.

Sometimes, instead of showing that a given statement p is true, we try to prove that it is false that is $\sim p$ is true. Such a *proof* is called **method of contradiction**.

A proof uses the hypothesis, axioms, definitions and so on, to reach a conclusion. Each step of the proof involves drawing intermediate conclusions. For the proof as a whole to be valid, each step of the proof must result in a valid, intermediate conclusion. So in this chapter we discuss the ***Rules of Inference***, which are brief **valid arguments*** used within a larger argument such as a proof. These rules provides the justification of the steps used to show that a conclusion follows logically from a set of hypothesis.

3.2. SEVERAL WAYS OF PROVING A THEOREM

There are several ways of proving a theorem. All of them are based on one or more rules of inference which are different form of arguments. Following are the most commonly used rules.

Table 3.1. Rules of Inference

	Rule of Inference	Tautology	Name of the Rule
1.	$\dfrac{p}{\therefore p \vee q}$	$p \to (p \vee q)$	Addition
2.	$\dfrac{p \wedge q}{\therefore p}$	$(p \wedge q) \to p$	Simplification
3.	p $\dfrac{p \to q}{\therefore q}$	$[p \wedge (p \to q)] \to q$	Modus pones
4.	$\sim q$ $\dfrac{p \to q}{\therefore \sim p}$	$[(\sim q) \wedge (p \to q)] \to \sim p$	Modus tollens
5.	$p \to q$ $\dfrac{q \to r}{\therefore p \to r}$	$[(p \to q) \wedge (q \to r)]$ $\to (p \to r)$	Hypothetical Syllogism (Chain Rule)
6.	$p \vee q$ $\dfrac{\sim p}{\therefore q}$	$(p \vee q) \wedge (\sim p) \to q$	Disjunctive Syllogism.

Now we will discuss some of the rules given in the table.

3.2.1. Law of Detachment (Modes Pones)

'Modes pones' is a latin term which means 'Method of affirmation'.

It is a tautology given in the table 3.2.

$$[(p \to q) \wedge p] \Rightarrow q$$

* An argument with premises (hypothesis) $p_1, p_2 .... p_r$ and conclusion q is said to be valid if and only if, the sentence $(p_1 \wedge p_2 \wedge ..... \wedge p_r) \to q$ is a tautology.

Table 3.2.

p	q	$p \to q$	$(p \to q) \wedge p$	$(p \to q) \wedge p \to q$
T	T	T	T	T
T	F	F	F	T
F	T	T	F	T
F	F	T	F	T

Modes pones states that *if both an implications and its hypothesis are known to be true, then the conclusion of this implication is true.*

This form of valid argument is called the law of detachment (Modes pones) because of conclusion q is detached from a premise $(p \to q)$.

EXAMPLE 3.1(*a*). Suppose that the implication 'If it rains today, then we will have rain bath' and its hypothesis is 'it is raining today,' are true. Then by Modes pones, it follows that the conclusion of the implication, "We will have rain bath," is true, as given below.

p : It rains today.

q : We will have rain bath.

then premises $(p \to q)$ and hypothesis is p

$\therefore$ $[(p \to q) \wedge p] \to q$ by Modes Pones

EXAMPLE 3.1(*b*). Consider the following argument:

If I work hard, then I will get first position in the examination.

I work hard.

Therefore, I get first position in the examination.

This is a **valid argument** by Modus pones as:

If p : I work hard

q : I get first position in the examination

In the symbolic form

$$\begin{array}{rl} & p \to q \\ & \underline{p} \\ \therefore & \underline{q} \end{array}$$

That is $(p \to q) \wedge p \Rightarrow q$ is a Tautology.

EXAMPLE 3.1(*C*). In the example 3.1(*b*) the following argument is not a valid argument.

If I work hard, then I will get first positioin in the examination

I get first position in the examination.

Therefore, I work hard.

p : I work hard

q : I will get first position in the examination

In symbolic form

$$\begin{array}{rl} & p \to q \\ & \underline{q} \\ \therefore & \underline{p} \end{array}$$

$(p \to q) \wedge q \to p$ is not a tautology given below

p	q	$p \to q$	$(p \to q) \wedge q$	$(p \to q) \wedge q \to p$
T	T	T	T	T
T	F	F	F	T
F	T	T	T	F
F	F	T	T	F

EXAMPLE 3.2. State which rule of inference is the basis of the following argument,' "It is below freezing now. Therefore, it is either below freezing or raining now".

Let p : It is below freezing now.

q : It is raining now.

therefore $p \to p \vee q$

Here **"addition"** is the **rule of inference.**

EXAMPLE 3.3. State which rule of inference is used in the argument :

If it rains today then we will not have a play today. If we do not have a play today, then we will have a play tomorrow. Therfore if it rains today then we will have a play tomorrow.

p : It rains today

q : We will not have a play today

r : We will have a play tomorrow.

hypothesis $(p \to q) \wedge (q \to r) \Rightarrow (p \to r)$

Here we use the rule of inference **'Hypothetical syllogism.'**

3.2.2. Law of Contraposition (or Modus tollens)

Modes tollens means method of **Denial**. It is given by tautology

$[(p \to q) \wedge \sim q] \to \sim p$

Consider the truth table (3.3)

Table 3.3.

p	q	$p \to q$	$\sim q$	$(p \to q \wedge \sim q$	$(p \to q) \wedge \sim q \Rightarrow \sim p$
T	T	T	F	F	T
T	F	F	T	F	T
F	T	T	F	F	T
F	F	T	T	T	T

3.2.3. Disjunctive Syllogism

It is given by the tautology $[(p \vee q) \wedge \sim p] \Rightarrow q$

Consider the example

EXAMPLE 3.4. Either Ram is a good boy or Rahul is a good boy. Ram is not a good boy therefore Rahul is a good boy.

Let p : Ram is a good boy

q : Rahul is a good boy

$[(p \vee q) \wedge (\sim p)] \to q$

is a tautology given as

p	q	$\sim p$	$(p \vee q)$	$(p \vee q) \wedge (\sim p)$	$[(p \vee q) \wedge (\sim p)] \Rightarrow q$
T	T	F	T	F	T
T	F	F	T	F	T
F	T	T	T	T	T
F	F	T	F	F	T

therefore, Disjunctive syllogism is a valid argument.

3.2.4. Hypothetical Syllogism

It is one of the most useful rule of inference and is given by the tautology It is also known as **Chain rule.**

$[(p \to q) \wedge (q \to r)] \Rightarrow p \to r$

Consider the truth table (3.4).

Table 3.4.

p	q	r	$p \to q$	$q \to r$	$(p \to q) \wedge (q \to r)$	$p \to r$	$(p \to q) \wedge (q \to r) \Rightarrow p \to r$
T	T	T	T	T	T	T	T
T	T	F	T	F	F	F	T
T	F	T	F	T	F	T	T
T	F	F	F	T	F	F	T
F	T	T	T	T	T	T	T
F	T	F	T	F	F	T	T
F	F	T	T	T	T	T	T
F	F	F	T	T	T	T	T

Therefore hypothetical syllogism is a valid argument.

EXAMPLE 3.5. If I will do my work, I will go to see movie.

If I will go to see movie, I will have popcorns.

Therefore if I will do my work, I will have popcorns.

Let p : I will do my work.

q : I will go to see movie.

r : I will have popcorn.

$$\begin{array}{l} p \to q \\ \underline{q \to r} \\ \Rightarrow p \to r. \end{array}$$

$(p \to q) \wedge (q \to r) \Rightarrow (p \to r)$

We observe that validity of an argument depends on its form but not on its contents. For example, consider the following arguments.

EXAMPLE 3.6. If a polygon is a square then it has 3 sides.

If a polygon has 3 sides, then it has 4 angles.

Therefore, if a polygon is a square, then it has 4 angles.

Now in the above argument, the premises are false but the conclusion is true, and the argument is true.

In symbolic form
p : A polygon is a square
q : A polygon has 3 sides
r : A polygon has 4 angles.

$$p \to q$$
$$q \to r$$
$$\therefore \quad p \to r$$

p	q	r	$p \to q$	$q \to r$	$p \to r$	$(p \to q) \wedge (q \to r) \Rightarrow (p \to r)$
F	F	F	T	T	T	T

EXAMPLE 3.7. If a polygon is a square than it has 4 sides.
If a polygon has 4 sides, then it has 3 angles.
Therefore, if a polygon is a square, then it has 3 angles.
Here the conclusion is false, but the argument is valid since it is of the form

3.2.4. Hypothetical Syllogism

p : Polygon is a square
q : Polygon has 4 sides
$\therefore$ r : Polygon has 3 angles

$(p \to q) \wedge (q \to r) \Rightarrow (p \to r)$

$p \to q$	$q \to r$	$(p \to q) \wedge (q \to r)$	$p \to r$	$(p \to q) \wedge (q \to r) \Rightarrow (p \to r)$
T	F	F	F	T

3.3. DIFFERENT METHODS OF PROOF

Let us now see how different forms of arguments can be put together to prove or disprove a statement.

3.3.1. Direct Proof

The direct proof is based entirely on modes pones.

A direct proof of $p \Rightarrow q$ is a logical valid argument in which we start with the assumption that hypothesis is true, that is, p is true.

A direct proof assumes that $p(x_1, x_2, \ldots x_n)$ is true and then using $p(x_1, x_2, \ldots x_n)$ as well as other axioms, definitions and previously derived theorems shows directly that $q(x_1, x_2, \ldots\ldots x_n)$ is true.

EXAMPLE 3.8. Give a direct proof of the statement,'

"The product of two odd integers is odd."

Let x and y be two odd integers then

$$x = 2n + 1 \qquad \text{for } n, m \in Z$$
$$y = 2m + 1$$
$$xy = (2n + 1)(2m + 1)$$
$$= 4mn + 2n + 2m + 1$$
$$= 2(mn + n + m) + 1$$
$$= 2p + 1 \qquad p \in Z$$

is an odd integer

$\therefore$ the result follows.

EXAMPLE 3.9. Give a direct proof the statement,

"The square of an even integer is an even integer".

Let x be an even integer then for some $n \in Z$

$$x = 2n$$

Squaring on both sides

$$\begin{aligned} x^2 &= (2n)^2 = 4n^2 = 2(2n)^2 \\ &= 2p \quad \text{where } p = 2n^2 \in Z \\ &= \text{even} \end{aligned}$$

$\therefore$ the result follows.

EXAMPLE 3.10. If $x \in R$ (set of real numbers) such that $x^2 = K$ then either $x = \sqrt{K}$ or $x = -\sqrt{K}$.

Let p and q are two propositions

$p : x \in R$ such that $x^2 = K$ and

$q : x = \sqrt{K}$ or $x = -\sqrt{K}$

Now $x^2 = K$

$$\Rightarrow \sqrt{x^2} = \pm\sqrt{K}$$

Therefore, p is true and $(p \Rightarrow q)$ is true, conclude that q is true by Rule of "***Modus Pones***".

EXAMPLE 3.11. Give a direct proof of the following statement.

For all real numbers x, x_1, x_2 and y,

If $\quad x = \max\{x_1, x_2\}$

and $\quad y \geq x$

then $\quad y \geq x_1$ and $y \geq x_2$

Now given $\quad y \geq x$

$$\Rightarrow y \geq \max\{x_1, x_2\}$$

Case I : If $\max\{x_1, x_2\} = x_1$

$\Rightarrow x_1 \geq x_2$

$\Rightarrow y \geq x_1$ as well as $y \geq x_2$

Case II : If $\max\{x_1, x_2\} = x_2$

$\Rightarrow x_2 \geq x_1$

$\Rightarrow y \geq x_2$ as well as $y \geq x_1$

Therefore $y \geq x_1$ and $y \geq x_2$.

3.3.2. Indirect Methods

In the direct method we prove the argument $p \Rightarrow q$ by directly using that p is true (*i.e.*, hypothesis is true) then proved that q is true.

Now we shall consider two round about methods for proving $p \Rightarrow q$ indirectly.

1. Proof by contrapositive.
2. Proof by contradiction.

Proof by Contrapositive

The method of contrapositive says $p \Rightarrow q$ is logicaly equivalent to its contrapositive $(\sim q \Rightarrow \sim p)$ *i.e.*,

$$(p \Rightarrow q) \equiv (\sim q \Rightarrow \sim p)$$

Consider the following truth table 3.5.

Table 3.5.

p	q	$p \rightarrow q$	$\sim p$	$\sim q$	$\sim q \rightarrow \sim p$
T	T	T	F	F	T
T	F	F	F	T	F
F	T	T	T	F	T
F	F	T	T	T	T

from the third and sixth column both $p \rightarrow q$ and $\sim q \rightarrow \sim p$ are equivalent.

In other words, in order to prove $p \Rightarrow q$ we assume that q is false and then show that p is false.

EXAMPLE 3.12. Prove that "If $x, y \in Z$ (set of integer) such that xy is odd then both x and y are odd, by proving its contrapositive.

As x and y both are integers then their product is also an integer *i.e.*, xy is an integer. Now

$P:$ xy is odd.

$q:$ x and y are odd.

Now we need to show

$$p \Rightarrow q$$

By using method of contrapositive we will prove $\sim q \Rightarrow \sim p$

Consider $\sim q$, *i.e.*, both x and y are not odd

Case I : x and y both are even

$\Rightarrow$ $x = 2n$ and $y = 2m$ for some $m, n \in Z$

$$\Rightarrow xy = (2n)(2m) = 4nm = 2(2mn) = 2p \quad \text{where } p = 2mn \in Z$$

$\Rightarrow$ xy is even

$\Rightarrow \sim p$ is true.

$\Rightarrow \sim q \Rightarrow \sim p$

Case II : One of x and y is even

Let x is even and y is odd

$\Rightarrow$ $x = 2n$ and $y = 2m + 1$ for some integer $m, n \in Z$

$$\Rightarrow xy = (2n)(2m + 1) = 2q \quad \text{where } q = n(2m + 1) \in Z$$

$\Rightarrow$ xy is even

$\Rightarrow$ $\sim p$ is true

$\Rightarrow$ $\sim q \Rightarrow \sim p$

Therefore, by method of contrapositives the result holds.

EXAMPLE 3.13. Prove that "If x and y are two integers such that x is even and y is odd then xy is even.

We will prove the result by direct method. By the given hypothesis

$p : x$ is even & y is odd.

$q : xy$ is even

$$\Rightarrow \quad x = 2n \text{ and } y = 2m + 1 \quad \text{for some } m, n \in Z$$

$$\Rightarrow \quad xy = (2n)(2m + 1)$$

$$= 4nm + 2n$$

$$= 2(2nm + n)$$

$$= 2p \qquad \text{where } p = (2mn + n) \in Z$$

$$= \text{even}$$

Therefore xy is even.

EXAMPLE 3.14. Prove that statement

"If x is an integer and x^2 is even then x is also even",

by proving its contrapositive.

Let p : x is an intger and x^2 is an even intger.

q : x is an even intger.

We need to prove $p \Rightarrow q$

By using contrapositive

$$\sim q \Rightarrow \sim p.$$

$\sim q$ says x is odd integer.

$$\Rightarrow x = 2n + 1,\ n \in Z$$

squaring on both sides

$$x^2 = (2n + 1)^2$$

$$= 4n^2 + 1 + 4n$$

$$= (4n^2 + 4n) + 1$$

$$= 2(2n^2 + 2n) + 1$$

$$= 2p + 1 \quad ; \quad p = 2n^2 + 2n \in Z$$

$\Rightarrow x^2$ is odd integer

Therefore $(\sim p)$ holds.

Therefore result is true by the method of contrapositive.

EXAMPLE 3.15. Give an indirect proof of the theorem "If $3n + 2$ is odd, then n is odd."

$p : 3n + 2$ is odd

$q : n$ is odd

we need to prove $p \Rightarrow q$

Consider the contrapositive.

$$\sim q \Rightarrow \sim p.$$

$\sim q : n$ is even

$$\Rightarrow n = 2k \quad , \quad k \in Z$$

Consider

$$3n + 2 = 3(2k) + 2$$

$$= 6p + 2$$

$$= 2(3k + 1)$$

$$= \text{even}$$

$\Rightarrow 3n + 2$ is even

$\Rightarrow \sim p$ holds

Therefore by method of contrapositive the result holds.

This result (Example 3.15.) can also be proved by another method called method of contradiction given as :

Method of Contradiction

This method is also called **reductio ad absurdum** in latin because it believes on reducing a given assumption to an absurdity.

For the argument $p \Rightarrow q$ we need to prove that if p is true then q is true. In this method we start by assuming that q is false (*i.e.*, $\sim q$ is true). Then by a logical argument we arrive at a situation where a statement is true as well as false that is we reach a contradiction $r \wedge \sim r$ for some statement r. This means that the truth of $\sim q$ implies a contradiction, a statement that is always false. This can only happen when $\sim q$ is false also. Therefore q must be true.

Now consider the example (3.15) again

we need to prove the theorem that

"If $3n + 2$ is odd, then n is odd".

$p : 3n + 2$ in odd

$q : n$ is odd

Now we use method of contradiction and we will prove the result.

Let $\sim q$ is true that is n is even

then $n = 2k$ for some integer k.

It follows that $3n + 2 = 3\,(2k) + 2 = 6k + 2 = 2\,(3k + 1)$

$\Rightarrow 3n + 2$ is even (since it is a multiple of 2)

Since the negation of the conclusion of the implication implies that the hypothesis is false, therefore the original implication is true.

EXAMPLE 3.16. Show that $\sqrt{3}$ is irrational.

Let us try and prove the given statement by contradiction. For this we begin by assuming that $\sqrt{3}$ is rational. This means that there exists positive integers x and y

Such that $\sqrt{3} = \dfrac{x}{y}$, where x and y are relatively prime.

This implies $x = \sqrt{3}\, y$

$\Rightarrow x^2 = 3y^2$

$\Rightarrow 3 | x^2$ (3 divides x^2)

$\Rightarrow 3 | x$ (3 divides x) ...(*i*)

Therefore, by definition, $x = 3p$ for some $p \in Z$

$\therefore \quad x^2 = 9p^2$

But $\quad x^2 = 3y^2$ also

so $\quad 9p^2 = 3y^2$

$\Rightarrow 3p^2 = y^2$

$\Rightarrow 3|y^2$ (3 divides y^2)

$\Rightarrow 3|y$ ...(*ii*)

(*i*) and (*ii*) implies that 3 divides both x and y which contradicts our assumption at x and y are relatively prime (that have no common factor).

Therefoere we conclude that our assumption is wrong that $\sqrt{3}$ is rational

So $\sqrt{3}$ is irrational.

EXAMPLE 3.17. Use the method of proof by contradiction to show that for $x \in R$, if $x^3 + 4x = 0$ then $x = 0$.

Let us assume that $x^3 + 4x = 0$ and $x \neq 0$

then $x(x^2 + 4x) = 0$ and $x = 0$.

$\Rightarrow x^2 + 4 = 0$

$\Rightarrow x^2 = -4$

But $x \in R$ and $x^2 = -4$ Not Possible.

Therefore our assumption is false.

Therefore the given statement is true.

EXAMPLE 3.18. Prove by method of contradiction the following statement :

"For all real numbers x and y, if $x + y \geq 2$ then either $x \geq 1$ or $y \geq 1$".

Solution : Let us assume that for $x, y \in R$,

$x + y \geq 2$ and the conclusion is not true.

Case I. $x < 1$ and $y < 1$

Adding both the inequalities

$x + y < 1$ Not possible

Case II. $x \geq 1$ and $y \geq 1$

Adding both the inequalities

$x + y \geq 2$

which is true.

Case III. $x \geq 1$ and $y < 1$

Now consider the given inequality

$\Rightarrow x + y \geq 2$

$\Rightarrow x \geq 2 - y$ ...(*i*)

As $y < 1$

$\Rightarrow -y > -1$

$\Rightarrow 2 - y > 2 - 1$

$\Rightarrow 2 - y > 1$

$\therefore$ (1) $\Rightarrow x \geq 1$ which is true

Therefore, the result is true for either $x \geq 1$ or $y \geq 1$.

EXAMPLE 3.19. From the following statement

"If n is an integer not divisible by 3, then $n^2 \equiv 1(\text{mod}3)$".

Let p : n is an integer not divisible by 3

$q : n^2 \equiv 1(\text{mod } 3)$.

Now $p \equiv p_1 \vee p_2$ where $p_1 : n \equiv 1 (\text{mod } 3)$ " and
$p_2 : n \equiv 2 (\text{mod } 3)$

Hence to show that $p \rightarrow q$ it can be shown that $p_1 \rightarrow q$ and $p_2 \rightarrow q$. It is easy to give direct proof of these two implications.

Case I. Let us suppose that p_1 is true.

$\Rightarrow n \equiv 1 (\text{mod } 3)$

$\Rightarrow 3 \mid (n-1)$ [3 divides $(n-1]$

$\Rightarrow n - 1 = 3k, k \in Z$

$\Rightarrow n = 3k + 1$

By squaring

$$n^2 = (3k+1)^2 = 9k^2 + 6k + 1 = 3(3k^2 + 2k) + 1$$

$\Rightarrow n^2 \equiv 1 (\text{mod } 3)$

Therefore $p_1 \Rightarrow q$ is true

Case II : Let us suppose that p_2 is true.

$\Rightarrow n \equiv 2$ (mod3)

$\Rightarrow 3 \mid (n-2)$ [3 divides $(n-2)$]

$\Rightarrow n - 2 = 3r$ $r \in Z$

$\Rightarrow n = 3r + 2$

By squaring

$$n^2 = (3r+2)^2 = 9r^2 + 12r + 4 = 3(3r^2 + 4r + 1) + 1$$

Hence $n^2 \equiv 1 (\text{mod } 3)$.

Therefore $p_2 \rightarrow q$ is true.

Since it has been shown that $p_1 \rightarrow q$ and $p_2 \rightarrow q$ are true, it can be concluded that $(p_1 \vee p_2) \rightarrow q$ is true.

As $p \equiv p_1 \vee p_2$

$\Rightarrow p \rightarrow q$ is true.

Therefore the result follows.

3.4. MATHEMATICAL INDUCTION

In this section, we will examine mathematical induction a technique for proving propositions over the positive integers. What does this principle really says? It says that if you can walk a few steps, say m steps, and if at each stage you can walk one more step, then you can walk any distance. Mathematical induction is used extensively to prove results about a large variety of discrete objects. For example, it is used to prove results about the complexity of algorithms, the correctness of certain types of computer programe, theorems about graphs and trees, as well as wide range of identities and inequalities.

An analogy : Mathematical induction is often useful in overcoming a problem such as this one. A proof by mathematical induction similar to knocking over a row or closely spaced dominos that are standing on end. To knock over the six dominos in fig. 3.1 all you need to do is push domino 1 to right. To be assured that they all will be knocked over, some work must be done ahead of time. The dominos must be positioned so that if any domino is pushed to the right, it will push the next domino in the line.

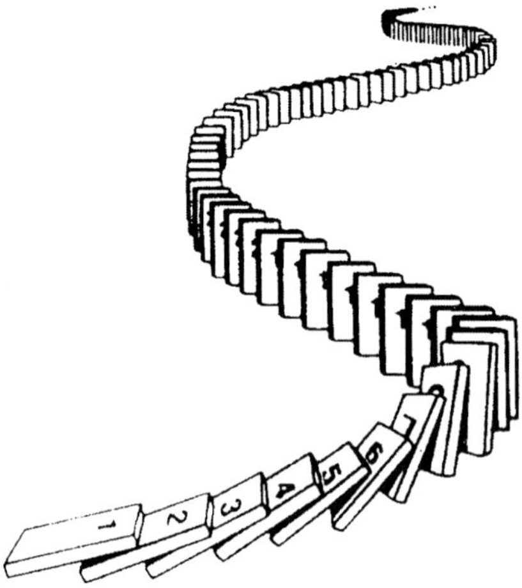

Fig. 3.1. Illustrating how mathematical induction works using dominos

Now imagine the propositions $p(1)$, $p(2)$, $p(3)$, ...to be an infinite line of dominos. Let's see if these propositions are in the same formation as the dominos were. First, we will focus on one specific point of the line : $p(99)$ and $p(100)$. We are not going to prove that either of these propositions is true, just that the truth of $p(99)$ implies the turth of $p(100)$. In terms of our analogy, if $p(99)$ is knocked over, it will knock over $p(100)$.

In proving p(99) $\Rightarrow p(100)$, we will use $p(99)$ as our premise. Consider the result "sum of the positive integers 1 to 100 is $\frac{100(100+1)}{2}$. We start by observing that the sum of the positive integers from 1 to 100 is (1 + 2 + ... + 99) + 100. That is, the sum of the positive integers from 1 to 100 equals the sum of the first ninety nine and 100. Now we apply our premise

$$p(99) : 1 + 2 + ... 99 = \frac{99(99+1)}{2}$$

then we consider

$$\begin{aligned} p(100) &= 1 + 2 + ... + 99 + 100 \\ &= (1 + 2 + ... + 99) + 100 \\ &= \frac{99}{2}(99 + 1) + 100 \\ &= 99 \times \frac{100}{2} + 100 \\ &= \frac{(100)(101)}{2} = \frac{100(100+1)}{2} \end{aligned}$$

and we get the success is proving the required result.

The validity of mathematical induction follows from the following fundamental axiom about the set of integers.

The well ordering property. Every non-empty set of non-negative integers has a least element.

Definition : Mathematical Induction

A proof by mathematical induction that $P(n)$ is true for every positive integer n consists of two steps :

1. **Basic step :** The proposition $p(1)$ is shown to be true.
2. **Inductive Step :** The implication $P(n) \rightarrow P(n + 1)$ is shown to be true for every positive integer n.

When we complete both steps of a proof by mathematical induction, we have proved that $P(n)$ is true for all positive integers n, that is, we have shown than $\forall\ n\ P(n)$ is true. It is to be noticed that in the proof by mathematical induction it is not assumed that $P(n)$ is true for all positive integers! It is only shown that if it is assumed that $P(n)$ is true then $P(n + 1)$ is also true.

EXAMPLE 3.20. Use mathematical induction to prove that the **sum of the first n odd positive integers is n^2.**

Let $p(n)$ is the proposition that sum of the first n odd positive integers is n^2.

Step 1. $p(1)$ says, the sum of the first one odd positive integer is 1^2 which is automatically true as the first odd positive integer is 1.

Step 2. Let us assume that $p(n)$ is true

i.e., sum of first n positive odd integer is n^2

i.e., $1 + 3 + 5 + ...(2n - 1) = n^2$.

Consider $p(n + 1)$, the sum of first $(n + 1)$ positive odd integers ;

$$\begin{aligned} 1 + 3 + 5 + ... + (2n - 1) + (2n + 1) &= n^2 + 2n + 1 \\ &= (n + 1)^2 \end{aligned}$$

This implies that $p(n + 1)$ is true.

Therefore by the principle of mathematical induction the result is true.

EXAMPLE 3.21. Prove the following result by the method of mathematical induction.

For all $n \geq 1$, $n^3 + 2n$ is a multiple of 3.

Let $p(n) : n^3 + 2n$ is a multiple of 3, $n \geq 1$

Basic step : $p(1) : 1^3 + 2 .1 = 3$ is a multiple of 3

$\therefore$ $p(1)$ is true.

Induction step : Assume that $p(n)$ is ture *i.e.*, $n \geq 1$ and $n^3 + 2n$ is a multiple of 3.

Consider $p(n + 1)$.

$$\begin{aligned} (n + 1)^3 + 2\,(n + 1) &= n^3 + 3n^2 + 3n + 1 + 2n + 2 \\ &= (n^3 + 2n) + 3n^2 + 3n + 3 \\ &= (n^3 + 2n) + 3(n^2 + n + 1) \end{aligned}$$

Now $n^3 + 2n$ is a multiple of 3 as $p(n)$ is true (By our hypothesis)

$(n + 1)^3 + 2(n + 1)$ is the sum of two multiples of 3; therefore, it is also a multiple of 3.

Thus by Principle of mathematical induction the result is true $\forall n$.

EXAMPLE 3.22. Use mathematical induction to prove that **$n^3 - n$ is divisible by 3 whenever n is a positive integer.**

Let $p(n)$: $(n^3 - n)$ is divisible by 3, $n \in Z^+$ (set of positive integers)

Basic Step :

$p(1) = (1^3 - 1) = 0$ is divisible by 3

Therefore, $p(1)$ is true.

$p(2) = 2^3 - 2 = 6$ is divisible by 3

Therefore, $p(2)$ is true.

Induction Step :

Now let us assume that $p(n)$ is true that is $(n^3 - n)$ is divisible by 3. We need to show $p(n + 1)$ is true, that is $[(n + 1)^2 - (n + 1)]$ is divisible by 3.

As $(n^3 - n)$ is divisible by 3 then by division algorithm there exists a positive integer k_1, such that

$$n^3 - n = 3k_1$$

Now consider $p(n + 1)$

$$\begin{aligned}(n + 1)^3 - (n + 1) &= (n^3 + 3n^2 + 3n + 1) - (n + 1)\\ &= n^3 + 3n^2 + 3n - n\\ &= (n^3 - n) + 3(n^2 + n)\end{aligned}$$

Now $(n^3 - n)$ is divisible by 3 as $p(n)$ is true.

$(n + 1)^3 - (n + 1)$ is the sum of two multiplies of 3, that is:

$$\begin{aligned}(n + 1)^3 - (n + 1) &= 3k_1 + 3k_2 & k_1 = n^3 - n \in Z\\ &= 3(k_1 + k_2) & k_2 = n^3 + n \in Z\end{aligned}$$

$\Rightarrow ((n + 1)^3 - (n + 1))$ is divisible by 3 therefore by principle of mathematical induction the result is true for $\forall n$.

EXAMPLE 3.23. Use mathematical induction to prove that

$$1 + 2 + 2^2 + ... + 2^n = 2^{n+1} - 1$$

for all non-negative integer n.

Solution : Let $p(n)$ be the predicate

$$1 + 2 + 2^2 + ...+ 2^n = 2^{n+1} - 1$$

Basic Step : $p(0)$

$1 = 2^1 - 1 = 1$

$\therefore p(0)$ is true.

Induction Step : Let us assume that the result is true for $n = k$,

that is, $p(k)$ is true that is

$$1 + 2 + 2^2 + ... + 2k = 2^{k+1} - 1 \text{ is true.}$$

Now to prove that $p(k + 1)$ is true. *i.e.*,

$$1 + 2 + 2^2 + 2^3 + ... + 2^k + 2^{k+1} = 2^{k+2} - 1$$

Consider $p(k + 1)$

$$\begin{aligned}1 + 2 + 2^2 + .. + 2^k &= 2^{k+1} - 1\\ 1 + 2 + 2^2 + ... + 2^k + 2^{k+1} &= 2^{k+1} - 1 + 2^{k+1}\\ &= 2.\ 2^{k+1} - 1\\ &= 2^{k+2} - 1\\ &= 2^{(k+1)+1} - 1\end{aligned}$$

$\therefore$ $p(k + 1)$ is true.

Therefore by principle of mathematical induction the result is true $\forall n$.

EXAMPLE 3.24. Use mathematical induction to prove that

$$1^2 + 2^2 + 3^2 + ... + n^2 = \frac{n}{6}(n + 1)(2n + 1) \; \forall \; n \in N$$

Solution : Let $p(n)$ is the predicate

$$1^2 + 2^2 + ... + n^2 = \frac{n}{6}(n + 1)(2n + 1)$$

Since we want to prove it for every $n \in N$, we take $n = 1$

Basic Step : $1^2 = 1 = \frac{1}{6}(1 + 1)(2.1 + 1) = \frac{2.3}{6} = 1$

Therefore $p(1)$ is true.

Induction Step : Let for an arbitracy $k \in N$, $p(k)$ is true

$$i.e., \; 1^2 + 2^2 + ... + k^2 = \frac{1}{6}k(k+1)(2k+1) \text{ is true.}$$

Consider $p(k + 1)$

$$\begin{aligned} p(k + 1) &= 1^2 + 2^2 + ... + k^2 + (k + 1)^2 \\ &= \frac{k(k+1)}{6}(2k + 1) + (k + 1)^2 \\ &= \frac{(k+1)\,[k(2k+1) + 6(k+1)]}{6} \\ &= \frac{1(k+1)\,[2k^2 + k + 6k + 6]}{6} \\ &= \frac{1}{6}(k + 1)\,[2k^2 + 3k + 4k + 6] \\ &= \frac{1}{6}(k + 1)(k + 2)(2k + 3) \end{aligned}$$

Therefore the result is true for $p(k + 1)$.

As both the conditions of the principle of mathematical induction hold. Therefore $p(n)$ is true for every $n \in N$.

EXAMPLE 3.25. Use mathematical induction to prove that $2^n < n!$ for every positive integer n with $n \geq 4$.

Solution. Let $p(n)$ be the predicate

$$2^n < n!$$

Basic step : $p(4)$, $2^4 = 16$.

$$4! = 4 \times 3 \times 2 \times 1 = 24$$

$\therefore$ $2^4 < 4!$

$\therefore$ $p(4)$ is true.

Induction step : Let us assume that $p(k)$ is true

$$\Rightarrow 2^k < k!\,,\; k \geq 4$$

Consider $p(k + 1)$

$$2^{k+1} = 2^k.2 < 2.\ k!$$

As $k \geq 4$

$k + 1 \geq 4 > 2$

$(k + 1) > 2$

$\therefore\ 2^{k+1} < (k + 1)\ k!$

$= (k + 1)\ !$

$\Rightarrow\ 2^{k+1} < (k + 1)!$

$\therefore\ p(k + 1)$ is true.

Therefore by principle of mathematical induction the result is true $\forall n$.

EXAMPLE 3.26. Show that $2^n > n^3$ for $n \geq 10$

Solution : Let $p(n)$ be the predicate $2^n > n^3$.

Basic Step : $p(10)$?

$2^{10} = 1024$

$10^3 = 1000$

$\therefore\ 2^{10} > 10^3$

$\therefore\ p(1)$ is true.

Induction Step : Let us assume that $p(k)$ is true for an arbitrary $k \geq 10$.

Therefore $2^k > k^3$

Consider $p(k + 1)$

$2^{k+1} = 2.2^k > 2.k^3$ By our assumption

$$> \left(1+\frac{1}{10}\right).k^3 \qquad ...(i)$$

as $$2 > \left(1+\frac{1}{10}\right)^3$$

As $k \geq 10 \Rightarrow \dfrac{1}{10} \geq \dfrac{1}{k}$

$$1+\frac{1}{10} \geq 1+\frac{1}{k}$$

$$\Rightarrow \left(1+\frac{1}{10}\right)^3 \geq \left(1+\frac{1}{k}\right)^3$$

$$\Rightarrow \left(1+\frac{1}{10}\right)^3 k^3 \geq \left(1+\frac{1}{k}\right)^3 k^3$$

$\therefore$ (1) $$\Rightarrow 2^{k+1} \geq \left(1+\frac{1}{k}\right)^3 k^3$$

$$= (k + 1)^3$$

Thus $p(k + 1)$ is true if $p(k)$ is true for $k \geq 10$.

Therefore by principle of Mathematical Induction the result is true.

EXAMPLE 3.27. Use principle of mathematical induction to show that for any integer n > 1 , $\frac{1}{\sqrt{1}}+\frac{1}{\sqrt{2}}+.....+\frac{1}{\sqrt{n}}>\sqrt{n}$.

Solution : Let $p(n)$ be the predicate

$$\frac{1}{\sqrt{1}}+\frac{1}{\sqrt{2}}+.....+\frac{1}{\sqrt{n}}>\sqrt{n}$$

Basic Step : $p(2)$

$$\frac{1}{\sqrt{1}}+\frac{1}{\sqrt{2}}>\sqrt{2}.$$

$$1+\frac{\sqrt{2}}{2}>\sqrt{2}$$

therefore $p(1)$ is true

Induction Step : Now let us assume that $p(k)$ is true for some $k \geq 2$. Then $p(k+1)$

$$\frac{1}{\sqrt{1}}+\frac{1}{\sqrt{2}}+....+\frac{1}{\sqrt{k}}+\frac{1}{\sqrt{k+1}}$$

$$>\frac{\sqrt{k(k+1)}+1}{\sqrt{k+1}} \qquad \text{[as } p(k) \text{ is true]}$$

As $\quad k+1>k \quad , \quad k>1$

$$\sqrt{k+1}>\sqrt{k}$$

$$\sqrt{k}\;\sqrt{k+1}>k$$

$$\therefore \quad \frac{1}{\sqrt{1}}+\frac{1}{\sqrt{2}}+...+\frac{1}{\sqrt{k+1}}>\frac{k+1}{\sqrt{k+1}}=\sqrt{k+1}$$

Hence $p(k+1)$ is true

therefore $p(n)$ is true $\forall n>1$.

EXAMPLE 3.28. Use mathematical induction to prove that

$$1+\frac{1}{4}+\frac{1}{9}+....+\frac{1}{n^2}\leq 2-\frac{1}{n}\;\forall n \in N.$$

Solution : Let $p(n)$ be the predicate

$$1+\frac{1}{4}+\frac{1}{9}+....+\frac{1}{n^2}\leq 2-\frac{1}{n}\quad \forall n \in N.$$

Basic Step : $p(1)$

$$1 \leq 2-\frac{1}{1}=1$$

$\therefore$ $p(1)$ is true.

Induction Step : Let us assume that $p(k)$ is true for $k \in N$.

$$\therefore \quad 1+\frac{1}{4}+\frac{1}{9}+....+\frac{1}{k^2}\leq 2-\frac{1}{k}$$

Consider $p(k+1)$

$$1+\frac{1}{4}+\frac{1}{9}+\ldots.+\frac{1}{k^2}+\frac{1}{(k+1)^2} \le \left(2-\frac{1}{k}\right)+\frac{1}{(k+1)^2} \quad \ldots(1)$$

(As $p(k)$ is true)

Now. $\left(2-\frac{1}{k}\right)+\frac{1}{(k+1)^2}+\frac{1}{k+1}-\frac{1}{k+1}$

$$= 2-\frac{1}{k+1}+\left[\frac{1}{(k+1)^2}+\frac{1}{k+1}-\frac{1}{k}\right]$$

$$= 2-\frac{1}{k+1}+\left[\frac{k+k(k+1)-(k+1)^2}{k(k+1)^2}\right]$$

$$= 2-\frac{1}{k+1}+\left(\frac{k^2+2k-k^2-2k-1}{k(k+1)^2}\right)$$

$$= 2-\frac{1}{k+1}-\frac{1}{k(k+1)^2}$$

$$= 2-\frac{1}{k+1} + \text{negative}$$

$$\le 2-\frac{1}{k+1}$$

Therefore, $1+\frac{1}{4}+\frac{1}{9}+\ldots+\frac{1}{(k+1)^2} \le 2-\frac{1}{k+1}$.

Therefore $p(k + 1)$ is true.

Thus by principle of mathematical induction $p(n)$ is true $\forall n \in N$.

3.5. PRINCIPLE OF STRONG MATHEMATICAL INDUCTION

Now let us look at the situation where the principle of mathematical induction does not work. Consider the Fibonacci Sequence of numbers 1, 1, 2, 3, 5, 8, 13, ... Each term in the sequence, from the third term on, is obtained by adding the previous two terms. So, if a_n is the nth term, then $a_1 = 1$, $a_2 = 1$ and $a_n = a_{n-1} + a_{n-2}$ $\forall\ n \ge 3$.

Suppose we want to show that $a_n < 2^n$ $\forall\ n \in N$

Using the principle of mathematical induction.

We use the predicate $p(n) : a_n < 2^n$.

Basic step : $p(1)$ is true.

Inductive Step : Let us assume that $p(k)$ *i.e.*, $a_k < 2^k$ is true for $k \in N$. we need to prove that $p(k + 1)$ is true.

$p(k + 1)$:

$$a_{k+1} < 2^{k+1}$$
$$a_k + a_{k-1} < 2^{k+1}$$

We know by our assumption that $a_k < 2^k$. but we don't know about a_{k-1}. So we cannot apply the principle of mathematical induction. In such a situation, a stronger, more powerful version of the principle of mathematical induction is to apply is given as:

Principle of Strong Mathematical Induction

$p(n)$ is true for all positive integer n if

1. **Basic Step :** The proposition $p(1)$ is shown to be true.
2. **Induction Step :** $[p(1) \wedge p(2) \wedge p(3) \wedge .... \wedge P(n)] \rightarrow p(n + 1)$ is true for every positive integer n.

Let us now go back to the Fibonacei sequence.

To use the strong form of the principle of mathematical induction, we take $n = 1$.

We have seen that $p(1)$ is true.

We also need to see if $p(2)$ is true.

This is because we have to use the relation

$a_n = a_{n-1} + a_{n-2}$ which is valid for $n \geq 3$.

Now we know $p(1)$ and $p(2)$ both are true then we go to next step. In step 2. we assume that $p(n)$ is true for $1 \leq n \leq k$, $k \geq 0$.

i.e., $a_n < 2^n$ $\quad 1 \leq n \leq k$

$p(k + 1)$:

$$\begin{aligned} a^{k+1} &= a_k + a_{k-1} && \text{[By Fibonacci sequence]} \\ &< 2^k + 2^{k-1} && \text{(By our assumption)} \\ &= 2^{k-1}(2 + 1) \\ &< 2^{k-1}.\ 2^2 \\ &= 2^k + 1 \end{aligned}$$

$\therefore$ $p(k + 1)$ is true.

Therefore $p(n)$ is true for all n.

From the above example it is clear that in case of strong mathematical induction we often need to check Basic Step for more than one value of n.

Both the form of mathematical induction is actually equivalent. This is because that one can be obtained from other. So we can use either form of mathematical induction.

EXAMPLE 3.29. Show that if n is an integer greater than 1, then n can be written as the product of primes.

Solution:

Let $p(n)$: If n is an integer greater than 1, then n is a product of primes.

Basic Step : $p(2)$: 2 is an integer greater than is 1 and 2 itself is a prime.

$p(3)$: 3 again itself is a prime and is greater than 1.

$p(4)$: $4 = 2 \times 2$ is the product of primes.

Induction Step :

Assume that $p(k)$ is true for all positive integers k with $k \leq n$.

We need to show that $p(n + 1)$ is true under this assumption.

Case I. $(n + 1)$ is a prime. Then $p(n + 1)$ is true automatically.

Case II. $(n + 1)$ is not a prime *i.e.*, $(n + 1)$ is a composite and can be written as product of two positive integers. *i.e.*, $(n + 1) = ab$ $\quad a$ and b are positive integer with. $2 \leq a \leq b < n + 1$

By our hypothesis a and b can be written as product of primes *i.e.*,

$$a = p_1 p_2 \ldots . p_r \qquad p_i \text{ are prime} \qquad 1 \le i \le r$$
$$b = q_1 q_2 \ldots . q_s \qquad q_j \text{ are prime} \qquad 1 \le j \le s$$
$$\therefore \quad n + 1 = ab$$
$$= (p_1 p_2 \ldots p_r)(q_1 q_2 \ldots q_s)$$

That is, $(n + 1)$ can be written as product of prime.

Thus $p(n + 1)$ is true.

Therefore, by principle of **strong mathematical induction** the result is true.

EXAMPLE 3.30. To prove that $\sum_{i=1}^{n} \frac{1}{\sqrt{i}} \le 2\sqrt{n} - 1 \ \forall \ n \in N$.

Which form of the principle of mathematical induction whould you use and why ? Also prove the inequality.

Solution : Here we will use the weak form of principle of mathematical induction as the assumption $p(k)$ is true is enough to prove that $p(k + 1)$ is true. We don't need to use $p(1)$, $p(2)$, ...$p(K - 1)$ also.

Basic Step : $p(1) = 1 \le 2 - 1 = 1$

$\therefore$ $p(1)$ is true.

Induction Step : Let us assume that $p(k)$ is true for some $k \in N$

Then $\frac{1}{\sqrt{1}} + \frac{1}{\sqrt{2}} + \ldots . + \frac{1}{\sqrt{k}} < 2\sqrt{k} - 1$

Consider,

$p(k + 1)$?

$$\frac{1}{\sqrt{1}} + \frac{1}{\sqrt{2}} + \ldots . + \frac{1}{\sqrt{k}} + \frac{1}{\sqrt{k+1}} < \left(2\sqrt{k} - 1\right) + \frac{1}{\sqrt{k+1}} < \left(2\sqrt{k} - 1\right) + \frac{1}{\sqrt{k+1}}$$

$$= \frac{2\sqrt{k}\,\sqrt{k+1} - \sqrt{k+1} + 1}{\sqrt{k+1}}$$

$$= \frac{\sqrt{k}\left(2\sqrt{k+1} - \sqrt{k}\right)}{(k+1)}$$

As $\quad k + 1 > k$

$$\Rightarrow \sqrt{k+1} > \sqrt{k}$$

$$\Rightarrow \frac{\sqrt{k}}{\sqrt{k+1}} < 1$$

$$\therefore \quad 1 + \frac{1}{\sqrt{2}} + \frac{1}{\sqrt{3}} + \ldots + \frac{1}{\sqrt{k+1}} < 2\sqrt{k+1} - \sqrt{k}$$

$$< 2\sqrt{k+1} - 1$$

$\therefore$ $p(k + 1)$ is true.

Therefore by principle of mathematical induction the result is true $\forall n$.

EXAMPLE 3.31. Use mathematical induction to prove that the sum of the first n odd positive integers is n^2.

Solution : Let $P(n)$ denote the proposition that the sum of the first n odd positive integers is n^2. We must first complete the basic step; that is, we must show that $P(1)$ is true. Then we must carry out the inductive step; that is, we must show that $P(n + 1)$ is true when $P(n)$ is assumed to be true.

Basic Step : P(1) states that the sum of the first one odd positive integers is 1^2. This is true as the sum of the first odd positive integer is 1.

Inductive Step : To complete the inductive step we must show that the proposition $P(n) \rightarrow P(n + 1)$ is true for every positive integer n. To do this, suppose that $P(n)$ is true for a positive integer n; that is,

$$1 + 3 + 5 + ... + (2n - 1) = n^2$$

Note : That the nth odd positive integer is $(2n - 1)$, since this integer is obtained by adding 2 a total of $n - 1$ times to 1. We must show that $P(n + 1)$ is true, assuming that $P(n)$ is true. Note that $P(n + 1)$ is the statement that

$$1 + 3 + 5 + ... + (2n - 1) + (2n + 1) = (n + 1)^2$$

$$\begin{aligned} \text{LHS} \quad & \Rightarrow 1 + 3 + 5 + ... + (2n - 1) + (2n + 1) \\ & = n^2 + 2n + 1 \qquad \text{[By our hypothesis]} \\ & = (n + 1)^2 \end{aligned}$$

Therefore, the result is true for (n + 1) also. Hence, by principle of mathematical Induction the result is true.

SOLVED PROBLEMS

PROBLEM 1: Give an argument using rules of inference to show that the conclusion follows from the hypothesis.

1. Hypotheses if Srishti can sing or kshitij can play, then I'll buy the compact disc. Srishti can sing. I'll buy the compact disc player. Conclusion : I'll buy the compact disc and the compact disc player.

Solution: p : Srishti can sing

q : Kshitij can play

r : I'll buy the compact disc

s : I'll buy the compact disc player

$$\begin{array}{l} (p \vee q) \rightarrow r \\ \quad p \\ \quad s \\ \hline \therefore (r \wedge s) \end{array}$$

Explanation: S is a hypothesis, so it holds. $((p \vee q) \rightarrow r) \wedge p$ implies r by Modus pones.

2. Hypothesis : Ken, a member of the Titans, can hit the ball a long way. Everyone who can hit the ball a long way can make a lot of money. Conclusion : Some member of the Titans can make a lot of money.

Solution. p : Member of the Titan

q : Can hit the ball a long way
r : Make lot of money

$$\begin{array}{ll} & p \to q \\ & q \to r \\ \hline \therefore & p \to r \quad \text{By transitive law} \end{array}$$

3. Hypothesis: Everyone in the discrete mathematics class loves proofs. Someone in the discrete mathematics class has never taken calculus. Conclusion: Someone who loves proofs has never taken calculus :

Solution: Let D be the discrete math class.

Let $P(x)$: x loves proofs.
$Q(x)$: x has never taken calculus

Then $\forall x, P(x)$ $(\forall x \in D$, that is $P(x))$

$$\begin{array}{l} \exists x,\ Q(x) \\ \hline \exists x\ (Q(x) \wedge P(x)), \end{array}$$

so someone loves proofs and have never taken calculus.
$\exists x \in D$, $P(x)$ is proven.

Mathematical Induction

PROBLEM 2. Prove that :

$$\frac{1}{2^2-1} + \frac{1}{3^2-1} + \ldots.. + \frac{1}{(n+1)^2-1} = \frac{3}{4} - \frac{1}{2(n+1)} - \frac{1}{2(n+2)}$$

i.e.,
$$\sum_{i=1}^{n} \frac{1}{(i+1)^2-1} = \frac{3}{4} - \frac{1}{2(n+1)} - \frac{1}{2(n+2)}$$

Solution: Basic Step: $\displaystyle\sum_{i=1}^{1} \frac{1}{(i+1)^2-1} - \frac{1}{2^2-1} = \frac{1}{3}$

RHS. $\dfrac{3}{4} - \dfrac{1}{4} - \dfrac{1}{6} = \dfrac{1}{3}$

Thus the result in true for $n = 1$.

Inductive Step : Let $k \in D$ be given. Assume $P(k)$ is true *i.e.,*

$$p(k) = \sum_{i=1}^{k} \frac{1}{(i+1)^2-1} = \frac{3}{4} - \frac{1}{2(k+1)} - \frac{1}{2(k+2)}$$

$$\therefore \quad p(k+1) = \sum_{i=1}^{k+1} \frac{1}{(i+1)^2-1}$$

$$= \sum_{i=1}^{k} \frac{1}{(i+1)^2-1} + \frac{1}{(k+2)^2-1}$$

$$= \frac{3}{4} - \frac{1}{2(k+1)} - \frac{1}{2(k+2)} + \frac{1}{(k+2)^2-1}$$

$$= \frac{3}{4} - \frac{1}{2(k+1)} - \frac{1}{2(k+2)} + \frac{1}{(k+1)(k+3)}$$

$$= \frac{3}{4} - \frac{1}{2(k+2)} - \frac{2-(k+3)}{2(k+1)(k+3)}$$

$$= \frac{3}{4} - \frac{1}{2(k+2)} + \frac{(-k-1)}{2(k+1)(k+3)}$$

$$= \frac{3}{4} - \frac{1}{2(k+2)} + \frac{1}{2(k+3)}$$

As $k \in D$ was arbitrary, the inductive step, $p(k) \rightarrow p(k+1)$, holds and alongwith the base case proves $\forall n \in D, p(n)$.

PROBLEM 3. Prove by principle of mathematical induction

$$1^2 - 2^2 + 3^2 - ......+(-1)^{n+1} n^2 = \frac{(-1)^{n+1} n(n+1)}{2}$$

Solution: Let $p(n) = 1^2 - 2^2 + 3^2 + ...+(-1)^{n+1} n^2 = \frac{(-1)^{n+1} n(n+1)}{2}$

$$= \sum_{i=1}^{n} (-1)^{i+1} i^2 = \frac{(-1)^{n+1} n(n+1)}{2}$$

Basic Step: $p(1) = \sum_{i=1}^{1} (-1)^{i+1} i^2 = (-1)^2 i^2 = 1$

R.H.S. $\frac{(-1)^2 \, 1(1+1)}{2} = 1$

Inductive Step: Let us assume that $p(k)$ is true, that is

$$\sum_{i=1}^{k} (-1)^{i+1} i^2 = \frac{(-1)^{k+1} k(k+1)}{2}$$

We will show that $p(k + 1)$ is true.

$$\sum_{i=1}^{k+1} (-1)^{i+1} i^2 = \sum_{i=1}^{k} (-1)^{i+1} i^2 + (-1)^{k+2} (k+1)^2$$

$$= \frac{(-1)^{k+1} k(k+1)}{2} + (-1)^{k+2} (k+1)^2$$

$$= (-1)^{k+1} (k+1) \left[\frac{k}{2} - (k+1)\right]$$

$$= (-1)^{k+1} (k+1) \left[\frac{-k-2}{2}\right]$$

$$= \frac{(-1)^{k+2} (k+1)(k+2)}{2}$$

As $k \in D$ is arbitrary, the inductive step is proven for all k and alongwith the basic case the result is true for all k.

PROBLEM 4. Use mathematical induction to show that if $E_1, E_2, \ldots\ldots, E_n$ are events, then

$$P(E_1 \cup E_2 \cup \ldots\ldots\ldots \cup E_n) \leq \sum_{i=1}^{n} P(E_i)$$

Solution : Basic step for $n = 1$. The result is true.

Inductive Step. Let the result is true for $n = k$. *i.e.*,

$$P(E_1 \cup E_2 \cup \ldots\ldots\ldots \cup E_k) \leq \sum_{i=1}^{k} P(E_i)$$

Then $\quad P(E_1 \cup E_2 \cup \ldots\ldots\ldots \cup E_k \cup E_{k+1})$

$$= P(E_1 \cup \ldots\ldots\ldots \cup E_k) + P(E_{k+1}) - P((E_1 \cup E_2 \ldots\ldots \cup E_k) \cap E_{k+1})$$

$$\leq \sum_{i=1}^{k} P(E_i) + P(E_{k+1}) - P(E_1 \cup \ldots\ldots\ldots E_k) \cap E_{k+1}) \leq \sum_{i=1}^{k+1} P(E_i)$$

As k is arbitrary thus the result is true for all n.

PROBLEM 5. Find the sum

$$1.2 + 2.3 + \ldots\ldots\ldots + (n-1)n$$

Solution : The sum $1.2 + 2.3 + \ldots\ldots\ldots + (n-1) \, . \, n$

$$= 2\left[\sum_{i=2}^{n} C(i, 2)\right]$$

Since each term is $C(i, 2) \, . \, 2$ and $C(i, 2) = \dfrac{i(i-1)}{2}$

So using $\displaystyle\sum_{1=k}^{n} C(i, k) = C(n+1, k+1)$

We have $\displaystyle\sum_{i=2}^{n} C(i, 2) = C(n+1, 3)$

So the required sum $= 2 \, C(n+1, 3)$

$$= 2 \, \frac{(n+1)(n)(n-1)}{3!} = \frac{n(n+1)(n-1)}{3}$$

EXERCISE 1

Convert each of the following arguments into the language of symbols and check their validity also.

1. Either the moon is cool or oxygen is a metal.
   The moon is cool.
   Therefore, oxygen is metal.
2. Either Srishti will play or Ribhav will practise karate.
   If Ribhave practises Karate, then Srishti studies.

Srishti does not study.

Therefore, Ribhav will practise Karte.

3. What rule of inference is used in each of the following arguments?
   (*a*) Kangaroos live in Australia and are marsupials. Therefore, kangaroos are marsupials.
   (*b*) It is either hotter than 100 degrees today or the pollution is dangerous. It is less than 100 degrees outside today. Therefore, the pollution is dangerous.
   (*c*) Lisa is an excellent swimmer. If Lisa is an excellent swimmer, then she can work as a lifeguard. Therefore, Lisa can work as a lifeguard.
   (*d*) Rahul will work at a computer company this summer. Therefore, this summer Rahul will work at a computer company or he will be a beach bum.
   (*e*) If I work all night on this homework, then I can answer all the exercises. If I answer all the exercises, I will get better understanding. Therefore, if I work all night on this homework, then I will have better understanting.
4. If Madhu is a 'President', she will be Head Girls in the School. If Madhu is Head Girl in the school; she wiil decide for annual function in the school.

   Therefore, if Madhu is a 'President', she will decide for annual function in the school.
5. Construct an argument using rules of inference to show that the hypotheses Raj works hard," "If Raj works hard, then he is a dull boy," and "If Raj is a dull boy, then he will not get the job" imply the conclusion "Raj will not get the job."
6. What rules of inference are used in the following famous argument? "All men are mortal. Socrates is a man. Therefore, Socrates is mortal."
7. Write down one example each of modus pones and modus tollens.
8. Show that $\sqrt{5}$ is irrational.
9. What rules of inference are used in the following argument? "No man is an island. Manhattan is an island. Therefore, Manhattan is not a man."
10. For each of the following sets of premises, what relevant conclusion or conclusions can be drawn? Explain the rules of inference used to obtain each conclusion from the premises.
    (*a*) "If I take the day off, it either rains or snows." "I took Tuesday off or I took Thursday off." "It was sunny on Tuesday." "It did not snow on Thursday."
    (*b*) "If I eat spicy foods, then I have strange dreams." "I have strange dreams if there is thunder while I sleep." " It did not have strange dreams."
    (*c*) "I am either clever or lucky." "I am not lucky." "If I am lucky, then I will win the lottery."
    (*d*) "Every computer science major has a personal computer." "Ralph does not have a personal computer". "Ann has a personal computer."
    (*e*) "What is good for corporations is good for the United States." "What is good for corporations is for you to buy lots of stuff."

(*f*) "All rodents gnaw their food." "Mice are rodents." "Rabbits do not gnaw their food." "Bats are not rodents."

11. Write down the contrapositive of the statement "If f is a $1-1$ function from a finite set X into itself, then f must be surjective".

12. For each of the following sets of premises, what relevant conclusion or conclusions can be drawn? Explain the rules of inference used to obtain each conclusion from the premises.
   (*a*) "If I play hockey, then I am sore the next day." "I use the whirlpool if I am sore." " I did not use the whirlpool."
   (*b*) "If I work, it is either sunny or partly sunny." "I worked last Monday or I worked last Friday." "It was not sunny on Tuesday". "It was not partly sunny on Friday."
   (*c*) "All insects have six legs." "Dragonflies are in sects." " Spiders do not have six legs." "Spiders eat dragonflies."
   (*d*) "Every student has an Internet account." "Aneesh does not have an internet account." " Srishti has an internet account."
   (*e*) "All foods that are healthy to eat do not taste good." "Apple is health to eat." "You only eat what tastes good." " You do not eat Apple." "Cheeseburgers are not healthy to eat."
   (*f*) "I am either dreaming or hallucinating." "I am not dreaming." "If I am hallucinating. I see elephants running down the road."

13. For each of the following arguments, explain which rules of inference are used for each step.
   (*a*) "Rekha, a student in this class, owns a red convertible. Everyone who owns a red convertible has gotten at least one spending ticket. Therefore, someone in this class has gotten a speeding ticket."
   (*b*) "Each of five roommates, Rahul, Aneesha, Priya, Soni and Ashish has taken a course in discrete mathematics. Every student who has taken a course in discrete mathematics can take a course in algorithms. Therefore, all five roommates can take a course in algorithms next year.
   (*c*) "All movies produced by John Sayles are wonderful. John Sayles produced a movie about coal miners. Therefore, there is a wonderful movie about coal miners".
   (*d*) "There is someone in this class who has been to USA. Everyone who goes to Franceisits the Niagrafalls. Therefore, someone in this class has visisted the Niagrafalls.

14. Determine whether each of the following arguments is valid. If an argument is correct, what rule of inference is being used ? If it is not, what fallacy occurs ?
   (*a*) If n is a real number such that $n > 1$, then $n^2 > 1$. Suppose that $n^2 > 1$. Then $n > 1$.
   (*b*) The number $\log_2 3$ is irrational if it is not the ratio of two integers. Therefore, since $\log_2 3$ cannot be written in the form a/b where a and b are integers, it is irrational.
   (c) If n is a real number with $n > 3$, then $n^2 > 9$, Suppose that $n^2 \leq 9$. Then $n \leq 3$.

(*d*) A positive integer is either a perfect square or it has an even number of positive integer divisors. Suppose that n is a positive integer that has an odd number of positive integer divisors. Then n is a perfect square.

(*e*) If n is a real number with $n > 2$, then $n^2 > 4$. Suppose that $n \leq 2$. Then $n^2 \leq 4$.

15. Prove the proposition $P(0)$, where $P(n)$ is the proposition "If n is a positive integer greater than 1, then $n^2 > n$. What kind of proof did you use ?
16. Let $P(n)$ be the proposition "If a and b are positive real numbers, then $(a + b)^n \geq a^n + b^n$." Prove that $P(1)$ is true. What kind of proof did you use ?
17. Prove that the square of an even number is an even number using
    (a) a direct proof
    (b) an indirect proof
    (c) a proof by contradiction
18. Prove that if n is an integer and $n^3 + 5$ is odd, then n is even using
    (*a*) an indirect proof
    (*b*) a proof by contradiction
19. Prove that if n is integer and $5n + 4$ is even, then n is even using.
    (*a*) an indirect proof
    (*b*) a proof by contradiction
20. Prove that the sum of two odd integers is even
21. Prove that the sum of two rational numbers is rational.
22. Prove that the sum of an irrational number and a rational number is irrational using a proof by contradiction.
23. Prove that the product of two rational number is rational.
24. Show that $\sqrt[3]{3}$ is irritational.
25. Prove that if x and y are real numbers, then max (x, y) + min$(x, y) = x + y$. (*Hint :* Use a proof by cases, with the two cases corresponding to $x \geq y$ and $x < y$, respectively).
26. Use a proof by cases to show that min (a, min (b, c)) = min (min(a, b), c) whenever a, b and c are real numbers.
27. Prove or disprove that if m and n are integers such that $mn = 1$, then either $m = 1$ and $n = 1$, or else $m = -1$ and $n = -1$.
28. Prove that $m^2 = n^2$ if and only if $m = n$ or $m = -n$.
29. Prove or disprove that $n^2 - 1$ is composite whenever n is a positive integer greater than 1.
30. Prove that if n is a positive integer, then n is odd if an only if $5n + 6$ is odd.

EXERCISE 2

1. Use mathematical induction to prove that

$$1 + \frac{1}{4} + \frac{1}{9} + \ldots + \frac{1}{n^2} \leq 2 - \frac{1}{n} \quad \forall n \in N$$

2. Use induction to prove that any integer $n \geq 2$ is either a prime or a product of primes.
3. Use the method of proof by contradiction to show that
   for $x \in R$, if $x^3 + 4x = 0$ then $x = 0$
4. Disprove the following statement :
   $(\forall\ x \in R)(\forall\ y \in R)\left[(x^2 = y^2) \Rightarrow (x = y)\right]$
5. Disprove the following statements by providing a suitable counter example :
   (*a*) $(x + y)^n = x^n + y^n\ \forall\ n \in N,\ \ x, y \in Z$
   (*b*) $\forall\ x \in Z,\ \ x \in N$
   (*c*) $f : Z \to Z$ is one-one if and only if f is onto.
6. Show that $1^3 + 2^3 + ... + n^3 = [n(n+1)/2]^2$ whenever n is a positive integer.
7. Prove that $1^2 + 3^2 + 5^2 + ... + (2n + 1)^2 = (n + 1)(2n + 1)(2n + 3)/3$ whenever n is a non-negative integer.
8. Prove that $1 \cdot 1! + 2 \cdot 2! + ... + n \cdot n! = (n + 1)! - 1$ whenever n is a positive integer.
9. Prove that $3n < n!$ whenever n is a positive integer greater than 6.
10. Show that $2n > n^2$ whenever n is an integer greater than 4.
11. Prove using mathematical induction that
    $$1 \cdot 2 + 2 \cdot 3 + ... + n(n + 1) = n(n + 1)(n + 2)/3$$
    whenever n is a positive integer.
12. Use mathematical induction to show that 3 divides $n^3 + 2n$ whenever n is a non-negative integer.
13. Use mathematical induction to show that 5 divides $n^5 - n$ whenever n is a non-negative integer.
14. Use mathematical induction to show that 6 divides $n^3 - n$ whenever n is a non-negative integer.
15. Use mathematical induction to prove that $\sum_{j=1}^{n} k^4 = n(n + 1)(2n + 1)(3n^2 + 3n - 1)/30$ whenever n is a positive integer.
16. Use mathematical induction to prove that if $A_1, A_2, ..., A_n$ are subsets of a universal set U, then
    $$\bigcup_{k=1}^{\overline{n}} A_k \subseteq \bigcap_{k=1}^{n} \overline{A_k}$$
17. Prove that if $A_1, A_2, ..., A_n$ and $B_1, B_2, ..., B_n$ are sets such that $A_k \subseteq B_k$ for $k = 1, 2, ..., n$, then
    (*a*) $\bigcup_{k=1}^{n} A_k \subseteq \bigcup_{k=1}^{n} B_k$
    (*b*) $\bigcap_{k=1}^{n} A_k \subseteq \bigcap_{k=1}^{n} B_k$

18. Prove that

$$1+\frac{1}{\sqrt{2}}+\frac{1}{\sqrt{3}}+\ldots+\frac{1}{\sqrt{n}}>2\left(\sqrt{n+1}-1\right)$$

19. Prove by induction :

$$1 + 2^3 + \ldots + n^3 = [n(n + 1)/2]^2 \text{ for } n \geq 1$$

20. Prove by induction :

$$1/(1 \cdot 2) + 1/(2 \cdot 3) + \ldots + 1/[n(n + 1)]$$
$$= n/(n + 1) \text{ for } n \leq 1$$

21. Prove by induction :

$$n^2 > n + 1 \text{ for } n \leq 2$$

22. Prove by induction for $n \leq 1$:

$$1 + 3^2 + 5^2 + \ldots + (2n + 1)^2 = (n + 1)(2n + 1)(2n + 3)/3$$

23. (*a*) Prove by induction that $2n + 1 \leq 2^n$ for $n \leq 3$.

    (*b*) Use the results of part (*a*) to prove by induction that $n^2 \geq 2^n$ for $n \leq 4$.

24. Let x be a real number such that $x \leq 1 + nx$.

$$(1 + x)^n \leq 1 + nx.$$

25. For $n \leq 1$, prove that $(x - y)$ is a factor of $(x^n - y^n)$. Hint : Add $0 = x^n y - x^n y$ to the term x^{n+1} and factor.

26. Prove by induction that, for any collection of n sets, $A_1, A_2, \ldots, A_n$, we have $(A_1 \cap A_2 \cap \ldots \cap A_n)' = A_1' \cup A_2' \cup \ldots \cup A'_n$.

EXERCISE 3

1. Applying the principle of mathematical induction, show that

(*i*) $1 + a + a^2 + \ldots. + a^n = \dfrac{1-a^{n+1}}{1-a}$ $(n \neq 1)$

(*ii*) $1 + 2 + 3 + \ldots + n = \dfrac{n(n+1)}{2}$

(*iii*) $1 + 2 + 2^2 + 2^3 + \ldots + 2^n n = 2^{n+1} - 1$

(*iv*) $2 + 2^2 + 2^3 + \ldots + 2^n = 2^{n+1} - 2$

(*v*) $1^2 + 3^2 + 5^2 + \ldots + (2^n - 1)^2 = \dfrac{n(2n-1)(2n+1)}{3}$

(*vi*) $2^n < n!$ $(n \geq 4)$

(*vii*) $1.1! + 2.2! + 3.3! + \ldots n.n! = (n + 1)! - 1$, where (factorial) $n! = 1.2.3. \ldots. n$.

(*viii*) $1.2 + 2.3 + \ldots + n(n + 1) = \dfrac{n(n+1)(n+2)}{3}$

(*ix*) $1^3 + 2^3 + \ldots + n^3 = (1 + 2 + 3 + \ldots + n)^2$.

(*x*) $\dfrac{1^2}{1.3}+\dfrac{2^2}{3.5}+\ldots+\dfrac{n^2}{(2n-1)(2n+1)}=\dfrac{n(n+1)}{2(2n+1)}$

2. Show that (by induction)
   (*i*) $n^4 - 4n^2$ is divisible by 3. $(n \geq 2)$
   (*ii*) $2^n \times 2^n - 1$ is divisible by 3. $(n \geq 1)$

3. Give an inductive definition for all the functions of the form $f(x) = x + \alpha$ $(\alpha \in N$; N is the universal).

4. Give an inductive definition of the well formed formulas of set theory incorporating the operations $\wedge$, $\vee$, and ~. (negation).

5. Let S stands for the set of natural numbers satisfying (*i*) the natural number n_0 is in S, and (*ii*) if the natural numbers n_0, $n_0 + 1$, $n_0 + 2$, ..., N are also in S, then the natural number $N + 1$ is also in S. Prove that N is the set of all natural numbers larger than or equal to n_0 $(N \geq n_0)$.

6. Among 100 students, 32 study mathematics, 20 study physics, 45 study biology, 15 study mathemtics and biology, 7 study mathematics, 10 study physics and biology, and 30 do not study any of the three subjects.
   (*a*) Find the number of students studying all three subjects.
   (*b*) Find the number of students studying exactly one of the three subjects.

7. It is known that at the university, 60 per cent of the professors play tennis, 50 percent of them play bridge, 70 per cent jogs, 20 per cent play tennis and bridge, 30 per cent play tennis and jog, and 40 percent play bridge and jog. If someone claimed that 20 percent of the professors jog and play bridge and tennis, would you believe this claim ? Why ?

8. Out of a total of 130 students, 60 are wearing hats to class, 51 are wearing scarves, and 30 are wearing both hats and scarves. Of the 54 student who are wearing sweaters, 26 are wearing hats, 21 are wearing scarves, and 12 are wearing both hats and scarves. Everyone wearing neither a hat nor a scarf is wearing gloves. Then, answer the following :
   (*a*) How many students are wearing gloves ?
   (*b*) How many students not wearing sweater are wearing hats but not scarves ?
   (M.C.A., VTU Karnataka, 1998)

9. (*a*) For any three sets A, B, and C, prove that $A - (B - C) = (A - B) - C$, if and only if $A \cap C = \phi$.
   (*b*) Using mathematical induction, prove that for an integer $n \geq 1.1^3 + 2^3 + 3^3 + ... + n^3 = (1 + 2 + 3 + ... + n)^2$.

10. Prove by induction
$$1 + 2^3 + ... + n^3 = [n(n + 1)/2]^2 \quad \text{for } n \geq 1$$

11. Prove by induction
$$1/(1{\cdot}2) + 1/(2{\cdot}3) + ... + 1/[n(n + 1)] = n/(n + 1) \quad \text{for } n \geq 1$$

12. Prove by induction
$$n^2 > n + 1 \quad \text{for } \; n \geq 2$$

13. Prove by induction for $n \geq 1$:
$$1 + 3^2 + 5^2 + ... + (2n + 1)^2 = (n + 1)(2n + 1)(2n + 3)/3$$

14. (*a*) Prove by induction that $2n + 1 \leq 2^n$ for $n \geq 3$.
    (*b*) Use the results of part(*a*) to prove by induction that, $n^2 \leq 2^n$ for $n \geq 4$.

15. Let x be a real number such that $x \geq -1$. Prove by induction that, for $n \geq 0$, $(1 + x)^n \geq 1 + nx$.

16. For $n \geq 1$, prove that $(x - y)$ is a factor of $(x^n - y^n)$.
[***Hint :*** Add $0 = x^n y - x^n y$ to the term $x^{n+1} - y^{n+1}$ and factor].

17. Prove by induction that, for any collection of n sets, $A_1, A_2, ..., A_n$. we have $(A_1 \cap A_2 \cap .... \cap A_n)' = A_1' \cup A_2' \cup ... \cup A_n'$.

18. (*a*) Use the principle of mathematical induction to prove that
$3 + 33 + ... + 33 ... 3_{n\text{-times}} = 10^{n+1} - 9n - 10)^{27}$
(*b*) In how many ways can five people be selected from a collection of Democrats, Republicans and independents?
(*c*) It is known that at the university 60% of the professors play tennis, 50% of them play bridge, 70% jog, 20% play tennis and bridge, 30% play tennis and jog, and 40% play bridge and jog. If someone claimed that 20% of the professors jog and play bridge and tennis would you believe this claim? Why? ***(End-term Exam., GGSIPU, 2001)***

19. (*a*) Prove using Mathematical induction that for $n \geq 4$
$n^3 < 3^n$

OR

Show that $\frac{1}{\sqrt{1}} + \frac{1}{\sqrt{2}} + ... + \frac{1}{\sqrt{n}} > \sqrt{n}$, for any positive integer $n > 1$.

(*b*) A function $A(m, n)$ is defined as follows :
(*i*) If $m = 0$ then $A(m, n) = n + 1$
(*ii*) If $m \neq 0$ but and $n = 0$ then $A(m, n) = A(m - 1, 1)$
(*iii*) If $m \neq 0$ and $n \neq 0$ then $A(m, n) = A(m - 1, A(m, n - 1))$ find the value of $A(1, 3)$ ***(Terminal Exam., GGSIPU, 2001)***

20. (*a*) Prove by induction principle that a tree with–n vertices has $(n–1)$ edges.
(*b*) Prove by induction principle that $2^n > n^3$ for $n \geq 10$
(End-term Exam., GGSIPU, 2002)

21. In how many ways can five people be selected from a collection of Democrats, Republicans & independents. ***(End-term Exam., GGSIPU, 2001)***

22. (*a*) State peono's axioms and prove by principle of mathematical induction that:

$$F_n < \left[1 + \frac{\sqrt{5}}{2}\right]^{n-1}, n = 1, 2, 3, ...$$

where fn is the nth Fibonacci number related by the recursive relation
$F_{n+1} = F_n + F_{n-1}$ $n \geq 2$ and $F_1 = F_2 = 1$.
(*b*) Explain the pigeonhole principle with suitable examples and hence show that in any set of eleven integers, there are two whose difference is divisible by 10.

EXERCISE 4

1. (*a*) Show that the following proposition is a tautology

   (*i*) $[(P \Rightarrow q) \wedge (q \Rightarrow r)] \Rightarrow (P \Rightarrow r)$

   (*ii*) $\sim(P \vee q) \vee [(\sim P) \wedge q] \vee P$

   (*b*) Establish the equivalence

   $P \Rightarrow q \vee r \equiv P \wedge \sim P \Rightarrow r \equiv P \wedge \sim r \Rightarrow q$

   Hence rewrite the following sentence in two different ways.

   If *n* is prime; then *n* is odd or *n* is 2.

   (*c*) Using laws and algebra of propositions, show that:

   (*i*) $(P \wedge q) \vee (P \wedge \sim q) \equiv P$

   (*ii*) $(P \Rightarrow q) \wedge (r \Rightarrow q) \equiv (P \vee r) \Rightarrow q$

   (*iii*) $\sim (P \vee (\sim P \wedge q)) \equiv \sim P \wedge q$

2. (*a*) Check the validity of the following argument.

   If I try hard and I have a talent, then I will become a scientist.

   If I become scientist then I will be happy. Therefore, if I will not try hard or I do not have talent.

   (*b*) Show that each of the following inference is fallacy.

   (*i*) If today is Anita's birthday, then today is February 16. Today is Feb. 16. Hence today is Anita's birthday.

   (*ii*) If the client is guilty, then he was at the scene of the crime. The client was at the scene of the crime. Hence the client is guilty.

   (*c*) Check the validity of the following arguments

   If *P* and *q* are true and *r* is false, find the truth values of the following

   (*i*) $(P \wedge \sim q) \Rightarrow r$ (*ii*) $(P \Leftrightarrow r) \Rightarrow r$

   (*iii*) $(P \vee q) \Leftrightarrow (P \Rightarrow \sim r)$

3. Show that the proposition $P \wedge - q$ and $(P \cup q) \wedge (-p) \wedge (-q)$ are contradiction.

4. Show that *t* is valid conclusion from the given promises

   $- p \wedge q,\ r \Rightarrow p,\ - r \Rightarrow s$ and $s \Rightarrow t$

5. (*a*) Define min terms and max terms obtained the disjunctive normal form (join of min terms) of the following boolean expression

   $(x_1 \wedge x^c_2) \vee (x_2 \wedge x^c_3) \vee (x_3 \wedge x^c_1)$

   where superscript '*c*' represents complement.

   (*b*) Determine the validity of the following argument with explanation :

   "If wages increase, then there will be inflection. The cost of living will not increase if there is no inflation wages will increase, therefore, the cost of living will not increase."

   (*c*) Without truth table prove that following is a tautology $P \wedge (P \vee Q) \rightarrow Q$.

6. Prove that following are all equivalence

   (*a*) $(p \wedge q) \vee (\sim P \wedge q) \Leftrightarrow q$

   (*b*) $p \rightarrow q \Leftrightarrow \sim q \rightarrow \sim p$

   (*c*) $p \vee q \Leftrightarrow q \vee p$

7. What common relation on *Z* are the transitive closures of the following relations :

   (*a*) *aSb* if and only if $a + 1 = a$

(b) aRb if and only if $|a - b| = 2$

8. (a) Show that
$((P \vee Q) \wedge \sim(\sim P \wedge (\sim Q \vee \sim R))) \vee (\sim P \wedge \sim Q) \vee (\sim P \wedge \sim R)$ is a Tautology.
(b) Prove that $(P \rightarrow Q) \Leftrightarrow (\sim P \vee Q)$.
9. Show that the truth values of following formulas are independent of their components.
(a) $(P \wedge (P \rightarrow Q)) \rightarrow Q$
(b) Check whether this statement $P \rightarrow Q, \sim P |\!\!- \sim Q$ is fallacy of a valid argument.

Answers to Exercise 1

1. p : The moon is cool
q : Oxygen is a metal

$$\begin{array}{l} p \vee q \\ p \\ \hline \therefore\ q \\ \hline \end{array}$$

Argument is $(p \vee q) \wedge p \Rightarrow q$
The argument is true if $(p \vee q) \wedge p \rightarrow q$ is a tautology.

p	q	$p \vee q$	$(p \vee q) \wedge p$	$(p \vee q) \wedge p \rightarrow q$
T	T	T	T	T
T	F	T	T	F
F	T	T	F	T
F	F	F	F	T

As the last column is not a tautology, Therefore it is not a valid argument.

2. p : Srishti will play
q : Ribhav will practise Karate
r : Srishti Studies

$$\begin{array}{l} p \vee q \\ p \rightarrow r \\ \sim r \\ \hline \therefore\ q \end{array}$$

The given argument is $(p \vee q) \wedge (q \rightarrow r) \wedge (\sim r) \Rightarrow q$

p	q	r	$p \vee q$	$q \rightarrow r$	$\sim r$	$(p \vee q) \wedge (q \rightarrow r) \wedge (\sim r)$	$(p \vee q) \wedge (q \rightarrow r) \wedge (\sim r) \rightarrow q$
T	T	T	T	T	F	F	T
T	T	F	T	F	T	F	T
T	F	T	T	T	F	F	T
T	F	F	T	T	T	T	(F)
F	T	T	T	T	F	F	T
F	T	F	T	F	T	F	T
F	F	T	F	T	F	F	T
F	F	F	F	T	T	F	T

Not a valid argument.

3(*a*) p : Kangaroos live in Australia.

q : Kangaroos are Marsupials.

$$\begin{array}{ll} & p \wedge q \\ \hline \therefore & p \\ \hline \end{array}$$

Argument : $(p \wedge q) \rightarrow p$

The Rule of Inference is **"SIMPLIFICATION"**

(*b*) p : It is hotter than 100 degrees today.

q : The pollution is dangerous.

$$\begin{array}{ll} & p \vee q \\ & \sim p \\ \hline \therefore & q \\ \hline \end{array}$$

Argument : $(p \vee q) \wedge (\sim p) \rightarrow q$

The rule of Inference is **"DISJUNCTIVE SYLLOGISM".**

(*c*) p : Lisa is an excellent swimmer.

q : Lisa can work as a lifeguard.

$$\begin{array}{ll} & p \\ & p \rightarrow q \\ \hline \therefore & q \\ \hline \end{array}$$

Argument : $[p \wedge (p \rightarrow q)] \rightarrow q$

The rule of Inference is **"MODUS PONES".**

(*d*) p : Rahul will work at a computer company this summer.

q : Rahul will be a beach bum.

$$\begin{array}{ll} & p \\ \hline \therefore & p \vee q \\ \hline \end{array}$$

Argument : $p \rightarrow p \vee q$

The rule of Inference is **"ADDITION".**

(*e*) p : Raj work all night on this homework.

q : I can answer all the exercise

r : I will get better understanding

$$\begin{array}{ll} & p \rightarrow q \\ & q \rightarrow r \\ \hline \therefore & p \rightarrow r \\ \hline \end{array}$$

Argument : $(p \rightarrow q) \wedge (q \rightarrow r) \Rightarrow (p \rightarrow r)$

The rule of Inference is **"HYPOTHETICAL SYLLOGISM".**

4. p : Madhu is a 'President'.

q : Madhu will be Head Girl in the School.

r : Madhu will decide for annual function in the school.

$$\begin{array}{ll} & p \rightarrow q \\ & q \rightarrow r \\ \hline \therefore & p \rightarrow r \\ \hline \end{array}$$

Here Argument is $(p \rightarrow q) \wedge (q \rightarrow r) \Rightarrow (p \rightarrow r)$

By the rule of **"Hypothetical Syllogism"** the given argument is a valid argument.

5. p : Raj works hard.
   q : Raj is a dull boy.
   r : Raj will not get the job.

$$\begin{array}{ll} & p \\ & p \to q \\ & q \to r \\ \hline \therefore & r \\ \hline \end{array}$$

Here Argument is $p \wedge (p \to q) \wedge (q \to r) \Rightarrow r$

p	q	r	$p \to q$	$q \to r$	$p \wedge (p \to q) \wedge (q \to r)$	$p \wedge (p \to q) \wedge (q \to r) \to r$
T	T	T	T	T	T	T
T	T	F	T	F	F	T
T	F	T	F	T	F	T
T	F	F	F	T	F	T
F	T	T	T	T	F	T
F	T	F	T	F	F	T
F	F	T	T	T	F	T
F	F	F	T	T	F	T

Valid argument.

6. p : Socrate
   q : Man
   r : Mortal
   $p \to q$: socrate is a man.
   $q \to r$: All men are mortal
   $p \to r$: Socrates is mortal

Argument : $(p \to q) \wedge (q \to r) \Rightarrow p \to r$

Therefore, by Rule **"Hypothetical Syllogism"**, the argument is valid.

9. Rule of **"Hypothetical Syllogism"**.

DO IT BY YOURSELF

1. Using direct proof technique, prove that for all real numbers d, $d1$, $d2$ and x, if $d = \min \{d1, d2\}$ and $x \leq d$, then $\leq d1$ and $x \leq d2$.
2. Using direct proof technique, prove that if x is a real number, then $x.0 = 0$. Assume that the following are previous theorems:
   If a, b, c are real numbers, then $b + 0 = b$; $a\ (b + c) = ab + ac$; If $a + b = a + c$, then $b = c$.
3. Using proof by contrapositive technique, prove that for all real numbers x and y, if $x + y \geq 2$, then either $x \geq 1$ or $y \geq 1$.
4. Using proof by contrapositive technique prove that if $xy = 0$, then either $x = 0$ or $y = 0$. Assume that if a, b and c are real numbers with $ab = ac$ and $a \neq 0$, then $b = c$.

5. Show, by giving a proof by contradiction, that if 100 balls are placed in nine boxes, some box contains 12 or more balls.
6. Show, by giving a proof by contradiction, that if 40 coins are distributed among nine bags so that each bag contains atleast one coin, atleast two bags contain the same number of coins.
7. Prove that there is no rational number p/q whose square is 2.
8. Let m and n be integers. Prove that $n^2 = m^2$ if and only if $m = n$ or $m = -n$
9. Let A and B be subsets of a universal set U, prove that $A \subseteq B$ iff $B \subseteq A$.
10. Prove the following using resolution.

    $a \vee b$

    $a \vee c$

    $c \vee d$

    $\therefore b \vee d$
11. State the Resolution principle. Apply it to find the validity of the following argument: "If the violinist plays the concerts, the prices will not too high. If the prices are not high, crowds will come. Therefore, if the violinist plays the concerts, crowds will come." Also, draw the Resolution tree.
12. Prove using induction:

    (*a*) $1.2 + 2.3 + \ldots.\ n(n + 1) = \dfrac{n(n+1)(n+2)}{3}$

    (*b*) $1^2 - 2^2 + 3^2 - 4^2 + \ldots.\ (-1)^{n-1}\, n^2 = (-1)^{n-1}\, \dfrac{n(n+1)}{2}$

    (*c*) $1^2 + 3^2 + 5^2 + \ldots.. + (2\,\text{n} - 1)^2 = \dfrac{n(2n-1)(2n+1)}{3}$

    (*d*) $1^3 + 2^3 + 3^3 + \ldots.. + n^3 = (1 + 2 + 3 + \ldots.. + n)^2$
13. Snow that:

    (*a*) $\dfrac{1^2}{1.3} + \dfrac{2^2}{3.5} + \ldots + \dfrac{n^2}{(2n-1)(2n+1)} = \dfrac{n(n+1)}{2\,(2n+1)}$

    (*b*) $\dfrac{1}{1.2} + \dfrac{1}{2.3} + \dfrac{1}{3.4} + \ldots + \dfrac{1}{n(n+1)} = \dfrac{n}{(n+1)}$

    (*c*) $\dfrac{1}{1.4} + \dfrac{1}{4.7} + \dfrac{1}{7.10} + \ldots + \dfrac{1}{(3n-2)(3n+1)} = \dfrac{n}{3n+1}$
14. Using induction, verify the inequalities:

    (*a*) $2^n \geq n^2$, $n = 4, 5 \ldots.$

    (*b*) $3^n + 7^n - 2$ is divisible by 8, for $n = 1, 2, \ldots.$
15. Show the postage of 24 cents or more can be achieved by using only 5-cent and 7-cent stamps.
16. Using complete induction, prove that postage of six cents or more can be achieved by using only 2-cents and 7-cents stamps.
17. Ackermann's function can be defined by the recurrence relation:

    $A\,(m, 0) = A\,(m - 1, 1)$, $m = 1, 2, \ldots\ldots,$

    $A\,(m, n) = A\,(m - 1, A(m, n - 1))$, $m = 1, 2, \ldots\ldots,$ $n = 1, 2, \ldots\ldots..,$

And the initial conditions $A(0, n) = n + 1, n = 0, 1, \ldots\ldots,$
Find the values of $A(1, 1)$ and $A(2, 3)$.

18. Find an explicit formula for minimum number of moves in which the n-disk Tower of Hanoi puzzle can be solved.
19. Find an explicit formula for the Fibonacci sequence
20. Solve the recurrence relation $d_n = 4(d_{n-1} - d_{n-2})$ subject to the initial conditions $d_0 = 1 = d_1$.
21. Find particular solution:
    (*a*) $a_r + a_{r-1} = 3r2^r$ (*b*) $a_r - 5a_{r-1} + 6a_{r-2} = 2^r + r$
    (*c*) $a_r + 5a_{r-1} + 6a_{r-2} = 3r^2 - 2r + 1$ (*d*) $a_r - 2a_{r-1} = 3.2^r$
    (*e*) $a_r - 4a_{r-1} + 4a_{r-2} = (r + 1).\ 2^r$
22. Find total solution:
    (*a*) $a_r - 5a_{r-1} + 6a_{r-2} = 1$ (*b*) $a_r - a_{r-1} + 7$
    (*c*) $a_r - 2a_{r-1} + a_{r-2} = 7$ (*d*) $a_r + 5a_{r-1} + 6a_{r-2} = 42.4^r$
23. Prove that the worst-time for binary search for input of size $n = \Theta(lg\ n)$
24. Solve the recurrence relation $a_n = 3a_{[n/2]} + n, n > 1$ in case n is a power of 2.

    Assume that $a_1 = 1$.
25. What is the worst case of Quick sort and prove that worst case running time for Quick Sort is $O(n^2)$.
26. Suppose that $\{a_n\}$ is an increasing sequence and that whenever m divides n, $a_n = c.a_{n/m} + d$, where c and d are real number satisfying $c > 1$ and $d > 0$, and m is an integer satisfying $m > 1$, show that $a_n = \Theta(n^{\log m^c})$
27. Solve the recurrence relation $t_n = 5t_{n-1}\ 8t_{n-2} + 4t_{n-3}$ for $n > 2$. Initial condition is $t_n = n$ for $n = 0, 1$ or 2.
28. Prove that worst case running time for selection sort is $\Theta(n^2)$ and best case running time is $\Theta(n)$.
29. Solve $T(n) = 2T\left(\frac{n}{2}\right) + \lg n$ exactly for n a power of 2. $T(1) = 1$. Express

    your answer as simply as possible using the Θ notation.
30. Solve $T(n) = 4\ T\left(\frac{n}{2}\right) + n$ exactly for n a power of 2. $T(1) = 1$.

❐❐❐

CHAPTER 4

COMBINATORICS

This Chapter Covers :

4.1. INTRODUCTION

Combinatorics deals with arrangements of objects according to some pattern and counting the number of ways it can be done. Mostly it deals with finite number of objects and finite number of ways of arranging them according to some pattern. Sometimes even infinite number of objects and infinite number of ways in which they can be arranged are considered.

EXAMPLE 4.1. In a test paper, there are ten true-false questions, In how many different ways can a student mark the test paper with one answer to each question?

Solution : Since each question can be attempted in two ways, there are altogether

$$2.\ 2.\ 2.\ 2.\ 2.\ 2.\ 2.\ 2.\ 2.\ 2. = 2^{10} = 1024 \text{ possibilities}$$

EXAMPLE 4.2. Consider the 26 letters of English alphabets.

Find the number of words of length 5.

Solution : The words can be formed as

1. All the letters are same *i.e.*, of the type $a\ a\ a\ a\ a$ etc.

Exactly one letter out of five is different $a\ a\ a\ a\ b$ etc.

Exactly two letters out of five are different $a\ a\ a\ b\ c$ etc.

Exactly three letters out of five are different $a\ a\ b\ c\ d$ etc.

Exactly four letters are different, *i.e.*, all the letters are different $a\ b\ c\ d\ e$ etc.

Therefore the words can be formed as

$a\ a\ a\ a\ a$, $a\ a\ a\ a\ b$, $a\ a\ a\ bc.$, ...$z\ z\ z\ z\ z$.

Clearly the number of words is $26 \times 26 \times 26 \times 26 \times 26$.

EXAMPLE 4.3. Consider the set of all positive integers. How many of them are prime?

Solution : As there are infinite numbers of prime numbers in a set of positive integers. Therefore answer is infinity.

Now we define two types of principles **Multiplication Principle** and **Addition Principle**.

Multiplication Principle

This principle can handle many situations which the simple permutation cannot. The formula for permutation *i.e.*, $P(n, r) = \dfrac{n!}{(n-r)!}$ has been derived using this principle. Suppose an experiment consists of a sequence of steps, say step 1, step 2, ...step r.

If one operation can be performed in m ways, and if corresponding to each of the m ways of performing this operation there are n ways of performing a second operation, then the number of ways of performing the two operations together is $m \times n$

Further suppose that step 1 can be performed in n_1 ways,

Step 2 can be performed in n_2 ways after step 1 has been performed

Step 3 can be performed in n_3 ways after step 2 has been performed

----

----

step r can be performed in n_r ways after step $(r-1)$ has been performed.

Then the number of ways the whole task can be performed is $n_1 \cdot n_2 \cdot \ldots n_k$.

For example : There are five different routes between place A and B and four different routes between place B and C. Find the number of ways a man can go form A to C *via*/B.

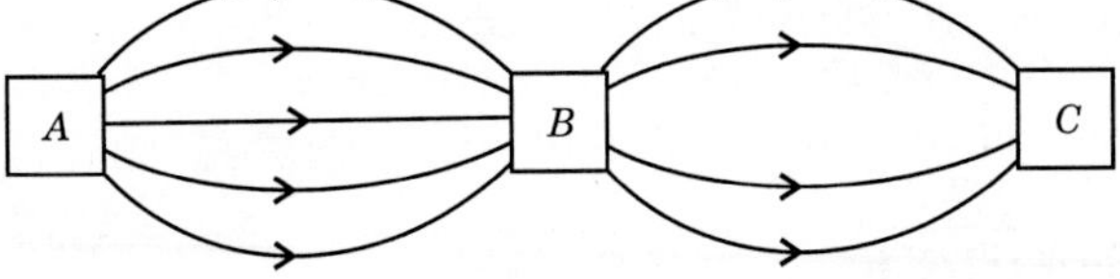

Fig. 4.1.

A person can go from A to B in 5 different ways and from B to C in four different ways. Hence, the number of ways a person can go from A to C is $5 \times 4 = 20$ ways.

Factorial

A factorial is the number obtained by multiplying all the positive integers less than or equal to a given positive integer. The factorial of a given integer n is usually written as n! or |n.

$n! = n \times (n-1) \times (n-2) \times \ldots\ldots 3 \times 2 \times 1$

By convention zero factorial (0 !) is taken to be unity.

Permutation

If r object is to be chosen from n, where $n \geq r$ and these r objects have to be arranged and the order in which these objects are arranged is important, such an arrangement is called a permutation of n objects taken r at a time. Permutation is denoted by ^nP_r or (n, r).

$$^nP_r = \frac{n!}{(n-r)!} \; r \leq n$$

Addition Principle

Suppose that a task consists of performing exactly one sub-task from among a collection of mutually exclusive (Disjoint or independent) subtasks, say sub-task 1, sub-task 2,.....sub-task k.

That is the task is performed if either the subtask 1 is performed or sub-task 2, is performed .... or sub-task k is performed. Further suppose that the task i can be performed in n_i ways, $i = 1, 2, \ldots k$ then the number of ways the task can be performed is $n_1 + n_2 + \ldots + n_k$.

EXAMPLE 4.4. Suppose that there are three companies C_1, C_2 and C_3. Suppose they want to select two persons, both from the same company for the post of president and vice president. In how many ways can this be done.

The company C_1 has 5 candidates and

The company C_2 has 4 candidates and

The company C_3 has 3 candidates for the respective posts.

Solution : Consider the first company C_1. There are 5 candidates for the post of president and vice presidents. Therefore choices for the post of president are 5 and rest of 4 can be chosen for the post of vice president. Therefore by multiplication principle there are $5 \times 4 = 20$ number of ways from the company C_1.

Similarly there are $4 \times 3 = 12$ number of ways the persons are selected from the company C_2 and there are $3 \times 2 = 6$ number of ways, the persons are selected from the company C_3.

If the persons are selected from the company C_1 then there is no chance for the companics C_2 and C_3. Similarly for other cases. Therefore by addition principle the numbers of selection is $20 + 12 + 6 = 38$.

4.2. PERMUTATIONS

Permutations are ordered arrangement of objects. Suppose if we are given n number of objects, permutation of them taken k at a time $(k \leq n)$ consists of arranging k of them on a line, the order in which they are arranged is important.

EXAMPLE 4.5. Let a, b, c and d be the set of four letters then (*i*) *ab, cd, bd* and *ad* are permutations of 4 letters taken 2 at a time.

and

(*ii*) *abc, acd, bcd, cda* are permutations of 4 letters taken 3 at a time.

EXAMPLE 4.6. The permutations of *a, b, c, d* taken 2 at a time are *ab, ba, ac, ca, ad, da, bc, cb, bd, db, cd, dc*. They are 12 in numbers. Here ab and ba are considered different even though they consist of same two objects, this is due to different arrangements of objects.

Suppose n and r are two positive integers with $r \leq n$. Then the number of permutations of n distinguishable objects taken r at a time is denoted by $\boldsymbol{p(n, r)}$, ^nP_r, $p_r^{\,n}$, $_nP_r$.

Theorem 4.1. The number of r- permutation from an n-set, where $o \leq r \leq n$, is given by:

$$P(n, r) = n(n-1)(n-2)\ldots(n-r+1) = \frac{n!}{(n-r)!}$$

To prove this theorem consider r boxes arranged in a line. Choose one object out of n and place it in the first box. This can be done in n ways. Then from the remaining $(n-1)$ objects choose one and place it in the second box. This can be done in $(n-1)$ ways. Therefore the first two boxes can be filled in $n(n-1)$ ways. Continuing like this till the rth box is filled, the number of ways of doing this is $n(n-1)(n-2)\ldots(n-r+1)$.

Thus we have

$$P(n, r) = n(n-1)(n-2)\ldots.(n-r+1)$$

$$= \frac{n(n-1)(n-2)\ldots..(n-r+1)(n-r)\ldots.2.1}{(n-r)\ldots\ldots.2.1}$$

$$= \frac{n!}{(n-r)!}$$

Permutations with Repetitions

In this case we will learn a method of finding the permutation of objects of which some are alike. Then the number of permutations of n objects of which n_1 are alike, n_2 are alike, and so on, is given by:

$$\frac{n!}{n_1!\,n_2!\ldots.n_r!}$$

EXAMPLE 4.7. How many 11 letter words can be formed using the letters of the word **INSTITUTION.**

Solution : Now we determine the number of permutations of 11 objects of which

the letter I is repeated thrice

the letter T is repeated thrice

the letter N is repeated twice.

and rest of letters come once.

Therefore by the above formula, the total number of required words are $\frac{11!}{3!\,3!\,2!}$.

Some Important Permutation Rules

(*i*) The total number of arrangements of n things taken r at a time, in which a particular thing always occurs $= r\left({}^{n-1}P_{r-1}\right)$

(*ii*) The total number of permutations of n distinct things taken r at a time in which a particular thing never occurs $= {}^{n-1}P_r$

Note : From (*i*) and (*ii*) above, we can conclude that the total number of permutations of n distinct things taken r at a time, ${}^nP_r$ will be equal to the sum of permutations from the above two cases.

i.e., $${}^nP_r = {}^{n-1}P_r + r\left({}^{n-1}P_{r-1}\right)$$

(*iii*) The number of permutations of n dissimilar things taken r at a time with repetitions $= n^r$.

If there are r things to be given to n persons, the total number of different ways in which this can be done $= n^r$

(*iv*) The number of permutations of n things taken all at a time, where p things are alike and of one kind. q others are alike and of another kind, r other are alike and of another kind, and so on is $\dfrac{n!}{p!\,q!\,r!...}$

e.g., In how many ways can the letters of the word 'SETS' be arranged?

Number of possible ways $= \dfrac{4!}{2!} = 12$

Circular Permutations

In the last section the permutation we have discussed is the linear permutation, that is, linear arrangement of objects. But there is a possibility, in which the objects are arranged along the circumference of a circle. On the circumference of a circle there is no fixed origin and hence the permutation $a\ b\ c\ d$, $b\ c\ d\ a$, $c\ d\ a\ b$, $d\ a\ b\ c$ will look exactly alike (see figure 4.1).

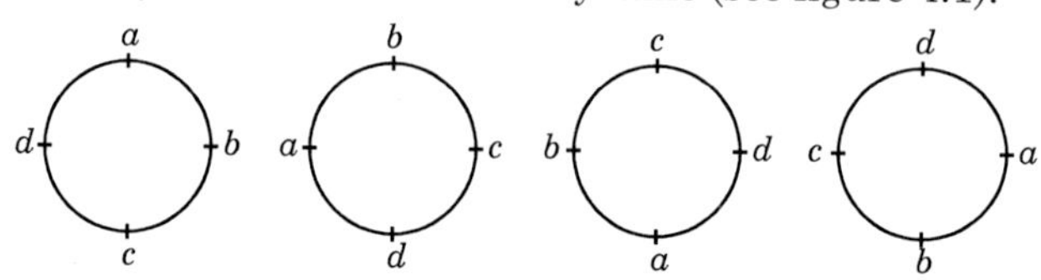

Fig. 4.2.

Thus the circular permutations are different only when relative order of object is changed.

If we consider all the $n!$ permutations of n objects, each permutation will be same as rest for $(n-1)$ obtained by the process of transferring the object at the first position to the last position repeatedly is linear permutation, *i.e.*, if arrangements are considered the same when one can be obtained from the other by rotation.

Hence, the position of one object is fixed and the remaining $(n-1)$ objects are arranged in all possible ways and the number of distance permutations, is $(n-1)!$.

Thus in circular permutations we will have exactly $\dfrac{n!}{n} = (n-1)!$.

Clockwise and Counter-Clockwise Permutations

There are two types of circular permutations depending on whether the clockwise and counter clockwise arrangements are distinguishable.

(*i*) The distinction between clockwise and anti-clockwise is ignored when a number of people have to be seated around a table so as not to have the same neighbours.

(*ii*) While making a necklace the clockwise arrangement ABCDE and the counter-clockwise arrangements AEDCB are not really different because the arrangement BCDE gives rise to the arrangement AEDCB if the necklace is turned over.

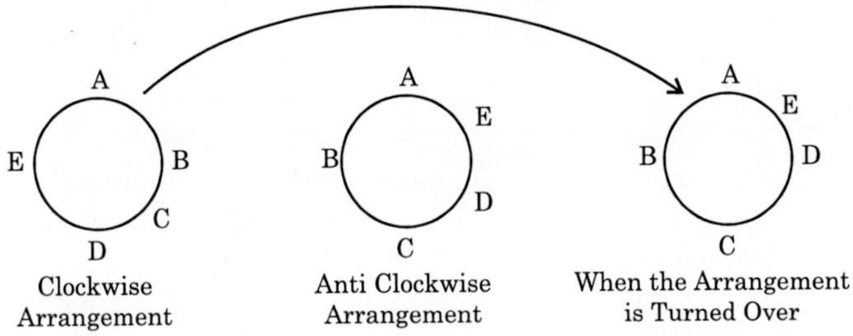

Fig. 4.3.

Thus the number of ways in which n beads of different colours can form a necklace $= \frac{1}{2} \times (n-1)!$

Theorem 4.2. The number of circular permutations of n things taken all at a time is $(n-1)!$

EXAMPLE 4.8. In how many ways ten people can sit around a table?

Solution : Clearly we need the circular permutation. Hence required total number of ways = 9!

$$= 9 \times 8 \times 7 \times 6 \times 5 \times 4 \times 3 \times 2 \times 1$$

EXAMPLE 4.9. Suppose there are 10 students in a class such that 5 are boys and 5 are girls. Suppose they are made to sit about a round table so that neither two boys nor two girls sit together. Find the number of circular arrangements.

Solution : Five boys are made to sit about a round table in $(5-1)! = 4!$ ways. One girl can sit between two boys. Therefore there are five positions (see fig (4.2) and hence they can be made to sit in 5! ways. Therefore by multiplication principle, the total number of ways of sitting arrangement is $4! \times 5! = 2800$.

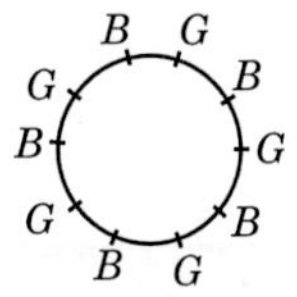

Fig. 4.4.

EXAMPLE 4.10. If there are 7 boys and 5 girls, how many circular arrangements are possible if the ladies do not sit adjacent to each other.

Solution : The boys can be sitted in circular arrangement in 6! ways. Now there are 5 girls and 7 seats (in between boys) so there are $P(7,5)$ number of ways, the girls can sit.

Therefore the total number of ways is:

$$= P(7, 5) \times 6!$$

$$= \frac{7!}{5!(7-5)!} \times 6!$$

$$= \frac{7 \times 6}{5!} \times 6!$$

= **252 ways.**

4.3. ORDERED SAMPLING

Consider a very familiar example of a bag containing n balls and we draw a ball from the bag with replacement is called **sampling with replacement.** In this case when the ball is replaced in the bag before drawing the next one. There are n different ways to choose each ball. Therefore using the principle of fundamental counting principle. We have

$$\underbrace{n \cdot n \ldots . n}_{r \text{ times}} = n^r$$

different ordered samples with replacement of size r.

On the other hand, consider the case when the ball is not replaced in the bag before taking out the next one, this case is called **sampling without replacement**. This is the case which corresponds to r permutation of the objects. Hence we have,

$$P(n, r) = \frac{n!}{(n-r)!} \qquad (n \geq r)$$

EXAMPLE 4.11. How many numbers between 10 and 99 have distinct digits?

Solution : There are 9.9 = 81 numbers between 10 and 99 that have distinct digits. (rest of the numbers are those which have 0 in the first position). As the first position can be filled by 9 ways (by any digit other than 0).

After filling the first position the second position can be filled up in 9 ways (by any of the 9 digits other than that occupying the first position, this will include 0 also).

EXAMPLE 4.12. In how many ways 3 cards can be chosen in succession from a pack of 52 ?

(*a*) with replacement.

(*b*) without replacement

Solution : (*a*) Since the card is replaced before the next card is chosen therefore for each card there are 52 choices. Hence we have 52 . 52 . 52

(*b*) Now this is the case when the first card is not replaced before choosing the second card. Therefore there are 51 different ways to choose the second card. Similarly there are 50 different ways to choose the third card. Hence we have $52 \times 51 \times 50$ ways to choose 3 cards by product rule.

EXAMPLE 4.13. How many licence plates can be made if each should have 3 letters with(*a*) no letter repeated (*b*) letters can be repeated ?

Solution : In a licence plate, for the first position there are 26 choices.

As there is no repitition of letters therefore there are 25 choices (as one choice is taken by first position).

Similarly there are 24 choices of letters for the third position.

Therefore the total number of ways

$$= 26 \times 25 \times 24$$

In case when the letters can be repeated there are number of choices

$$= 26 \times 26 \times 26$$

4.4. COMBINATIONS

Suppose there are n distinct objects and we want a selection of k objects, $k \le n$, and the order of the objects in the selection does not matter. This is called a combination of n things taken k at a time. It is denoted by:

$$C(n, r), {}^nC_r, {}_nC_r, \binom{n}{r}, C_r^n.$$

Therefore we can say in combination is the fact that **only choice is involved but not ordering.** In set theory $C(n,r)$ is the number of subsets of size r from a set containing n elements.

Theorem 4.3. The number of r-combinations from a n-set, where $0 \le k \le n$ is given by

$$C(n, r) = \frac{n!}{r!\,(n-r)!}$$

Theorem 4.4. Relationship between the permutation and combination can be expressed as

$$C(n, r) = \frac{p(n, r)}{r!}$$

Theorem 4.5. $C(n, r) = C(n, n - r)$

Some Important Combination Rules

(*i*) ${}^nC_r = {}^nC_{n-r}$ **Note :** ${}^nC_0 = {}^nC_n = 1$

(*ii*) ${}^{n+1}C_r = {}^nC_r + {}^nC_{r-1}$

(*iii*) ${}^{n+1}C_{r+1} = {}^nC_{r+1} + {}^{n-1}C_r + {}^{n-1}C_{r-1}$

(*iv*) The total number of combinations of 'n' distinct things taken some or all at a time $= {}^nC_0 + {}^nC_1 + {}^nC_2 + \dots {}^nC_n = 2^n$

(*v*) The number of combinations of 'n' things taken r at a time in which 'p' particular things will always occur $= {}^{n-p}C_{r-p}$.

(*vi*) The number of combinations of n things taken r at a time in which p particular things never occur $= {}^{n-p}C_r$.

(*vii*) The number of ways of dividing (partitioning) n distinct things into r distinct groups (where some groups may be empty) $= r^n$.

If the partitions are not distinguishable, then the total number of ways $= \frac{r^n}{n!}$

(*viii*) The number of ways of partitioning n non-distinct things into r distinct groups (all groups non-empty) is ${}^{n-1}C_{r-1}$.

(*ix*) The number of ways of partitioning n non-distinct things into r distinct groups (some groups may be empty) is ${}^{n+r-1}C_{r-1}$.

(*x*) The number of ways in which $(m + n)$ things can be divided into two groups containing m and n things respectively $= {}^{m+n}C_n = \frac{(m+n)!}{m! \times n!}$.

(*xi*) If n things are divided into two groups each containing m things then the number of ways to divide the n things are $\frac{(2m)!}{(m!)^2}$

(*xii*) The number of ways to divide n things into different groups, one containing p things, another q things and so on is $\frac{(p+q+r+....)!}{p! \times q! \times r! \times ....}$ where $n = p + q + r$ ....

(*xiii*) Suppose out of n things, p are alike and of one kind, q are alike and of another kind, r are alike and of a third kind and the rest t things are different, the total numbers of selections that can be made $= (p + 1)(q + 1)(r + 1)\, 2^t - 1$.

Note : The p like things can be selected in $(p + 1)$ ways. q like things in $(q + 1)$ and r like things in $(r + 1)$ ways. The t different things can be selected in 2^t ways. This includes the case in which all are rejected. Hence total number of selection is $(p + 1)(q + 1)(r + 1)\, 2^t - 1$.

EXAMPLE 4.14. Prove that, $C(n, r) = C(n - 1, r) + C(n - 1, r - 1)$

Solution :

$$\begin{aligned} C(n, r) &= \frac{n!}{r!\,(n-r)!} \\ &= \frac{n!}{r!\,(n-r)!}\left(\frac{n-r}{n} + \frac{r}{n}\right) \\ &= \frac{(n-1)!}{r!\,(n-r-1)!} + \frac{(n-1)!}{(r-1)!\,(n-r)!} \\ &= C(n-1, r) + C(n-1, r-1). \end{aligned}$$

EXAMPLE 4.15. Evaluate. $C(7, 2)$, $C(7, 5)$, $C(9,0)$ and $C(9, 3)$.

$$\begin{aligned} C(7, 2) &= \frac{7!}{2!\ (7-2)!} = \frac{7 \times 6}{2!} = 21 \\ C(7, 5) &= C(7, 7-5) \end{aligned}$$

$$= C(7, 2)$$
$$= 21$$

$$C(9, 0) = \frac{9!}{0!\,9!} = 1$$

Therefore, $\quad C(n, 0) = 1 \quad \text{for } n \in N.$

$$C(9, 3) = \frac{9!}{3!\,(9-3)!}$$
$$= \frac{9!}{3!\,6!}$$
$$= \frac{9 \times 8 \times 7}{3 \times 2 \times 1} = 84$$

Theorem 4.6. The number of ways to choose r objects from a collection of n objects with repetition allowed is $C(n + r - 1, r)$.

EXAMPLE 4.16. A man wants to buy some pet birds. The pet store sells parrots, sparrows and mainas. How many different selections are possible if the man wants to take home 7 birds.

Solution : Here $r = 7$ and $n = 3$

Implies the no. of objects are repeated

Hence the number of possible selection of

$$\text{pet birds} = C(3 + 7 - 1, 7)$$
$$= C(9, 7)$$
$$= \frac{9!}{7!\,2!}$$
$$= 36.$$

EXAMPLE 4.17. Let a_r = the number of ways of selecting r objects from n objects with unlimited repetitions. Prove that

$$(1 + z + z^2 + ... + z^n) = C(n + r - 1, r)$$

Solution : As each object is selected as many times as we with, hence let a_r will be equal to the co-efficient of z^r in

$$(1 + z + z^2 + ... + ...)^n$$

As $(1 + z + z^2 + z^3 + ...)^n = \left(\frac{1}{1-z}\right)^n$

$$= (1 - z)^{-n}$$

Hence, we have

$$a_r = \frac{(-1)^r (-n)\,(-n-1)....(-n-r+1)}{r!} = \text{coefficient of } z^n$$
$$= \frac{(n+r-1)!}{r!\,(n-1)!}$$
$$= C(n + r - 1, r)$$

EXAMPLE 4.18. In how many ways 11 apples can be distributed among the three children so that the first child gets 5 apples and then rest two children get 3 apples each

Solution : $\frac{11!}{5!\,3!\,3!}$ ways.

EXAMPLE 4.19. Find ${}^6P_4$ and ${}^{25}C_{22}$

Solution :

$$
\begin{aligned}
{}^6P_4 &= \frac{6!}{(6-4)!} = \frac{6!}{2!} \\
&= 6 \times 5 \times 4 \times 3 = 360 \\
{}^{25}C_{22} &= \frac{25!}{(25-22)!\,22!} \\
&= \frac{25!}{3!22!} \\
&= \frac{25 \times 24 \times 23}{3 \times 2} = 2300
\end{aligned}
$$

EXAMPLE 4.20. How many three digit numbers can be formed using 1, 2, 3, 4, 5?

Solution : Number of three digit numbers can be formed

= Number of permutations of 5 things taken 3 at a time

$= {}^5P_3 = 5 \times 4 \times 3 = 60$

EXAMPLE 4.21. How many eight letter words can be formed from the letters of the words "COURTESY" beginning with C and ending with Y?

Solution : The first place will always be taken by C and last place always by Y. The remaining six places can be filled by the remaining six letters in ${}^6P_6$ ways.

$\therefore$ Total number of words $= 1 \times 1 \times {}^6P_6 = 6! = 720$

EXAMPLE 4.22. In how many ways can the seven letters A, B, C, D, E, F, and G be arranged so that B and C are always together ?

Solution : Considering B and C as one, the six things can be arranged in ${}^6P_6$ ways. B and C can be arranged among themselves in 2 ways *i.e.*, BC and CB.

Total number of arrangements $= {}^6P_6 \times 2 = 2 \times 720 = 1440$ ways.

EXAMPLE 4.23. In how many ways can 7 Englishmen and 7 American sit down at a round table such that no 2 Americans being in consecutive positions ?

Solution : Putting 1 Englishman in a fixed position, the remaining 6 can be arranged in 6! = 720 ways.

For each such arrangement, there are 7 positions for the 7 Americans and they can be arranged in 7! ways.

Total each such arrangement $= 7! \times 6! = 3628800$

EXAMPLE 4.24. What is the number of ways to assign the numbers (1 to 20) to the 25 rooms of 5 by 5 grid having 15 rooms empty ?

Solution : There are 25 rooms.

One of 1, 2, ... 20 is placed in these 25 rooms in 25 ways.

After placing 1st number 2nd can be placed in remaining 24 rooms in 24 ways.

3rd number in 23 rooms in 23 ways.

20th number of 6 rooms in 6 ways.

∴ Number of ways = $6 \times 7 \times 8 \times \ldots \times 23 \times 24 \times 25$

EXAMPLE 4.25. How many numbers divisible by 2 and lying between 50,000 to 70000 can be formed from the digits 3, 4, 5, 6, 7, 8, 9, no digit being repeated in any numbers ?

Solution : Number is divisible by 2 ∴ 4, 6, or 8 can be in units place

This can be done in 3 ways.

The number is between 50000 to 70000

∴ Ten thousands place can be filled in by 5 or 6 in 2 ways, as we require number between 50,000 to 70,000.

After filling two places, remaining three places can be filled by remaining 5s number in $^5P_3 = \frac{5!}{2!} = 3 \times 4 \times 5 = 60$ ways .

∴ Total numbers that can be formed = $3 \times 2 \times 60 = 360$

EXAMPLE 4.26. A committee of 2 hawkers and 3 shopkeepers is to be formed from 7 hawkers and 10 shopkeepers. Find the number of ways in which this can be done if a particular shopkeeper included and a particular hawker is excluded.

Solution : A particular shopkeepers is included

∴ We have to choose 2 from remaining 9 in $^9C_2 = \frac{9!}{7! \times 2!}$

$$= \frac{8 \times 9}{2} = 36 \text{ ways}$$

1 hawker is excluded

∴ Choose 2 hawker from 6 in $^6C_2 = \frac{6!}{4! \times 2!} = \frac{5 \times 6}{2} = 15$ ways

Total number of ways = $36 \times 15 = 540$.

EXAMPLE 4.27. In how many ways can a team of 11 cricketers be chosen from 6 bowlers, 4 wicket keepers and 11 batsmen to give a majority of batsmen if at least 4 bowlers are to be included and there is one wicket keeper ?

Solution : 1 wicket keeper from 4 can be selected in $^4C_1 = \frac{4!}{3! \times 1!} = 4$ ways

If 4 bowlers are chosen then remaining 6 batsmen can be chosen in $^{11}C_6$.

$$^6C_4 \times {}^{11}C_6 = \frac{6!}{4! \times 2!} \times \frac{11!}{6! \times 5!}$$

$$= \frac{5 \times 6}{2} \times \frac{11 \times 10 \times 9 \times 8 \times 7}{5 \times 4 \times 3 \times 2}$$

$$= 15 \times 14 \times 33 = 6930$$

If we choose 5 bowlers then we have to choose 5 batsmen.

∴ There is no majority.

∴ Total number of ways = 4 × 6930 = 27720.

EXAMPLE 4.28. In how many ways can 4 cards be selected from a pack of cards so as to include at least one diamond.

Solution : Four cards can be selected from 52 cards in $^{52}C_4$ ways.

We need to find the figure if at least one card is diamond. which is $^{52}C_4$ – (No diamond)

$$= {}^{52}C_4 - {}^{39}C_4$$

EXAMPLE 4.29. In how many ways can 24 identical white balls and 19 black balls be arranged in a row so that no 2 black balls are together ?

Solution : First arrange 24 white balls in a row. This can be done in 1 way (since they are identical).

Now there are 25 places for the 19 black balls and so the places can be filled in $^{25}C_{19}$ ways $= \dfrac{25!}{3!\,22!}$ ways.

$$= \frac{25 \times 24 \times 23}{3 \times 2} = 2300 \text{ ways}$$

EXAMPLE 4.30. There are 3 copies of a 4 different books. In how many ways can they be arranged on a shelf ?

Solution : Since there are 3 things alike in each set of 4.

$$\text{Number of arrangements} = \frac{12!}{3!\,3!\,3!\,3!} = \frac{12!}{(3!)^4}$$

4.5. THE BINOMIAL COEFFICIENTS AND PASCAL'S TRIANGLE

The symbol $\binom{n}{r}$ where r and n are positive integers with $r \le n$, is given as

$$C(n, r) = {}^nC_r = \binom{n}{r}$$

$$= \frac{n!}{r!\,(n-r)!}$$

But $\quad r = n - (n - r)$

hence we have the following important relation :

$$C(n, r) = C(n, n - r)$$

as $\quad C(n, n-r) = \dfrac{n!}{(n-r)!\,(n - n + r)!}$

$$= \frac{n!}{(n-r)!\,r!}$$

In other words, if $a + b = n$ then $\binom{n}{a} = \binom{n}{b}$

The numbers $\binom{n}{r}$ are called the binomial co-efficients since they appear as the co-efficients in the expansion of $(x + y)^n$ as

$$(x + y)^n = x^n + {}^nC_1\, x^{n-1}\, y + {}^nC_2\, x^{n-2}\, y^2 + \ldots + y^n$$

$$= \sum_{k=0}^{n} {}^nC_k\, x^{n-k} y^k$$

The co-efficients of the successive powers of $x + y$ can be arranged in a triangular array of numbers, called **Pascal's triangle**, as given below:

$$(x + y)^0 = 1$$
$$(x + y)^1 = x + y$$
$$(x + y)^2 = x^2 + 2xy + y^2$$
$$(x + y)^3 = x^3 + 3x^2y + 3xy^2 + y^3$$
$$(x + y)^4 = x^4 + 4x^3y + 6x^2y^2 + 4xy^3 + y^4$$
$$(x + y)^5 = x^5 + 5x^4\, y + 10\, x^3\, y^2 + 10\, x^2\, y^3 + 5\, xy^4 + y^5$$
$$(x + y)^6 = x^6 + 6x^5y + 15x^4\, y^2 + 20\, x^3\, y^3 + 15\, x^2\, y^4 + 6xy^5 + y^6$$
$$\ldots\ldots = \ldots\ldots$$

In other way, considering the coefficients we will have.

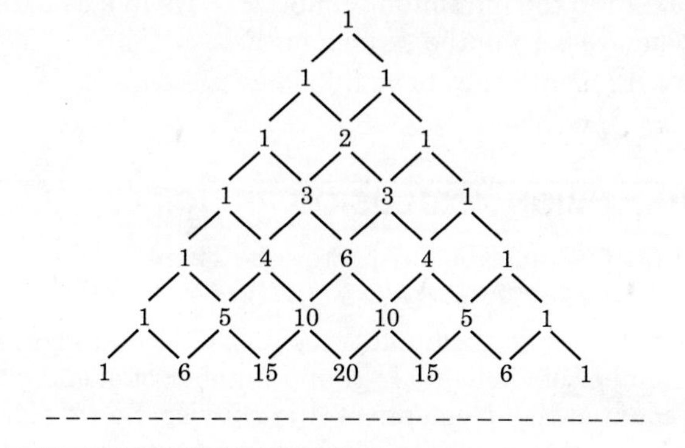

Fig. 4.5.

The numbers in Pascal's triangle have the following intersecting properties:

(1) The first number and the last number in each row is 1.

(2) Every other number in the array can be obtained by adding the two numbers appearing directly above it.

For example, $4 = 1 + 3$, $10 = 6 + 4$, $15 = 5 + 10$

Since the numbers appearing in Pascal's triangle are binomial co-efficients, property (2) of Pascal's triangle comes from the theorem given below:

$C(n + 1, r) = C(n, r - 1) + C(n, r)$ Proved earlier.

4.6. THE PIGEONHOLE PRINCIPLE

The pigeonhole principle states that if there are more pigeons than pigeonholes, then there must be at least one pigeonhole with at least two pigeons in it.

Definition : If $k + 1$ or more objects (pigeons) are placed into k boxes (pigeonholes) then there is at least one box containing two or more of the objects.

Generalised Pigeonhole Principle

If n objects (pigeons) are placed into k $(n > k)$ boxes (pigeonholes), then there is at least one box containing atleast $\lceil n / k \rceil$ objects or $\left[\frac{n-1}{k}\right] + 1$.

That is if n pigeonholes are occupied by $mn + 1$ or more pigeons, where m is a positive integer then atleast one pigeonhole is occupied by $m + 1$ or more pigeons.

EXAMPLE 4.31. Show that if 20 persons are selected for presenting a cultural programme, then one may select a subset of 3 so that all 3 would be able to present their programmes on the same day of the week.

Solution : Now we assign each person to the day of the week on which he (or she) would present his programme. Now we have given 20 persons (pigeons) need to assign to 7 days (pigeonhole). Hence according to pigeonhole principle at least $\left\lceil\frac{20}{7}\right\rceil$ or $\left[\frac{(20-1)}{7}\right] + 1 = 2 + 1 = 3$ of the persons will be able to present their programmes on the same day of the week.

EXAMPLE 4.32. Find the minimum number of boys in a community to be sure that five of them are born in the same month.

Hint : Here $n = 12$ months are to be filled *i.e., pigconholes.*

$k = 5$ boys *i.e.*, pigeons.

4.7. THE INCLUSION -EXCLUSION PRINCIPLE

Theorem 4.7 : Let A and B be any finite sets. Then

$$n(A \cup B) = n(A) + n(B) - n(A \cap B)$$

In other words, to find the number $n(A \cup B)$, that number of elements in $A \cup B$, we add number of elements in A and number of elements in B. By doing this we have counted the elements of $A \cap B$ twice. So we substract $n(A \cap B)$ from the resultant. It is explained in the fig. 4.4.

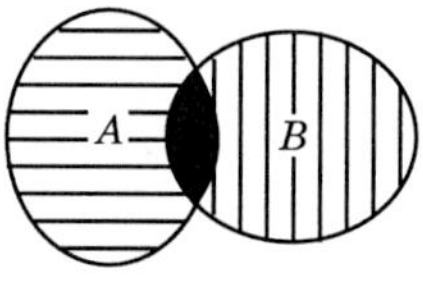

Fig. 4.6.

The shaded portion is repeated twice

Theorem 4.8 : For any finite set A, B, C

$$n(A \cup B \cup C) = n(A) + n(B) + n(C) - n(A \cap B) - n(A \cap C) - n(B \cap C) + n(A \cap B \cap C)$$

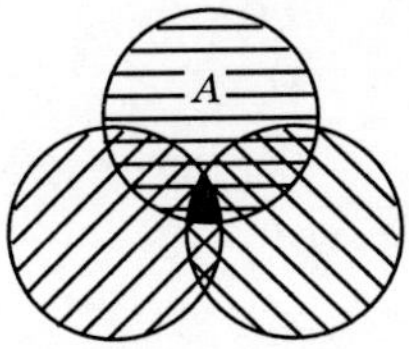

Fig. 4.7.

Theorem 4.9 : $n(A_1 \cup A_2 \cup ... \cup A_m) = t_1 - t_2 + t_3 ......+(-1)^{m-1}t_m$

$$= [n(A_1) + n(A_2) + ... + n(A_m)] - [n(A_1 \cap A_2) + n(A_2 \cap A_3) + ... + n(A_1 \cap A_m)] + [n(A_1 \cap A_2 \cap A_3) + n(A_1 \cap A_2 \cap A_4) + ... +]....$$

where $A_1, A_2, ... A_n$ are finite sets.

EXAMPLE 4.33. Out of 200 students, 50 of them take the course mathematics, 140 of them take the course economics & 24 of them take both the couses. Since both courses have scheduled examinations for the following day, only those students who are not taken any of these courses will be able to go to see movie. How many students will be able to go to see movie.

Solution : Let A be the set of students take mathematics and B be the set of students take economics.

$$\therefore \quad n(A) = 50$$
$$n(B) = 140$$
$$n(A \cap B) = 24$$
$$n(A \cup B) = n(A) + n(B) - n(A \cap B)$$
$$= 50 + 140 - 24$$
$$= 166$$

So 166 students cannot go to see the movie as they have to give the examination of mathematics or economics.

Therefore, the number of students can go to see the movie = 200 – 166 = 34

Probability

The probability of happening of an event B when another event A is known to have already happened is called conditional probability and is denoted by $P(B/A)$ is given as

$$P(B/A) = \frac{P(A \cap B)}{P(A)}, \quad \text{where } P(A) \neq 0$$

Similarly, $P(A/B) = \frac{P(A \cap B)}{P(B)}$, where $P(B) \neq 0$

In the above example,

$$P(A \cap B) = \frac{13}{52} \times \frac{12}{51}$$

$$P(A) = \frac{13}{52}$$

$$\therefore \quad P(B/A) = \frac{P(A \cap B)}{P(A)}$$

$$= \frac{13}{52} \times \frac{12}{51} \times \frac{52}{13}$$

$$= \frac{12}{51}$$

Note : $P(A \cap B) = P(A) \cdot P(B/A)$
$= P(B) \cdot P(A/B)$

EXAMPLE 4.34. Find the probability that the birthday of a child is a Saturday or Sunday.

Solution : S = {Sun, Mon, Tue, Wed, Thur, Fri, Sat}

$\therefore \quad n(S) = 7$

Let A be the event that the child is born on a Sunday or a Saturday.

$\therefore \quad n(A) = 2$

$$P(A) = \frac{2}{7}$$

EXAMPLE 4.35. Find the probability of getting a multiple of 2 in the throw of a die.

Solution : $S = \{1, 2, 3, 4, 5, 6\}$ $\therefore n(S) = 6$

$A = \{2, 4, 6\}$ $\therefore n(A) = 3$

$$P(A) = \frac{3}{6} = \frac{1}{2}$$

EXAMPLE 4.36. Find the probability that the sum of the score is odd in a throw of two dice.

Solution : $S = \{(1, 1), (1, 2)\ (1, 3), (1, 4), (1, 5), (1, 6),$
$(2, 1), (2, 2), (2, 3), (2, 4), (2, 5), (2, 6),$
$(3, 1), (3, 2), (3, 3), (3, 4), (3, 5)\ (3, 6),$
$(4, 1), (4, 2), (4, 3), (4, 4), (4, 5), (4, 6),$
$(5, 1), (5, 2), (5, 3), (5, 4), (5, 5), (5, 6),$
$(6, 1), (6, 2), (6, 3), (6, 4), (6, 5)\ (6, 6)\}$

$\therefore \quad n(S) = 36$

Let A be the event that the score on the two dice is odd.

$\therefore \quad A = \{(1, 2)\ (1, 4)\ (1, 6)\ (2, 1)\ (2, 3)\ (2, 5)\ (3, 2)\ (3, 4)$
$(3, 6)\ (4, 1)\ (4, 3)\ (4, 3)\ (4, 5)\ (5, 2)\ (5, 4)\ (5, 6)\ (6, 1)$
$(6, 3)\ (6, 5)\}$

$\therefore \quad n(A) = 19$

$$\therefore \quad P(A) = \frac{19}{36} = \frac{1}{2}$$

EXAMPLE 4.37. A committee of 5 employees is to be chosen from 6 men and 4 women. Find the probability that the committee contains exactly 2 women.

Solution : 5 Students can be selected from 10 in $^{10}C_5$ ways.

$\therefore$ Total number of possibilities $= {}^{10}C_5 = \dfrac{10!}{5!5!}$

$$= \frac{10 \times 9 \times 8 \times 7 \times 6}{5 \times 4 \times 3 \times 2} = 252$$

Let X be the event that the committee includes exactly 2 women and 3 men. The two women can be selected in $^4C_2$ ways and the 3 men can be selected in $^6C_3$ ways.

$\therefore \qquad n(A) = {}^4C_2 \times {}^6C_3 = 6 \times 20 = 120$

$\therefore \qquad P(A) = \dfrac{120}{252} = \dfrac{10}{21}$

EXAMPLE 4.38. A bag contains 7 red, 5 blue, 4 white and 4 black balls. Find the probability that a ball drawn at random is red or white.

Solution : A ball can be selected from 20 in $^{20}C_1 = 20$ ways.

Let A be an event that the ball drawn is red or white.

$\therefore \qquad n(A) = {}^7C_1 + {}^4C_1 = 7 + 4 = 11$

$\therefore \qquad P(A) = \dfrac{11}{20}$

EXAMPLE 4.39. In a shooting competition, the probability of hitting the target by A is $\dfrac{2}{5}$, by B is $\dfrac{2}{3}$ and by C is $\dfrac{3}{5}$. If all of them fire independently at the same target, calculate the probability that only one of them will hit the target.

Solution : $P(A) = \dfrac{2}{5} \qquad P(A') = \dfrac{3}{5}$

$P(B) = \dfrac{2}{3} \qquad P(B') = \dfrac{1}{3}$

$P(C) = \dfrac{3}{5} \qquad P(C') = \dfrac{2}{5}$

Probability that only one of them hits the target.

= Probability that A hits the target but not B and C.
+ Probability that B hits the target but not A and C.
+ Probability that C hits the target but not A and B.

$= P(A \cap B' \cap C') + P(A' \cap B \cap C') + P(A' \cap B' \cap C)$

$$= \frac{2}{5} \times \frac{1}{3} \times \frac{2}{5} + \frac{2}{3} \times \frac{3}{5} \times \frac{2}{5} + \frac{3}{5} \times \frac{1}{3} \times \frac{3}{5}$$

$$= \frac{4}{75} + \frac{12}{75} + \frac{9}{75}$$

$$= \frac{25}{75} = \frac{1}{3}$$

EXAMPLE 4.40. A box contains 6 white balls and 3 black balls, and another box contains 4 white balls and 5 black balls. Find the probability that a ball selected from one of the box again selected at random is a white ball?

Solution : Let A be the event that the first box is selected, B be the event that the second box is selected and C be the event that a white ball is selected.

$$P(A) = \frac{1}{2} \quad \text{and} \quad P(B) = \frac{1}{2}$$

$$P(C/A) = \frac{6}{9} \qquad P(C/B) = \frac{4}{9}$$

The probability of getting the first box and getting a white ball

$$= P(A \cap C) = P(A) \cdot P(C/A)$$

$$= \left(\frac{1}{2}\right)\left(\frac{4}{9}\right) = \frac{6}{18}$$

The probability of getting the second box and getting a white ball

$$= P(B \cap C) = P(B) \cdot P(B/C)$$

$$= \left(\frac{1}{2}\right)\left(\frac{4}{9}\right) = \frac{4}{18}$$

$\therefore$ The event $(A \cap C)$ and $(B \cap C)$ are mutually exclusive events.

$$\therefore \quad \text{P}(A \cap \text{C}) \cup (B \cap C) = \frac{6}{18} + \frac{4}{18} = \frac{10}{18} = \frac{5}{9}$$

Alternatively Probability of selecting any bag is $\frac{1}{2}$.

Probability of getting a white ball will be $= \frac{1}{2} \times \frac{{}^6C_1}{{}^9C_1} + \frac{1}{2} \times \frac{{}^4C_1}{{}^9C_1} = \frac{5}{9}$

EXAMPLE 4.41. In a garden, 40% of the flowers are roses and the rest are carnations. If 25% of the roses and 10% of the carnations are red, find the probability that a red flower selected at random is a rose.

Solution : Suppose there are 100 flowers.

Number of roses = 40 and number of carnations = 60

25% of 40 = 10 roses are red and 10% of 60 = 6 carnations are red.

Let A be the event that the flower is red and B be the event that the flower is a rose.

$\therefore A \cap B$ is the event that the flower is a red rose.

number of red flowers = $n(A) = 10 + 6 = 16$

$$\therefore \qquad P(A) = \frac{16}{100}$$

$$n(A \cap B) = 10 \quad \therefore P(A \cap B) = \frac{10}{100}$$

$P(B/A)$ = probability that a selected flower is a rose, red in colour.

$$P(B/A) = \frac{P(A \cap B)}{P(A)} = \frac{10/100}{16/100} = \frac{10}{16} = \frac{5}{8}$$

EXAMPLE 4.42. Two sisters X and Y appeared for an interview. The probability of selection of X is $\frac{1}{5}$ and that of Y is $\frac{2}{5}$. Find the probability that at least one of them is selected.

Solution : Let A be the event that the X is selected and B be the event that Y is selected.

$$\therefore \qquad P(A) = \frac{1}{5} \qquad\qquad P(B) = \frac{2}{7}$$

P(at least one of them is selected)

$$= 1 - P(\text{No one is selected})$$

$$= 1 - P(\overline{A})\, P(\overline{B})$$

$$= 1 - \frac{4}{\cancel{5}} \cdot \frac{\cancel{5}}{7} = \frac{3}{7}$$

SOLVED PROBLEMS

PROBLEM 1. How many ways can you arrange 7 different books, so that a specific book is on the third place ?

Ans. Remove the third book. Now you can arbitarily arrange the 6 books in 6! ways.

PROBLEM 2. In how many ways can you take 3 marbles out of a box with 15 different marbles ?

Ans. Taking a subset of three elements out a set of 15 elements can be done in $C(15, 3)$ ways.

PROBLEM 3. In a firm are 20 men and 10 women. In how many ways can you have a commitee with 3 men and 2 women ?

Ans. First we choose the 3 men. This can be done in $C(20, 3)$ ways. Then we choose the 2 women. This can be done in $C(10, 2)$ ways. The committee can be assembled in $C(20, 3) . C(10, 2)$ ways.

PROBLEM 4. In how many ways can you take 5 cards, with at least 2 aces, out of a game of 52 cards ?

Ans. First consider 5 cards, with exactly 2 aces.

For the two aces we have $C(4, 2)$ possibilities and for the three other cards we have $C(48, 3)$ possibilities. The 5 cards can be choosen in $C(4, 2) . C(48, 3)$ ways.

Analogous 5 cards, with exactly 3 aces can be chosen in $C(4, 3) . C(48, 2)$ ways. 5 cards, with exactly 4 aces can be chose in 48 ways.

$$\text{Total} = C(4, 2) . C(48, 3) + C(4, 3) + C(48, 3) + 48.$$

PROBLEM 5. In how many ways can you split a group of 15 persons in 4 persons and 11 persons ?

Ans. It is sufficient to choose 4 persons to split the group. This can be done in $C(15, 4)$ ways.

PROBLEM 6. How many diagonals are there in a convex n-polygon ?

Ans. From each angular point we can count $(n - 3)$ diagonal and since there are

n angular points we have counted $n(n-3)$ diagonals. But now we have counted twice each diagonal. Hence there are $n(n-3)/2$ diagonals.

PROBLEM 7. How many numbers consisting of 3 digits, can you make with the digits 0, 1, 2, 3, 4 ?

Ans. There are 4 possibilities for the first digit of the number.

There are 5 possibilities for the second figure of the number.

There are 5 possibilities for the third figure of the number.

$$\text{Total} : (4)(5)(5) = 100$$

PROBLEM 8. How many subsets are the there in a set of 10 elements ?

Ans. We can construct a subset of the set by deciding for each of the ten elements, if it belongs to the subset or not. So for each element there are 2 possibilities.

So total possibilities $= 2^{10}$.

PROBLEM 9. In how many ways can you arrange m identical stones into k piles so that each pile has at least 1 stone in it.

Ans. By removing one stone from each pile, this is the number of ways you can arrange $(m\text{-}k)$ identical stones into k (possibly empty) piles. Now, consider the k piles as a numbered set.

Write on each stone the number of a chosen pile and order the stones accordingly. The numbered stones constitute a combination with repetition of k elements (the numbers) choose $m\text{-}k$ (the stones). This can be done in

$$C(k, m\text{-}k) = C(m\text{-}1, m\text{-}k) \text{ ways.}$$

PROBLEM 10. How many strictly positive integer solutions (x, y, z) are there, such that $x + y + z = 100$.

Ans. This is the same problem as in how many ways can you arrange 100 identical stones into 3 piles so that each pile has at least 1 stone in it.

So the answer to this problems is:

$$C(99, 97) = 4851 \text{ ways.}$$

PROBLEM 11. How many terms are contained in $(a + b + c)^{20}$?

Ans. All terms can be written as A $a^p\, b^q\, c^r$ with $p + q + r = 20$.

The number of terms is the number of solutions of the equation $p + q + r = 20$ with p, q, r as positive integer unknowns.

Now regard (p, q, r) as three ordered elements.

Point 20 times one of these elements and order these elements in the same order as the given elements.

This corresponds with one solution of $p + q + r = 20$ and it is a combination with repetition of 3 elements choose 20.

The number of terms is the number of such combinations.

$$C(3, 20) = C(22, 20) = C(22, 2) = 231.$$

PROBLEM 12. The term $A \,.\, a^{10}\, b^3\, c^7$ is contained in $(a + b + c)^{20}$. Calculate A.

Ans. The number of terms $a^{10}\, b^3\, c^7$ is the number of permutations with repetition of the elements. a(10 times), b(3 times) and c(7 times).

This number is $\dfrac{20!}{10!\,3!\,7!}$.

PROBLEM 13. Consider a group of 191 students, of which 10 are taking French, Jermen and Sanskrit; 36 are taking French and Jerman; 20 are taking French and Sanskrit; 18 are taking Jerman and Sanskrit; 65 are taking French; 76 are taking Jerman and 63 are taking Sanskrit.

(*a*) How many are taking Jerman and neither French and Sanskrit ?

(*b*) How many are taking French or Jerman (or Botes) ?

(*c*) How many are taking none of three subjects?

Solution: 191 students;

10 F, J, S,

20 F, S,

18 J, S,

65 F

76 J

63 S

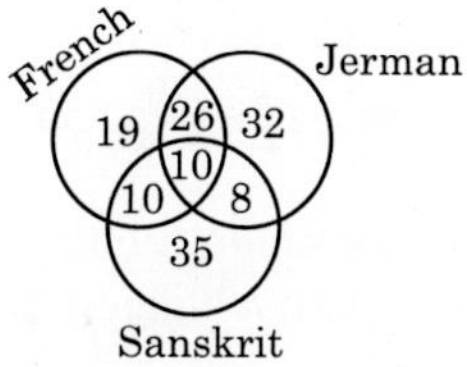

(*a*) How many are taking only Jerman ?

We can see by the picture that 32 students are taking only Jerman. We can also figure this out by looking at the numbers; 76 are in Jerman, 36 are in French and Jerman and 18 are in Jerman and Sanskrit; but 10 of these are also in French, so they are counted as part of the 36 in French and sanskrit. So :

$$76 - 36 - (18 - 10) = \mathbf{32}$$

(*b*) We want the union of French and Jerman : Using the picture, this is anyone is either the French or the business circle :

$$19 + 26 + 32 + 10 + 10 + 8 = 105$$

Also,
$$(F \cup J) = (F + J) - (F \cap J)$$
$$= 65 + 76 - 36$$
$$= 105$$

(*c*) From part (*b*) 105 are taking for J. So all we need to do is figure out the number of people taking only Sanskrit and the answer we are looking for is

$$191 - |F \cup J| - |S - (F \cap J)|$$

63 students take sanskrit, but 20 of these are also in French and another (18–10) of these take business. So $63 - 20 - 8 = 35$. $191 - 105 - 35 = 51$ students. We can also add up the numbers in the diagram and subtract this number from 191.

PROBLEM 14. Let T_n denotes the highest power of 2 that divides n. Show that

$$T_{mn} = T_m + T_n \text{ for all } m, n \geq 1$$

Solution: T_n is the highest power of two that divides n.

$$\frac{n}{2^{T_n}} = C_1$$

$$\frac{m}{2^{T_m}} = C_2$$

where C_1 and C_2 are odd integers, because they cannot be divided by 2 again,

$$\frac{n}{2^{T_n}} \cdot \frac{m}{2^{T_m}} = \frac{nm}{2^{T_m + T_n}} = C_1 C_2$$

Since an odd number times another odd number is odd, C_1C_2 is also add, so $T_m + T_n$ is in fact the highest power of 2 that divides mn. So $T_{mn} = T_m + T_n$.

PROBLEM 15. How many terms are there in the expansion of

$$(x + y)\ (a + b + c)\ (e + f + g)\ (h + i)\ ?$$

Solution : $(x + y)\ (a + b + c)\ (e + f + g)\ (h + i)$

Each term in each polynomial will be multiplied by every term in the other polynomials.

So $\quad (2)\ (3)\ (3)\ (2) = 36$ terms.

PROBLEM 16. How many subsets of a $(2n + 1)$ elements set have n elements or less ?

Solution : A set with $(2n + 1)$ elements has 2^{2n+1} subsets. A subset S has n elements or less if $\overline{S}$ has more than n elements. So *half* of the subsets have n or less elements.

$$\frac{1}{2}(2^{2n+1}) = \frac{1}{2}(2^{2n})\ (2) = 2^{2n} \text{ subsets.}$$

PROBLEM 17. Show that the product of any positive integer and its $k - 1$ successors is divisible by $k!$

Solution : The product of n and its $(k - 1)$ successors :

$$n(n + 1)\ (n + 2)\ (n + 3)\ ....\ (n + k - 1)$$

There are k terms. We want to show that $\dfrac{n(n+1)(n+2).....(n+k-1)}{k!}$ is a whole number, and

$$\frac{n(n+1)(n+2).....(n+k-1)(n-k)!}{(n-k)!\,k!} = c(n + k - 1,\ k)$$

which is an integer. So the claim is true.

PROBLEM 18. Four telephonesets are randomly selected from a lot of 100 telephonesets among which 10 are defective. Find the probability of obtaining at most one defective telephoneset.

Hint : P(at most one defective telephone set)

$$P(\text{no defective}) + P(1 \text{ defective}) = \frac{C(90,4) + 10.\,C(90,3)}{C(100,4)} = 95\%$$

EXERCISE

1. If ${}^nC_{10} = {}^nC_{15}$, find ${}^{27}C_n$
2. If ${}^nC_6 : {}^{(n-3)}C_3 = 33 : 4$, find n
3. In how many ways can the letters of the word "ARRANGE" be rearranged? In how many arrangements the two R's do not come together? How many arrangements are possible in which neither the two R's nor the two A's are allowed to come together?
4. From 4 teachers and 8 students, in how many ways can a team of six be chosen (*i*) to include exactly one teacher; (*ii*) to include atleast one student ?
5. Evaluate ${}^{15}C_3 + {}^{15}C_6 - {}^{15}C_7$

6. How many triangles can be formed by joining 12 points, 7 of which are collinear ?
7. In how many ways can the letters of the words "COMMITTEE" be arranged so that the vowels may never be separated?
8. How many numbers between 4000 and 9000 can be formed by using the digits 2, 4, 7 and 9 if each digit may be repeated ?
9. 8 chairs are numbered from 1 to 8. Two women first choose two chairs from those marked 1 to 4 and 3 chairs from the remaining. Find the number of possible arrangements.
10. Srishti hosted a party to which he invited 20 persons. In how many ways can he and the guests be seated at a circular table? In how many ways of these will two particular persons be seated on either side of the host?
11. There are 6 pairs of gloves of different sizes. In how many ways can you choose one for the left hand one for the right hand such that they are not of the same pair ?
12. In how many ways can you put 9 coins of different value into 2 packets ?
13. From 3 mangoes, 4 apples and 2 oranges, how many selections of fruits can be made, taking at least one of each kind?
14. From 6 gentlemen and 4 ladies a committee of 5 is to be formed. In how many ways can this be done, if the committee is to include at least one lady?
15. A candidate is required to answer 6 out of 10 questions divided into 2 groups each containing 5 questions. He is not permitted to attempt more than 4 from each group. In how many ways can he make the choice ?
16. In how many ways can you choose 8 cards out of a pack of 52 and have all suits present and have 2 spade.
17. Given 5 different green kite, 4 different blue kite, 3 different red kite, how many combinations of kite can be chosen taking atleast 1 green and 2 blue kite ?
18. How many numbers greater than a million can be formed with the digits 2, 3, 0, 3, 4, 2 and 3 ?
19. In a railway compartment there are 2 rows of seats facing each other with accommodation for 5 in each. 4 wish to sit facing forward and 3 facing towards the rear while 3 others are indifferent. In how many ways can the 10 passengers be seated?
20. Find the chance of throwing more than 15 in one throw with 3 dice.
21. In a race where 12 horses are running, the chance that horse A will win is $\frac{1}{6}$, that B will win is $\frac{1}{10}$ and that C will win is $\frac{1}{8}$. Assuming that a dead heat is impossible, find the chance that one of them will win.
22. Two balls are to drawn from a bag containing 7 red and 9 white balls; find the chance that they will both be red.

23. Two cards are drawn from a pack of well shuffled cards. Find the probability that one is a club and the other is an Ace.

24. There are two boxes, one contains 8 cubes and 6 spheres and the other contains 6 cubes and 12 spheres. One object is transferred at random from the first box to the second. Find the probability that an object selected from the second box is a cube.

25. There are two bags, one of which contains 5 red and 7 white balls and the other 3 red and 12 white balls. A ball is to be drawn from one or other of the two bags; find the chance of drawing a red ball.

26. (*a*) A computer password consists of a letter of the alphabet followed by 3 or 4 digits. Find the total number of passwords that can be formed and the number of passwords in which no digits is being repeated.

(*b*) How many solutions are there of $x + y + z = 17$ subject to the constraints $x \geq 1, y \geq 2, z \geq 3$.

(*c*) In how many ways can a committee of 10 is to be selected consisting of at least 4 women and 4 men from 9 women and 8 men if

(*i*) Ms X refuses to work with Mr. Y.

(*ii*) Ms X and *Y* insist to work together.

27. (*a*) There are *n* circles drawn in a plane such that each circle intersects every other circle in exactly two points and no 3 circle, meet in the same point. If a_n denotes the region created by these *n* circles, find a recurrence relation for a_n.

(*b*) Find the general solution of the following relation

$2a_n - 7a_{n-1} + 3a_{n-2} = 2^n$

(*c*) Find the sequence $\{y_x\}$ having the generating function *G* given by:

$$G(x) = \frac{3}{1-x} + \frac{1}{1-2x}$$

28. Solve recurrence relation

$a_n - 8a_{n-1} + 21a_{n-2} - 18a_{n-3} = 0$

29. How many arrangement can be made of the letter of the word.

(*a*) APPLE

(*b*) COMMERCE

(*c*) PROGRAMMING

Answers

1. ${}^nC_r = {}^nC_{n-r}$

$\therefore {}^nC_{10} = {}^nC_n - 15 \quad \therefore n - 15 = 10 \quad \therefore n = 25$

$${}^{27}C_n = {}^{27}C_{25} = \frac{27 \times 26}{1 \times 2} = 351$$

2. $$\frac{{}^nC_6}{{}^{(n-3)}C_3} = \frac{33}{4}$$

$$\frac{n!}{6!(n-6)!} \times \frac{(n-3-3)!3!}{(n-3)!} = \frac{33}{4}$$

$$\frac{n!}{6!(n-6)!} \times \frac{(n-6)!\,3!}{(n-3)!} = \frac{33}{4}$$

$$\frac{n \times (n-1)\,(n-2)\,(n-3)!}{6 \times 5 \times 4 \times 3!\,(n-6)!} \times \frac{(n-6)!\,3!}{(n-3)!} = \frac{33}{4}$$

$$\frac{n(n-1)\,(n-2)}{6 \times 5 \times 4} = \frac{33}{4}$$

$n(n-1)\,(n-2) = 30 \times 33 = 11 \times 10 \times 9$

$\therefore\ n = 11$

3. (*i*) We have 7 letters, of which there are two R's two A's and the rest unlike

$\therefore$ Total number of arrangements $= \frac{33}{4} = 1260$

(*ii*) To find the arrangements in which two R's come together, consider teh two R's as one letter and find the arrangement of 6 things in which two A's are present.

$\therefore$ Number of arrangements $= \frac{6!}{2!} = 360$

$\therefore$ Number of arrangements in which R's do not come together

(*iii*) a. Number of arrangements in which two R's are together = 360

b. Number of arrangements in which two A's are together = 360

c. Number of arrangements in which two R's and two As are together.

$=$ No. of ways of arranging 5 things taken all $= {}^5P_5 = 5! = 120$

$\therefore$ No. of arrangements in which neither the two R's nor the two A's are together $= 1260 - [(360 + 360 - 120)] = 1260 - (600)) = 660$

4. (*i*) 1 teacher and 5 students can be chosen in ${}^4C_1 \times {}^8C_5 = 224$ ways

(*ii*) If no teacher is chosen then 6 students can be chosen from 8 in

$${}^8C_6 = \frac{8!}{6!\,6!}$$

$= 28$ ways.

Number of ways of choosing 6 form 4 teachers and 8 students

i.e., from $12 = {}^{12}C_6 = \frac{12!}{6!\,6!} = 924$

Total number of ways such that at least one teachers is included

$= 924 - 28 = 896.$

5. ${}^{15}C_8 + {}^{15}C_9 - {}^{15}C_6 - {}^{15}C_7$

$= ({}^{15}C_8 + {}^{15}C_9) - ({}^{15}C_6 + {}^{15}C_7) \ldots ({}^nC_r + {}^nC_{r+1}) = {}^{n+1}C_{r+1})$

$= {}^{16}C_9 - {}^{16}C_7 = {}^{16}C_9 - {}^{16}C_9 = 0 \ldots ({}^nC_r = {}^nC_{n-r})$

6. 3 points out of 12 can be selected in 12C3 ways, out of which 7C3 triangles are impossible $\therefore$ Number of triangles formed $= {}^{12}C_3 - {}^7C_3 = 220 - 35 = 185$

7. Considering all vowels *i.e.*, *a*, *e*, *i*, o as one, the eight letters are to be arranged in eight places in $^8P_8 = 8!$ ways
   The 4 vowels can be arranged themselves in $^4P_4 = 4!$ way
   $\therefore$ Total number of arrangements = $8! \times 4!$

8. The first place can be taken by 4 or 7 in 2 ways
   The remaining three places can be taken by the four numbers in $4 \times 4 \times 4 = 64$ ways.

9. 2 persons can be arranged in 4 in $^4P_2 \times {}^6P_3 = 4 \times 3 \times 6 \times 5 \times 4 = 1440$

11. 21 persons can be arranged in $(21 - 1)! = 20!$ ways.
    Treating the host and the two persons as one, we can arrange the 19 persons in $(19 - 1)! = 18!$ ways.
    But the two can be arranged on either side of the cost in 2 ways.
    So total number of arrangements = $2 \times 18!$

12. Number of choice = $^6C_1 \times {}^5C_1 = 6 \times 5 = 30$
    (After choosing 1 pair for left hand, there are only 5 pairs left for choice for the right hand).

13. Each coin can go into one of the two packets.
    So, total ways = 29. Hence, [4].

14. Number of ways in which mangoes can be selected = 23. But this also includes, the case where all three mangoes are not selected Hence, number of ways in which at least mango is selected = $(2^3 - 1)$
    Similarly, Number of ways in which apples and oranges are selected is (24 – 1) and $(2^2 - 1)$
    $\therefore$ Total number of selections = $(2^3 - 1)(2^4 - 1)(2^2 - 1) = 315$.

DO IT BY YOURSELF

1. U = {1, 2, 3, ..., 8, 9}, and A = {1, 2, 5, 6}, B = {2, 5, 7}, C = {1, 3, 5 , 7, 9} Find and make Venn diagram of each of the following:
   (*a*) $A \cap B$ and $A \cap C$ (*b*) $A \cup B$ and $B \cup C$
   (*c*) A^C and C^C (*d*) $A \setminus B$ and $A \setminus C$
   (*e*) $A \oplus B$ and $A \oplus C$
   (*f*) $(A \cup C) \setminus B$, $(A \cup B)^C$, $(B \oplus C) \setminus A$
2. Write all set identities and define Boolean sum of two sets.
3. The formula $A \setminus B = A \cap B^C$ defines the difference operation in terms of the operations of intersection and complement. Find a formula that defines the union $A \cup B$ in terms of the operations of intersection and complement.
4. Draw a Venn diagram of sets *A*, *B*, *C* where $A \subseteq B$, sets *B* and *C* are disjoint, but *A* and *C* have elements in common.
5. Let A, B, C be subsets of $\cup$. Given that
   $$A \cap B = A \cap C$$
   $$\bar{A} \cap B = \bar{A} \cap C$$
   Is it necessary that $B = C$? Justify your answer.

6. Give that

$$(A \cap C) \subseteq (B \cap C)$$
$$(A \cap \bar{C}) \subseteq (B \cap \bar{C})$$

Shown that $A \subseteq B$.

7. What can you say about the sets P and Q if
   (*a*) $P \cap Q = P$? (*b*) $P \cup Q = P$?
   (*c*) $P \oplus Q = P$? (*d*) $P \cap Q = P \cup Q$?
8. Let $A \subseteq B$ and $C \subseteq D$. Is it always the case that $(A \cup C) \subseteq (B \cup D)$? Is it always the case that $(A \cap C) \subseteq (B \cap D)$?
9. Let $W \subset X$ and $Y \subset Z$. Is it always the case that $(W \cup Y) \subset (X \cup Z)$? Is it always the case that $(W \cup Y) \subset (X \cup Z)$?
10. (*a*) Given that $A \cup B = A \cup C$, is it necessary that $B = C$?
    (*b*) Given that $A \cap B = A \cap C$, is it necessary that $B = C$?
    (*c*) Given that $A \oplus B = A \oplus C$, is it necessary that B = C?
    Justify your answers?
11. Let A, B, C, sets. Under what condition is each of the following statements are true.
    (*a*) $(A - B) \cup (A - C) = A$ (*b*) $(A - B) \cup (A - C) = \phi$
    (*c*) $(A - B) \cap (A - C) = \phi$ (*d*) $(A - B) \oplus (A - C) = \phi$
12. Let A, B be two sets.
    (*a*) Given that $A - B = B$, What can be said about A and B?
    (*b*) Given that $A - B = B - A$, What can be said about A and B?
13. Let A, B, C be arbitrary sets. Shows that $(A - B) - C = A - (A \cup C)$.
14. Let A, B, C be arbitrary sets. Shows that $(A - B) - C = (A - C) - B$.
15. Let A, B, C be arbitrary sets. Shows that $(A - B) - C = (A - C) - (B - C)$.
16. Out of 200 students, 50 of them take the discrete Mathematics, 140 of them take the course Economics, and 24 of them take both courses. Since both courses have scheduled examination s for the following day, only students who are not in either one of these courses will be able to go to the party the night before. How many students will be at the party. Suppose that 60 of the 200 are underclass students. Among the underclass students, 20 of them take Discrete Mathematics, 45 of them take Economics, and 16 of them take both. How many upper-class students will be at the party?
17. Thirty cars were assembled in a factory. The options available were a radio, an air conditioner, and white-wall tires. It is known that 15 of the cars have radios, 8 of them have air conditioners, and 6 of them have white-wall tires. Moreover, 3 of them have all three options. Atleast how many cars do not have any options at all.
18. Determine the number of integers between 1 and 300 that are divisible by any of the integers 2, 3, 5 and 7.
19. A survey on a sample of 25 new cars being sold at a local auto dealer was conducted to see which of three popular options air-conditioning (A) radio (R), and power windows (W) were already installed. The survey found.
    - 15 had air conditioning,
    - 12 had radio

- 11 had power windows
- 5 had air conditioning and power windows,
- 9 had air conditioning and radio
- 4 had radio and power windows,
- 3 had all three options.

Find the number of cars that had

(*a*) Only power windows

(*b*) Only air-conditioning

(*c*) Only radio

(*d*) Radio and power windows but not air conditioning

(*e*) Air conditioning and radio, but not power windows

(*f*) Only one of the option

(*g*) At least one option

(*h*) None of the options.

20. Show that the set of all rational numbers is countably infinite.
21. Show that the set of all real numbers is uncountably infinite.
22. Prove that if a set is countably infinite, then its power set is not countably infinite.
23. If A and B are disjoint sets S.T. $|A \cup B| = |A|$, what can we say about B?
24. There is a set of five distinct computer science books, and two distinct art books.

    (*a*) In how many ways can these books be arranged on a shelf?

    (*b*) In how many ways these books be arranged on a shelf if all books of the same discipline are grouped together.

    (*c*) In how many ways these books be arranged on a shelf if the two art books are not together?
25. (*a*) How many eight-bit strings contain exactly three 0's?

    (*b*) How many eight-bit string contain three 0's in a row and five 1's?

    (*c*) How many eight-bit string contain atleast two 0's in a row?
26. (*a*) Find the next row of Pascal's triangle given the row:

    1 7 21 35 35 21 7 1

    (*b*) Use the Binomial theorem to show that $\Sigma^n\ k = 0\ 2^k\ C\ (n, k) = 3^n$.
27. It is possible to interconnect five processors so that exactly two processors are directly connected to an identical number of processors? Explain.
28. Find the minimum number of the students needed to guarantee that five of them belong to the same group (Bhangra, Rock, Classical).
29. A sample of 80 cars owned revealed that 24 owned station wagons and 62 owned cars that are not station wagon. Find the number k of people who owned both a station wagon and some other car.
30. A student must take five classes from three areas of study. Numerous classes are offered in each discipline, but the student cannot take more than two classes in any given area.

    (*a*) Using the Pigeonhole principle, show that the student will take at least two classes in one area.

    (*b*) Using the inclusion-exclusion principle, show that the student will have to take at least one class in each area.

MULTIPLE CHOICE QUESTIONS

1. The number of numbers from 1 to 100 that are divisible by 2 or 3 is:
(*a*) 52 (*b*) 80 (*c*) 67 (*d*) 21

2. Five letter words are to be formed from the given 10 different letters. The numbers of words having at least one letter repeated is:
(*a*) $10^5 - 10\,P_5$ (*b*) $5^{10} - 10\,P_4$ (*c*) $5^{10} - 10\,P_5$ (*d*) $10\,P_4 - 10^5$

3. How many numbers of 7 digits can be formed with the digits 0, 2, 2, 5, 6, 6, 6
(*a*) 420 (*b*) 60 (*c*) 340 (*d*) None

4. Find the number of permutations of the letters in the word POSSESSION
(*a*) 76500 (*b*) 67500 (*c*) 75600 (*d*) 56700

5. A student is to answer 10 out of 13 questions on an exam. How many choices he has:
(*a*) 786 (*b*) 286 (*c*) 986 (*d*) 386

6. Find the minimum number of students that can be admitted to a college so that there are at least 15 students from one of the 50 states
(*a*) 701 (*b*) 702 (*c*) 703 (*d*) None

7. Bhaskar has 6 friends. In how many ways can he invite one or more of them at a dinner?
(*a*) 62 (*b*) 63 (*c*) 64 (*d*) None of these

8. There are three identical red balls and four identical blue balls in a bag. Three balls are drawn. The number of different color combinations is:
(*a*) 20 (*b*) 30 (*c*) 40 (*d*) 50

9. The number of ways in which a team of 11 players can be selected from 22 players including 2 of them and excluding 4 of them *i*
(*a*) ${}^{22}C_6$ (*b*) ${}^{16}C_9$ (*c*) ${}^{22}C_{11}$ (*d*) None of these

10. If ${}^{n}C_{15} = {}^{n}C_9$ then n is equal to:
(*a*) 6 (*b*) 12 (*c*) 24 (*d*) None

Answers

1. (*c*)	2. (*a*)	3. (*c*)	4. (*c*)	5. (*b*)	6. (*a*)
7. (*b*)	8. (*b*)	9. (*b*)	10. (*c*)		

❐❐❐

CHAPTER 5

RELATION

This Chapter Covers :

5.1. INTRODUCTION

Relationship between elements of sets occur in many way. In daily life we deal with the relationships such as those in business, relatives like between mother and son, father and son. In mathematics we study relationship between functions, numbers, equations etc. Even computer languages and programs may have certain relationship defined between them, like relation between the two computer languages *C* and *C*++ and in computer programs different variables are related. For example, function A "calls" function B is a program, here, "calls" is a relation. Therefore the relation is nothing but the link between the members of different sets. The most direct way to express a relationship between two sets is to use ordered pairs made up of elements from each set. Therefore, the set of ordered pairs are called binary relations.

5.2. BINARY RELATION

Let A and B be two sets. A binary relation from the set A to the set B is a subset of $A \times B$. In other words, a binary relation R from A to B is a set of ordered pairs where the first element is from the set A and the second element is from the set B

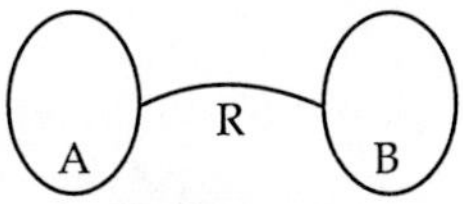

Fig. 5.1.

$R = \{(x, y) \mid x \in A, y \in B\}$

Two elements x and y are related by the relation R is denoted by $(x, y) \in R$ or xRy. If x is not related to y by the relation R then the notation is $(x, y) \notin R$ or $x \not R y$. Therefore binary relation represent relationship between the elements of two sets.

EXAMPLE 5.1. Let $A = \{1, 2, 3\}$, $B = \{4, 5\}$

$A \times B = \{(1, 4), (1, 5), (2, 4), (2, 5), (3, 4), (3, 5)\}$ then the set $\{(1, 4), (2, 5), (3, 4)\}$ is a relation from A into B.

Example 5.2. Let A be the set of all cities and let B let the set of all states in a country. Define a relation R by specifying that aRb if the city a is in the state b.

EXAMPLE 5.3. Let A be the set of integers. Define a relation R on A is a subset of $A \times A$ such that for x, y in A, x is related to y if x divides y.

i.e., xRy if $x \mid y$.

for instance $2R4$ but $2 \not R 5$.

EXAMPLE 5.4. Let $A = \{1, 2, 3, 4\}$. Define the following relation R on A :

aRb *iff* $a < b$

$R = \{(1, 2), (1, 3), (1, 4), (2, 4), (3, 4), (2, 3)\}$

EXAMPLE 5.5. Let A = The set of all lines in plane. Define the following relation R on A :

$l_1 R l_2$ *iff* $l_1 \mid\mid l_2$ where l_1 and l_2 are lines in the planes, *i.e.*, l_1 is related to l_2 by relation R if l_1 is parallel to l_2.

Theorem 5.1. There are 2^{n^2} relations on a set A with n elements.

Proof : A relation on a set A is a subset of $A \times A$. Since $A \times A$ has n^2 elements when A has n elements. A set with m elements has 2^m subsets therefore there are 2^{n^2} subsets of $A \times A$. Thus there are 2^{n^2} relations on a set with n elements.

[Readers are required to verify the above theorem on a set of three elements]

Theorem 5.2. Let R and S be relations from A to B.

If $R(a) = S(a)$ $\forall$ $a \in A$ then $R = S$.

If R is a relation from a set A to a set B then

Domain of R is the set of elements in A that are related to some elements in B *i.e.*, the set of all first elements in the ordered pairs that make up R. Doman of R is denoted by Dom(R).

Range of R

Range of the relation R is the set of elements in B that are second elements of the ordered pair in R.

EXAMPLE 5.6. Consider

$$A = \{4, 5, 6\}$$
$$B = \{a, b\}$$
$$R = \{(4, a), (5, a), (6, a)\}$$

then

$$\text{Domain}(R) = \{4, 5, 6\} = A$$
$$\text{Range}(R) = \{a\}$$

EXAMPLE 5.7. Let $A = \{1, 2, 3, 4, 5\}$

for $a, b \in A$

aRb iff $a < b$

$$R = \{(1, 2), (1, 3), (1, 4), (1, 5), (2, 3), (2, 4), (2, 5), (3, 4), (3, 5), (4, 5)\}$$

then

$$\text{Domain}(R) = \{1, 2, 3, 4\}$$
$$\text{Range}(R) = \{2, 3, 4, 5\}$$

5.3. *R*-RELATIVE SET

If R is a relation from the set A to B and $x \in A$ we define $R(x)$, the R-Relative set of x, to be the set of all y in B with the property that x is related to y. Thus

$$R(x) = \{y \in B \mid xRy\}$$

EXAMPLE 5.7. Let $A = \{1, 2, 3, 4\}$

and $R = \{(1, 1), (1, 2), (2, 3), (3, 1), (4, 3), (3, 2)\}$

then

$$R(1) = \{1, 2\}$$
$$R(2) = \{3\}$$

If $A_1 = \{3, 4\} \subset A$

then $R(A_1) = \{1, 2, 3\}$

If $A_1 \subseteq A$ then $R(A_1)$, the **R-Relative set of A_1** is the set of all y in B with the property that xRy for some $x \in A_1$ *i.e.*,

$$R(A_1) = \{y \in B \mid xRy \text{ for some } x \in A_1\}$$

Theorem 5.2. Let A and B be two sets and R be the relation from the set A to B. Let A_1 and A_2 be the subsets of A, then

(*a*) If $A_1 \subseteq A_2$ then $R(A_1) \subseteq R(A_2)$

(*b*) $R(A_1 \cup A_2) = R(A_1) \cup R(A_2)$

(c) $R(A_1 \cap A_2) \subseteq R(A_1) \cap R(A_2)$

Proof : Basic of proof of the theorem is that $A \subseteq B$ if for any arbitrary element in A is in B.

(a) : Let $y \in R(A_1)$ then xRy for some $x \in A_1$

and $A_1 \subseteq A_2$

$\Rightarrow x \in A_2$

$\Rightarrow \quad x\,Ry$ for some $x \in A_2$

$\Rightarrow y \in R(A_2)$

$\Rightarrow R(A_1) \subseteq R(A_2)$

Therefore,

If $\quad A_1 \subseteq A_2 \quad$ then $\quad R(A_1) \subseteq R(A_2)$

(*b*) Let $\quad y \in R(A_1 \cup A_2)$

then xRy for some $x \in A_1 \cup A_2$

$\Rightarrow x \in A_1 \quad$ or $\quad x \in A_2 \quad$ or $\quad x \in A_1 \cap A_2$

If $\quad x \in A_1$ and $xRy \Rightarrow y \in R(A_1) \Rightarrow y \in R(A_1) \cup R(A_2)$

If $\quad x \in A_2$ and $xRy \Rightarrow y \in R(A_2) \Rightarrow y \in R(A_1) \cup R(A_2)$

If $\quad x \in A_1 \cap A_2$ and $xRy \Rightarrow y \in R(A_1)$ and $y \in R(A_2) \Rightarrow y \in R(A_1) \cup R(A_2)$

$\Rightarrow R(A_1 \cup A_2) \subseteq R(A_1) \cup R(A_2)$ ...(*i*)

Conversely Let $y \in R(A_1) \cup R(A_2)$

$\Rightarrow y \in R(A_1)$	or $y \in R(A_2)$	or $y \in R(A_1) \cap R(A_2)$
$\Downarrow$	$\Downarrow$	$\Downarrow$
xRy, for some $x \in A_1$	xRy, for some $x \in A_2$	xRy, for some $x \in A_1 \cap A_2$
$\Downarrow$	$\Downarrow$	$\Downarrow$
$y \in R(A_1 \cup A_2)$	$y \in R(A_1 \cup A_2)$	$y \in R(A_1 \cup A_2)$

$\Rightarrow R(A_1) \cup R(A_2) \subseteq R(A_1 \cup A_2) \quad \Rightarrow y \in R(A_1) \cap R(A_2)$ ...(*ii*)

Therefore, (1) and (2) $\Rightarrow$

$$R(A_1 \cup A_2) = R(A_1) \cup R(A_2).$$

(*c*) Let $\quad y \in R(A_1 \cap A_2)$

$\Rightarrow xRy,\ x \in A_1 \cap A_2$

$\Rightarrow xRy,\ x \in A_1$ and $x \in A_2$

$\Rightarrow xRy,\ x \in A_1 \Rightarrow y \in R(A_1)$

$xRy,\ x \in A_2 \Rightarrow y \in R(A_2)$

Therefore, $y \in R(A_1 \cap A_2) \Rightarrow y \in R(A_1) \cap R(A_2)$

As y is arbitary $\Rightarrow R(A_1 \cap A_2) \subseteq R(A_1) \cap R(A_2)$.

Converse of the above result is not true *i.e.*, equality does not holds. The following example will give better understanding.

EXAMPLE 5.8. Consider an example of $R(A_1 \cap A_2) \neq R(A_1) \cap R(A_2)$

Let $A = Z$, set of integers

Let "R", "$\leq$" is the given relation on the let A

$$A_1 = \{0, 1, 2\}$$
$$A_2 = \{9, 13\}$$

then $R(A_1)$ consists of all integers n such that $0 \leq n$, $1 \leq n$, $2 \leq n$.

$\Rightarrow R(A_1) = \{0, 1, 2, ...\}$

Similarly $\quad R(A_2) = \{9, 10, 11, 12, 13, ...\}$

So $\quad R(A_1) \cap R(A_2) = \{9, 10, 11, 12, 13, ...\}$

whereas $\quad A_1 \cap A_2 = \phi$

$$\Rightarrow R(A_1 \cap A_2) = \phi$$
$$\Rightarrow R(A_1 \cap A_2) \neq R(A_1) \cap R(A_2).$$

5.4. COMPOSITION OF RELATIONS

Let R be a relation from a set A into a set B and S be a relation from set B into a set C. The composition of R and S, written as RS or RoS is the set of pairs of the form $(a,c) \in A \times C$ where $(a,c) \in RS$ if and only if $b \in B$ such that $(a, b) \in R$ and $(b, c) \in S$.

EXAMPLE 5.9. Let $A = \{1, 3, 5\}$

Let R be the relation such that xRy iff $y = x + 2$
and S be the relation such that xRy iff $x < y$

(*a*) Find RS
(*b*) Find SR
(*c*) Illustrate RS and SR *via.,* a diagram.
(*d*) Is the relation (set) RS equal to the relation SR ? Why?

Solution : $R = \{(1,3) , (3,5)\}$
$S = \{(1,3) , (1,5) , (3,5)\},$

(*a*) $RS = \{(1,5)\}$
(*b*) $SR = \{(1,5)\}$

(*c*)

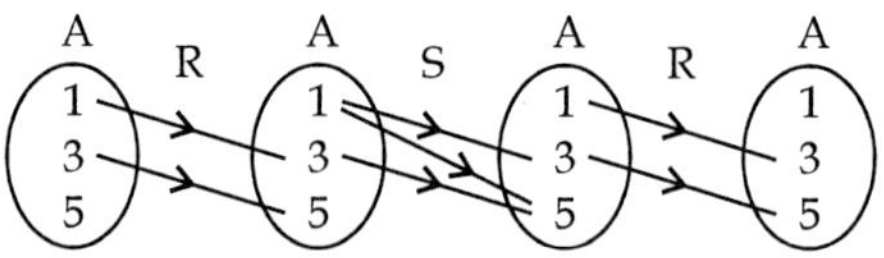

Fig. 5.2.

(*d*) $\therefore$ The relations $RS = SR$.
Relations are same as the elements of the set.

5.5. REPRESENTATION OF RELATION GRAPHICALLY

We can represent a relation on a finite set A with a diagram called a directed graph or digraph. To do this, we represent elements of the set A as a vertex or a node by a small circle. If the ordered pair (x, y) is a related pair *ie* if xRy then connect the two vertices by an arrow (starts from x towards y) called directed graph or diagraph. An edge of a graph is also referred to as an arc, a line or a branch.

EXAMPLE 5.10. Let $A = \{1,2,3,4\}$

$R = \{(1, 1), (1, 2), (2, 1), (2, 2), (2, 3), (2, 4), (3, 4), (4, 4)\}$

Graphical representation of the given relation is given by:

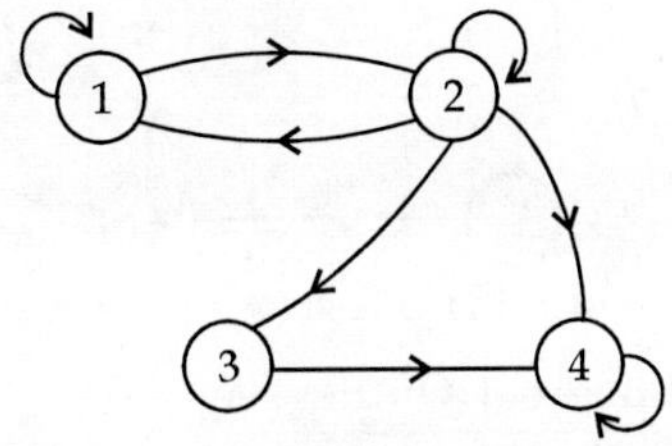

Fig. 5.3.

Arrows between the vertices 1 & 2 indicate that 1 is related 2 and 2 is related to 1 etc. self loop indicates the element is related to itself. Therefore following are the conclusions from the graph given by the figure (5.3)

1. Every element is related to itself.
2. 1 is related to 2 and 2 is related to 1
3. 2 is related to 3 and 2 is related to 4
4. 3 is related to 4.

If R is a relation on a set A and a is any element in A then **degree of a vertex** is the number of edges incident on that vertex. **The degree of a vertex** v is denoted by $d(v)$.

Indegree of a Vertex

Let R is a relation on a set A and $a \in A$ then indegree of the vertex a is the number of $b \in A$ such that $(b,a) \in R$, *ie* number of edges coming towards a.

Outdegree of a Vertex

Outdegree of a vertex of the vertex a is the number of $b \in A$ such that $(a, b) \in R$, *i.e.*, the number of edges going towards a.

EXAMPLE 5.11. Let $A = \{1,2,3,4\}$

$$R = \{(1, 1), (1, 2), (2, 1), (2, 2), (3, 3), (2, 3), (2, 4), (3, 4)\}$$

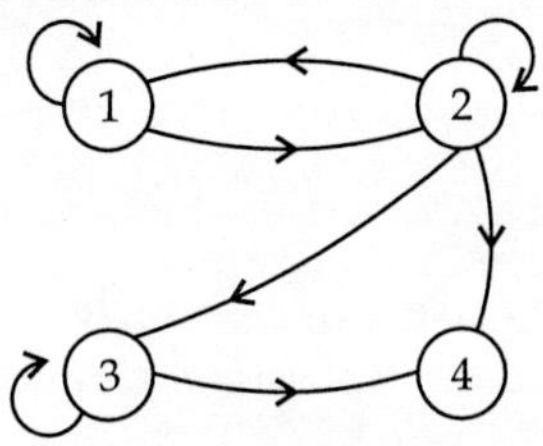

Fig. 5.4.

Indegree and outdegree of all the vertices are given by the following table:

Vertices	Indegree	Outdegree
1	2	2
2	2	2
3	2	2
4	2	0

EXAMPLE 5.12. Consider the set

$$A = \{x_1, x_2, x_3, x_4\}$$

$$\text{and } R = \{x_1, x_3), (x_3, x_4), (x_2, x_4)\}$$

Find indegree and outdegree of each vertex.

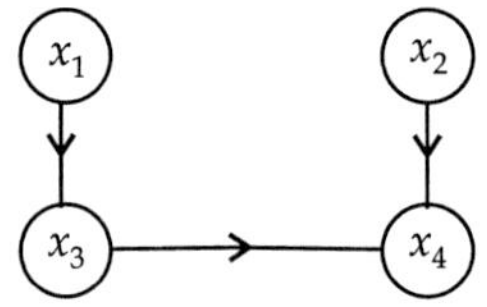

Indegree & outdegree of all the vertices are given by the following table:

Vertices	Indegree	Outdegree
x_1	0	1
x_2	0	1
x_3	1	1
x_4	1	0

5.6. MATRIX REPRESENTATION OF A RELATION

There are many ways to represent a relation between finite sets. One way is to list its ordered pairs. In this section we uses matrices to represent relations and the corresponding matrix is called Boolean matrix, adjacency matrix or zero one matrix.

Let A and B be the two sets with dimension m & n respectively.

Let R is a relation from A to B.

We can represent R by using $m \times n$ matrix $M_R = [m_{ij}]$ and this matrix is called adjacency matrix.

Adjacency Matrix

Adjacency Matrix for R is defined as:

$$m_{ij} = \begin{cases} 1 & if \quad (a_i, b_j) \in R \\ 0 & if \quad (a_i, b_j) \notin R. \end{cases}$$

If A and B are the finite sets with n, m number of elements respectively and R is a relation from A to B then M_R (matrix representation of R) is a matrix of order $m \times n$.

EXAMPLE 5.13. Let $A = \{4,5,6\}$

$B = \{r,s\}$

and let $R = \{(4,r), (5,s), (6,r)\}$

As R is a relation from the set A to the set B.

then M_R is a matrix of order 3×2 and is given by:

$$M_R = \begin{bmatrix} 1 & 0 \\ 0 & 1 \\ 1 & 0 \end{bmatrix}$$

Here elements of A represents row & element of B represents columns.

In the example (5.12), a matrix representation if the relation R is given. In the following example matrix representation is given & we need to give relation set.

EXAMPLE 5.14. For the given adjacency matrix define the relation R.

$$M_R = \begin{bmatrix} 1 & 0 & 0 & 1 \\ 0 & 1 & 1 & 0 \\ 1 & 0 & 1 & 1 \end{bmatrix}$$

As the matrix M_R is of the order 3 × 4 so if R is a relation from the set A to B then set, A and B will have 3 and 4 elements respectively.

Let $A = \{a_1, a_2, a_3\}$

$B = \{b_1, b_2, b_3, b_4\}$

Now $(a_i, b_j) \in R$ if and only if $m_{ij} = 1$

$\therefore \quad R = \{(a_1, b_1), (a_1, b_4), (a_2, b_2), (a_2, b_3), (a_3, b_1), (a_3,b_3), (a_3,b_4)\}$

EXAMPLE 5.15. Let $A = \{a,b,c,d\}$ and let R be a relation on the set A that has the matrix representation given as

$$M_R = \begin{bmatrix} 1 & 0 & 0 & 0 \\ 0 & 1 & 0 & 0 \\ 1 & 1 & 1 & 0 \\ 0 & 1 & 0 & 1 \end{bmatrix}$$

Construct the diagraph of R and list indegree and outdegree of all the vertices.

From the matrix it is clear that

$$R = \{(a, a), (b, b), (c, a), (c, b), (c, c), (d, b), (d, d)\}$$

The Graphical representation of the given matrix is as follows :

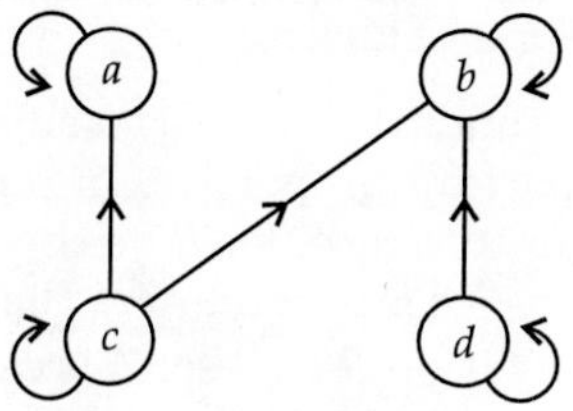

Fig. 5.5.

In the matrix notation the indegree represents the number of 1's in columns and outdegree represents the number of 1's in rows. Graphically indegree and outdegree explained earlier.

Vertices	a	b	c	d
Indegree	2	3	1	1
Outdegree	1	1	3	2

EXAMPLE 5.16. Consider the following relations on the set of integers

$R_1 = \{(a,b) \mid a \le b\}$

$R_2 = \{(a,b) \mid a = b\}$

$R_3 = \{(a,b) \mid a > b\}$

$R_4 = \{(a,b) \mid a = b+1\}$

$R_5 = \{(a,b) \mid a + b \le 3\}$

Which of these relations contain each of the pairs (1, 1), (1, 2), (2, 1), (1, 1), (2,2) ?

The ordered pair (1,1) is in R_1, R_2, R_5

The ordered pair (1,2) is in R_1, R_5

The ordered pair (2,1) is in R_3, R_4, R_5

The ordered pair (1,–1) is in R_3, R_5

The ordered pair (2,2) is in R_1, R_2

5.7. *N*-ARY RELATIONS

In the previous section we have discussed binary relation that is relation between two sets. This section includes relationship among elements from more than two sets.

Definition : Let $A_1, A_2, ... A_n$ be sets. An n-ary relation on these sets is a subset of $A_1 \times A_2 \times ..... \times A_n$. The sets $A_1, A_2, ... A_n$ are called the domains of the relations and n is called its degree.

EXAMPLE 5.17. Let R be the relation consisting of triplets (a, b, c) where a, b and c are integers with $b = a + 1$, $c = b + 1$. Then $(1, 2, 3) \in R$ but $(2,4,9) \notin R$.

The degree of this relation is 3. Its domains are all equal to the set of integers.

EXAMPLE 5.18. Let V be a n-dimensional vector space over the field K.

Let $v_i \in V$

$$V = (v_1, v_2, ...v_n)$$

then R is the any defined relation among v_i's.

5.8. PROPERTIES OF RELATIONS

There are several properties that are used to classify relations on a set. Here we will introduced some important properties. In some relations an element is always related to itself. For instance, let R be the relation on the set of all students consisting of pairs $(x, y) \in R$ if x and y are the student of same college.

Reflexive Relation. A relation R on a set A is called reflexive if $(a,a) \in R$ for every element $a \in A$ that is a relation on A is **reflexive** if every element of A is related to itself.

EXAMPLE 5.19. A binary relation R on the set A such that $(a,b) \in R$ if $a \mid b$. Here R is a reflexive relation as $a \mid a$, *i.e.*, every element divides itself.

EXAMPLE 5.20. If some students are in a class then $(a, b) \in R$ if a nominates b for the post of president. Then R is reflexive if everyone nominates himself or herself and suppose some of the students are of lack of confidence then R is not reflexive.

The adjacency matrix which is a square matrix can be used to determine whether the relation has certain properties.

R is reflexive iff $(a_i, a_i) \in R$ for $i = 1,2, ...n$. Hence R is reflexive if and only if $m_{ii} = 1$ for $i = 1, 2, ..n$. In other words, R is reflexive if all the elements on the main diagonal of M_R is equal to 1, as shown in figure 5.6.

$$\begin{bmatrix} 1 & & & & \\ & 1 & & & \\ & & 1 \ddots & & \\ & & & 1 & \\ & & & & 1 \end{bmatrix}$$

Fig. 5.6.

The adjacency matrix for reflexive relation.

Symmetric Relation

A relation R is said to be symmetric if $(a, b) \in R$ whenever $(b, a) \in R$. *i.e.,* whenever aRb then bRa.

That is a relation is symmetric if and only if a is related to b implies that b is related to a.

Again the adjacency matrix can be used to determine whether the given matrix is a symmetric or not. As in symmetric relation if $(a_i, a_j) \in R$ then $(a_j, a_i) \in R$ *i.e.* $m_{ij} = 1$ then $m_{ji} = 1$, moreover if $a_{ji} = 0$ then $a_{ij} = 0$, *i.e.,* $M_R = M_R^T$ $\Rightarrow M_R$ should be symmetric.

EXAMPLE 5.21. Consider the ordered pair $(x, y) \in R$ if x is a brother of y is a symmetric relation as $(y, x) \in R$, *i.e.* y is also a brother of x.

EXAMPLE 5.22. Let $A = \{a, b, c, d\}$

$$R = \{xRy \mid (x, y) \in A \;\&\; x = y\}$$

then R is a symmetric relation.

Note here R is reflexive also.

Asymmetric Relation

If the relation R is not a symmetric relation then R is called, asymmetric relation for example, ($R_,$ '$\subseteq$'), ($R_,$ '$\le$'), ($R_,$ '$\ge$'), ($R_,$ '/').

That is in the asymmetric relation aRb then $b \not R a$. In matrix representation the corresponding matrix should not be symmetric.

So in the adjacency matrix if $a_{ij} = 1$, then $a_{ji} = 0$

Note: *If R is asymmetric, it follows that* $a_{ii} = 0 \forall_j$ *that is main diagonal of the matrix* M_R *consists entirely of 0's.*

Anti-symmetric Relation

A relation R on a set A is said to be anti-symmetric if aRb & bRa only if $a = b$ for $a, b \in A$.

In other words, (a, b) is in R implies (b, a) is in R, unless $a = b$.

Same as above adjancy matrix can be used to find whether the given relation is a antisymmetric or not.

The matrix $M_R = [m_{ij}]$ of an anti-symmetric relation R satisfies the property that for $i \ne j$ then $m_{ij} = 0$ or $m_{ji} = 0$

[As $m_i \, R \, m_j \Rightarrow m_{ij} = 1$ & $m_i \ne m_j$ then $m_j \not R \, m_i \Rightarrow m_{ji} = 0$]

Example 5.23. The example (5.19) is $a \; R \; b$ if $a \mid b$ is a anti-symmetric relation.

EXAMPLE 5.23. Let "R, ≤" be the given relation a set

$A = \{1, 2, 3, 4\}$ then $a, b \in A \; \forall \; a, b \in A$

$$\left.\begin{array}{l} aRb \Rightarrow a \le b \\ \& \, bRa \Rightarrow b \le a \end{array}\right] \Rightarrow a = b$$

Implies the given relation is an antisymmetric relation.

EXAMPLE 5.24. Suppose A be a set of tests to be performed on a patient in the hospital then if a and b are two tests then a is related to b if the test a must perform before b, then the test b must not performed before the set a. In other words $a \; R \; b$ and a $\ne b$ then $b \not R \; a$. This implies that R is an anti-symmetric relation.

EXAMPLE 5.25. Check whether the following relations are anti-symmetric or not.

(*a*) $R = \{(1, 1), (2, 2)\}$

(*b*) $R = \{(1, 1), (2, 2), (1, 2)\}$

(*c*) $R = \{(1, 1), (2, 2), (1, 2), (2, 3)\}$

(*d*) $R = \{(1, 1), (2, 2), (1, 2), (2, 1)\}$

(*e*) Let X be the set of lines in the plane

Let $R = \{(l_1, l_2) \mid l_1 \perp l_2, l_1, l_2 \in X\}$

(*f*) Let X be the set of members in a family & let R be the relation on the set X such that R is a brother relation that is for $a, b \in R$.

$a\ R\ b$ iff a is brother of b.

Answer : (*a*), (*b*), (*c*) are the anti-symmetric relation.

(*d*) is not anti-symmetric as $1\ R\ 2$ and $2\ R\ 1$ but $1 \neq 2$

Similarly (*e*) and (*f*) are not anti-symmetric relations.

EXAMPLE 5.26. Consider the relation given by the example 5.23 on the set $A = \{1, 2, 3, 4\}$ & R, $\leq$

Therefore, $R = \{(1, 1), (1, 2), (1, 3), (1, 4), (2, 2), (2, 3), (2, 4), (3, 3), (3, 4), (4, 4)\}$

The corresponding matrix of the relation R is given by M_R

$$M_R = \begin{bmatrix} 1 & 1 & 1 & 1 \\ 0 & 1 & 1 & 1 \\ 0 & 0 & 1 & 1 \\ 0 & 0 & 0 & 1 \end{bmatrix}$$

By using boolean addition & multiplication, *i.e.*,

$$\begin{aligned} 1 + 1 &= 1 \\ 1 + 0 &= 1 \\ 0 + 1 &= 1 \\ 0 + 0 &= 0 \\ 1 \cdot 1 &= 1 \\ 1 \cdot 0 &= 0 \\ 0 \cdot 1 &= 0 \\ 0 \cdot 0 &= 0 \end{aligned}$$

$$M_R{}^2 = M_R \odot M_R$$

$$= \begin{bmatrix} 1 & 1 & 1 & 1 \\ 0 & 1 & 1 & 1 \\ 0 & 0 & 1 & 1 \\ 0 & 0 & 0 & 1 \end{bmatrix} \odot \begin{bmatrix} 1 & 1 & 1 & 1 \\ 0 & 1 & 1 & 1 \\ 0 & 0 & 1 & 1 \\ 0 & 0 & 0 & 1 \end{bmatrix}$$

$$= \begin{bmatrix} 1 & 1 & 1 & 1 \\ 0 & 1 & 1 & 1 \\ 0 & 0 & 1 & 1 \\ 0 & 0 & 0 & 1 \end{bmatrix} = M_R$$

Therefore R is transitive.

EXAMPLE 5.27. Consider the set $\{1, 2, 3, 4\}$ & define a relation R, '/' over a set A such that xRy if and only if x divides y.

Here $R = \{(1, 1), (1, 2), (1, 3), (1, 4), (2, 2), (2, 4), (3, 3), (4, 4)\}$
It's matrix representation is given as:

$$M_R = \begin{bmatrix} 1 & 1 & 1 & 1 \\ 0 & 1 & 0 & 1 \\ 0 & 0 & 1 & 0 \\ 0 & 0 & 0 & 1 \end{bmatrix}$$

$$M_R^2 = M_R \odot M_R$$

$$= \begin{bmatrix} 1 & 1 & 1 & 1 \\ 0 & 1 & 0 & 1 \\ 0 & 0 & 1 & 0 \\ 0 & 0 & 0 & 1 \end{bmatrix} \odot \begin{bmatrix} 1 & 1 & 1 & 1 \\ 0 & 1 & 0 & 1 \\ 0 & 0 & 1 & 0 \\ 0 & 0 & 0 & 1 \end{bmatrix} = \begin{bmatrix} 1 & 1 & 1 & 1 \\ 0 & 1 & 0 & 1 \\ 0 & 0 & 1 & 0 \\ 0 & 0 & 0 & 1 \end{bmatrix}$$

$\therefore$ R is transitive.

Transitive Relation

A relation R on a set A is called transitive if whenever $(a, b) \in R$ and $(b, c) \in R$
then $(a, c) \in R$ for $a, b, c \in A$.

Again in matrix from a relation R is transitive iff its matrix $M_R = [m_{ij}]$ has the property if $m_{ij} = 1$ and $m_{jk} = 1$ then $m_{ik} = 1$. The transitivity of R means that if $(M_R)^2$ has 1 in any position then M_R must have 1 in the same position. Thus, in particular, if $M_R^2 = M_R$ then R is transitive and converse is not true for example.

EXAMPLE 5.28. Let R be the relation consisting of all pairs (x, y) of students at your school that x has taken more credits than y. Suppose that x is related to y and y is related to z. This means that x has taken more credits than y and y has taken more credits than z. This implies that x has taken more credits than z that is x is related to z. It means that R has transitive property.

EXAMPLE 5.29. The example (5.19) is $a \, R \, b$ if $a \mid b$ is a transitive relations as if $a \mid b$ and $b \mid c$ this implies that $a \mid c$.

Proof. As if $a \mid b$ then $\exists$ an integer say x ; $b = ax$

and if $b \mid c$ then $\exists$ an integer say y ; $c = yb$

$= y(ax)$

$= (xy)\, a$

$= za$, where z is an integer

[As product of two integers is an integer]

$\Rightarrow a \mid c$

EXAMPLE 5.30. Example of a relation which is not transitive.

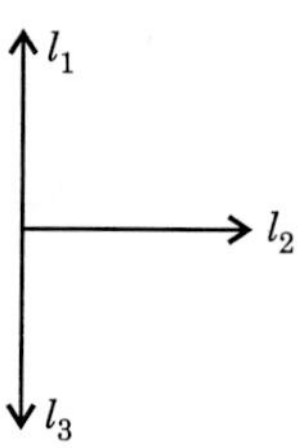

Let X-Set of lines in the plane & Let R be such that $\ell_1 \, R \, \ell_2$ of $\ell_1 \perp \ell_2$ then the relation is not perpendicular for example, if

$\ell_1 \perp \ell_2$ and $\ell_2 \perp \ell_3$

then $\ell_1 \mid\mid \ell_3$ that is not perpendicular

$\therefore$ $\ell_1 \not R \, \ell_3$

$\therefore$ the relation is not transitive.

Equivalence Relation

A relation *R* on a set *A* is called an equivalence relation if and only if it is reflexive, symmetric and transitive.

EXAMPLE 5.31. If $A = \{1, 2, 3, 4, 5, 6\}$ and 'R, =' is a relation on the set A then R is an equivalence relation.

EXAMPLE 5.32. Let Z^* be the set of non-zero integers. One of the most basic equivalence relations in mathematics is the relation R on $Z^* \times Z^*$ defined by

$$(a, b)\ R\ (c, d) \text{ if and only if } ad = bc.$$

R is *Reflexive as* $(a, b)\ R(a, b)$

as $\quad ab = ab$

$$\frac{a}{b} = \frac{a}{b}$$

Therefore the relation R is reflexive.

Symmetric Let $(a, b)\ R(c, d)$

$$\Rightarrow \quad ad = bc$$

$$\Rightarrow \quad \frac{a}{b} = \frac{c}{d}$$

$$\Rightarrow \quad \frac{b}{a} = \frac{d}{c}$$

$$\Rightarrow \quad (b, a)\ R(c, d)$$

Therefore, the relation R is symmetric.

Transitive Let $(a, b)\ R(c, d)$ and $(c, d)\ R(e, f)$

then we need to show that $(a, b)\ R(e, f)$

Now $(a, b)\ R(c, d)$

$$\Rightarrow \quad ad = bc$$

$$\Rightarrow \quad \frac{a}{b} = \frac{c}{d} \qquad \ldots(i)$$

and $(c, d)\ R(e, f)$

$$\Rightarrow \quad cf = de$$

$$\Rightarrow \quad \frac{c}{d} = \frac{e}{f} \qquad \ldots(ii)$$

$$(i) \text{ and } (ii) \quad \Rightarrow \quad \frac{a}{b} = \frac{e}{f}$$

$$\Rightarrow \quad (a, b)\ R(e, f)$$

Therefore the relation R is transitive.

Hence R is an equivalence relation.

$$M_R = \begin{bmatrix} 1 & 1 & 1 \\ 0 & 0 & 1 \\ 0 & 0 & 1 \end{bmatrix}$$

$$M_R^2 = \begin{bmatrix} 1 & 1 & 1 \\ 0 & 0 & 1 \\ 0 & 0 & 1 \end{bmatrix} \begin{bmatrix} 1 & 1 & 1 \\ 0 & 0 & 1 \\ 0 & 0 & 1 \end{bmatrix}$$

$$= \begin{bmatrix} 1 & 1 & 1 \\ 0 & 0 & 1 \\ 0 & 0 & 1 \end{bmatrix} = M_R$$

$\Rightarrow R$ is transitive

5.9. PARTIAL ORDER RELATION

A relation R on a set A is said to be a **partial order relation** if it satisfies **reflexivity, anti-symmetry** and **transitivity** and the corresponding set A is called **partial ordered set** or **poset**.

EXAMPLE 5.33. The set of natural numbers form a poset under the usual notation, '≤'. Similarly the set of integers, set of rationals and set of real numbers also form posets under usual, '≤'.

EXAMPLE 5.34. Let X be the collection of sets $A,B,C,$....then X together with the relation '⊆' form a poset.

Reflexivity is satisfied as every set is a subset of itself that is $A \subseteq A$.

Antisymmetry is also satisfied If $A \subseteq B$ and $B \subseteq A$ then $A = B$

In the case of Transitivity $A \subseteq B$ and $B \subseteq C$ then A is also a subset of C.

Therefore X under the relation R is a poset and R is a partial order relation.

EXAMPLE 5.35. The example (5.18) is aRb if $a \mid b$ is a partial order relation.

EXAMPLE 5.36. The set of positive integers under the relation 'R, <' is not a poset as reflexivity does holds.

Theorem 5.3. If R be a partial order on a set A and let R^{-1} be the inverse relation of R then R^{-1} is also partial order relation.

Proof. Reflexivity : $aR^{-1}a$ as $aRa \ \forall \ a \in A$

$\Rightarrow \ R^{-1}$ is reflexive.

Antisymmetry : Let $a \ R^{-1} \ b$ and $b \ R^{-1} \ a$

$\Rightarrow \ b \ R \ a$ and $a \ R \ b$

as R is partial order relation this implies $a = b$

Therefore, R^{-1} is anti-symmetric

Transitive : Let $a \ R^{-1} \ b$ and $b \ R^{-1} \ c$

$\Rightarrow \ b \ R \ a$ and $c \ R \ b$

$\Rightarrow \ c \ R \ b$ and $b \ R \ a$

As R is transitive this implies that $c \ R \ a$

$\Rightarrow \ a \ R^{-1} \ c$.

$\Rightarrow \ R^{-1}$ is transitive.

Therefore R^{-1} is a partial order relation if R is a partial order relation.

The poset (A, R^{-1}) is called the **dual of the poset** (A, R) and the partial order relation R^{-1} is called the **dual partial order relation of R.** For instance '≥' and '≤' both are partial order relations and are dual of each other.

If (A, R) is a poset that is the relation R is a partial order relation on the set A then the elements a and b of the set A are said to be **comparable** if $a \ R \ b$ or $b \ R \ a$.

Totally Ordered Set

If A is a poset in which every two members are comparable then the set A is called **totally ordered set or toset or chain.**

The example (5.23) the set A is a toset where as in the example (5.18) , the corresponding set is not a toset.

Theorem 5.4. Show that a non empty subset S of a poset P is a poset. If P is a chain then so is S.

Solution : Let $a, b, c \in S \Rightarrow a, b, c \in P$ and P is a poset so all the there conditions of posets are satisfied in P & so in S. This implies that S is a poset. Similarly we can prove for chain also.

EXAMPLE 5.37. Give the graphical representations of the given relation :

Let $A = \{ 1, 2, 3, 4, 6\}$ and 'R' is a relation on the set A such that $a\ R\ b$ if $a \mid b$.

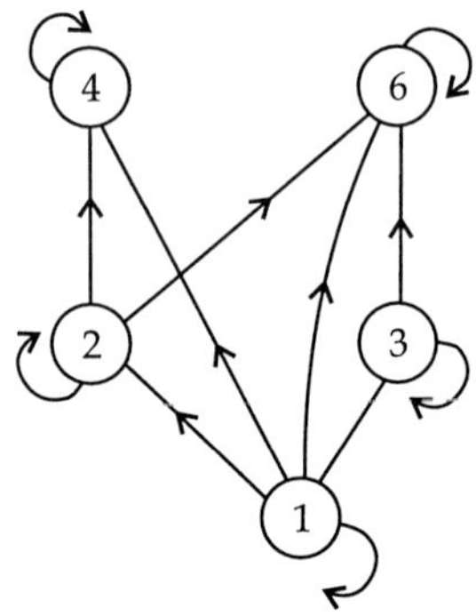

Fig. 5.7.

Geometrical Representation of a poset (Hasse diagrams)

In the section (5.5) we have discussed the graphical representation of a given relation. Many edges in the directed graph for a finite poset do not have to be shown since they must be present. For instance consider the above example (5.33).

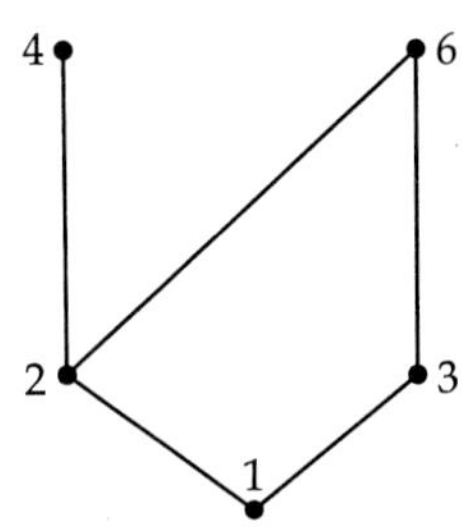

Fig. 5.8.

Since this relation (Division) is a partial ordering, it is reflexive and its directed graph has loops at all vertices. Consequently, there is no need to show these loops since they must be present. As shown in the figure (5.8). Because a partial ordering is a transitive relation, again there is no need to show those edges that must be present because of transitivity as in figure the edges (1, 4), (1, 6) are not present and as if we assume that all edges are pointed "upward", we do not have to show the directions of the edges. Such geometrical representation of a partial order relation is called **Hasse diagram** or **ordering diagram** which has the following properties :

(1) Each vertex of A must be related to itself , so arrows from a vertex to itself are not necessary.

(2) If vertex b appears above vertex a and if vertex a is connected to vertex b by an edge then $a\ R\ b$, so direction arrows are not necessary.

(3) If vertex c is above vertex a and if c is connected to a by a sequence of edges then $a\ R\ c$.

(4) The vertices (or nodes) are denoted by points rather than by a "circles"

5.10. UNION AND INTERSECTION OF RELATIONS

By the above discussion it is clear that relation is nothing but a set. So we can apply some set operation on them. The Boolean operations join & meet can be used to find the matrix representing the union and the intersection of two relation.

Let R_1 and R_2 be the two relations on a set A represented by the Boolean matrix M_{R_1} and M_{R_2} [As discussed before] respectively. Then relation R representing the union of these relations be equal to the join to two matrixes M_{R_1} and M_{R_2}

i.e., $$M_R = M_{R_1 \cup R_2} = M_{R_1} \vee M_{R_2}$$

and By the property of join, the resultant matrix will have 1 in the position where M_{R_1} or M_{R_2} has a 1.

Similary relation say S representing the intersection of there two relations equal to the meet of two natives M_{R_1} and M_{R_2}

i.e., $$M_S = M_{R_1 \cap R_2} = M_{R_1 \wedge R_2}$$

and By the property of meet the resultant matrix will have 1 in the position where both M_{R_1} and M_{R_2} have a 1.

Composition of Relations

Composition of relations can be found by using the Boolean product of corresponding matrixes.

In particular, suppose R is a relation from A to B and S is a relation from B to C. Suppose that A, B and C have m, n and p elements respectively. Let Boolean matrices for S oR, R and S be $M_{SoR} = [t_{ij}]$, $M_R = [r_{ij}]$ and $M_S = [s_{ij}]$ resp. therefore these matrixes have sizes mxp, mxn and n x p resp.

The ordered pair (a_i, c_j) belongs to SoR iff there is an element b_k such that $(a_i , b_k) \in R$ and $(b_k, c_j) \in S$.

That is $t_{ij} = 1$ iff $r_{ik} = s_{kj} = 1$ for some k

EXAMPLE 5.38. Suppose that the relations R_1 and R_2 on a set A are represented by the matrices

$$M_{R_1} = \begin{bmatrix} 1 & 0 & 1 \\ 1 & 1 & 0 \\ 0 & 0 & 1 \end{bmatrix}$$

and $$M_{R_2} = \begin{bmatrix} 1 & 0 & 0 \\ 1 & 0 & 1 \\ 1 & 1 & 1 \end{bmatrix}$$

then relation $R = R_1 \cup R_2$ is represented by the matrix M_R is given by:

$$M_R = M_{R_1 \cup R_2} = M_{R_1} \vee M_{R_2}$$

$$= \begin{bmatrix} 1 & 0 & 1 \\ 1 & 1 & 1 \\ 1 & 1 & 1 \end{bmatrix}$$

and the relation $S = R_1 \cap R_2$ is represented by the matrix M_S is given by:

$$M_S = M_{R_1 \cap R_2} = M_{R_1} \wedge M_{R_2}$$

$$= \begin{bmatrix} 1 & 0 & 0 \\ 1 & 0 & 0 \\ 0 & 0 & 1 \end{bmatrix}$$

[Readers are now required to verify the above result graphically].

Suppose that R is a relation on a set A . A path of length n in R from a to b is a finite sequence of n + 1 element (not necessarily distinct) $X : a, x_1, x_2, \ldots x_{n-1}, b$, starting from a and ending with b such that

$$a\ R\ x_1,\ x_1\ R\ x_2,\ \ldots\ldots,\ x_{n-1}\ Rb.$$

The length of a path is the number of edges in the path where the vertices need not be distinct. A path that begins and ends at the same vertex is called a cycle.

Consider the following graph

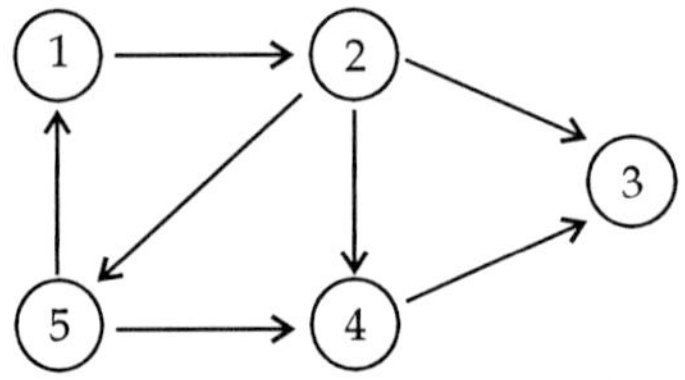

Fig. 5.9.

Consider the sequence

$X_1 : 1, 2, 4, 3$

is a path of length 3 from vertex 1 to vertex 3,

Consider the another sequence

$X_2 : 1, 2, 5, 4, 3$

is a path of length 4 from the vertex 1 to the vertex 3.

Path in a relation R can be used to define new relations that are quite useful.

If n is a fixed positive integer, we define a relation R^n on A as follows :

$x\ R^n\ y$ means that there is a path of length n from, x to y in R. We can define a relation from x to y of some length and is denoted by R^∞.

The relation R^∞ is called the **connectivity relation** for R. That is R^∞ is a relation from x to y such that length of the path does not matter.

EXAMPLE 5.39. Find the matrix representing the relation $R_1 \odot R_2$ where the R_1 and R_2 are:

$$R_1 = \{(a, a), (a, c), (b, a), (b, b)\}$$
$$R_2 = \{(a, b), (b, c), (c, a), (c, c)\}$$

The matrix representing the relations R_1 and R_2

$$M_{R_1} = \begin{bmatrix} 1 & 0 & 1 \\ 1 & 1 & 0 \\ 0 & 0 & 0 \end{bmatrix}$$

$$M_{R_2} = \begin{bmatrix} 0 & 1 & 0 \\ 0 & 0 & 1 \\ 1 & 0 & 1 \end{bmatrix}$$

$$M_{R_1 \cdot R_2} = M_{R_1} \odot M_{R_2}$$

$$= \begin{bmatrix} 1 & 1 & 1 \\ 0 & 1 & 1 \\ 0 & 0 & 0 \end{bmatrix} \text{ [By using the boolean product[}$$

EXAMPLE 5.40. Let $A = \{1, 2, 3, 4, 5\}$ and

$$R = \{(1, 1), (1, 2), (2, 3), (3, 5), (3, 4), (4, 5)\}$$

Compute (*a*) R^2 (*b*) R^∞

Solution : Consider the matrix represented of R

$$M_R = \begin{bmatrix} 1 & 1 & 0 & 0 & 0 \\ 0 & 0 & 1 & 0 & 0 \\ 0 & 0 & 0 & 1 & 1 \\ 0 & 0 & 0 & 0 & 1 \\ 0 & 0 & 0 & 0 & 0 \end{bmatrix}$$

$$M_{R^2} = M_R \odot M_R = \begin{bmatrix} 1 & 1 & 0 & 0 & 0 \\ 0 & 0 & 1 & 0 & 0 \\ 0 & 0 & 0 & 1 & 1 \\ 0 & 0 & 0 & 0 & 1 \\ 0 & 0 & 0 & 0 & 0 \end{bmatrix} \odot \begin{bmatrix} 1 & 1 & 0 & 0 & 0 \\ 0 & 0 & 1 & 0 & 0 \\ 0 & 0 & 0 & 1 & 1 \\ 0 & 0 & 0 & 0 & 1 \\ 0 & 0 & 0 & 0 & 0 \end{bmatrix}$$

$$= \begin{bmatrix} 1 & 1 & 1 & 0 & 0 \\ 0 & 0 & 0 & 1 & 1 \\ 0 & 0 & 0 & 0 & 1 \\ 0 & 0 & 0 & 0 & 0 \\ 0 & 0 & 0 & 0 & 0 \end{bmatrix}$$

Therefore, $\quad R^2 = \{(a, a), (a, b), (a, c), (b, e), (b, d), (c, e)\}$

Further computation R^∞ can be computed

R^∞ = {(a, a), (a, b), (a, c), (a, d), (a, e), (b, b), (b, c), (b, d), (b, e), (c, c) (c, d), (c, e), (d, e)}

Readers are required to verify the above result by using diagraphs.

5.11. CLOSURES OF RELATIONS

In the previous sections we studies relations and one important operation on relations, namely composition. This operation enabled us to generate new relations from previously known relations. After that we studied some key properties of the relations. In this section we wish to consider the situation of constructing a new relation R^+ from previously known relation R such that R^+ contains R. This relation is called the closure of R.

In general let R be a relation on a set A. R may or may not have some property say p, such as reflexivity, symmetry or transitivity. If there is a relation R^+ with property p containing R such that R^+ is a subset of every relation with property p containing R, then R^+ is called closure of R with property p. Here we will show how reflexive, symmetric and transitive closures of relations can be found.

5.12. DIVIDE AND CONQUER RELATIONS

Recurrence relations of the following type are called divide and conguer relations.

$$a_n = \frac{c.a_n}{d} + f(n)$$

where C : constant

n is a non-negative Integer power of d (*i.e.*, $n = d^k$, $k \geq 0$) and $f(n)$ is a function of n.

EXAMPLE 5.41. Solve divide and congress relation

$$a_n = 2.\frac{an}{2} + (n-1) \quad n_1 \geq 2, a_1 = 0$$

Solution : We can solve this d and c recurrance relation for $n = 2k$

(*i.e.*, $d = 2$ given and $k \geq 1$)

$$\boxed{a_n = 2\frac{an}{2} + (n-1)} \quad ...(i)$$

Replace n by $\frac{n}{2}$ in (i) and then multiply by ,

$$2\cdot\frac{an}{2} = \left[2\cdot\frac{a_n}{4} + \left(\frac{n}{2} - 1\right)\right]$$

Replace n by $\frac{n}{4}$ in (1) and then multiply by 2^2,

$$2^2\,\frac{a_n}{4} = 2^2\left[2\,\frac{a_n}{8} + \left(\frac{n}{4} - 1\right)\right]$$

Replace n by $\frac{n}{8}$ in (i) and then multiply by 2^3,

$$2^3 \frac{a_n}{8} = 2^3 \left[2 \frac{a_n}{16} + \left(\frac{n}{8} - 1 \right) \right]$$

$$2^{k-1} \frac{a_n}{2^{k-1}} = 2^{k-1} \left[2 \frac{a_n}{2^k} + \left(\frac{n}{2^{k-1}} - 1 \right) \right]$$

on adding the above 'k^{th}' relations we obtain

$$a_n = (n-1) + (n-2) + \ldots + (n - 2^{k-1}) + 2^k \frac{a_n}{2^k}$$

Note:

$$\frac{a_n}{2^k} = a \frac{2^k}{2^k} \quad (\because n = 2^k \text{ we assumed in beginning})$$

$$\frac{a_n}{2^k} = a_1 \text{ (but } a_1 = 0 \text{ given in beginning)}$$

$$\frac{a_n}{2^k} = 0$$

$$\therefore \quad a_n = (n-1) + (n-2) + \ldots + (n - 2^{k-1}) + 2^k[0]$$

$$a_n = (n-1) + (n-2) + \ldots + (n - 2^{k-1})$$

$$= n \,.\, k - (1 + 2 + \ldots + 2^{k-1})$$

$$= n.k - \left(\frac{1 - 2^k}{1 - 2} \right) = nk + 1 - 2\text{k}$$

$$= (nk - 2k + 1) \quad \because 2^k = n$$

$$\Rightarrow \quad k \log 2 = \log n$$

$$\Rightarrow \quad k = \frac{\log n}{\log 2}$$

$$\Rightarrow \quad k = \log_2 n$$

$$\because \quad a_n = n \log_2 n - 2 \log_2{}^n + 1$$

$$\boxed{a_n = n \log_2{}^n - n + 1} \text{ as } \boxed{2^{\log_2{}^n} = n} \textbf{ Ans.}$$

EXAMPLE 5.42. Solve the D and C relation

$$a_n = 7 \frac{a_n}{3} + 2n$$

$$n = 3k$$

$$a_1 = \frac{5}{2}$$

Solution : $a_n = 7 \frac{a_n}{3} + 2n$

Replace n by $\frac{n}{3}$ and multiply by 7 in equation (*i*)

$$7 \cdot \frac{a_n}{3} = 7\left[7\frac{a_n}{3} + 2\frac{n}{3}\right]$$

Replace n by $\frac{n}{3^2}$ and multiply by 7^2 in 1.

$$7^2\frac{a_n}{a} = 7^2\left[7\frac{a_n}{3 \times 3^2} + 2 \cdot \frac{n}{3^2}\right]$$

$$7^2\frac{a_n}{a} = 7^2\left[7\frac{a_n}{27} + 2\ \frac{n}{9}\right]$$

$$7^{k-1}\frac{a_n}{3^{k-1}} = 7^{k-1}\left[7\frac{a_n}{3^k} + 2\frac{n}{3^{k-1}}\right]$$

adding above k relations

$$a_n = 2^n\left[1 + \frac{7}{3} + \frac{7^2}{3^2} + \ldots \frac{7^{k-1}}{3^{k-1}}\right] + \frac{7^k}{3^k}\frac{a_n}{3^k}$$

Note that:

at $\qquad n = 3^k,\ a\frac{n}{3^k} \equiv a\frac{\not{n}}{\not{n}} \equiv a_1 = \frac{5}{2}$ given

$$\therefore \qquad a_n = 2n\left[\frac{(7/3)^k - 1}{(7/3 - 1)}\right] + \frac{7^k}{3^k} \cdot \frac{5}{2}$$

$$\Rightarrow \qquad a_n = 2n\left[\frac{(7/3)^k - 1}{4/3}\right] + \frac{7^k}{3^k} \cdot \frac{5}{2} \qquad \ldots(ii)$$

as $\qquad n = 3^k$

$$k = \frac{\log n}{\log 3} \equiv \log_3 n$$

so put $k = \log_3^n$ in equation (2) we obtain

$$\Rightarrow \qquad a_n = 2n\left[\frac{(7/3)^{\log_3^n} - 1}{4/3}\right] + \frac{5}{2} \cdot \left(\frac{7}{3}\right)^{\log_3^n}$$

$$\Rightarrow \qquad a_n = \frac{\not{2}n \times 3}{\not{4}}\left[(7/3)^{\log_3 n} - 1\right] + \frac{5}{2} \cdot \left(\frac{7}{3}\right)^{\log_3^n}$$

$$\Rightarrow \qquad a_n = \frac{3n}{2}\left[(7/3)^{\log_3^n} - 1\right] + \frac{5}{2}\left[\frac{7}{3}\right]^{\log_3^n}$$

$$\Rightarrow \qquad a_n = \frac{3n}{2}\left[\frac{7^{\log_3^n}}{3^{\log_3^n}} - 1\right] + \frac{5}{2}\left[\frac{7^{\log_3^n}}{3^{\log_3^n}}\right]$$

but $3^{\log_3^n} = n$

$$a_n = \frac{3n}{2}\left[\frac{7^{\log_3 n}}{n} - 1\right] + \frac{5}{2}\left[\frac{7^{\log_3 n}}{n}\right]$$

$$a_n = \frac{3n}{2}\left[\frac{7^{\log_3 n} - n}{n}\right] + \frac{5}{2n}\,7^{\log_3^n}$$

Verification $$a_1 = \frac{3}{2}\left[7^{\log_3^1} - 1\right] + \frac{5}{2.1}\,7^{\log_3^n}$$

$$a_1 = \frac{3}{2}[1-1] + \frac{5}{2}\cdot 1 = 0 + \frac{5}{2} = \frac{5}{2} \quad \textbf{Proved.}$$

Reflexive Closure

Let R be the relation on a set $A = \{a, b, c\}$ such that $R = \{ (a, a), (a, b) , (b, a) , (a, c)\}$. R is not a reflexive relation. Now we produce a reflexive relation containing R that is as small as possible. This can be done by adding (b, b) and (c, c) to R and then the resultant relation $R^+ = \{(a, a) , (b, b) , (c, c) , (a, b), (b, a) , (a, c)\}$. Now R^+ is reflexive containing R and R^+ is contained within every reflexive relation that contains R, it is called *reflexive closure* of R. Therefore from the above example it is clear that for a given relation R on a set A, the reflexive closure of R can be formed by adding to R all pairs of the form (a, a) with $a \in A$, not already in R. That is reflexive closure of R equals $R \cup \Delta$, where $\Delta = \{(a, a) \mid a \in A\}$.

EXAMPLE 5.43. What is the reflexive closure of the relation

$R = \{(X, Y) \mid X \subset Y\}$.

Solution : By the above discussion the reflexive closure of R is $R \cup \Delta = \{(X, Y) \mid X \subset Y) \cup \{(X, X)\} = \{(X, Y) \mid X \subseteq Y)]$

EXAMPLE 5.44. Find reflexive closure of the relation

$R = \{(a, b) \mid a > b\}$ on the set of integers.

Solution : $$R^+ = R \cup \Delta = \{(a, b) \mid a > b\} \cup \{(a, a) \mid (a \in Z\}$$
$$= \{(a, b) \mid a \geq b\}$$

Symmetric Closure

Same as in the above example the symmetric closure of a relation R can be constructed by adding all ordered pairs of the form (b, a) whenever (a, b) is in the relation that are not already present in R. By doing this we can produce a relation that a symmetric, contains R and that is contained in any symmetric relation that contains R. Therefore the symmetric closure a relation can be constructed as

$$R^+ = R \cup R^{-1} \text{ where } R^{-1} = \{(b, a) \mid (a, b) \in R\}$$

EXAMPLE 5.45. Let $R = \{(1, 1) , (1, 2) , (2, 2) , (2, 3) , (2, 1) , (3, 1) , (3, 2)\}$. What is the symmetric closure of R.

Now R is not a symmetric relation because the order pair $(3, 1) \in R$ but $(1, 3) \notin R$. Therefore we can find symmetric closure of R by adding the ordered pair $(1, 3)$ to R and the resultant relation R^+ will be symmetric closure of R and $R^+ = R \cup R^{-1}$

$$R^{-1} = \{(1, 1), (2, 1), (2, 2), (3, 2), (1, 2), (1, 3), (2, 3)\}$$
$$R \cup R^{-1} = \{(1, 1)\ (1, 2), (2, 2), (2, 3), (2, 1), (3, 1), (3, 2), (1, 3)\}$$

EXAMPLE 5.46. What is the symmetic closure of the relation $R = \{(a, b) \mid a < b]$ on the set of positive integers.

$$R^+ = R \cup R^{-1} = \{(a, b) \mid a < b) \cup \{(b, a) \mid a < b\}$$
$$= \{(a, b) \mid a \neq b\}.$$

Transitive Closure

Consider a telephone network in which the main office x is connected to and can communicate to, individuals y and z. Both y and z can communicate to another person, t ; however, the main office cannot communicate with t. Assume that communication is only one way. This program can be described by the relation $R = \{(x, y), (x, z), (y, t), (z, t)\}$. Now suppose we want to change the program such that the main office x can communicate with person t and still maintain the previous system, *i.e.*, most economical system. This can be done by finding the smallest relation R^+ which contains R as a subset and which is transitive,

i.e., $R^+ = \{(x, y), (x, z), (y, t), (z, t), (x, t)\}$.

Therefore, we can define the transitive closure as follows :

Let A be a set and R be a relation on a set A. The transitive closure of R, denoted by R^+ is the smallest relation which contains R as subset and which is transitive.

EXAMPLE 5.47. Let $A = \{1, 2, 3, 4\}$. Consider the successor relation on A, *i.e.*,

$$R = \{(1, 2), (2, 3), (3, 4)\}$$

Now we analyse that the given relation R is not a transitive relation because the order pairs (1, 3) and (2, 4) are not there. So we can find the transitive relation R by adding these two ordered pairs in R.

$$R \cup \{(1, 3), (2, 4)\} = \{(1, 2), (2, 3), (3, 4), (1, 3), (2, 4)\}$$

Is this relation is transitive ?

Again by inspection (1, 4) is not there as $1R2$ and $2R4$ then 1 should relate to 4 under the relation R.

$$\therefore \qquad R^+ = \{(1, 2), (2, 3), (3, 4), (1, 3), (2, 4), (1, 4)\}$$

It is difficult to find a transitive closure by this procedure. From the above example it is clear that in the first step. If $(a, b) \in R$ and $(b, c) \in R$ then $(a, c) \in R^+$. This condition is exactly the membership requirement for the pair (a, c) to be in the composition $RR = R^2$. So every element in R^2 must be an element in R^+. Therefore, R^+ contains at leat $R \cup R^2$. In particular for the above example

$$R = \{(1, 2), (2, 3), (3, 4)\}$$
$$R^2 = \{(1, 3), (2, 4)\}$$

and we have

$$R \cup R^2 = \{(1, 2), (2, 3), (3, 4), (1, 3), (2, 4)\}$$

and still we require the pair (1, 4). As $(1, 3) \in S^2$ & $(3, 4) \in S$ & the composition $S^2S = S^3$ produces (1, 4). This implies $R^3 \subseteq R^+$. This process must be continued untill the resulting relation is transitive. Here $R^+ = R \cup R^2 \cup R^3$.

Theorem 5.5. If R is a relation on a set A and if dim $A = n$, then transtive closure R^+ of R is:

$$R^+ = R \cup R^2 \cup R^3 \cup .... \cup R^n.$$

Let's now consider the matrix analogue of the transitive closure.

Thereom 5.6. Let $M_R{}^+$ be the matrix of R^+. The transitive closure of R, which is a relation on a set of n elements. Then $M_{R^+} = M_R + M_{R^2} + ... + M_{R^n}$, where addition is done using Boolean arithmetic.

EXAMPLE 5.48. Consider the relation $R = \{(1, 4), (2, 1), (2, 2), (2, 3), (3, 2), (4, 3), (4, 5), (5, 1)\}$ on the set $A = \{1, 2, 3, 4, 5\}$. The matrix M_R of the relation R is:

$$M_R = \begin{bmatrix} 0 & 0 & 0 & 1 & 0 \\ 1 & 1 & 1 & 0 & 0 \\ 0 & 1 & 0 & 0 & 0 \\ 0 & 0 & 1 & 0 & 1 \\ 1 & 0 & 0 & 0 & 0 \end{bmatrix}$$

Recall R_2, R_3, ... can be determined through computing the matrices $M_R{}^2$, $M_R{}^3$ ..... Here

$$M_R{}^2 = M_R \odot M_R$$

$$= \begin{bmatrix} 0 & 0 & 0 & 1 & 0 \\ 1 & 1 & 1 & 0 & 0 \\ 0 & 1 & 0 & 0 & 0 \\ 0 & 0 & 1 & 0 & 1 \\ 1 & 0 & 0 & 0 & 0 \end{bmatrix} \odot \begin{bmatrix} 0 & 0 & 0 & 1 & 0 \\ 1 & 1 & 1 & 0 & 0 \\ 0 & 1 & 0 & 0 & 0 \\ 0 & 0 & 1 & 0 & 1 \\ 1 & 0 & 0 & 0 & 0 \end{bmatrix}$$

$$= \begin{bmatrix} 0 & 0 & 1 & 0 & 1 \\ 1 & 1 & 1 & 0 & 0 \\ 1 & 1 & 1 & 0 & 0 \\ 1 & 1 & 0 & 0 & 0 \\ 0 & 0 & 0 & 1 & 0 \end{bmatrix}$$

$$M_R{}^3 = M_R{}^2 \odot M_R$$

$$= \begin{bmatrix} 0 & 0 & 1 & 0 & 1 \\ 1 & 1 & 1 & 0 & 0 \\ 1 & 1 & 1 & 0 & 0 \\ 1 & 1 & 0 & 0 & 0 \\ 0 & 0 & 0 & 1 & 0 \end{bmatrix} \odot \begin{bmatrix} 0 & 0 & 0 & 1 & 0 \\ 1 & 1 & 1 & 0 & 0 \\ 0 & 1 & 0 & 0 & 0 \\ 0 & 0 & 1 & 0 & 1 \\ 1 & 0 & 0 & 0 & 0 \end{bmatrix}$$

$$= \begin{bmatrix} 1 & 1 & 0 & 0 & 0 \\ 1 & 1 & 1 & 0 & 0 \\ 1 & 1 & 1 & 1 & 0 \\ 1 & 1 & 1 & 1 & 0 \\ 0 & 0 & 1 & 0 & 1 \end{bmatrix}$$

$$M_R^4 = M_R^3 \odot M_R$$

$$= \begin{bmatrix} 1 & 1 & 0 & 0 & 0 \\ 1 & 1 & 1 & 0 & 0 \\ 1 & 1 & 1 & 1 & 0 \\ 1 & 1 & 1 & 1 & 0 \\ 0 & 0 & 1 & 0 & 1 \end{bmatrix} \odot \begin{bmatrix} 0 & 0 & 0 & 1 & 0 \\ 1 & 1 & 1 & 0 & 0 \\ 0 & 1 & 0 & 0 & 0 \\ 0 & 0 & 1 & 0 & 1 \\ 1 & 0 & 0 & 0 & 0 \end{bmatrix}$$

$$= \begin{bmatrix} 1 & 1 & 1 & 1 & 0 \\ 1 & 1 & 1 & 1 & 0 \\ 1 & 1 & 1 & 1 & 1 \\ 1 & 1 & 1 & 1 & 1 \\ 1 & 1 & 0 & 0 & 0 \end{bmatrix}$$

$$M_R^5 = M_R^4 \odot M_R$$

$$= \begin{bmatrix} 1 & 1 & 1 & 1 & 0 \\ 1 & 1 & 1 & 1 & 0 \\ 1 & 1 & 1 & 1 & 1 \\ 1 & 1 & 1 & 1 & 1 \\ 1 & 1 & 0 & 0 & 0 \end{bmatrix} \odot \begin{bmatrix} 0 & 0 & 0 & 1 & 0 \\ 1 & 1 & 1 & 0 & 0 \\ 0 & 1 & 0 & 0 & 0 \\ 0 & 0 & 1 & 0 & 1 \\ 1 & 0 & 0 & 0 & 0 \end{bmatrix}$$

$$= \begin{bmatrix} 1 & 1 & 1 & 1 & 1 \\ 1 & 1 & 1 & 1 & 1 \\ 1 & 1 & 1 & 1 & 1 \\ 1 & 1 & 1 & 1 & 1 \\ 1 & 1 & 1 & 1 & 0 \end{bmatrix}$$

$\therefore \quad M_R^+ = M_R + M_R^2 + M_R^3 + M_R^4 + M_R^5$

Using Boolean Addition.

5.13. WARSHALL'S ALGORITHM

Let G be a directed graph of n vertices say $V_1, V_2, ... V_n$. Let R be a relation on the set $\{V_1, V_2, ... V_n\}$ with the adjacency matrix M of G.

The matrix of transitive closure, *i.e.*, M^+, can be computed by the equation $M^+ = M + M^2 + ... + M^n$.

By using polynomial evaluation method one can compute $M^+ = M(1 + M(1 + M(1 + M)))$.

For example if $n = 3$, $R^+ = R(1 + R(1 + R))$.

which is much more efficient than calculating the powers of the adjacency matrix M.

Warshall's algorithm makes use of the fact that if M is a relation matrix, $M + M = M$ due to the fact that $1 + 1 = 1$ is Boolean arithmetic.

First we define n-square Boolean matrices $M_0, M_1, ... M_n$ as given below.

Let m_k (i, j) denote the m_{ij} entry of the matrix M_k. Then

$m_k(i, j) = 1$ if there is a simple path from V_i to V_j which does not use any other vertices except possible $V_1, V_2, \ldots V_k$.

0 otherwise.

$m_0(i, j) = 1$ If there is an edge from V_i to V_j (*i.e.* at initial stage).

$m_1(i, j) = 1$ If there is a simple path from V_i to V_j which does not pass through any other vertex except V_1.

$m_2(i, j) = 1$ If there is a simple path from V_i to V_j which does not pass through any other vertex except V_1 & V_2.

--- . .------------

--- . .------------

$m_k(i, j) = 1$ If there is a simple path from V_i to V_j which does not pass through any other vertex except $V_1, V_2, \ldots, V_k$.

Warshall's Algorithm

Let M be a known relation matrix and let M^+ be its transitive closure matrix which is to be computed.

Step 1 : $T = M$ (Initialize M)

If $a(i, j) = 0$ then set

$m(i, j) = 0$ else $m(i, j) = 1$. (End of Loop)

Step 2 : For $k = 1$ to n DO

For $i = 1$ to n DO

For $j = 1$ to n DO

$m(i, j) = m(i, j) + m(i, k) \cdot m(k, j)$

Step 3 : Terminate with $M^+ = T$.

Modification of Warshall's Algorithm

In the above discussion the matrix M tells us whether or not there are paths between the vertices. Now we want to find a matrix say P which tells us the length of the shortest path between the vertices or more exactly, a matrix $P = P(i, j)$ where

$$p_{ij} = \text{length of shortest path from } V_i \text{ to } V_j.$$

Now we define a sequence of matrices

where $(P_r(i, j)$, the ij entry of $P_0, P_1, \ldots, P_n$. is defined as follows :

$P_r(i, j)$ = The smaller of the length of the preceding path from V_i to V_j or the sum of the lengths of the preceding paths from V_i to V_k and from V_k to V_j.

i.e.,

$$P_r(i, j) = \min\left(P_{r-1}(i, j), (P_{r-1}(i, k) + P_{r-1}(k, j))\right)$$

The initial matrix P_0 is the same as weighted matrix M_0 except that each 0 is M_0 is replaced by ∞. The final matrix P_n will be the desired matrix T.

EXAMPLE 5.49.

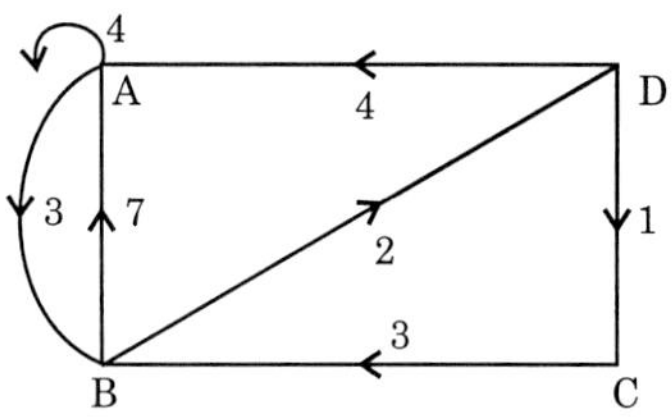

$$P_0 = \begin{array}{c} \\ A \\ B \\ C \\ D \end{array} \begin{array}{c} \begin{array}{cccc} A & B & C & D \end{array} \\ \begin{pmatrix} 4 & 3 & \infty & \infty \\ 7 & \infty & \infty & 2 \\ \infty & 3 & \infty & \infty \\ 4 & \infty & 1 & \infty \end{pmatrix} \end{array} \qquad M_0 = \begin{array}{c} \\ A \\ B \\ C \\ D \end{array} \begin{array}{c} \begin{array}{cccc} A & B & C & D \end{array} \\ \begin{pmatrix} 1 & 1 & 0 & 0 \\ 1 & 0 & 0 & 1 \\ 0 & 1 & 0 & 0 \\ 1 & 0 & 1 & 0 \end{pmatrix} \end{array}$$

It is to be noted in the initial matrix we give weight to the edge *i.e.*, $w_{ik} = \infty$. If there is no edge (path) from i to j.

$$P_1 = \begin{array}{c} \\ A \\ B \\ C \\ D \end{array} \begin{array}{c} \begin{array}{cccc} A & B & C & D \end{array} \\ \begin{pmatrix} 4 & 3 & \infty & \infty \\ 7 & \textcircled{10} & \infty & 2 \\ \infty & 3 & \infty & \infty \\ 4 & \textcircled{7} & 1 & \infty \end{pmatrix} \end{array} \qquad M_0 = \begin{array}{c} \\ A \\ B \\ C \\ D \end{array} \begin{array}{c} \begin{array}{cccc} A & B & C & D \end{array} \\ \begin{pmatrix} 1 & 1 & 0 & 0 \\ 1 & 1 & 0 & 1 \\ 0 & 1 & 0 & 0 \\ 1 & 1 & 1 & 0 \end{pmatrix} \end{array}$$

Taking A is the Centre

$$\begin{aligned} P_1(2, 2) &= \min\big(P_0(2,2), (P_0(2,1) + P_0(1,2))\big) \\ &= \min(\infty, 7+3) \\ &= \min(\infty, 10) = 10 \end{aligned}$$

$$\begin{aligned} P_1(4, 2) &= \min\big(P_0(4,2), (P_0(4,1) + P_0(1,2))\big) \\ &= \min(\infty, 4+3) \\ &= \min(\infty, 7) = 7 \end{aligned}$$

$$P_2 = \begin{array}{c} \\ A \\ B \\ C \\ D \end{array} \begin{array}{c} \begin{array}{cccc} A & B & C & D \end{array} \\ \begin{pmatrix} 4 & 3 & \infty & \textcircled{5} \\ 7 & 10 & \infty & 2 \\ \textcircled{10} & 3 & \infty & 5 \\ 4 & 7 & 1 & \textcircled{9} \end{pmatrix} \end{array} \qquad M_1 = \begin{array}{c} \\ A \\ B \\ C \\ D \end{array} \begin{array}{c} \begin{array}{cccc} A & B & C & D \end{array} \\ \begin{pmatrix} 1 & 1 & 0 & 0 \\ 1 & 1 & 0 & 1 \\ 1 & 1 & 0 & 1 \\ 1 & 1 & 1 & 1 \end{pmatrix} \end{array}$$

Taking A & B as the centre

$$\begin{aligned} P_2(1, 4) &= \min\big(P_1(1,4), (P(1,2) + P(2,4)\big) \\ &= \min(\infty, 3+2) \\ &= 5 \end{aligned}$$

$$\begin{aligned} P_2(3, 1) &= \min\big(P_1(3,1), (P_1(3,2) + P_1(2,1)\big) \\ &= \min\ (\infty,\ 3+7) \\ &= \min(\infty, 10) \\ &= 10 \end{aligned}$$

$$\begin{aligned} P_2(4, 4) &= \min\big(P_1(4,4), (P_1(4,2) + P_1(2,4)\big) \\ &= \min(\infty, 7+2) \\ &= \min(\infty, 9) \\ &= 9 \end{aligned}$$

Taking A , B & C are the centres

$$P_3 = \begin{array}{c} \\ A \\ B \\ C \\ D \end{array}\begin{array}{c} \begin{array}{cccc} A & B & C & D \end{array} \\ \begin{pmatrix} 4 & 3 & \infty & 5 \\ 7 & 10 & \infty & 2 \\ 10 & 3 & \infty & 5 \\ 4 & \textcircled{4} & 1 & \textcircled{6} \end{pmatrix} \end{array} \qquad M_3 = \begin{array}{c} \\ A \\ B \\ C \\ D \end{array}\begin{array}{c} \begin{array}{cccc} A & B & C & D \end{array} \\ \begin{pmatrix} 1 & 1 & 0 & 0 \\ 1 & 1 & 0 & 1 \\ 1 & 1 & 0 & 1 \\ 1 & 1 & 1 & 1 \end{pmatrix} \end{array}$$

$$\begin{aligned} P_3(4, 2) &= \min(P_2(4, 2), (P_2(4, 3) + P(3, 2)) \\ &= \min(7,\ 1+3) \\ &= 4. \\ P_3(4, 4) &= \min(P_3(4, 4), (P_3(4, 2) + P_3(2, 4)) \\ &= \min(9,\ \ 4+2) \\ &= 6 \end{aligned}$$

A, B, C, D all the centres

$$P_4 = \begin{array}{c} \\ A \\ B \\ C \\ D \end{array}\begin{array}{c} \begin{array}{cccc} A & B & C & D \end{array} \\ \begin{pmatrix} 4 & 3 & \textcircled{6} & 5 \\ \textcircled{6} & \textcircled{6} & \textcircled{3} & 2 \\ \textcircled{9} & 3 & \textcircled{6} & 5 \\ 4 & 4 & 1 & 6 \end{pmatrix} \end{array} \qquad M_3 = (m_{ij}) \begin{array}{l} i = 1 \text{ to } 4 \\ j = 1 \text{ to } 4 \end{array}$$

$$\begin{aligned} P_4(1, 3) &= \min(P_3(1, 3), (P_3(1, 4) + P_3(4, 3)) \\ &= \min(\infty,\ 5+1) \\ &= 6 \\ P_4(2, 1) &= \min(P_3(2, 1), (P_3(2, 4) + P_3(4, 1)) = \min(7,\ 2+4) = 6 \\ P_4(2, 2) &= \min(P_3(2, 2), (P_3(2, 4) + P_3(4, 2)) = \min\ (10,\ 2+4) = 6 \\ P_4(2, 3) &= \min(P_3(2, 3), (P_3(2, 4) + P_3(4, 3)) = \min(\infty,\ 2+1) = 3 \\ P_4(3, 1) &= \min(P_3(3, 1), (P_3(3, 4) + P_3(4, 1)) = \min(10,\ 5+4) = 9 \\ P_4(3, 3) &= \min(P_3(3, 3), (P_3(3, 4) + P_3(4, 3)) = \min(\infty,\ 5+1) = 6 \end{aligned}$$

The last matrix P_4 gives the shortest path matrix.

EXERCISE 1

1. Consider the following five relations on the set
   $A = \{ 1, 2, 3, 4\}$
   $R_1 = \{(1, 1), (1, 2), (2, 3), (1, 3), (4, 4)$
   $R_2 = \phi$, the empty relation
   $R_3 = A \times A$, the universal relation
   R_4 = Relation $\leq$ (less than or equal) on the set of integers
   R_5 = Relation $\perp$ (perpendicular) on the set L of lines in the plane.
   R_6 = Relation $||$ (parallel) on the set L of lines in the plane.
   Determine which of the relations are reflexive, symmetric, Transitive.
2. Given : $A = \{a, b, c\}$ and $B = \{a, b\}$
   Find (*a*) $A \times B$
   (*b*) $B \times A$
   (*c*) $A \times A$

3. $A = \{a, b, c, d\}$ and $B = \{1, 2, 3)$. Let R be the following relations from A to B :

$$R = \{(a, 1), (a, 3), (b, 2), (b, 3)), (c, 1), (c, 3), (d, 1), (d, 2), (d, 3)\}.$$

(*a*) Give matrix representation of the following relatipn.

(*b*) Find R^{-1}

(*c*) Find R^2

(*d*) Give graphical representation of R.

Let R and S be the following relations on $A = \{1, 2, 3\}$

$$R = \{(1, 1), (1, 2), (2, 3), (3, 1), (3, 3)\}$$

$$S = \{(1, 2), (1, 3), (2, 1), (3, 3)\}$$

Find(*a*) $R \cup S$

R^C

$R \cap S$

(*b*) $R^{-1} \cup R$

(*c*) $S^2 = S \circ S$

4. Consider a set $A = \{1, 2, 3\}$ and the relation R on A defined by:

$$R = \{(1, 1), (1, 2), (2, 3), (3, 3), (1, 3)\}$$

Find :(*a*) Reflexive closure

(*b*) Symmetric closure

(*c*) Transitive closure of R.

5. Consider the set Z of integers and an integer $n > 1$ then define a congruence relation R on Z as "x is congruent to y modulo n, written

$$x \equiv y \pmod{n}$$

if $(x - y)$ is divisible by n.

Check whether or not this relation R is an equavalence relation or not.

6. Let R be the relation on the positive integer N defined by the equation $x + 3y = 12$ that is

$R = \{(x, y) \mid x + 3y = 12\}$

Write : (*a*) R as the set of ordered pairs.

(*b*) Is R is an equivalence relation and give reason also.

(*c*) Find R^{-1} also.

(*d*) Find the composition Relation $R \circ R$.

7. Prove that if R is an equivalence relation then R^{-1} is also an equavalence relation

8. Let $A = \{1, 2, 3, 4\}$

Give example of a relation in A which is:

(*i*) Equivalence relation.

(*ii*) Partial order relation.

(*iii*) Reflexive, not symmetric.

(*iv*) Symmetric, not reflexive.

(*v*) Symmetric, transitive not symmetric.

(*vi*) Transitive but not reflexive or symmetric.

9. Draw the diagraph of the relation defined on $\{a_1, a_2, a_3, a_4\}$ by the matrix

	a_1	a_2	a_3	a_4
a_1	0	1	1	1
a_2	0	0	1	1
a_3	1	1	0	0
a_4	1	0	0	0

10. Let $A = \{a, b, c\}$ and let $P(A)$ be the power set of A.. Define the relation R on $P(A)$ as follows : If S and T are subsets of A, then SRT if $S \subset T$. Draw a diagraph for this relation. Determine its reflexivity, transitivity, symmetry and anti-symmetry.

11. Let R be the relation on the let $\{a_1, a_2, a_3, a_4\}$ defined by the following matrix M:

$$\begin{pmatrix} 0 & 0 & 0 & 1 \\ 1 & 0 & 0 & 0 \\ 1 & 0 & 0 & 0 \\ 0 & 1 & 0 & 1 \end{pmatrix}$$

(*a*) Draw the diagraph of the relation R defined by M.

(*b*) Find $M \times M$

(*c*) Find digraph of R o R

(*d*) Find the matrix for R o R and compare it with $M \times M$.

Let R be a relation on a set A.

Define transitive closure of A.

Find the transitive closure of each of the following relations.

(*a*) $A = \{a, b, c, d, e\}$ and $R = \{a, b), (b, c), (c, d), (d, a), (a, e)\}$

(*b*) $A = \{a, b, c, d, e, f\}$

where R is the relation given by

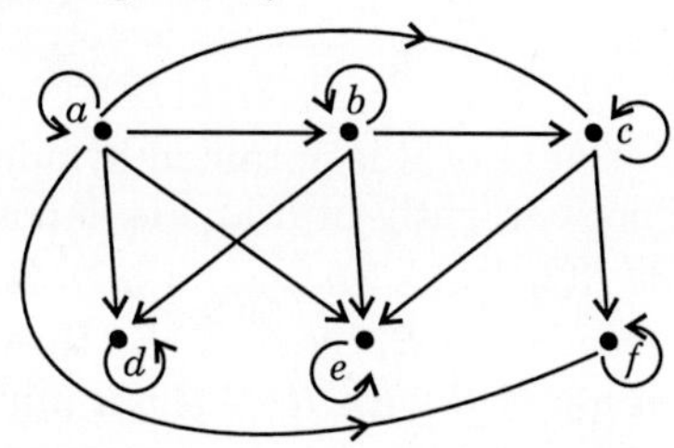

12. Decide which of the following diagraphs defines an equivalence relation.

(*a*)
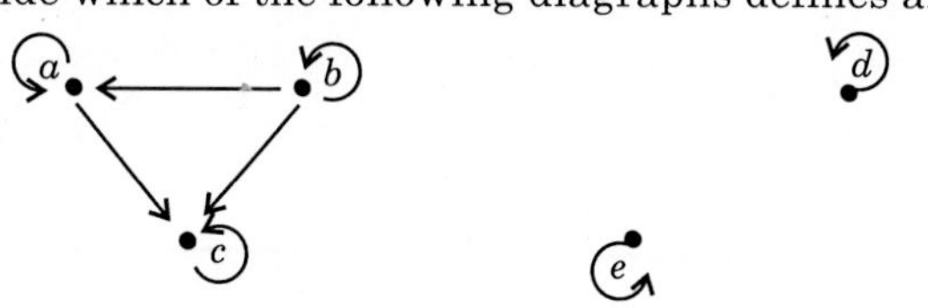

(*b*)
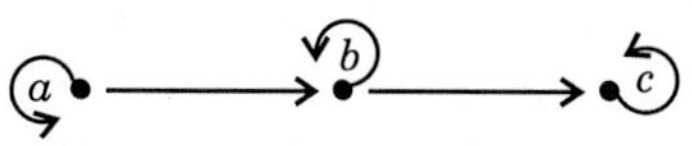

13. Draw the Hasse diagram for each of the following partially ordered sets.

(*a*) The power set of $\{a, b, c, d\}$ under inclusion.

(*b*) The set {1, 2, 3, ...12} under $x \, R \, y$ whenever $x \mid y$.
(*c*) The set {1,2, 3, 4, 5} under $x \, R \, y$ whenever $x \geq y$.

14. Let X be the set of all four-bit strings (*e.g.*, 0011, 0101, 1000). Define a relation R on X as $s_1 R s_2$ if some substring of s_1 of length 2 is equal to some substring of s_2 of length 2. Examples : 0111 R 1010 (because both 0111 and 1010 contain 01). 1110 $\not R$ 0001 (because 1110 and 0001 do not share a common substring of length 2). Is this relation reflexive, symmetric, antisymmetric, transitive and/or a partial order?
15. Tell whether the relation given by the following matrix is reflexive, symmetric, transitive, antisymmetric, a partial order and/or an equivalence relation.

$$\begin{array}{c} \\ w \\ x \\ y \\ z \end{array} \begin{array}{c} \begin{array}{cccc} w & x & y & z \end{array} \\ \begin{pmatrix} 1 & 0 & 1 & 0 \\ 0 & 0 & 0 & 0 \\ 1 & 0 & 1 & 0 \\ 0 & 0 & 0 & 1 \end{pmatrix} \end{array}$$

Answers for Exercise 1

1. R_2, R_3, R_4, R_6 are reflexive relations
   R_2, R_3, R_5 are symmetric
   R_1, R_3, R_6, R_2 are transitive.
2. $A \times B = \{(a, a), (a, b), (b, a), (b, b), (c, a), (c, b)\}$
   $B \times A = \{(a, a), (a, b), (a, c), (b, a), (b, b), (b, c)\}$
   $A \times A = \{(a, a), (a, b), (a, c), (b, a), (b, b), (b, c), (c, a), (c, b), (c,c)\}$
5. (*a*) As the reflexive closure of R is obtained by adding all diagonal pairs of $A \times A$ to R which are not carrently in R and is denoted by Reflexive (R^+)
   Reflexive $(R^+) = \{(1, 1), (2, 2), (1, 2), (1, 2), (2, 3), (3, 3), (1, 3)\}$
   (*b*) The symmetric closure of R is obtained by adding all the pairs in R^{-1} to R which are not currently in R and is denoted by symmetric (R^+).
   Symmetric $(R^+) = \{(2, 1), (3, 2), (3, 1), (1, 1), (1, 2), (2, 3), (3, 3), (1, 3)\}$.
   (*c*) The transitive closure on R. Since A has three elements, is obtained by taking the union of R with $R^2 = R \text{ o } R$ and $R^3 = R \text{ o } R \text{ o } R$. and is denoted by transitive (R^+).

   Note that

   $$R^2 = R \text{ o } R = \{(1, 1), (1, 2), (2, 3), (3, 3), (1, 3)\} = R$$
   $$R^3 = (R \text{ o } R) \text{ o } R = R \text{ o } R = R$$
   $$\therefore \text{ transitive } (R^+) = R \cup R^2 \cup R^3 = R$$
6. Equivalence relation.
7. (*a*) $R = \{9, 1), (6, 2), (3, 3)\}$
   (*b*) R is not an equavalence relation as reflexive does not exists.
   (*c*) $R^{-1} = \{(1, 9), (2, 6), (3, 3)\}$
   (*d*) $R \text{ o } R = \{(3, 3)\}$

15. This relation is reflexive since $\forall x \in D$, xRx.

This relation is symmetric since $xRy \to yRx$.

This relation is NOT antisymmetric. $\forall (x, y) \in R$,

$$xRy \to yRx, \text{ even when } x \neq y.$$

This relation is NOT transitive. Example :

$$\{1010, 0110\} \in R\,; \quad \{0110, 1111\} \in R$$

But $\{1010, 1111\} \notin R$

Since R is not transitive it is NOT a partial order.

xRy means $\exists T \in S$ such that $(x, y) \in T$.

This function is necessarily reflexive and symmetric, but not necessarily transitive.

Since $xRy \to (x, y) \in T,\ T \subset S$

$yRz \to (y, z) \in T,\ T \subset S$

Since the subset of S are not pair wise disjoint, it is not necessarily so that $S = T$. So xRz does not necessarily hold.

EXERCISE 2

1. Let $A = \{2, 3\}$ and $B = \{a, b, c, d\}$
   (*a*) Find the elements of $A \times B$.
   (b) Represent the set $A \times B$ by diagram.
2. Consider the ordered triple {2, 3, 5}. Represent the ordered triple as an ordered pair.
3. Let R be the relation from $A = \{2, 3, 4\}$ to $B = \{3, 6, 7, 8, 9, 10\}$ defined by "x divides y".
   (*a*) Write R on a coordinate diagram $A \times B$.
   (*b*) Obtain the inverse relation R^{-1}.
4. In the light of the set definition of an ordered pair, if $(a, b) = (c, d)$, then show that $a = c$ and $b = d$.
5. Which of the following relations are equivalence relations (R defines the relation on the set of integers).
   (a) aRb if and only if $a > b$.
   (b) aRb if and only if $a + b = 8$
   (c) aRb if and only if $a + b < 8$.
6. Let $A = \{a, b, c, d\}$
   $R = \{(a, a), (a, b), (b, a), (b, b), (b, c), (b, d), (c, d), (d, a)$
   Sketch the diagram of R.
7. Let A be a set of people and let
   $R = \{(x, y) \in A \times A : \text{such that } x \text{ is a sister of } y\}$
   Determine whether R is a symmetric relation.
8. Show whether the following relations are transitive.
   $R_1 = \{(1, 1)\},\ R_2 = \{(1, 2), (2, 2)\}$
   $R_3 = \{(1, 2), (2, 3), (1, 3), (2, 1)\}$.

9. Let $S = \{1, 2, 3, 4\}$. A relation R on S is defined by:
$R = \{(1, 2), (4, 3), (2, 2), (2, 1), (3, 1)$.
Find a relation $R_1 \supseteq R$ such that R_1 is transitive.

10. Let $R = \{(1, 2), (3, 4), (2, 3)\}$ and $S = (2, 5), (4, 6), (3, 1), (1, 3)\}$. Determine RoS, and SoR, and answer whether RoS equals to SoR. Also, determine $Ro(SoR)$, and $RoRoR$.

11. Let R and S be two relations on a set of positive integers Z.
$R = \{(x, 3x) : x \in Z\}$.
$S = \{(x, 7x) : x \in Z\}$; Find RoS, and $RoRoR$.

12. Let R be the relation defined by xRy if and only if $|x - y|$ is even. Show that R is an equivalence relation.

13. Let R be the relation defined on the integers by aRb of $a - b$ is even. Show that R is an equivalence relation, and determine the equivalence classes.

14. Partition the set $A = \{1, 2, 3, 4, 5, 6\}$ by $\{\{1, 2\}, \{3\}, \{4, 5\}, \{6\}\}$. List the symmetric relation determined by this relation.

15. Find all partitions of $S = \{x_0, x_1, x_2, x_3\}$.

16. Let $S = (a, b, c, d\}$ and R_1, R_2, R_3, R_4, R_5 and R_6 be the following relations on S :
$R_1 = \{(a, a), (a, b)\}$
$R_2 = \{(a, a), (a, c), (b, c), (d, a)\}$
$R_3 = \{(a, c), (b, d)\}$
$R_4 = \{(a, a), (b, b)\}$
$R_5 = \text{W} \times \text{W}$
$R_6 = \phi$
Determine whether or not each of above relations is (*i*) symmetric, (*ii*) transitive.

17. Consider the following relation given by the diagraph

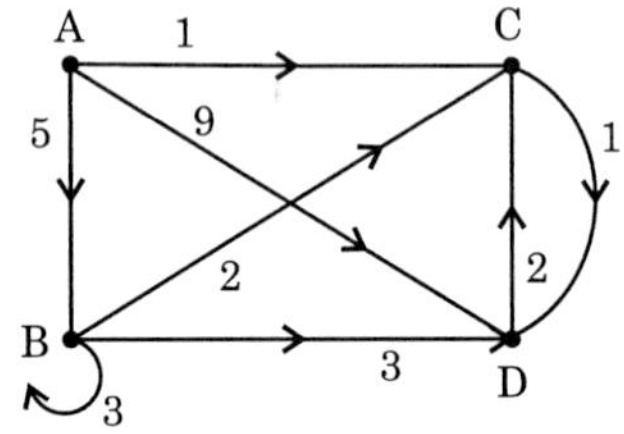

Apply Warshall's algorithm and find all shortest paths.

18. Given $X = \{1, 2, 3, 4, 5\}$. R is a relation on X given by
$R = \{(1, 1), (2, 5), (3, 1), (4, 2), (5, 4)\}$
Find the relations ROR^{-1} and R^{-1}OR. ***(End-term Exam., GGSIPU, 2001)***

19. (*a*) Let $A = \{0, 1, 2, 3\}$ and $R = \{(0, 0), (1, 1), (2, 2), (3, 3), (1, 2), (2, 1), (3, 2), (2, 3), (3, 1), (1, 3)\}$
Show that R is an equivalence relation and hence find the partition of A.
(*b*) Find the transitive closure of the relation R defined on set
$X = \{1, 2, 3, 4\}$ and give $R = \{(1, 2), (2, 3), (3, 4)\}$
(End-term Exam., GGSIPU, 2001)

20. (*a*) Give various properties of binary relations.

    (*b*) If $X = \{1, 2 .....7\}$ and

    $R = \{(x, y) \mid x - y \text{ is divisible by } 3\}$,

    show that R is an equivalence relation. Also, determine the equivalence class of 4. ***(Terminal Exam., GGSIPU, 2001)***

21. State which of the following diagraphs represent a poset, a well ordered set and find the lub and glb for ever pair of vertices.

    (End-term Exam., GGSIPU, 2001)

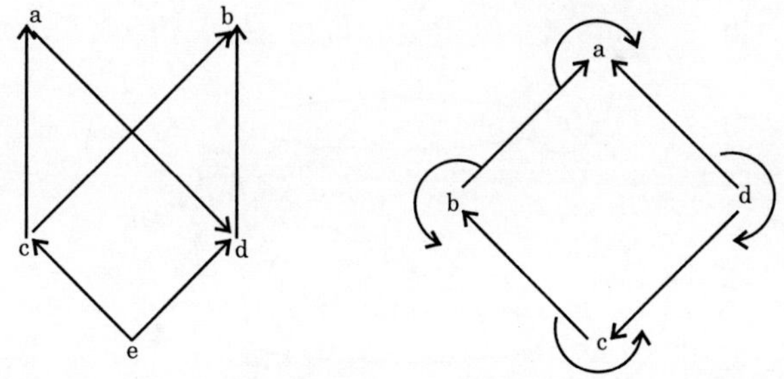

22. (*a*) What do you mean by transitive closure of relation R defined in set A. Given the set-$A = \{1, 2, 3\}$ and relation $R = \{(1,2), (2, 3), (3, 3)\}$. Find transitive closure of relation R.

    (*b*) Define a lattice. Show that the set of factors of 24 forms a lattice under the relation $\leq$ of divisibility *i.e.*, $a \leq b$ if and only if a divides b. Draw the hasse diagram of lattice. Find the values of $3 \wedge 4$ and $3 \vee 4$.

    (End-term Exam., GGSIPU, 2001)

Answers for Exercise 2

1.
$$\begin{aligned} A &= \{2, 3\} \;\&\; B = \{a, b, c, d\} \\ A \times B &= \{(a, b) \mid a \in A, b \in B\} \\ &= \{2, 3\} \times \{a, b, c, d\} \\ &= \{(2, a), (2, b), (2, c), (2, d), (3, a), (3, b), (3, c), (3, d)\} \end{aligned}$$

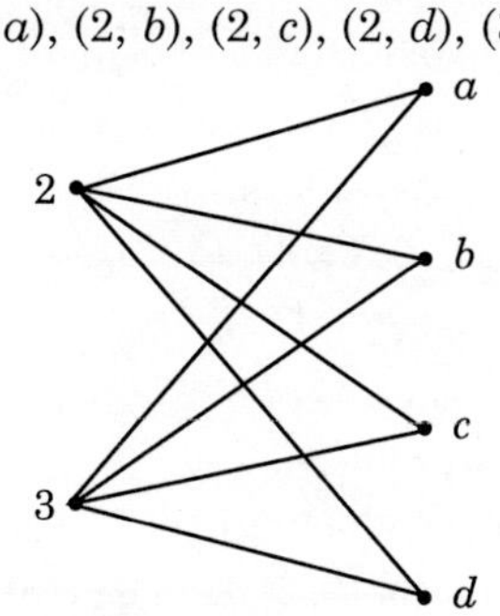

2. Hint
$$\begin{aligned} A &= \{2, 3, 5\} \\ A \times A &= \{(a, b) \mid a \in A, b \in A\} \end{aligned}$$

3. (*a*)
$$A = \{2, 3, 4\}, B = \{3, 6, 7, 8, 9, 10\}$$
$$R \subseteq A \times B$$

$$R = \{(a, b) \mid a \text{ divides } b,\ a \in A \ \&\ b \in B\}$$
$$= \{(2,6), (2,8), (2, 10), (3,3), (3, 6), (3, 9), (4, 8)\}$$
$$(b)\ R^{-1} = \{(b, a) \mid b \text{ is divisible by } a,\ b \in B \ \&\ a \in A\}$$
$$= \{(6, 2), (8, 2), (10, 2), (3, 3), (6, 3), (9, 3), (8, 4)\}$$

5. (*a*) & (*c*) are not equivalence relation as Reflexivity does not hold.

(*b*) Not an equivalence relation as transitivity, and reflexivity does not hold.

6. Let $A = \{a, b, c, d\}$

$$R = \{(a, a), (a, b), (b,a), (b, b), (b,c), (b, d), (c, d), (d, a)\}$$

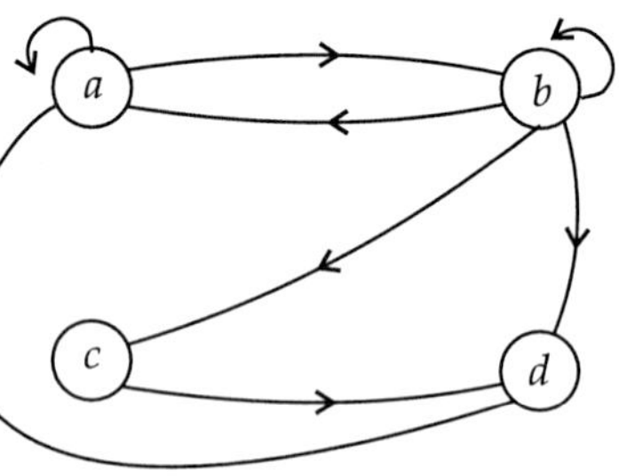

7. A = Set of people

$$R = \{(x, y) \in A \times A : x \text{ is sister of } y\}$$

R is not symmetric relation as if y is a brother of x then $x\,R\,y$ but $y \not R\, x$.

8. $R_1 = \{(1,1)\}$ is a transitive relation

$R_2 = \{(1,2), (2, 2)\}$ is a transitive relation

$R_3 = \{(1, 2), (2, 3), (1, 3), (2, 1)\}$

Not a transitive relation as $1R2$ & $2R1$ but $1 \not R\, 1$.

9. $S = \{1, 2, 3, 4\}$

$$R = \{(1, 2), (2, 2), (2, 1), (4, 3), (3, 1)\}$$

We need to find

$R_1 \supseteq R$ such that R_1 is transitive means we need to give transitive closure of R.

$$R^2 = \{(1, 1), (4, 1), (3, 2)\}$$
$$R \cup R^2 = \{(1, 2, (2, 2), (2, 1), (4, 3), (3, 1), (4, 1), (1, 1), (3, 2)\}$$
$$R^3 = \{(4, 2)\}$$
$$R \cup R^2 \cup R^3 = \{(1, 2), (2, 2), (2, 1), (4, 3), (3, 1), (4, 1), (1, 1), (3, 2), (4, 2)\}$$

$\therefore \quad R_1$ = Transitive closure of R

$$= \{(1, 1), (1, 2), (2, 2), (2, 1), (4, 3), (3, 1), (3, 2), (4, 1), (4,2)\}$$

10. $R = \{(1, 2), (3, 4), (2, 3)\}$

$$S = \{(2, 5), (4, 6), (3, 1), (1, 3)\}$$

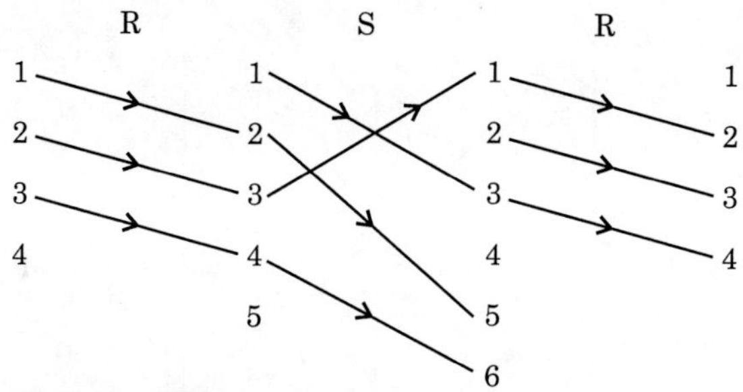

$$\therefore \quad RoS = \{(1, 5), (2, 1), (3, 6)\}$$

$$SoR = \{(1, 4), (3, 2)\}$$

$$RoS \neq SoR$$

$$RoSoR = \{ (2, 2)\}$$

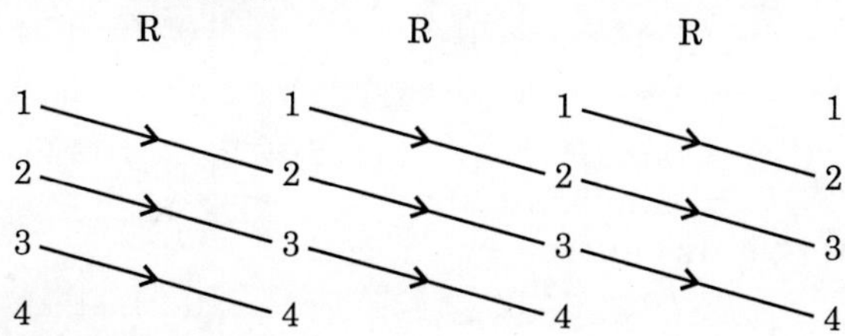

$$RoRoR = \{ (1, 4)\}$$

12. $x \, R \, y$ if and only if $|x - y|$ is even

Reflexivity : $x \, R \, x$ as $|x - x| = |0| = 0$ is even

Symmetry : $x \, R \, y \Rightarrow |x - y|$ is even

$y \, R \, x \Rightarrow |y - x|$ is even True

$\Rightarrow \quad x \, R \, y$ if and only if $y \, R \, x$.

Transitivity : Let $x \, R \, y$ and $y \, R \, z$

$\Rightarrow \quad |x - y|$ is even

$\Rightarrow \quad |y - z|$ is even

Consider $|x - z| = |x - y + y - z|$ is even [as even plus even]

16. $R_1 = \{(a, a), (a, b)\}$ Not Symmetric but transitive.

$R_2 = \{ (a, a), (a, c), (b, c), (d, a)\}$ Not symmetric but transitive.

$R_3 = \{ (a, c), (b, d)\}$ Transitive but not symmetric.

$R_4 = \{ (a, a), (b, b)\}$ Symmetric as well as transitive.

$R_5 = W \times W$ Symmetric as well as transitive.

$R_6 = \phi$ Symmetric as well as transitive.

17.

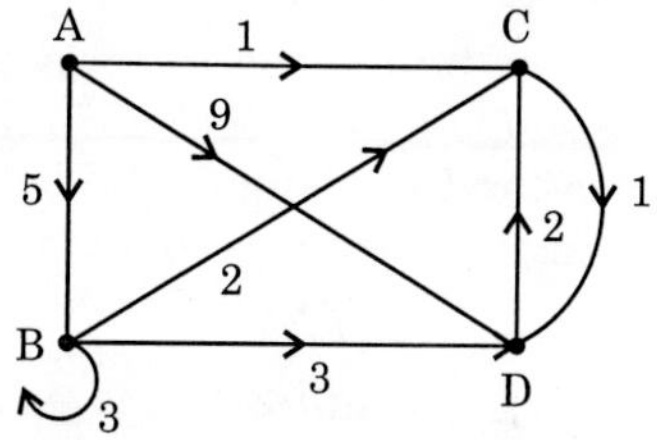

$$P_0 = \begin{pmatrix} \infty & 5 & 1 & 9 \\ \infty & 3 & 2 & 3 \\ \infty & \infty & \infty & 1 \\ \infty & \infty & 2 & \infty \end{pmatrix} \qquad M_0 = \begin{pmatrix} 0 & 1 & 1 & 1 \\ 0 & 1 & 1 & 1 \\ 0 & 0 & 0 & 1 \\ 0 & 0 & 1 & 0 \end{pmatrix}$$

Taking *A* as a centre

$$P_1 = \begin{pmatrix} \infty & 5 & 1 & 9 \\ \infty & 3 & 2 & 3 \\ \infty & \infty & \infty & 1 \\ \infty & \infty & 2 & \infty \end{pmatrix} \qquad M_1 = \begin{pmatrix} 0 & 1 & 1 & 1 \\ 0 & 1 & 1 & 1 \\ 0 & 0 & 0 & 1 \\ 0 & 0 & 1 & 0 \end{pmatrix}$$

Taking *A* and *B* as a centre

$$P_2 = \begin{pmatrix} \infty & 5 & 1 & \textcircled{8} \\ \infty & 3 & 2 & 3 \\ \infty & \infty & \infty & 1 \\ \infty & \infty & 2 & \infty \end{pmatrix} \qquad M_2 = \begin{pmatrix} 0 & 1 & 1 & 1 \\ 0 & 1 & 1 & 1 \\ 0 & 0 & 0 & 1 \\ 0 & 0 & 1 & 0 \end{pmatrix}$$

$$\begin{aligned} P_2(1, 4) &= \min (P_1(1, 4), P_1 (1, 2) + P_1 (2, 4)) \\ &= \min (9, 5 + 3) = 8 \end{aligned}$$

Taking *A*, *B*, *C* as centre

$$P_3 = \begin{pmatrix} \infty & 5 & 1 & \textcircled{2} \\ \infty & 3 & 2 & 3 \\ \infty & \infty & \infty & 1 \\ \infty & \infty & 2 & \textcircled{3} \end{pmatrix} \qquad M_3 = \begin{pmatrix} 0 & 1 & 1 & 1 \\ 0 & 1 & 1 & 1 \\ 0 & 0 & 0 & 1 \\ 0 & 0 & 1 & 1 \end{pmatrix}$$

$$\begin{aligned} P_3(1, 4) &= \min (P_2(1, 4), P_2 (1, 3) + P_2(3, 4)) = \min (8, 1 + 1) = 2 \\ P_3(4, 4) &= \min (P_2(4, 4), P_2(4, 3) + P_2(3, 4)) = \min (\infty, 2 + 1) = 3 \end{aligned}$$

Taking *A*, *B*, *C*, *D* as centres

$$= \begin{pmatrix} \infty & 5 & 1 & 2 \\ \infty & 3 & 2 & 3 \\ \infty & \infty & \textcircled{4} & 1 \\ \infty & \infty & 2 & 3 \end{pmatrix} \qquad M_4 = \begin{pmatrix} 0 & 1 & 1 & 1 \\ 0 & 1 & 1 & 1 \\ 0 & 0 & 1 & 1 \\ 0 & 0 & 1 & 1 \end{pmatrix}$$

$$\begin{aligned} P_4 (3, 3) &= \min (P_3 (3, 3), P_3 (3, 4) + P_3 (4, 3)) \\ &= \min (\infty, 1 + 3) = 4 \end{aligned}$$

EXERCISE 3

1. (*a*) Show that the functions f and g, both defined from NXN to N given by $f(x, y) = x + y$ and $g(x, y) = xy$ are onto but not one-to-one.

   (*b*) If $f: A \to B$ and $g : B \to C$ be one-one onto functions then show that gof is also one-one onto and $(gof)^{-1} = f^{1}\, og^{-1}$.

   (*c*) What do you understand by recursively defined function. Determine whether the following function is recursively defined function :

$$f(n) = \begin{cases} 1 \text{ if } n = 1 \\ 1 + g(n/2) \text{ if } n \text{ is even} \\ g(3n - 1) \text{ if } n \text{ is odd and } n > 1 \end{cases}$$

2. (*a*) If R and S are equivalence relations on the set A then show that
   (*i*) R^{-1} is an equivalence relation.
   (*ii*) $R \cap S$ is an equivalence relaton.
   (*b*) Show that the relation 'is congruent modulo 4' on the set of integer {0, 1, 2, 3, ...10} is an equivalence relation.
   (*c*) Let R be a relation on a Set $A = \{1, 2, 3, 4, 5\}$ whose matrix is given below.

$$M = \begin{vmatrix} 1 & 1 & 1 & 1 & 0 \\ 0 & 0 & 0 & 0 & 0 \\ 1 & 1 & 0 & 0 & 0 \\ 0 & 0 & 0 & 0 & 0 \\ 1 & 1 & 1 & 1 & 1 \end{vmatrix}$$

   Draw the directed graph of R.
3. If R be a relation in the set of integers Z defined by $R = \{(x, y) :: x \in Z, y \in Z, (x - y)$ is multiple of $3\}$ Show that it is an equivalence relation. What is equivalence class of 0 ?
4. Show that the relation "less than or equal to" on the set of integers is a partial order.
5. (*a*) Let R be the relation on the positive integers N defined by the equation $x + 3y = 12$, *i.e.*, $R = \{(x, y); x + 3y = 12\}$ Find (*i*) domain of R (*ii*) range of R (*iii*) R^{-1} and (*iv*) the composition relation RoR.
   (*b*) Show that the relation $(x, y)\, R\, (a, b) \Leftrightarrow x^2 + y^2 = a^2 + b^2$ is an equivalence relation on the plane and describe the equivalence classes.
   (*c*) Show that every semi-group homomorphism induces a congruence relation and *vice-versa*.
6. (*a*) Let $f : R \to R$ and $g : R \to R$, where R is the set of real numbers. Find whether these functions are injective, surjective and bijective.
   (*b*) Show that $f(x, y) = x^y$ is a primitive recursive-function.
   (*c*) Explain reflexive and symmetric closure with example.
7. If $\{\{a, c, e\}, \{b, d, f\}\}$ is partition of the set $A = \{a, b, c, d, e, f\}$, determine the correspondency equivalence relation R.
8. If a set A has n elements, how many function are there from A to A?
9. (*a*) Give an example of a relation which is neither reflexive nor irreflexive.
   (*b*) Give an example of a relation which is both symmetric and anti-symmetric.
10. Write a short note on relation matrix. Give a relation matrix for the given graph.

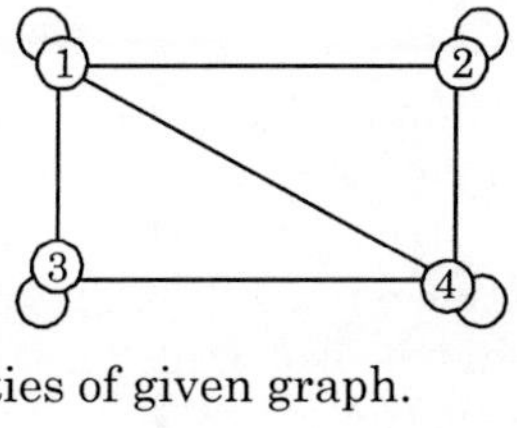

and explain the properties of given graph.

DO IT BY YOURSELF

1. Let A = {a, b,}, and B = {4, 5, 6}, List the elements in AXB, BXA, AXA and BXB.
2. If B = {0, 3, 6, 9.....}, give a partition of B containing
   (*a*) two infinite subsets (*b*) three infinite subsets.
3. If A has three elements and B has $n \geq 1$ elements, use mathematical induction to prove that $|AXB| = 3n$.
4. If A has n elements and B has m elements, how many different relations are there from A to B?
5. Let relation R on the set {1, 2, 3, 4, 5} be defined by the rule $(x, y) \in R$ if 3 divides $x - y$.
   (*a*) List the elements of R and R^{-1}.
   (*b*) Find the domain and range of R and R^{-1}.
   (*c*) Draw directed graph.
   (*d*) List all properties of the relation and explain.
6. Let X be a nonempty set. Define a relation on (X), the power set of X as $(A, B) \in R$ if $A \subseteq B$, is this relation reflexive symmetric, antisymmetric, transitive, and/or a partial order?
7. Let $R1$ and $R2$ be the relations on {1, 2, 3, 4} given by
   $R1$ = {(1, 1), (1, 2), (3, 4), (4, 2)}
   $R2$ = {(1, 1), (2, 1), (3, 1), (4, 4), (2, 2)}
   List the elements of $R1$ o $R1$ and $R2$ o $R1$
8. What is wrong with the following argument, which supposedly shows that my relation R on X that is symmetric and transitive is reflexive?
   Let $x \in X$. Using symmetry, we have (x, y) and (y, x) both in R. Since (x, y), $(y, x) \in R$, by Transitivity we have $(x, x) \in R$. Therefore R is reflexive.
9. Define the relation R on the set of positive integers by $(x, y) \in R$ if the greatest common divisor of x and y is 1. Determine whether R is reflexive, symmetric, antisymmetric, transitive and a partial order.
10. Show that if R is an equivalence relation on X, then domain R = range $R = X$.
11. Consider the following five relations:
    (*a*) Relation $\leq$ (less than or equal) on the set of Z of integers.
    (*b*) Set inclusion $\subseteq$ on a collection C of sets.
    (*c*) Relation $\perp$ (perpendicular) on the set L of lines in the plane.
    (*d*) Relation $\parallel$ (parallel) on the set L of lines in the plane.
    (*e*) Relation $\mid$ of divisibility on the set N of positive integers.
    Determine which of the above relations are reflexive, symmetric, antisymmetric, transitive, and/or a partial order
12. Let A be a set of books.
    (*a*) Let R be a binary relation on A.S.T. (a, b) is in R if book a costs more and contain fewer pages than book b. In general, is R reflexive? Symmetric, Antisymmetric? Transitive?
    (*b*) Let S be a binary relation on A.S.T. (a, b) is in S if book a costs more or contain fewer pages than book b. In general is S reflexive? Symmetric, Antisymmetric? Transitive?

13. Let P be the set of all people. Let R be a binary relation on P.S.T. (a,b) is in R if a is a brother of b. (Disregard half-brothers and fraternity brothers.) Is R reflexive? Symmetric? Antisymmetric? Transitive? An equivalence relation? A partial ordering relation?
14. Let A be set with 10 distinct elements
    (*a*) How many different binary relations on A are there?
    (*b*) How many of them are reflexive?
    (*c*) How many of them are symmetric?
    (*d*) How many of them are reflexive and symmetric?
    (*e*) How many of them are total ordering relations?
15. (*a*) Let R and S be binary relations from A to B. Is it true that $(R \cup S)^{-1} = R^{-1} \cup S^{-1}$?
    (*b*) Let R be a binary relation on A. If R reflexive, is inverse/converse of R necessarily reflexive? If R is symmetric, is its inverse necessarily symmetric? If R is transitive, is its inverse necessarily transitive?
16. Let $A = \{1, 2, 3, 4\}$, and let $R = \{(1, 2), (2, 3), (3, 4), (2, 1)\}$. Find the reflexive, symmetric and transitive closure of R.
17. (*a*) Let $f(x) = x + 2$, $g(x) = x - 2$ and $h(x) = 3x$ for $x \in R$. Where R is a set of real numbers. Find $g \circ f$, $f \circ g$, $f \circ f$, h, $h \circ g$, $h \circ f$, $f \circ h \circ g$.
    (*b*) Let $f : R \to R$ be given by $f(x) = -x^2$ and $g : R_+ \to R_+$ be given by $g(x) = \sqrt{x}$.
    Where R_+ is the set of non-negative real numbers and R is the set of real numbers. Find $f \circ g$. Is $g \circ f$ defined?
18. Consider functions $f : A \to B$ and $g : B \to C$. Prove that following
    (*a*) If f and g are one-to-one, then composition function $g \circ f$ is one-to-one.
    (*b*) If f and g are onto functions, then $g \circ f$ is an onto function
19. If $A = \{1, 2, \ldots\ldots, n\}$ show that any function from A to A which is one-to-one must also be onto, and conversely.
20. Prove that the function $f = \{(1, a), (2, c), (3, b)\}$ from $X = \{1, 2, 3\}$ to $Y = \{a, b, c\}$ is bijection and hence find inverse.
21. (*a*) Give an example of a function that is one-to-one but not onto.
    (*b*) Give an example of a function that is onto but not one-to-one.
    (*c*) Give an example of a function that is neither one-to-one nor onto.
22. Define characteristic function and write all properties of characteristic function.
23. Using properties of characteristic function, show that
    (*a*) $A \cap (B \cup C) = (A \cap B) \cup (A \cap C)$
    (*b*) $\neg\neg A = A$
24. Find the orbit and cycles of the set S = {1, 2, 3, 4, 5, 6} under the following permutations:

(*a*) $\begin{bmatrix} 1 & 2 & 3 & 4 & 5 & 6 \\ 2 & 1 & 5 & 6 & 4 & 3 \end{bmatrix}$ (*b*) $\begin{bmatrix} 1 & 2 & 3 & 4 & 5 & 6 \\ 2 & 3 & 4 & 5 & 6 & 1 \end{bmatrix}$

(*c*) $\begin{bmatrix} 1 & 2 & 3 & 4 & 5 & 6 \\ 2 & 3 & 1 & 6 & 5 & 4 \end{bmatrix}$

25. Express the following permutations as product of disjoint cycles.
(*a*) (1, 2, 3) (2, 3, 4) (4, 5, 6) (6, 7) (*b*) (1, 2) (1, 3) (1, 4) (1, 5) (1, 6)
(*c*) (2, 4) (2, 6) (2, 8) (1, 3) (1, 7)
26. Find out which of the following permutations are even (odd) permutations
(*a*) (1, 2, 3) (1, 2) (*b*) (1, 2, 3, 4, 5) (1, 2, 3) (4, 5)
(*c*) (1, 2) (1, 4) (1, 5, 3)
27. Express the following permutations as product of transpositions
(*a*) (1, 2, 3, ...*n*) (*b*) (1, 2, 3) (4, 5, 7, 6)
(*c*) (1, 2) (3, 4, 5)
28. Find the orbits of all elements under the permutation

$$\begin{bmatrix} 1 & 2 & 3 & 4 & 5 & 6 & 7 & 8 & 9 \\ 2 & 1 & 3 & 6 & 4 & 5 & 9 & 7 & 8 \end{bmatrix}_{\text{in } S9}$$

29. (*a*) Let $A = \begin{bmatrix} 0 & 1 & 0 \\ 1 & 0 & 1 \\ 1 & 0 & 0 \end{bmatrix}$ and $B = \begin{bmatrix} 1 & 0 & 0 \\ 1 & 0 & 0 \\ 0 & 1 & 1 \end{bmatrix}$ be Boolean matrices.

Find the Boolean matrices : 1. A + B 2. AB 3. BA 4. A^2 5. B^2.

(*b*) Let $A = \begin{bmatrix} 1 & 0 & 0 \\ 0 & 0 & 1 \\ 1 & 1 & 0 \end{bmatrix}$ and $B = \begin{bmatrix} 0 & 1 & 1 \\ 1 & 0 & 0 \\ 0 & 1 & 0 \end{bmatrix}$ be Boolean matrices.

Find the Boolean matrices : 1. A + B 2. AB 3. BA 4. A^2 5. B^2.

30. Show that any two disjoint permutations commute *i.e.*, If *f* and *g* are any two disjoint permutations of set *S*, then $f \circ g = g \circ f$.

MULTIPLE CHOICE QUESTIONS

1. The number of binary relations on asset with *n* elements is:
(*a*) 2^n (*b*) n^2 (*c*) 2^{n^2} (*d*) None
2. The number of equivalence relations of the set {1, 2, 3, 4} is:
(*a*) 15 (*b*) 16 (*c*) 17 (*d*) None
3. Suppose *A* is finite set with *n* elements. The number of elements in the largest equivalence relation of *A* is:
(*a*) *n* (*b*) n^2 (*c*) $n + 1$ (*d*) 1
4. The 'subset' relation on a set of sets is:
(*a*) An equivalence relation (*b*) Reflexive and antisymmetric
(*c*) A partial ordering (*d*) None of these
5. *R* is transitive if
(*a*) $a\,R\,b$ and $b\,R\,c \Rightarrow a\,R\,c$ (*b*) $c\,R\,a$ and $b\,R\,a \Rightarrow c\,R\,b$
(*c*) $a\,R\,c$ and $b\,R\,c \Rightarrow a\,R\,c$ (*d*) None of these
6. Let $X = \{1, 2, 3, 4\}$ and $R = \{(x, y)\} : x > y\}$ the which will be ordered pairs of *R*
(*a*) {(1, 2), (1, 3), (2, 3), (1, 4), (2, 4), (3, 4)}
(*b*) {(2, 1), (3, 1), (3, 2), (4, 1), (4, 2), (4, 3)}
(*c*) {(1, 2), (1, 3), (1, 4), (2, 3), (2, 4), (3, 3), (3, 4)}
(*d*) None of these.

7. If R^{-1} and 5^{-1} are the inverse of the relation R and S respectively, then which is true.
   (*a*) $(R_0 S)^{-1} = R^{-1}{}_0 S^{-1}$
   (*b*) $(R_0 S)^{-1} = S^{-1}{}_0 R^{-1}$
   (*c*) Both are true
   (*d*) None of these
8. Transitive closure of R can be given by:
   (*a*) $R^+ = R \cup R^{-1}$
   (*b*) $R^+ = R \cup R^2 \cup R^3 \ldots R^4$
   (*c*) $R^+ = R \cup \Delta$
   (*d*) None of these.
9. If $f(x) = x^2$ and $g(x) = \sin(x)$ then the value of go $f(x)$ is:
   (*a*) $\sin x^2$
   (*b*) $\sin^2 x$
   (*c*) $x^{\sin x}$
   (*d*) None of these
10. A function f : $R \rightarrow R$ given by $f(x) = e^x$ is:
   (*a*) one - one
   (*b*) one-one into
   (*c*) one-one onto
   (*d*) None of these.

Answers

1. (*c*)	2. (*a*)	3. (*b*)	4. (*c*)	5. (*a*)	6. (*b*)
7. (*b*)	8. (*b*)	9. (*a*)	10. (*b*)		

❐❐❐

CHAPTER 6

LATTICES

This Chapter Covers :

6.1. INTRODUCTION

In the previous chapter we have discussed various types of relations that can be defined on a set. In particular we have done partial order relations which will finally lead us to the difinition of lattice. But before that we need to strengthen our base about the elements of poset.

In the next section we will discuss bounds on a poset and the basic theorems which lead to the definition of lattices.

6.2. BOUNDS OF A POSET

A non-empty set A together with the binary relation R is said to be poset or partially ordered set if R is reflexive, Anti-symmetric and transitive. For instance, the set of positive integers forms a poset under the usual R. Each poset can be represented by the help of diagram called Hasse diagram. It is to be noted that in some books author uses the relation '$\leq$' instead of 'R'.

EXAMPLE 6.1. If $X = \{1, 2, 3\}$, then power set of X, the power set, $P(X) = \{\phi, \{1\}, \{2\}, \{3\}, \{1, 2\}, \{1, 3\}, \{2, 3\}, \{1, 2, 3\}\}$ under $\subseteq$ relation is given by the fig 6.1.

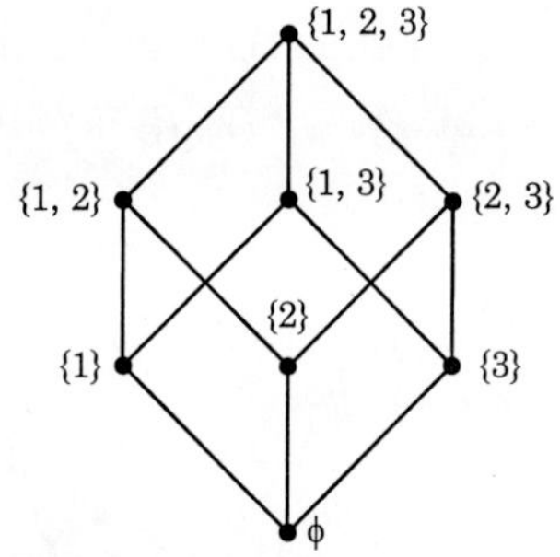

Fig. 6.1.

Now we discuss the bounds of elements of a set.

Lower bound. Let (X, R) be a poset. Let $a, b \in X$ then $c \in X$ is said to be lower bound of a, b if $c\ R\ a$ and $c\ R\ b$ *For example* In the example 6.1, ϕ is the lower bound of {1}, {2}etc.

Upper bound. Let $a, b \in X$ where X is a poset under the relation R then an element $d \in X$ is called upper bound of a and b if $a\ R\ d$ and $b\ R\ d$.

In the example 6.1, {1} is the upper bound of ϕ, {1, 2} is the upper bound of {1} and {2}.

Greatest Lower Bound (*GLB*) or Infimum. An element $g \in X$ is called the greatest lower bound (*GLB*) of $a, b \in X$ if and only if

(*i*) $g\ Ra$ and gRb *i.e.*, g is the lower bound of a and b.

(*ii*) If there exists one element $g' \in L$ such that g' is also a lower bound of a and b then $g'\ Rg$ that is if g' is also a lower bound of a,b then g' is the lower bound of g also *i.e.*, g is the greatest of all lower bounds of a and b.

EXAMPLE 6.2. In the example 6.1 the lower bound of {1, 2} are {1, 2}, {2}, ϕ and lower bounds of {2, 3} are {2, 3}, {2}, ϕ,.Therefore the lower bounds of {1, 2} and {2, 3} are {2} and ϕ. and the greatest lower bound of {1, 2} and {2, 3} is {2}.

Least Upper Bound (LUB) or Supremum

Let (X, R) be a poset. An element $l \in X$ is called least upper bound of a and b in X if and only if.

(*i*) aRl and bRl *i.e.*, l is the upper bound of a and b.

(*ii*) If $\exists$ an element $l' \in X$ such that aRl' and bRl' then lRl'.

That is if l' is another upper bound of a and b then l' is also the upper bound of l. Thus l is the least upper bound of a and b.

EXAMPLE 6.3. In the example 6.1 consider the elements {1} and {3}.

Upper bounds of {1} are {1}, {1, 2}, {1, 3}, {1, 2, 3}

Upper bounds of {3} are {3}, {2, 3}, {1, 3}, {1, 2, 3}

Therefore the upper bounds of {1} and {3} together are {1, 3} and {1, 2, 3}.

Then least upper bound of {1} and {3} is {1, 3}.

Greatest Element. An element m in a poset (X,R) is called the greatest (maximal) element if $\forall\ a \in X$, aRm.

EXAMPLE 6.4. In the example 6.1 the greatest element is {1, 2, 3}.

Least element. An element l in a poset (X, R) is called the least (minimal) element if $\forall\ a \in X$, lRa.

EXAMPLE 6.5. In the example 6.1 the ϕ is the least element.

EXAMPLE 6.6. Let $X = \{1, 3, 5, 7, 15, 21, 35, 105\}$ and let R be the relation '/ ' (divides) on the set X then X is the poset.

To determine the $L.U.B$ of 3 and 7. We look for all $x \in X$ such that $3|x$ and $7|x$ Both $x = 21$ and $x = 105$ work, This implies 21 and 105 are the upper bounds [see fig. 6.2] of 3 and 7 both.

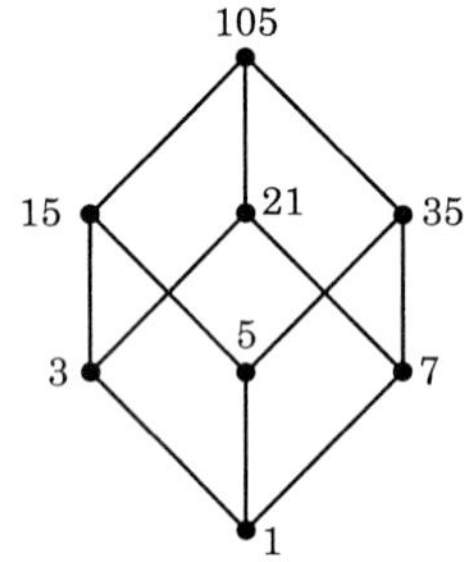

Fig. 6.2.

Now 21/105 implies 21 is the $l.u.b$ of (3, 5)
Similarly LUB of 3 and 5 is 15,
The greatest element of X is 105 as $a/105\ \forall\ a \in X$
The least element of X is 1 as $1/a\ \forall\ a \in X$
[Readers are required to find the GLB and LUB of 15 and 35].

EXAMPLE 6.7. $D_n = \{x : x \mid n\ \forall\ x \in N\}$

Consider $D_{30} = \{1,2,3,5,6,10,15,30\}$

(*a*) Find all the lower bounds of 10 and 15.
(*b*) Determine GLB of 10 and 15.
(*c*) Find all the upper bounds of 10 and 15.
(*d*) Determine LUB of 10 and 15.
(*e*) Find greatest element of D_{30}.
(*f*) Find least element of D_{30}.

Solution : See Fig. 6.3.

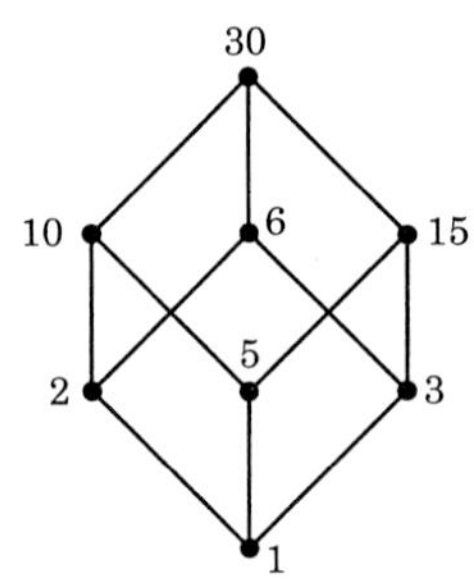

Fig. 6.3.

(*a*) The lower bounds of 10 are 10, 2, 5, 1.
The lower bounds of 15 are 15, 5, 3, 1.
Therefore, the lower bounds of 10 & 15 are 5, 1.

(*b*) The greatest lower bound of 10 & 15 are 5.

(*c*) Upper bounds of 10 are 10, 30.
Upper bounds of 15 are 15, 30.
Therefore, the upper bounds of 10 & 15 are 30.

(*d*) As there is only one upper bound of 10 and 15 which is the least upper bound.

(*e*) The greatest element of D_{30} is 30.

(*f*) The least element of D_{30} is 1.

Join and Meet in a Poset

Let X be a poset under an ordering '≤' or R. Let $a, b \in X$ we define
Join $(a, b) = a \vee b = LUB$ of a,b
Meat $(a, b) = a \wedge b = GLB$ of a, b

Theorem 6.1. Let X be a poset under the relation R and let a and b be the elements in X then

(*a*) If a and b have LUB then this LUB is unique.
(*b*) If a and b have GLB then this GLB is unique.

Proof. (*a*) Let l_1 and l_2 be the two LUB of a and b in X then by definition of LUB

$a \leq l_1$ and $b \leq l_1$ (as l_1 and l_2 are upper bounds of a and b)
$a \leq l_2$ and $b \leq l_2$
Further since l_1 is the LUB and l_2 is the upper bound then $l_1 \; R \; l_2$ ...(*i*)
Again since l_2 is the LUB and l_1 is the upper bound then $l_2 \; R \; l_1$ ...(*ii*)
then (1) and (2) implies $l_1 = l_2$.
Similarly part (*b*) of the theorem can be proved.

Theorem 6.2. Both the join and meet operations are commutative.

Proof. The proof of the theorem follows from the definition of LUB and GLB.

Theorem 6.3. Both the join and meat operations are associative.

Proof. Let X be a poset and let $a, b, c \in X$.
We need to show that $a \vee (b \vee c) = (a \vee b) \vee c$
Let $a \vee (b \vee c) = t$ and $(a \vee b) \vee c = r$
As t is the join of a and $(b \vee c)$
$\Rightarrow a \leq_R t$ and $b \vee c \leq_R t$.
As $b \vee c$ is the LUB of b and c therefore $b \leq_R b \vee c$ and $c \leq_R b \vee c$
$\Rightarrow b \leq_R t$ and $c \leq_R t$
As $a \vee b$ is the LUB of a and b and $a \leq_R t$ and $b \leq_R t$
$\Rightarrow a \vee b \leq_R t$
which together with $c \leq_R t$ leads to $(a \vee b) \vee c \leq_R t$
Therefore, we have $r \leq t$

Conversely. Now we show $t \leq r$.

As r is the join of $(a \vee b)$ and c

$\Rightarrow (a \vee b \leq_R r$ and $c \leq_R r$.

As $a \vee b$ is the LUB of a and b

Therefore, $a \leq_R a \vee b$ and $b \leq_R a \vee b$.

$\Rightarrow a \leq r$ and $b \leq r$.

As $b \vee c$ is the LUB of b and c and $b \leq_R r$, and $c \leq_R r$

$\Rightarrow b \vee c \leq_R r$.

which together with $a \leq_R r$ leads to $a \vee (b \vee c) \leq_R r$.

$\Rightarrow t \leq_R r$

$\therefore \quad t = r$

$\Rightarrow$ Associatively law holds.

Similarly the result for the meat operation holds.

Theorem 6.4. For every a in the poset (X, R)

$a \vee a = a$ and $a \wedge a = a$.

Readers are required to do the proof

[Hint use the definition of meet and join in lattice of GLB and LUB].

Theroem 6.5. For any a and b in A,

$$a \vee (a \wedge b) = a$$
$$a \wedge (a \vee b) = a$$

Proof. Since $a \vee (a \wedge b)$ is the join of a and $(a \wedge b)$,

means $a \vee (a \wedge b)$ is the LUB of a and $(a \wedge b)$.

that is $a \vee (a \wedge b)$ is the upper bound of a & $(a \wedge b)$

$\Rightarrow \qquad a \leq_R a \vee (a \wedge b)$ ...(1)

Since $a \leq_R a$ and $a \wedge b \leq_R a$

[*i.e.*, a is the upper bound of a and $a \wedge b$]

therefore $a \vee (a \wedge b) \leq_R a \vee a = a$

$\Rightarrow a \vee (a \wedge b) \leq_R a$ ...(2)

therefore (1) and (2) implies.

$$a = a \vee (a \wedge b)$$

Again by principle of duality

$$a = a \wedge (a \vee b).$$

6.3. LATTICE

In the previous section we have discussed the properties of join and meet as LUB and GLB respectively. Now we define a particular type of poset called lattice.

A lattice is a poset $(L, \leq_R)$ in which every subset $\{a,b\}$ consisting of two elements has a LUB and GLB in L and is denoted by $(L, \leq_R, \vee, \wedge)$.

$$LUB\ \{a,b\} = a \vee b$$
$$GLB\ \{a,b\} = a \wedge \text{b}$$

EXAMPLE 6.8. Let S be any set, then $L = P(S)$ under '$\subseteq$' as a partial order relation is a lattice where $A \vee B = A \cup B$ and $A \wedge B = A \cap B$, For instance let $S = \{1,2,3\}$

$P(S) = \{\phi,\{1\},\{2\}, \{3\}, \{1,2\}, \{1,3\}, \{2,3\},\{1,2,3\}\}$

then $(P(s), \subseteq, \cup, \cap)$ is a lattice. [Discussed the fig in Example 6.1]

EXAMPLE 6.9. Consider $(Z^+, \leq_R)$ where $a, b \in Z^+$ and $a \leq_R b$ if $a|b$ then $(Z^+, \leq_R, \vee, \wedge)$ is a lattice where

$$a \vee b = LCM(a,b)$$
$$a \wedge b = GCD(a,b)$$

EXAMPLE 6.10. consider the following fig. 6.4 of posets and check whether they are lattice or not.

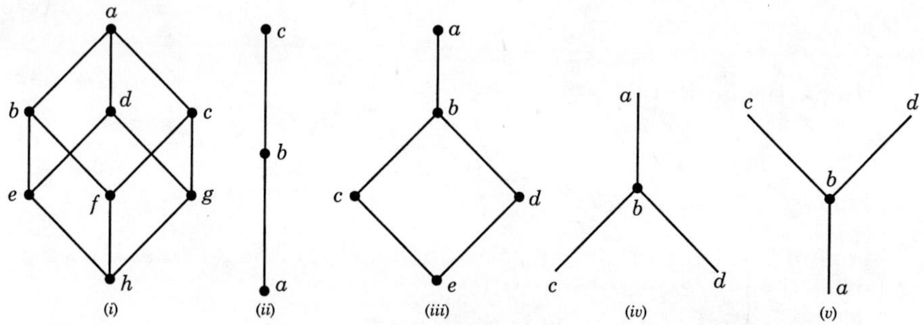

Fig. 6.4.

Solution : The posets given by figure (*i*), (*ii*) & (*iii*) are lattices whereas (*iv*) and (*v*) are not lattices.

EXAMPLE 6.11. Example of a poset which is not a lattice.

The poset {2, 3, 4, 6} under divisibility is not a lattice as $4 \vee 6$ does not exist.

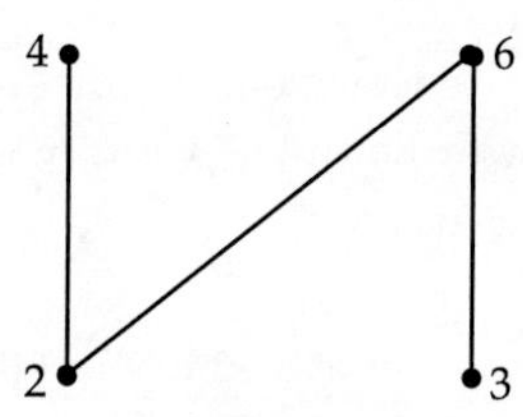

Fig. 6.5.

EXAMPLE 6.12. D_n where n is a positive integer is always be a lattice under the relation divisibility.

Theorem 6.6. If L is any lattice, then for any $a,b,c \in L$ the following holds.

(*i*) $a \wedge a = a$ Idempotunt law

$a \vee a = a$

(*ii*) $a \wedge (b \wedge c) = (a \wedge b) \wedge c$ Associativity law

$a \vee (b \vee c) = (a \vee b) \vee c$

(*iii*) $a \wedge b \leq_R a, \ a \wedge b \leq_R b$

$a \leq_R a \vee b, \ b \leq_R a \vee b$

(iv) If $a \leq b \Leftrightarrow a \wedge b = a$

$\Leftrightarrow a \vee b = b$

(v) If 0 is the universal lower bound and 1 is the universal upper bound then

$$0 \wedge a = 0, \ 0 \vee a = a$$
$$1 \wedge a = a, \ 1 \vee a = 1$$

(vi) $a \wedge (a \vee b) = a$

$a \vee (a \wedge b) = a$ Absorption property

(vii) $a \leq_R b$ and $c \leq_R d$

$\Rightarrow a \wedge c \leq_R b \wedge d$

$\Rightarrow a \vee c \leq_R b \vee d$

In particular if

$a \leq b$

$\Rightarrow$ $a \wedge x \leq b \wedge x \ \forall x \in L$

and $a \vee x \leq b \vee x$

Theorem 6.7. Let $(L, \leq_R)$ be a poset and let $(L, \geq_R)$ be its dual. If $(L, \leq_R)$ is a lattice then $(L, \geq_R)$ is also a lattice.

That is, **dual of a lattice is a lattice.**

Proof. Let $x, y \in L$, as L is a lattice therefore

Supremum $\{x,y\}$ exists in L

Supremum $\{x, y\} = x \vee y$, that is $x \vee y$ is the LUB of x and y in L then

$x \leq_R x \vee y$ and $y \leq_R x \vee y$

$\Rightarrow x \vee y \geq_{R'} x$ and $x \vee y \geq_{R'} y$

$\Rightarrow x \vee y$ is the lower bound of x and y in $(L, \geq_{R'})$.

Now need to show $x \vee y$ is the greatest lower bound of x and y in $(L, \geq_{R'})$.

Let $z \in L$ is another lower bound of x and y in $(L, \geq_{R'})$ then

$z \geq_{R'} x$ and $z \geq_{R'} y$

$\Rightarrow x \leq_R z$ and $y \leq_R z$

implies z is the upper bound of x and y in $(L, \leq_R)$.

As $(L, \leq_R)$ is a lattice therefore $x \vee y$ exists.

therefore $x \vee y \leq_R z$ [As $x \vee y$ is the LUB]

$\Rightarrow z \leq_R x \vee y$

As z is asbitrary therefore $x \vee y$ is the greatest lower bound of x and y in $(L, \geq_R)$.

EXAMPLE 6.13. If $(L, \subset, \cup, \cap)$ is a lattice then $(L, \supseteq, \cap, \cup)$ is also a lattice.

Solution : Let $S = \{a, b, c\}$ and

$P(s)$ = Power sets of S

$= \{\phi, \{a\}, \{b\}, \{c\}, \{a, b\}, \{a, c\}, \{b, c\}, \{a, b, c\}\}$

is a Lattice $(L, \subseteq, \cup, \cap)$ given by the figure 6.6.

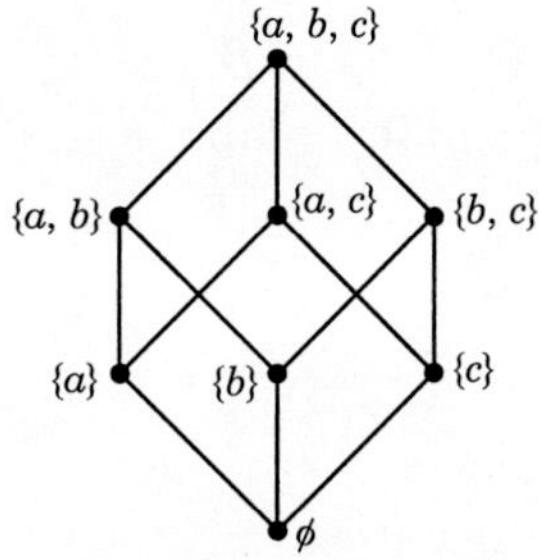

Fig. 6.6.

Taking its reverse image *i.e.* 'R', '$\supseteq$' superset.

Consider the $(L, \supseteq, \cup, \cap)$

In that case $\{a, b, c\}$ will be the least element and ϕ is the greatest element.

That is by taking reverse image

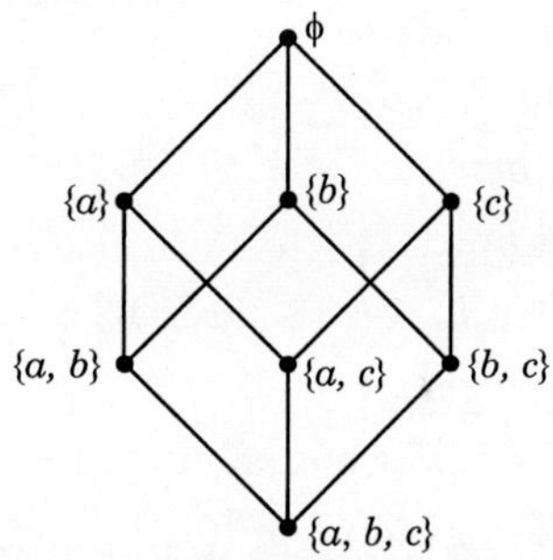

Fig. 6.7.

By figure it is easy to verify that If $(L, \supseteq, \cup, \cap)$ is also a lattice.

Theorem 6.8. If $(L_1, \leq_R)$ and $(L_2, \leq_R)$ are two lattices then $(L_1 \times L_2, \leq_R)$ is also a lattice.

That is **product of two lattices is a lattice.**

Proof. Let L_1 and L_2 be two lattices implies L_1 and L_2 are posets.

$L_1 \times L_2 = \{(a, b) \mid a \in L_1 \text{ and } b \in L_2\}$ is a poset under the relation $\leq_R$ defined by:

$$(a_1,b_1) \leq_R (a_2,b_2)$$
$$\Leftrightarrow a_1 \leq_R a_2 \text{ in } L_1 \text{ and } b_1 \leq_R b_2 \text{ in } L_2$$

Reflexivity :

$(a,b) \leq_R (a, b)$ as $a\ R\ a$ in L_1 as $b\ R\ b$ in L_2

reflexivity holds in L_1 and L_2 both.

Antisymmetry :

If $(a_1, b_1)\ R\ (a_2, b_2)$ then $(a_2, b_2)\ R\ (a_1, b_1)$

If $(a_1, b_1)\ \not R\ (a_2, b_2)$

then

$(a_1, b_1) \neq (a_2, b_2)$

Transitivity :

Let $(a, b) \leq_R (c, d)$ and $(c, d) \leq_R (e, f)$

$\Rightarrow a \leq_R e$ and in L_1 and $b \leq_R f$ in L_2

As transitivity holds in L_1 and L_2 both.

$\Rightarrow a \leq_R c$ and $c \leq_R e$ in L_1

and $b \leq_R d$ and $d \leq_R f$ in L_2

$\Rightarrow (a, b) \leq_R (e, f)$

$\Rightarrow L_1 \times L_2$ is a poset.

Now we need to show that *$L_1 \times L_2$ form a lattice.*

Let (a_1,b_1) and (a_2,b_2) are any two elements of $L_1 \times L_2$ then $a_1, a_2 \in L_1$ and $b_1,b_2 \in L_2$.

As L_1 and L_2 are lattices then $\{a_1,a_2\}$ and $\{b_1,b_2\}$ have *LUB* and *GLB* in L_1 and L_2.

Let $a_1 \wedge a_2 = \text{Inf } \{a_1,a_2\}$

$\Rightarrow a_1 \wedge a_2 \leq_R a_1$ and $a_1 \wedge a_2 \leq_R a_2$

Let $b_1 \wedge b_2 = \text{Inf } \{b_1,b_2\}$

$\Rightarrow b_1 \wedge b_2 \leq_R b_1$ and $b_1 \wedge b_2 \leq_R b_2$

$\Rightarrow (a_1 \wedge a_2, b_1 \wedge b_2) \leq_R (a_1,b_1)$

and$(a_1 \wedge a_2, b_1 \wedge b_2) \leq_R (a_2,b_2)$

$\Rightarrow (a_1 \wedge a_2, b_1 \wedge b_2)$ is the lower bound of $\{(a_1, b_1), (a_2,b_2)\}$ in $L_1 \times L_2$.

Let (c, d) is any other lower bound of $\{(a_1,b_1), (a_2,b_2)\}$ in $L_1 \times L_2$.

$\Rightarrow (c, d) \leq_R (a_1,b_1)$ and $(c, d) \leq_R (a_2, b_2)$

$\Rightarrow c \leq_R a_1$ and $c \leq_R a_2$

$\Rightarrow c$ is the lower bound of $\{a_1, a_2\}$ in L_1

$d \leq_R b_1$ and $d \leq_R b_2 \Rightarrow d$ is the lower bound of $\{b_1,b_2\}$ in L_2.

$\Rightarrow c \leq_R a_1 \wedge a_2 = \text{Inf}\{ a_1,a_2\}$ in L_1

and $d \leq_R b_1 \wedge b_2 = \text{Inf } \{b_1,b_2\}$ in L_2

$\Rightarrow (c, d) \leq (a_1 \wedge a_2 , b_1 \wedge b_2\}$ in $L_1 \times L_2$

As (c, d) is arbitrary

$\Rightarrow (a_1 \wedge a_2, b_1 \wedge b_2)$ is the *GLB* of $\{(a_1,b_1)\ (a_2,b_2)\}$ in $L_1 \times L_2$

Similarly by duality

$(a_1 \vee a_2, b_1 \vee b_2)$ is the *LUB* of $\{(a_1, b_1)\ (a_2,b_2)\}$ in $L_1 \times L_2$.

$\Rightarrow L_1 \times L_2$ is a lattice with

LUB of$\{(a_1,b_1), (a_2,b_2)\} = (a_1 \vee a_2, b_1 \vee b_2)$

GLB $\{(a_1,b,), (a_2, b_2)\} = (a_1 \wedge a_2, b_1 \wedge b_2)$

EXAMPLE 6.14. Let L_1 and L_2 be lattices as shown in the fig 6.8 then $L = L_1 \times L_2$ is also a lattice.

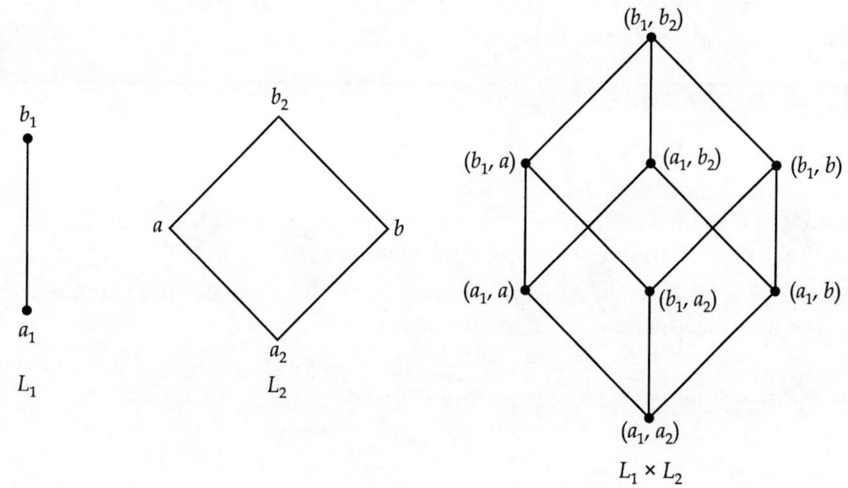

Fig. 6.8.

Theorem 6.9. In any lattice the distributive inequalities hold. That is If $(L, \leq)$ is a lattice then for $a, b, c \in L$:

(*i*) $a \wedge (b \vee c) \geq (a \wedge b) \vee (a \wedge c)$

(*ii*) $a \vee (b \wedge c) \geq (a \vee b) \wedge (a \vee c)$

hold for any a, b, c.

Proof. (*i*) $a \wedge b \leq a$ and $a \wedge b \leq b$ [As $a \wedge b$ is the *GLB* of a and b]

Now $a \wedge b \leq b \leq b \vee c$ [as $b \vee c$ is the upper bound of b and c]

$\Rightarrow a \wedge b$ is lower bound of $\{a, b \vee c\}$

$\Rightarrow a \wedge b \leq a \wedge (b \vee c)$ ...(*i*)

[As $a \wedge (b \vee c)$ is the *GLB* $\{a, b \vee c\}$].

similarly $a \wedge c \leq a \wedge (b \vee c)$ ...(*ii*)

(*i*) and (*ii*) implies that $a \wedge (b \vee c)$ is an upper bound of $\{a \wedge b, a \wedge c\}$

$\Rightarrow (a \wedge b) \vee (a \wedge c) \leq a \wedge (b \vee c)$

[As $(a \wedge b) \vee (a \wedge c)$ is the *LUB* of $(a \wedge b)$ and $(a \wedge c)$

Second inequality can be proved by duality.

Theorem 6.10. In any lattice L

$$(x \wedge y) \vee (y \wedge z) \vee (z \wedge x) \leq (x \vee y) \wedge (y \vee z) \wedge (z \vee x) \quad \forall\, x, y, z \in L.$$

Proof. As $x \wedge y \leq x \leq x \vee y$

$x \wedge y \leq y \leq y \vee z$

$x \wedge y \leq x \leq z \vee x$

$\Rightarrow x \wedge y$ is the lower bound of $x \vee y$, $y \vee z$, $z \vee x$

$\Rightarrow x \wedge y \leq (x \vee y) \wedge (y \vee z) \wedge (z \vee x)$

[as *R.H.S.* is the *GLB* of $x \vee y$, $y \vee z$, $z \vee x$]

Similarly $y \wedge z \leq (x \vee y) \wedge (y \vee z) \wedge (z \vee x)$

Similarly $z \wedge x \leq (x \vee y) \wedge (y \vee z) \wedge (z \vee x)$

Implies $(x \vee y) \wedge (y \vee z) \wedge (z \vee x)$ is the upper bound of $x \wedge y$, $y \wedge z$, $z \wedge x$.

$\Rightarrow (x \wedge y) \vee (y \wedge z) \vee (z \wedge x) \leq (x \vee y) \wedge (y \vee z) \wedge (z \vee x)$
[as L.H.S. is the LUB of $x \wedge y$, $y \wedge z$, $z \wedge x$]

6.4. SUBLATTICES

A non-empty subset K of a lattice L is called a sublattice of L if $a, b \in K$
$\Rightarrow a \wedge b$ and $a \vee b \in K$

EXAMPLE 6.15. The lattice D_n of all positive divisors of $n \in N$ is a sublattice of the lattice z^+ under the relation divisibility.

EXAMLE 6.16. Let $(z^+, /)$ is a lattice then $X = \{2,3,4,6\}$ under the same relation is not a sublattice as $4 \vee 6$ does not exist in X.

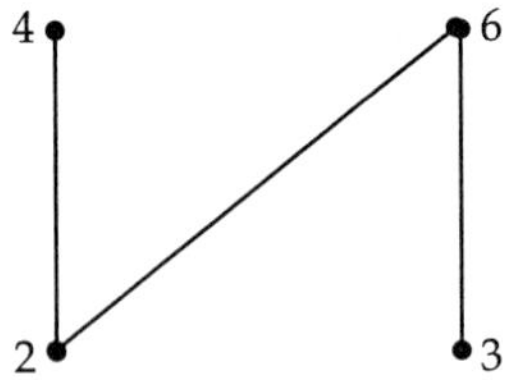

Fig. 6.9.

Theorem 6.11. Intersection of two sublattices is a sublattice.

Proof. Let L be any lattice under the relation R.

Let K_1 and K_2 are sublattices of L.

We need to show that $K_1 \cap K_2$ is a sublattice of L.

Let $x, y \in K_1 \cap K_2$

$\Rightarrow x, y \in K_1$ and $x, y \in K_2$ as K_1 and K_2 are sublattices.

Therefore, $x \wedge y \in K_1$ and $x \vee y \in K_1$

Similarly, $x \wedge y \in K_2$ and $x \vee y \in K_2$

Therefore, by definition of intersection

$$x \wedge y \in K_1 \cap K_2 \qquad x \vee y \in K_1 \cap K_2$$

As K_1 and K_2 are arbitrary. Therefore, $K_1 \cap K_2$ is a sublattice.

Theorem 6.12. Union of two sublattices may not be a sublattice.

Proof. We will prove the theorem by taking an example.

Consider the lattice

$L = \{1, 2, 3, 5, 6, 10, 15, 30\}$ factors of 30 under divisibility.

then $K_1 = \{1,2\}$ and $K_2 = \{1,5\}$ are sublattices of L.

But $K_1 \cup K_2 = \{1, 2, 5\}$ is not a sublattice

As $2,5 \in K_1 \cup K_2$

But $2 \vee 5 = 10 \notin K_1 \cup K_2$.

Theorem 6.13. Let L be a lattice then for every a and b in L

(*i*) $a \vee b = b$ if and only if $a \leq b$.

(*ii*) $a \wedge b = a$ if and only if $a \leq b$.

Proof. (*i*) Let $a \vee b = b$

$a \leq a \vee b$ [As $a \vee b$ is the upper bound of a and b]
but $a \vee b = b$ (given)
$\Rightarrow a \leq b$
Conversely,
Let $\quad a \leq b$
$\quad a \vee b \leq a \leq b$ [As $a \vee b$ is the $L \cup B$ of a and b]
Also, $\quad b \leq a \vee b$
[As $a \vee b$ is the upper bound of b]
$\Rightarrow \quad a \vee b = b$

Second part of the theorem can be proved similarly.

Theorem 6.14. Let L be a lattice then for all a,b,c, d in L
$(a \wedge b) \vee (c \wedge d) \leq (a \vee c) \wedge (b \vee d)$ holds true in L.

6.5. SPECIAL TYPES OF LATTICES

There are the lattices which have some extra (additional) properties. By using these special type of lattices we may then define an algebraic system which is even more structured.

1. Complete Lattice

A lattice is said to be complete if its each non-empty subsets possesses a Least Upper Bound (LUB) and a Greatest Lower Bound (GLB).Therefore it is to be noted that every finite lattice must be complete which must have a least element (0) and a greatest element (1) and these two elements 0 and 1 are called **universal lower bound** and **universal upper bound** respectively.

If $(L, R, \wedge, \vee)$ be the lattice ; $L = \{a, a_2, ---, a_n\}$ then

$$\bigwedge_{i=1}^{n} a_i = 0 \quad ; \quad \bigvee_{i=1}^{n} a_i = 1$$

The universal upper bound and universal lower bound have the following properties

$$a \vee 0 = a, \qquad a \wedge 0 = 0$$
$$a \vee 1 = 1, \qquad a \wedge 1 = a$$

The real interval $[0,1]$ with usual $\leq$ forms a complete lattice.
The lattice $(z, \leq)$ is not a complete lattice as the subset $S = \{x \in z \mid x > 0\}$ does not have an upper bound in z.

2. Bounded Lattice

A lattice L is called a bounded lattice if it has a least element 0 and a greatest element 1.

EXAMPLE 6.17. Let Z be the set of all positive integers and let $\{Z, R\}$ be a lattice under the partial order relation of divisibilily is not a bounded lattice as it does not possess a greatest element, although, it has the least element 1.

Theorem. 6.15. Dual of a complete lattice is complete.
(Readers are required to prove the above theorem).

EXAMPLE 6.18. The lattice $(z, \leq)$ under the partial order relation "$\leq$" is not bounded as it has neither a least element nor a greatest element.

EXAMPLE 6.19. The lattice $(L, \subseteq)$ where $L = P(s)$ = power set of the set $S = \{a,b,c\}$ is a bounded lattice. Its least element is ϕ and its greatest element is S.

Complement of an Element in a Lattice

Let L be a bounded lattice $(L, \wedge, \vee, 0,1)$ and let $a \in L$. The element $a' \in L$ is called complement of a if

$$a \vee a' = 1 \qquad \text{and} \qquad a \wedge a' = 0$$

For example, in the fig. 6.10 complement of $\{1\}$ is $\{2\}$ and complement of $\{2\}$ is $\{1\}$ and complement ϕ is $\{1, 2\}$ and complement of $\{1, 2\}$ is ϕ.

3. Complemented Lattice

A lattice $\{L, R, \wedge,\vee,0,1\}$ is called complemented if it is bounded and if every element in L has at least one complement, therefore the operation complement exists in L. Thus in notation we can write complemented lattice as $\{L, R, \wedge, \vee, 0, 1, '\}$

For example $\{ P(s), \subseteq\}$ where $S = \{1,2\}$ is the complemented lattice.

$$P(s) = \{\phi, \{1\}, \{2\}, \{1,2\}\}$$

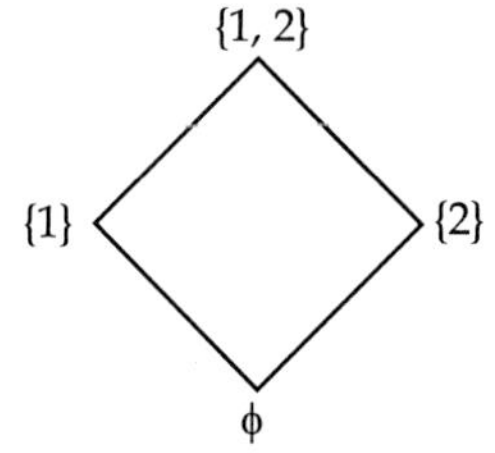

Fig. 6.10.

EXAMPLE 6.20. Consider the following fig 6.11.

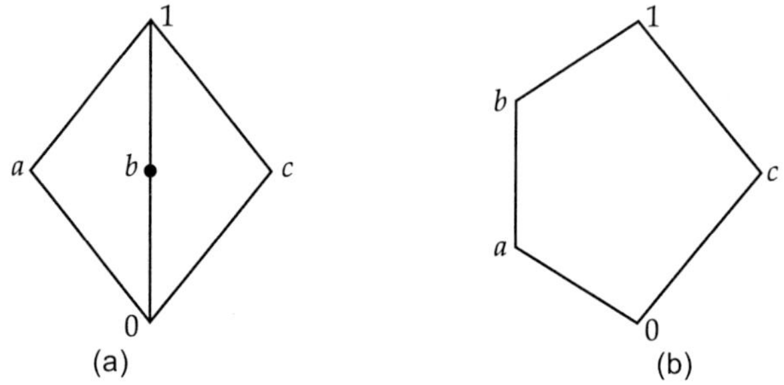

Fig. 6.11.

In fig. 6.11(a) complement of b is a and c both as:

$b \vee a = 1$, $b \wedge a = 0$ and

$b \vee c = 1$, $b \wedge c = 0$

In fig 6.11 (b) the element c has 2 complements a and b as:

$$c \vee a = 1, \qquad c \wedge a = 0$$
$$c \vee b = 1, \qquad c \wedge b = 0$$

EXAMPLE 6.21. Let L be the lattice given by the following fig 6.12. Check whether the following lattice is a complemented lattice or not.

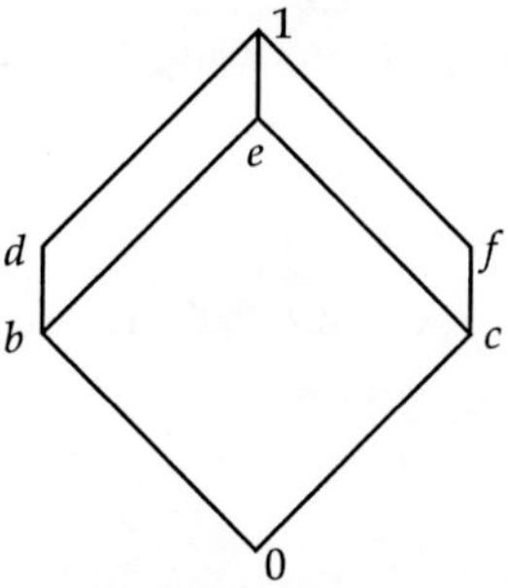

Fig. 6.12.

Solution : A lattice is complemented if every element in it has complement.

0 & 1 are complements of each other

as $0 \vee 1 = 1$

$0 \wedge 1 = 1$

Similarly, b & f are complements of each other.

d & c are complements of each other.

e & a are complements of each other.

Therefore, every element 0, a, b, c, d, f, e, 1 has complements in L.

Therefore, L is complemented lattice.

Theorem 6.16. Two bounded lattices A and B are complemented if and only if $A \times B$ is complemented.

Proof. Let A and B be two complemented lattices and suppose 0 & 1 and $0'$ & $1'$ are the universal bounds of A and B respectively.

Then $(0,0')$ and $(1,1')$ will be least and greatest elements of $A \times B$ (Discussed earlier).

Let $(a,b) \in A \times B$ be any element then $a \in A$ and $b \in B$.

As A and B are complemented lattices therefore, $\exists\ a' \in A$ and $b' \in B$ such that

$$a \wedge a' = 0 \text{ and } a \vee a' = 1$$
$$b \wedge b' = 0' \text{ and } b \vee b' = 1'$$

Now
$$(a,b) \wedge (a',b') = (a \wedge a', b \wedge b') = (0,0')$$
$$(a,b) \vee (a', b') = (a \vee a', b \vee b') = (1,1')$$

this implies (a',b') is complement of (a, b) in $A \times B$. As (a, b) is arbitrary in $A \times B$ therefore $A \times B$ is complemented.

Conversely Let $A \times B$ is complemented

Let $(a,b) \in A \times B$ then $\exists\ (a',b') \in A \times B$ such that

$$(a,b) \wedge (a', b') = (0,0')$$
$$(a,b) \vee (a',b') = (1,1')$$
$$\Rightarrow \quad (a \wedge a', b \wedge b') = (0,0')$$
$$(a \vee a', b \wedge b') = (1,1')$$

$\Rightarrow a \wedge a' = 0$ and $b \wedge b' = 0'$

$a \vee a' = 1$ and $b \vee b' = 1$

$\Rightarrow$ a' is complement of a in A

and b' is complement of b in B

As a and b are arbitrary in A and B respectively.

Therefore, A and B are complemented.

EXAMPLE 6.22. Consider the example 6.14. and verify the above theorem.

Theorem 6.17. Dual of a complemented lattice is complemented.

Proof. Let (L,R) be a complemented lattice with 0 and 1 as least and greatest elements respectively.

Let (L,R') be the dual of (L,R) then 1 and 0 are its least and greatest elements respectively.

Let a $\in L$ be any element.

As (L,R) is complemented $\exists$ $a' \in L$ and such that

$a \wedge a' = 0$ and $a \vee a' = 1$ in L

$\Rightarrow 0 = GLB\ \{a,a'\}$ in L

$\Rightarrow$ o $R\ a$ and o Ra' in (L, R)

$\Rightarrow a\ R'$ o and $a'\ R'$ in (L,R')

$\Rightarrow$ o is the upper bound of $\{a, a'\}$ in $(L,R'\}$

Let $y \in (L, R')$ be another upper bound of $\{a, a'\}$

then $a\ R'\ y$ and $a'\ R'\ y$ in (L,R')

$\Rightarrow y\ R\ a$ and $y\ R\ a'$ in (L, R)

$\Rightarrow y$ is the lower bound of $\{a,a'\}$ in (L, R)

as o is the GLB of $\{a,a'\}$ in (L, R)

$\Rightarrow y\ R$ o in (L, R)

$\Rightarrow$ o $R'\ y$ in (L, R')

As y is arbitrary, o is the least upper bound of $\{a,a'\}$ in (L, R').

Similarly we can prove that

1 is the GLB of $\{a,a'\}$ in (L, R')

$\Rightarrow$ (L, R') is complemented.

4. Iso lattices L_1 and L_2 are called isomorphic lattice if there is a bijection from L_1 to L_2 *i.e.*, $f : L_1 \to L_2$ such that

$$\left.\begin{aligned} f(a \wedge b) &= f(a) \wedge f(b) \\ f(a \vee b) &= f(a) \vee f(b) \end{aligned}\right\} \text{ for every } a, b \in L_1$$

EXAMPLE 6.23. Determine whether the lattice shown are isomorphic

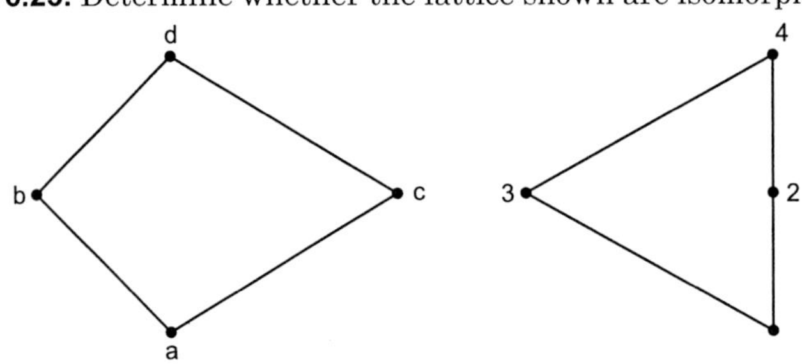

Fig. 6.13.

Solution : Consider the mapping $f = \{(a, 1), (b, 2), (c, 3), (d, 4)\}$

for example $f(b \wedge c) \equiv f(a) \equiv 1$ also we have $f(b) \wedge f(c) \equiv 2 \wedge 3 \equiv 1$

5. Pentagonal lattice: Least element of pantagonal is p and greatest element of pentagonal is q.

as $a \wedge b = p$ and $a \vee b = q$

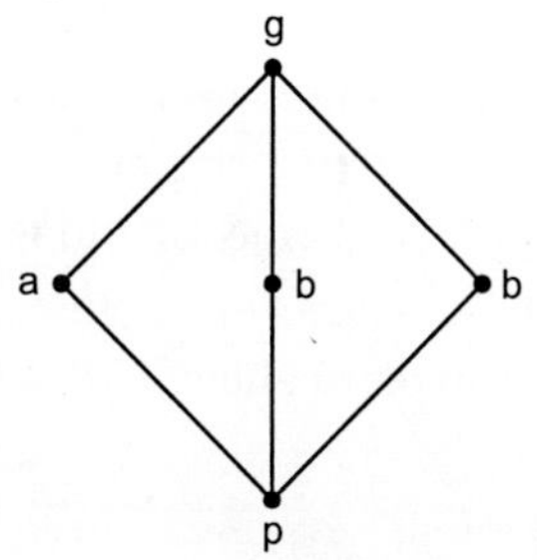

Fig. 6.14. Complementary property

Now as $\bar{b} = a$ and $\bar{a} = b$

$$b \wedge c = p \text{ and } b \vee c = q$$

$\therefore$ $\bar{b} = c$ and $\bar{c} = b$

So b has two complement a and 6, also complement may not be unique.

So we observed that

$$\bar{b} = a \text{ and } \bar{b} = c \text{ but } a \neq c$$

EXAMPLE 6.24. Consider the lattice under divisibility as $D_6 = \{1, 2, 3, 6\}$

Solution: In this least element is 1 and greatest element is 6.

Now,

$$1 \wedge x = 1 \equiv \text{HCF of 1 and } x$$

$$1 \vee x = 6 \equiv \text{LCM of 1 and } x$$

here, $x = 6$ so

$$\left.\begin{array}{l}\bar{1} = 6 \\ \bar{6} = 1\end{array}\right\} \text{ so } \left.\begin{array}{l}1 \wedge 6 = 1 \\ 1 \vee 6 = 6\end{array}\right\}$$

so both are complement of each other.

Now for 2 and 3

$$2 \wedge x = 1 \equiv \text{HCF of 2 and } x$$

$$2 \vee x = 6 \equiv \text{LCM of 2 and } x$$

so both 2 and 3 are complement of each other.

6.6. HOMOMORPHISMS

Let L and M be lattices.

A mapping $\theta : L \to M$ is called meet homomorphisms if

$\theta(a \wedge b) = \theta(a) \wedge \theta(b)$

θ is called join homomorphism if $\theta(a \vee b) = \theta(a) \vee \theta(b)$.

θ is called homomorphisms if θ is both meet as well as join homomorphisms. Thus in the homomorphism both the operations of meet and join are preserved. If the mapping θ is bijective *i.e.*, 1-1 and onto then θ is called isomorphism.

Then $L \cong M$ [L is isomorphic to M].

Theorem 6.18. Any meet (join) homomorphism preserves order. That is if $a \leq b$ then $\theta(a) \leq \theta(b)$ if θ is homomorphism.

Proof. Let $\theta : L \to M$ be a meet homomorphism then for $a, b \in L$,

$$\theta(a \wedge b) = \theta(a) \wedge \theta(b).$$

Let $a \leq b$ in L, Now we need to show $\theta(a) \leq \theta(b)$ in M.

Now $a \leq b$

$$\Rightarrow a = a \wedge b$$

$$\Rightarrow \theta(a) = \theta(a \wedge b) = \theta(a) \wedge \theta(b) \leq \theta(b)$$

$$\Rightarrow \theta(a) \leq \theta(b)$$

Now applying duality result, the result for join operation can be proved.

This leads us to conclude that it is possible to represent two isomorphic lattices by the same diagram by simply replacing the nods with the images.

Theorem 6.19. If L_1, L_2, M_1, M_2 are lattices such that $L_1 \cong M_1$ and $L_2 \cong M_2$ then $L_1 \times L_2 \cong M_1 \times M_2 \cong M_2 \times M_1$

Proof. As $L_1 \cong M_1$ and $L_2 \cong M_2$

Define a mapping $f : L_1 \to M_1$ such that f is 1-1, onto homomorphism.

Define a mapping $g : L_2 \to M_2$ such that g is 1-1, onto homomorphism.

Define $\theta : L_1 \times L_2 \to M_1 \times M_2$

$$\theta(a, b) = (f(a), g(b))$$

then

$$\theta(a, b) = \theta(c, d)$$

$$\Leftrightarrow (f(a), g(b)) = (f(c), g(d))$$

$$\Leftrightarrow f(a) = f(c) \text{ and } g(b) = g(d)$$

$$\Leftrightarrow a = c \text{ and } b = d \text{ as } f \text{ and } g \text{ are 1-1}$$

$$\Leftrightarrow (a, b) = (c, d)$$

$\Rightarrow \theta$ is well defined 1-1 mapping

Ontoness

Let $(m_1\ m_2) \in M_1 \times M_2$

Since $m_1 \in M_1$ and $m_2 \in M_2$ and f and g are onto mapping therefore

$\exists\ l_1 \in L_1$ and $l_2 \in L_2$ such that

$$f(l_1) = m_1 \text{ and } g(l_2) = m_2$$

and

$$\theta((l_1, l_2)) = (f(l_1), g(l_2)) = (m_1, m_2)$$

for any element in $M_1 \times M_2$ we can find corresponding element in $L_1 \times L_2$.

$\Rightarrow \theta$ is onto.

Homomorphism

Let $(a, b), (c, d) \in L_1 \times L_2$

Consider

$$\theta((a, b) \wedge (c, d)) = \theta(a \wedge c, b \wedge d)$$
$$= (f(a \wedge c), g(b \wedge d))$$
$$= (f(a) \wedge f(c), g(b) \wedge g(d)) \text{ [As } f \text{ and } g \text{ are homomorphism]}$$
$$= (f(a), g(b)) \wedge (f(c), g(d))$$
$$= \theta(a, b) \wedge \theta(c, d)$$

$\Rightarrow \theta$ is meet homomorphism.

By applying duality

$$\theta((a, b) \vee (c, d)) = \theta(a, b) \vee \theta(c, d)$$

$\Rightarrow \theta$ is join homomorphism.

$$\Rightarrow L_1 \times L_2 \cong M_1 \times M_2 \quad ...(i)$$

Now to show $M_1 \times M_2 \cong M_2 \times M_1$

Define $\phi : M_1 \times M_2 \to M_2 \times M_1$

$$\phi(x_1, x_2) = (x_2, x_1)$$

Let $\phi(x_1, x_2) = \phi(y_1, y_2)$

$$\Leftrightarrow (x_2, x_1) = (y_2, y_1)$$
$$\Leftrightarrow x_2 = y_2, x_1 = y_1$$
$$\Leftrightarrow (x_1, x_2) = (y_1, y_2)$$

$\Rightarrow \phi$ is well defined, 1-1 mapping

Homomorphism $\phi((x_1,x_2) \wedge (y_1,y_2)) = \phi(x_1 \wedge y_1, x_2 \wedge y_2)$

$$= (x_2 \wedge y_2, x_1 \wedge y_1)$$
$$= (x_2, x_1) \wedge (y_2, y_1)$$
$$= \phi(x_1, x_2) \wedge \phi(y_1, y_2)$$

By applying duality

$$\phi((x_1, x_2) \vee (y_1, y_2)) = \phi(x_1, x_2) \vee \phi(y_1, y_2)$$

$$\Rightarrow M_1 \times M_2 \cong M_2 \times M_1 \quad ...(ii)$$

$\therefore$ (*i*) and (*ii*) $\Rightarrow$

$$L_1 \times L_2 \cong M_1 \times M_2 \cong M_2 \times M_1$$

Theorem 6.20. Homormorphic image of a complemented lattice is complemented.

6.7. DISTRIBUTIVE LATTICE

A lattice $(L, \wedge, \vee)$ is said to be distributive lattice if for any $a, b, c \in L$ it satisfies the following distributive properties

(*i*) $a \wedge (b \vee c) = (a \wedge b) \vee (a \wedge c)$

(*ii*) $a \vee (b \wedge c) = (a \vee b) \wedge (a \vee c)$

That is, in a distributive lattice the meet operation ($\wedge$) distributes over the join operation ($\vee$) and the join operation distributes over the meet operation.

If a lattice is not distributive then it is called **non-distributive**.

EXAMPLE 6.25. Let S be any set $P(S)$ is the power set of S then $(L = P(S), \subseteq, \cup, \cap)$ is the distributive lattice because of $X, Y, Z \in L$

$$X \cap (Y \cup Z) = (X \cap Y) \cup (X \cap Z)$$
$$X \cup (Y \cap Z) = (X \cup Y) \cap (X \cap Z)$$

EXAMPLE 6.26. Check whether the lattices given by the following fig 6.15 (*a*), (*b*), (*c*) are distributed.

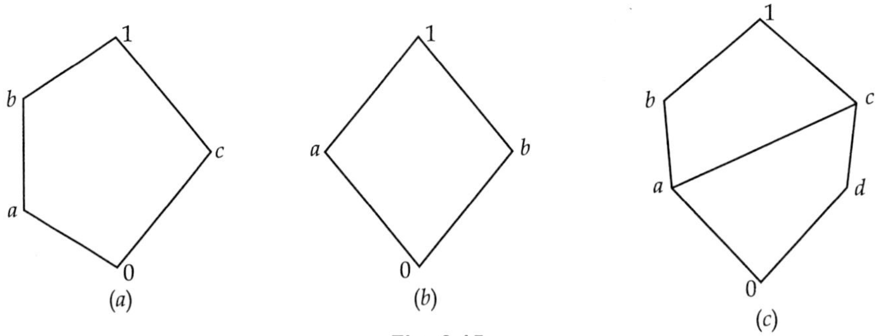

Fig. 6.15.

The lattice given by (*a*) is non distributive as:

$$b \wedge (a \vee c) \neq (b \wedge a) \vee (b \wedge c)$$

Now $b \wedge (a \vee c) = b \wedge 1 = b$

$$(b \wedge a) \vee (b \wedge c) = a \vee 0 = a$$

Similarly,

The lattice given by (*b*) is distributive.

The lattice given by (*c*) is also distributive.

Theorem 6.21.

(*i*) If the meet operation is distributive over the join operation in a lattice, then the join operation is also distributive over the meet operation.

(*ii*) If the join operation is distributive over the meet operation in a lattice then the meet operation is also distributive over the join operation.

Proof. (*i*) Let meet operation is distributive over the join operation in a lattice L and $a, b, c \in L$ such that

$$a \wedge (b \vee c) = (a \wedge b) \vee (a \wedge c)$$

Consider

$$\begin{aligned}(a \vee b) \wedge (a \vee c) &= [(a \vee b) \wedge a] \vee [(a \vee b) \wedge c]\\ &= [(a \wedge a) \vee (a \wedge b)] \vee [(c \wedge a) \vee (b \wedge c)]\\ &= [a \vee (a \wedge b)] \vee [(c \wedge a) \vee (b \wedge c)]\\ &= a \vee (a \wedge c) \vee (b \wedge c)\\ &= a \vee (b \wedge c)\end{aligned}$$

Similarly the 2nd part of the theorem can be proved by using ϕ duality.

Theorem 6.22. If L be a bounded distributive lattice, and an element in L possesses a complement, then this is unique.

Proof. Let $a \in L$ be any element

Let a has two complements say a' and a'' in L.

$$\Rightarrow a \wedge a' = 0 \text{ and } a \vee a' = 1$$

and $a \wedge a'' = 0$ and $a \vee a'' = 1$

Now
$$\begin{aligned} a' &= a' \wedge 1 \\ &= a' \wedge (a \vee a'') && \text{[as } a'' \text{ is the complement of } a] \\ &= (a' \wedge a) \vee (a' \wedge a'') && \text{[By distributivily]} \\ &= 0 \vee (a' \wedge a'') \\ &= (a \wedge a') \vee (a' \wedge a'') && [\text{as } a \wedge a'' = 0] \\ &= (a \vee a') \wedge a'' \\ &= 1 \wedge a'' = a''. \end{aligned}$$

$$\Rightarrow \quad a' = a''$$

Therefore in a distributive lattice the complement element is unique.

Theorem 6.23. Dual of a distributive lattice is distributive.

Theorem 6.24. A lattice L is distributive then

$$(a \vee b) \wedge (b \vee c) \wedge (c \vee a) = (a \wedge b) \vee (b \wedge c) \vee (c \wedge a) \ \forall \ a, b, c \in L$$

Proof. Let L be a distributive lattice.

Consider. $(a \vee b) \wedge (b \vee c) \wedge (c \vee a)$

$$\begin{aligned} &= [a \wedge ((b \vee c) \wedge (c \vee a))] \vee [b \wedge [(b \vee c) \wedge (c \vee a))] \\ &= [(a \wedge (c \vee a)) \wedge (b \vee c)] \vee [(b \wedge (b \vee c)] \wedge (c \vee a)] \\ & \qquad \text{(By associativity)} \\ &= [a \wedge (b \vee c)] \vee [b \wedge (c \vee a)] \text{ [By Idempotent law]} \\ &= (a \wedge b) \vee (a \wedge c) \vee (b \wedge c) \vee (b \wedge a) \\ &= (a \wedge b) \vee (a \wedge c) \vee (b \wedge c) \end{aligned}$$

Theorem 6.25. Sublattice of a distributive lattice is distributive.

Proof. Let L be a distributive lattice and let S be a sublattice of L.

let $a, b, c \in S \subseteq L$

$\Rightarrow a, b, c \in L$

$\Rightarrow \quad a \wedge (b \vee c) = (a \wedge b) \vee (a \wedge c)$

and $\quad a \vee (b \wedge c) = (a \vee b) \wedge (a \vee c)$

and as S is closed under join and meet operation therefore these equations hold in S.

Theorem 6.26. Homorphic image of a distributive lattice is distributive.

Proof. Let $\theta : L \rightarrow M$ be an homomorphism and let L is a distributive lattice.

We need to show that M is distributive

Let $x, y, z \in M$

as θ is onto $\exists \ a, b, c \in L$ such that

$$\theta(a) = x, \theta(b) = y, \theta(c) = z$$

$$\begin{aligned} x \wedge (y \vee z) &= \theta(a) \wedge (\theta(b) \vee \theta(c)) \\ &= \theta(a) \wedge ((\theta(b \vee c)) \end{aligned}$$

$$= \theta(a) \wedge \theta(b \vee c)$$
$$= \theta(a \wedge (b \vee c))$$
$$= \theta(a \wedge b) \vee (a \wedge c) \qquad \text{As } L \text{ is distributive}$$
$$= \theta(a \wedge b) \vee \theta(a \wedge c) \qquad \text{(as } \theta \text{ is homomorphism)}$$
$$= (\theta(a) \wedge \theta(b)) \vee (\theta(a) \wedge \theta(c))$$
$$= (x \wedge y) \vee (x \wedge z)$$

$\Rightarrow M$ is distributive.

Theorem 6.27. If L and M are two distributive lattices then $L \times M$ is also distributive.

Proof. Let (a_1,b_1) , (a_2,b_2) , $(a_3, b_3) \in L \times M$

where $a_1, a_2, a_3 \in L$ and $b_1, b_2, b_3 \in$ M

Consider. $(a_1, b_1) \wedge ((a_2, b_2) \vee (a_3, b_3))$

$$= (a_1,b_1) \wedge ((a_2 \vee a_3, b_2 \vee b_3))$$
$$= (a_1, b_1) \wedge ((a_2 \vee a_3), (b_2 \vee b_3))$$
$$= (a_1 \wedge (a_2 \vee a_3) , b_1 \wedge (b_2 \vee b_3))$$
$$= ((a_1 \wedge a_2) \vee (a_1 \wedge a_3) , (b_1 \wedge b_2) \vee (b_1 \wedge b_3))$$
$$= (a_1 \wedge a_2, b_1 \wedge b_2) \vee (a_1 \wedge a_3, b_1 \wedge b_3)$$
$$= ((a_1, b_1) \wedge (a_2, b_2)) \vee ((a_1, b_1) \wedge (a_3,b_3))$$

Similarly by duality

$$(a_1, b_1) \vee ((a_2, b_2) \wedge (a_3, b_3)) = [(a_1, b_1) \vee (a_2, b_2)] \wedge (a_1, b_1) \vee (a_3, b_3)$$

$\Rightarrow L \times M$ is distributive.

EXERCISE 1

1. Define the term lattice and give two example of lattices which are not poset. Give reason also.
2. Let X be any set and L be a lattice.
   Let T = set of functions from $X \to L$.
   Define a relation R on T by
   $$f \, R \, g \Leftrightarrow f(x) \; R \; g(x) \; \forall \; x \in X, f, g \in T$$
   Show that T is a lattice.
3. Show that idempotent laws follow from the absorption law.
4. Show that in a poset greatest (least) element if exists is unique.
5. Show that non-empty intersection of two sublattices is a sublattice.
6. Draw the diagram of the lattice of factors of 100 under divisibility. check whether it is a distributive lattice or not.
7. Show that sublattices of a lattice from a complete lattice under $\subseteq$.
8. If L is a poset with five or less elements and also has least and greatest elements then show that L is a lattice.

9. Consider the set T in problem (2) then show that T is distributive if L is distributive.

10. If L is a lattice in which

    $(x \vee y) \wedge [z \vee (x \wedge y)] = (x \wedge y) \vee (y \wedge z) \vee (z \wedge x) \ \forall\ x, y, z$

    then show that L is distributive.

11. Explain why the partially ordered set given in the following figure are not lattices.

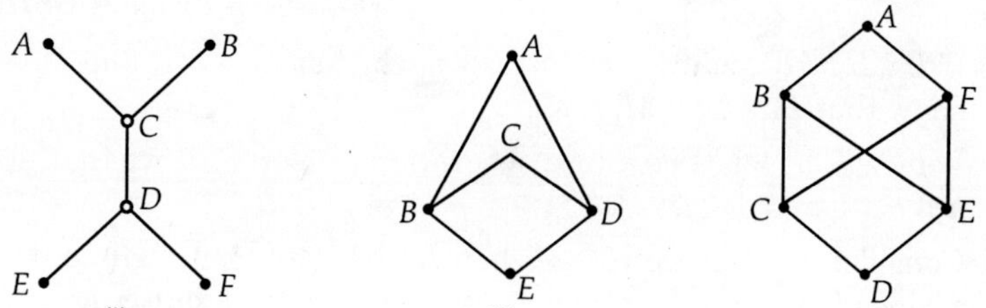

(i) (ii) (iii)

Fig. 1.

12. Show that the operations of meet and join on a lattice are commutative, associative and idempotent.

13. Let the sets S_0, S_1, ... S_7 be given by:

    $S_0 = \{a, b, c, d, e, f\}$
    $S_1 = \{a, b, c, d, e\}$
    $S_2 = \{a, b, c, e, f\}$
    $S_3 = \{a, b, c, e\}$
    $S_4 = \{a, b, c\}$
    $S_5 = \{a, b\}$
    $S_6 = \{a, c\}$
    $S_7 = \{a\}$

    Draw the diagram of $< L, \subseteq >$ where $L = \{ S_0, S_1, S_2, .. S_7\}$. Give the operations of meet and join.

14. Show that every interval of a lattice is a sublattice.

15. Find all the sublattices of the lattice D_{12}.

16. Check whether the following are lattices or not. If so give comments on distributive lattice.

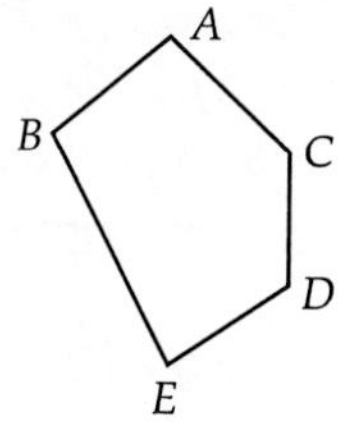

Fig. 2

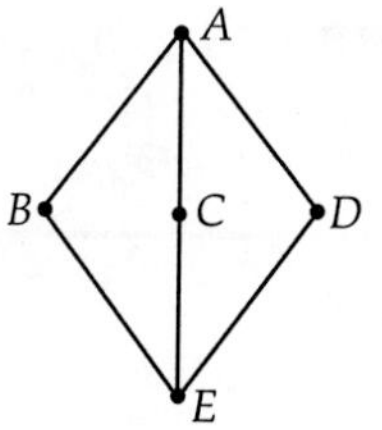

Fig. 3.

17. Find the complements of every element of the lattice $(S_n, D >$ for $n = 30$

18. Show that in a lattice with two or more elements no element is its own complement.

19. Show that in a complemented, distributive lattice
$a \leq b \Leftrightarrow ab' = 0 \Leftrightarrow a' + b = 1 \Leftrightarrow b' \leq a'$

20. Let $(L, >)$ be a lattice. Show that, for every a, b and c in L, $a \wedge (b \vee c) > (a \wedge b)$ $\vee (a \wedge c)$ ***(End-term Exam., GGSIPU, 2001)***

21. Show that a lattice is distributive iff.
$$(a \wedge b) \vee (b \wedge c) \vee (c \wedge a) = (a \vee b) \wedge (b \vee c) \wedge (c \vee a)$$
(End-term Exam., GGSIPU, 2001)

22. (*a*) If L_1, L_2, M_1 and M_2 are lattices such that $L_1 \cong M_1$, and $L_2 \cong M_2$ then show that $L_1 \times L_2 \cong M_1 \times M_2$

    (*b*) Let $(L, \leq)$ be a lattice. State absorption property. Prove that $A \wedge C \leq B \wedge d$ if $a \leq b$ and $c \leq d$.

    (*c*) Consider the Lattice $(L, \leq)$ when $L = \{1, 2, 4, 5, 10, 20\}$ and $a \leq b \Leftrightarrow$ a divides b. Is $4 \leq 10$? Let $S = \{1, 2, 4, 5\}$, is S a sub lattice of L.
(Terminal Exam., GGSIPU, 2001)

Hints and Answers for Exercise

3. We have $a \wedge (a \vee b) = a$ ...(1)
and $a \vee (a \wedge b) = a$ ...(2)
Take $b = a \wedge b$ in (1) and we get
$$a \wedge (a \vee (a \wedge b)) = a \wedge a \quad \text{(by 2)}$$
$$= a$$
Similarly we can show $a \vee a = a$

4. ***Hint:*** Let's take the poset has two greatest elements say a and a' then prove the contradiction.

5. ***Hint:*** $D_{100} = \{1, 2, 4, 5, 10, 20, 25, 50, 100\}$

11. The partial ordered set given by figure (*i*) is not a lattice as $A \vee B$ does not exist. But (*ii*) and (*iii*) are lattices.

16. Posets given in both the figures are lattices but lattice in figure 3 is not distributive whereas lattice given by fig. 2 is distributive.

As figure 2.

$$B \vee (C \wedge D) = B \vee D = A \qquad B \vee (A \wedge E) = B \vee (E)$$
$$(B \vee C) \wedge (B \vee D) = A \wedge A = A \qquad (B \vee A) \wedge (B \vee E) = A \wedge B = B$$
$$B \vee (C \wedge D) = (B \vee C) \wedge (B \vee D) \quad \Rightarrow \quad B \vee (A \wedge E) = (B \vee A) \wedge (B \vee E)$$
and
$$D \vee (C \wedge A) = D \vee C = C \qquad B \vee (E \wedge D) = B \vee E = B$$
$$(D \vee C) \wedge (D \vee A) = C \wedge A = C \qquad (B \vee E) \wedge (B \vee D) = B \wedge A = B$$
$$D \vee (C \wedge A) = (D \vee C) \wedge (D \vee A) \quad \Rightarrow \quad B \vee (E \wedge D) = (B \vee E) \wedge (B \vee D)$$

In figure 3.

$$B \vee (C \wedge D) = B \vee E = B \qquad (B \vee C) \wedge (B \vee D) = A \wedge A = A$$
that is
$$B \vee (C \wedge D) \neq (B \vee C) \wedge (B \vee D)$$

EXERCISE 2

1. (*a*) Define Posets and show that $(A \times B, \leq)$ is a poset if $(A, \leq)$ and $(B, \leq)$ are posets with partial order $\leq$ defined by:

$$(a, b)\,(a', b') \text{ if } a \leq a' \text{ in } A \text{ and } b \leq b' \text{ in } B.$$

   (*b*) Let $A = \{1, 3, 9, 27, 81\}$, draw the Hasse diagram of the poset $(A, 1)$.

   (*c*) Determine whether the posets represented by Hasse diagram in the following figures have greatest element, least element, minimal element and maximal element.

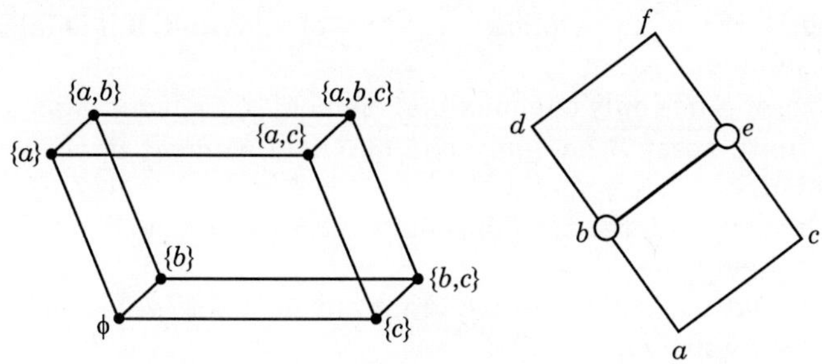

Fig. 1.

2. (*a*) Define isomorphic ordered sets.
   If $A = \{1, 2, 4, 8\}$ with divisibility and $B = \{0, 1, 2, 3\}$ with usual $\leq$ as partial order. Consider $f : A \to B$ defined by:
   $F(1) = 0, f(2) = 1, f(4) = 2, f(8) = 3$. Show that f is an isomorphism from A and B.

   (*b*) Describe the following with at least two examples
   (*i*) Bounded lattices; (*ii*) Distributive lattices.

3. (*a*) $A = \{2, 4, 6, 8, 12, 18, 24, 36, 72\}$ with the partial order of divisibility.

   (*b*) $A = [2, 3, 4, 6, 12, 18, 24, 36]$ with the partial order of divisibility.

4. Let P be relation on the Power Set $P(S)$ of a finite Set S of cardinality n define P by $(A, B) \in P$ *iff* $A \cap B = \phi$.

   (*a*) Consider the specific case $n = 3$ and determine the cardinality of P.

   (*b*) What is the cardinality of P for an arbitrary $n7$. Express answer in terms n.

DO IT BY YOURSELF

1. Prove that power set of a finite set under subset relation is a lattice. Define meet, join and complement operation in general for this lattice.
2. Prove that set of all divisors of a positive integer under division relation is a lattice. Define meet, join and complement operation in general for this lattice.
3. Show that a lattice with three or fewer elements in a chain.
4. Show that a chain of three or more elements is not complemented.
5. Show that in abounded distributive lattice, the elements that have complements form a sub lattice.

6. Consider the set Q of rational number with the usual $\leq$, and consider the subset D of Q defined by.
   $D = \{x: x \in Q \text{ and } B < x^3 < 15\}$
   (*a*) Is D bounded above or below?
   (*b*) Find upper bound and lower bound of D.
   (*c*) Do sup (D) and inf (D) exist?
7. Write the dual of each statement:
   (*a*) $(a \wedge b) \vee \text{c} = (a \vee b) \wedge (c \vee a)$ (*b*) $(a \wedge b) \vee \text{a} = a \wedge (b \vee a)$
8. Draw the Hasse diagrams of the factors of 6, 8, 12, 15, 18, 20, 24, 25, 30, 32, 36, 40, 48 all ordered by divisibility.
9. State whether each of the following is true of false and, if it is false, give a counter example.
   (*a*) If a poset S has only one maximal element a, then a is a last element.
   (*b*) If a finite poset S has only one maximal element a, then a is last element.
   (*c*) If a linearly ordered set S has only one maximal element a, then a is the last element.
10. Show that a lattice is distributive iff for any elements a, b, c in the lattice $(a \vee b) \wedge c \leq a \vee (b \vee c)$.
11. Let (A, $\leq$) be a distributive lattice. Show that, if
    $a \wedge x = a \wedge y$ and $a \vee x = a \vee y$ for some a, then $x = y$.
12. Show that a lattice $(A, \leq)$ is distributive iff for any elements a, b, c in A, $(a \wedge b) \vee (b \wedge c) \wedge (c \wedge a) = (a \vee b) \wedge (b \wedge c) \wedge (c \wedge a)$.
13. Let a and b be two elements in a lattice $(A, \leq)$. Show that $a \wedge b = b$ iff $a \wedge b = a$
14. Let a, b, c be elements in a lattice $(A, \leq)$. Show that if $a \leq b$, then $\wedge (b \wedge c) \leq b \wedge (a \wedge c)$.
15. Let a, b, c be elements in a lattice $(A, \leq)$. Show that
    $a \vee (b \wedge c) \leq (a \vee b) \wedge (a \vee c)$
    $(a \wedge b) \vee (a \vee c) \leq a \wedge (b \vee c)$
16. Show that if $(L_1, \leq_R)$ and $(L_2, \leq_R)$ are two lattices, then $(L_1 \times L_2, \leq_R)$ is also a lattice.
17. Prove that greatest element (if exists) is unique maximal element.
18. Prove that Absorption law for lattices:
    $a \wedge (a \wedge b) = a$, $a \vee (a \vee b) = a$
19. Show that:
    $a \leq b$, $c \leq d \Rightarrow a \wedge c \leq b \wedge d$
    $a \vee c, \leq b \vee d$
    In particular, show that $a \leq b \Rightarrow a \wedge x \leq b \wedge x$
    $a \vee x, \leq b \vee x$ for $\forall x \in L$
20. Prove that if L is a bounded distributive lattice, then complements are unique if they exist.
21. Show that two lattices L and M are distributive iff LXM is distributive.
22. Show that in any lattice L, for any elements a, b, c,
    $(a \wedge b) \vee (b \wedge c) \vee (c \wedge a) \leq (a \vee b) \wedge (b \wedge c) \wedge (c \wedge a)$

23. Prove that in any lattice, the distributive inequalities,
$a \wedge (b \vee c) \geq (a \wedge b) \vee (a \wedge c)$
$a \vee (b \vee c) \leq (a \wedge b) \wedge (a \vee c)$
hold for any a, b, c.

24. Prove that dual of a lattice is a lattice, i.e. if (L, R) is a lattice and $(L, \bar{R})$ is its dual, then for all x, y in L, Supremum (w.r.t. $\bar{R}$) and infimum (w.r.t. $\bar{R}$) exists.

MULTIPLE CHOICE QUESTIONS

1. Let A = {2, 3, 4, 6, 8, 24, 48} with partial ordering of divisibility. Determine all the maximal and minimal elements of A.
(*a*) maximal = {48} and minimal = {2}
(*b*) maximal = {24, 48} and minimal = {2, 3}
(*c*) maximal = {48} and minimal = {2, 3}
(*d*) maximal = {24, 48} and minimal = {2}

2. Let D_{100} = {1, 2, 4, 5, 10, 20, 25, 50, 100} and let the relation ≤ be the relation / (divider) be a partial ordering on D_{100} [shown in Fig. 1.]. Determine the *GLB* and *LUB* of B B where B = {10, 20}
(*a*) *GLB*(*B*) = {10} and *LUB*(*B*) = {20, 50}
(*b*) *GLB* (*B*) = {10} and *LUB*(*B*) = {20}
(*c*) *GLB* (*B*) = {100} and *LUB*(*B*) = {1}
(*d*) *GLB* (*B*) = {1} and *LUB*(*B*) = {4, 25}

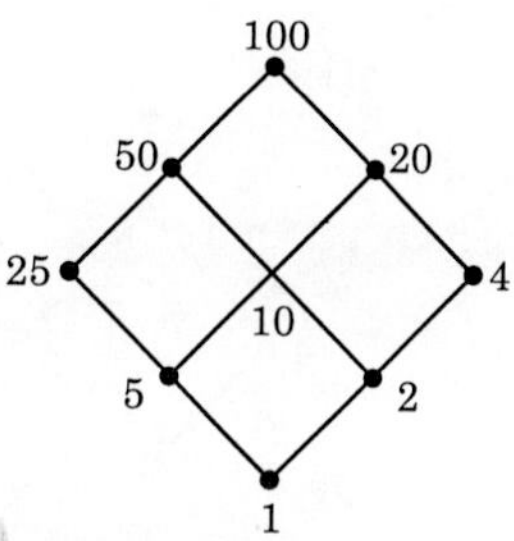

Fig. 1.

3. Determine which of the posets shown in Fig. 2. an lattices.

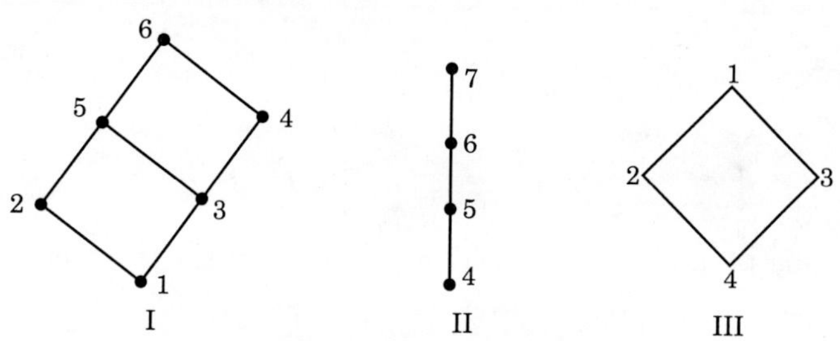

Fig. 2.

(*a*) I and III only (*b*) I and II but not II
(*c*) II and III but not I (*d*) All are lattices

4. A binary relation ≤ on a non-empty set A is called a partial order on A if the relation is
(*a*) Reflexive but not transitive
(*c*) Reflexive anti symmetric and tansitive
(*d*) none of these

5. The binary relation ≤ on the set of all positive integers I^+ defined by $x \leq y$ *iff* x divides y is a
(*a*) Partial ordered relation but not a total ordered relation
(*b*) Total ordered relation but not a partial ordered relation

(*c*) Total ordered relation
(*d*) Partial ordered relation

6. The lattice ($P(s)$, $\leq$) is bounded with greatest and least elements equal to
(*a*) 0 and 1 respectively
(*b*) ϕ and S respectively
(*c*) S and ϕ respectively
(*d*) None of these

7. Let S be the ordered set in Fig. 3. Suppose $A = \{1, 2, 3, 4, 5\}$ is isomorphic to S and $f = \{(a, 1), (b, 3), (c, 5), (d, 2), (e, 4)\}$ is a similarity mapping from 5 to A. Draw the Hasse diagram of A.

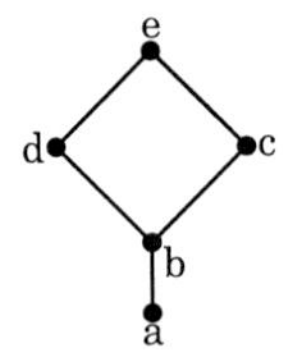

Fig. 3.

(*a*)

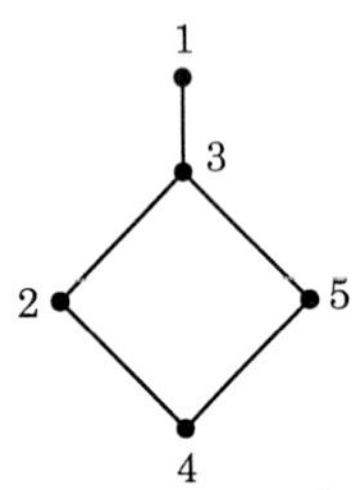

(*b*)

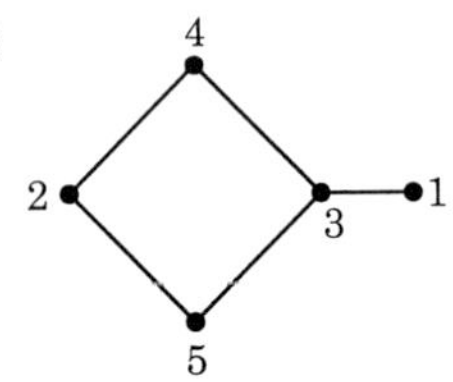

(*c*)

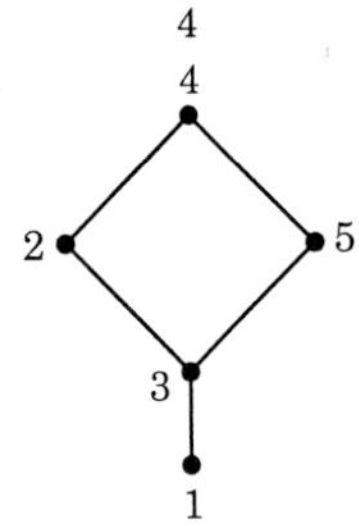

(*d*)

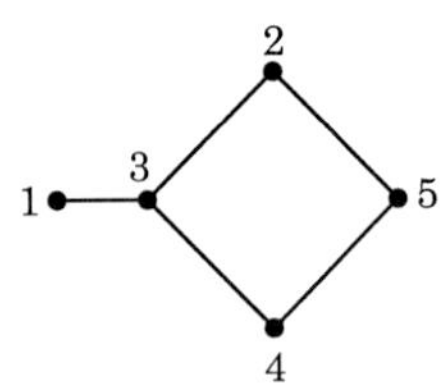

8. Determine the greatest and least elements, if the exist, of the poset given in Fig. 4.

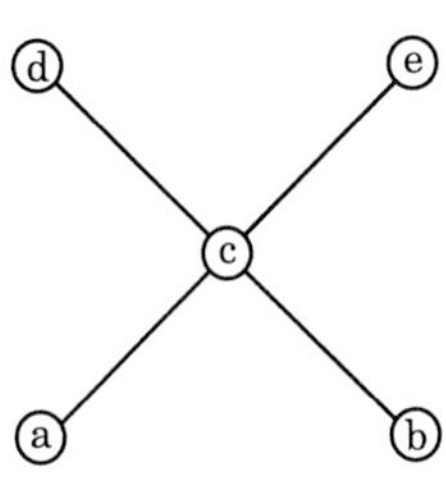

Fig. 4.

(*a*) greatest element = $\{d, e\}$, least element = $\{a\}$
(*b*) greatest element donot exists, least element do not exists
(*c*) greatest element = $\{d\}$, least element do not exists

(*d*) greatest element do not exists, least element = $\{a, b\}$

9. What is $(S, \leq)$?
   (*a*) *A* Poset but not Boolean Algebra
   (*b*) A Poset
   (*c*) A Boolean Algebra
   (*d*) A Boolean Algebra but not Poset
10. The number of complement of an element in a bounded lattice, if it exists is,
   (*a*) 1 (*b*) 2
   (*c*) 3 (*d*) 3
   (*e*) any positive integer

Answers

1. (*c*)	2. (*b*)	3. (*d*)	4. (*b*)	5. (*a*)	6. (*c*)
7. (*c*)	8. (*b*)	9. (*b*)	10. (*a*)		

❐❐❐

CHAPTER 7

BOOLEAN ALGEBRA

This Chapter Covers :

7.1. INTRODUCTION

In the previous chapter we have done lattices and theorem based on lattice and type of lattices also as complemented lattice and distributive lattice. A lattice which is complemented and distributive both is called boolean lattice. Since every element in a boolean lattice has a unique compliment, so we can define a unary operation complement on the lattice (Considering that L is a boolean lattice). That is if $(L, R, \vee, \wedge)$ be a boolean lattice then we define an algebraic system $(L, R, \vee, \wedge, ')$ where $\vee$, $\wedge$ and $'$ are the join, meet and complementation operations respectively. This algebraic system defined by boolean lattice is called a boolean algebra.

As in digital computer, there are only two possibilities written as 0 and 1 for smallest, indivisible object (*i.e.*, bits). All programs and data are ultimately reducible to combination of bits. A variety of devices have been used in digital computers to store bits. Electronic circuits allow these storage devices to a circuit is transmitted to another part of the circuit as a volatge. [Thus two voltage levels are needed. For example, a high voltage can communicate 1 and a low voltage can communicate 0.

7.2. BOOLEAN ALGEBRA

Definition : A complemented distributive lattice is called a Boolean lattice. Since complement is unique in a Boolean lattice so we can regard Boolean lattice as Boolean algebra with two binary operation $\vee$ and $\wedge$ and one unary operation ','. Therefore by Boolean Algebra we mean a system consisting of a non empty set L together with two binary operation $\wedge$ and $\vee$ and a unary operation, satisfying $\forall\ a, b, c \in L$

(1) $a \wedge a = a$
$a \vee a = a$

(2) $a \wedge b = b \wedge a$, $a \vee b = b \vee a$ — Commutative law

(3) $a \wedge (b \wedge c) = (a \wedge b) \wedge c$, $a \vee (b \vee c) = (a \vee b) \vee c$ — Associative law

(4) $a \wedge (b \vee c) = (a \wedge b) \vee (a \wedge c)$, $a \vee (b \wedge c) = (a \vee b) \wedge (a \vee c)$ — Distributive law

(5) $a \wedge (a \vee b) = a$, $a \vee (a \wedge b) = a$ — Indempotent law

(6) $\forall\ a \in L\ \exists\ a' \in L :\ a \vee a' = 1$
$a \wedge a' = 0$

Where 1 and 0 are the elements satisfying

$$0 \leq x \leq 1\ \forall\ x \in L$$

i.e., 1 is the universal upper bound and 0 is the universal lower bound.

EXAMPLE 7.1. Let K be a non-empty set. Consider the power set of K, *i.e.*, $P(K)$ then the lattice $(P(K), \subseteq, \cup, \cap, ')$ is a boolean lattice.

where K itself is universal upper bound and ϕ is the universal lower bound.

EXAMPLE 7.2. Let

$L = D_{30}$ = Set of all divisers of 30.
= {1, 2, 3, 5, 6, 10, 15, 30} under divisibility.

Then L is a boolean lattice. Verify it by the fig. 7.1.

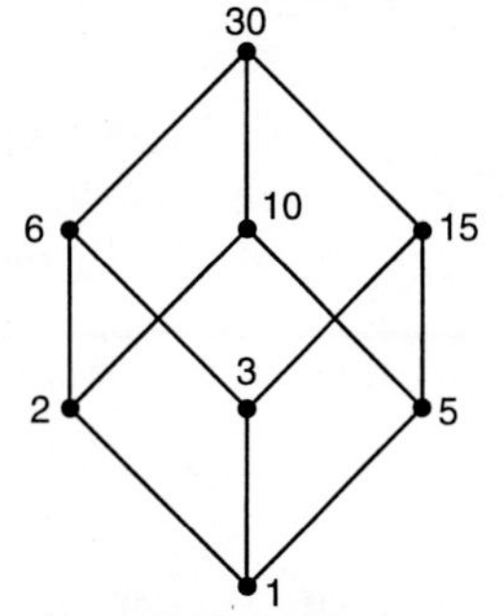

Fig. 7.1.

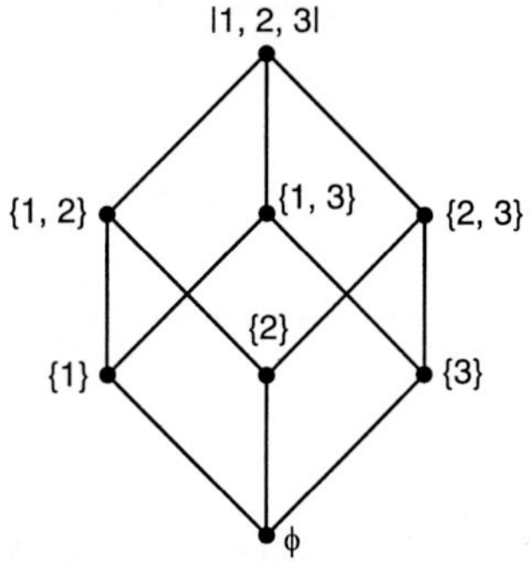

Fig. 7.2.

By the fig 7.1 it is clear that the D_{30} is isomorphic to $(P(S), \subseteq)$ where $S = \{1, 2, 3\}$. See fig 7.2.

$$P(S) = \{\phi, \{1\}, \{2\}, \{3\}, \{1, 2\}, \{1, 3\}, \{2,3\}, \{1, 2, 3\}\}$$

Therefore the result holds if we define a mapping $\theta : D_{30} \to P(s)$ such that

$$\begin{aligned} \theta(1) &= \phi \\ \theta(2) &= \{1\} \\ \theta(3) &= \{2\} \\ \theta(5) &= \{3\} \\ \theta(6) &= \{1,2\} \\ \theta(10) &= \{1,3\} \\ \theta(15) &= \{2,3\} \\ \theta(30) &= \{1, 2, 3\}, \end{aligned}$$

EXAMPLE 7.3. Let $L = \{0, x, y, 1\}$. Define $\vee$, $\wedge$ and $'$ by the follwoing table [Readers are advised to recall truth tables of chapter 2]

Table 7.1.(*a*)

$\vee$	0	x	y	1
0	0	x	x	1
x	x	x	1	1
y	y	1	y	1
1	1	1	1	1

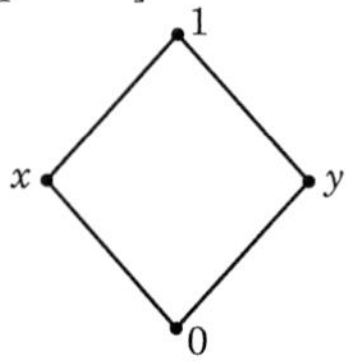

Fig. 7.3.

Table 7.1(*b*)

$\wedge$	0	x	y	1
0	0	0	0	0
x	0	x	0	x
y	0	0	y	y
1	0	x	y	1

Table 7.1(*b*)

$'$	
0	1
x	y
y	x
1	0

Readers can verify the results of the tables 7.1.(*a*), (*b*), (*c*) by using figure (7.3).

Theorem 7.1. For any two elements a and b in a boolean algebra.

$$\overline{a \vee b} = \bar{a} \wedge \bar{b}$$

$$\overline{a \wedge b} = \bar{a} \vee \bar{b}$$

That is $a \vee b$ is the complement of $\bar{a} \wedge \bar{b}$

and $a \wedge b$ is the complement of $\bar{a} \vee \bar{b}$

Proof. (*i*) Now we need to show that $a \vee b$ is the complement of $\bar{a} \wedge \bar{b}$. Therefore by the definition of complement we have to show the following results

$$(a \vee b) \vee \left(\bar{a} \wedge \bar{b}\right) = 1$$

$$(a \vee b) \wedge \left(\bar{a} \wedge \bar{b}\right) = 0$$

Now consider

$$(a \vee b) \vee \left(\bar{a} \wedge \bar{b}\right) = \left((a \vee b) \vee \bar{a}\right) \wedge \left((a \vee b) \vee \bar{b}\right) \text{ [By Distributive law]}$$

$= \left(a \vee \left(b \vee \bar{a}\right)\right) \wedge \left(a \vee \left(b \vee \bar{b}\right)\right)$ [By associativty law]

$= \left(a \vee \left(\bar{a} \vee b\right)\right) \wedge (a \vee 1)$ [By commutative law and complement law]

$= \left(\left(a \vee \bar{a}\right) \vee b\right) \wedge 1$

$= (1 \vee b) \wedge 1$

$1 \wedge 1 = 1$

Consider $(a \vee b) \wedge \left(\bar{a} \wedge \bar{b}\right) = \left[a \wedge \left(\bar{a} \wedge \bar{b}\right)\right] \vee \left[b \wedge \left(\bar{a} \wedge \bar{b}\right)\right]$ (Distributive)

$= \left[\left(a \wedge \bar{a}\right) \wedge \bar{b}\right] \vee \left[\left(b \wedge \bar{b}\right) \wedge \bar{a}\right]$ (Associative)

$= \left(0 \wedge \bar{b}\right) \vee \left(0 \wedge \bar{a}\right)$ (Complement)

$= 0 \vee 0 = 0$

Thus $\bar{a} \wedge \bar{b}$ is the complement of $a \vee b$ *i.e.*,

$$\overline{a \vee b} = \bar{a} \wedge \bar{b}$$

By the principle of duality

$$\overline{a \wedge b} = \bar{a} \vee \bar{b}.$$

Theorem 7.2. Homomorphic image of a Boolean algebra is a Boolean algebra.

Theorem 7.3. Dual of a Boolean algebra is a Boolean algebra.

Theorem 7.4. L and M are the Boolean algebras if and only if $L \times M$ is a boolean algebra.

[**Hint.** In the previous chapter we have done results for complemented and distributive lattices that is for boolean lattices].

Theorem 7.5. In a Boolean algebra, the following results hold.

(*i*) $(a')' = a$

(*ii*) $a \le b \Leftrightarrow a' \ge b'$

(*iii*) $a \le b \Leftrightarrow a \wedge b' = 0 \Leftrightarrow a' \vee b = 1$

(*i*) To show $(a')' = a$

That is a the complement of a'

Let x is the complement of a'

$(a')' = x$ then

$a' \wedge x = 0$ and $a' \vee x = 1$

and also $a \wedge a' = 0$ and $a \vee a' = 1$ [as a' is the complement of a]

$\Rightarrow a \wedge a' = a' \wedge x$ and $a \vee a' = a' \vee x$

$\Rightarrow a \wedge a' = x \wedge a'$ and $a \vee a' = x \vee a'$

$\Rightarrow a = x$

$\Rightarrow (a')' = a$

(*ii*) $a \le b \Rightarrow a = a \wedge b$

$\Rightarrow a' = (a \wedge b)' = a' \vee b'$

$\Rightarrow b' \le a'$

Let $b' \leq a' \Rightarrow b'' \geq a'' \Rightarrow b \geq a$

(*iii*) $a \leq b \Rightarrow a \wedge b' \leq b \wedge b' \Rightarrow a \wedge b' \leq 0$ ...(*i*)

$\Rightarrow 0 \leq a \wedge b'$ ...(*ii*)

[as 0 is the universal lower bound]

$\therefore$ (*i*) and (*ii*) $\Rightarrow a \wedge b' = 0$

Again let $a \wedge b' = 0$

then $a = a \wedge 1 = a \wedge (b \vee b') = (a \wedge b) \vee (a \wedge b')$

$= (a \wedge b) \vee 0$

$= a \wedge b$

$\Rightarrow a \leq a \wedge b \leq b$

$\Rightarrow a \leq b$

another part of result is proved by duality.

7.3. BOOLEAN EXPRESSIONS AND BOOLEAN FUNCTIONS

In chapter 2, we have learnt how a compound statement can be formed by combining simple statement or atomic statement, $p_1, p_2, ...p_n$ with the help of logical connectives $\wedge$, $\vee$ and $'$ also in chapter 2 we have discussed disjunctive normal form, conjunctive normal form, principle disjunctive normal form, principle conjunctive normal form. While expressing circuits mathematically, we identify each circuit in terms of some Boolean variables. Each of these variables repsesents either a simple switch or an input to some electronic switch.

Boolean Expression

Let L, $(\vee, \wedge, ', 0, 1)$ be a Boolean algebra. A boolean expression in variables $x_1, x_2, ...x_k$ (say), each taking their values in the set L is defined as follows :

(*i*) Each of the variables $x_1, x_2,...x_k$ as well as the elements 0 and 1 of the boolean algebra L are Boolean expressions or Boolean polynomial.

(*ii*) If x_1 and x_2 are previously defined Boolean expressions then $x_1 \wedge x_2$, $x_1 \vee x_2$ are also Boolean expressions, for example, $x_1 \wedge x_2'$, $x_1 \wedge x_2 \wedge x_1'$ etc. are Boolean expressions. That is disjunction, conjunction or complementation of any previousely defined Boolean expression is a Boolean expression.

Boolean function

Any function specifying these Boolean expressions is called a Boolean function. Thus if $f(x,y) = x \vee y$ then f is a boolean function and $x \vee y$ is the Boolean expression or the value of the function f.

EXAMPLE 7.4. Let $A = \{0,1\}$ and $f : A^3 \rightarrow A$ be the function defined by $f(x, y, z) = x \wedge (y \vee z)$. Give all the functional values of f :

Consider the table

x	y	z	$y \vee z$	$f = x \wedge (y \vee z)$
1	1	1	1	1
1	1	0	1	1
1	0	1	1	1
1	0	0	0	0
0	1	1	1	0
0	1	0	1	0
0	0	1	1	0
0	0	0	0	0

∴ functional values of f are

$$f(1, 1, 1) = 1$$
$$f(1, 1, 0) = 1$$
$$f(1, 0, 1,) = 1$$
$$f(1, 0, 0) = 0$$
$$f(0, 1, 1,) = 0$$
$$f(0, 1, 0,) = 0$$
$$f(0, 0, 1) = 0$$
$$f(0, 0, 0) = 0$$

Now let us look converse of the above result *i.e.*, if the functional values are given how to find the boolean function. As every function is equal to a function in its *DN* form. So we give the function in *DN* form as explained below and will solve it further.

The maximum number of minterms in the expression of f can be $2^3 = 8$ and in that case the function is in the *PDNF* (*i.e.*, complete normal form). But that will happen when all the eight values of $f(x, y, z)$ are equal to 1. Since in this case all the values are not equal to 1 therefore it can be considered that some of the minterns are missing. Thus we look at those values which give us the value of $f(x, y, z)$ equal to 1.

$$\therefore \quad f(x, y, z) = (x \wedge y \wedge z) \vee (x \wedge y \wedge z') \vee (x \wedge y' \wedge z)$$

EXAMPLE 7.5. Obtain a *DN* form for the given boolean function.

$$f(x, y, z) = (x' \wedge y) \vee (x \wedge z)$$

Solution : We can write

$$x' \wedge y = x' \wedge y \wedge 1 \qquad \text{(Identity law)}$$
$$= (x' \wedge y) \wedge (z \vee z') \qquad \text{(Complementation law)}$$
$$= (x' \wedge y \wedge z) \vee (x' \wedge y \wedge z') \qquad \text{(Distributive law)}$$

Also,

$$x \wedge z = (x \wedge z) \wedge 1$$
$$= (x \wedge z) \wedge (y \vee y')$$
$$= (x \wedge z \wedge y) \vee (x \wedge z \wedge y')$$

$$\therefore f(x, y, z) = (x' \wedge y \wedge z) \vee (x' \wedge y \wedge z') \vee (x \wedge z \wedge y) \vee (x \wedge z \wedge y')$$

in the disjunctive Normal form.

EXAMPLE 7.6. Obtain the disjunctive normal form of the Boolean expression

$$f(x, y, z) = (x \vee y)' \vee (x' \wedge z)$$

Solution : $f(x, y, z) = (x \vee y)' \vee (x' \wedge z)$

$= (x' \wedge y') \vee (x' \wedge z)$ [De Morgan's law]

$= (x' \wedge y' \wedge 1) \vee (x' \wedge z \wedge 1)$ [as 1 is the universal upper bound]

$= [(x' \wedge y') \wedge (z \vee z')] \vee [(x' \wedge z) \wedge (y \vee y')]$

$= (x' \wedge y' \wedge z) \vee (x' \wedge y' \wedge z') \vee (x' \wedge z \wedge y) \vee (x' \wedge z \wedge y')$

is the required form.

7.4. CIRCUITS

One of the major applications of Boolean algebra is to the electrical network consisting of switches that involves two state of devices. The simplest example of such a device being an ordinary on-off switch. The switch can assume two states 'closed' or 'open' (ON or OFF). In the first case the current flows and in the second case the current does not flow.

The basic elements of the circuits are called gates (arrangement of switches). Each type of gates implements a Boolean operation. In this section we define three types of gates. Using these gates and applying the properties of Boolean algebra we design circuits in the simplest form. The circuits that we will study give output that depends only on the input and not on the current state of the circuit.

There are two basic ways in which switches are generally inter-connected. These are 'in series' and 'in parallel'. The switches are arranged in series is nothing but using **AND gate** and the switches are arrenged in parallel is **OR gate**. They are reperented by the fig. 7.4(*a*) and 7.4(*b*) and there is another gate that we will study is invertal that is related to complement of element (switches) is given in fig. 7.4(*c*).

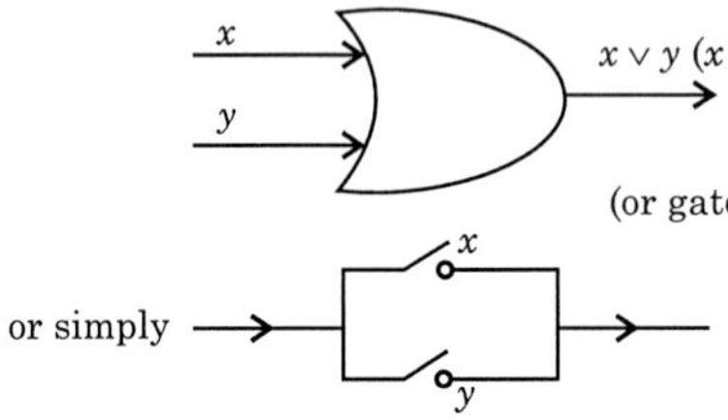

Fig. 7.4(a)

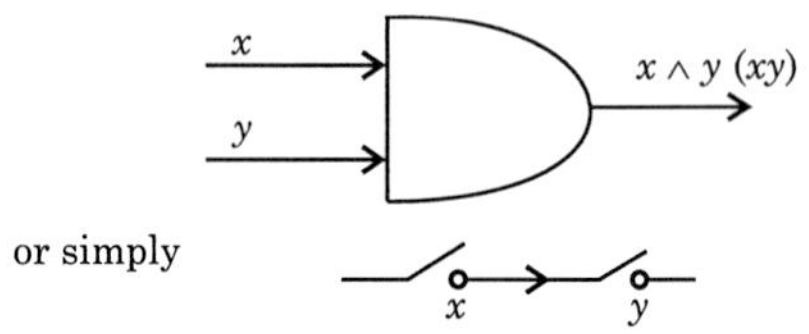

Fig. 7.4(b)

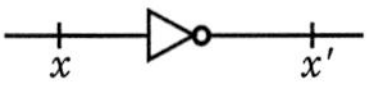

Fig. 7.4.(c)

The input to the circuit (gates) is shown on the left side of the entering element and output is shown on the right side, leaving the element. Closed state of the circuit means current is flowing and open state means current is not flowing. From the fig. 7.4(*a*) and 7.4(*b*) it is clear that two switches a and b are said to be connected 'in series' or by **AND gate**, if the current can pass only when both are in the closed state and current does not flow if anyone or both are open see fig 7.4(*b*).

Two switches a and b are said to be connected '**in parallel**' or by '**OR gate**' if current flows when any one or both are closed and current does not pass when both are open. see fig 7.4(a)

If a switch x is open then we denote x' for closed switch and is reprsented by the third gate that is **inverter.**

As we have done in chapter 2, the compound statements can be constructed by using simple statement. Similarly using the same logic we can make some compound circuits. We will do this with the help of Boolean algebra.

Let 0 denotes open circuit (current does not pass)

and 1 denotes closed circuit (current passes)

therefore we can consider the Boolean algebra as circuit system $\{\{0, 1\}, \vee, \wedge, '\}$.

7.5. REPRESENTATION OF CIRCUITS

Now we consider how a boolean function can represent a circuit and how a circuit can be represented by a Boolean function. We use the basic as given in the section (7.4).

EXAMPLE 7.7. Draw the circuit represented by the boolean function

$$f(x, y, z) = (x \vee y) \wedge z$$

Solution. $x \vee y$ [OR gate] are represented by the parallel circuits

$\therefore$ $x \vee y$ is given as

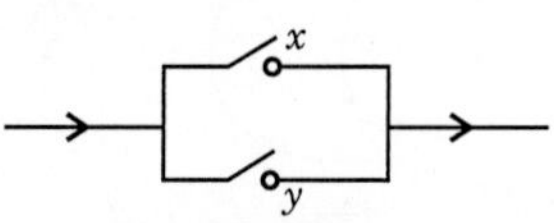

Fig. 7.5(a)

Now $(x \vee y) \wedge z$ is given by (AND gate) or switches in series therefore the required circuit is:

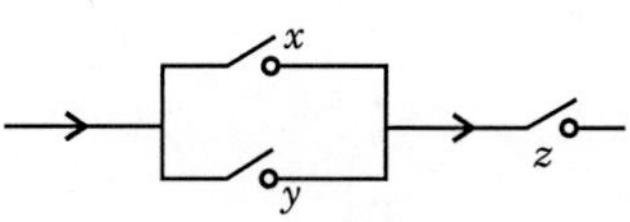

Fig. 7.5(b)

ALITER

This circuit can be given in terms of gates.

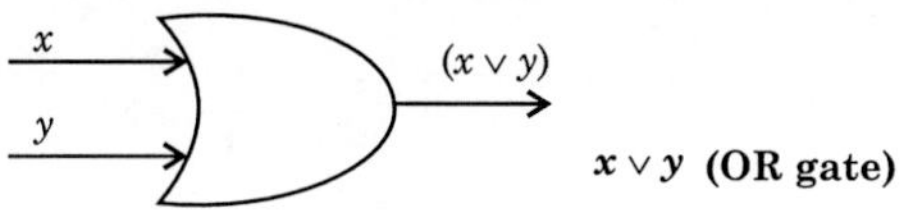

Fig. 7.5(c)

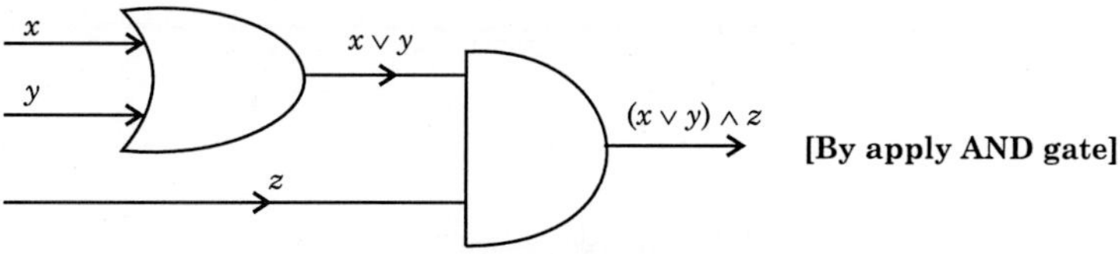

Fig. 7.5(d)

EXAMPLE 7.8. Now readers are required to draw the circuit represented by the Boolean function $f = x \vee (y \wedge z)$ and give the condition that when the current will flow.

Solution : By applying the rule of precedence

$(y \wedge z)$

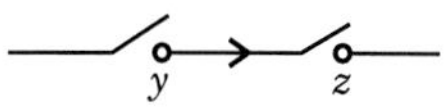

Fig. 7.6(a)

$x \vee (y \wedge z)$

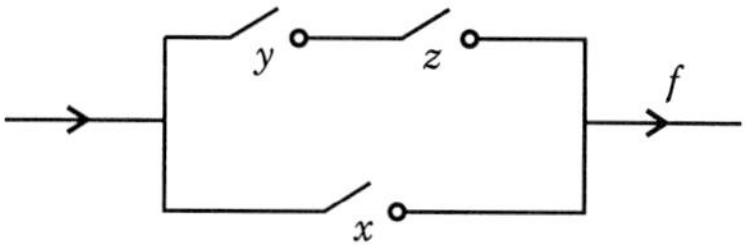

Fig. 7.6(b)

ALITER

The required circuit can be given in terms of gates

$(y \wedge z)$

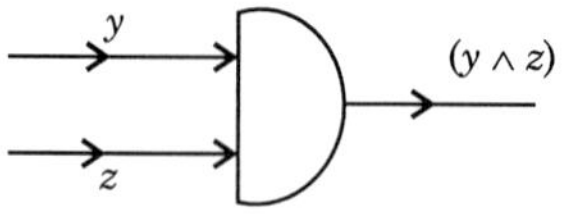

(AND gate)

Fig. 7.6(c)

$x \vee (y \wedge z)$

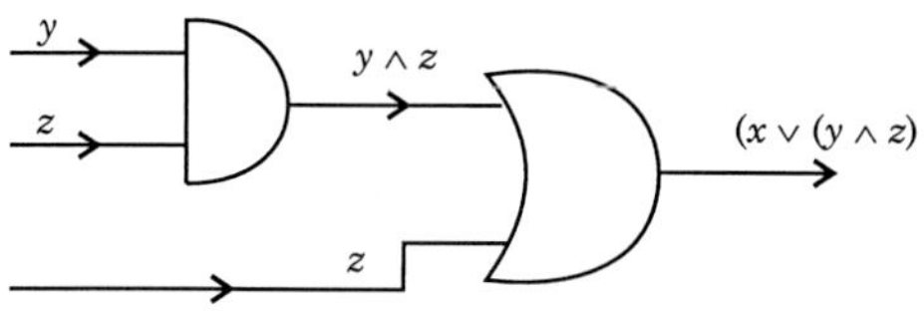

(OR GATE)

Fig. 7.6 (d)

EXAMPLE 7.9. Draw the circuit represented by the following Boolean function.

$$f : xy + \bar{x}\, y$$

Solution : $f(x, y, z) = xy + \bar{x}\, y = (x \wedge y) \vee (\bar{x} \wedge y)$

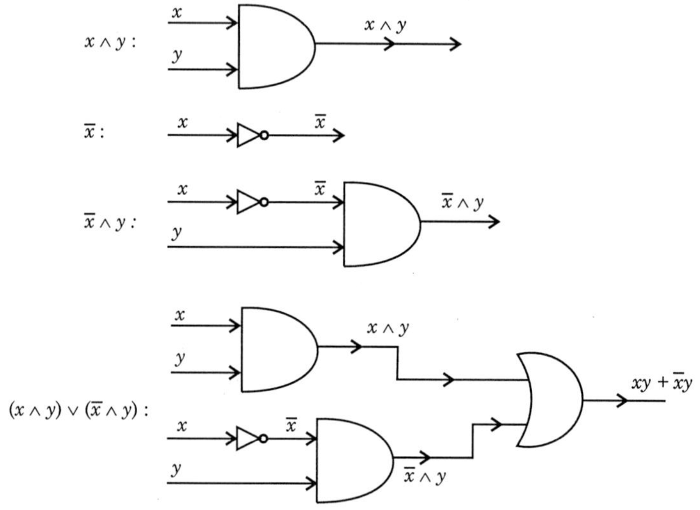

Fig. 7.7.

EXAMPLE 7.10. Draw the circuit which realises the function

$$f(x, y, z, t) = x \wedge [(y \vee t') \vee (z \vee (x \vee t \vee z'))] \wedge y$$

Solution : It is to be noticed that while drawing a circuit one should use rule of precedence

$y \vee t'$ $\qquad$ $x \vee t \vee z'$

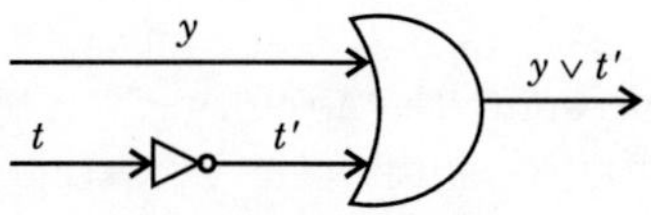

Fig. 7.8(a)

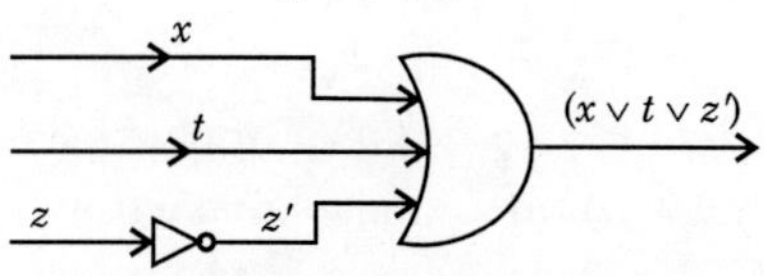

Fig. 7.8(b)

$z \wedge (x \vee t \vee z')$

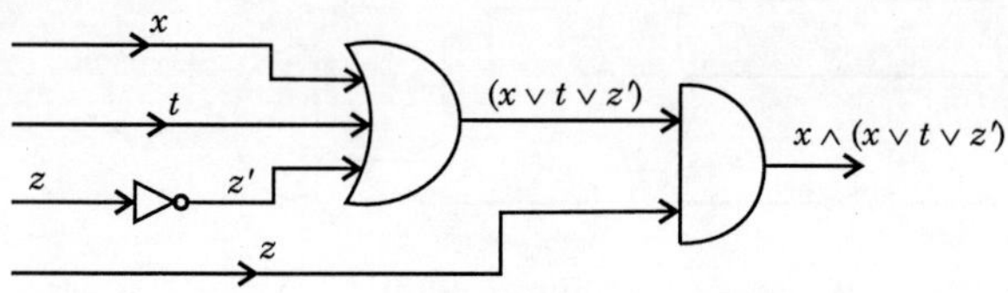

Fig. 7.8(c)

By (*a*) and (*c*)

$$g = (y \vee t') \vee (z \wedge (x \vee t \vee z'))$$

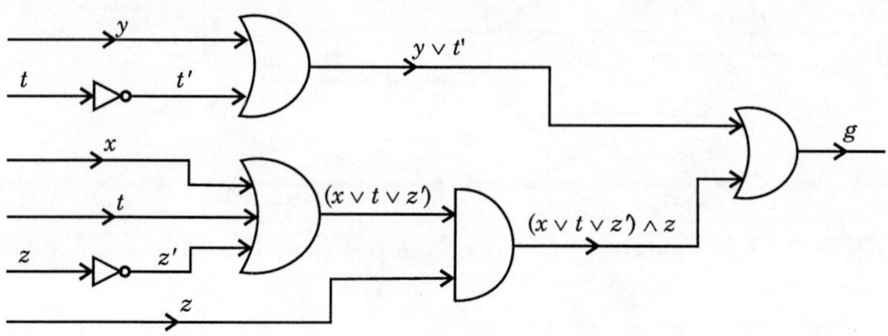

Fig. 7.8 (d)

$f : x \wedge g \wedge y$

Using AND gate we get

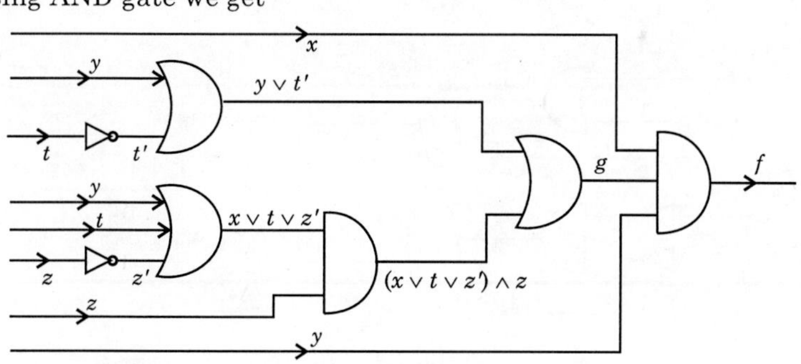

Fig. 7.8(e)

is the required circuit diagram of the given function.

EXAMPLE 7.11. Find the functon *f* that represents the circuit.

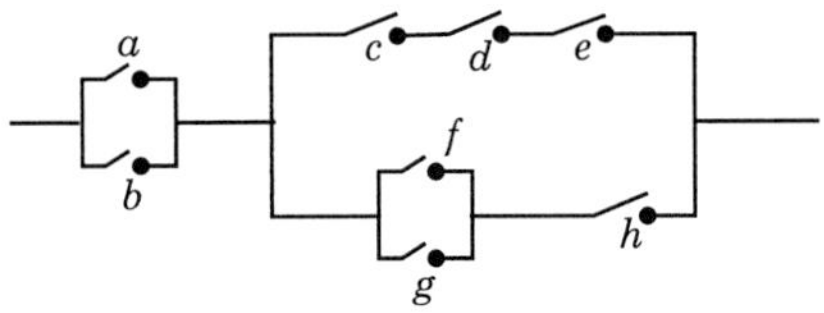

Fig. 7.9.

EXAMPLE 7.12. Find the function f that represents the circuits given by the fig. 7.10 (a) (b) and also give value [boolean expression] in the blocks marked as $\boxed{?}$.

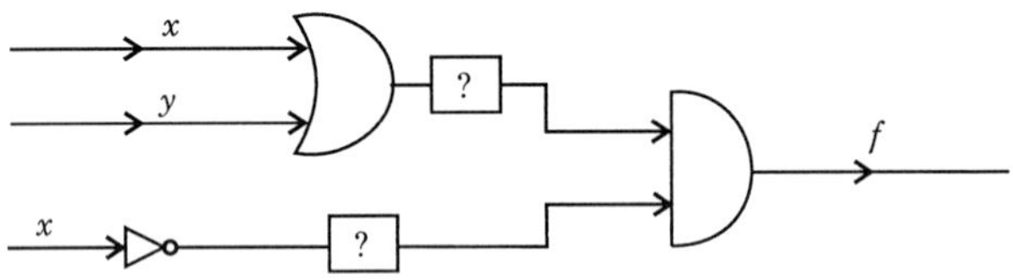

Fig. 7.10(a) **Ans.** $(x + y)\,\overline{x}$

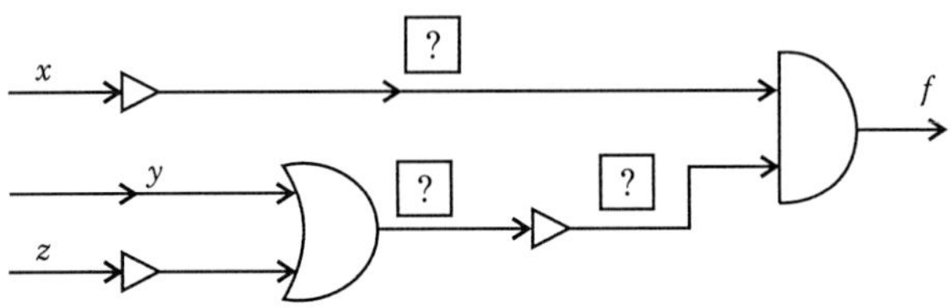

Fig. 7.10(b) **Ans.** $\overline{x}(\overline{y} + \overline{z})$

EXAMPLE 7.13. Construct circuits that produce the following outputs and their complements.

(i) $P(x, y, z) = (x + y + z)\ (\overline{x}\ \overline{y}\ \overline{z})$

(ii) $P(a, b, c, d, e, f, g, h) = (a \vee b) \wedge ((c \wedge d \wedge e) \vee ((f \vee g) \wedge h))$

(iii) $P(x, y, z) = xy + xz + yz$

Solution :

(i)

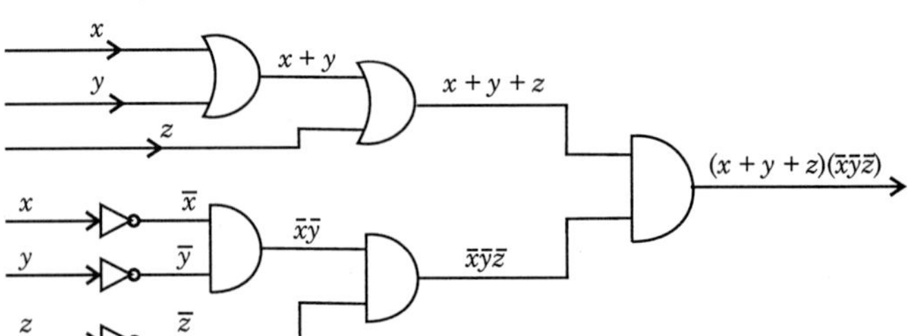

Fig. 7.11 (a).

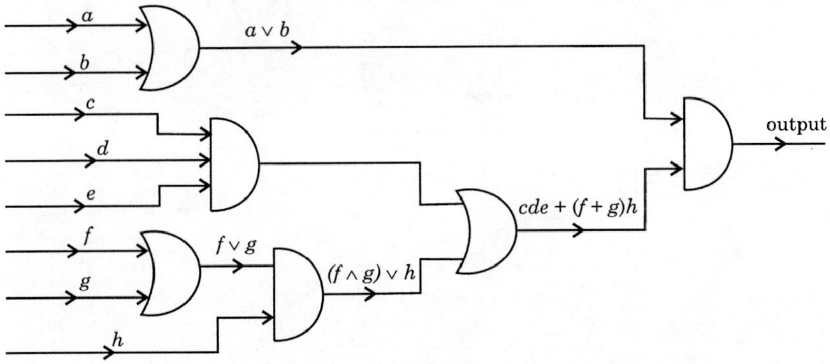

Fig. 7.11(b).

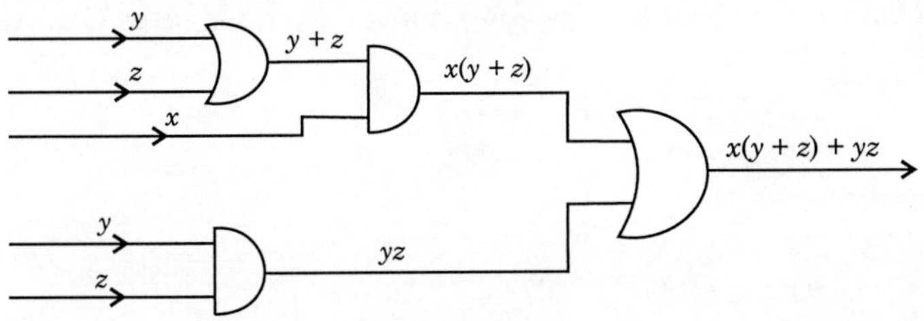

Fig. 7.11(c).

7.6. SIMPLIFICATION OF CIRCUITS

The efficiency of a circuit depends on the number and arrangement of its gates. Simplification of a circuit would normally mean the least complicated circuit with minimum cost and best result, *i.e.*, economical circuit. The simplification of circuits would mean lesser number of switches, which we achieve by using different properties of Boolean algebra.

EXAMPLE 7.14. Simplify the circuit given by the function :

$$f(x, y, z) = x\,y\,z + x\,\overline{y}\,z$$

The circuit diagram of given function is:

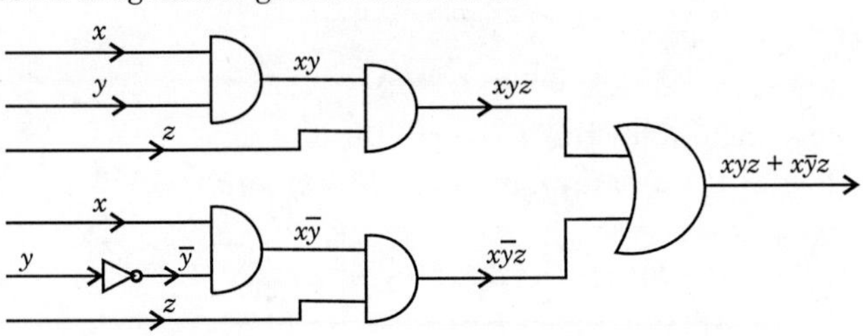

Fig. 7.12 (a).

As in the above function switches x and z are repeated

Now simplify

$$f(x, y, z) = x(y + \overline{y})z$$
$$= x(1)z = xz$$

Therefore new resultant circuit in the simplified form is:

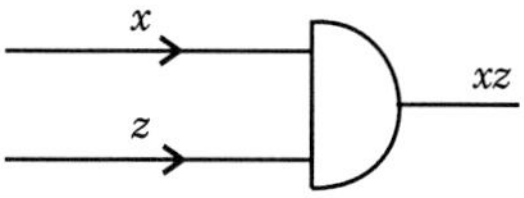

Fig. 7.12 (b).

EXAMPLE 7.15. Find the output of the given circuit.

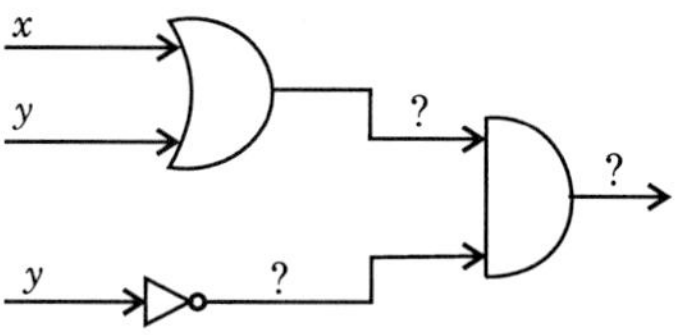

Fig. 7.13(a)

Solution : The given circuit is the combination of three gates OR, AND and invertor.

Therefore

$$(x \vee y) \wedge \overline{y}$$

on simplifying

$$(x \wedge \overline{y}) \vee (y \wedge \overline{y})$$
$$(x \wedge \overline{y}) \vee 0 = x \wedge \overline{y}$$

Therefore the simplified circuit is:

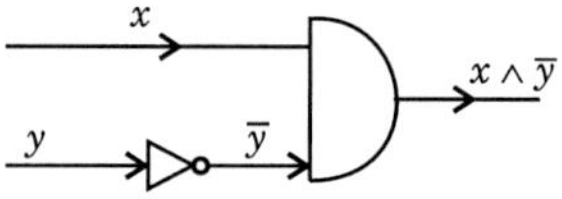

Fig. 7.13(b)

EXAMPLE 7.16. Simplify the circuit

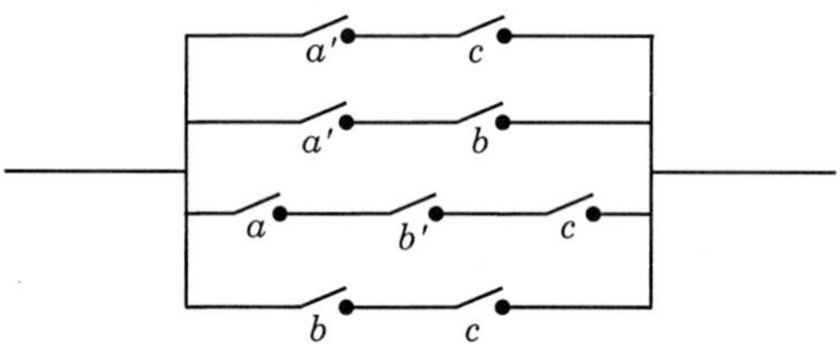

Fig. 7.14(a)

The Boolean function for the given circuit is:

$$(a' \wedge c) \vee (a' \wedge b) \vee (a \wedge b' \wedge c) \vee (b \wedge c)$$

which is equal to

$$(a' \wedge b) \vee [a' \vee (a \wedge b') \vee b] \wedge c$$
$$= (a' \wedge b) \vee [a' \vee (a \wedge b') \vee 1 \wedge b] \wedge c$$
$$= (a' \wedge b) \vee [a' \vee (a \wedge b') \vee (a \vee a') \wedge b] \wedge c$$
$$= (a' \wedge b) \vee [a' \vee (a \wedge b') \vee (a \wedge b) \vee (a' \wedge b)] \wedge c$$
$$= (a' \wedge b) \vee [a' \vee (a \wedge (b' \vee b)) \vee (a' \wedge b)] \wedge c$$
$$= (a' \wedge b) \vee [a' \vee (a \wedge 1) \vee (a' \wedge b)] \wedge c$$
$$= (a' \wedge b) \vee [a' \vee a \vee (a' \wedge b)] \wedge c$$

$= (a' \wedge b) \vee ((1 \vee (a' \wedge b)) \wedge c)$
$= (a' \wedge b) \vee ((1 \wedge c) \vee (a' \wedge b) \wedge c)$
$= ((a' \wedge b) \vee c) \vee ((a' \wedge b) \wedge c)$
$= (a' \wedge b) \vee c$

∴ Simplified circuit

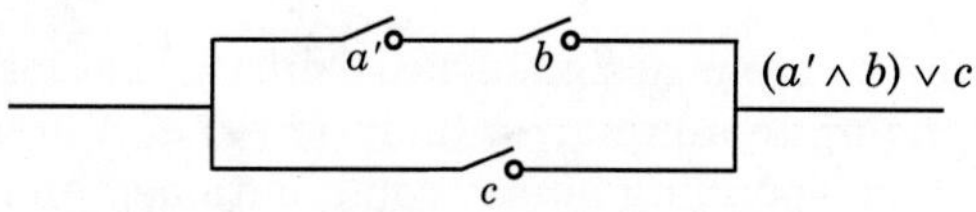

Fig. 7.14(b)

Readers can make the circuit diagram by the use of gates.

EXAMPLE 7.17. Give output of the gives circuit and also simplify if

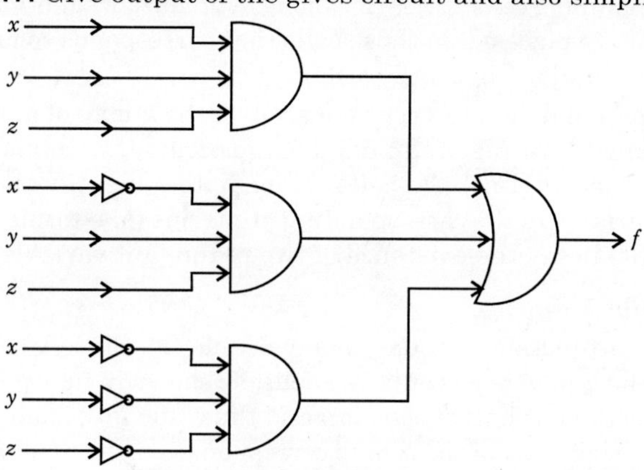

Fig. 7.14(c)

$f : (x \wedge y \wedge z) \vee (\overline{x} \wedge y \wedge z) \vee (\overline{x} \wedge \overline{y} \wedge \overline{z})$
$[(x \vee \overline{x}) \wedge (y \wedge z)] \vee (\overline{x} \wedge \overline{y} \wedge \overline{z})$
$[1 \wedge (y \wedge z)] \vee (\overline{x} \wedge \overline{y} \wedge \overline{z}]$

$(y \wedge z) \vee (\overline{x} \wedge \overline{y} \wedge \overline{z})$ is the simplified form

∴ the simplified circuit is

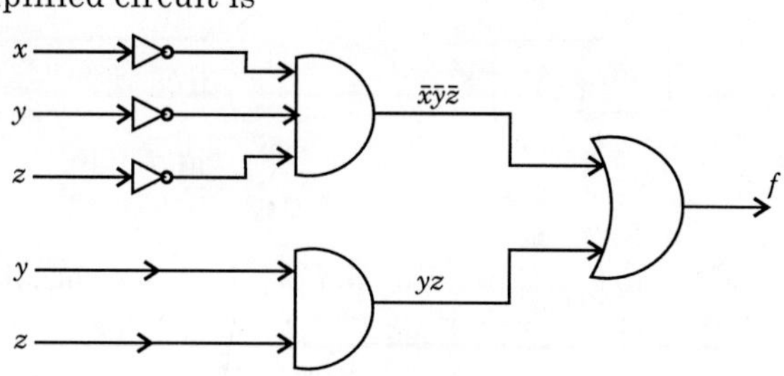

Fig. 7.14(d).

7.7. KARNAUGH MAP

A karnaugh map (or k-map for short) is a graphical method for representing the truth table of a Boolean function.

If n is the number of variables, then k-map has 2^n cells one for each of the 2^n minterms (maxterms) and, therefore, one for each now of the truth table. Moreover, these cells one arranged such that the minterm (maxterms) are displayed in a geometric pattern so that logically adjacent terms are visibly apparent.

Minimisation, therefore, can be accomplished by recognising basic pattern on the k-map.

Each cell of the k-map is assigned a coordinate that forms an n-bit binary number corresponding to a minterm (maxterms) of n literals. Two cells are adjacent if their corresponding binary combinations differ by only one bit. If a function F is expressed in canonical Sum-Of-Products (SOP) form, then we complete the k-map of F by placing a 1 in those cells whose coordinates corresponds to minterms that occur is the canonical form and placing a 0 in each of the remaining cells. If F is in canonical product of sums form (conjunctive normal form), we place a 0 in those cells that corresponds to maxterms of the canonical form and a 1 in all the others.

If the functional value 1 occurs in a cell of the k-map of a function F, then the cell in called a 1-cells of F ; if a 0 value occurs, the cell is called a o-cell. Hence, two adjacent 1-cells (o-cells) immediately identify the presence of a redundant variable in the corresponding minterms (maxterms) ; four adjacent 1-cells (o-cells) imply the existence of two redundant variables ; and so on.

One-Variable Maps

Consider a function F of only one variable (x). Since x can take on only two values, the k-map needs only two cells, as shown in figure 1(a). The region where $x = 0$ is the left half of the map and the value $F(0)$ should be entered in that cell. The right half of the k-map corresponds to $x = 1$, and $F(1)$ is centered in that cell. In the fig 7.15(b) the locations of minterms are indicated so that a 1 is entered in the left cell if the minterms m_0 appears in the canonical form of F and a 1 is entered in the right cell if the minterm m_1 appears.

Higher order k-maps can be obtained by successively doubling the size of the 1- variable k-map. This can be done for example by simply "folding out" the larger map from the smaller.

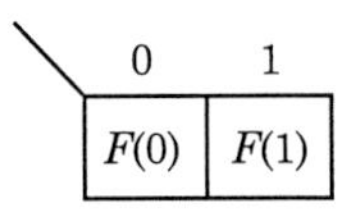

Fig. 7.15(a)

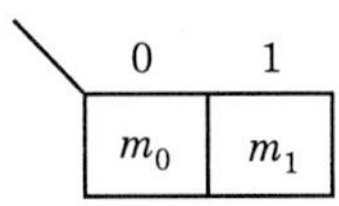

Fig. 7.15(b)

Two-variable Maps

A 2- variable k-map corresponding to $F(x_1, x_2)$ can be obtained by the fold-out procedure as shown in fig. 7.16(a) and 7.16(b).

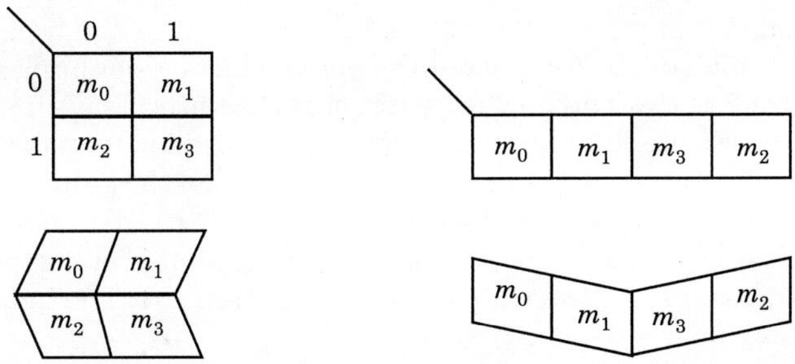

(a) (b)

Fig. 7.16.

The cells correspond to the minterms $m_o = x_1' x_2'$, $m_1 = x_1' x_2$, $m_2 = x_1 x_2'$, $m_3 = x_1 x_2$.

This particular arrangement forces minterms differing in only one variable to be adjacent. Figure 7.16(a), and 7.16(b) depict general form of 2 variable k-maps and figure 7.16(b) shows the k-map for $F(x_1, x_2) = \Sigma\ (2, 3)$.

The representation of $F(x_1, x_2)$ in fig. 7.16(a), & (b) is useful as each conveys the information that certain minterms are logically adjacent to others.

It should be noted that in the 2-variable k-map ; Every minterm is adjacent to the two other minterms. Similarly in the 3-variable k-map every minterms will adjacent to 3 other minterms and in 4-variable k-map every minterms will adjacent to 4 other minterms. Therefore in general in n-variable k-map minterm is adjacent to n other minterms.

Three-variable Map

The 3-variable k-map of figure 7.17(a) is produced by a fold out from the 2-variable map as given by the figure 7.17(b).

In the fig 7.17(b) the cell assigned to $m_5 = x_1\ x_2'\ x_3$ corresponds to row 1 and column 01 because binary 101 is equal to decimal 5. In other words, m_5 appears at the intersection of the row marked x_1 and column belonging to x_2' x_3.

The fig 7.17(c) shows the k-map for

$F = \Sigma(0, 3, 5, 7)$

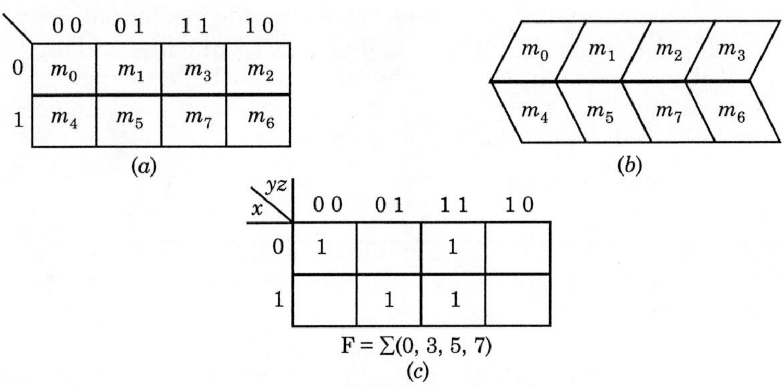

Fig. 7.17.

Four Variable Maps

Four variable k-maps can be obtained as indicated in the figure 7.18(a) from the 3 variable map by folding out the map and again doubling its size as shown in figure 7.18(b). The new cells represents minterms with subscripts that are 2^3 greater than their unfolded counterparts. Again, we have indicated in figure 7.18(c) the regions where the variables take on the value 1. To verify the location of minterms, let us consider $m_{13} = x_1\, x_2\, x'_3\, x_4$.

Since decimal 13 is equivalent to binary 1101, m_3 is located at the intersection of the row with heading 11 and the column with heading 01. The figure 7.18(c) shows the k-map for $F = \Sigma(1, 6, 7, 9, 12, 15)$.

wx \ yz	y'z'	y'z	y z	y z'
w'x'	m_0	m_1	m_3	m_2
w'x	m_4	m_5	m_7	m_6
w x	m_{12}	m_{13}	m_{15}	m_{14}
w x'	m_8	m_9	m_{11}	m_{10}

(a)

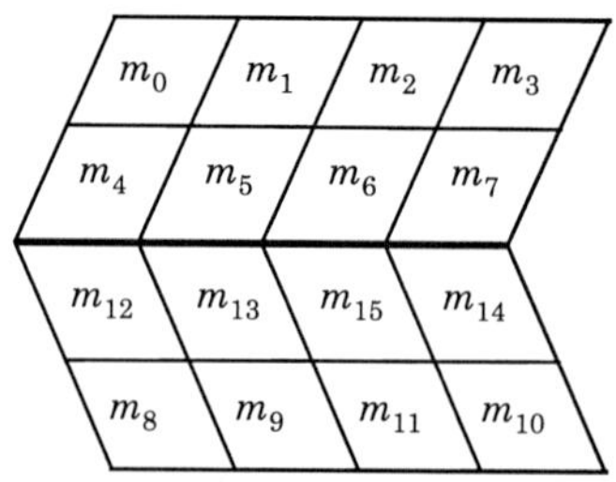

(b)

wx \ yz	y'z'	y'z	y z	y z'
w'x'		1		
w'x			1	1
w x	1		1	
w x'		1		

(c)

Fig. 7.18 (a).

Every minterm is the 4-variable k-map is adjacent to four other minterms because both the column and row headings follow the cyclic code sequence {0 0, 01, 11, 10}. Minterms m_o and m_2, m_4 and m_6, m_{12} and m_{14} and m_8 & m_{10} and adjacent if we view right edge as identified with the left edge. Likewise minterms m_o and m_8, m_1 and m_0, m_3 and m_{11}, m_2 and m_{10} are adjacent by identifying the top edge with the bottom edge.

Simplifying Karnaugh Map

From the postulates of Boolean algebra, it follows that the sum of two minterms in adjacent squares can be simplified to a single AND terms consisting of only two literals. To clarify this, consider the sum of two adjacent sequence such as m_5 and m_7.

$$m_5 + m_7 = xy'z + xyz = xz(y' + y)$$
$$= xz$$

Here the two squares differ by the variable y, which can be removed when the sum of the two minterms in formed. Thus any two minterms in adjacent squares that are O Red together will cause a removal of the different variable. The following example explains the procedure for minimising a Boolean function with a map.

Karnaugh Map

Each square represents one minterm. Since any Boolean function can be expressed as a sum of minterms, it follows that a Boolean function is recognized graphically in map from area enclosed by those squares whose minterms are included in the function.

Two Variable Map

$x \backslash y$	y'	y
x'	$x'y'$	$x'y$
x	xy'	xy

Four minterms for two variables $[2^2 = 4]$

Three Variable Map

Eight minterms for three variables $[2^3 = 8]$

$x \backslash yz$	$y'z'$ 0 0	$y'z$ 01	yz 11	yz' 10
0	$x'y'z'$	$x'y'z$	$x'yz$	$x'yz'$
1	$xy'z'$	$xy'z$	xyz	xyz'

Four Variable Map

$wx \backslash yz$	00	01	11	10
00	$w'x'y'z'$	$w'x'y'z'$	$w'x'yz$	$w'xyz'$
01	$w'xy'z'$	$w'xy'z$	$w'xyz$	$w'xyz'$
11	$wxy'z'$	$wxy'z$	$wxyz$	$wxyz'$
10	$wx'y'z'$	$wx'y'z$	$wx'yz$	$wx'yz'$

Four Binary Variables – 2^4 – minterms.

EXAMPLE 7.18. Simplify the boolean f^n

(*i*) $F = x'yz + x'yz' + xy'z' + xy'z$

$x \backslash yz$	$y'z'$ 0 0	$y'z$ 0 1	yz 1 1	yz' 1 0
x' 0	m_0	m_1	1 m_2	1 m_3
x' 1	1 m_4	1 m_5	m_6	m_7

$$\begin{aligned} f(x, y, z) &= x'yz + x'yz' + xy'z' + xy'z \\ &= x'y(z + z') + xy'(z' + z') \\ &= x'y + xy' \end{aligned}$$

(*ii*) $F = x'yz + xy'z' + xyz + xyz'$

$x \backslash yz$	$y'z'$ 00	$y'z$ 01	yz 11	yz' 10
x' 0			1	
x' 1	1		1	1

$$\begin{aligned} f(x, y, z) &= x'yz + xyz + xy'z' + xyz' \\ &= (x' + x)\,yz + x(y' + y)\,z' \\ &= yz + xz' \end{aligned}$$

(*iii*) $F = A'C + A'B + AB'C + BC$

$= A'C(B + B') + A'B(C + C') + AB'C + BC(A + A')$

As there is only a pair of 2 so multiply *e.g.* $A'C$, *i.e.*,

$$\begin{aligned} A'C &= (A'C)\,.\,1 = (A'C)\,(1 + 0) \\ &= (A'C)\,(B + B') = A'CB + A'CB' \end{aligned}$$

$A \backslash BC$	$B'C'$	$B'C$	BC	BC'
A'		1	1	1
A		1	1	

Applying the prucedure same as above we will get

$$F = C + A'B$$

(*iv*) $F(x, y, z) = \Sigma(0, 2, 4, 5\ 6)$

$x \backslash yz$	$y'z'$	$y'z$	yz	yz'
x'	$x'y'z'$	$x'y'z$	$x'yz$	$x'yz'$
x	$xy'z'$	$xy'z$	xyz	xyz'

Now in the form of Binary Notation

000	001	011	010
= 0	= 1	= 3	= 2
100	101	111	110
= 4	= 5	= 7	= 6

$$\begin{aligned} e.g.\ 001 &= 0 \times 2^2 + 0 \times 2' + 1 \times 2^0 = 1 \\ 011 &= 0 \times 2^2 + 1 \times 2' + 1 \times 2^0 = 3 \end{aligned}$$

So. (1) reduces to:

$x \backslash yz$	00	01	11	10
0	0	1	3	2
1	4	5	7	6

Now $\quad F(x, y, z) = \Sigma(0, 2, 4, 5, 6)$

x \ yz	00	01	11	10
0	1			1
1	1	1		1

Applying the procedure same as above.

So $F = z' + xy'$

Similarly, for the Four variable map

wx \ yz	y'z'	y'z	y z	y z'
w'x'	0	1	3	2
w'x	4	5	7	6
w x	12	13	15	14
w x'	8	9	11	10

EXAMPLE 7.19. Simplify the boolean expression

(*i*) $F(w, x, y, z) = \Sigma(0, 1, 2, 4, 5, 6, 8, 9, 12, 13, 14)$

wxyz *y′z′* *y′z* *yz* *yz′*

wx \ yz	y'z'	y'z	y z	y z'
w'x'	1	1		1
w'x	1	1		1
w x	1	1		1
w x'	1	1		

$$f = y' + w'z' + xz'$$

(*ii*) $F = A'B'C' + B'CD' + A'BCD' + AB'C'$

$= A'B'C'(D + D') + B'CD'(A + A') + A'BCD' + AB'C'(D + D')$

wx \ yz	C'D'	C'D	CD	CD'
w'x'	1	1		1
w'x				1
w x				
w x'	1	1		1

Ans. $F = B'C' + A'CD' + B'D'$

EXAMPLE 7.20. Simplify the Boolean function

$$f(w,x,y,z) = \Sigma(1, 3, 7, 11, 15)$$

and don't care conditions

$$f(w,x,y,z) = \Sigma(0, 2, 5)$$

wx \ yz	y'z'	y'z	y z	y z'
w'x'		1	1	
w'x			1	
w x			1	
w x'			1	

Ans. $f = yz + w'x'z'$

$= (y + w'x')z$

EXAMPLE 7.21. Obtain the simplified expressions in sum of product for the following Boolean functions.

$$F(x, y, z) = \Sigma(2, 3, 6, 7)$$

x \ yz	0 0	0 1	1 1	1 0
0			1	1
1		1	1	1

Ans. $f(x, y, z) = y$

EXAMPLE 7.22. Show the sum-of-product for the Karnaugh map given below:

(*a*)

	$\bar{z}\,\bar{w}$	$\bar{z}\,w$	$z\,w$	$z\,\bar{w}$
$\bar{x}\,\bar{y}$	0	0	0	0
$\bar{x}\,y$	0	0	0	1
$x\,y$	1	1	1	1
$x\,\bar{y}$	1	1	1	1

Ans. $f(x, y, z, w) = x + y\,\overline{zw}$

(*b*)

	$\bar{z}\,\bar{w}$	$\bar{z}\,w$	$z\,w$	$z\,\bar{w}$
$\bar{x}\,\bar{y}$	1	1	1	1
$\bar{x}\,y$	1	1	1	0
$x\,y$	1	0	0	0
$x\,\bar{y}$	1	0	0	0

Ans. $f(x, y, z, w) = \bar{z}\,\bar{w} + \bar{x}\,\bar{y} + \bar{x}\,w$

SOLVED PROBLEMS

PROBLEM 1. Write the Boolean expression that represents the combinatorial circuit, write the logic table and write the output of each gate symbolically.

Solution:

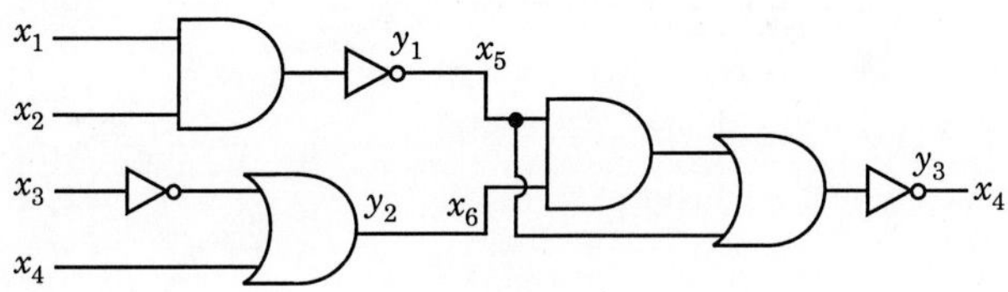

Fig. 1

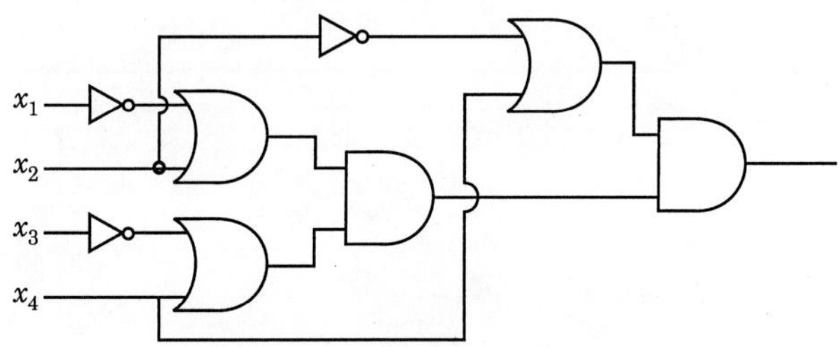

Fig. 2

PROBLEM 2. Show that the combinatorial circuits to following are equivalent.

Solution:

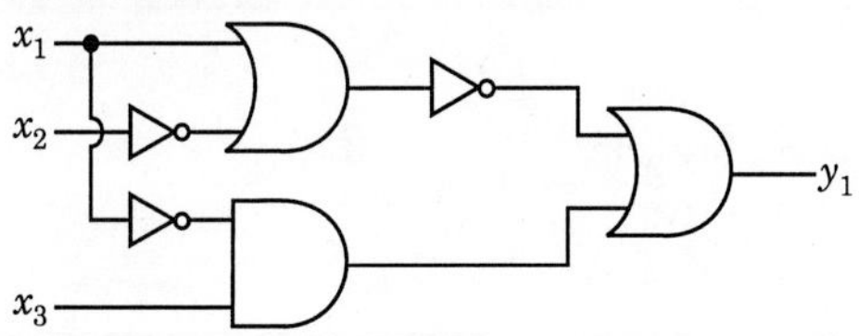

Fig. 3.

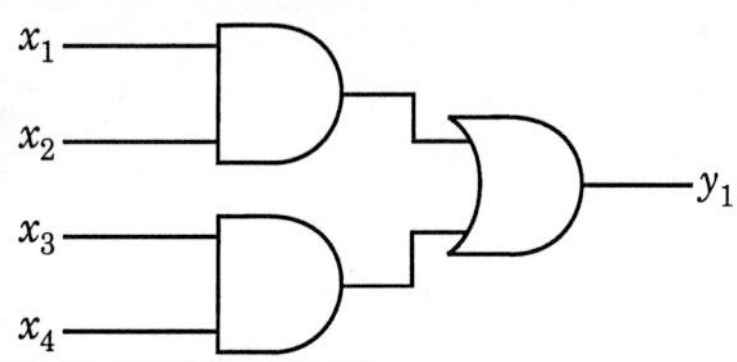

Fig. 4.

PROBLEM 3. Write the dual of each statement and also prove it.

(*a*) If $x + y = x + z$ and $x' + y = x' + z$ then $y = z$

(*b*) $x = 0$ if and only if $y = xy' + x'y$ for all y.

Solution: (*a*) *Dual* : if $xy = xz$ and $x'y = x'z$ then $y = z$

Proof: Assume $xy = xz$ and $x'y = x'z$

then
$$\begin{aligned} y &= y \,.\, 1 = y(x + x') = yx + yx' = xy + x'y \\ &= xz + x'z \\ &= zx + zx' \\ &= z(x + x') \\ &= z \,.\, 1 \\ &= z \end{aligned}$$

(*b*) *Dual* : $x = 1$ if and only if $y = (x + y')\,(x' + y)\ \forall y$

To prove this, we must recall the definition of q if and only if p or $p \leftrightarrow q$.

This is logically equivalent to $(p \to q) \wedge (q \to p)$

So we must prove that

$x = 1 \to y = (x + y')\,(x' + y)$ and then prove that $y = (x + y')\,(x' + y) \to x = 1$

Assume $x = 1$. Then
$$\begin{aligned} (x + y')\,(x' + y) &= (1 + y')\,(1' + y) \\ &= (1 + y')\,(0 + y) \\ &= 1(y) = y. \end{aligned}$$

So $x = 1 \to y = (x + y')\,(x' + y)$

Assume $y = (x + y')\,(x' + y)$ then
$$\begin{aligned} x &= x + 0 \\ &= x + yy' \\ &= x + (x + y')\,(x' + y)\,y' \qquad \text{(given)} \\ &= x + (xx' + xy + y'x' + y'y)y' \\ &= x + (xy + x'y')\,y' \\ &= x + (xyy' + x'y'y') \\ &= x + x + x'y' \\ &= x + x'y' \end{aligned}$$

In addition,
$$\begin{aligned} x = x + 0 &= x + yy' \\ &= x + y[(x + y')'\,(x' + y)'] \\ &= x + y[x'y + xy'] \\ &= x + x'y \end{aligned}$$

So we have $x = x + x'y'$ and $x = x + x'y$

Since
$$\begin{aligned} x = x + x &= x + x'y' + x + x'y \\ &= x + x'y' + x'y \\ &= x + x'\,(y' + y) \\ &= x + x' = 1 \end{aligned}$$

Since $x = 1 \to (x + y')\,(x' + y) = y$ and $y = (x + y')\,(x' + y) \to x = 1$ thus the statement.

PROBLEM 4. Find the disjunctive normal form of each functions and draw the combinatorial circuit corresponding to the disjunctive normal form.

(*a*)

x	y	z	$f(x, y, z)$
1	1	1	0
1	1	0	0
1	0	1	0
1	0	0	1
0	1	1	1
0	1	0	1
0	0	1	1
0	0	0	0

(*b*)

w	x	y	z	$f(w, x, y, z)$
1	1	1	1	1
1	1	1	0	0
1	1	0	1	1
1	1	0	0	0
1	0	1	1	0
1	0	1	0	0
1	0	0	1	0
1	0	0	0	1
0	1	1	1	1
0	1	1	0	0
0	1	0	1	0
0	1	0	0	0
0	0	1	1	1
0	0	1	0	0
0	0	0	1	0
0	0	0	0	0

Solution: (*a*)

(*b*)

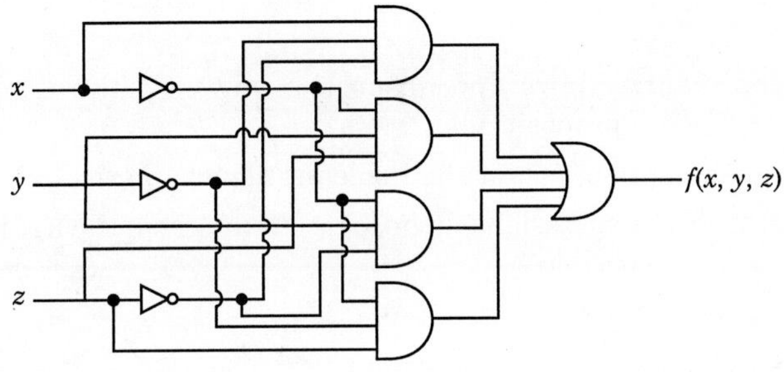

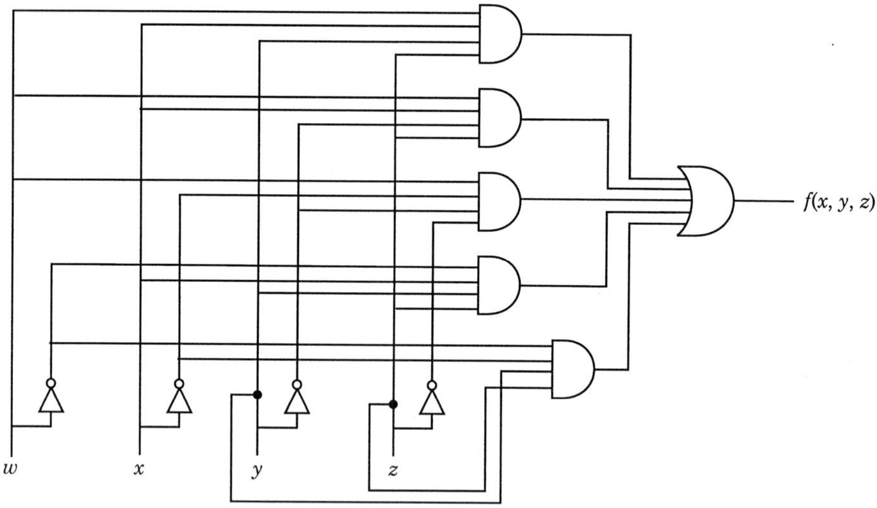

PROBLEM 5. Write $x \downarrow y$ in terms of $\uparrow$

Solution:

$$x \downarrow y = \overline{x \vee y}$$
$$= \overline{x}\,\overline{y}$$
$$= \overline{\overline{x} \uparrow \overline{y}}$$
$$= (x \uparrow y) \uparrow (x \uparrow y)$$
$$= (x \uparrow x) \uparrow (y \uparrow y) \uparrow (x \uparrow x) \uparrow (y \uparrow y)$$

So
$$x \downarrow y = (x \uparrow x) \uparrow (y \uparrow y) \uparrow (x \uparrow x) \uparrow (y \uparrow y)$$

EXERCISE 1

1. Verify the identity laws and absorption laws for the Boolean algebra $(S, \wedge, \vee, \sim, T, F)$ of propositions.
2. Verify the absorption law for the Boolean Algebra $(P(x), \cup, \cap, ', \phi, X)$
3. Find the Boolean expression for the logic circuit given the figures below and simplify them if possible.

(a)

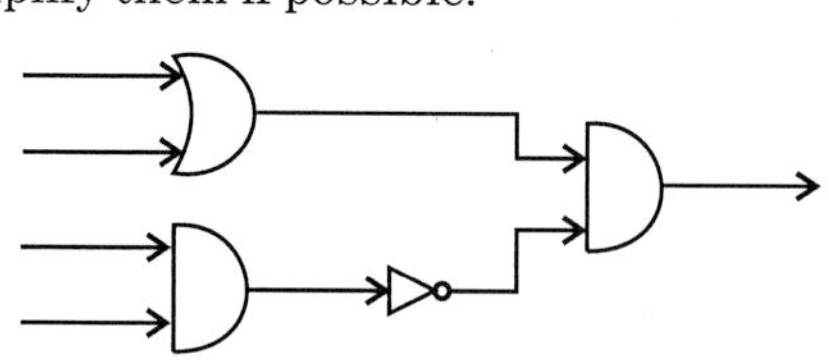

(b)

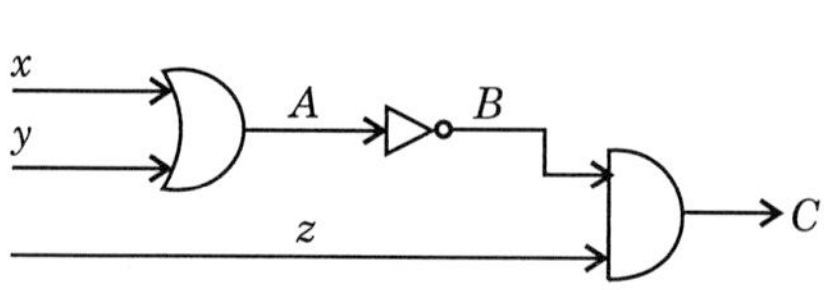

(c) 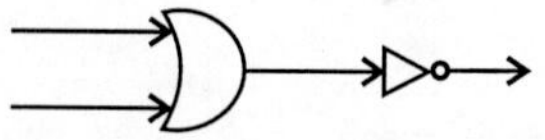

(d) 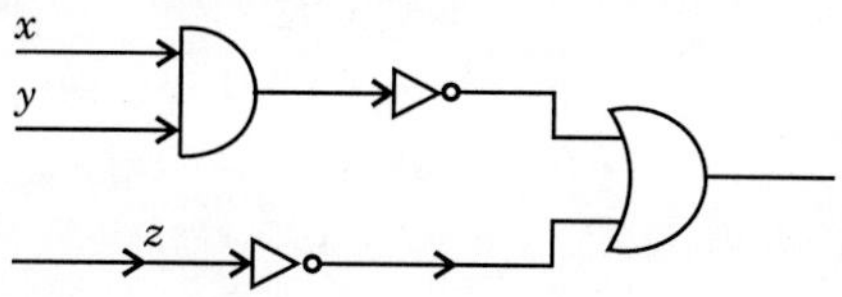

(e) 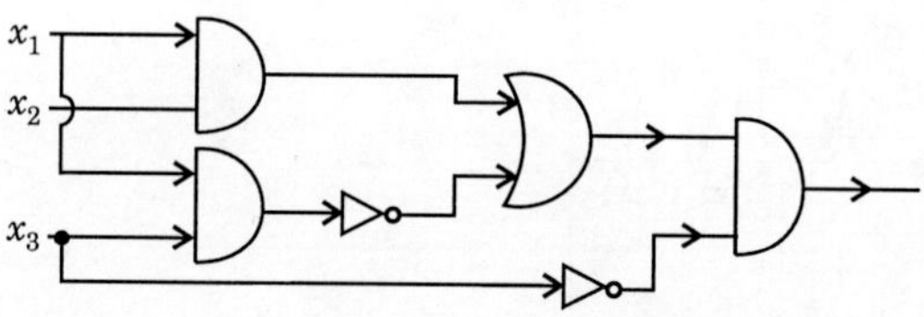

4. Construct the logic circuit represented by the Boolean expression $(x_1' \wedge x_2) \vee (x_1 \vee x_3)$, where $x_i (1 \le i \le 3)$ are assume to be the inputs.
5. Given the expression $(x_1' \vee (x_2 \wedge x'_3)) \wedge (x_2 \vee x'_4)$, find the corresponding circuit where $x_i(1 \le 1 \le 4)$ are consider to be input to the circuits.

In the following exercises 6–10 find the value of the Boolean Expression for

$$x_1 = 1, x_2 = 1, x_3 = 0, x_4 = 1$$

6. $\overline{x_1 \wedge x_2}$
7. $\left((x_1 \wedge (x_2 \vee (x_1 \wedge \overline{x_2}))\right) \vee \left(x_1 \wedge \overline{x_2}\right) \vee \overline{\left(x_1 \wedge \overline{x_3}\right)}$
8. $x_1 \vee \left(\overline{x_2} \wedge x_3\right)$
9. $\left((x_1 \wedge x_2) \vee (x_1 \wedge \overline{x_4})\right) \vee \overline{\left((\overline{x_1 \vee x_3}) \wedge (\overline{x_2} \vee x_3)\right)} \vee \left(x_1 \wedge \overline{x_3}\right)$
10. $x_1 \wedge \overline{x_2} \vee x_3$
11. Let $S = \{1, 2, 3, 6\}$. Define

$$x + y = lcm\,(x, y)$$
$$x \cdot y = gcd\,(x, y)$$
$$x' = \frac{6}{x}$$

for $x, y \in S$(*lcm* and *gcd* denote respectively, the least common multiple and the greatest common divisor).

Show that $(S, +, \cdot, ')$ is a Boolean algebra.

12. Write the dual of each statement
    (*a*) $(x + y)\,(x + 1) = x + xy + y$
    (*b*) $(x' + y)' = xy$
    (*c*) If $x + y = x + z$ and $x' + y = x' + z$ then $y = z$
13. Find the disjunctive normal form of each function and draw the combinatorial circuit corresponding to the disjunctive normal form and give the simplified form also.

(a)

x	y	$f(x, y)$
1	1	1
1	0	0
0	1	1
0	0	1

(b)

x	y	$f(x, y)$
1	1	0
1	0	1
0	1	0
0	0	1

(c)

x	y	z	$f(x, y, z)$
1	1	1	1
1	1	0	1
1	0	1	0
1	0	0	1
0	1	1	0
0	1	0	0
0	0	1	1
0	0	0	1

(d)

x	y	z	$f(x, y, z)$
1	1	1	1
1	1	0	1
1	0	1	1
1	0	0	0
0	1	1	0
0	1	0	1
0	0	1	1
0	0	0	1

14. Find the Disjunctive Normal form for each of the following functions.

(a) $f(x, y) = \left(x \vee y\right) \wedge \left(\bar{x} \vee \bar{y}\right)$

(b) $f(x, y, z) = x \vee y \wedge x$

(c) $f(x, y, z) = \left(\bar{x}\, y + \overline{xz}\right) \left(\overline{x \vee yz}\right)$

(d) $f(x, y, z) = x + \left(\bar{y} + (x\bar{y} + (x\bar{y} + x\bar{z})\right)$

15. Find the conjunctive Normal form for each of the functions given in the problem (14).

16. Simplify the following Boolean functions by using Karnaugh Map.

(i) $f(x, y, z) = \Sigma\,(3, 4, 6, 7)$

(ii) $f(x, y, z) = \Sigma\,(0, 2, 4, 5, 6)$

(iii) $X'Z + X'Y + XY'Z + YZ$

(iv) $f(x, y, z, w) = \Sigma(1, 3, 7, 11, 15)$

Answers for Exercise 1

6. 0

7. 1

8. 1

9. 1

10. 0

13. (*a*) $y + x'y'$
    (*b*) y'
    (*c*) $xyz + y'$
    (*d*) $xy + x'y' + xy'z + x'yz'$

16. (*i*) z'
    (*ii*) $z' + xy'$
    (*iii*) $z + x'y$
    (*iv*) $z' + wy'$

EXERCISE 2

1. Define Boolean algebra. Give two examples and justify your answer also.

2. Write the dual of each Boolean equation
   1. $(a \cdot 1) \cdot (0 + a') = 0$
   2. $a + ab' = a + b$
   3. $a(a' + b) = ab$
   4. $(a + 1)(a + 0) = a$
   5. $a(a' + b) = ab$

3. Consider the Boolean algebra D_{210}
   (*a*) List its elements and draw its diagram.
   (b) Is $X = \{1, 2, 3, 6\}$ a sublattice of D_{210} ?
   (c) Is $X = \{1, 2, 3, 6\}$ a subalgebra of D_{210} ..... ?
   (*d*) Find the number of subalgebras of D_{210}

4. Prove that the following are equivalent in a Boolean algebra.
   (*i*) $x + y = y$ (*ii*) $x\,y = x$
   (*iii*) $x' + y = 1$ (*iv*) $x\,y' = 0$

5. Reduce the following Boolean products to either 0 or a fundamental product.
   $xyx'z$
   $xyzy$
   $xyz'yz$

6. Consider the following lattices :
   (*a*) D_{20} (*b*) D_{36}
   (c) D_{100} (*d*) D_{130}
   Which of them are Boolean algebras and what are their atoms.

7. Express in the principle disjunctive normal form.
   (The following Booleanm function)
   (*i*) $f(a, b, c) = (a' + b)' + a' c$
   (*ii*) $f(x, y, z) = y\ (x + y\ z)'$
   (*iii*) $f(x, y, z) = x(x\ y' + x'y + y'z)$
   (*iv*) $f(x, y, z) = (x + y' z)\ (y + z')$
   (*v*) $f(x, y, z) = (x' y)' (x' + xyz')$
   (*vi*) $f(x, y) = (x\ y')' (x + y)'$
8. Give the output function f for an **AND gate** with inputs x, y, z that is give the function $f(x, y, z)$
   (1) $x = 1\ 1\ 1\ 0\ 0\ 1$
   $y = 1\ 0\ 0\ 1\ 0\ 0$
   $z = 1\ 1\ 0\ 0\ 1\ 1$

   (2) $x = 0\ 0\ 1\ 1\ 1\ 1\ 1\ 1$
   $y = 1\ 1\ 1\ 1\ 1\ 1\ 0\ 0$
   $z = 1\ 1\ 0\ 0\ 0\ 0\ 1\ 1$
9. Find the truth table T for the Boolean expression $f = f(x, y, z)$ where
   (*a*) $f = x\,z + x'\,y$ (*b*) $f = x\,y\,z' + x' + y\,z'$
10. Use Karnaugh map to obtain the following functions in the simplified form as sum of product
    (*a*) $x'z' + y'z' + yz' + xyz$ (*b*) $xy + x'y'z' + x'yz'$
    (*c*) $F(x, y, z, t) = \Sigma(7, 13, 14, 15)$ (*d*) $F(x, y, z, t) = \Sigma(2, 3, 12, 13, 14, 15)$
11. Using Karnaugh map find a minimal sum of product expression of the following function
    $F(A, B, C, D) = \Sigma(0, 1, 4, 5, 9, 11)$ ***(End-term Exam., GGSIPU, 2001)***
12. (*a*) Define: Distributive lattice, complemented lattice and Boolean algebra.
    (*b*) Use karnaugh map procedure to find the minimal sum of product form of the following Boolean expression.

$$\overline{X}_1\overline{X}_2X_3X_4 + X_1X_2X_3\overline{X}_4 + \overline{X}_1X_2X_3X_4 + \overline{X}_1X_2X_3\overline{X}_4 + X_1X_2\overline{X}_3\overline{X}_4 + \overline{X}_1\overline{X}_2X_3\overline{X}_4 + X_1\overline{X}_2\overline{X}_3X_4 + X_1\overline{X}_2X_3\overline{X}_4$$

(End-term Exam., GGSIPU, 2002)

Answers for Exercise 2

5. Hint :
   (*a*) $x\,y\,x'\,z = x\,x'\,y\,z$ (Commutature law)
   $= 0\,y\,z = 0$
   (*b*), (*c*), (*d*) same as (*a*).

9. (i)

x	y	z	$f = x \wedge y \wedge z$
1	1	1	1
1	0	1	0
1	0	0	0
0	1	0	0
0	0	1	0
1	0	1	0

$f = 100000$

(ii)

x	y	z	$f = x \wedge y \wedge z$
0	1	1	0
0	1	1	0
1	1	0	0
1	1	0	0
1	1	0	0
1	1	0	0
1	0	1	0
1	0	1	0

10. (a)

x	y	z	xz	x'	$x'y$	$x_2 + x'y$
1	1	1	1	0	0	1
1	1	0	0	0	0	0
1	0	1	1	0	0	1
1	0	0	0	0	0	0
0	1	1	0	1	1	1
0	1	0	0	1	1	1
0	0	1	0	1	0	0
0	0	0	0	1	0	0

(b) Same as (a)

DO IT BY YOURSELF

1. Let $E(x_1, x_2, x_3) = \left(\overline{x_1} \vee \overline{x_2}\right) \vee \left(\overline{x_1} \wedge x_2\right) \vee (x_2 \wedge x_3)$ be a Boolean expression over the two-valued Boolean algebra. Write $E(x_1, x_2, x_3)$ in both disjunctive and conjunctive normal forms.
2. Let $E(x_1, x_2, x_3) = (x_1 \vee x_2) \vee (x_1 \wedge x_3)$ be a Boolean expression over the two-valued Boolean algebra. Write $E(x_1, x_2, x_3)$ in both disjunctive and conjunctive normal forms.
3. Simplify the following Boolean expressions:

   (a) $(a \wedge b) \vee (a \wedge b \wedge c) \vee (b \wedge c)$ (b) $(a \wedge b) \vee (a \wedge \overline{b} \wedge c) \vee (b \wedge c)$

   (c) $(a \wedge b) \vee (\overline{a} \wedge b \wedge \overline{c}) \vee (b \wedge c)$ (d) $((a \wedge \overline{b}) \vee c) \wedge (a \wedge \overline{b}) \wedge c$
4. Show that Boolean Algebra is self dual.

5. Express each Boolean expression E (x, y, z) as a sum-of-products and then in its complete sum of-product form:
   (a) $E = x\,(xy' + x'\,y + y'z)$ (b) $E = (x + y'\,z)\,(\,y + z')$
   (c) $E = (x' + y)' + y'z$ (d) $E = (x'\,y)'\,(x' + xyz')$
   (e) $E = (x + y)\,(xy')'$ (f) $E = y(x + yz)'$
6. Use a Karnaugh map to find a minimal sum of each Boolean expression:
   (a) $E = xy + x'\,y + x'\,y'$ (b) $E = x + x'\,yz + xy'z'$
   (c) $E = y'z + y'z't' + z't$ (d) $E = y'\,zt + xzt' + xy'z'$

MULTIPLE CHOICE QUESTIONS

1. In Boolean Algebra the dual of $a.(b + c) = (a\,.\,b) + (a\,.\,c)$ is
   (a) $a\,.\,b + a\,.\,c = (a.\,b) + (a\,.\,c)$ (b) $a + (b\,.\,c) = (a + b)\,.\,(a + c)$
   (c) $(b + c)\,.\,a = (a\,.\,c) + (a\,.\,b)$ (d) None of these
2. In Boolean Algebra $a\,.\,a = a$ is known as
   (a) De-Morgan's law (b) Absorption law
   (c) Idempotent law (d) Associative law
3. In Boolean Algebra, which of the following statements is true for $x, y \in B$.
   (a) $(x.\,y)' = x'\,.\,y'$ (b) $(x.\,y)' = x' - y'$
   (c) $(x.\,y)' = x' + y'$ (d) $(x.\,y)' = x'\,/\,y'$
4. In Boolean Algebra, which of the following statements is true for $x, y \in B$ then value of $a' + a\,.\,b$ is equal to
   (a) $a + b'$ (b) $a' + b'$ (c) $a' + b$ (d) $a - b'$
5. Simplified form of the switching function $F(x, y) = x + y\,.\,y$ is
   (a) x . y (b) x (c) y (d) $x + y$
6. The Boolean expression $x\,.\,y' + x.\,y + x\,.\,y' + x'y$ is equivalent to
   (a) 0 (b) 1 (c) $x\,y$ (d) None of these
7. Complete D.N.F of a Boolean function in two variable p and q is
   (a) $p\,.\,q + p' + q + p\,.\,q' + p'\,.\,q'$ (b) $p\,.\,q' + p'\,.\,q$
   (c) $p\,.\,q$ (d) $p\,.\,p + q\,.\,q + p'\,.\,q + p\,.\,q'$
8. The complement of Boolean function $f(x, y) = x'\,y + xy' + x'\,y'$ is
   (a) $x\,.\,y$ (b) $x + y(c)$ $(x\,.\,y)'$ (d) $x'\,.\,y'$
9. Which of the following statement is true in Boolean algebra, when $a, b \in B$
   (a) $(x + y)' = x\,.\,y'$ (b) $(x + y)' = x + y'$
   (c) $(x + y)' = x\,.\,y$ (d) $(x + y)' = x + y$
10. In Boolean algebra, the dual of $a\,.\,a' = 0$
    (a) $a + a' = 1$ (b) $a'\,.\,a = 0$ (c) $a\,.\,a' = 1$ (d) $a' + a = 0$

Answers

1. (b)	2. (c)	3. (c)	4. (b)	5. (b)	6. (b)
7. (a)	8. (a)	9. (a)	10. (a)		

❒❒❒

CHAPTER 8

RECURRENCE RELATION

This Chapter Covers :

8.1. INTRODUCTION

A recurrence relation is an equation that expresses a given problem for n objects in terms of the same problem posed for less than n objects. For example, let us consider the most famous example of Fibonacci series proposed by Fibonacci in (1170-1250).

Consider an example of a pair of rabbits, one male and one female, are left on an island. These rabbits begin breeding at the end of two months and produce a pair of rabbits of opposite sex at the end of each month thereafter. The problem is to determine the number of pairs of rabbits after n months assuming no rabbit dies on this island.

Let a_n denotes the number of pairs of rabbits after n months. Therefore, $a_1 = 1$. As the pair does not breed in the second month also, therefore $a_2 = 1$.

To find the number of pairs of rabbits after n months, the number of pairs after $(n - 1)$ months must be added. But the new borns come from the pairs atleast two month old that is from the pairs that already existed after $(n - 2)$ months. Therefore the given sequence meets the condition.

$$a_n = a_{n-1} + a_{n-2} \quad \text{and if } n \geq 3., \; a_1 = 1 \text{ and } a_2 = 1$$

This sequence is called Fibonacci sequence and a_n are called Fibonacci numbers. The series obtained from the above relation is called Fibonacci series. There are various ways to solve the recurrence relation as substitution method, back tracking method. But in some cases it is easy to find the explicit formula for the given recurrence relation. As it gives directly the result at the rth stage. In this chapter we will discuss the method to find the explicit formula.

8.2. RECURRENCE RELATION

Let $\{a_n : n \geq 0\}$ be a sequence of real or complex numbers. A recurrence relation is an expression of the form.

$$a_n = F(a_{n-1}, a_{n-2}, ... a_0, g(n)) \qquad ...(i)$$

where $g(n)$ is a function of variable n and 'F' is a function of some of the variables $a_{n-1}, a_{n-2}, .. a_0, g(n)$

The expression (1) is equivalent to the linear differential equation with constant co-efficients, whose solution consists of two parts homogeneous solution or complementary solution and particular solution. Therefore the solution for (1) involves two parts homogeneous solution and particular solutions. The homogeneous solution, which satisfies the difference equation when the right hand side of the equation is set to 0 and the particular solution which satisfies the difference equation with $g(n)$ on the right hand side.

Following are some examples of homogeneous and non-homogeneous recurrence relations or difference equations.

EXAMPLE 8.1.

1. $a_n = 5a_{n-2} + 3\, a_{n-1}$ is a homogeneous equation
2. $a_n = 5a_{n-1} + n^2$ is a non-homogeneous equation.
3. $a_n = 2a_{n-2} + 2^n$ is a non-homogeneous equation.
4. $a_n = \sqrt{a_{n-1}} + a_{n-1}^2$ is a homogeneous equation.
5. $a_n = a_{n-1} + a_{n-2} + a_{n-3} + ... + a_1 + a_0$ is a homogeneous equation.
6. $a_n = a^3_{n-1} + a_{n-2} + a_{n-3}\, a_{n-4}$ is a homogenous equation.
7. $a_n = \sin a_{n-2} + \cos a_{n-1} + \sin a_{n-3} + ... + e^n$ is a non-homogeneous equation

From the above example we can say that a recurrence relation is said to be **homogeneous** if it has no term that depends only on the variable n, otherwise the recurrence relation is said to be **Non-homogeneous** recurrence relation.

Order of the Recurrence Relation

The order of the recurrence relation

$$a_n = F(a_{n-1}, a_{n-2}, ... a_{n-k}, g(n))$$

is k, where a_n depends on one or more of the previous k terms and k is the smallest such integer.

We do not define an order for recurrence relation of the form $a_n = F(a_{n-1}, a_{n-2}, ..., n)$ that depends on each of its previous terms.

Therefore if we can compute the nth term of a sequence from the preceding k terms, but not from the preceding $k - 1$ terms then we define the order to be k.

Now we try to understand the order of Recurrence relation by considering the following examples.

EXAMPLE 8.2. Consider the example (8.1) again and find the order of each given recurrence relation.

Solution :

1. $a_n = 5\, a_{n-2} + 3a_{n-1}$
   The order is 2
2. $a_n = 5a_{n-1} + n^2$
   The order is 1.
3. $a_n = 2a_{n-2} + 2^n$
   The order is 2
4. $a_n = \sqrt{a_{n-1}} + {a_{n-2}}^2$
   The order is 2
5. $a_n = a_{n-1} + a_{n-2} + a_{n-3} + ... + a_1 + a_0$
   The order is n
6. $a_n = a^3{}_{n-1} + a_{n-2} + a_{n-3}\, a_{n-4}$
   The order is 4.
7. $a_n = \sin a_{n-2} + \cos a_{n-1} + \sin a_{n-3} + ... + e^n$
   The given recurrence relation has no order.

Degree of the Recurrence Relation

The degree of the Recurrence relation is the degree of F in (1), treating as a polynomial in its variables excluding n.

It is to be noticed that if F is not a polynomial in its variables, no degree is assigned to the recurrence relation. Same as in differential equation a recurrence relation of degree one is called **Linear Recurrence Relation** a recurrence relation of degree two is called **quadratic recurrence relation** and so on.

EXAMPLE 8.3. Consider the following recurrence relation and order as well as degree of each recurrence relation.

$$a_n = \sin a_{n-1} + \cos a_{n-2} + \sin a_{n-3} + ...+ e^n$$

(Order and degree does not exist)

$$a_n = \sin a_{n-1}$$

$$= a_{n-1} - \frac{a_{n-1}^3}{3!} + \frac{a_{n-1}^5}{5!} + ... = f(a_{n-1})$$

Order 1 but degree does not exist & $f(a_{n-1})$ is not a polynominal & is an infinite series.

$$a_n = a_{n-1}\, a_{n-2}^3$$

⇒ degree is 4 and order is 2

EXAMPLE 8.4. Consider again the example 8.1

In this example the recurrence relation 1, 2, 3, 5 has degree 1. and 4, 7 has no degree and 6 has degree 3.

EXAMPLE 8.5. The general solution to $a_n = a_{n-1}$ is $a_n = k$ where k is any constant but if in addition $a_0 = 1$ then the solution is $a_n = 1, n \geq 0$.

This we have done by back tracking method. In this method we consider each of the recurrence relation in terms of previous term as given below.

$$a_n = a_{n-1} \quad \ldots(i)$$

Replacing $n \to n-1$

$$a_{n-1} = a_{n-2}$$
$$\Rightarrow a_n = a_{n-2}$$

Now by (*i*)

$$a_{n-2} = a_{n-3}$$
$$\Rightarrow a_n = a_{n-3}$$

Continuing like this we will get

$$a_n = a_0 = k \quad \text{and if} \quad k = 1 \quad \text{then } a_n = 1$$

EXAMPLE 8.6. The general solution to $a_n = a_{n-1} + 1$ is

$$a_n = c + n$$

where c is any constant.

As Let

$$\begin{aligned} a_0 &= c \\ a_1 &= a_0 + 1 = c + 1 \\ a_2 &= a_1 + 1 = c + 2 \\ a_3 &= a_2 + 1 = c + 3 \\ .. &= \ldots\ldots \\ a_n &= a_{n-1} + 1 \\ &= c + (n-1) + 1 \\ &= c + n \end{aligned}$$

Thus $a_n = C + n$

EXAMPLE 8.7. The general solution $a_n = k\, a_{n-1}$ can be found by the method of back tracking as given below:

$$\begin{aligned} a_n &= k\, a_{n-1} \\ a_n &= k(k\, a_{n-2}) \\ &= k^2 a_{n-2} \\ &= k^2(k\, a_{n-3}) \\ &= k^3 a_{n-3} \\ &\text{- - - -} \\ &= k^n a_0 \\ &= a_0 k^n \qquad n \geq 0. \end{aligned}$$

8.3. LINEAR HOMOGENEOUS RECURRENCE RELATION

The most general linear homogeneous equations with constant coefficient is

$$a_n = k_1 a_{n-1} + k_2 a_{n-2} + \ldots + k_r a_{n-r}, \; n \geq r \quad \ldots(i)$$

where all k_i's are constants

The characterstic equation or auxiliary equation of the linear homogeneous recurrence of order r is given by

$$z^r - k_1 z^{r-1} - k_2 z^{r-2} - \dots - k_{r-1} z - k_r = 0 \qquad \dots (ii)$$

The **roots** of the characterstic equation (ii) are called characterstic roots of (i). The **multiplicity** of the characterstic root of (i) is the greatest integer m such that $(z - a)^m$ is a factor of the characterstic polynomial of (i).

Solve the following recurrence using substitution method.

(a) $a_n = a_{n-1} + 3n^2 + 3n + 1$ for $n \geq 1$ where $a_0 = 1$

Ans. $a_n = n(n^2 + 4n + 2) + 1$

(b) $a_n = a_{n-1} + n^2$ for $n \geq 1$ where $a_0 = 7$

Ans. $a_n = \dfrac{n(n+1)(2n+1)}{6} + 7$

(c) $a_n \mp 2a_{n-1} + n$ for $n \geq 1$ where $a_0 = 1$ **Ans.** $a_n = 3.2^n - n - 2$

(d) $a_n = 2a_{n-1} + n$ for $n \geq 1$ where $a_0 = 1$ **Ans.** $a_n = 2^n - 1$

(e) $a_n = a_{n-1} + 2n + 1$ for $n \geq 1$ where $a_0 = 1$ **Ans.** $a_n = (n + 1)^2$

(f) $a_n = a_{n-1} + n^3$ for $n \geq 1$ and $a_0 = 5$ **Ans.** $a_n = \dfrac{n^2 (n+1)^2}{4} + 5$

(g) $a_n = a_{n-1} + \lfloor \sqrt{n} \rfloor$ for $n \geq 1$ where $a_0 = 1$ **Ans.** $a_n = \dfrac{n^2 (4n^2 - 3x + 5)}{6} + 1$

(h) $a_n = 2\dfrac{an}{2} + n - 1$ for r $n \geq 2$ where $a_1 = 0$

Ans. $a_n = n\dfrac{n-1}{n} + n\log_2(n)$

EXAMPLE 8.8. Consider the recurrence relation

(i) $\qquad U_{n+2} = 2U_n - U_{n-2} \qquad n \geq 2$

Its characterstic equation is given by

$$z^{n+2} = 2z^n - z^{n-2}$$

that is

$$z^4 = 2z^{2-1}$$

$$z^4 - 2z^2 + 1 = 0.$$

$$(z^2 - 1)^2 = 0$$

$\therefore z = \pm 1$ are the roots, each of multiplicity 2.

(ii) $\qquad a_n = 16\, a_{n-2}$

Its characterstic equation is

$$z^n = 16\, z^{n-2}$$

$$i.e.\ z^2 = 16$$

$$\Rightarrow z^2 - 16 = 0$$

$\Rightarrow$ z = ± 4 are the roots each of multiplicity 2.

Following theorem gives us the method to solve recurrence relation by finding explicit formula.

Theorem 8.1. (a) If the characterstic equation

$$x^r - k_1 x^{r-1} - k_2 x^{r-2} - \dots\dots\dots - k_{r-1} x - k_r = 0$$

of the recurrence relation

$$a_n - k_1 a_{n-1} - k_2 a_{n-2} - \ldots\ldots - k_r a_{n-r} = 0, \quad n \geq r$$

has roots say $\alpha_1, \alpha_2, \ldots. \alpha_r$ then explicit formula for the sequence is given as follows

Case I : If all roots are different

$$\alpha_i \neq \alpha_j \qquad i \neq j$$

$$a_n = C_1 \alpha_1^n + C_2 \alpha_2^n + C_3 \alpha_3^n + \ldots\ldots + C_r \alpha_r^n$$

Case II : If some of the roots are same. Say $\alpha_1 = \alpha_2 = \ldots\ldots \alpha_p = \alpha$, $p \leq r$ then,

$$a_n = \left(C_1 + C_2 n + C_3 n^2 + \ldots + C_p n^{p-1}\right)\alpha^n + C_{p+1}\alpha_{p+1}^n + \ldots\ldots + C_r \alpha_r^n$$

Case III : If all roots are same say $\alpha_1 = \alpha_2 = \ldots\ldots \alpha_r = \alpha$ then,

$$a_n = \left(C_1 + C_2 n + C_3 n^2 + \ldots\ldots + C_r n^{r-1}\right)\alpha^n$$

Following theorem is the particular case of the theorem (8.1) if $r = 2$.

Theorem 8.2.

(*a*) If the characterstic equation $x^2 - k_1 x - k_2 = 0$ of the recurrence relation $a_n = k_1 a_{n-1} + k_2 a_{n-2}$ has two distinct roots say α_1 and α_2 then

$$a_n = A\alpha_1{}^n + B\,\alpha_2{}^n$$

where A and B depend on initial conditions, is the explicit formula for the sequence.

(*b*) If the characterstic equation $x^2 - k_1 x - k_2 = 0$ has a single root say α, then the explicit formula is $a_n = (A + B^n)\,\alpha^n$

where A and B depend on the initial conditions.

EXAMPLE 8.9. Find an explicit formula for the sequence defined by:

$C_n = 5C_{n-1} - 6\,C_{n-2}$ with the initial conditions $C_1 = 2$ and $C_3 = 1$

Solution : Now $C_n = 5C_{n-1} - 6\,C_{n-2}$ is the recurrence relation of degree 2. It's characterstic equation is given as:

$$x^2 - 5x + 6 = 0$$

$$\Rightarrow x^2 - 3x - 2x + 6 = 0$$

$$\Rightarrow x(x-3) - 2(x-3) = 0$$

$$\Rightarrow (x-2)(x-3) = 0$$

$\Rightarrow x = 2$ and $x = 3$ are the roots with multiplicity 1.

Therefore by the above theorem

$$C_n = A(2)^n + B(3)^n$$

Now applying initial conditions we will get

$C_1 = 2$

$$\Rightarrow \quad 2 = A(2) + B(3)$$

$$\Rightarrow 2A + 3B = 2 \qquad \ldots(i)$$

$C_3 = 1$

$$\Rightarrow \quad 1 = A(2)^3 + B(3)^3$$

$$1 = 8A + 27\,B \qquad \ldots(ii)$$

By solving (*i*) and (*ii*) we get

$$A = \frac{3}{10} \text{ and } B = \frac{7}{15}$$

∴ the explicit formula for the given recurrence relation is

$$C_n = \frac{3}{10}(2)^n + \frac{7}{15}(3)^n$$

EXAMPLE 8.10. Solve the recurrence relation $d_n = 2\,d_{n-1} - d_{n-2}$ with the initial conditions.

$$d_1 = 1.5,\ d_2 = 3$$

Solution : Consider the given recurrence relation

$$d_n = 2d_{n-1} - d_{n-2}$$

Its characterstic equation is

$$z^2 - 2z + 1 = 0$$

$$(z-1)^2 = 0$$

∴ Its explicit formula is

$$d_n = (c_o + c_1 n)1^n$$

$$d_n = c_o + c_1 n$$

$$\Rightarrow d_1 = c_o + c_1 \qquad \Rightarrow 1.5 = c_o + c_1$$

$$\Rightarrow d_2 = c_o + 2c_1 \qquad \Rightarrow 3 = c_o + 2c_1$$

$$\Rightarrow c_1 = 1.5 \quad \text{and} \quad c_o = 0$$

$$\Rightarrow d_n = c_1 n$$

$$\Rightarrow \boxed{d_n = 1.5n}$$

EXAMPLE 8.11. Consider the recurrence relation

$$a_{n+2} = 2a_{n+1} + 4a_n - 8a_{n-1}$$

Find explicit formula.

Solution : Its characterstic polynominal in the variable Z is:

$$p(Z) = Z^3 - 2Z^2 - 4Z + 8$$

$$p(Z) = (Z-2)^2 (Z+2)$$

∴ Its characterstic roots are $Z = 2$ and $Z = -2$.

$Z = 2$ has multiplicity 2

$Z = -2$ has multiplicity 1

∴ Solution is

$$a_n = (c_0 + c_1 n)2^n + c_3(-2)^n.$$

where c_0, c_1 and c_2 are constants

EXAMPLE 8.12. Solve the recurrence

$$a_n = 4\,a_{n-2}$$

where (1) $a_0 = 4$ and $a_1 = 6$

(2) $a_0 = 6$ and $a_2 = 20$

(3) $a_1 = 6$ and $a_2 = 20$.

Solution : The characterstic equation of the given recurrence relation

$$a_n = 4\,a_{n-2} \quad \text{is}$$

$$Z^2 - 4 = 0$$

Its roots are $Z = \pm 2$

∴ By theorem (8.1).

the general solution is of the form

$$a_n = A(2)^n + B(-2)^n \text{ where } A \text{ and } B \text{ are constants}$$

(1) Now if $a_0 = 4$ and $a_1 = 6$

$\Rightarrow a_0 = A + B$ and $a_1 = 2(A - B)$

$\Rightarrow 4 = A + B$ and $6 = 2(A - B)$

Solving $8 + 6 = 4A$

$$\Rightarrow \boxed{A = \frac{7}{2}}$$

$$B = 4 - \frac{7}{2} = \frac{1}{2}$$

$$\Rightarrow \boxed{B = \frac{1}{2}}$$

∴ the solution is $a_n = \frac{7}{2} 2^n + \frac{1}{2}(-2)^n$

$$a_n = 7 \cdot 2^{n-1} - (-2)^{n-1}$$
$$= 2^{n-1}[7 + (-1)^n]$$

$$\Rightarrow a_n = \begin{cases} 2^{n-1} \times 8 & \text{if } n \text{ even} \\ 2^{n-1} \times 6 & \text{if } n \text{ odd} \end{cases}$$

$$\Rightarrow a_n = \begin{cases} 2^{n+2} & \text{if } n \text{ even} \\ 3(2^n) & \text{if } n \text{ odd} \end{cases}$$

(2) $a_0 = 6$ and $a_2 = 20$

$\Rightarrow A + B = 6$ and $20 = 4(A + B) \Rightarrow A + B = 5$ (Not possible)

Therefore, Inconsistent, that is, no solution.

(3) $a_1 = 6, a_2 = 20$

$$a_n = A\,2^n + B(-2)^n$$

$a_1 = A - 2B \quad \Rightarrow 6 = A - 2B$

$a_2 = 4A + 4B \quad \Rightarrow 5 = A + B$

$\Rightarrow B = 1$ and $A = 4$

∴ Solution is $a_n = 4(2)^n + (-2)^n.$

$$= 2^n[4 + (-1)^n]$$

$$a_n = 5 \times 2^n \quad \text{If } n - \text{even}$$
$$3 \times 2^n \quad \text{if } n \text{ odd.}$$

8.4. NON-HOMOGENEOUS RECURRENCE RELATION

The recurrence relation of the form

$$a_n - k_1 a_{n-1} - k_2 a_{n-2} \ldots - k_r a_{n-r} = g(n)$$

is called **non homogeneous recurrence relation** where $g(n)$ is a function of n. As we have discussed earlier that we should consider the particular solution also. The theorem (8.3) gives the general method for determining the particular solutions.

Solve the following *Homogenous recurrence* using the characteristics roots method.

1. $a_n = 6a_{n-1} - 8\, a_{n-2}$ for $n \geq 2$ where $a_0 = 2$ and $a_1 = 5$

Ans. $a_n = \frac{4^n}{2} + \frac{3}{2} 2^n$

2. $a_n = 3a_{n-1} + 4a_{n-2}$ for $n \geq 2$ where $a_0 = 1$ and $a_1 = 1$

Ans. $a_n = \frac{4^n}{5} + \frac{4}{5}(-1)^n$

3. $a_n = 4a_{n-1} - 4a_{n-2}$ for $n \geq 2$ where $a_0 = 5$ and $a_1 = 8$

Ans. $a_n = (5 - n)2^n$

4. $a_n = 5\, a_{n-1} - 6a_{n-2}$ for $n \geq 2$ where $a_0 = 0$ and $a_1 = 1$

Ans. $a_n = 3^n - 2^n$

5. $a_n = 2\, a_{n-1} - a_{n-2}$ for $n \geq 2$ where $a_0 = 1$ and $a_1 = 1$

Ans. $a_n = 1$

Theorem 8.3. Suppose that $\{a_n\}$ satisfies the linear nonhomogeneous recurrence relation

$$a_n - k_1\, a_{n-1} - k_2\, a_{n-2} - .... - k_r a_{n-r} = F(n)$$

where $k_1, k_2, ..., k_r$ are the real numbers and

$$F(n) = \left(\alpha_p\, n^p + \alpha_{p-1}\, n^{p-1} + ... + \alpha_1 n + \alpha_0\right) S^n$$

where $\alpha_0, \alpha_1, ..., \alpha_p$ and S are real numbers.

Case I : When S is not a root of the characterstic equation then there is a particular solution of the form

$$\left(\beta_p\, n^p + \beta_{p-1}\, n^{p-1} + ... + \beta_1 n + \beta_0\right) S^n$$

Case II : When S is the root of the characterstic equation and its multiplicity is m, then there is a particular solution of the form.

$$n^m \left(\beta_p\, n^p + \beta_{p-1}\, n^{p-1} + ... + \beta_1 n + \beta_0\right) S^n$$

where $\beta_1, \beta_2, ..., \beta_p$ are real numbers and the value of β_i depend on the initial conditions.

EXAMPLE 8.13. $C_n - 5C_{n-1} + 6C_{n-2} = 3n^2 - 2\, n + 1$ ...(*i*)

The solution for the left hand side part is already done in example (8.9). For the another part (that is Right hand side) By applying theorem (8.3), the general form of the solution is

$P_1\, n^2 + P_2\, n + P_3$

where P_1, P_2 and P_3 are constants to be determined. Substituting this expression in equation (1) and equating the coefficients and calculate the values of P_1, P_2 and P_3

Therefore,

$$(P_1n^2 + P_2n + P_3) - 5(P_1(n-1)^2 + P_2(n-1) + P_3) + 6(P_1(n-2)^2 + P_2(n-2) + P_3) = 3n^2 - 2n + 1$$

$$\Rightarrow (2P_1)n^2 + (2P_2 - 14\,P_1)\,n + (2P_3 - 7P_2 + 19P_1) = 3n^2 - 2n + 1$$

Now equating the co-efficients of power of n

$$2P_1 = 3$$

$$2P_2 - 14P_1 = -2$$

$$2P_3 - 7P_2 + 19P_1 = 1$$

$$\Rightarrow \quad P_1 = \frac{3}{2}$$

$$P_2 = \frac{19}{2}$$

$$P_3 = \frac{39}{2}$$

∴ Particular solution of the given recurrence relation is

$$\frac{3}{2}n^2 + \frac{19}{2}n + \frac{39}{2}$$

Now, by example (8.9) the explicity formula is

$$A(2)^n + B(3)^n$$

Therefore the complete solution is

$$a_n = \text{Explicit formula} + \text{Particular solution}$$

$$a_n = A(2)^n + B(3)^n + \left(\frac{3}{2}\right)n^2 + \left(\frac{19}{2}\right)n + \left(\frac{39}{2}\right)$$

EXAMPLE 8.14. Consider the recurrence relation.

$$C_n - 5\,C_{n-1} + 6\,C_{n-2} = 5\,n2^n \qquad ...(i)$$

Solution: The solution for the left hand part is already done is example (8.9). Now to find particular solution, compare the function on right hand side with the function $F(n)$ in the theorem (8.3)

i.e., $\quad F(n) = 5\,n\,2^n$

Here 2 is the root of the characterstic equation with multiplicity 1 therefore the particular solution is

$$n(P_1\,n + P_0)2^n$$

Substituting this value in equation (1) we will get

$$n(p_1n + P_0)2^n - 5(n-1)\,(P_1(n-1) + P_0)2^{n-1} + 6(n-2)(P_1(n-2) + P_0)\,2^{n-2} = 5n2^n$$

Divide both the sides by 2^{n-2}

$$n(p_1n + P_0)2^2 - 5(n-1)(P_1(n-1) + P_0)2 + 6(n-2)(P_1(n-2) + P_0) = 20n$$

By solving

$$-4P_0n + (14\,P_1 - 11P_0) = 20\,n$$

Now compare cofficients of powers of n, we will get

$$-4P_0 = 20 \quad \text{and} \quad 14P_1 - 11P_0 = 0$$

$$\boxed{P_0 = -5} \quad \text{and} \quad 14P_1 = -55$$

$$\boxed{P_1 = \frac{-55}{14}}$$

Thus, the Particular solution is:

$$-n\left(\frac{55}{14}n + 5\right)2^n$$

The Explicit formula is

$$C_n = \text{(Complementary function)} + \text{(Particular solution)}$$

$$C_n = A(2)^n + B(3)^n - n2^n\left(5 + \frac{55}{14}n\right)$$

EXAMPLE 8.15. Find the particular solution for the following recurrence relation.

$$a_n - 6a_{n-1} + 9a_{n-2} = 3^n \qquad \ldots(i)$$

Solution : The auxillary equation is given by:

$$x^2 - 6x + 9 = 0$$

$$\Rightarrow \quad (x-3)^2 = 0$$

$$\Rightarrow \quad x = 3, 3$$

Here $x = 3$ is the root of the auxillary equation of multiplicity 2. Therefore by applying the theorem (8.3), the particular solution will have the form

$$n^2 P\, 3^n$$

Substituting this value in equation (*i*) we will get

$$n^2 P\, 3^n - 6(n-1)^2 P\, 3^{n-1} - 9(n-2)^2 P\, 3^{n-2} = 3^n$$

$$n^2 P\, 3^n - 6(n^2 - 2n + 1) P\, 3^{n-1} - 9(n^2 - 4n + 4) P\, 3^{n-2} = 3^n$$

Divide both the sides by 3^{n-2}, we will get

$$9n^2 P - 18P(n^2 - 2n + 1) + 9P(n^2 - 4n + 4) = 9$$

$$\Rightarrow -18P + 36P = 9$$

$$\Rightarrow \quad 18P = 9$$

$$\Rightarrow \quad \boxed{P = \frac{1}{2}}$$

Therefore, the particular solution of the given recurrence relation will be:

$$\boxed{\frac{n^2}{2}3^n}$$

EXAMPLE 8.16. Find the explicit formula for the given recurrence relation

$$a_r - 7a_{r-1} + 10a_{r-2} = 2r^2 + 2,$$

with the initial conditions $a_0 = 0$, $a_1 = 1$

Solution: The given recurrence relation is:

$$a_r - 7a_{r-1} + 10a_{r-2} = 2r^2 + 2$$

Characterstic equation is in the variable x is given as:

$$x^2 - 7x + 10 = 0$$

$$\Rightarrow \quad x^2 - 5x - 2x + 10 = 0$$
$$\Rightarrow \quad x(x-5) - 2(x-2) = 0$$
$$\Rightarrow \quad (x-2)(x-5) = 0$$
$$\Rightarrow \quad x = 2, 5$$

The characterstic roots are 2 & 5 each of multiplicity 1. Consider the function on RHS of the given equation

$$F(r) = 2r^2 + 2$$
$$= 2(r^2 + 1)$$

As 2 is the root of the characterstic equation with multiplicity 1. Therefore the particular solution is given as

$$r\left(P_2r^2 + P_1r + P_0\right)2$$

Now substitute this value in the given recurrence relation.

$$2r(P_2\, r^2 + P_1\, r + P_0) - 14\,(r-1)\,[P_2(r-1)^2 + P_1(r-1) + P_0] + 20(r-2)$$
$$[P_2(r-2)^2 + P_1(r-2) + \mathrm{P}_0] = 2(r^2 + 1)$$

By Solving

$$(4P_2)r^3 + (4P_1 - 39\,P_2)r^2 + (99\,P_2 - 26P_1 + 8P_0)r + (59\,P_2 + 33\,P_1 + 17\,P_0)$$
$$= r^2 + 1$$

Compairing the coefficients of power of r, we will get

$$\boxed{P_2 = 0}$$
$$4P_1 - 39P_2 = 1$$
$$\Rightarrow \quad \boxed{P_1 = 1/4}$$
$$-26 \times \frac{1}{4} + 8P_0 = 0$$
$$\Rightarrow \quad P_0 = \frac{-26}{32}$$
$$\Rightarrow \quad \boxed{P_0 = \frac{-13}{16}}$$

The required explicit formula is:

$$a_n = A\,2^r + B\,5^r + 2r\left[\frac{1}{4}r - \frac{13}{16}\right]$$

Solve the following Non-homogeneous recurrence relations using the characteristics roots method:

1. $a_n = 2a_{n-1} + 3\,a_{n-2}$ for $+\,n^2$ for $n \geq 2$ where $a_0 = 7$ and $a_1 = 3$

Ans. $a_n = \dfrac{19}{4}(-1)^n + \left(\dfrac{27}{8}\right)3^n - \dfrac{9}{8} - n - \dfrac{n^2}{4}$

2. $a_n = n^3 - 2\,a_{n-2} - a_{n-2} - a_{n-2}$ for $n \geq 2$ where $a_0 = 5$ and $a_1 = 3$

Ans. $a_n = \left(\dfrac{47}{8} - \dfrac{73}{8}n\right)(-1)^n - \dfrac{7}{8} - \dfrac{3}{8}n + \dfrac{3}{4}n^2 + \dfrac{n^3}{4}$

3. $a_n = 6\, a_{n-1} - 8\, a_{n-2} + n\, 4^n$ for $+ n^2$ for $n \geq 2$
where $a_0 = 1$ and $a_1 = 3$

Ans. $a_n = \dfrac{4^n}{2} + \dfrac{2^n}{2} + (n^2 - n)\, 4^n$

4. $a_n = 4\, a_{n-2} + n^2 - 3n + 1$ for $n \geq 2$

Ans. $a_n = \dfrac{189}{108} 2^n - \dfrac{13}{108}(-2)^n - \dfrac{n^2}{3} - \dfrac{7}{9} n - \dfrac{17}{27}$

8.5. GENERATING FUNCTION

For many practical purposes, it is essentially very important specially to engineering and science students to have an alternative way of representation of numbers or functions. In this section, we introduce another alternative way of representing sequences (or numeric functions).

Definition : (Generating function) The generating function for the sequence of real (or complex) numbers, $\{a_0, a_1, ...a_n, ...\}$ is given by the power series

$$F(z) = \sum_{k=0}^{\infty} a_k z^k$$

$$= a_0 + a_1 z + a_2 z^2 + ... + a_n z^n + ...$$

Thus, the $(n + 1)^{th}$ term a_n of the sequence $\{a_n\}_{n \geq 0}$ is simply the co-efficient of z^n in $F(z)$.

Thus generating function identifies the different terms of a sequence by different powers of the symbol z.

Thus A functional form of a series associated with a sequence is a generating function.

Thus the associated power series of the constant sequence $\{a, a, ...\}$ is

$$a - az + az^2 - az^3 + ... = a\,[1 - z + z^2 - z^3 + ...]$$

$$= a\,(1 + z)^{-1}$$

Thus $(1 + z)^{-1}$ is the generating function of the constant sequence $\{a, a, ...\}$.

Algebra of Generating Function

Let $a = <a_n>$ and $b = <b_n>$ and $c = <c_n>$ be sequences and $G_1(z)$, $G_2(z)$ and $G_3(z)$ be the corresponding generating functions, then

1. For $b = \alpha\, a$,

$$G_2(z) = \alpha\, G_1(z)$$

2. Corresponding to $c = a + b$, we have

$$G_3(z) = G_1(z) + G_2(z)$$

3. Let b be a sequence such that $b_r = \alpha^r a_r$ for some constant α. Then

$$G_2(z) = a_0 + \alpha a_1 z + \alpha^2 a_2 z^2 + ... + \alpha^r a_r z^r + ...$$

$$= a_0 + a_1(\alpha z) + a_2(\alpha z)^2 + ... + a_r(\alpha z)^r + ...$$

$$= G_1(\alpha z)$$

4. For the convolution of two sequences $c = a + b$ where

$$c_r = a_0\, b_r + a_1\, b_{r-1} + ... + a_{r-1}\, b_1 + a_r\, b_0$$

is the co-efficient of z^r in the product

$$(a_0 + a_1 z + \ldots\ldots + a_r z^r + \ldots\ldots)\ (b_0 + b_1 z + \ldots\ldots + b_r z^r + \ldots\ldots)$$

then the generating function of c is

$$G_3(z) = G_1(z)\ G_2(z)$$

More generally, let $G(z)$ be the generating function of the geometric progression $\{ar^n\}_{n \geq 0}$, *i.e.*,

$$G(z) = a + (ar)z + (ar^2)\ z^2 + \ldots\ldots$$

Then $$G(z) - a = rz\ [a + (ar)z + (ar^2)\ z^2 + \ldots\ldots] = rz\ G(z)$$

$$\Rightarrow \qquad G(z) = \frac{a}{1 - rz}$$

EXAMPLE 8.17. The generating function for the finite geometric progression

$$\{a, ar, ar^2, \ldots\ldots, ar^{k-1}\} \text{ is } \frac{a(1 - r^k z^k)}{(1 - rz)}$$

Solution: Let $G(z)$ be the generating function of the geometric progresssioin

$$\{a, ar, \ldots\ldots, ar^{k-1}\}$$

then $$G(z) = a + (ar)z + \ldots\ldots + (ar^{k-1})\ z^{k-1}$$

$$G(z) = a\ [1 + rz + r^2 z^2 + \ldots\ldots + r^{k-1} z^{k-1})$$

$$= \frac{a(1 - (rz)^k)}{(1 - (rz))} \quad \text{(sum of GP.)}$$

EXAMPLE 8.18. The generating function for the sequence of bionomial co-efficients

$$\{C(k, 0), C(k, 1)a, C(k, 2)a^2, \ldots\ldots\} \text{ is } (1 + az)^k.$$

Solution : Let $G(z)$ be the generating function of the sequence of bionomial co-efficients

$$\{C(k, 0), C(k, 1)a, C(k, 2)a^2, \ldots\ldots\}$$

$$G(z) = C(k, 0) + C(k, 1)a + C(k, 2)a^2 + \ldots\ldots$$

$$= 1 + ka + \frac{k(k-1)}{2} a^2 +$$

$$= (1 + az)^k$$

EXAMPLE 8.19. Determine the number of subsets of a set of n elements, $n \geq 0$.

Solution : Let a_n denotes the number of subsets that a set of n elements can have. Then

$$a_n = 2\ a_{n-1} \quad \text{if} \quad n \geq 1 \quad \text{and} \quad a_0 = 1$$

Let $G(z)$ be the generating function of the sequence $\{a_n\}_{n \geq 0}$. So

$$G(z) = \sum_{n=0}^{\infty} a_n z^n \qquad \ldots(i)$$

$$= 1 + \sum_{n=1}^{\infty} a_n z^n$$

$$= 1 + 2\sum_{n=1}^{\infty} a_{n-1} z^n$$

$$= 1 + 2z\sum_{n=0}^{\infty} a_n z^n$$

$$= 1 + 2z \,.\, G(z)$$

$$\Rightarrow \qquad G(z) = \frac{1}{1-2z} = \sum_{n=0}^{\infty} 2^n z^n \quad ....(ii) \quad \text{[By Bionomial Theorem]}$$

Comparing (*i*) and (*ii*)

$$a_n = 2^n,\ n \geq 0$$

Thus the number of subsets of a set of n elements is 2^n $\forall n$.

Let us now look at some general results which provides connection between the generating functions of various sequences, terms of which are related in some manner to each other.

Following theorem can help us to obtain the generating function of the product of two sequence if the generating functions of individual sequences are known.

Theorem. If $G(z)$ is the generating function for the sequence $\{a_n\}_{n \geq 0}$ and $G_2(z)$ is the generating function for the sequence $\{b_n\}_{n \geq 0}$, then $G_1(z) \times G_2(z)$ is the generating function for the sequence $\{c_n\}_{n \geq 0}$ where

$$c_n = \sum_{k=0}^{n} a_k\, b_{n-k},\quad n \geq 0$$

Proof :

$$G_1(z) \times G_2(z) = \left(\sum_{i=0}^{\infty} a_i z^i\right)\left(\sum_{j=0}^{\infty} b_j z^j\right)$$

$$= a_0\left(\sum_{i=0}^{\infty} b_j z^j\right) + a_1\left(\sum_{j=0}^{\infty} b_j z^{j+1}\right) + \,.....$$

$$= \sum_{i=0}^{\infty} a_i \sum_{j=0}^{\infty} b_j z^{i+j}$$

$$= \sum_{n=0}^{n}\left(\sum_{j=0}^{n} a_j b_{n-j}\right) z^n$$

Theorem. Suppose that the sequence $\{a_n\}_{n \geq 0}$ has the generating function $G(z)$. Then the generating function $F(z)$ (say) for the sequence $\{b_n\}_{n \geq 0}$, where $b_n = a_n - a_{n-1}$, for $n \geq 1$ and $b_0 = a_0$, is given by

$$F(z) = (1 - z)\, G(z).$$

Proof : By definition, the generating function for the sequence $\{b_n\}$ is:

$$F(z) = \sum_{n=0}^{\infty} b_n z^n$$

$$= b_0 + \sum_{n=1}^{\infty} b_n z^n$$

$$= a_0 + \sum_{n=1}^{\infty} (a_n - a_{n-1}) z^n$$

$$= a_0 + \sum_{n=1}^{\infty} a_n z^n - z \sum_{n=1}^{\infty} a_{n-1} z^{n-1}$$

$$= a_0 + [G(z) - a_0] - z \, . \, G(z)$$

$$= (1 - z) \, G(z)$$

Theorem. If the generating function for the sequence a_r is known then the generating function for the sequence a_{r+1}, a_{r+2} can be find easily.

Proof : Let $G(z)$ be the generating function for the sequence having the general term a_r, then we have

$$G(z) = a_0 + a_1 z + a_2 z^2 + \ldots\ldots. + a_r z^r + \ldots.. \quad \ldots(i)$$

$$\frac{G(z) - a_0}{z} = a_0 + a_2 z + a_3 z^2 + \ldots\ldots. + a_r z^{r-1} + a_{r+1} z^r + \text{- - -}$$

Thus $\frac{G(z) - a_0}{z}$ is the generating function for the sequence a_{r+1}.

Now consider $G(z)$ again then

$$\frac{G(z) - a_0 - a_1 z}{z^2} = a_2 + a_3 z + a_4 z^2 + \ldots\ldots$$

is the generating function for a_{r+2}.

Table 8.1. Useful Generating Functions

General term of the sequence a_r	*Generating function G(z)*
1	$\frac{1}{(1-z)}$
$r + 1$	$\frac{1}{(1-z)^2}$
r	$\frac{z}{(1-z)^2}$
$r(r + 1)$	$\frac{2z}{(1-z)^3}$
k^r, k is constant	$\frac{1}{1-zk}$
$c(n, k)$	$(1 + z)^n$
$c(n, k)\, a^k$	$(1 + az)^n$
$\frac{1}{k!}$	e^z.

$\frac{(-1)^{k+1}}{k}$	$\ln(1+z)$
a_r	$G(z)$
a_{r+1}	$\frac{G(z)-a_0}{z}$
a_{r+2}	$\frac{G(z)-a_0-a_1z}{z^2}$
a_{r+3}	$\frac{G(z)-a_0-a_1z-a_2z^2}{z^3}$
a_{r+n}	$\frac{G(z)-a_0-a_1z \ldots\ldots a_{n-1}z^{n-1}}{z^n}$

EXAMPLE 8.20. Find the generating function $G(z)$ (say) for the sequence in the arithmetic progression $\{a, a+d, a+2d, \ldots..\}$

Solution: The generating function $G(z)$ for the given sequence is

$$\begin{aligned} G(z) &= a + (a+d)z + (a+2d)z^2 + \ldots\ldots \\ &= (a + az + az^2 + \ldots\ldots) + (dz + 2dz^2 + \ldots\ldots) \\ &= (a + az + az^2 + \ldots\ldots) + (dz)(1 + 2z + 3z^2 + \ldots\ldots) \\ &= a(1 + z + z^2 + \ldots\ldots) + (dz)(1 + 2z + 3z^2 + \ldots\ldots) \\ &= a(1-z)^{-1} + (dz)(1-z)^{-2} \\ &= \{a + (d-a)z\}(1-z)^{-2} \end{aligned}$$

EXAMPLE 8.21. If $G(z)$ is the generating function for the sequence $\{a_n\}_{n \geq 0}$. Show that the generating function $F(z)$ for the sequence $\{b_n\}$ of its partial sums *viz.* $b_n = \sum_{i=0}^{n} a_i$, $(n \geq 0)$ is given by $F(z) = \frac{G(z)}{1-z}$.

EXAMPLE 8.22. The generating function of the sequence $(3^0, 3^1, 3^2, \ldots\ldots 3^r, \ldots\ldots)$ is

$$\begin{aligned} G(z) &= 3^0 + 3^1 z + 3^2 z^2 + \ldots\ldots + 3^r z^r + \ldots\ldots \\ &= 1 + 3z + (3z)^2 + \ldots\ldots + (3z)^r \\ &= \frac{1}{1-3z} \end{aligned}$$

EXAMPLE 8.23. Suppose the sequence $< a_n >$, $0 \leq n \leq \infty$, has generating function $G(z)$

(*i*) What sequence is generated by the function

$$F(z) = (1+z)\, f(z)\ ?$$

(*ii*) What sequence is generated by the function

$$H(z) = \frac{f(z)}{1+z}\ ?$$

Solution: (*i*) Here $G(z) = a_0 + a_1 z + a_2 z^2 + \ldots\ldots + a_r z^r + \ldots\ldots$

$$F(z) = (1+z)\, G(z)$$

$$= (1 + z)(a_0 + a_1 z + \ldots\ldots + a_r z^r + \ldots\ldots)$$
$$= a_0 + (a_0 + a_1)z + (a_1 + a_2)z^2 + \ldots\ldots + (a_{r-1} + a_r)z^r + \ldots\ldots$$

Thus the sequence corresponding to $F(z)$, say $< b_n >$ is

$$b_i = \begin{cases} a_0, & i = 0 \\ a_{i-1} + a_i, & i \geq 1 \end{cases}$$

(ii) $$H(z) = \frac{f(z)}{1+z}$$
$$= (a_0 + a_1 z + \ldots\ldots + a_r z^r + \ldots\ldots) + (1 + (-1)z + (-1)^2 z^2 + \ldots\ldots)$$
$$= a_0 + (a_1 + (-1)a_0)z + (a_2 + (-1)a_1 + (-1)^2 a_0)z^2 + \ldots\ldots$$
$$+ \ldots\ldots + (a_r + a_{r-1}(-1) + a_{r-2}(-1)^2 + \ldots\ldots$$
$$+ a_1(-1)^{r-1} + a_0(-1)^r)z^r + \ldots\ldots$$

Thus if $<c_n>$ is the sequence corresponding to the generating function $H(x)$. Then

$$c_r = a_r + a_{r-1}(-1) + a_{r-2}(-1)^2 + \ldots\ldots + a_1(-1)^{r-1} + a_0(-1)^r, \; r \geq 0$$

8.5.1. Solution of Recurrence relation by the method of Generating Function

Generating functions can be used to find the explicit formula for the recurrence relation. In the last section, on solving a recurrence relation, we have got the solution in the form of sequence. Instead of solving in the form of a sequence expression for the value of the sequence can easily be obtained once the generating functions are known.

EXAMPLE 8.24. Solve the recurrence relation

$$a_r - 3a_{r-1} + 2a_{r-2} = 0, \quad r \geq 2 \qquad \ldots(i)$$

by the generating function method with initial conditions $a_0 = 2$ and $a_1 = 3$.

Solution: Let $G(z)$ be the generating function of the sequence $<a_n>$, $n \geq 0$

Multiplying each side of (1) by z^r, we get

$$a_r z^r - 3a_{r-1} z^r + 2a_{r-2} z^r = 0$$

Taking summation on both the sides keeping $r = 2$ to ∞, we obtain

$$\sum_{r=2}^{\infty} a_r z^r - 3\sum_{r=2}^{\infty} a_{r-1} z^r + 2\sum_{r=2}^{\infty} a_{r-2} z^r = 0$$

$$(G(z) - a_0 - a_1 z) - 3z(G(z) - a_0) + 2z^2 G(z) = 0$$
$$(2z^2 - 3z + 1)\, G(z) - a_0 - a_1 z + 3a_0 z = 0$$

Now using the given conditions *i.e.*, $a_0 = 2$, $a_1 = 3$, we get

$$(2z^2 - 3z + 1)\, G(z) - 2 - 3z + 6z = 0$$

$$G(z) = \frac{2 - 3z}{2z^2 - 3z + 1} = \frac{1}{1-z} + \frac{1}{1-2z}$$
$$= (1 + z + z^2 + \ldots\ldots) + (1 + 2z + 2^2 z^2 + \ldots\ldots)$$
$$= (1 + 2^0)\,(1 + 2^1)z + (1 + 2^2)z^2 + \ldots\ldots$$

Thus, we have

$$a_r = 1 + 2^r$$

EXAMPLE 8.25. Solve the recurrence relation

$$a_r - 5a_{r-1} + 6a_{r-2} = 2, \ r \geq 2$$

by the generating function method with the boundary conditions $a_0 = 1$ and $a_1 = 1$.

Solution: Let $G(z)$ be the generating function of the sequence $< a_n >$, $n \geq 0$. Multiplying each side of the given equation by z^r and summing for all $r \geq 2$, we obtain

$$\sum_{r=2}^{\infty} a_r z^r - 5\sum_{r=2}^{\infty} a_{r-1} z^r + 6\sum_{r=2}^{\infty} a_{r-2} z^r = 2\sum_{r=2}^{\infty} z^r$$

$$(G(z) - a_0 - a_1 z) - 5z(G(z) - a_0) + 6z^2\, G(z) = 2z^2 \sum_{r=2}^{\infty} z^{r-2}$$

Applying initial value conditions we will get

$$(1 - 5z + 6z^2)\, G(z) - 1 + 4z = \frac{2z^2}{1-z}$$

$$(6z^2 - 5z + 1)\, G(z) = \frac{2z^2}{1-z} + 1 - 4z$$

$$G(z) = \frac{2z^2}{(1-z)(1-2z)(1-3z)} + \frac{1-4z}{(1-2z)(1-3z)}$$

$$= \frac{6z^2 - 5z + 1}{(1-z)(1-2z)(1-3z)} = \frac{1}{1-z}$$

Consequently we have $a_0 = 1$.

EXAMPLE 8.26. Solve the recurrence relation by the method of generating function.

$$a_r - 3a_{r-1} = 2, \ r \geq 1, \ a_0 = 1$$

Solution. Multiplying both the sides by z^r and do summations from 1 to ∞ we will get

$$\sum_{r=1}^{\infty} a_r z^r - 3\sum_{r=1}^{\infty} a_{r-1} z^r = 2\sum_{r=1}^{\infty} z^r$$

Let $G(z) = \sum_{r=0}^{\infty} a_r z^r$ be the generating function of the sequence $< a_n >$, $n \geq 0$.

Then we have

$$(G(z) - a_0) - 3z\, G(z) = 2\,(1 + z + z^2 + z^3 + \ldots\ldots)$$

$$(1 - 3z)\, G(z) - a_0 = 2(1 - z)^{-1}$$

Applying initial value condition

$$-3z\, G(z) = \frac{2}{(1-z)}$$

$$G(z) = \frac{2}{-3z(1-z)}$$

$$G(z) = \frac{-2}{3z}\ (1 - z)^{-1}$$

$$= \frac{-2}{3z}\ [1 + z + z^2 + z^3 + \ldots\ldots]$$

$$= \frac{-2}{3}\left[\frac{1}{z} + 1 + z + z^3 + \ldots\ldots\right]$$

Consequently $a_r = \frac{-2}{3} z^{r-1}$.

EXAMPLE 8.27. Solve the recurrence relation

$$a_{r+2} - 3a_{r+1} + 2a_r = 0$$

by the method of generating functions with the initial conditions $a_0 = 2$ and $a_1 = 3$.

Solution: Multiplying the given equation by z^r and summing from $r = 0$ to $r \to \infty$, we have

$$\sum_{r=0}^{\infty} a_{r+2}\, z^r - 3\sum_{r=0}^{\infty} a_{r+1}\, z^r + 2\sum_{r=0}^{\infty} a_r\, z^r = 0$$

$$(a_0 + a_3 z + a_4 z^2 + \ldots\ldots) - 3(a_1 + a_2 z + a_3 z^2 + \ldots\ldots)$$
$$+ 2(a_0 + a_1 z + a_2 z^2 + \ldots\ldots) = 0$$

$$\left(\frac{G(z) - a_0 - a_1 z}{z^2}\right) - 3\left(\frac{G(z) - a_0}{z}\right) + 2G(z) = 0$$

Applying the initial conditions we will have

$$\left(\frac{G(z) - 2 - 3z}{z^2}\right) - 3\left(\frac{G(z) - 2}{z}\right) + 2G(z) = 0$$

Solving we will get

$$G(z) = \frac{2 - 3z}{1 - 3z + 2z^2}$$

$$= \frac{2 - 3z}{(1 - z)(1 - 2z)} = \frac{1}{1 - z} + \frac{1}{1 - 2z}$$

Consequently $a_r = 1 + 2^r$.

EXAMPLE 8.28. Solve the recurrence relation

$$a_r - 7a_{r-1} + 10a_{r-2} = 0$$

by the method of generating functions with the initial conditions $a_0 = 3$ and $a_1 = 3$.

Solution: Multiplying the given recurrence relation z^r and summing from 0 to ∞.

$$\sum_{r=2}^{\infty} a_r z^r - 7\sum_{r=2}^{\infty} a_{r-1} z^r + 10\sum_{r=2}^{\infty} a_{r-2} z^r = 0$$

$$(G(z) - a_0 - a_1 z) - 7z(G(z) - a_0) + 10z^2\, G(z) = 0$$

Now, put $a_0 = 3$ and $a_1 = 3$

$$G(z) = \frac{3+24z}{10z^2 - 7z + 1}$$

$$G(z) = \frac{3+24z}{(5z-1)(2z-1)}$$

$$G(z) = \frac{13}{1-5z} - \frac{10}{1-2z}$$

Consequently $a_r = 13(5)^r - 10(2)^r$.

EXAMPLE 8.29. Solve the recurrence relation

$$a_r - 2a_{r-1} + a_{r-2} = 2^r$$

by the method of generating functions with the initial conditions $a_0 = 2$ and $a_1 = 1$.

Solution. Multiplying the given recurrence relation by z^r and summing from 2 to ∞.

$$\sum_{r=2}^{\infty} a_r z^r - 2\sum_{r=2}^{\infty} a_{r-1} z^r + \sum_{r=2}^{\infty} a_{r-2} z^r = \sum_{r=2}^{\infty} 2^r z^r$$

$$\left(\frac{G(z) - a_0 - a_1 z}{z^2}\right) - 2\left(\frac{G(z) - a_0}{z}\right) + G(z) = \frac{1}{1-2z}$$

Applying the initial conditions, we will get on solving

$$(z^2 - 2z + 1)\, G(z) = 2 + 3z + \frac{z^2}{1-2z}$$

$$G(z) = \frac{2}{(1-z)^2} + \frac{3z}{(1-z)^2} + \frac{z^2}{(1-2z)(1-z)^2}$$

$$= \frac{1}{(1-z)^2} + \frac{3z}{(1-z)^2} + \frac{1}{1-2z}$$

Consequently $\quad a_r = (r + 1) + 3r + 2^r = 1 + 4r + 2^r$.

8.6. ACKERMANN'S FUNCTION

Ackermann's function can be defined by the recurrence relations.

$$A(m, 0) = A(m-1, 1) \qquad m = 1, 2, \ldots\ldots.$$

$$A(m, n) = A(m-1, A(m, n-1)) \qquad m = 1, 2, \ldots\ldots.;\ n = 1, 2, \ldots\ldots..$$

and the initial conditions

$$A(0, n) = n + 1, \quad n = 0, 1, \ldots\ldots.$$

EXAMPLE 8.30. Show that $A(1, n) = n + 2,\ n = 0, 1, \ldots\ldots..$

Solution: We use the principle of Mathematical induction

Basic Step : $\quad A(1, 0) = A(0, 1) = 2$

Induction Step :

$$A(1, n+1) = A(0, A(1, n))$$
$$= A(0, n+2)$$
$$= n + 3 = (n+1) + 2$$

n is arbitrary.

Thus the result is true for all n.

EXAMPLE 8.31. Use induction to show that

$$A(2, n) = 3 + 2n, \quad n = 0, 1, \ldots\ldots$$

Solution: Basic Step :

$$A(2, 0) = A(1, 1) = 3 \qquad \text{(As } A(1, n)\ n + 2 \text{ proved earlier)}$$

Inductive Step :

$$\begin{aligned} A(2, n+1) &= A(1, A(2, n)) \\ &= A(1, 3 + 2n) \qquad \text{(By inductive assumption0} \\ &= 3 + 2n + 2 \\ &= 2n + 5 \\ &= 2(n+1) + 3 \end{aligned}$$

As n is arbitrary.

Thus the result is true for all n.

EXAMPLE 8.32. Prove that $A(m, n) > n$ for all $m \geq 0$, $n \geq 0$ by induction on m. The inductive step will use induction on n.

Solution : Basic Step : $m = 0$ $A(0, n) = n + 1 > n \quad \forall n \geq 0$

Inductive Step : Let us assume that

$$A(m, n) > n \quad \forall\, n \geq 0$$

Now we prove $A(m+1, n) > n \quad \forall n \geq 0$

Basics ($n = 0$) $A(m+1, n) = A(m+1, 0) = A(m, 1) > 1$

$$> 0 = n \qquad \text{(By our assumption)}$$

Inductive

$$\begin{aligned} A(m+1, n+1) &= A(m, A(m+1, n)) \\ &> A(m+1, n) \qquad \text{(By original assumptions)} \\ &> n \qquad \text{(By present inductive assumptions)} \end{aligned}$$

Now $A(m+1, n+1) > A(m+1, n) \geq n + 1$ and that completes the current inductive step. Therefore $A(m+1, n) > n \quad \forall n \geq 0$.

Thus the result is true.

EXAMPLE 8.33. Solve the recurrence relation $a_n = 2^n a_{n-1}$, $n > 0$, with initial condition $a_0 = 1$.

Solution :

$$\begin{aligned} a_n &= 2^n a_{n-1} \\ &= 2^n \cdot 2^{n-1} a_{n-2} \\ &= 2^n \cdot 2^{n-1} \cdot 2^{n-2} \cdot a_{n-3} \\ &\ldots \quad \ldots \quad \ldots \\ &\ldots \quad \ldots \quad \ldots \\ &= 2^n\, 2^{n-1}\, 2^{n-2} \ldots\ldots\, 2^1 a_0, \ a_0 = 1 \\ &= 2^{[n + (n-1) + (n-2) + \ldots\ldots + 1]} \\ &= 2^{\frac{n(n+1)}{2}}. \end{aligned}$$

EXAMPLE 8.34. Solve the recurrence relation given below:

$$a_n = 7a_{n-1} - 10a_{n-2}, \ a_0 = 5, a_1 = 16$$

Solution : $a_n - 7a_{n-1} + 10a_{n-2} = 0$

Characteristic equation

$$t^2 - 7t + 10 = 0$$

$$(t-2)(t-5)=0$$
$$t = 2, 5$$
So $$a_n = A(2)^n + B(5)^n \quad \text{...(*)}$$
$$a_0 = 5$$
$$\Rightarrow \quad 5 = A + B \quad \text{...}(i)$$
$$a_1 = 16$$
$$\Rightarrow \quad 16 = 2A + 5B \quad \text{...}(ii)$$

Solving (*i*) and (*ii*) for A and B and substitute the values of A and B in *, we get

$$a_n = 3.2^n + 2.5^n.$$

EXAMPLE 8.35. Solve the recurrence relation

$$a_n = 2a_{n-1} + 8a_{n-2}$$

with the initial conditions $a_0 = 4$, $a_1 = 10$.

Solution : $$a_n = 2a_{n-1} + 8a_{n-2}, \quad a_0 = 4, a_1 = 10$$

Characteristic equation is

$$t^2 - 2t - 8 = 0$$
$$(t+2)(t-4) = 0$$
$$t = (-2, 4)$$
$$\Rightarrow \quad a_n = A.(-2)^n + B(4)^n$$
$$a_0 = 4 = A + B \quad \text{...}(i)$$
$$\Rightarrow \quad a_1 = 10 = -2A + 4B \quad \text{...}(ii)$$

Solving (*i*) and (*ii*) we get $A = 1, B = 3$

So $$a_n = (-2)^n + 3(4)^n.$$

EXAMPLE 8.36. Solve the recurrence relation for the initial conditions given:

$$c_n = 2 + \sum_{i=1}^{n-1} c_{i,} \quad n \ge 2; \quad c_1 = 1$$

Solution: Given : $$c_n = 2 + \sum_{i=1}^{n-1} c_i$$

$$\Rightarrow \quad c_{n+1} = 2 + \sum_{i=1}^{n} c_i$$

Consider $$c_{n+1} - c_n = \sum_{i=1}^{n} c_i + \sum_{i=1}^{n-1} c_i = c_n$$

$$\Rightarrow \quad c_{n+1} = 2c_n, \quad n \ge 2$$
or
$$c_{n+1} = 2 \,.\, 2 \,.\, c_{n-1}$$
$$= 2^3\, c_{n-2}$$
$$\cdots \quad \cdots \quad \cdots$$
$$\cdots \quad \cdots \quad \cdots$$
$$= 2^{n-1}\, c_2$$
$$= 3 \,.\, 2^{n-1}, \quad n \ge 2 \qquad [\text{As } c_2 = 2c_1 = 2]$$
$$\Rightarrow \quad c_n = 3 \,.\, 2^{n-1}, \quad n \ge 2$$

EXERCISE 1

1. Define recurrence relation, order and degree of the recurrence relation. Explain it with the help of at least two example.
2. Give difference between homogeneous and non-homogeneous recurrence relation. Explain it by giving at least three differences.
3. What is the characterstic equation and characteristic roots of given recurrence relation.
4. $a_r - 7a_{r-1} + 10a_{r-2} = 0$, given that $a_0 = 0$ and $a_1 = 4$.
5. $a_r - 4a_{r-1} + 4a_{r-2} = 0$, given that $a_0 = 2$ and $a_1 = 9$.
6. $a_r - 2a_{r-1} + 2a_{r-2} - a_{r-3} = 0$, given that $a_0 = 2$, $a_1 = 1$ and $a_2 = 4$.
7. $a_r - a_{r-1} - a_{r-2} = 0$, given that $a_0 = 1$ and $a_1 = 1$.
8. $a_r + 6a_{r-1} + 9a_{r-2} = 9$, given that $a_0 = 0$ and $a_1 = 1$.
9. $a_r - 7a_{r-1} + 10a_{r-2} = 2$, given that $a_0 = 0$ and $a_1 = 1$.
10. Consider the problems 4 to 9 and solve them.
11. Solve the following recurrence relation
    $a_r + 6a_{r-1} + 12a_{r-2} + 8a_{r-3} = 0$
12. Consider the problem 8 and give the general solution if right hand side is replaced by $3r^2 - 2r + 1$.
13. Solve the recurrence relation
    $$a_r - 5a_{r-1} + 6a_{r-2} = 1$$
    With the initial conditions $a_0 = 0$ & $a_1 = 1$
14. Give the explicit formula for the difference equation given as:
    $$a_r + a_{r-1} = 5r\,4^r$$
15. Given that $a_0 = 0$, $a_1 = 1$, $a_2 = 4$ and $a_3 = 12$ satisfy the recurrence relation
    $$a_r + C_1a_{r-1} + C_2a_{r-2} = 0$$
16. Give the explicit formula for the difference equation given as:
    $$f(r) = 3r^2$$
    where $f(r)$ is given in the left hand side of problem 11.
17. Consider the recurrence relation
    $$a_r = a_{r-1} - a_{r-2}$$
    (*a*) Solve the recurrence relation, given that $a_1 = 1$ and $a_2 = 0$.
    (*b*) Can you solve the recurrence relation if it is given that $a_0 = 0$ and $a_3 = 0$?
    (*c*) Repear part (*b*) if it is given that $a_0 = 1$ and $a_3 = 2$
18. Determine the particular solution for the the difference equation
    $$a_r - 3a_{r-1} + 2a_{r-2} = 4^r$$
19. Solve the following recurrence relation and give explicity formula
    $$a_r - 2a_{r-1} = 5r$$
20. Determine the particular solution for the the difference equation
    $$a_r - 4a_{r-1} + 4a_{r-2} = r3^r$$
21. Solve the recurrence relation 8 to 10 by the method of backtracking.

22. (*a*) Find the general solution of recurrence relation

$T(k) - 3T(k-1) - 4T(k-2) = 4^k$

(*B*) Solve as completely as possible the following recurrence relation :

$T(n) = 3T(\lfloor n/2 \rfloor)$,

$T(0) = 0$ ***(End-term Exam., GGSIPU, 2001)***

23. Give an explicit formula for Fibonacci sequence, and solve it. ***(Terminal Exam., GGSIPU, 2001)***

24. Solve the recurrence relation :

$a_n + 5a_{n-1} + 6\,a_{n-2} = 3\,n^2 - 2n + 1$

Describe binary search algorithm also. ***(Terminal Exam., GGSIPU, 2001)***

25. (*a*) Solve the recurrence relation:

$a_n = a_{n-1} + a_{n-2};\ n \geq 2$ with $a_0 = 1$ and $a_1 = 1$

(*b*) Explain the meaning of recursively defined functions. When a recursive function is said to be well defined. Find the value of Acbermann Function $A(1, 3)$. ***(End-term Exam., GGSIPU, 2002)***

26. Solve the following recurrence relation

(*i*) $T(k) + 3kT(k-1) = 0\ T(0) = 1$

(*ii*) $nS(n) - S(n-1) = 0\ S(0) = 1$

27. Solve the recurrence relation using generating function

$S(n) - 2S(n-1) - 3S(n-2) = 0\ n \geq 2$ with $S(0) = 3$ and $S(1) = 1$

28. What do you mean by sum rule and product rule ?

29. (*a*) Solve the recurrence relation

$A_r + 3A_{r-1} + 2A_{r-2} = f(r)$

where $f(r) =$ 1, $r = 2$

0 otherwise

With boundary condition $A_0 = A_1 = 0$.

EXERCISE 2

Tell whether or not each recurrence relation in exercises 1—10 is a linear homogeneous recurrence relation with constant co-efficients. Give the order of each linear homogeneous recurrence relation with constant co-efficients.

1. $a_n = -5a_{n-1}$
2. $a_n = 3n\,a_{n-2} - a_{n-1}$
3. $a_n = 2n\,a_{n-2}$
4. $a_n = a_{n-1} + 5n$
5. $a_n = 7a_{n-2} - 9a_{n-3}$
6. $a_n = a_{n-1} + 1 + 2^{n-1}$
7. $a_n = (\log 2n)a_{n-1} - [\log(n-1)]a_{n-2}$
8. $a_n = 6a_{n-1} - 7a_{n-3}$
9. $a_n = -a^2_{n-2} - 5a_{n-5}$
10. $a_n = -a_{n-1} - 7a_{n-2} - 3a_{n-3}$

Answers

1. linear, order 1.
2. Not, co-efficient not constant
3. Not, co-efficient not constant
4. linear, non Homogeneous
5. Yest order 3
6. No.
7. Coefficients are not constants
8. Yest order 3
9. Not linear
10. Yes, order 3.

MULTIPLE CHOICE QUESTIONS

1. The recurrence relation $a_r = 7\, a_{r-1} + r^3$ is
   (*a*) Non homogeneous of order 1 and degree 3
   (*b*) Non homogeneous of order 1 and degree 1
   (*c*) Homogeneous of order 2 and degree 1
   (*d*) Homogeneous or order 1 and degree 1
2. The characteristics roots of the characteristics equation $(x-2)\,(x-3)^4 = 0$ are
   (*a*) $x = 2$ of multiplicity 1 and $x = 3$ of multiplicity 4
   (*b*) $x = 2$ and $x = 3$
   (*c*) $x = 2$
   (*d*) $x = 3$
3. For a given sequence 3, 7, 11, 15, 19, 23, ... The recussive formula is given by
   (*a*) $d_1 = 2\,n + 1,\ dn\ = 2n + 6$ (*b*) $d_1 = 2\,n,\ dn\ = 2\,(n+1) + 3$
   (*c*) $d_1 = 3,\ dn = d_{n-1} + 4$ (*d*) $d_1 = 3\,n,\ dn\ = 2\,d_{n+1} + 3$
4. Find the recurrence relation and give initial conditions for the the number of bit strings of length n that do not contain the pattern 11.
   (*a*) $C_n = 2\,C_{n-1} + 3\,C_{n-2},\ n \geq 2,\ C_1 = 2,\ C_2 = 3$
   (*b*) $C_n = 3\,C_{n-2} + 2\,C_{n-3},\ n \geq 3,\ C_1 = 3,\ C_2 = 2$
   (*c*) $C_n = 2\,C_{n-2} + C_{n-3},\ n \geq 2\ C_1 = 2,\ C_2 = 3$
   (*d*) $C_n = C_{n-1} + C_{n-2},\ n \geq 3\ C_1 = 2,\ C_2 = 3$
5. Find the generating function for the sequence $1, a, a^2, a^3$ ... where a is fixed constant
   (*a*) $\dfrac{a}{1+ax}$ (*b*) $\dfrac{a}{1-ax}$ (*c*) $\dfrac{a}{1-ax}$ (*d*) $\dfrac{1}{1+ax}$
6. Let $F(x)$ and $G(x)$ are two generating functions. What will be the value of $F(r).\ G(x)$ [Multiplicaiton of two generating functions]
   (*a*) $\sum\limits_{k=0}^{\infty} (a_k + b_k)\, x^k$ (*b*) $\sum\limits_{k=0}^{\infty} \left[\sum\limits_{i=0}^{k} a_i\, b_{k-1}\right] x^k$
   (*c*) $\sum\limits_{k=0}^{\infty} (a_k - b_k)\, x^{k+1}$ (*d*) $\sum\limits_{k=0}^{\infty} \left[\sum\limits_{i=0}^{k} a_{i-1}\, b_k\right] x^{k+1}$

7. If the general term of sequence a_r is a_{r+2} then generating function $G(t)$ is given by

(a) $\dfrac{G(t) - a_0 - a_1 t}{t_2}$ (b) $\dfrac{G(t) - a_0^2 - a_1 t}{t+1}$

(c) $\dfrac{G(t) - a_0 - a_1 t^2}{t+2}$ (d) $\dfrac{G(t) - a_0 - a_1^2 t}{t^2}$

8. Let $a_{r+4} + 4\, a_{r+3} + 8\, a_{r+2} + 8\, a_{r+1} + 4\, a_r = 0$ is a recurrence relation and ithas complex root of multiplicity 2 then solution has form of

(a) $(A_1 + A_2 r)\ \alpha_1^r + (A_3 + A_4\, r)\ \alpha_2^r$ where $\alpha_{1,\,2} = -1 \pm i$

(b) $(A_1 r + A_2)\ \alpha_1^r + (A_3 r + A_4)\ \alpha_2^r$ where $\alpha_{1,\,2} = -1 \pm i$

(c) $(A_1 + A_2 r + A_3\, r^2 + A_4\, r^3)\ \alpha^r$ where $\alpha = -1 \pm i$

(d) $A_1\ \alpha_1^r + A_2\ \ \alpha_2^r$ where $\alpha_{1,\,2} = -1 \pm i$

9. Tell which one is not a linear homogeneous recurrence relation with constant coefficient

(a) $a_n = 7\, a_{n-2} - 9\, a_{n-3}$ (b) $a_n = a_{n-2} - 1\ 2^{n-1}$

(c) $a_n = -5\, a_{n-1}$

(d) $a_n = (\log 2n)\, a_{n-1} - [\,\log (n-1)]\, a_{n-2}$

10. In A ckermann's function if the value of m and n are not equal to zero, then the recurence relation is given by

(a) $A(m, n) = A(m-1, 1)$ () (b) $A(m, n) = A(m-1, A\,(m, n-1))$

(c) $A\,(m, n) = A\,(m-1, n+1)$ (d) $A\,(m, n) = m + n + 1$

Answers

1. (b)	2. (a)	3. (c)	4. (d)	5. (b)	6. (b)
7. (a)	8. (b)	9. (d)	10. (b).		

❐❐❐

CHAPTER 9

GRAPH THEORY

This Chapter Covers :

9.1. INTRODUCTION

Throughout in this book we are dealing with set and their elements and the relation between two or more sets. We have done some graphical representation of relations. Graph we have done are set of vertices and edges. So graph gives us the geometrical knowledge of a relation.

The basic idea of graph theory was introduced by great mathematician **Leonhard Euler** who used graph to solve the famous **Konigsberg bridge problem** which we will discuss in this chapter. We will also discuss the three utility problems. One of the application of the graph theory is in the field of circuit system, where by using graph we can say whether the circuit is correct or not. Graph theory is also useful in networking system.

In this chapter we do some special types of graph and their applications.

9.2. GRAPH THEORY

A graph is nothing but the set of vertices and set of edges. Let V be the set of vertices and E be a given set of edges, so that each element of E corresponds to (v_i, v_j) for $v_i, v_j \in V$. Then a graph G can be given as the set $G = (V, E)$. For example consider the graph G given by figure (9.1)

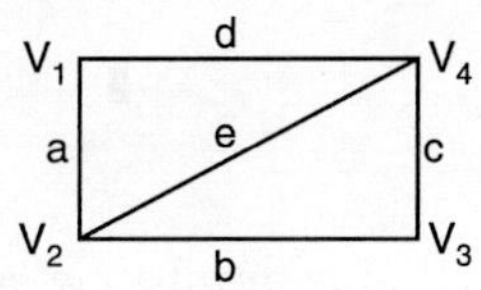

Fig. 9.1.

$$V = \{v_1, v_2, v_3, v_4\}$$
$$E = \{a, b, c, d, e\}$$

a is the edge incident at (v_1, v_2)
b is the edge incident at (v_2, v_3)
c is the edge incident at $(v_3, v_4,)$
d is the edge incident at (v_4, v_1)
e is the edge incident at (v_2, v_4)

The elements of V are called vertices and the elements of E are called edges.

If each element of E (set of edges) corresponds to an unordered pair (u, v) for $u, v \in V$ then we say that the linear graph is ***unoriented*** or ***undirected***. Otherwise the graph G is said to be an ***oriented graph***. or ***directed.***

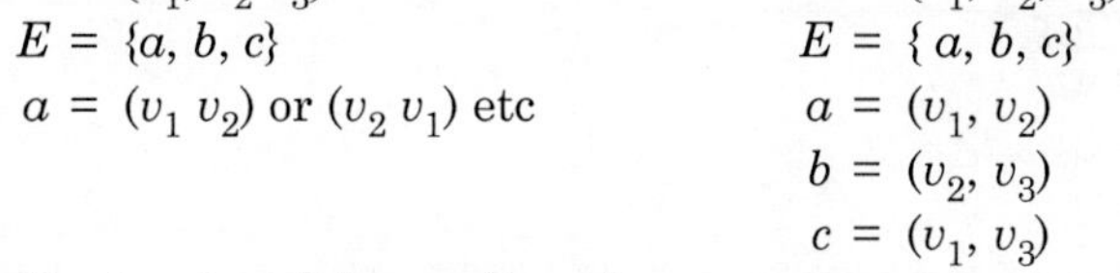

Fig. 9.2.

$V = \{v_1, v_2\ v_3\}$
$E = \{a, b, c\}$
Here $a = (v_1\ v_2)$ or $(v_2\ v_1)$ etc

$V = \{v_1, v_2, v_3\}$
$E = \{\ a, b, c\}$
$a = (v_1, v_2)$
$b = (v_2, v_3)$
$c = (v_1, v_3)$

Degree of a Vertex in a Linear Graph

By degree of a vertex in a linear graph we mean the number of edges including self-loop which are incident at the vertex v. Degree of the vertex v is denoted by $d(v)$.

$$d(v) = n_e + 2n_l$$

Where n_e is the number of edges incident at the vertex v and n_l is the number of self loops incident at the vertex v.

Self-loop. An edge $a \in E$ which corresponds to (v,v) for $v \in V$ is called a self-loop.

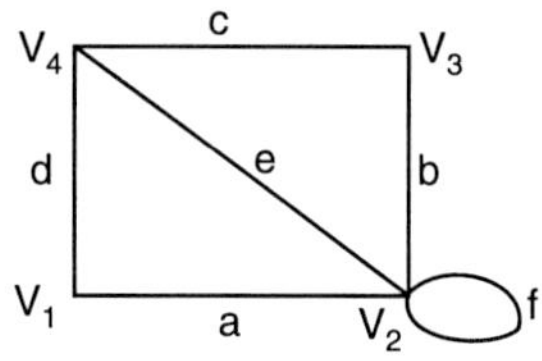

Fig. 9.3.

Here f is the self-loop incident on v_2

degree $(v_1) = 2$

degree $(v_2) = 3 + 2 = 5$

degree $(v_3) = 2$

degree $(v_4) = 3$.

EXAMPLE 9.1. Consider the following graphs given by the fig. 9.4 (*a*) and (*b*) and also give the degree of each vertex.

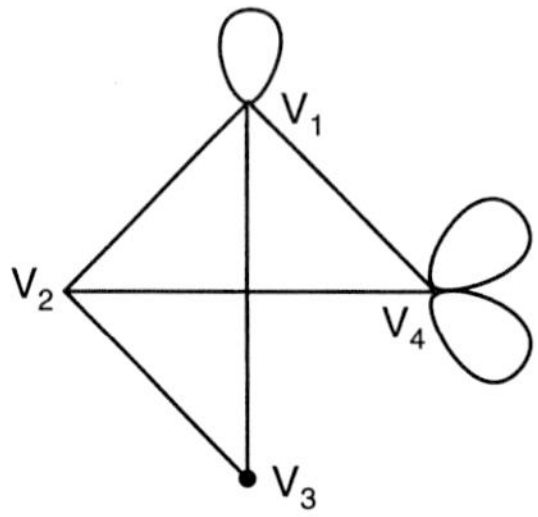

Fig. 9.4. (a)

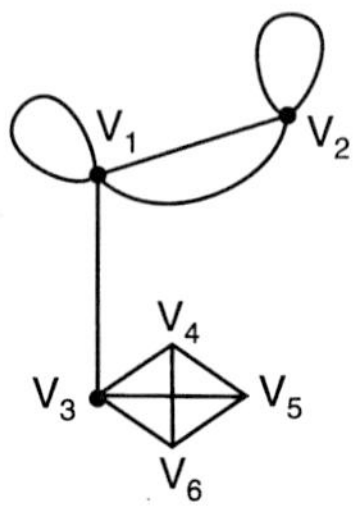

Fig. 9.4(b)

Solution: Consider the fig. 9.4 (*a*) Recall the definition of degree of a vertex in a given graph.Then

degree $(v_1) = 3 + 2 = 5$

degree $(v_2) = 3$

degree $(v_3) = 2$

degree $(v_4) = 2 + 4 = 6$

Similarly in the fig. 9.4 (*b*)

degree $(v_1) = 3 + 2 = 5$

degree $(v_2) = 2 + 2 = 4$

degree $(v_3) = 4$

degree $(v_4) = 3$

degree $(v_5) = 3$

degree $(v_6) = 3$

Theorem 9.1: Given a linear graph $G = (V, E)$ and $v \in V$ then $\sum_{v \varepsilon G} d(v) = 2n_e$

where n_e is the number of edges in the graph G.

Proof. Take any edge a in the graph G such that a corresponds to (u, v). Now the edge a contributes degree 1 to the vertex u and degree 1 to the vertex v if $u \neq v$.

a contributes degree 2 if $u = v$.

That is we will count the edge a twice.

$\therefore \sum \deg(v) = 2n_e$

That is sum of degree of all the vertices of a graph is same as twice the number of edges in the graph.

Theorem 9.2: The number of vertices of odd degree in a linear graph is even.

Proof. Let G be a linear graph then by above theorem

$$\sum_{v \in G} d(v) = 2\, n_e$$

where n_e is the number of edges in G

$$\Rightarrow \sum_1 d(v) + \sum_2 d(v) = 2\, n_e$$

where $\sum_1$ represents sum of vertices of even degree and

$\sum_2$ represents sum of vertices of odd degree

$$\Rightarrow \sum_2 d(v) = 2n_e - \sum_1 d(v)$$

$$= \text{even}$$

Therefore the result holds.

By the above theorem by looking at the graph we can tell whether the graph is possible or not. Consider the following example

EXAMPLE 9.2. Show that whether you can draw a linear graph in which degree of each vertex is given.

(*i*) 2, 3, 2, 3

(*ii*) 1, 3, 4, 5

Using the above theorem the graph 1 is possible as number of vertices of odd degree are even but the graph 2 is not possible.

One can ask the question that "what does mean by possibility of a graph"?

By the practical point of view graph is useful in many construction and paths etc. For this consider the following example

Konigsberg Bridge Problem

The **Konigsberg bridge problem** is the best known example in graph theory. It is given as follows :

There are four lands which are connected by seven bridges. The problem is that a person has to start with any of the four lands and cross each bridge once and only once and travel each land and back to the starting one.

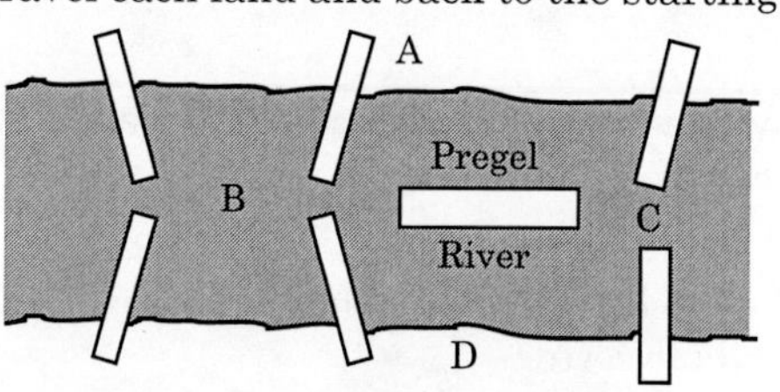

Fig. 9.5. Konigsberg Bridge

In terms of vertices and edges the above problem can be given as by fig. 9.6.

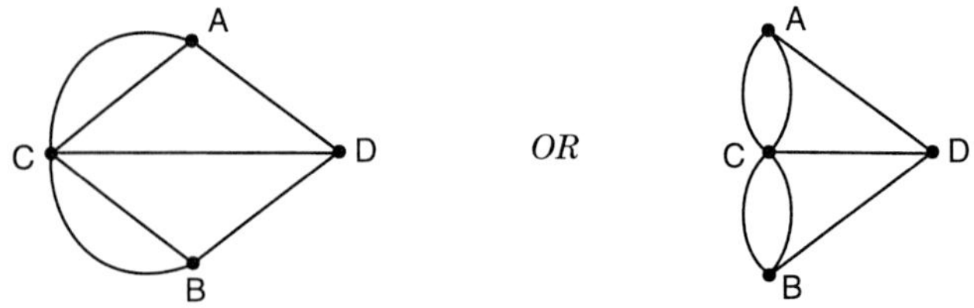

Fig. 9.6.

The answer of the above problem is a very tedious job or is a hit and trial method. But by using the above theorem we can say that it is not possible to draw a graph from the given constraints as the number of vertices of odd degree is odd.

Similarly we can consider the another example

Three Utilities Problem

There are three houses say H_1, H_2, H_3 with three facilities gas, water and electricity respectively. We want each facility to each house and there is connections by pipes between each house and facility, such that it should be done without crossing any pipe.

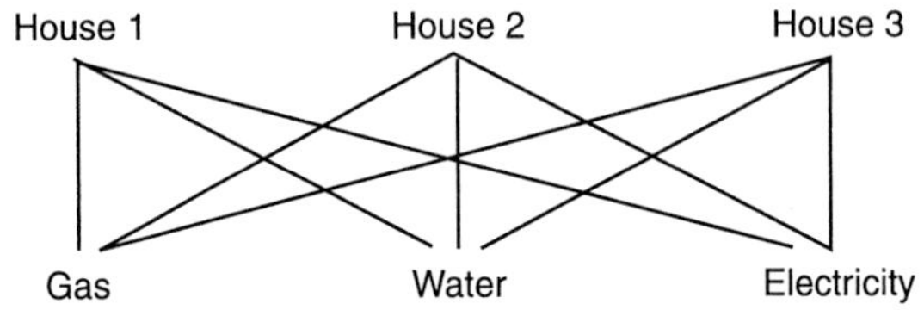

Fig. 9.7(a)

By the graph 9.7(*b*), it is not possible as there can not be an edge between H_3 and *G*.

One can interpret the result by means of degree of vertices.

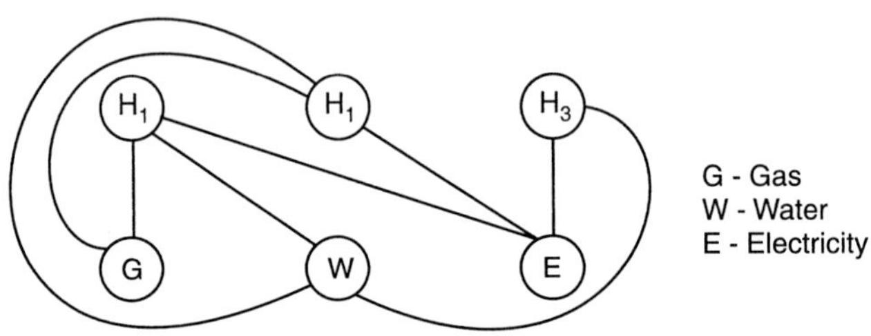

Fig. 9.7(b)

Now we discuss some types of vertices in a graph.

Isolated Vertex

The vertex having no incident edge is called isolated vertex. In the fig. 9.8, v_1 is the isolated vertex.

Pendant Vertex

A vertex with degree 1 is called pendant vertex. In the gives fig. (9.8) the vertex v_7 is the pendant vertex.

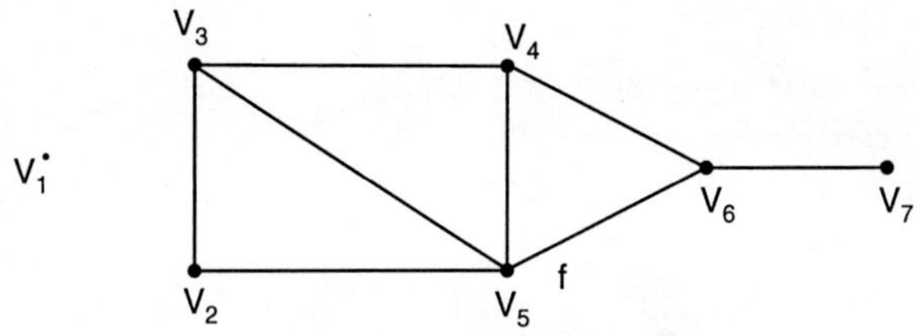

Fig. 9.8.

EXAMPLE 9.3. Consider the graph given by the fig 9.9(*a*) to 9.9(*e*) find

(*a*) Degree of each vertex

(*b*) Find all isolated vertices (if any)

(*c*) Find pendant vertex.

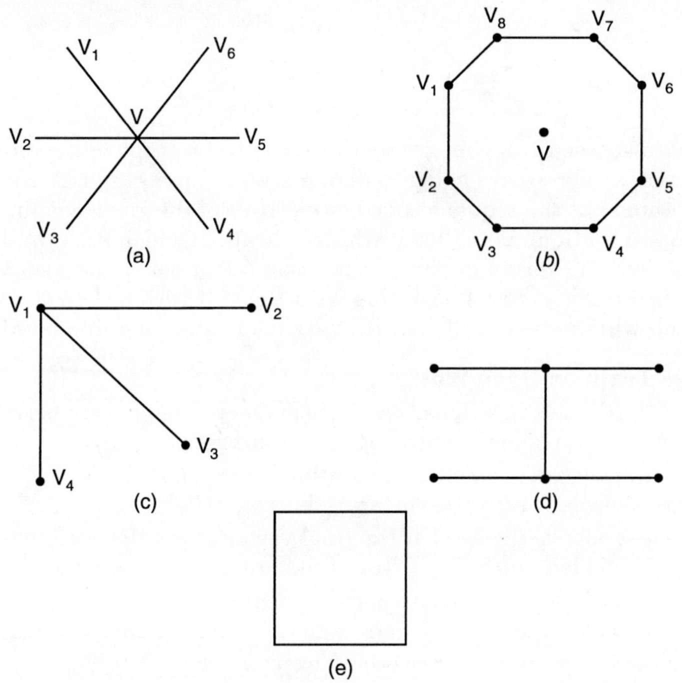

Fig. 9.9.

Solution : Consider the fig. 9.9(*a*)

degree $(v) = 6$

degree (v_1) = degree (v_2) = ... = degree $(v_6) = 1$

Therefore v_1, v_2, v_3, v_4, v_5, v_6, are all pendant vertices and graph has no isolated vertex.

Consider the fig. 9.9(*b*)

degree $(v) = 0$

degree (v_1) = degree (v_2) = ... = degree $(v_6) = 2$

Therefore the vertex v is an isolated vertex and the graph has no pendant vertex.

Consider the fig. 9.9(*c*)

degree (v_1) = 3

degree(v_2) = degree (v_3) = degree (v_4) = 1

$\therefore$ v_2, v_3, v_4 are isolated vertices.

Similarly the reader can do for rest of the graphs given by the fig. 9.9(*d*) and 9.9(*e*).

Null Graph

A graph *G* with no edge is called null graph. For instance see figure 9.10.

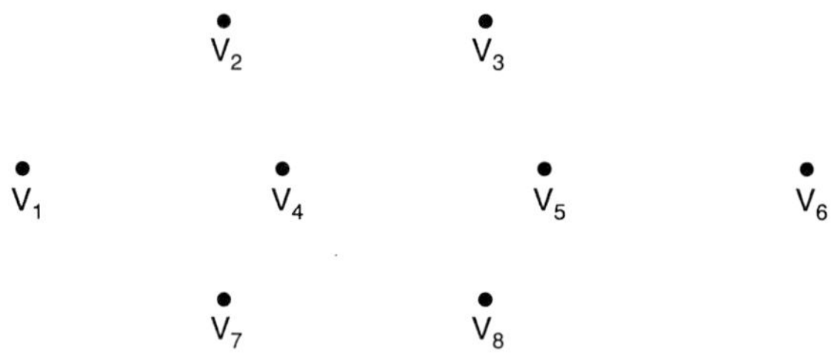

Fig. 9.10.

We have discussed graph, degree of the vertices in a graph and some special type of vertices. Now we are going to define a **walk** on a graph. A **walk** on a graph is nothing but the sequence of edges/vertices of the given graph. A walk can be an **open walk** and is called **path** *i.e.*, a walk in which the initial vertex is different from the vertex at the end position (*i.e.* a point at which we stop otherwise. If the initial vertex coincide with the end vertex then the walk is called **closed walk** or is called **Circuit.**. We elaborate our definition as follows.

Open Edge Train or Open Walk

A sequence of edges in a linear graph form an open edge train between the vertices u and v if the following properties are satisfied.

(1) The first edge is incident at the initial vertex *u* and the other end of the edge is incident with the vertex of the succeeding edge.

(2) The last edge is incident at the final vertex *v* and the other end of the edge is incident with the vertex of the proceeding edge.

(3) All the remaining edges have ends incident at the vertex of the preceding edge and the other end incident with the vertex of the succeeding edge.

(4) Every edge in the sequence occurs exactly once.

For example

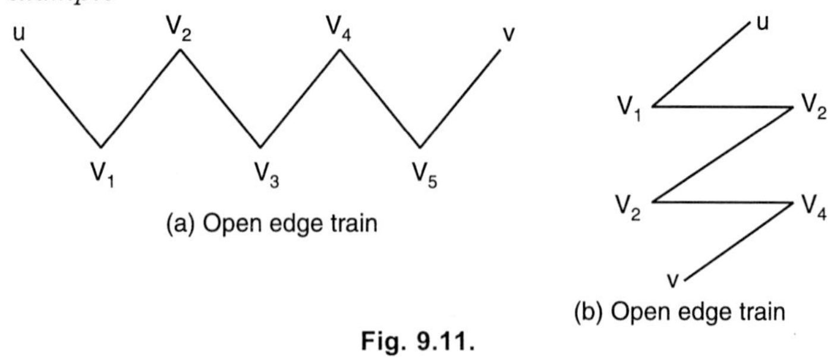

Fig. 9.11.

Consider the figure 9.12

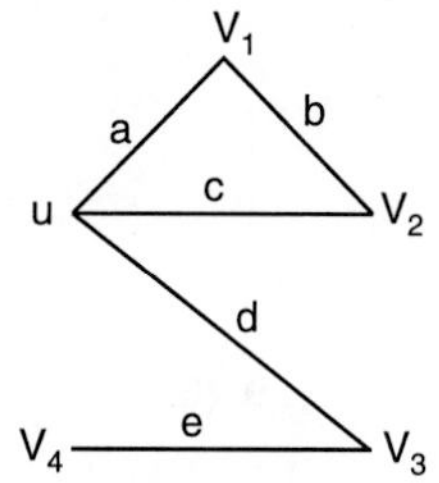

Fig. 9.12.

a, b, c, d, e is not an open edge train as the vertex *u* appears twice, but *b, c, d, e* is an open edge train.

An open edge train is also called path.

A closed walk or a Circuit

A graph in which no vertex (except the initial and final vertex) appears more than once is called a circuit or a closed walk or a closed edge train. For example,

A circuit is also called a **cycle, elementary cycle, circular path.**

Consider the graphs given by the figure 9.13.

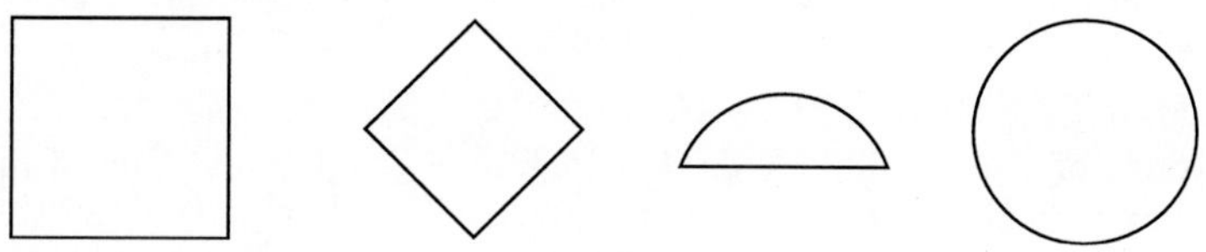

Fig. 9.13.

In electrical engineering a circuit is sometimes referred to as a loop-not to be confused with self-loop.

In the fig. 9.12 the walk *a b c* is the circuit.

9.3. OPERATIONS ON GRAPHS

As we have discussed that graph is nothing but the set of vertices and edges. Therefore the same operations of sets are applicable on the graph also. So in this section we will discuss some of these operations.

Let G_1 and G_2 be two graphs as given below:

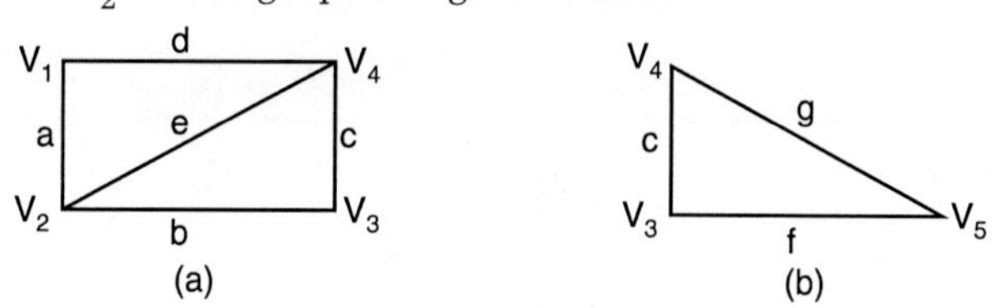

Fig. 9.14.

Union of Two Graphs

The union of graphs G_1 and G_2 that is given as

$$G = G_1 \cup G_2$$

is a linear graph consisting of edges either in G_1 or in G_2 or in both as given is th fig. 9.15.

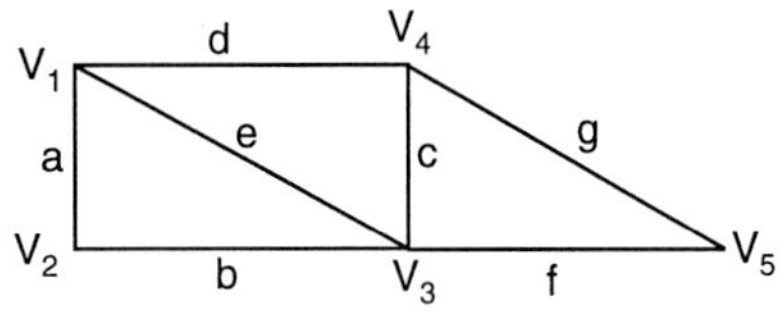

Fig. 9.15.

Intersection of Two Graphs

The intersection of two graphs G_1 and G_2

$$G = G_1 \cap G_2$$

is a linear graph consisting of all the edges which are in both G_1 and G_2 see fig. (9.16).

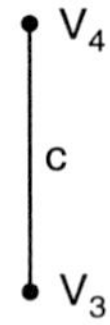

Fig. 9.16.

Difference of Two Graphs ($G_1 - G_2$)

The difference of two graphs G_1 & G_2 is a linear graph consisting of all the edges which are in G_1 but not G_2 as in the fig. 9.17.

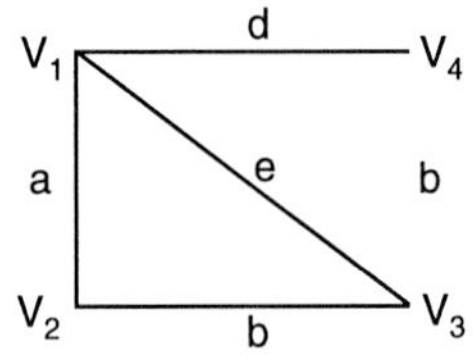

Fig. 9.17.

4. Addition of Two Graph $G_1 \oplus G_2$

The addition of two graphs G_1 and G_2 is a linear graph which includes the edges either in G_1 or in G_2 but not in both that is a graph which is obtained from $G_1 \cup G_2$ after deleting the common edges. That is:

$$G \oplus G_2 = (G_1 \cup G_2) - (G_1 \cap G_2)$$

is given gy fig. (9.18).

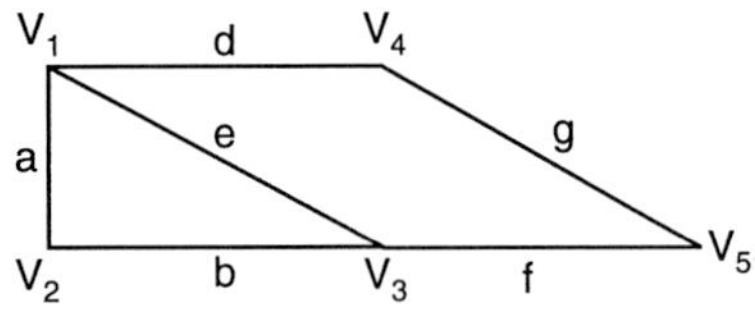

Fig. 9.18.

EXAMPLE 9.4. Consider the following given graphs G_1 and G_2 given by fig. 9.19(*a*) and 9.19(*b*) and compute (1) $G_1 \cup G_2$

(2) $G_1 \cap G_2$

(3) $G_1 \sim G_2$

(4) $G_1 \oplus G_2$

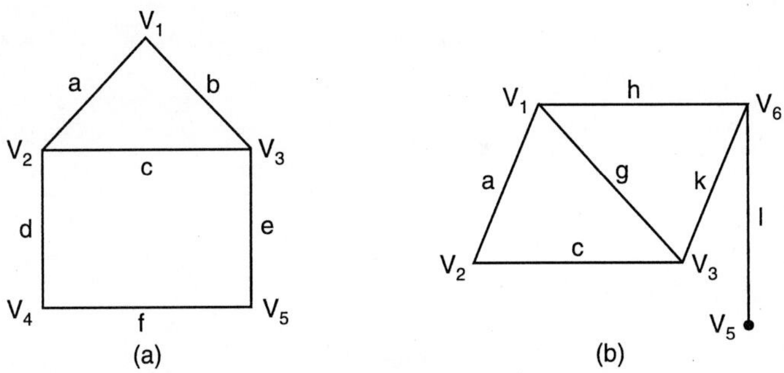

Fig. 9.19.

Solution :

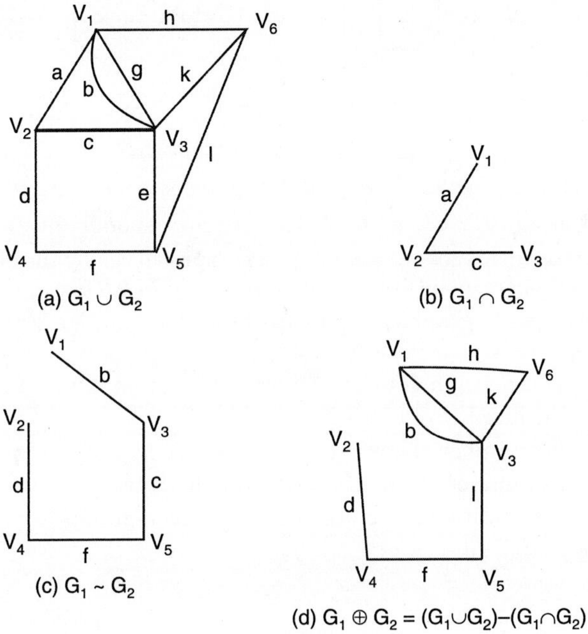

Fig. 9.20.

9.4. ISOMORPHIC GRAPHS

In the chapter of relation we have discussed the equivalence relations and also the relation can be represented by a graph so the same property (equivalence) is also applicable in the graph theory and is called **isomorphisms**. Therefore let G and G' be the two graphs then these two graphs are said to be isomorphic to each other if there is a one-to-one correspondence (that is one-one and onto) between their vertices and between their edges such that the incidence relationship is preserved. In other words, suppose that an edge e is incident on vertices v_1 and v_2 in a graph G then the corresponding edge e' in G' must be incident on the vertices v'_1 and v_2' that corresponds to v_1 and v_2. For example, fig. (9.21 and 9.22).

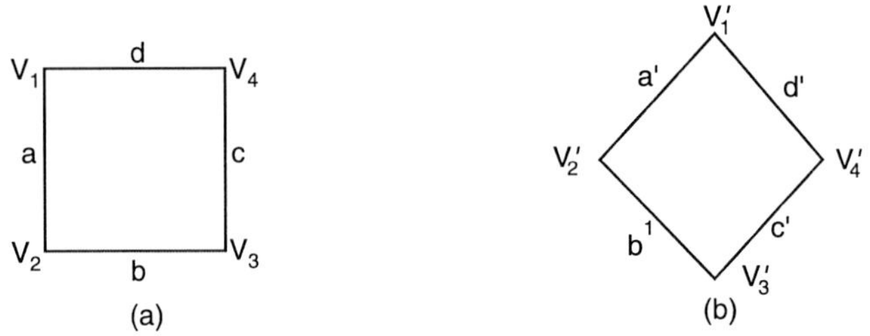

Fig. 9.21.

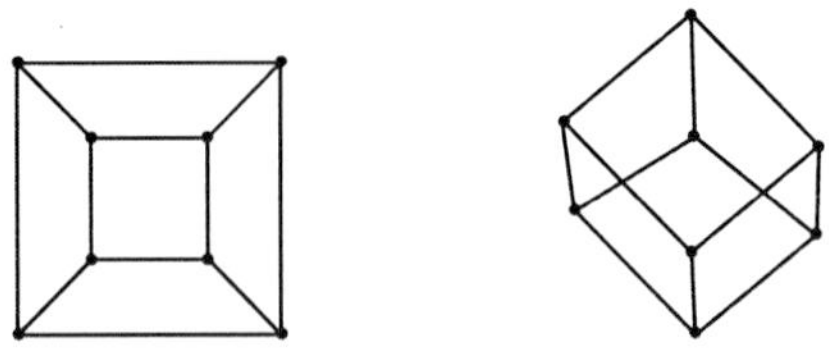

(a) (b)

Fig. 9.22. (Isomorphic graphs)

In the fig. 9.21 the vertices of 9.21(*b*) are corresponding to the vertices of 9.21(*a*) that is V_1', V_2', V_3', V_4' in 9.21(*b*) are corresponding to V_1, V_2, V_3, V_4 in 9.21(*a*), also the incidence relationship is also preserved by the edges. *i.e.*, a', b', c', d' in 9.21(*b*) corresponds to the edges a, b, c, d in 9.21(*a*).

Similarly the two graphs given by the fig. 9.22(*a*) and 9.22(*b*) are isomorphic to each other.

Therefore for a graph G to be isomorphic we must have

1. The same number of vertices.
2. The same number of edges.
3. An equal number of vertices with a given degree.

If one of the condition is violated then the given graphs are not isomorphic.

EXAMPLE 9.5. Non isomorphic graphs.

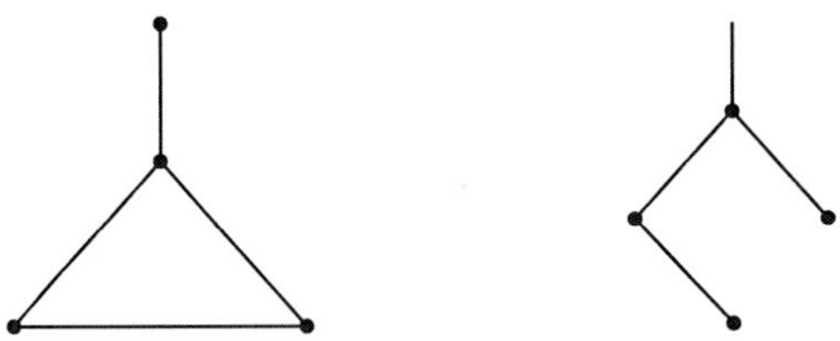

(a) (b)

Fig. 9.23.

As the third condition is violated in the above graphs therefore the graphs given by fig. 9.23. are not isomorphic to each other as the degree of vertices of even degree in 9.23(*a*) is not same in 9.23(*b*)

From the above discussion we conclude that the following definition of isomorphic graphs.

Isomorphic Graphs

The simple graphs $G_1 = (V_1, E_1)$ and $G_2 = (V_2, E_2)$ are isomorphic if there is a one-to-one and onto function f from V_1 to V_2 with the property that a and b are adjacent in G_1 if and only if $f(a)$ and $f(b)$ are adjacent in G_2, for all a and b in V_1. Such a function f is called isomorphism.

EXAMPLE 9.6.

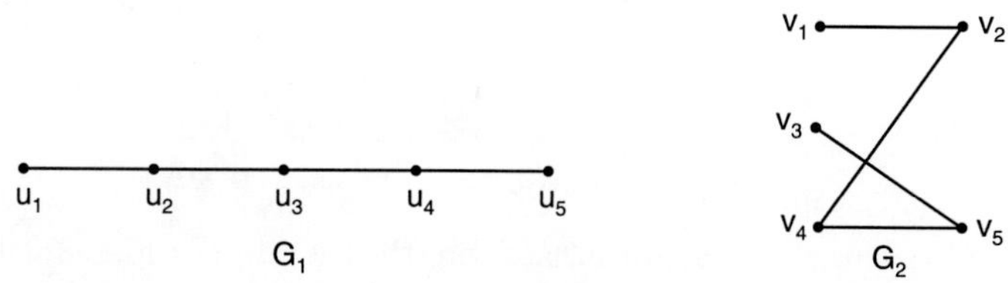

Fig. 9.24.

The two graphs G_1 & G_2 are isomorphic by the given isomorphism

$$f(u_1) = v_1$$
$$f(u_2) = v_2$$
$$f(u_3) = v_4$$
$$f(u_3) = v_5$$
$$f(u_5) = v_3$$

EXAMPLE 9.7.

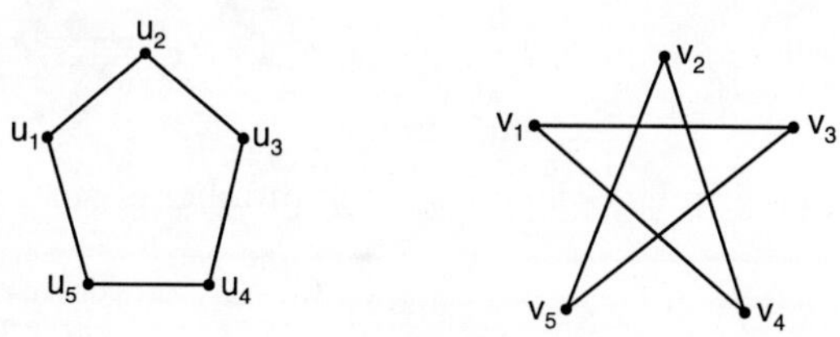

Fig. 9.25.

The above two graphs are isomorphic and the isomorphism is given as follows.

$$f(u_1) = v_1$$
$$f(u_2) = v_3$$
$$f(u_5) = v_4$$
$$f(u_3) = v_5$$
$$f(u_4) = v_2$$

EXAMPLE 9.8.

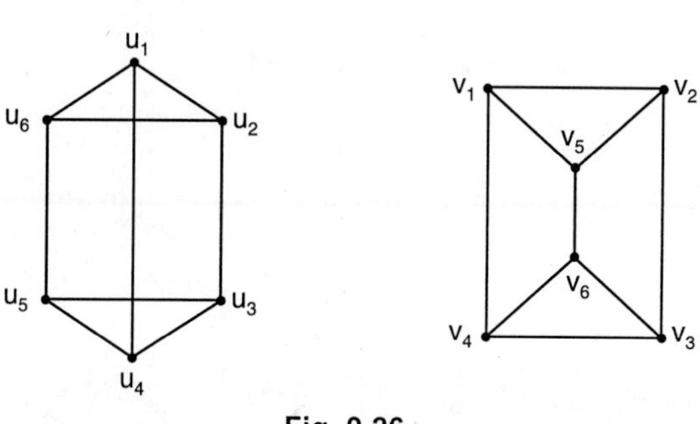

Fig. 9.26.

The above two graphs are isomorphic & the isomorphism is given as follows:

$$f(u_1) = v_5$$
$$f(u_2) = v_2$$
$$f(u_3) = v_3$$
$$f(u_4) = v_6$$
$$f(u_5) = v_4$$
$$f(u_6) = v_1$$

However, G_2 has a simple circuit of length 3 while G_1 does not have a simple circuit of length 3.

Since the existence of a simple circuit of length 3 is an isomorphic invariant so G_1 and G_2 are not isomorphic.

EXAMPLE 9.9. Check whether the following graphs are isomorphic or not.

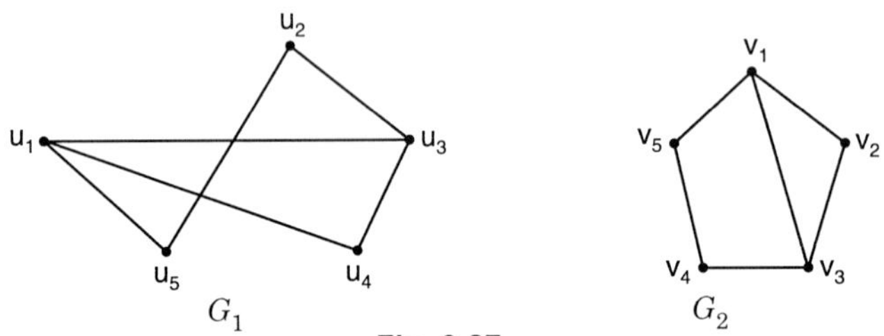

Fig. 9.27.

In the above given graphs G_1 and G_2 the of number of vertices in G_1 & G_2 are same *i.e.*, 5.

The number of edges in G_1 and G_2 are same *i.e.*, 6

The number of vertices of degree 3 in G_1 is 2 as well as in G_2 is also 2.

The number of vertices of degree 2 in G_1 is 3 as well as in G_2 is also 3.

Also the simple graph of length 3 in G_1 is 1 and also in G_2 is 1.

Also the simple graph of length 4 in G_1 is 1 and in G_2 is 1.

$\therefore$ G_1 and G_2 both are isomorphic.

and the mapping is given by:

$$f(u_1) = v_1$$
$$f(u_3) = v_3$$
$$f(u_4) = v_2$$
$$f(u_2) = v_4$$
$$f(u_5) = v_5$$

Paths and Isomorphism

There are several ways that path and circuits can help determine whether two graphs are isomorphic. For example, the existence of a simple circuit of particular length $K(K > 2)$ is a useful invariant that can be used to show that two graphs are not isomorphic. In addition, paths can be used to construct mappings that may be isomorphic.

EXAMPLE 9.10. Check whether the given graphs in Fig. 9.28 are isomorphic or not.

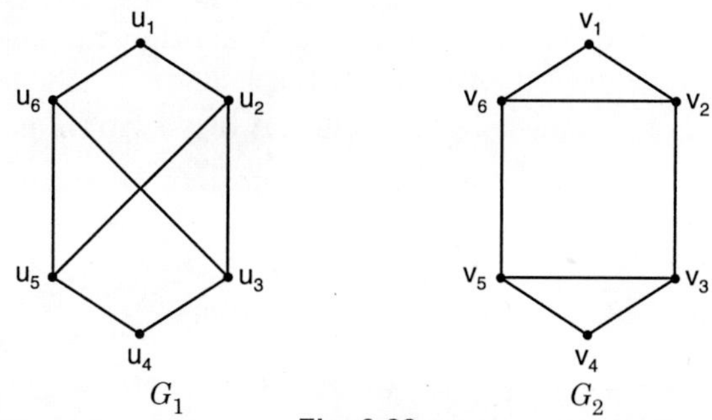

Fig. 9.28.

Both the given graphs say G_1 and G_2 have 6 vertices and eight edges. Each has four vertices of degree 3 and two vertices of degree 2.

So, the three conditions,

number of vertices

number of vertices of same degree

all agree for these two graphs.

Connected Graph

A linear graph G is said to be connected if between any two vertices of G there is at least one path connecting these two vertices. (See fig. 9.29).

For example :

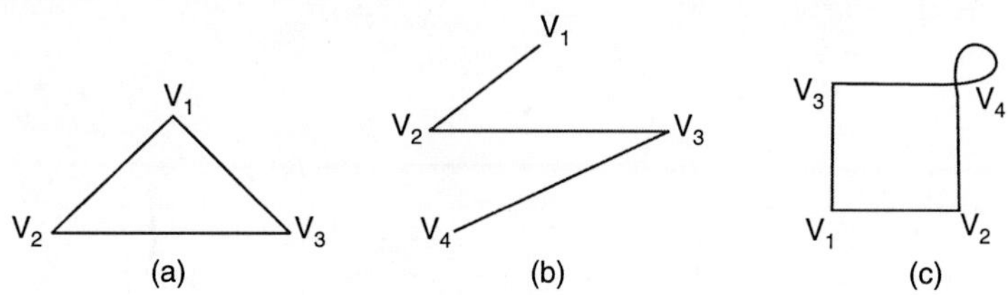

Fig. 9.29.

Theorem : If a graph (connected or disconnected) has exactly two vertices of odd degree, there must be a path joining there two vertices.

Theorem : A graph with n vertices and k components (disjoint sub-graphs of the given graph) can have at most $^1/_2\ (n - k)\ (n - k + 1)$ edges.

Euler Graphs

Euler Path :

An euler path in a graph G be a path that traverse every edge exactly once and only once.

9.5. BIPARTITE GRAPHS

Sometimes a graph has the property that its vertex set can be divided into two disjoint subsets such that each edge connects a vertex in one of these subsets to a vertex in the other subset.

Def^n. A simple graph G is called bipartite if its vertex set V can be partitioned into two disjoint nonempty sets V_1 and V_2 such that every edge in the graph connects a vertex in V_1 and a vertex in V_2 so that no edge in G connects either two vertices in V_1 or two vertices in V_2.

For example Consider **C_4 → cycle having 4 vertices**

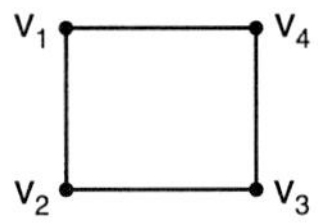

Fig. 9.30.

Fig. 9.31.

The vertex set S can be partitioned into two set $S_1 = \{V_1, V_3\}$ & $S_2 = \{V_2, V_4\}$ and every edge of C_4 connects a vertex in S_1 & a vertex in S_2.

EXAMPLE 9.10. Consider the example of a cycle of length 6, *i.e.*, C_6

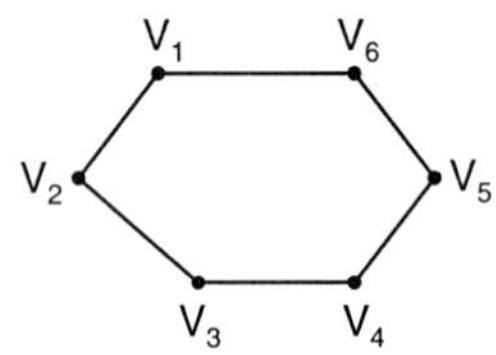

Fig. 9.32.

Here C_6 is bipartite as show in the figure, since its vertex set can be partitioned into two set $S_1 = \{V_1, V_3, V_5\}$ and $S_2 = \{V_2, V_4, V_6\}$ and every edge of C_6 connects a vertex in S_1 & a vertex in S_2.

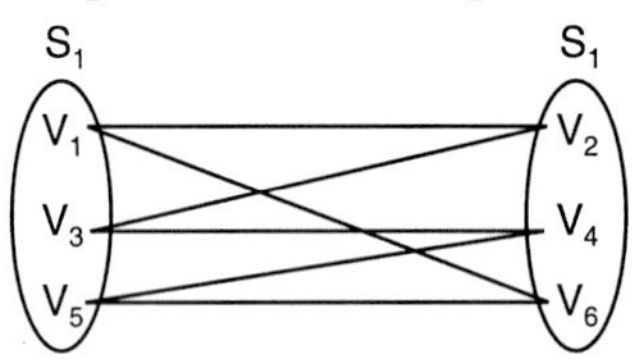

Fig. 9.33.

EXAMPLE 9.11. Determine whether the following graphs are bipartite or not.

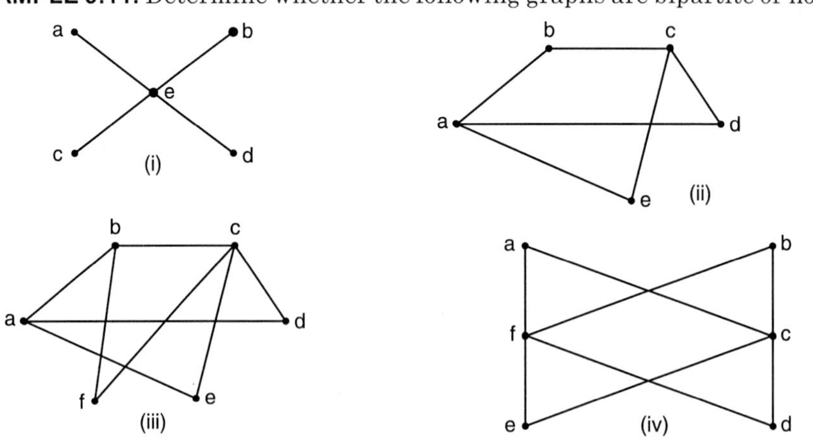

Fig. 9.34.

Solution : The graphs given by the figure 9.34(*i*) & (*iv*) are bipartite as the vertex set V can be divided into sets say V_1 & V_2 such in figure 9.34(*i*)

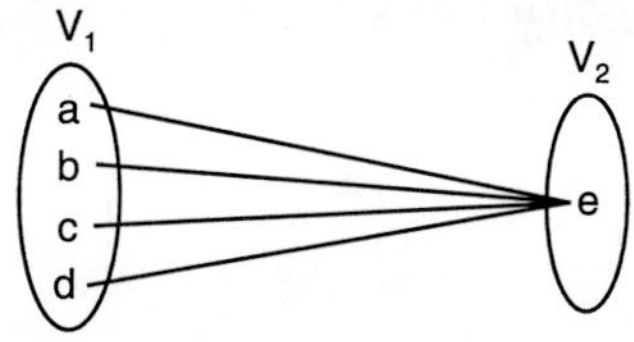

for fig. (iv)

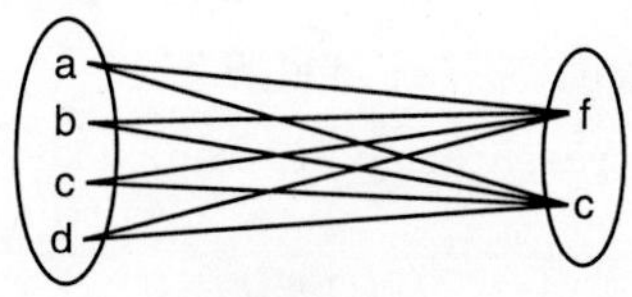

Fig. 9.35.

Complete Bipartite Graph

The complete bipartite graph $k_{m,n}$ is the graph that has its vertex set partitioned into two subsets of m and n vertices respectively. There is an edge between two vertices if and only if one vertex is in the first subset and the other vertex is in the second subset. Following are the examples of the some complete bipartite graphs.

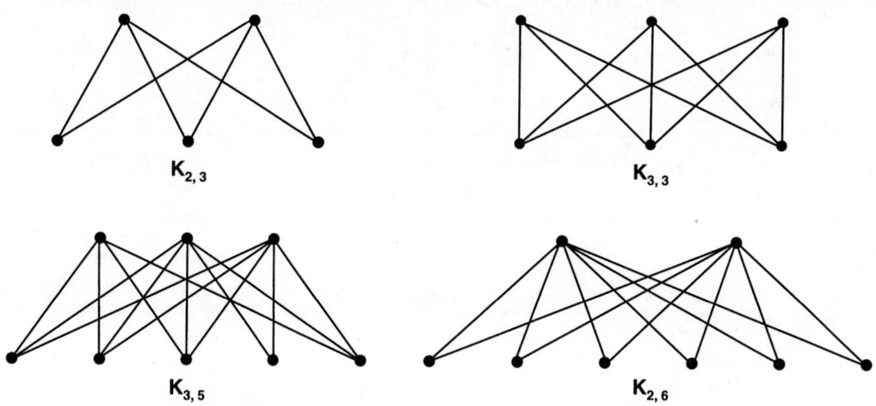

Fig. 9.36.

9.6. MATRIX REPRESENTATION OF LINEAR GRAPH

In the previous chapters we have discussed representation of sets and relations. As every relation can be given in terms of matrices.

In this section we will discuss the matrix representation of linear graph, which will be useful in defining different properties of graph theory more conveniently. Here we will consider a linear graph G in the form of matrix where the vertices form rows and edges are written as columns.

Exhaustive Incidence Matrix

It is denoted by A_e given as

$$A_e = (A_{pq})$$

$$A_{pq} = \begin{cases} 1 & \text{if the edge } q \text{ incident at the vertex } P \\ 0 & \text{otherwise.} \end{cases}$$

where A_{pq} is the (p, q) entry of A_e.

EXAMPLE 9.12.

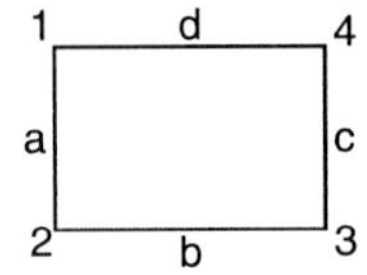

Fig. 9.37.

then its matrix representation is given by the matrix A_e given as follows :

$$A_e = \begin{matrix} & a & b & c & d \\ 1 & 1 & 0 & 0 & 1 \\ 2 & 1 & 1 & 0 & 0 \\ 3 & 0 & 1 & 1 & 0 \\ 4 & 0 & 0 & 1 & 1 \end{matrix}$$

9.7. INCIDENCE SET

Consider a linear graph G. An incidence set at the vertex v is denoted by $s(v)$ is the set of all edges incident at v. Consider the graph given by figure 9.38.

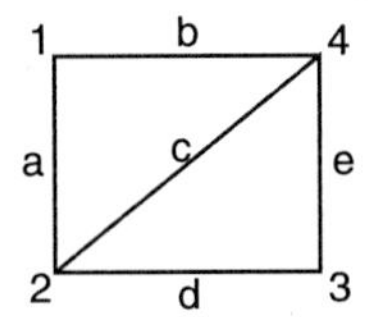

Fig. 9.38.

$$S(1) = \{a, b\}$$
$$S(2) = \{a, c, d\}$$
$$S(3) = \{d, e\}$$
$$S(4) = \{b,. c, e\}$$

Theorem 9.3 : Given two incident sets $S(v_1)$ and $S(v_2)$, the ring sum $S(v_1) \oplus S(v_2)$ is not an incident set.

Proof : We will prove the theorem with the help of one example. Consider the graph given by the figure 9.39.

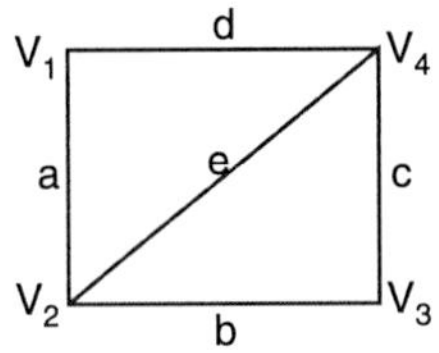

Fig. 9.39.

$$S(v_1) = \{a, d\}$$
$$S(v_2) = \{a, e, b\}$$

$$S(v_3) = \{b, c\}$$
$$S(v_4) = \{d, e, c\}$$
$$S(v_1) \oplus S(v_2) = \{b, d, e\}$$

is not an incident set as the edge named 'a' is absent in $S(v_2) \oplus S(v_4)$.

$$S(v_2) \oplus S(v_4) = \{a, b, c, d\}$$

is not an incident set as the edge named 'e' is absent in $s(v_2) \oplus s(v_4)$.

EXAMPLE 9.13. Consider the graph given by the figure 9.40. Find

$$S(5) \oplus S(3)$$
$$S(1) \oplus S(4)$$

and check whether the resultant set is an incidence set or not.

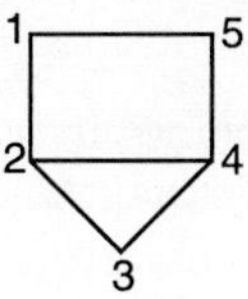

Fig. 9.40.

Linearly Independent

The collection of sets S_1, S_2 ..., S_n is said to be linearly independent if it contains no subcollection which is linearly dependent.

EXAMPLE 9.14. Consider the graph given in the figure 9.41. Find

$$S_1 \oplus S_2$$
$$S_1 \oplus S_2 \oplus S_3$$
$$S_1 \oplus S_2 \oplus S_3 \oplus S_4$$

and comment on their linearly independent property.

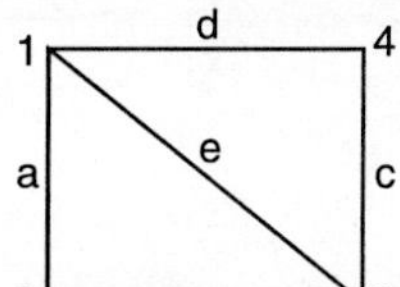

Fig. 9.41.

$$S_1 = \{a, d, e\}$$
$$S_2 = \{a, b\}$$
$$S_3 = \{b, c, e\}$$
$$S_4 = \{c, d\}$$
$$S_1 \oplus S_2 = \{d, e, b\}$$
$$S_1 \oplus S_2 \oplus S_3 = \{c, d\}$$
$$S_1 \oplus S_2 \oplus S_3 \oplus S_4 = \phi$$

All the resultant three sets are linearly independent.

Theorem 9.3 : Let G be a connected graph of n vertices then it has almost (n–1) linearly independent incident sets.

EXAMPLE 9.15. Consider the incidence matrix of the graph given by fig. 9.41.

$$A_e = \begin{array}{c} \\ 1 \\ 2 \\ 3 \\ 4 \end{array} \begin{array}{c} \begin{array}{cccc} a & b & c & d \end{array} \\ \begin{bmatrix} 1 & 0 & 0 & 1 \\ 1 & 1 & 0 & 0 \\ 0 & 1 & 1 & 0 \\ 0 & 0 & 1 & 1 \end{bmatrix} \end{array}$$

The incidence set corresponding to the exhaustive matrix is the set of edges incident on that vertex. For example, the incidence set for the above matrix is given as

$$S_1 = S(1) = \{a, d\}$$
$$S_2 = S(2) = \{a, b\}$$
$$S_3 = S(3) = \{b, c\}$$
$$S_4 = S(4) = \{c, d\}$$

Since each edge has two end points in a linear graph therefore there is exactly two *i*'s in each column of an exhaustive matrix.

Since we are considering only linear graph that does not contain self-loops and it is possible to produce self-loops when two vertices are made to coincide. So whenever self loops are produced by some operation on a linear graph, we will delete them, so that the resultant graph will have no self loop. For this we always use addition modulo 2 operation ($\oplus_2$).

Addition Modulo : 2 ($a \oplus_2 b$) add a and b and divide it by 2 the result will be the remainder].

Definition (Multi-graph)

Let $G = (V, E)$, where V is the set of all vertices of G and E is the set of all edges. G is said to be a multi-graph if it can be represented as a set of marked points (v) with the set of edges (E) between the points where there is no restriction on the number of edges from one point to another

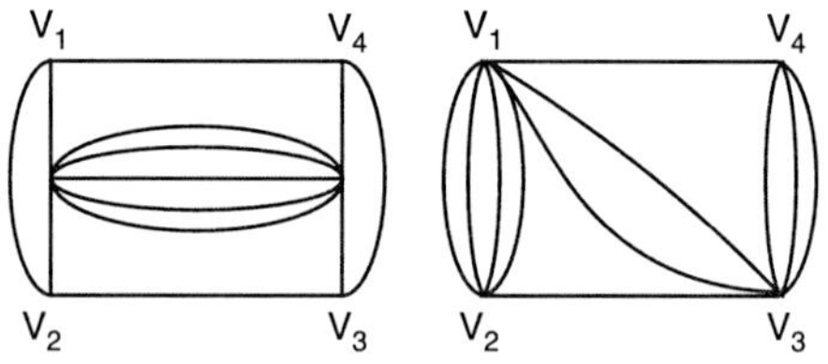

Fig. 9.42.

Also consider a graphical representation of a highway map in which an edge between two cities correspond to a lane in a highway between cities

Since there are often multiway highways between pair of cities, this representation yields a multigraph.

Simple Path : A path is said to be simple if it does not include the same edge twice.

For example :

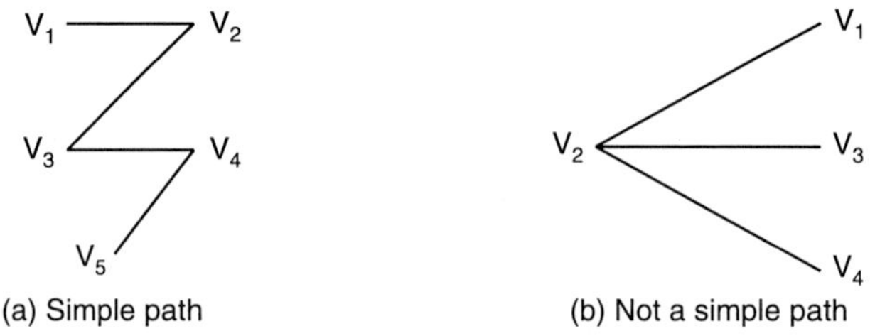

Fig. 9.43.

Elementary Path. A path is said to be elementary if it does not meet the same vertex twice. For example

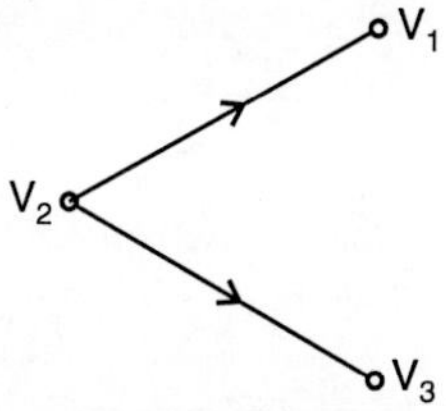

(a) Elementary path

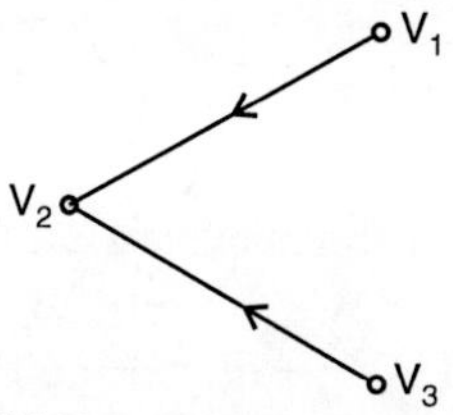

(b) Non-elementary path

Fig. 9.44.

Simple Circuit. It does not include the same edge twice.

Elementary Circuit. An elementary circuit does not meet the same vertex twice.

Now consider the following example

EXAMPLE 9.18. Consider the linear graph given as follows :

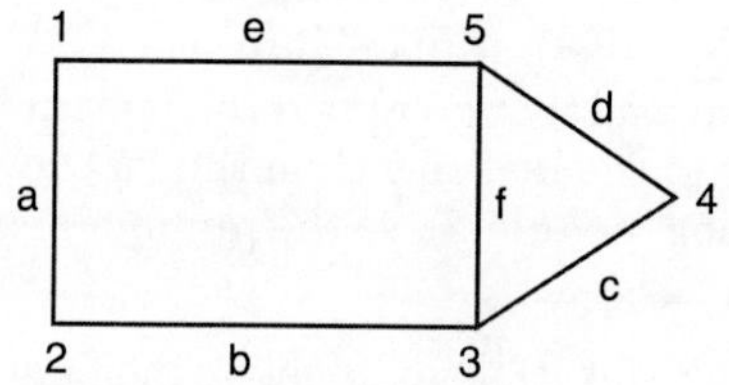

Fig. 9.45(a)

The exhaustive incidence matrix corresponding to the graph G is:

$$A_e = \begin{matrix} & a & b & c & d & e & f \\ 1 & 1 & 0 & 0 & 0 & 1 & 0 \\ 2 & 1 & 1 & 0 & 0 & 0 & 0 \\ 3 & 0 & 1 & 1 & 0 & 0 & 1 \\ 4 & 0 & 0 & 1 & 1 & 0 & 0 \\ 5 & 0 & 0 & 0 & 1 & 1 & 1 \end{matrix}$$

Let us combine the two vertices in a given graph *i.e.*, if $3 = 4$ then c will be deleted. By coinciding vertex 3 and 4, we make a self-loop by edge c so the resultant exhaustive incidence matrix A_e is given as:

$$\begin{matrix} & a & b & d & e & f \\ 1 & 1 & 0 & 0 & 1 & 0 \\ 2 & 1 & 1 & 0 & 0 & 0 \\ 3,4 & 0 & 1 & 1 & 0 & 1 \\ 5 & 0 & 0 & 1 & 1 & 1 \end{matrix}$$

It is to be noted that by applying the operations addition modulo 2 in the above matrix for the rows corresponding to the vertices numbered 3 and 4, the column c will have all zero therefore there is no harm in deleting the columns corresponding to the edge c.

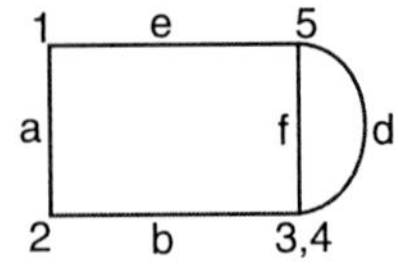

Fig. 9.45(b).

From the above example it is clear that if we add i^{th} row to the jth row of $A_e = [a_{pq}]$ then three cases arise

Case I. $a_{ir} + a_{jr} = 1$ if either a_{ir} or a_{jr} is 1.

Case II. $a_{ir} + a_{jr} = 0$ if both a_{ir} and a_{jr} are 0.

Case III. $a_{ir} + a_{rj} = 0$ if both a_{ir} and a_{rj} are 1.

Rank of the Exhaustive Incidence Matrix

The rank of matrix A is the number of linearly independent rows in A.

Theorem 9.4 : Rank of $A_e = R(A_e) \leq n_v - 1$ where n_v is the number of vertices in G

Rank of a matrix = Number of non-zero rows in a matrix in the echelon form
= Number of non-zero columns of matrix in the echelon form
(Row rank = column rank = Rank).

Proof : Since there are exactly two non-zero elements in each column of A_e so by adding all the rows to the last row then the last row becomes row of zeroes, (as we are doing addition modulo-2), so the number of non zero rows in A_e is n_v– 1 Therefore, rank of $A_e \leq n_v - 1$.

Theorem 9.5 : The rank of A_e of an exhaustive incident matrix A_e of a connected linear graph consisting of N_v vertices is $n_v - 1$.

Proof. Let G be a connected linear graph of n_v vertices and let A_e be the exhaustive incidence matrix, we can interchange rows (columns) in A_e so that 1 occurs in the first row and first column of A_e we will get

$$\begin{bmatrix} 1 & 0 & \cdots & \cdots & 0 \\ 0 & & & & \\ 0 & A_{e_1} & & & \\ 1 & & & & \\ \vdots & & & & \\ 0 & & & & \end{bmatrix}$$

Apply addition modulo 2 in the above matrix then the resultant matrix is:

$$\begin{bmatrix} 1 & 0 & \cdots & 0 \\ 0 & & & \\ 0 & A_{e_1} & & \\ \vdots & & & \\ 0 & & & \end{bmatrix}$$

Where A_{e_1} is exhaustive incidence matrix of some linear graph G' which is obtained from G by shorting an edge (or coinciding two vertices) and G' is connected (shorting an edge in a connected linear graph produces a connected graph). So A_{e_1} has same property as A_e. Also A_{e_1} has exactly two 1's in every column, so by the same procedure we have adopted above.

$$A_{e_1} = \begin{bmatrix} 1 & \cdots & \cdots & \cdots \\ 0 & 1 & \cdots & \cdots \\ \vdots & \vdots & A_{e_2} & \\ 1 & \vdots & & \end{bmatrix}$$

Where A_{e_2} is exhaustive incidence matrix of some linear graph G'' which is obtained from G'. Here G'' is connected. So by repeating the process we can get matrix of the form after k steps.

$$A_{e_2} = \begin{bmatrix} 1 & 0 & \cdots & \cdots & \cdots & \cdots \\ 0 & 1 & 0 & \cdots & \cdots & \cdots \\ \vdots & 0 & 1 & 0 & \cdots & \cdots \\ \vdots & \vdots & \vdots & 1 & 0 & \cdots \\ \vdots & \vdots & \vdots & \ddots & & \\ \vdots & \vdots & \vdots & & \ddots & A_{e_k} \end{bmatrix}$$

Since we are not taking self-loops so A_{e_k} has only one row of 0 (zero). Repeating the process n times we have

$$A_e = \begin{bmatrix} 1 & 0 & \cdots & \cdots & \cdots & \cdots \\ 0 & 1 & 0 & \cdots & \cdots & \cdots \\ \vdots & 0 & 1 & 0 & \cdots & \cdots \\ \vdots & \vdots & 0 & & & \\ 0 & 0 & 0 & 0 & 0 & 0 \end{bmatrix}_{nth\ \text{row}}$$

Here last row contains 0 only and since elementary row operation does not change the rank of matrix therefore, there exists atleast one non-singular sub-matrix of order $n_v - 1$.

∴ Rank of an exhaustive incidence matrix A_e of a connected linear graph with n_v vertices is $n_v - 1$ *i.e.*, $\mathcal{R}(A_e) = n_v - 1$.

Theorem 9.5 : Rank of a linear graph is equal to the rank of the exhaustive incidence matrix.

Proof. Case(1). If the graph G is connected of n vertices, then by previous theorem, rank of exhaustive incidence matrix is $n-1$ and we know that the rank of a connected linear graph consisting of n vertices is $n-1$,.

So for connected linear graph result is true.

Case (*ii*). If graph G is Separated. Let G has p mutually connected subgraphs say $g_1, g_2 \ldots g_p$. So each graph can be written in the form of an exhaustive incidence matrix. Let n_{v_r} be the number of vertices in g_r and n_{e_r} is the number of edges in g_r where $r = 1, 2, ..p$. So, by arranging the rows and columns, we can put them in such a way that first n_{v_1} rows represent vertices in g_1, next n_{v_2} rows represent vertices in g_2 and last n_{v_p} rows represent vertices in g_p.

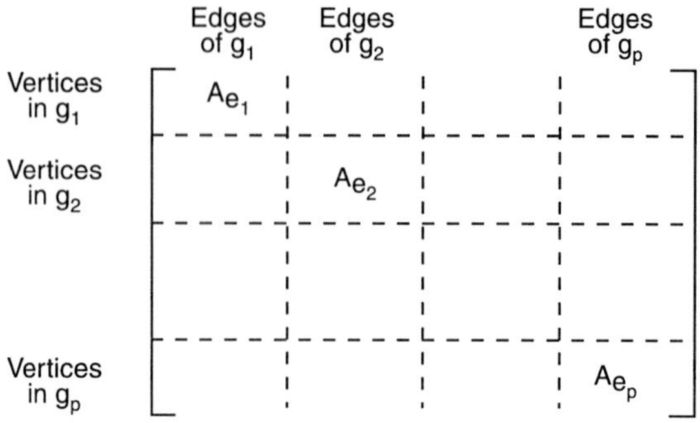

Similarly, first n_{e_1} columns represent edges in g_1, next n_{e_2} columns represent edges in g_2 and last n_{e_p} columns represents edges in g_p.

Where A_{e_x} is an exhaustive incidence matrix of g_r where $r = 1 \ldots, p$ then by using elementary row operation on $A_{e_1}, \ldots, A_{e_p}$ we obtain an exhaustive incident matrix of the form.

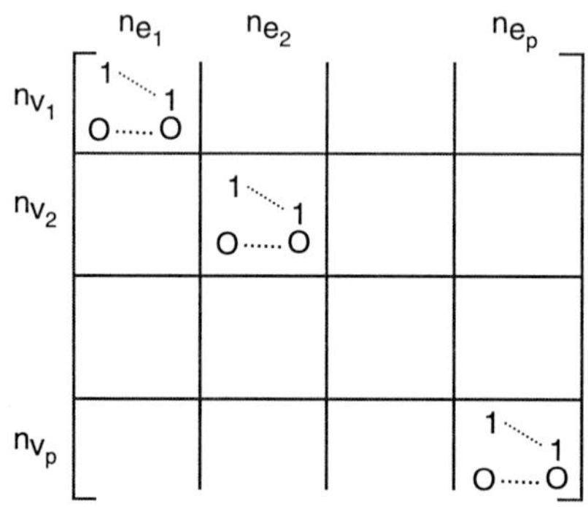

So the rank of A_{e_r} is $n_{v_r} - 1$ [As A_{e_r} is the matrix for connected graph.

$$\begin{aligned}
\text{Therefore, the rank of } A_e &= P(A_{e_1}) + P(A_{e_2}) + \ldots. + P(A_{e_p}) \\
&= (n_{v_1} - 1) + (n_{v_2} - 1) + \ldots + (n_{v_p} - 1) \\
&= (n_{v_1} + n_{v_2} + \ldots.. + n_{v_p}) - \text{p} \\
&= n - p
\end{aligned}$$

Where n is number of vertices in G and p is the number of maximally connected subgraphs of G.

Hence, rank of an exhaustive incidence matrix is equal to the rank of a linear graph.

9.8. EULER GRAPHS

By discussing **Konigsberg Bridge problem** Euler gave us the theory in which we say it is possible to find a closed walk running through every edge of G exactly once. Such a walk is called Euler path and a graph that consists of an Euler path is called an Euler graph.

Since the Euler path contains all the edges of the graph, therefore Eular graph is always connected.

Euler Path. Euler path in a graph G is a path that traverse every edge exactly once and only once.

Euler Circuit. Euler circuit in a graph G is a circuit that traverse each edge exactly once and only once.

Theorem 9.6 : A given connected graph G is an Euler graph if and only if all the vertices of G are of even degree.

Proof : Let us assume that the connected graph G having all the vertices of even degree. Now we construct a walk starting from any arbitrary vertex v of G and move through all the edges of G such that all the edges should traverse exactly once. Continue this process as far as possible. Since degree of each vertex in G is even therefore we exit from the same vertex as we enter. Therefore at last we reach the vertex v, which must be an end vertex of our walk. If in this closed walk we traced all the edges of G then G is an Euler graph otherwise consider a new graph obtained from G by deleting all the edges of our previous walk. Let it be G'. Also the degree of all the vertices of G' are of even degree. Again starting from any vertex u of G' construct a walk in the graph G'. Since all the vertices of G' are of even degree, this walk will terminate at u. But this walk in G' can be combined with previous walk to form a new walk which starts and end with the vertex v and has more edge than the previous walk. Continuing this process until we obtain a closed walk that traverses all the edges of G. Thus G is an Euler graph.

Conversely : Let G is an Euler graph. Therefore it contains a closed walk that always meets a vertex v. This closed walk goes through two new edges incident on v – with one we entered v and another we exited v. This is true for all the vertices of G. Thus if G is an Euler graph the degree of each vertex is even.

For example :

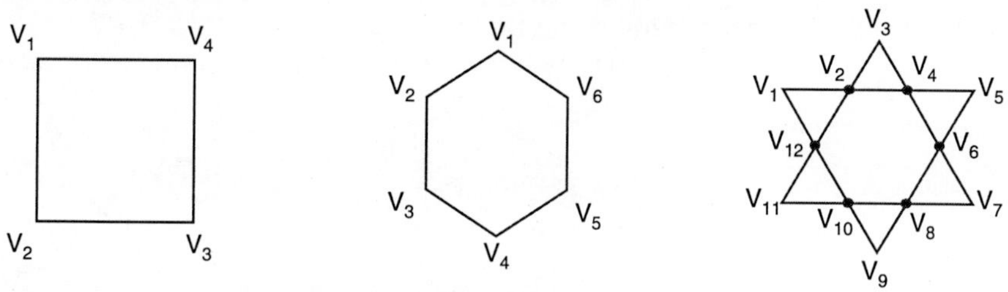

Fig. 9.46(a).

The graphs given by the figure 9.46(a) all are Euler graphs.

From the above theorem it is clear that a graph G having a vertex of odd degree does not possess an Euler circuit.

Now consider an example given by the fig. 9.46(*b*)

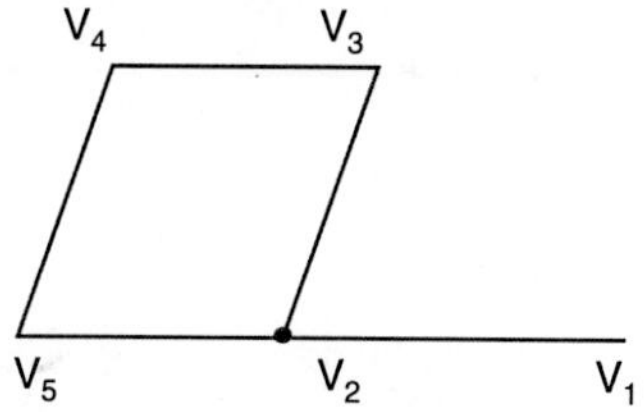

Fig. 9.46(b)

In the above example we start with the vertex v_1 we cannot find an Euler circuit. Similar case is with the vertex v_2.

Theorem 9.7 : For a graph G having more than two vertices of odd degree there cannot be an euler path.

For example, consider the graph given by the figure 9.47

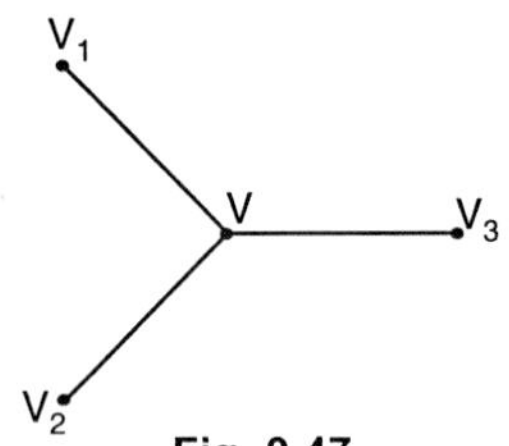

Fig. 9.47.

In the above graph we have three vertices of odd degree say v_1, v_2, v_3. It is clear from the figure that we cannot find a path in which each edge should traverse exactly once.

Theorem 9.8 : A connected graph G is an Euler graph if and only if it can be decomposed into circuits.

Proof. Let us suppose that G be a graph that can be decomposed into circuits that is G is a union of edge-disjoint-circuits.

Since degree of each vertex in a circuit is two therefore degree of every vertex in G is even therefore G is an Euler Graph.

Conversely : Let the graph G is an Euler graph. Consider a vertex v_1 of G. As degree of each vertex in G is even therefore there are at least two edges incident on v_1 say e_1 and e_2. Let e_1 be the edge between v_1 and v_2. Since vertex v_2 is also of even degree, it must have at least two edges incident on v_2 say e_3 and e_4. Let e_3 is between v_2 and v_3. Continuing this process we arrive at a vertex that has previously been traversed, thus forming a circuit C_1. Let us remove C_1 from G. Then all the vertices in the remaining graph say G' must also be of even degree. From G' remove another circuit in exactly the same way as we removed c_1 from G. Continue this process till no edges are left.

Hence the theorem.

EXAMPLE 9.17. Consider the following Euler graph

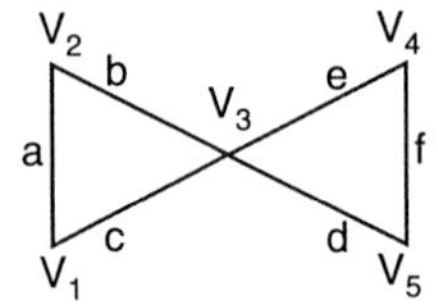

Fig. 9.48.

Let $\quad G = (V, E)$ be the given graph

where $\quad V = \{v_1, v_2, v_3, v_4, v_5\}$

$\quad E = \{a, b, c, d, e, f\}$

Starting from the vertex v_1 we traversed through the circuit c_1 as $a\ b\ c$. Now Remove c_1 from G, we will have the resultant graph G' as:

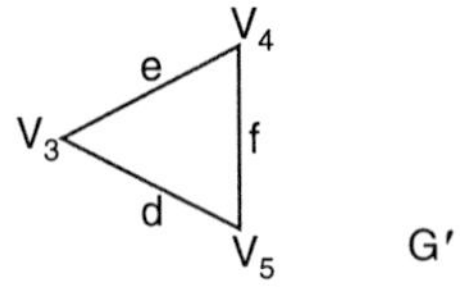

Fig. 9.49.

Again applying the same procedure above starting from v_3 we will get a circuit c_2 as *e f d*. By removing *e f d* from G' we are left with no edge.

EXAMPLE 9.18. Determine whether the given directed graphs has an Eular Circuit/Path or not if so construct and Euler Circuit/Path.

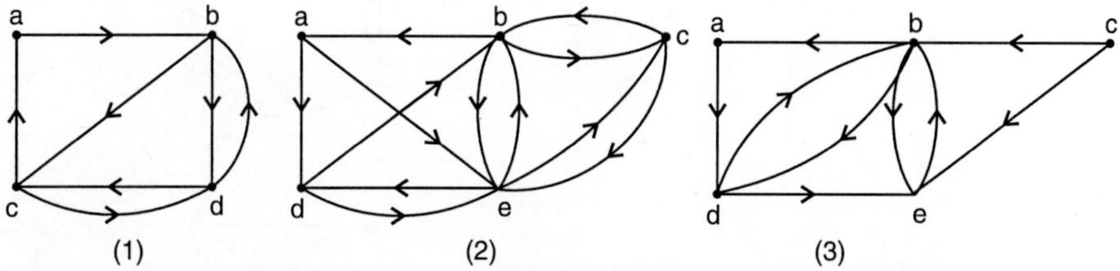

Fig. 9.50.

Solution: In graph (1) & (3), the degree of each vertex is even

The Euler circuit in (1) is *c a b c d b d c*

The Euler circuit in (3) is *c b a d b d e b e c*.

For Euler Path the no. of vertices of odd degree must be even.

(1) No vertex of odd degree therefore, Euler path by removing last vertex in the Euler circuit *c a b c d b d*

(2) The number of vertices of odd degree are 2. Therefore, the Euler path exists.

(3) *c b a d b d e b* e

EXAMPLE 9.19. Which of the undirected graphs given below have an Euler circuit? Of those that do not, which have an Euler Path.

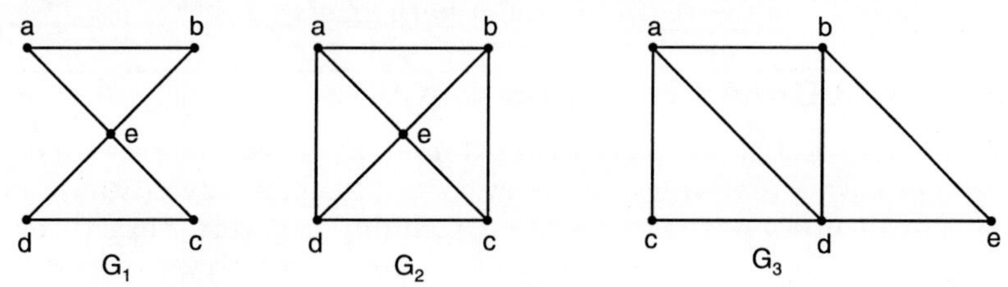

Fig. 9.51.

The graph G_1 has Euler circuit *a, b, e, d, c, e, a*. While G_1 & G_2 does not posses Euler circuit.

Therfore, the graph G_1 has Euler path *a b e d c e* and the graph G_3 has Euler path *a c d b e d a b* but G_2 does not posses any Euler path.

EXAMPLE 9.20. Determine whether each graph has an Euler circuit. Construct such a circuit when one exists.

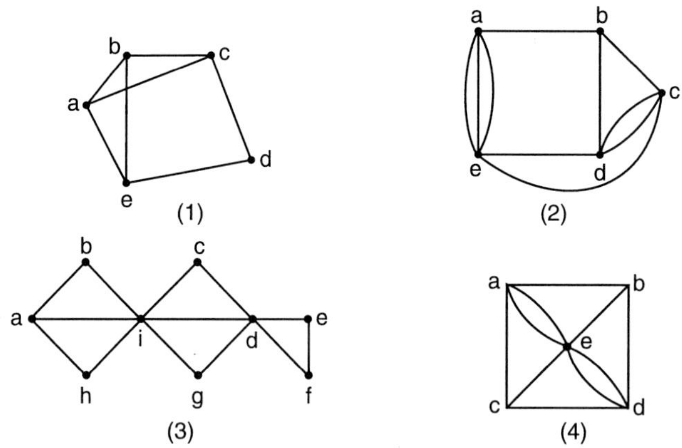

Fig. 9.52.

Solution: For the graph to have Euler circuit the degree of each vertex should be even.

Therefore all the four graph given above doesnot posses Euler circuit.

But for Eular path the number of vertices of odd degree should be even

in (1) The number of vertices of odd degree are 4.

in (2) The number of vertices of odd degree are 2.

in (3) The number of vertices of odd degree are 3.

in (4) The number of vertices of odd degree are 2.

therefore the in figure (3) No. Euler path exists

9.9. HAMILTONIAN PATH AND CIRCUITS

In the previous section we have discussed Euler path and Euler circuits having property of a walk that traverses every edge of the graph exactly once.

A **Hamiltonian path** is a path that contains each vertex exactly once.

A **Hamiltonian circuit** is a circuit that contains each vertex exactly once except for the first vertex which is also the last vertex.

For example, see fig. 9.53.

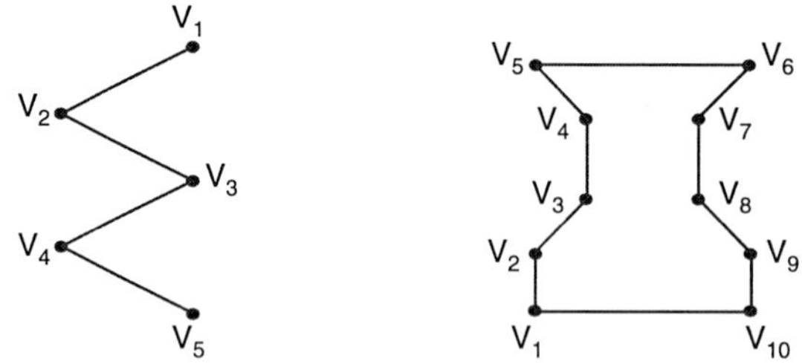

Fig. 9.53.

The fig. 9.53(*a*) represents the Hamiltonian path $\{v_1, v_2, v_3, v_4, v_5\}$ while the fig. 9.53(*b*) represents Hamiltonian the circuit $\{v_1, v_2, v_3, v_4, v_5, v_6, v_7, v_8, v_9, v_{10}, v_1\}$.

A circuit in a connected graph G is said to be Hamiltonian if it includes every vertex of G. Hence a Hamiltonian circuit in a graph of n vertices consists of exactly n edges.

If we remove any one edge from a **Hamiltonian circuit**, we are left with a path called **Hamiltonian path.** For example, consider the Hamiltonian circuit given by the fig. 9.53(*b*)

$c_1 : \{v_1, v_2, v_3, v_4, v_5, v_6, v_7, v_8, v_9, v_{10}, v_1\}$

If we remove one of the edge from the given circuit say the edge (v_1, v_{10}) then we are left with a path $\{v_1, v_2, v_3, v_4, v_5, v_6, v_7, v_8, v_9, v_{10}\}$ in which we travel along all the vertices exactly once is the **Hamiltonian path.** Since a Hamiltonian path is a sub-graph of a Hamiltonian circuit which we get by removing an edge from the circuit. Therefore every graph that has a Hamiltonian circuit also has a Hamiltonian path.

Therefore the length of a Hamiltonian path (if exists) in a connected graph of n vertices and $(n–1)$ edges.

Complete Graph

A simple graph in which there exists an edge between every pair of vertices is called a complete graph. For example see fig. 9.54.

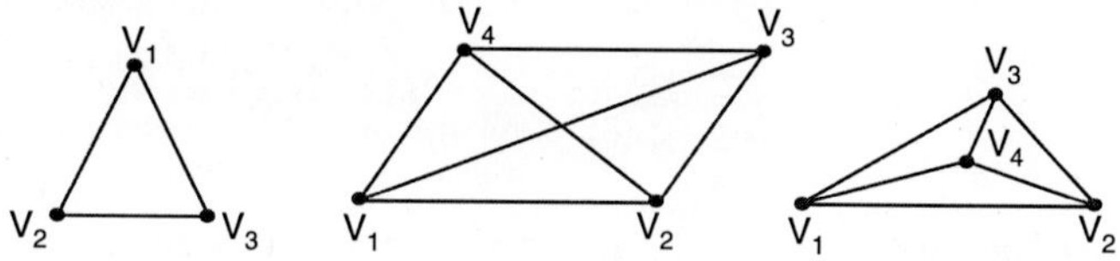

Fig. 9.54.

Since every vertex is joined with every other vertex through an edge, therefore in a complete graph G of n vertices the degree of each vertex is $n– 1$. See in the fig. (9.54) and verify it.

Theorem 9.9 : Let G be a graph of n vertices then G has a Hamiltonian path if for any two vertices u and v of G

$$\deg u + \deg v \geq n$$

Proof : The given theorem is true if the graph G is a connected graph. Let's first prove that G is connected.

On the contrary, suppose G is not connected, that is, it has two (or more) components.

Suppose v_1 is a vertex in one component which has n_1 vertices and v_2 be the vertex in another component which has n_2 vertices. Now degree of v_1 is $n_1 -1$ and degree of v_2 is $n_2 - 1$. Hence

$$\text{degree } v_1 + \deg v_2 = n_1 + n_2 - 2$$
$$< n - 1$$

This contradicts our assumption.

Hence G is connected.

The converse of the above term is not always true *i.e.* condition above is sufficient not necessary.

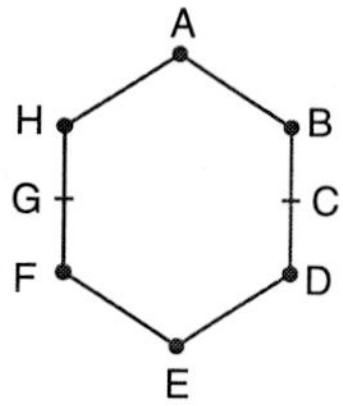

Fig. 9.55.

No. of vertices = 8

and each vertex has deg. 2

So deg u + deg v = 4 for non adjacent vertices u and v.

But from the term

deg u + deg $v \geq n$

But $4 \geq 8$ Not true.

But the graph is Hamiltonian.

Hence the converse is not true.

Theorem 9.10 : Let G be a graph of n vertices where $n \geq 3$. G has a Hamiltonian circuit if each vertex has degree greater than or equal to $^n/_2$.

Proof. Let v_1 and v_2 be two vertices of G

then degree v_1 + degree $v_2 \geq n$ (by above theorem)

$$\Leftrightarrow \text{degree } v_1 + \text{degree } v_2 \geq \frac{n}{2} + \frac{n}{2}$$

By applying comparison test in binomial expansion for $n = 1$

degree $v_1 \geq {}^n/_2$ and degree $v_2 \geq {}^n/_2$

As v_1 and v_2 are arbitrary elements, therefore degree of each vertex in Hamiltonian circuit is greater than or equal to $^n/_2$.

Theorem 9.11 : Let G be a graph of n vertices and m edges then G has a Hamiltonian circuit if

$$m \geq \frac{1}{2}(n^2 - 3n + 6)$$

Proof. Suppose that u and v are any two vertices of G that are adjacent. Let H be a graph produced by eliminating u and v from G along with any edge that have u and v as end point.

Then H has $(n - 2)$ vertices and $(m$ – deg u – deg $v)$ number of edges.

The maximum number of edges that H could possibly have is $^{n-2}C_2$. This happens when there is an edge connecting every distinct pair of vertices.

Thus the number of edges of H is at most $^{n-2}C_2$ *i.e.,*

$$^{n-2}C_2 = \frac{(n-2)!}{2!(n-4)!}$$

$$= \frac{(n-2)(n-3)}{2}$$

$$= \frac{n^2 - 5n + 6}{2} \text{ edges in } G.$$

Thus we have

$$m - \text{degree } u - \text{degree } v \geq \frac{1}{2}(n^2 - 5n + 6)$$

or
$$\text{deg } u + \text{deg } v \geq m - \frac{1}{2}(n^2 - 5n + 6)$$

$$\geq \frac{1}{2}(n^2 - 3n + 6) - \frac{1}{2}(n^2 - 5n + 6)$$

$$= n.$$

$\therefore$ deg u + deg $v \geq n$

Thus by previous theorems G has Hamiltonian circuit.

Theorem 9.12 : A directed complete graph always possesses a Hamiltonian Path.

Proof. Let G be a complete graph of n vertices then degree of each vertices of G must have degree $(n-1)$.

Let u and v be two vertices of G

Therefore,

$$\begin{aligned} \text{degree } u + \text{degree } v &= (n-1) + (n-1) \\ &= 2(n-1) \\ &\geq (n-1) \end{aligned}$$

Therefore G possesses a Hamiltonian Path.

EXAMPLE 9.21. Determine whether the given graph has a Hamiltonian circuit if so find a such circuit. If it does not given an argument to show why no such circuit exists.

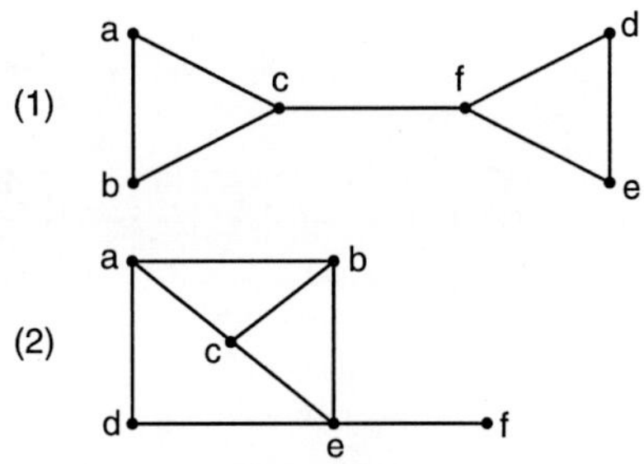

Fig. 9.56.

In a connected graph with n vertices where $n \geq 3$, then G has a Hamiltonian circuit if the

$$d(v) > \frac{n}{2} \qquad \forall\, v \in V.$$

(1) deg. $a, b, d, e = 2 \quad \frac{6}{2}$ $\Rightarrow$ Not posses Hamiltonics cirucit.

(2) As deg (f) $= 1 \not> \frac{6}{2}$ $\Rightarrow$ Not posses Hamitonia circuit.

9.10. CUT-SET IN A LINEAR GRAPH

Let G be a linear graph. The set S of edges in G is a Cut-Set if deletion of edges in S from G reduces the rank of G exactly by one and no non-empty proper subset of S has this property.

EXAMPLE 9.22. Consider the graph given by the figure 9.57(a) and 9.57(b). In the figure 9.57(a) the rank of the graph is 3. By deleting the edges a & b. The rank of the resultant graph is reduced by 1 that is now rank of the resultant graph is 2.

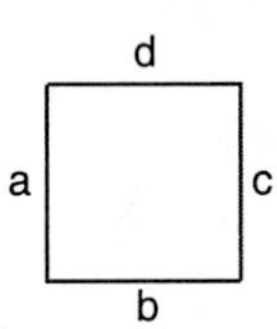

Rank = 3
Deleting $\{a, b\}$

By deleting the edges a and b

We have

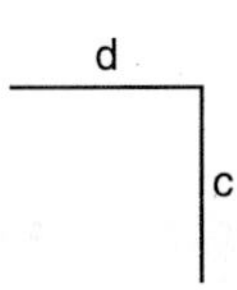

Rank = 2

Fig. 9.57(a)

In the figure 9.57(*b*). If we delete the edges *a* & *b* from the graph we will get the resultant graph of rank 2 therefore {*a*, *b*} is a cut-set.

So {*a*, *b*} is a cut-set

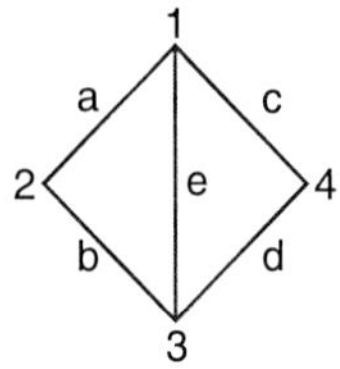

Rank = 3
Deleting {*a*, *b*}

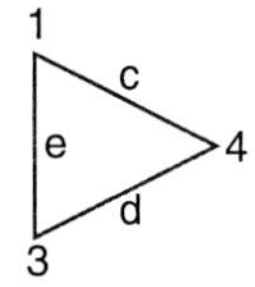

Rank = 2
{*a*. *b*} are cut-sets
{*c*, *d*}

Fig. 9.57(b)

Again consider the graph given by the figure 9.57(*b*) and delete the edges *a*, *d* and *e*, then the resultant graph is a separated graph as shown in the figure 9.57(*c*).

Rank of separated graph = n – {no. of maximally connected sub-graph}

$$S = \{a, e, d\} \text{ is a cut-set}$$

Fig. 9.57(c)

Rank of linear graph is $n_v - 1$.

EXAMPLE 9.23. Consider the graph given by the figure 9.58 and find a set which is cut set and a set which is not a cut-set.

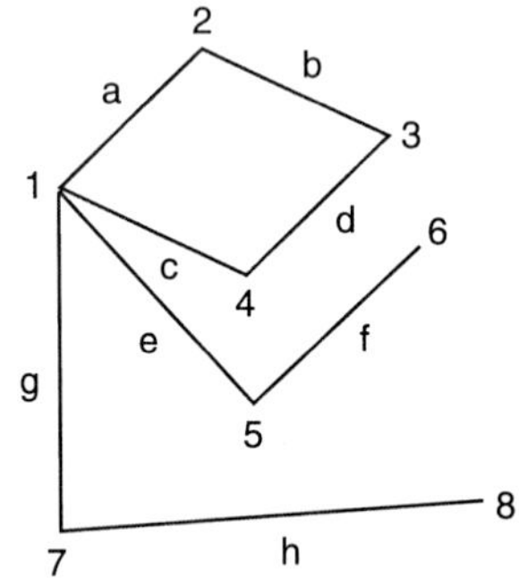

Fig. 9.58.

Now rank of the given graph is 7. If we delete the edges a & b from the graph then we will get a resultant graph as shown in the figure 9.59. Whose rank is 6. That is the rank of the resultant graph is reduced by exactly 1, therefore the set {a, b} is a cut set. But if we delete the edge say *a* only then rank of the graph will remain same therefore the set {*a*} is not a cut-set

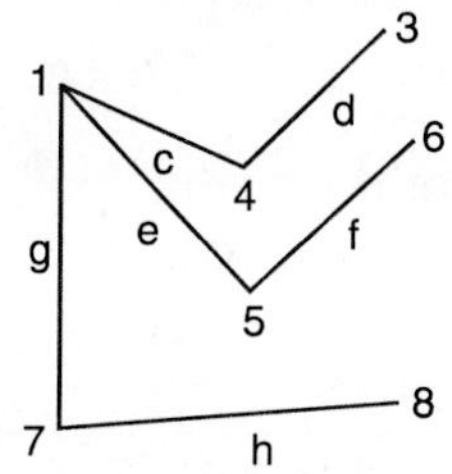

Fig. 9.59.

Theorem 9.13 : Let Ω_c be the set of vertices in a complete linear graph G and let Ω_1 be a non empty proper subset of Ω_c *i.e.*, $\Omega_1 \subset \Omega_c$

$$\overline{\Omega_1} = \Omega_c - \Omega_1 \text{ then}$$

$$E(\Omega_1 \times \overline{\Omega_1}) \text{ is a cut set}$$

We will prove the theorem by taking one example first let's define.

***E* (*A* × *B*) :** Let G be a linear graph of n vertices, let A and B be sub-graphs of G then $E(A \times B)$ is the set of edges joining the vertices of A to the vertices of B. Now consider the following example to understand the theorem (9.13).

EXAMPLE 9.24. Consider the graph given by the figure (9.60).

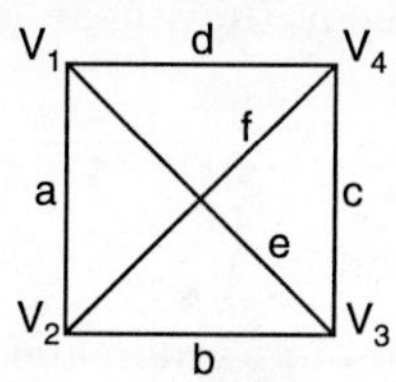

Fig. 9.60.

Here
$$\Omega_c = \{v_1, v_2, v_3, v_4\}$$
$$\Omega_1 = \{v_1, v_2\}$$
$$\overline{\Omega_1} = \{v_3, v_4\}.$$
$$E(\Omega \times \overline{\Omega_1}) = \{v_1, v_2\} \times \{v_3, v_4\}$$
$$= \{(v_1, v_3), (v_1, v_4), (v_2, v_3), (v_2, v_4)\}$$
$$= \{e, d, b, f\}$$

Therefore, rank of the resultant graph = 4 – (number separated sub-graph)
$$= 4 - 2 = 2$$

Now, delete the edges of the set $E(\Omega \times \overline{\Omega_1})$ from the given graph then the resultant graph will be as given in figure (9.61). The graph is a separated graph.

Hence, $E(\Omega \times \overline{\Omega_1})$ is a cut-set.

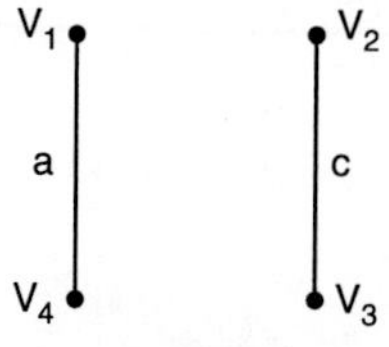

Fig. 9.61.

Proof :

$$\Omega_c = \Omega_1 \cup \overline{\Omega_1}$$

$$E(\Omega_c \times \Omega_c) = \{(\Omega_1 \cup \overline{\Omega_1}) \times (\Omega_1 \cup \overline{\Omega_1})$$

$$= E(\Omega_1 \times \Omega_1) \cup E(\Omega_1 \times \overline{\Omega_1}) \cup E(\overline{\Omega_1} \times \Omega_1) \cup E(\overline{\Omega_1} \times \overline{\Omega_1})$$

$$= E(\Omega_1 \times \Omega_1) \cup E(\Omega_1 \times \overline{\Omega_1}) \cup E(\overline{\Omega_1} \times \overline{\Omega_1})$$

Consider an edge e of G with the end of vertices p and q.

Then

Case (*i*) $p \in \Omega_1, q \in \Omega_1$ then $e \in \Omega_1 \times \Omega_1$

(*ii*) $p \in \Omega_1, q \in \overline{\Omega_1}$ then $e \in \Omega_1 \times \overline{\Omega_1}$

(*iii*) $p \in \overline{\Omega_1}, q \in \Omega_1$ then $e \in \overline{\Omega_1} \times \Omega_1$

(*iv*) $p \in \overline{\Omega_1}, q \in \overline{\Omega_1}$ then $e \in \overline{\Omega_1} \times \overline{\Omega_1}$

$$E(\Omega_c \times \Omega_c) - E(\Omega_1 \times \overline{\Omega_1}) = E(\Omega_1 \times \Omega_1) \oplus E(\overline{\Omega_1} \times \overline{\Omega_1})$$

$$E(\Omega_c \times \Omega_c) - E(\Omega_1 \times \overline{\Omega_1}) = E(\Omega_1 \times \Omega_1) \cup E(\overline{\Omega_1} \times \overline{\Omega_1})$$

$E(\Omega_1 \times \Omega_1)$ and $E(\overline{\Omega_1} \times \overline{\Omega_1})$ are both connected sets.

Also we observe that if we take a non empty proper subset of $E(\Omega_1 \times \overline{\Omega_1})$ say S_1 deletion of edges of S_1 keeps the linear graph $E(\Omega_1 \times \Omega_1)$, $E(\overline{\Omega_1} \times \overline{\Omega_1})$ connected.

Hence, the deletion of edges in $E(\Omega_1 \times \overline{\Omega_1})$ from $E(\Omega_c \times \Omega_c)$ lowers the rank by 1. Hence, $E(\Omega_1 \times \overline{\Omega_1})$ is a cut - set.

Theorem 9.14 : Let G be a complete graph and Ω_c be the set of vertices of G and let Ω_1 and Ω_2 be the two non empty proper sub-sets of Ω_c then $E(\Omega_1 \times \overline{\Omega_1})$ and $E(\Omega_2 \times \overline{\Omega_2})$ are cut - sets and ring sum of the cut - set is a cut - set *i.e.*, if $S_1 = E(\Omega_1 \times \overline{\Omega_1})$ and $S_2 = E(\Omega_2 \times \overline{\Omega_2})$.

Then $S_1 \oplus S_2$ is a cut - set.

Proof : We will prove the theorem by taking an example.

Consider a pentagon

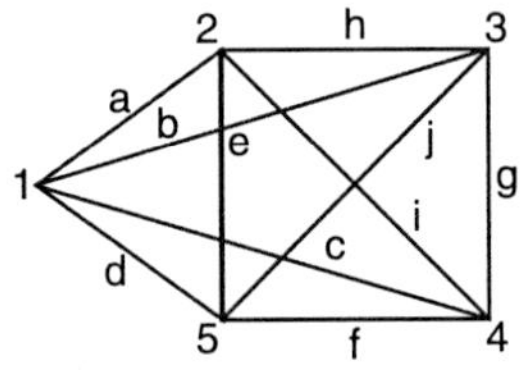

Fig. 9.62.

Let

$$\Omega_1 = \{1, 2, 3\}$$

$$\overline{\Omega_1} = \{4, 5\}$$

$$\Omega_2 = \{1, 2, 5\}$$

$$\overline{\Omega_2} = \{3, 4\}$$

$$E(\Omega_1 \times \overline{\Omega_1}) = \{c, d, i, e, g, j\}$$

$$E(\Omega_2 \times \overline{\Omega_2}) = \{b, c, h, i, j, f\}$$

Deleting the edges of S_1 from G given by fig 9.63(a) and deleting the edges of S_2 from G given by the fig. 9.63(b) and it is clear from the figure (9.63) that S_1 & S_2 both are cut set.

$$s_1 \oplus s_2 = \{d, e, g, b, h, f\}$$

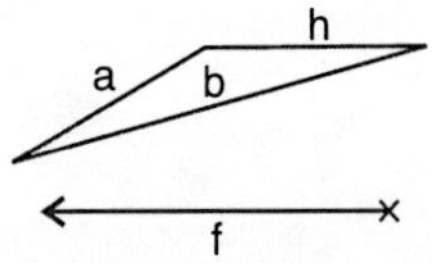

Fig. 9.63(a)

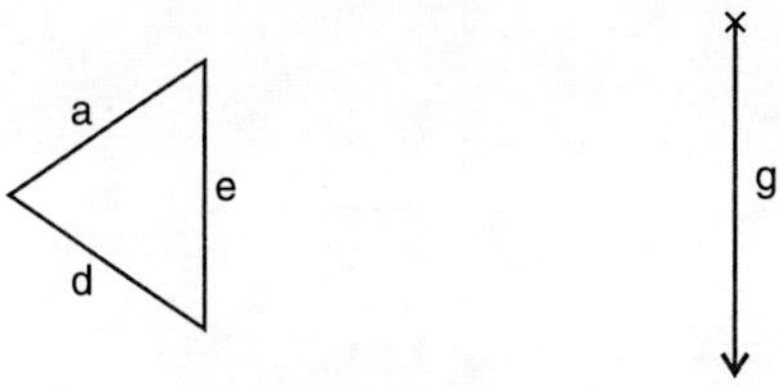

Fig. 9.63(b)

Now deleting the edges of $S_1 \oplus S_2$ from the given graph we will get the resultant graph given by the fig. 9.63(c).

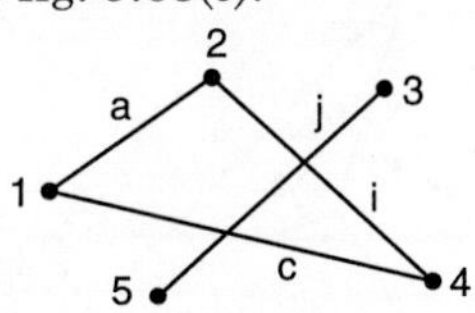

Fig. 9.63(c)

This is seperated graph having two maximally connected sub-graphs. Therefore the rank of the resultant graph given in the figure 9.63(c) is (5 – 2) = 3.

Therefore, $S_1 \oplus S_2$ is a cut-set.

Theorem 9.15 : Let s_1 and s_2 be the two cut sets and let G be a connected graph then the ring sum of s_1 and s_2 is a cut - set.

Proof : Follows from the theorem 9.14.

Theorem 9.16 : In a linear graph G, any incident set can be written as ring sum of all other incident sets.

Linearly Dependent

Let s_1 s_2, ... , s_n be the sets of a linear graph. Then there are said to be linearly dependent if $s_1 \oplus s_2 \oplus ... \oplus s_n = \phi$.

9.11. TREES

Now in this section we will discuss special type of graphs called tree which are most important in the application of graphs. A tree is a connected graph without any circuits.

Definition : Let G be a connected linear graph of n vertices, a subgraph of G of n vertices and $(n-1)$ edges which is connected is called a tree.

Properties of Tree

The tree is characterised by any three of the following properties

(1) It is connected.

(2) It has n vertices.

(3) It has $(n-1)$ edges.

(4) It has no circuit.

For example 9.24 :

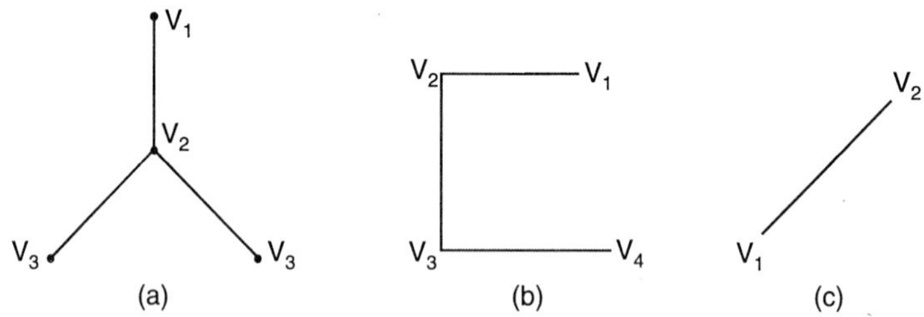

Fig. 9.64.

EXAMPLE 9.25. Examples of graphs which are not a tree are given by the figure (9.46).

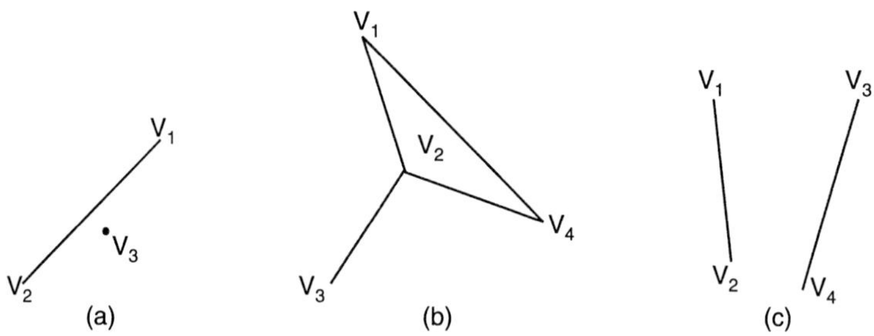

Fig. 9.65.

Theorem 9.17 : A tree of a connected graph has no circuit.

Proof. Suppose a tree has circuits, then by deleting an edge in a circuit, the resultant graph is left to be connected.

However, the no of edges is reduced by 1. Thus, there will be a connected graph of n vertices and $n-2$ edges.

Now by shorting an edge, again the resultant graph consist of $n-1$ vertices and $n-3$ edges.

Hence, by continuing the shorting of edges, we will have a connected graph of two vertices and no edges.

It is a contradiction [as G is connected].

Hence, a tree of a connected graph has no circuit.

EXAMPLE 9.26. G is connected graph having

(1) n vertices

(2) $n-1$ edges

(3) no circuit

Then G forms a tree (by defintion)

Path

A path in a linear graph G from a vertex u to a vertex v is an open edge train, the degree of whose vertices excluding u and v is 2 and degree of u and v is 1.

EXAMPLE 9.27. Open edge train which is a path

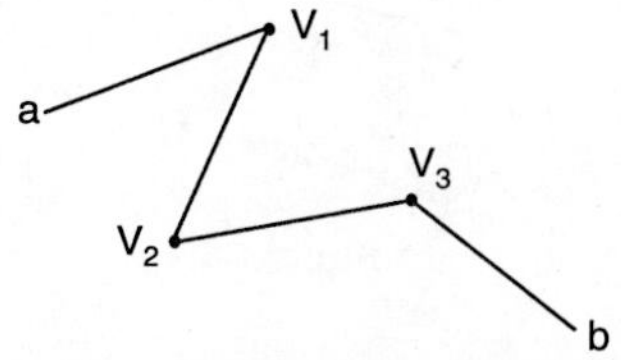

Fig. 9.66.

Open edge train which is not a path

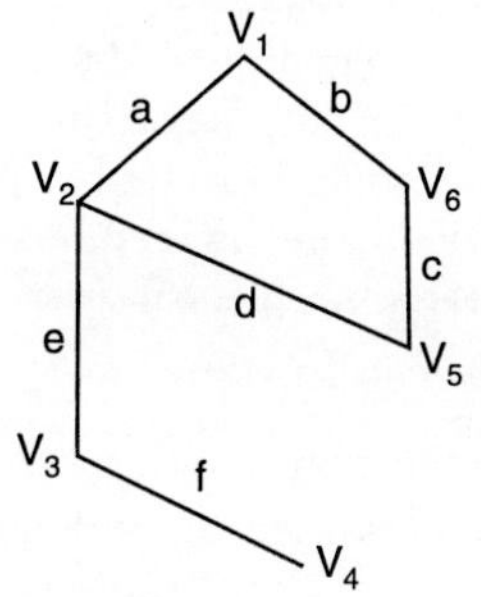

Fig. 9.67.

Theorem 9.18 : If a linear graph has exactly one path between any two vertices. The linear graph is a tree and conversely.

Proof : Let us assume that linear graph G has exactly one path between any two vertices

$\Rightarrow G$ is connected

Obviously it has no circuit

Therefore G has the following properties

(1) It has two vertices

(2) It has one edge.

(3) It has no circuit.

(4) G is connected

$\Rightarrow G$ is a tree.

Conversely, let G is a tree

(1) G has n vertices.

(2) G has $n - 1$ edges.

(3) G has no circuit.

(4) G is connected.

Now by deleting an edge from G we are left with $n - 1$ vertices and vertices $n - 2$ edges. Continuing like this, we'll have 2 vertices and 1 edge and obviously between two vertex there is a path. Therefore, the resultant graphs has exactly one path between any two vertices.

Theorem 9.19 : A tree with n vertices has $(n - 1)$ edges.

Proof. We will prove the theorem by the method of mathematical induction.

Basic step : for $n = 1, 2, 3$ the result is true see fig. 9.68.

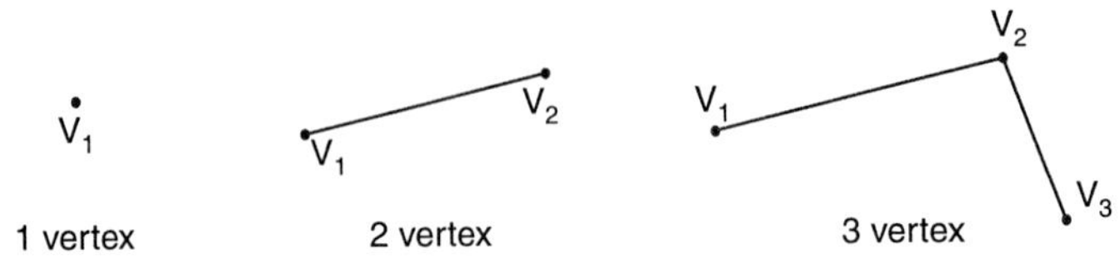

Fig. 9.68.

Inductive Step :

Let us assume that the result is true for $n = k$ *i.e.*, graph H has k vertices say $v_1, v_2, v_3 \ldots v_k$ and $(k - 1)$ edges say $e_1, e_2, .. e_{k-1}$.

Now we will prove it for $n = k + 1$.

Now adding a vertex v to the graph H. If this is connected to only one of the vertices $v_1\ v_2 \ldots v_k$ by means of an edge say e then we are done. It cannot connect to atleast two of vertices otherwise there will be a circuit in the resultant graph that will violate the property of tree, as tree has no circuit.

Therefore the result is true for $n = k + 1$

Hence, by induction the result is true.

Theorem 9.20 : Any connected graph with n vertices and (n – 1) edges is a tree.

Proof : Let the given graph G is with the following properties.

G is connected.

G has n vertices

G has $(n - 1)$ edges

and let us assume that

G is not a tree.

Now by deleting an edge we are left with the resultant graph say G' with $(n - 1)$ vertices and $(n - 2)$ edges. Again delete an edge from the graph G' and now again the resultant graph will have $(n - 2)$ vertices and $(n - 3)$ edges, continuing this process of reducing the edges, at last we are left with a graph with 2 vertices and no edge which is not possible as G is connected. Therefore our assumption is wrong.

Theorem 9.21 : A graph G with n vertices and $(n - 1)$ edges and no circuits in connected.

9.12. CHORD

For a tree t in a linear graph, an edge that is not in t is called a chord. For example consider the graph given by the figure (9.69*a*). One of the tree is given by fig. 9.69(*b*) for which the corresponding chords are b and d.

(a) (b) Here, d, b are chords of this rate

Fig. 9.69.

$$s = \{b, d\}$$

s is set of chords.

Remark. If we consider a connected linear graph of n_v vertices and n_e edges, then tree consists of $n_v - 1$ edges.

Therefore, there are $n_e - n_v + 1$ chords.

Fundamental Circuits

If we insert an edge *e* from set of chords in a tree then $t \cup (e)$ will have exactly one circuit and this circuit is called **fundamental circuit**.

Since there are $n_e - n_v + 1$ chords in a tree *t* and introduction of each chord in tree gives rise to exactly one circuit.

Therefore, there are $\boldsymbol{n_e - n_v + 1}$ fundamental circuits.

Fundamental Circuit Matrix is the matrix having fundamental circuits

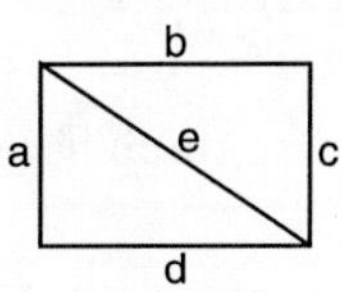

Fig. 9.70.

$$B_e = \begin{matrix} & a & b & c & d & e \\ C_2 & 0 & 1 & 1 & 0 & 1 \\ C_1 & 1 & 0 & 0 & 1 & 1 \end{matrix}$$

$$B_e = \begin{matrix} & b & d & ; & a & c & e \\ C_1 & 1 & 0 & ; & 1 & 0 & 1 \\ C_2 & 0 & 1 & ; & 0 & 1 & 1 \end{matrix}$$

Fundamental Circuit Matrix is

$$B_e = [I : B_f]$$

where *I* is the identity matrix gives chords in the corresponding circuit *B* & B_f is the matrix that gives the edges (branches) of the the tree *t* in the corresponding circuits. Therefore, now consider the following theorem.

Theorem 9.21 : In a fundamental circuit matrix $[I : B_f]$, of a connected linear graph, the edges corresponding to the columns of B_f forms a tree.

Proof. Let *G* be a connected linear graph of n_v vertices and n_e edges. Let *A* be an incidence matrix in *G*, we can write $A = [A_1, A_2]$.

Where columns of A_1 correspond to the columns of *I*, the identity matrix and the columns of A_2 correspond to the column of B_f.

Also $$AB_f^T = 0$$

so
$$(A_1 \; A_2)\begin{pmatrix} I_t \\ B_f \end{pmatrix} = 0$$

$$A_1 + A_2 B_f^t = 0$$
$$A_1 = -A_2 B_f^t$$

Now
$$A = [A_1 \; A_2]$$
$$= [-A_2 \;\; B_f^t A_2]$$
$$= A_2[-B_f^t \;\; I] = \qquad A_2 M$$

$$\begin{bmatrix} R_1 \\ R_2 \\ : \\ R_{n-1} \end{bmatrix} = \begin{bmatrix} r_1 \\ r_2 \\ : \\ r_{n-1} \end{bmatrix} M$$

Since $p(A)$ is number of rows in A.

$\Rightarrow A_2$ must be non-singular.

$\therefore$ The edge corresponding to columns of A_2 forms a tree.

Since the columns of B_f correspond to columns of A_2. Therefore, the edge corresponding to columns of B_f also forms a tree.

Connected Graph

A linear graph is said to be connected if between any two vertices of G there is a path connecting these two vertices.

Euler Graph

A linear graph is said to be Euler graph if degree of each vertex is even.

Complete Graph

A linear graph G is said to be complete if there exists an edge between every pair of vertices in G.

For example consider the following graphs given by the figure 9.71.

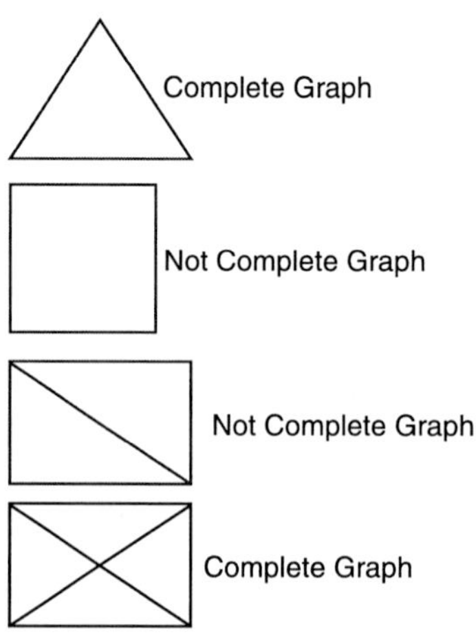

Fig. 9.71.

A complete graph for n vertices is denoted by K_n. The graphs K_n, for $n =$ 1, 2, 3, 4, 5, 6 are given as follows :

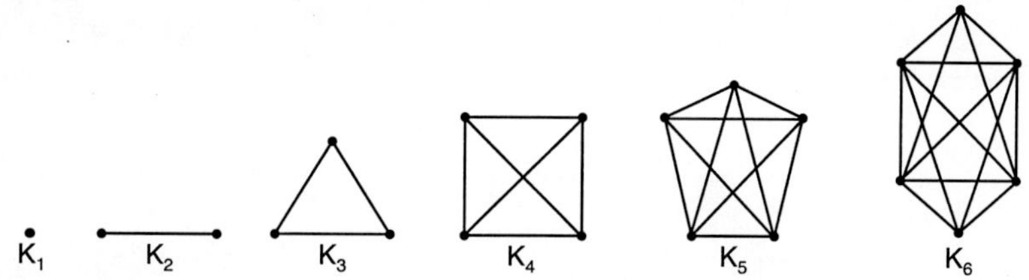

Fig. 9.72.

Wheels

We obtain the wheel W_n when we add an additional vertex to the cycle C_n, for $n \geq 3$, and connect this new vertex to each of the n vertices in C_n, by new edges.

The wheels W_3, W_4, W_5 and W_6 are given as follows :

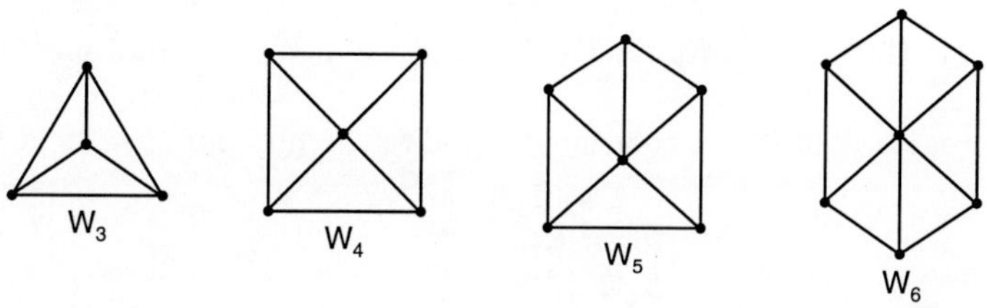

Fig. 9.73.

Theorem 9.22 : A complete graph is always a connected graph but converse is not true.

In order to verify the theorem (9.22), consider the following graph given by figure 9.74.

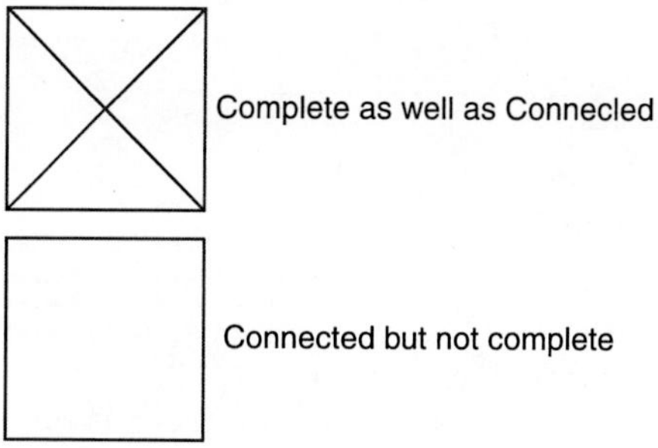

Fig. 9.74.

Sylverter's Law : Let $H = (h_{ij})_{p \times e}$

and $K = (k_{ij})_{q \times e}$ be the two

matrices where $p \leq e$, $q \leq e$

and $P(H) = p$ and suppose $HK = 0$

then $P(k)$ is atmost $(e - p)$

Proof : Since $p(H) = p$, we can write $H = (H_1 \; H_2)$

where H_1 is of order $p \times (e - p)$
and H_2 is of order $p \times p$

By partitioning $$K = \begin{pmatrix} K_1 \\ K_2 \end{pmatrix}$$

We have $$HK = 0$$

$$\Rightarrow (H_1 \ \ H_2)\begin{pmatrix} K_1 \\ K_2 \end{pmatrix} = 0$$

$$\Rightarrow H_1 K_1 + H_2 K_2 = 0$$

$$\Rightarrow H_2 K_2 = - H_1 K_1$$

Pre-multiplying above eqn by H_2^{-1}

$$K_2 = -H_2^{-1} H_1 K_1$$

[H_2 is non- singular]

$$= L K_1$$

Where $$L = - H_2^{-1} H_1$$

This shows that rows of K_2 can be obtained as linear combination of rows of K_1.

Therefore, Rank of K can be atmost equal to the number of rows of K_1

$$(K = [K_1 \ \ K_2]$$
$$= [K_1 \ \ LK_1]$$
$$= K_1[I \ \ L]$$

Since H_1 is order $p \times (e - p)$

$\therefore$ K_1 must be $(e - p) \times q$ matrix

$\therefore$ Rank of K is atmost $(e - p)$.

Odd or Even Vertex

Vertex of a graph is called odd or even depending on whether its degree is even or odd, that is,

If deg $v = 2n$

then v is an odd vertex.

If deg $v = 2n + 1$

then v is an even vertex.

Bipartite Graphs

If the vertex set V of a graph G can be partitioned into two non-empty disjoint subsets X and Y in such a way that each edge of G has one end in X and one end in Y then G is called Bipartite. The partition $V = X \cup Y$ is called bipartition of G. For example consider the following graph in which the vertex set V is partitioned into X & Y such that $V = X \cup Y$.

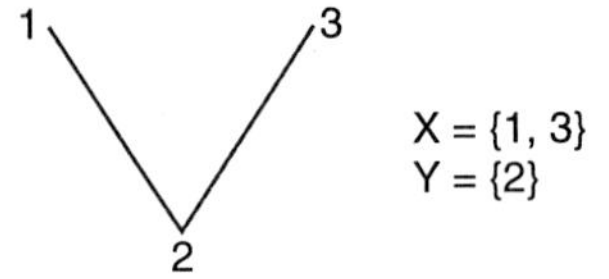

Fig. 9.75.

Subgraph

If G and H are two graphs with vertex sets $V(H)$, $V(G)$ and edge set $E(H)$ and $E(G)$ respectively such that $V(H) \leq V(G)$ and $E(H) \leq E(G)$ then we call H as a **subgraph** of G or G is a **supergraph** of H.

If $V(H) \subset V(G)$ and $E(H) \subset E(G)$ *then* H is a proper subgraph of G and if $V(H) = V(G)$ then we say H is a **spanning subgraph** of G.

Fig. 9.76.

Here, H is spanning subgraph of G as number of vertices is same but number of edges are not same.

Distance Between Two Vertices

For any two vertices u and v connected by a path in G, we define distance between u and v denoted by $d(u, v)$ to be the length of the shortest $u - v$ path.

Remark : $d(u, v) = \infty$ if there is no path connecting them.

Eccentricity of V

Suppose V be a vertex set of graph G and $v \in V$ then eccentricity of v is denoted by $e(v)$ and it is given by:

$$e(v) = \max \{d(u, v) : u \in V, u \neq v\}$$

Radius of G

The radius of graph G is denoted by rad(G) & is given as:

$$\text{rad}(G) = \min \{e(v) : v \in V\}$$

Diameter of G

The diameter of a graph G is denoted by dia(G) and is given as:

$$\text{dia}(G) = \text{Max}\{e(v) : v \in V\}$$

Centre Point

A point v is a centre point if

$$e(v) = \text{rad}(G)$$

EXAMPLE 9.28. Consider the graph given by the figure 9.77. Find distance of each vertex of G from the vertex V_6 and also find diameter of G and radius of G. Also find the eccentricity of the vertex.

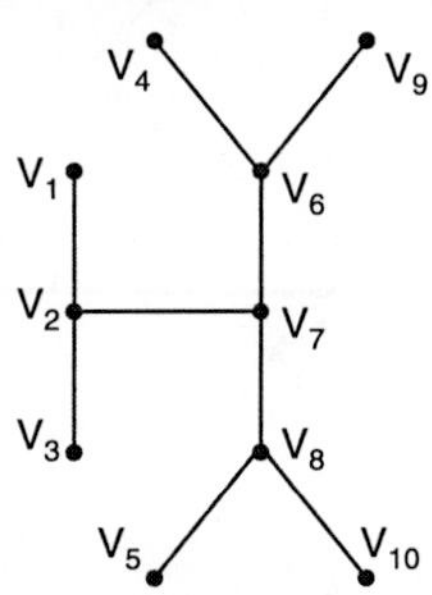

Fig. 9.77.

$$d(v_1, v_6) = 3$$
$$d(v_2, v_6) = 2$$
$$d(v_3, v_6) = 3$$
$$d(v_4, v_6) = 1$$
$$d(v_5, v_6) = 3$$
$$d(v_7, v_6) = 1$$
$$d(v_8, v_6) = 2$$
$$d(v_9, v_6) = 1$$
$$d(v_{10}, v_6) = 3$$

$e(v_6) = \max \{d(u, v_6) : u \in V, v_6 \in V\}$

$e(v_6) = 3$

Similarly
$$e(v_1) = 4,$$
$$e(v_2) = 3$$
$$e(v_3) = 4$$
$$e(v_4) = 4$$
$$e(v_5) = 4$$
$$e(v_6) = 3$$
$$e(v_7) = 2$$
$$e(v_8) = 3$$
$$e(v_9) = 4$$
$$e(v_{10}) = 4$$

Radius of the graph $G = \min \{e(v) : v \in V\} = 2$

Diameter of the graph $G = \max \{e(v) : v \in V\} = 4$

Centre point $= v_7$.

Spanning Tree

Let G be a graph. A subgraph H of G is called a spanning subgraph of G if the vertex set of H is same as vertex set of G.

A **spanning Tree** of a graph G is a spanning subgraph of G which is a tree.

Remark. Complete graph G of n vertices has n^{n-2} different spanning trees.

EXAMPLE 9.29. In the following graph having four vertices there are $4^2 = 16$ spanning trees.

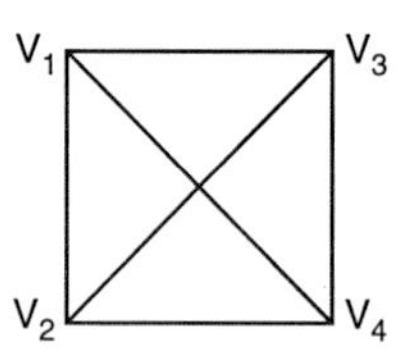

Fig. 9.78.

We have discussed the rank of a linear graph as graph can be given in terms of matrices, therefore, its rank exists. For a connected graph of n vertices the rank is $(n - 1)$ and nullity $(e-n + 1)$. For a graph having k components (seperated subgraphs) must have rank $(n - k)$ and nullity $(e - n + k)$ where e is the number of edges in the given graph G. Now one can ask the question what's the use of Rank and Nullity in a graph G.

As we defined edges, vertices and components in a graph G. The rank and nullity give the inter relationship among all three as:

Rank $r = n - k$ = number of branches (edges) in any spanning tree (or forest) of G.

nullity u = $e - n + k$ = number of chords in G.

rank + nullity = number of edges in G.

Weighted Graph

A weighted graph is a graph G in which each edge e has been assigned a real number $w(e)$. This $w(e)$ is called weight of e. If H is a subgraph of a weighted graph then $w(H)$ is the sum of weights $w(e_1)$, $w(e_2)$, ... , $w(e_k)$ where each e_j is edge of H.

$w(H) = w(e_1) + w(e_2) + ... + w(e_k)$.

9.13. ALGORITHM

We try to find subgraph of weighted graph with the minimum or maximum weights. We look at a connected spanning subgraph of G which should not have a cycle that is we look at a spanning tree of G and finally we look for a spanning tree with minimum weights. Such a tree is called ***minimal spanning tree*** or ***optimal tree.***

Kruskal's Algorithm

In kruskal's algorithm, we choose an edge of G which has smallest weight among the edges of G which are not a loop. This algorithm gives an sub-graph of G such that this is a minimum spanning tree of G.

Step 1: Choose e_1 an edge of G such that $w(e_1)$ is smallest among all and e_1 is not a loop.

Step 2 : If edges e_1, e_2, ..., e_i have been chosen, then choose an edge $e_i + 1$ not already chosen such that

(*i*) the induced graph $G\{e_1, .., e_{i+1}\}$ is acyclic

(*ii*) $w(e_{i+1})$ is as small as possible.

Step 3 : If G has n vertices, stop after $(n-1)$ edges have been chosen,. Otherwise repeat step 2.

Result : Let G be a weighted connected graph in which weights of each edge are non-negative numbers. Let T be a subgraph of G obtained by Kruskal's algorithm, then T is a minimal spanning tree of G.

EXAMPLE 9.30. Given below is a weighted graph G with 5 vertices. Apply kruskal's. Algorithm to determine a minimal spanning tree.

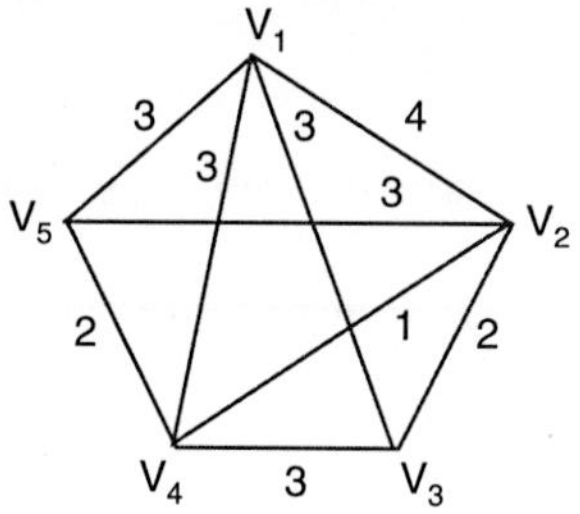

Fig. 9.79.

Step 1. Choose e_1 as $v_2\ v_4$ obviously, it is not a loop.

Applying step 2 we will get

Step 2. $e_2 = v_2\ v_3$

$e_3 = v_4\ v_5$

$e_4 = v_5\ v_1$

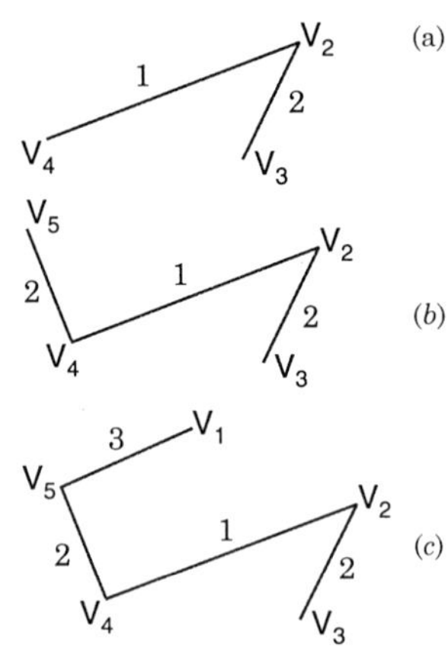

Fig. 9.80.

From the above example it is clear that we can find another spanning tree also. Readers are required to find a spanning tree other than T.

9.14. PRIM'S ALGORITHM

Now consider the another algorithm to find. In the following algorithm is given by Prim so called Prim's algorithm in which we do not require listing of all the edges but which starts with an arbitrary edge.

In this algorithm we choose a vertex first and choose an edge with smallest weight incident on that vertex.

Step 1: Choose any vertex v_1 of G.

Step 2 : Choose an edge $e_1 = v_1\ v_2$ of G *s.t.* $v_2 \neq v_1$ and e_1 has smallest weight among the edges of G incident with v_1.

Step3 : If edges $e_1, e_2 \ldots, e_i$ have been chosen involving

$$v_1\ v_2 \ldots v_{i+t}$$

Choose an edge $\quad e_{i+1} = v_j\ v_k$ where

$$v_j \in \{v_1 \ldots v_{i+1}\}$$
$$v_k \notin \{v_1 \ldots v_{i+1}\}$$

such that e_{i+1} has smallest weight among the edges of G where precisely one end in $\{v_1, \ldots v_{i+1}\}$.

Step 4 : If G has n vertices then stop after $(n-1)$ edges have been chosen, otherwise go to step 3.

Consider the example

EXAMPLE 9.31. Start with the vertex v_1 choose an edge e_1 say $(v_1\ v_3)$. Now consider edges start with v_1 and v_3 choose the $(v_3\ v_2)$ of weight 2. Similary doing this at last we will get a spanning tree of minimum weight say T.

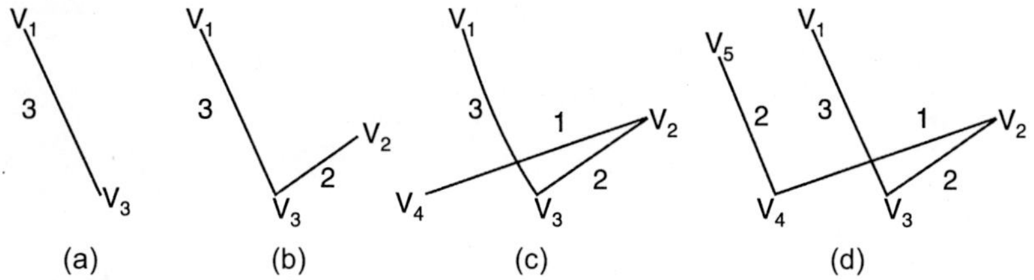

Fig. 9.81.

EXAMPLE 9.32. In the above problem we have taken the initial vertex v_1. The readers are required to take another vertex say v_4 and find the minimal spanning tree.

9.15. BREADTH-FIRST SEARCH

The general idea behind a breadth-first search begining at a starting vertex S as follows. First we process the starting vertex S. Then we process all the neighbours of S. Then we process all the neighbours of S. And so on. Naturally we need to keep track of the neighbours of a vertex and we need to guarantee that no vertex is processed twice.

Breadth First Search Algorithim (BFS)

This algorithm is used to find a path which uses least number of edges, that is the shortest path.

Step 1 : Label vertex S with 0, set $i = 0$.

Step 2 : Find all unlabelled vertices in G which are adjacent to vertices labelled i. If there are no. such vertices then t is not connected to S. If there are such vertices label them $i + 1$.

Step 3 : If t is labelled, go to step 4, if not increase i to $i + 1$ and go to step 2.

Step 4 : The length of shortest path from S to t is $i + 1$, then stop.

Once the length of shortest path is found from BFS, we use the **back-tracking algorithm**, to find the shortest path from s to t. This algorithm uses the label $\lambda(v)$ which are generated in BFS algorithm.

Back Tracking algorithm for shortest path

Step 1 : Set $i = \lambda(t)$ and assign $v_i = t$

Step 2 : Find a vertex u adjacent to v_i and with $\lambda(u) = \text{i} - 1$
Assign $v_{i-1} = u$

Step 3 : If $i = 1$, then stop, if not then decrease i to $i - 1$ and go to step 2.

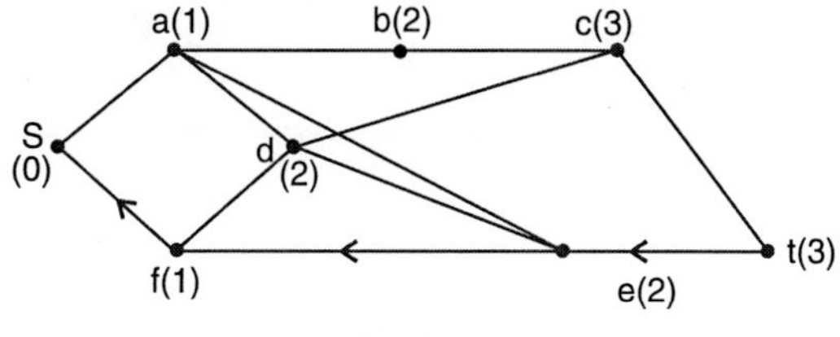

Fig. 9.82.

Length of shortest path = 3

Applying Back Tracking algorithm

$$i = \lambda(t) \Rightarrow \lambda(t) = 3$$

$$i = 3, v_3 = t.$$

Choose e adjacent to $v_3 = t$ with

$$\lambda(e) = 2 \text{ and } v_2 = e.$$

Choose f adjacent to v_2 with

$$\lambda(f) = 1 \text{ and } v_1 = f$$

Shortest path can be $t\ e\ f\ s$
$t\ e\ a\ s$
$c\ d\ f\ s$
$c\ d\ a\ s$
$c\ b\ a\ s$

9.16. DEPTH-FIRST SEARCH (DFS)

As the name suggests in the DFS we move along depth and then do backtracking.

The general idea behind a depth first search beginning at a starting vertex S is as follows.

First we process the starting vertex S. Then we process each vertex N along a path P which begins at S that is, we process a neighbour of S, then a neighbour of a neighbour of S and so on.

After coming to a "dead end", *i.e.*, to the vertex at which we cannot process further we back track on the path P until we continue along another path P' and so on.

Depth-first search algorithm

Step 1 : Label vertex S with 0 and set $i = 0$

Step 2 : Find an unlabelled vertex in G which is adjacent to the vertex in G labelled i. If there is no such vertex then t is not connected to S and if there such vertex exists label them $i + 1$.

Step 3 : If t is labelled go to step 4 if not increase i to $i + 1$ and go to step 2.

Step 4 : Do backtracking

Backtracking :

Start with the vertex t and move to the vertex adjacent to t then find the another path as we have done in step 2. First we reach the end vertex (t) of the new path then do back tracking again and the process should continue till the initial vertex S reached.

Consider the example done in the *BFS* algorithm.

EXAMPLE 9.33.

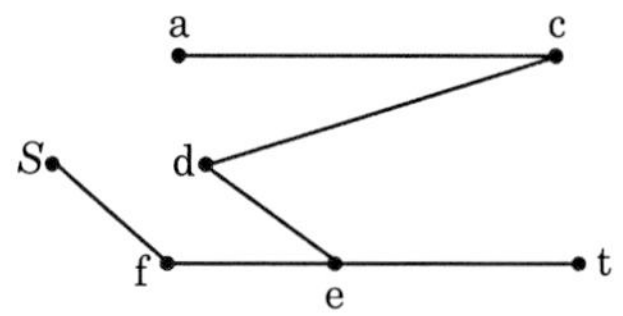

Fig. 9.83.

We start with the initial vertex S

Step 1. S •

Step 2.

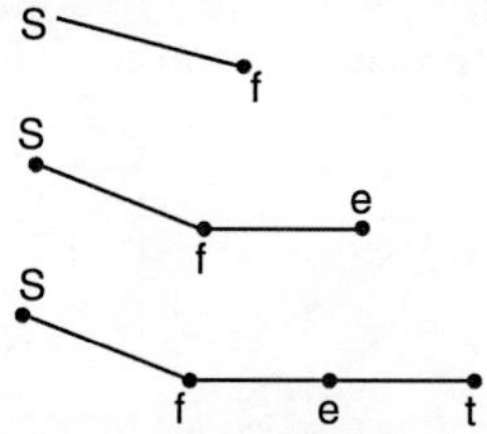

Fig. 9.84.

Step 3. G0 to step 4 *i.e.,* backtracking starting with t

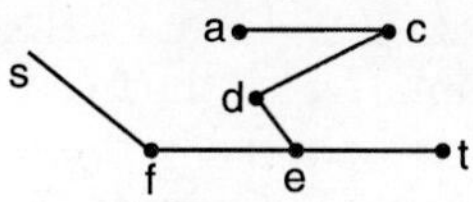

Fig. 9.85.

As the number of edges are (i– 1) then stop.

EXAMPLE 9.34. Apply DFS to find a spanning tree for the following graph.

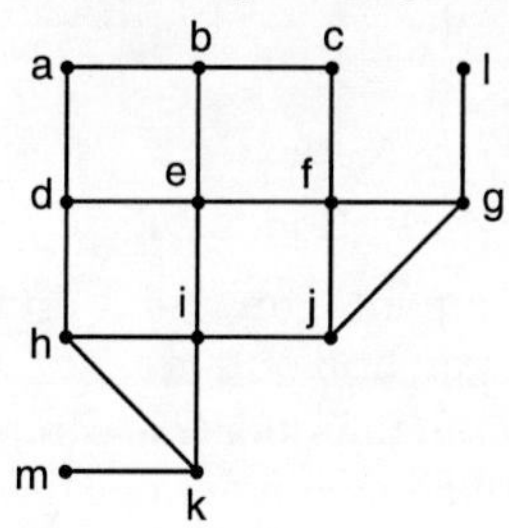

Fig. 9.86.

Solution : Let Source vertex is f

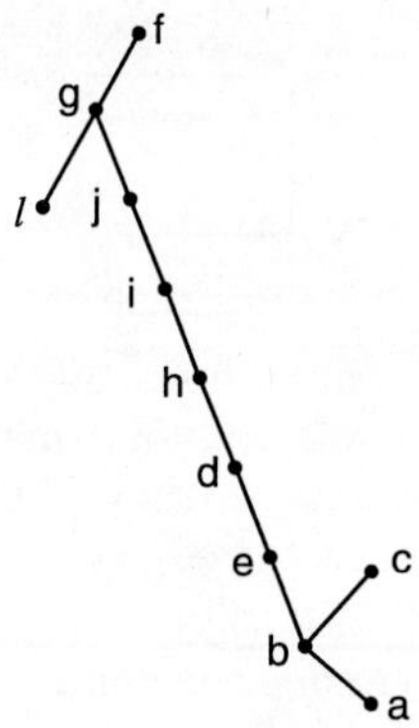

Fig. 9.87.

9.17. PLANAR GRAPHS

A graph G is said to be a **planar graph** if there exists some geometric representation of G which can draw a plane such that no two of its edges intersect. A graph that cannot be drawn on a plane without a crossover between its edges is called **non-planar graph**.

The geometric representation of a graph on any surface such that no edges intersect is called **embedding.** For example,

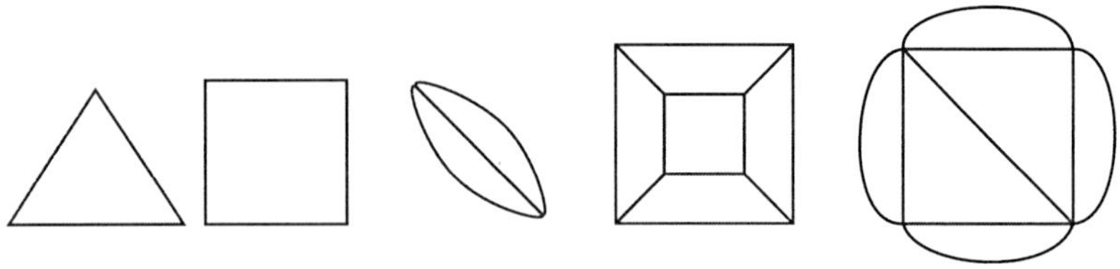

Fig. 9.88. Planar graphs

Theorem 9.23 : A complete graph of five vertices is non-planar.

Proof. Let us consider a graph G with the five vertices v_1, v_2, v_3, v_4, v_5. As G is a complete graph of five vertices therefore there must be a circuit from $v_1 \rightarrow v_2 \rightarrow v_3 \rightarrow v_4 \rightarrow v_5 \rightarrow v_1$ as show in the figure 9.89.

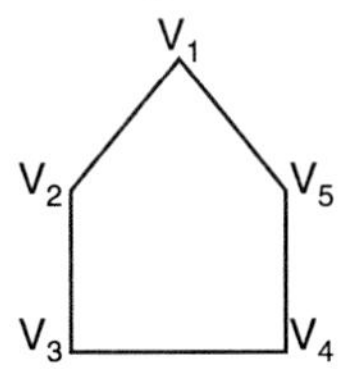

Fig. 9.89.

The resultant circuit is a pentagon and it divides the plane paper into two parts that is inside of pentagon and outside.

Since v_2 must be connected to v_5 (as G is complete) by the edge (v_2, v_5) then this edge can be drawn outside or inside. Let us draw this edge (v_2, v_5) inside the pentagon as shown in figure 9.90.

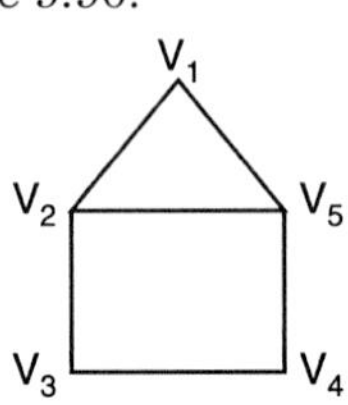

Fig. 9.90.

Now as G is complete there must be an edge from v_1 to v_3 and an edge from v_1 to v_4. As we can not draw these two edges inside the pentagon otherwise they will intersect from the edge (v_2, v_5). So we draw the edge (v_1, v_3) and (v_1, v_4) outside the pentagon As shown in fig. 9.91.

Now we draw edge (v_2, v_4) inside the pentagon and the resultant graph is shown in the figure 9.91.

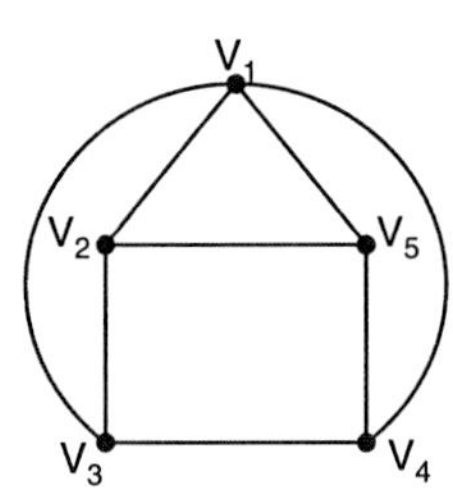

Fig. 9.91.

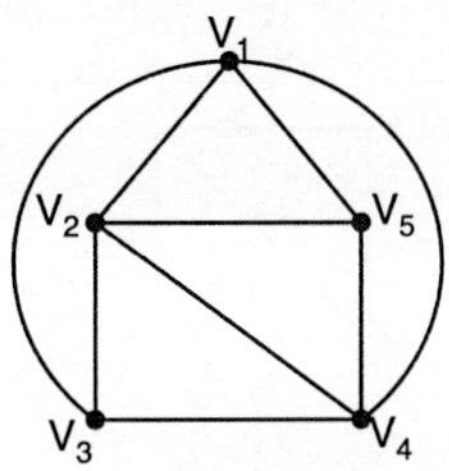

Fig. 9.92.

Now we cannot draw edge (v_3, v_5) without intersecting any edge. Therefore the complete graph with five vertices is non-planar.

EXAMPLE 9.35. Is K_4 is planar

Solution: A graph is called planar if it can be drawn in the plane without any edge crossing. Such a drawing is called planar representation of the graph.

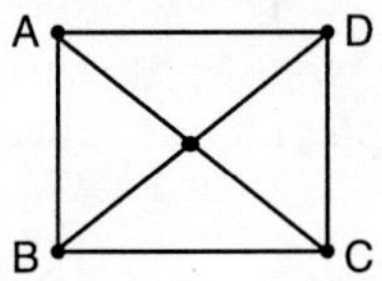

Planar Representation

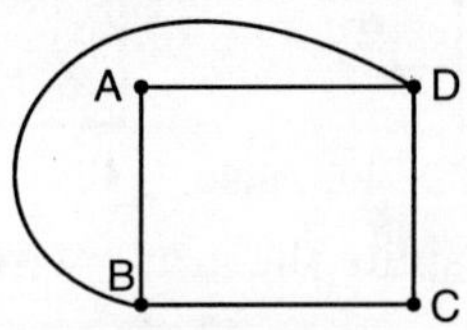

Fig. 9.93.

EXAMPLE 9.36. The given graph G is planner

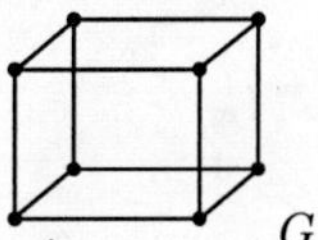

and its planar representation is:

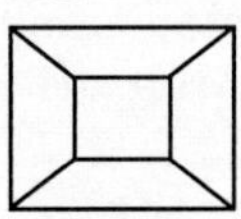

Fig. 9.94.

EXAMPLE 9.37. $K_{3, 3}$ a complete Bipartite graph is not planar.

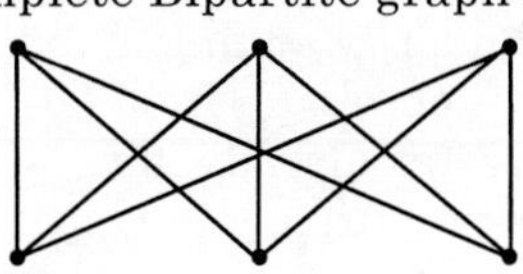

Fig. 9.95.

EXAMPLE 9.38. Is K_4 is planar.

Solution : A graph is called planar if it can be drawn in the plane without any edge crossing. Such a drawing is called planar representation of the graph.

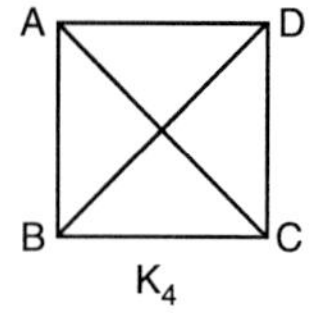

Its Planar representation is:

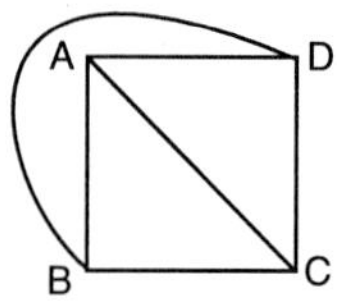

Fig. 9.95(a)

Euler's Formula

A planar representation of a graph splits the plane into regions, including an unbounded region. For example, planar representation of the graph given in example 9.36 *i.e.*,

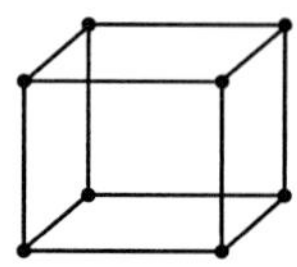

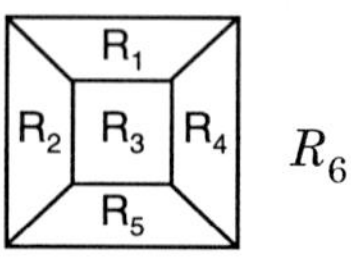

Fig. 9.96.

Its planar representation splits the graph with 6 regions say R_1, R_2, R_3, R_4, R_5, R_6.

Euler give the relationship among the number of regions, the number of vertices and the number of edges in the planar graph.

Theorem 9.24 : Euler's Formula : Let G be a connected planar simple graph with $\boldsymbol{e}$ edges and $\boldsymbol{v}$ vertices. Let r be the number of regions in a planar representation of G. Then $\boldsymbol{r = e - v + 2}$.

Proof : First, we specify a planar representation of G. We will prove the theorem by constructing a sequence of sub-graphs $G_1, G_2, \ldots G_e = G$, successively adding an edge at each stage. Now Arbitrarily pick one edge of G to obtain G_1 as shown in the fig. 9.97.

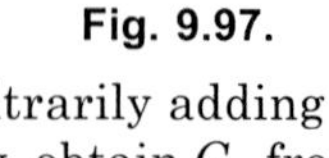

Fig. 9.97.

Obtain G_2 from G_1 by arbitrarily adding an edge that is incident with a vertex already in G_1. Similarly, obtain G_n from G_{n-1} by arbitrarily adding an edge that is incident with a vertex already in G_{n-1}, adding the other vertex incident with this edge if it is not already in G_{n-1}.

This construction is possible as G is connected. G is obtained after e edge are added.

Let r_n, e_n and v_n represent the number of regions, edges and vertices of the planar representation of G_n.

Now we will prove the result by **method of Induction.**

Basic Step : $n = 1$

$$r_1 = e_1 - v_1 + 2 \text{ is true for } G_1$$

as $$e_1 = 1, v_1 = 2 \ \& \ r_1 = 1.$$

This is shown as

Fig. 9.98(a)

Inductive Step : Let the result is true for $n = k$ *i.e.,*

$$r_k = e_k - v_k + 2 \text{ is true for } G_k.$$

Let (a_{k+1}, b_{k+1}) be the edge that is added to G_k to obtain G_{k+1}.

Two cases arises :

Case I : both a_{k+1}, b_{k+1} are already in G_k as shown in fig. 9.98

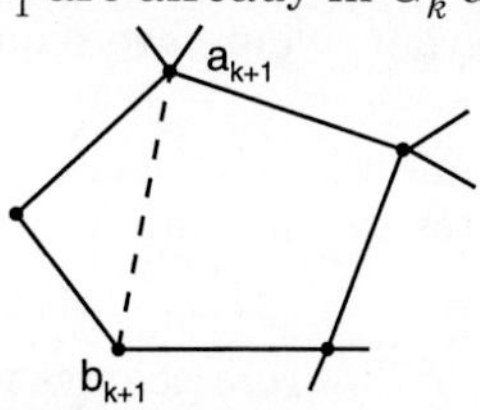

Fig. 9.98.

These two vertices must be on the boundary of a common region R otherwise it would be impossible to add the edge (a_{k+1}, b_{k+1}) to G_k without edge-crossing.

The addition of this new edge splits R into two regions.

Consequently

$$r_{n+1} = r_n + 1$$
$$e_{n+1} = e_n + 1$$
$$v_{n+1} = v_n$$

Thus, $$r_{n+1} = e_{n+1} - v_{n+1} + 2$$
$$r_{n+1} = e_{n+1} - v_n + 2$$

Hence, the result is true.

Case II : One of the two vertices of the new edge is not already in G_k. Suppose that a_{k+1} is in G_k but b_{k+1} is not. Adding of this new edge does not produce any new region, since b_{n+1} must be in a region that has a_{k+1} on its boundary, consequently

$$r_{k+1} = r_k$$
$$e_{k+1} = e_k + 1$$
$$v_{k+1} = v_k + 1$$

Now, $$r_{k+1} = e_{k+1} - v_{k+1} + 2$$
$$r_k = (e_k + 1) - (v_k + 1) + 2 \quad \text{True.}$$

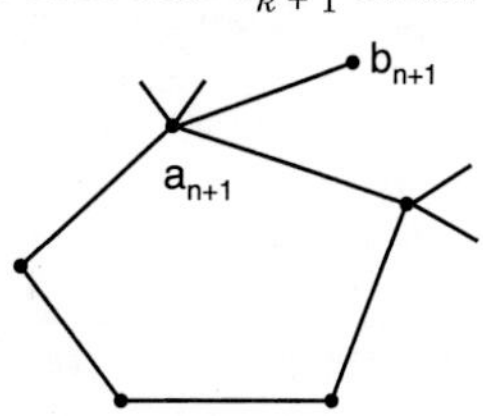

Fig. 9.99.

Therefore, by principle of induction the result is true for all n.

EXAMPLE 9.39. Suppose that a connected planar simple graph has 20 vertices, each of degree 3. Into how many regions does a representation of this planar graph split the plane ?

Solution: This graph has 20 vertices each of degree 3

Therefore, $3v = 3 \times 20 = 60$, which is twice the number of edges

i.e., $2e = 60$

$\Rightarrow \quad e = 30$

$\therefore$ Total no. of edges = 30

$\therefore$ By Euler's Formula.

$$r = e - v + 2 = 30 - 20 + 2 = 12$$

Thus, the graph must have (12) regions.

9.18. COLOURING

Colouring of a graph constitutes the colouring the vertices or edges of that graph.

Suppose we have a graph G of n vertices and it is required to paint its vertices such that no two adjacent vertices have the same colour. Problem here arises that "How many different colours are required".

That is **colouring** all the vertices of a graph is the property that no two adjacent vertices have the same colour. Proper colouring (or Best job in colouring) in a graph is one that requires the minimum number of colour.

Chromatic number of *G*

A graph G that requires k different colours for its proper colouring and no less that exactly k different colours) is called ***k*-chromatic graph** and the number k is called the **chromatic number of *G***. Consider the following figures the different colouring of the graphs are given in figure 9.100.

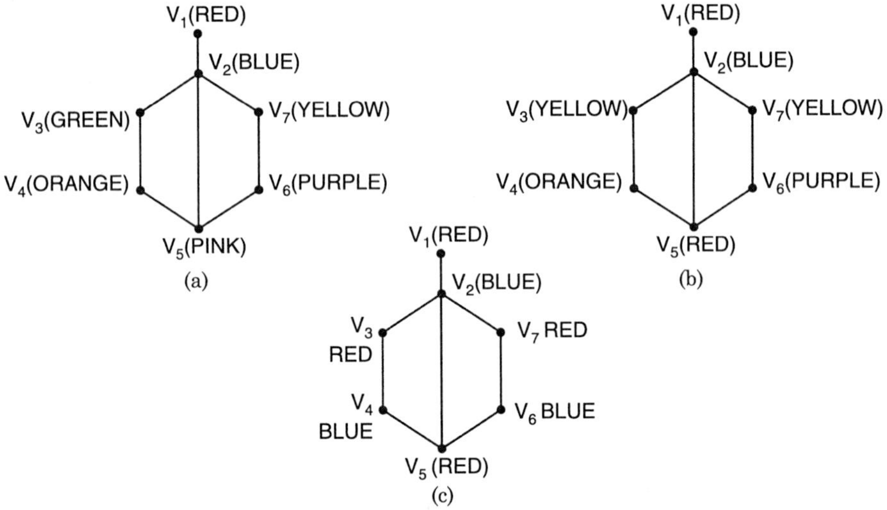

Fig. 9.100. Chromatic graph

In figure 9.100, it is shown that in the graph(*c*) that the exactly 3 colours are required to colour the graph.

Some Useful Results in Colouring a Graph

1. A graph consisting of only isolated vertices is 1- chromatic.
2. A graph with one or more edges (not self loop) is atleast 2- chromatic

For example :

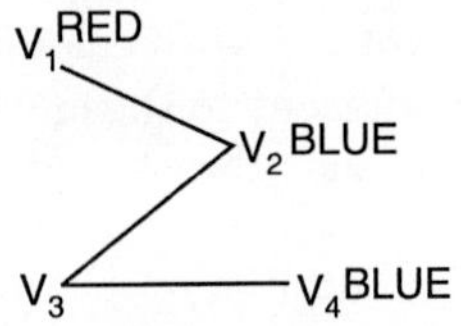

Fig. 9.101.

3. A complete graph with n vertices is n chromatic as all its vertices are adjacent.

For example

Consider a triangle which is a complete graph with three vertices is 3 chromatic as shown in the figure 9.102.

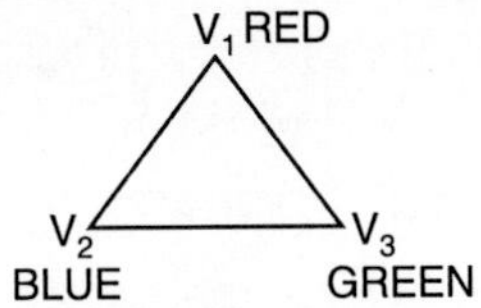

Fig. 9.102.

For another example consider a complete graph with 4 vertices this graph must be 4 chromatic.

Suppose we colour v_1 by red colour then we can not colour v_4, v_2, v_3 by the same red colour as all are adjacent vertices. Similarly for other vertices also. Therefore every vertex in the above graph must have different colours.

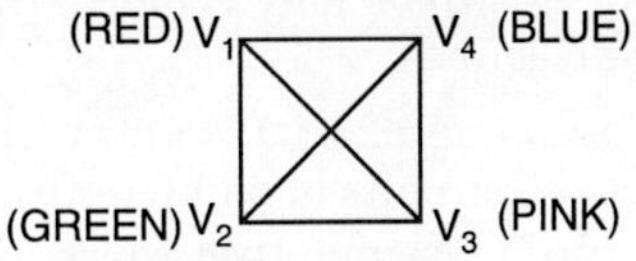

Fig. 9.103.

A graph consisting of one circuit with $n \geq 3$ vertices is 2 chromatic if n is even and is 3 chromatic if n is odd for instance consider the sequence of n vertices $x_1, x_2, ... x_n$ and give colour red to the odd vertices (that is the vertex at the odd numbering position) and the **green** colour to the **even vertices**. If n is even, no adjacent vertices will have the same colour. Therefore the graph is 2 chromatic.

But if n is odd, the nth and the first vertex will have the adjacent and will have the same colour which gives the requirement of the third colour.

For example

Consider the figure 9.104. It is clear froms the figure 9.104(a) that if n(the number of vertices in a graph is odd then G is 3-chromatic graph. In the figure 9.104(b) the graph having even number of vertices is 2-chromatic.

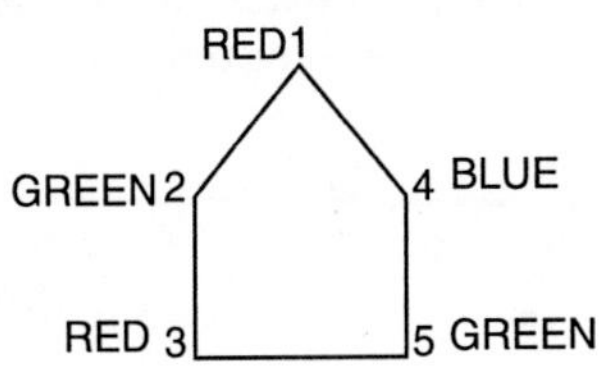

Fig. 9.104(a)

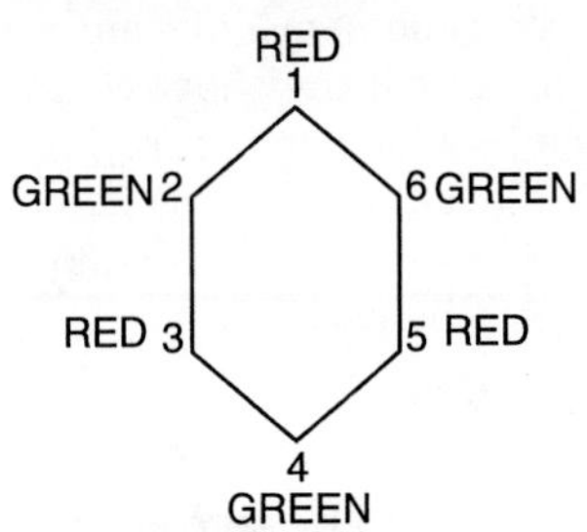

Fig. 9.104(b)

Theorem 9.24 : Every tree with two or more vertices is 2-chromatic.

Proof : Consider a tree T. Select any vertex v in T as a rooted vertex.

Colour the vertex v with the colour 1.

Colour all the vertices adjacent to v with colour 2.

Now colour all the vertices adjacent to these vertices (already coloured by colour 2) by colour 1 and continue this process till the colouring of all vertices is completed.

Now in T, all the vertices at odd distances from v have colour 2 and the vertices at even distances from v have colour 1.

Therefore at last the graph (tree) is fully coloured by the two colours.

Therefore the tree with two or more vertices is 2 chromatic. See Fig. 9.105.

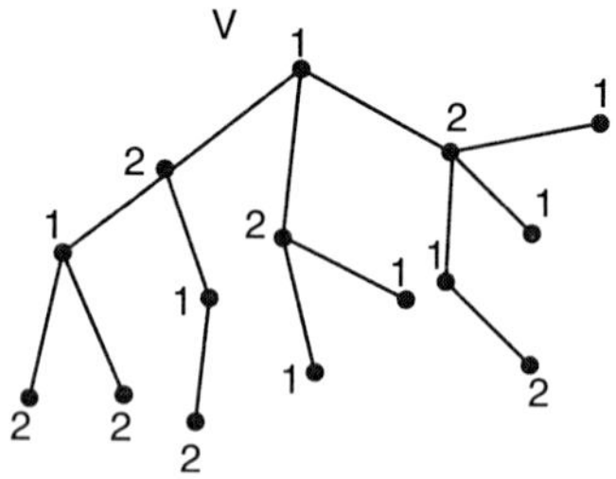

Fig. 9.105.

It is to be noted that in the figure (9.105), 1 stands for colour 1 & 2 stands for colour 2.

Theorem 9.25 : A graph with at least one edge is 2-chromatic if and only if it has no circuits of odd length.

Proof : Assume that there exists a graph G such that it has circuits of even lengths. Let T be the spanning tree of G. Start to properly colour the tree T using two colours. Add the chord to T in a step-by-step manner. The end vertices of every chord which are replaced are differently coloured (because G does not possess circuits of odd length). Thus G is coloured with two colours, with no adjacent vertices having the same colour. Thus G is 2-chromatic.

Conversely : Let G has a circuit of odd length then as discussed above at least three colours would have been necessary to paint G.

Four Colour Theorem

We were discussing the proper colouring of the given G. Now let us consider the proper colouring of regions in a planar graph.

Just as in the case of vertices of a graph that no two adjacent vertices have the same colour. Similarly the planer graphs are said to be properly coloured if no two contiguous or adjacent regions have the same colour. (Two regions are said to be same it they have common edge between them). For example, consider the atlas in which the maps having different countries are coloured such that no two adjacent countries should have same colour.

Same as in the case of colouring of the vertices we require minimum number of colours to be used.

Theorem 9.26 : The Four Colour Theorem : Every planer graph has a chromatic number not greater than four.

Theorem 9.27 : The five colour Theorem : The vertices of every planer graph can be properly coloured with five colours.

Proof : We will prove the result by principle of mathematical induction. Let $P(n)$ be the statement that every planer graph with n vertices can be properly coloured with five colours.

Basic Step : For $n \leq 5$ that the vertices of all graphs with 1, 2, 3, 4, or 5 vertices can be properly coloured with five colours.

Induction Step : Let us assume that the result is true for the graph with $(n-1)$ vertices that is vertices of every planer graph with $(n-1)$ vertices can be properly coloured with five colours.

Now we prove that any planer graph with n vertices can be properly coloured with five colours.

Consider the Planer graph G with n Vertices

Since G is planer, it must have at least one vertex with degree five or less. Let this vertex is v. Remove this vertex v from the graph G and the resultant graph will be a graph G' (say) with $(n-1)$ vertices. Therefore by our hypothesis this graph G' is properly coloured with five colours.

Now introduce this vertex v to G'. If degree of v is less than 5 *i.e.*, 1, 2, 3 or 4, then there is no problem and the result is proved. But if the vertex v has degree 5 and all the five colours have been used in colouring the vertices adjacent to v as shown in fig. 9.106.

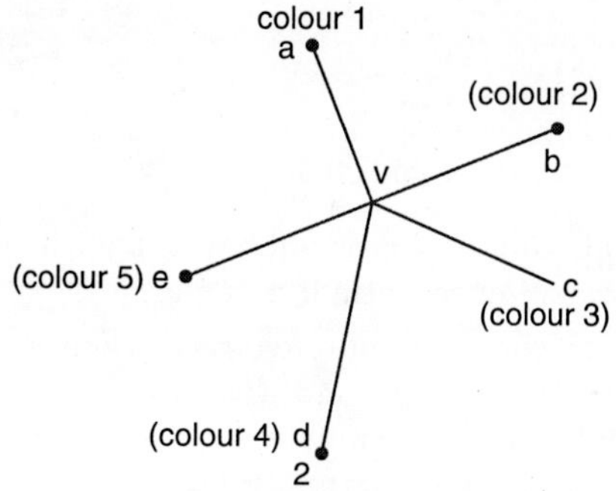

Fig. 9.106.

Now suppose there is a path in G' between b and d coloured alternately with colours 2 and 4 (as in the Fig. 9.107).

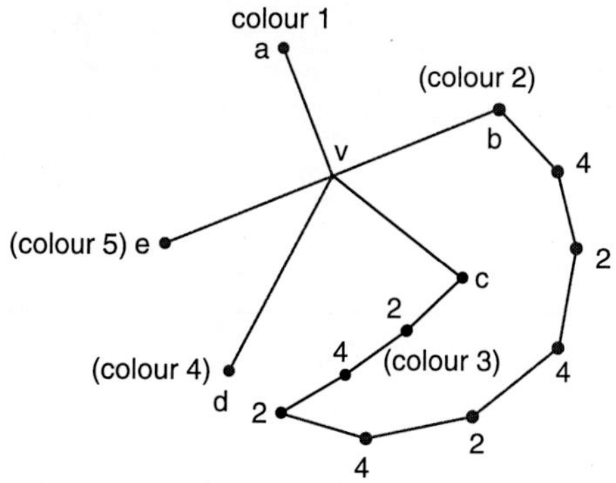

Fig. 9.107.

Remark : Here the numerals 4 and 2 stand for the respective colours.

Then the similar path between a and c, coloured alternately with colours 1 and 3 cannot exist, otherwise these two paths intersects and cause G to be non-planer. If there is no path between a and c coloured alternately with colours 1 and 3, starting from the vertex a we can interchange colours 1 and 3 of all vertices connected to a through vertices of alternating colours 1 and 3. By this we colour the vertex a by the colour 3 and yet keep G' properly coloured. Since vertex c is still with colour 3, we are left with colour 1 in our hand which we will give to vertex v and hence the theorem is proved.

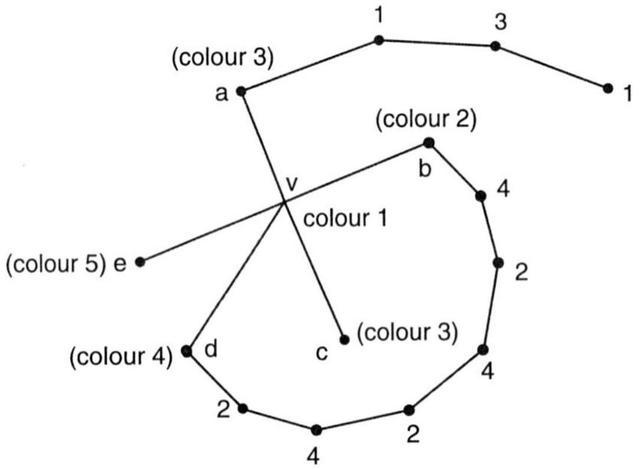

Fig. 9.108(a).

Theorem 9.28. A graph consisting of one circuit with $n \geq 3$ vertices is 2-chromatic if n is even & is 3-chromatic if n is odd.

Proof : Consider a graph consiting of one circuit with n number of vertices $x_1, x_2, \ldots x_n$. Give colour **RED** to the odd vertices (*i.e.*, the vertex at the odd numbering position) and **GREEN** colour to even vertices.

If n is even, no adjacent vertex will have the same colour $\therefore$ the graph is 2 chromatic.

If n is odd, the nth and the 1st vertex will have the adjacent vertex and will have the same colour. Which gives the requirement of the third colour.

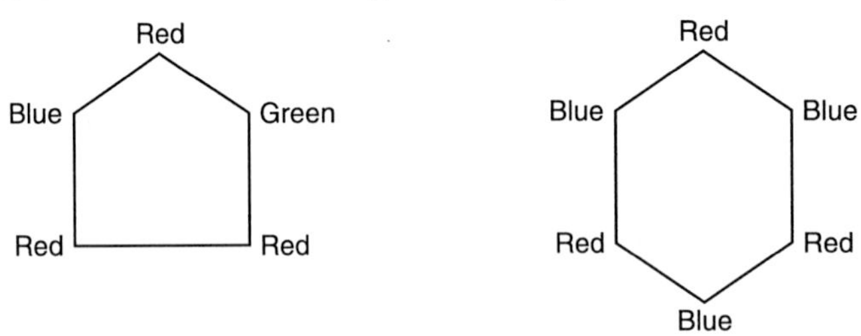

Fig. 1.109.

SOLVED PROBLEMS

PROBLEM 1. Show that each graph given below has a path form a to a that passes through each edge exactly one time by finding such a path by inspection.

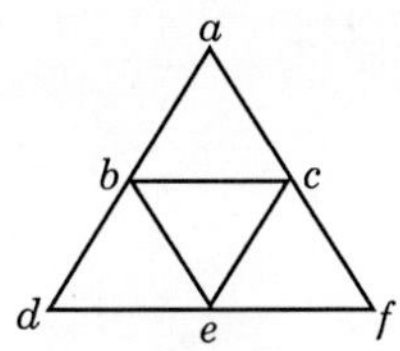

(G_1)

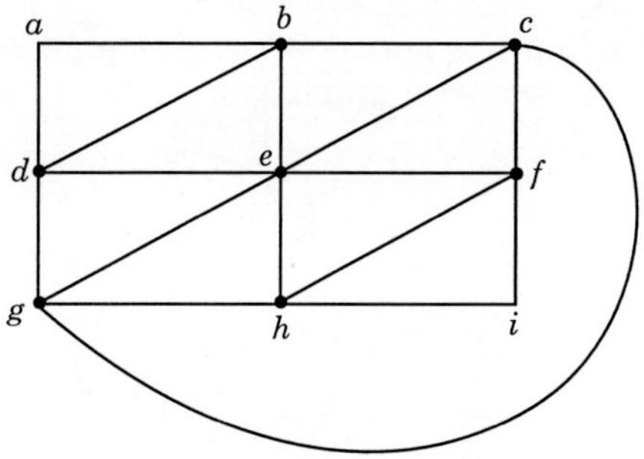

(G_2)

In G_1

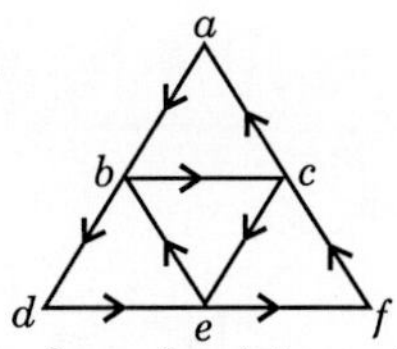

($a\ b\ d\ e\ b\ c\ e\ f\ c\ a$) is an Euler cycle.

In G_2

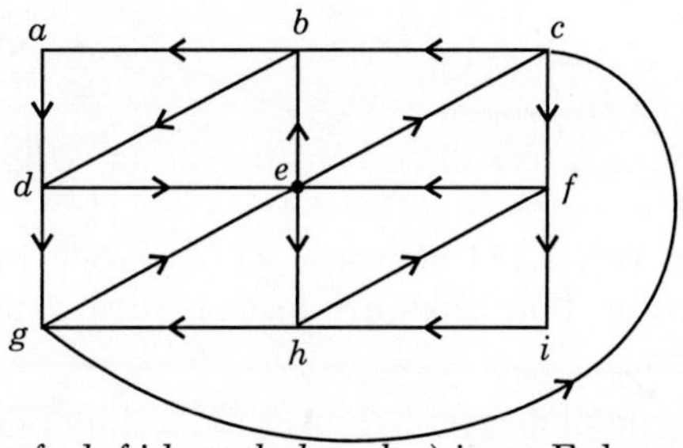

($a\ d\ e\ c\ f\ e\ h\ f\ i\ h\ g\ e\ b\ d\ g\ c\ b\ a$) is an Euler path.

PROBLEM 2. Consider the following graph. The vertices represent offices. An edge connects two offices if there is a communication link between the two. Notice that any office can communicate with any other either directly through a communication link or by having others relay the message.

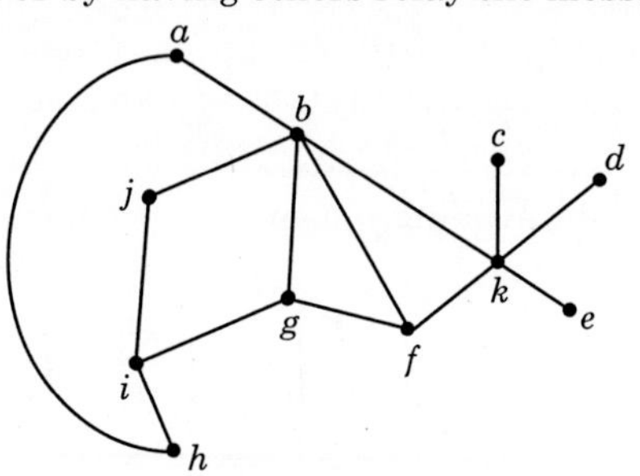

Fig. 9.110.

PROBLEM 3. What is the maximum number of communication links that can be broken with communication among all offices still possible ? Also show a configuration in which the maximum number of communication links are broken with communication among all offices still possible.

Solution: If we break edges (b, k) and edge (g, f), the graph is still connected and the resultant graph is:

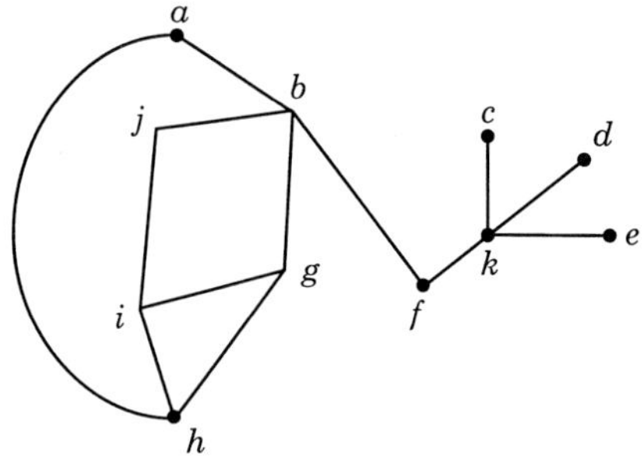

Fig. 9.111.

We can now still break (g, i) and (i, j) and (a, h) keeping connectedness. So the resultant graph is now

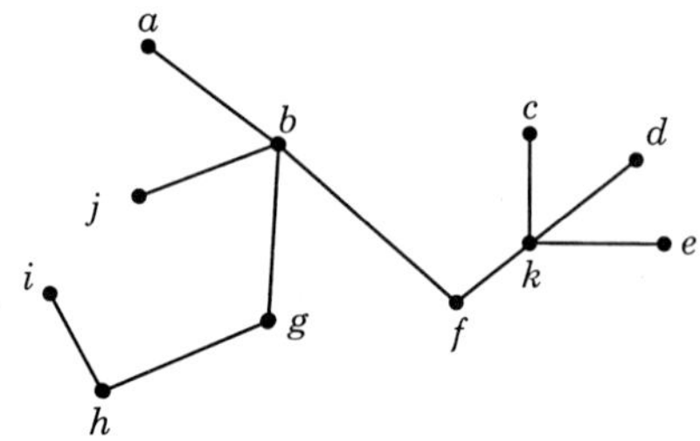

Now we cannot break the graph further otherwise the resultant graph will be a seperable graph.

Thus the maximum number of communication links that can be broken with communication among all offices still possible is **5**.

Thus the graph given in Fig. 9.111 shows a configuration in which the maximum number of communication links are broken with communication among all offices still possible.

PROBLEM 4. When does the complete graph k_n contain an Euler cycle ?

Solution: The complete graph k_n must have every vertex connected to every other vertex. So each vertex has degree $n - 1$. In order to have an Euler graph, therefore, n must be odd.

PROBLEM 5. When does the complete bipartite graph $k_{m, n}$ contain an Euler cycle ?

Solution: In the complete bipartite graph m, n, one set of vertices must be connected to every other vertex in the other set and vice versa. So m and n must both be even.

PROBLEM 6. Solve the travelling salesperson problem for the graph.

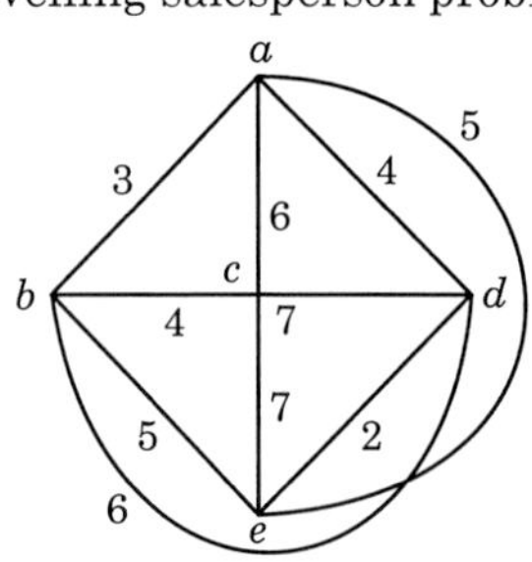

Solution: (e, d, a, b, c).

PROBLEM 6. Consider the graph given by the following matrix.

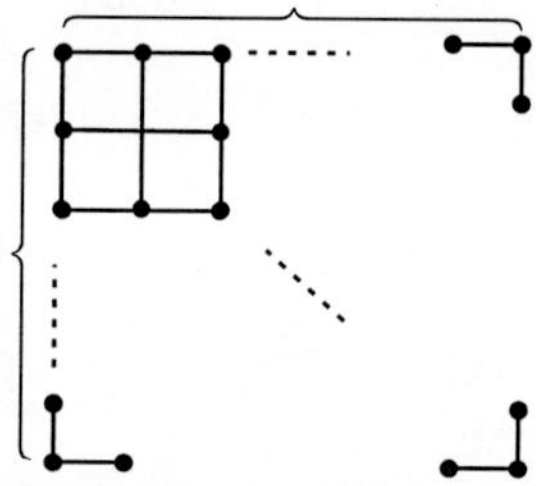

Modify the graph by inserting an edge between the vertex in row i, column 1 and the vertex in row i, column m, for $i = 1, \ldots\ldots, n$. show that the resulting graph always has a Hamiltanian cycle.

PROBLEM 7. Find the length of a shortest path between each pair of vertices in the weighted graph.

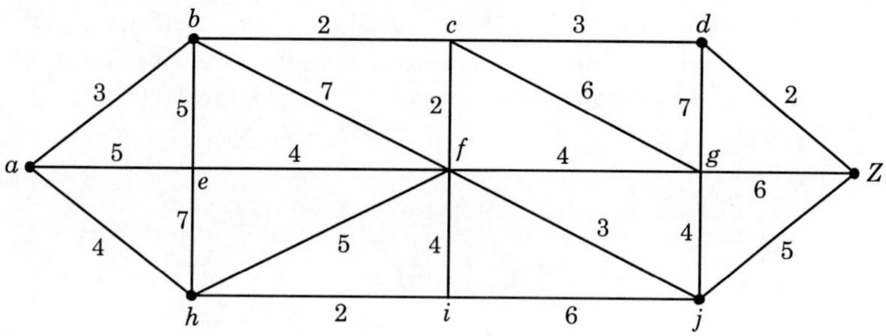

1. a, g
2. a, z
3. h, d

Solution: 1. length 11, path (a, b, c, g)
2. Length 10, path (a, b, c, d, z)
3. Length 10, path (h, f, c, d)

PROBLEM 8. Draw the graph represented by the 7×7 matrix whose ij^{th} entry is 1 if $i + 1$ divides $j + 1$ or $j + 1$ divides $i + 1$, $i \pm j$; whose ij^{th} entry is 2 if $i = j$ and whose ij^{th} entry is 0 otherwise.

Solution: Matrix is

$$\begin{array}{c} \\ 1 \\ 2 \\ 3 \\ 4 \\ 5 \\ 6 \\ 7 \end{array} \begin{array}{c} \begin{array}{ccccccc} 1 & 2 & 3 & 4 & 5 & 6 & 7 \end{array} \\ \begin{bmatrix} 2 & 0 & 1 & 0 & 1 & 0 & 1 \\ 0 & 2 & 0 & 0 & 1 & 0 & 0 \\ 1 & 0 & 2 & 0 & 0 & 0 & 1 \\ 0 & 0 & 0 & 2 & 0 & 0 & 0 \\ 1 & 1 & 0 & 0 & 2 & 0 & 0 \\ 0 & 0 & 0 & 0 & 0 & 2 & 0 \\ 1 & 0 & 1 & 0 & 0 & 0 & 2 \end{bmatrix} \end{array}$$

So the required graph is

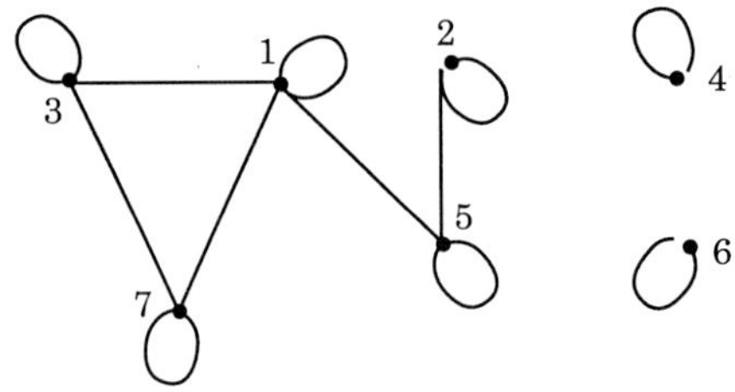

PROBLEM 9. If $A = \begin{pmatrix} 0 & 1 & 1 & 1 & 1 \\ 1 & 0 & 1 & 1 & 1 \\ 1 & 1 & 0 & 1 & 1 \\ 1 & 1 & 1 & 0 & 1 \\ 1 & 1 & 1 & 1 & 0 \end{pmatrix}$

Let n be a positive integer. It can be easily varified that all the diagonal elements of A^n are equal and all the off-diagonal elements of A^n are equal.

Let d_n be the common value of the diagonal elements of A^n and let a_n be the common value of the off diagonal elements of A^n. Then show that:

$$d_{n+1} = 4\, a_n,\ a_{n+1} = d_n + 3a_n,\ a_{n+1} = 3a_n + 4a_{n-1}$$

Solution : We know

$$A^{n+1} = \begin{pmatrix} d_{n+1} & a_{n+1} & \cdots & a_{n+1} \\ \vdots & d_{n+1} & & \vdots \\ \vdots & \vdots & & \vdots \\ a_{n+1} & a_{n+1} & & d_{n+1} \end{pmatrix}$$

$$A^{n+1} = \begin{pmatrix} d_n & a_n & \ldots & a_n \\ a_n & d_n & & a_n \\ \vdots & & & \\ a_n & \ldots & \ldots & d_n \end{pmatrix}_{5\times 5} \begin{pmatrix} 0 & 1 & 1 & 1 & 1 \\ 1 & 0 & 1 & 1 & 1 \\ 1 & 1 & 0 & 1 & 1 \\ 1 & 1 & 1 & 0 & 1 \\ 1 & 1 & 1 & 1 & 0 \end{pmatrix}$$

So
$$d_{n+1} = 4a_n$$
$$a_{n+1} = d_n + 3a_n$$
$$a_{n+1} = 3a_n + 4a_{n-1}$$

PROBLEM 10. Determine whether the graphs G_1 and G_2 are isomorphic. If the graphs are isomorphic, find functions f and g; otherwise, give an invariant that the graphs do not share.

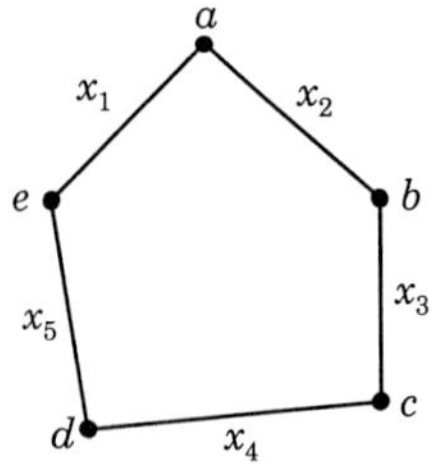

(G_1)

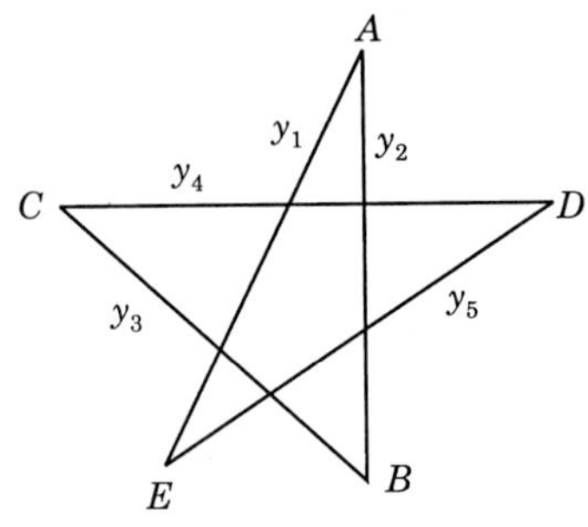

(G_2)

Solution : Graphs G_1 and G_2 are isomorphic if there is a one-to-one, onto function f from the vertices of G_1 to the vertices of G_2 and a one-to-one, onto function g from the edges of G_1 to the edges of G_2, so that an edge e is incident on v and w in G_1 if and only if the edge $g(e)$ is incident on $f(v)$ and $f(w)$ in G_2. The pair of functions f and g is called an isomorphism of G_1 onto G_2. So an isomorphism for the graphs G_1 and G_2 is defined by:

$$f(a) = A$$
$$f(b) = B$$
$$f(c) = C$$
$$f(d) = D$$
$$f(e) = E$$
$$g(x_i) = y_i,\ i = 1, ... 5.$$

Note : If we define a relation R on a set of graphs by the rule G_1RG_2 if G_1 and G_2 are isomorphic, R is an equivalence relation.

PROBLEM 11.

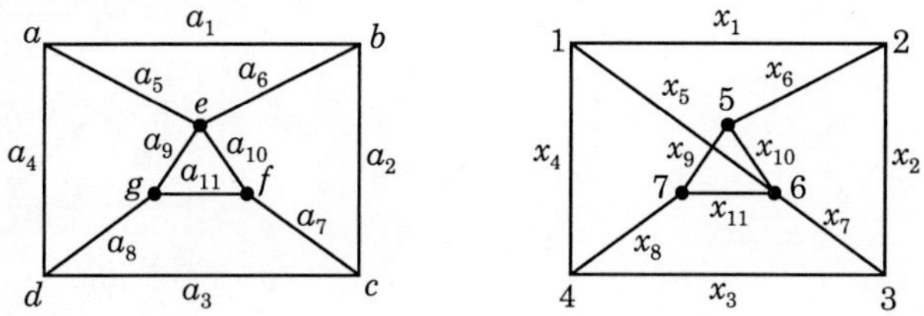

$$f(a) = 1$$
$$f(b) = 2$$
$$f(c) = 3$$
$$f(d) = 4$$
$$f(e) = 5$$
$$f(f) = 6$$
$$f(g) = 7$$
$$g(a_i) = g(x_i),\ i = 1, ..., 11.$$

EXERCISE 1

1. Show that the sum of indegrees of all the nodes of a simple digraph is equal to the sum of outdegrees of all its nodes and that this sum is equal to the number of edges of the graph.
2. Draw all possible simple digraphs having three nodes. Show that there is only one digraph with no edges, one with one edge, four with two edges, four with there edges, four with four edges, one with five edges, and one with six edges. Assume that there are no loops and that graphs which are isomorphic are not distinguishable.

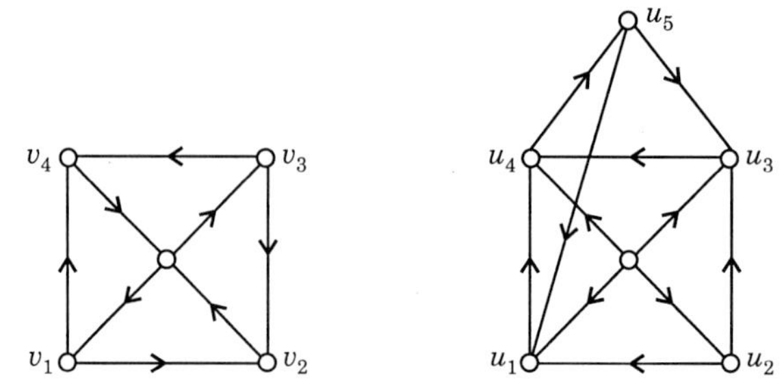

(a) (b)

Fig. 9.111(i).

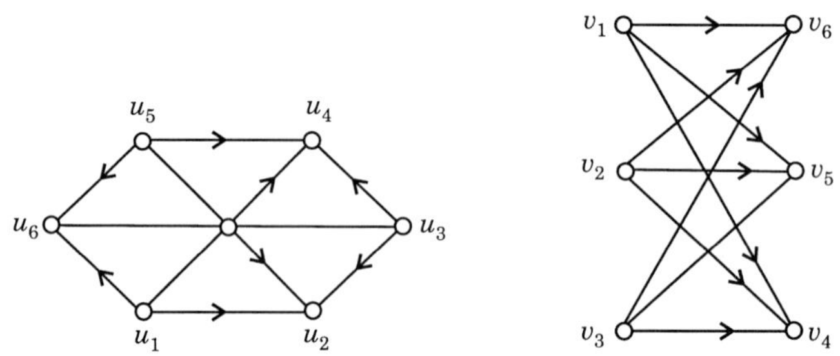

(a) (b)

Fig. 9.111(ii).

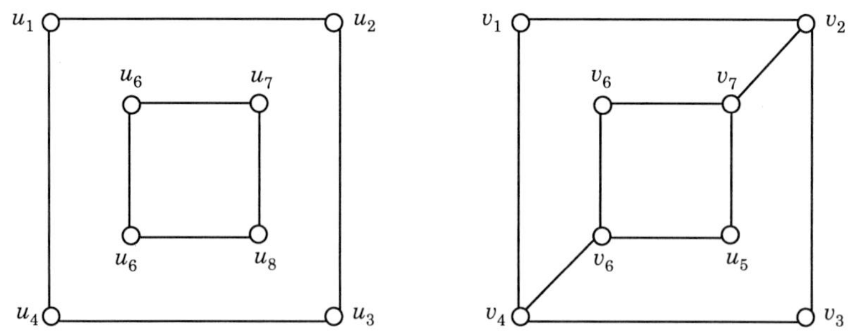

Fig. 9.112.

3. Show that the digraphs in Fig. 9.112 are not isomorphic.
4. A smiple digraph $G = (V, E)$ is said to be complete if every node is adjacent to all other nodes of the graph. Show that a complete digraph with n nodes has the maximum number of edges *viz.*, $n(n-1)$ edges, assuming that there are no loops.
5. The complement of a simple digraph $G = (V, E)$ is the digraph $\overline{G} = (V, \overline{E})$ where $\overline{E} = V \times V - E$. Find the complements of the graphs given in Prob. 2.
6. Give three different elementary paths from v_1 to v_3 for the digraph given in Fig. 9.113 What is the shortest distance between v_1 and v_3? Is there any cycle in the graph? Is the digraph transitive? In case it is not transitive, find the transitive closure of the digraph.

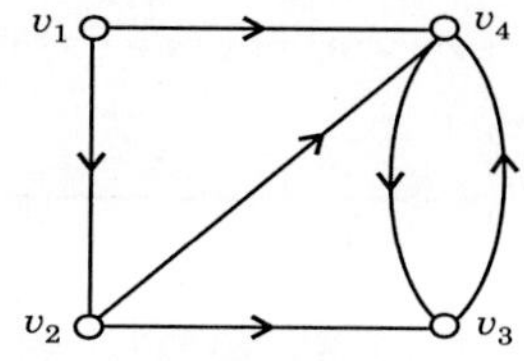

Fig. 9.113.

7. Find all the indegrees and outdegrees of the nodes of the graph given in Fig. 9.113. Give all the elementary cycles of this graph. Obtain an acyclic digraph by deleting one edge of the given digraph. List all the nodes which are reachable from another node of the digraph.

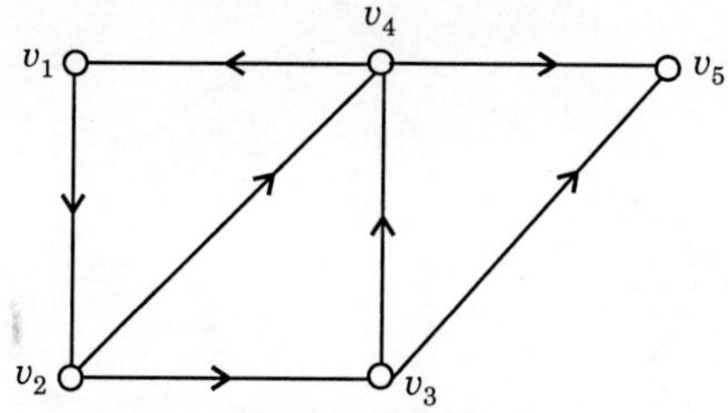

Fig. 9.114.

8. The diameter of a simple digraph $G = (V, E)$ is given by δ, where $\delta = \max d(u, v)\ u, v \in V$

9. Obtain the adjacency matrix A of the digraph given in Fig. 9.115. Find the elementary paths of lengths 1 and 2 from v_1 to v_4. Show that there is also a simple path of length 4 from v_1 to v_4. Verify the results by calculating A^2, A^3, and A^4.

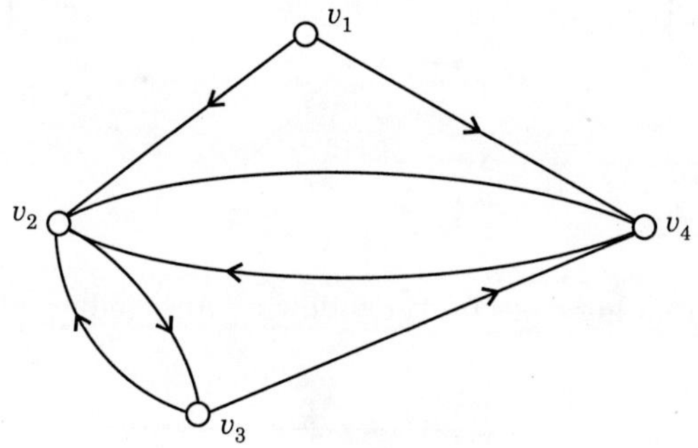

Fig. 9.115.

10. For the digraph of Fig. 9.115. determine A', AA'. Interpret the entries of the matrix $A \wedge A'$. (A' is the transpose of A).

11. For a simple digraph $G = (V, E)$ whose adjacency matrix is denoted by A, its distance matrix is given by:

    $d_{ij} = \infty$ if $(v_i, v_j) \notin E$
    $d_{ii} = 0$ for all $i = 1, 2, \ldots, n$
    $d_{ij} = k$ where k is the smallest integer for which $a_{ij}^{(k)} \neq 0$

12. Show by means of an example that a simple digraph in which exactly one node has indegree 0 and every other node has indegree 1 is not necessarily a directed tree.

EXERCISE 2

1. A tree has $2n$ vertices of degree 1, $3n$ vertices of degree 3, and n vertices of degree 4. Determine the number of vertices and edges in the tree.
2. A tree has two vertices of degree 2, one vertex of degree 3, and three vertices of degree 4. Draw its graph, and show how many vertices of degree 1 does it have ?
3. Give a binary tree representation of the following directed tree* (remember that every tree can be uniquely represented by a binary tree).

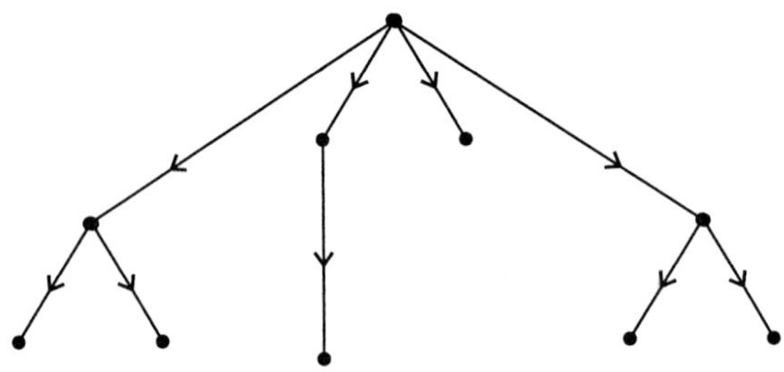

Directed tree

Fig. 9.116.

4. Construct differnent number of (as many as possible) undirected spanning trees for the following graphs (Fig. 9.117) (vertices are distinct).

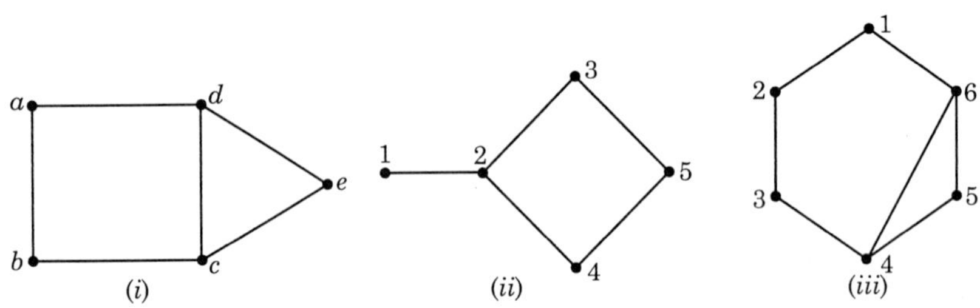

Fig. 9.117.

5. Obtain all spanning trees for the following directed graphs.

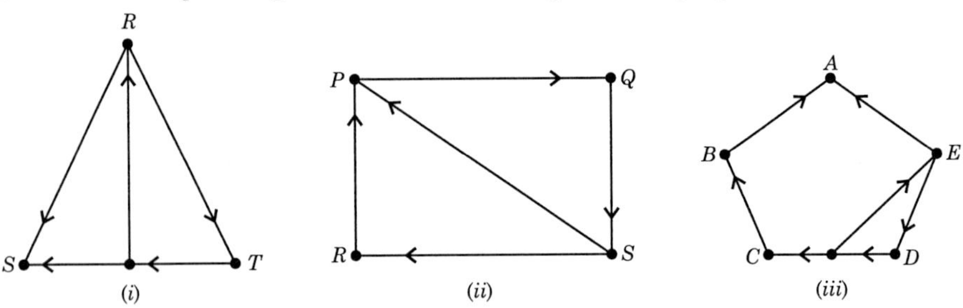

Fig. 9.118.

6. Using Kruskal's algorithm, find a minimal spanning tree for the following graphs in fig. 9.119.

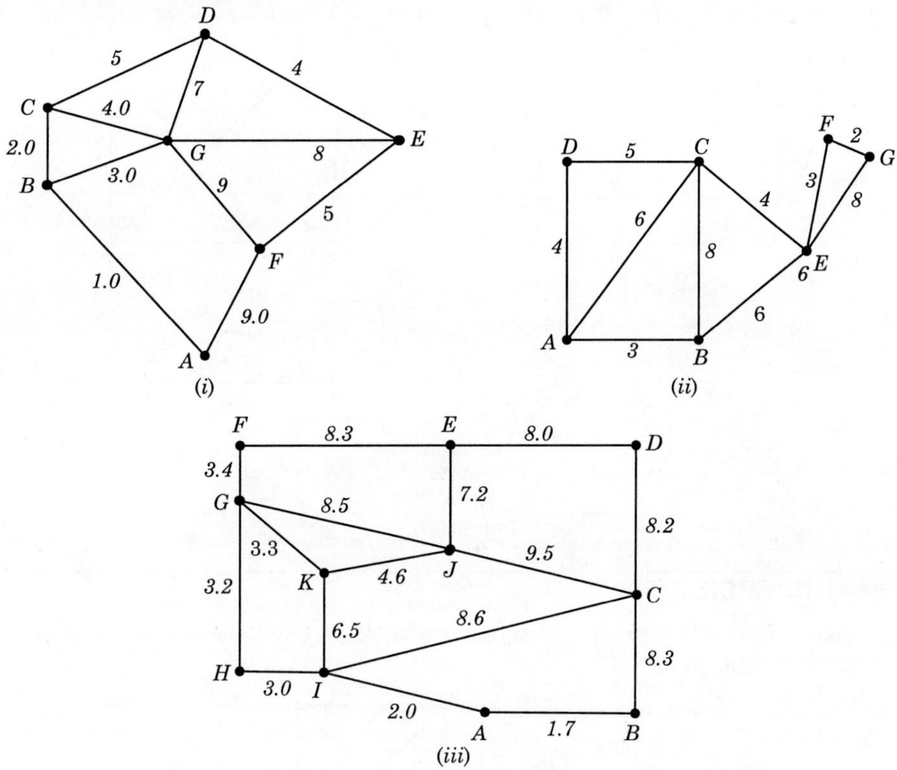

Fig. 9.119.

7. Use Prism's algorithm to determine the minimal spanning tree for the graphs (*i*) —(*iii*) in the foregoing problem number 6.

8. The following table represents distances between eight places. Apply Kruskal's algorithm to determine the minimal spanning tree (vertices represent the places).

	GKP	*KS*	*DE*	*BA*	*LKO*	*FA*	*AZ*	*ZA*
GKP	0	70	122	31	280	71	136	64
KS	70	0	53	98	171	118	164	17
DE	122	13	0	150	223	161	207	60
BA	31	98	150	0	93	64	123	94
LKO	280	171	223	93	0	156	205	175
FA	71	118	161	64	156	0	67	102
AZ	136	164	207	123	205	67	0	148
ZA	64	17	60	94	175	102	148	0

9. Use Kruskal's algorithm to determne a minimum spanning tree for the graphs in Fig. 9.120.

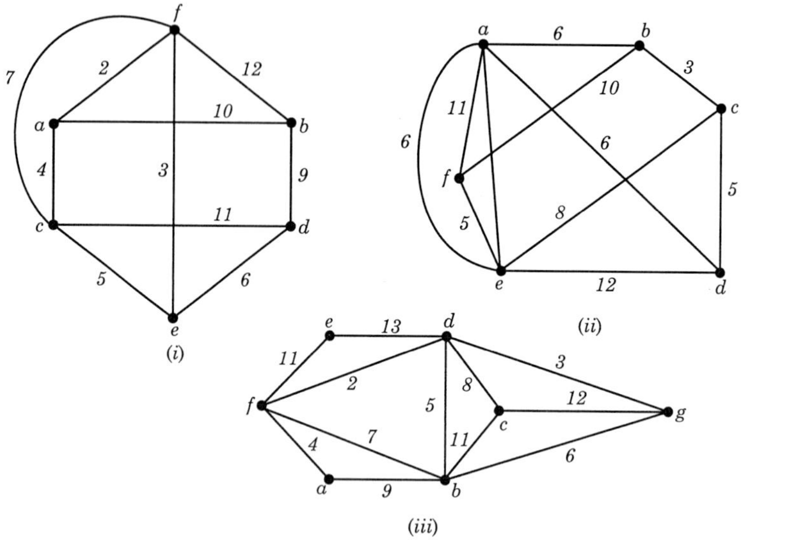

Fig. 9.120.

EXERCISE 3

1. Determine for the graphs given in figure 9.121 (*i*) edge connectivity, and (*ii*) vertex connectivity.

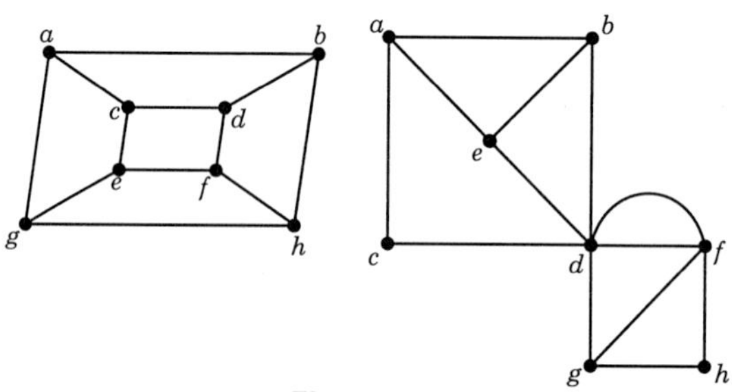

Fig. 9.121.

2. Tell whether each of the following graphs possess an Eulerian path. In case it has, determine it, and if not, give your view in brief.

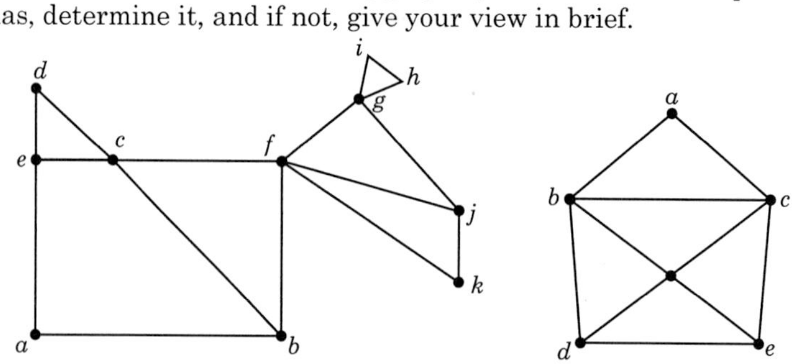

Fig. 9.122.

3. In Fig. 9.123 tell whether the graphs possess the Euler circuit and Euler path or neither of the two.

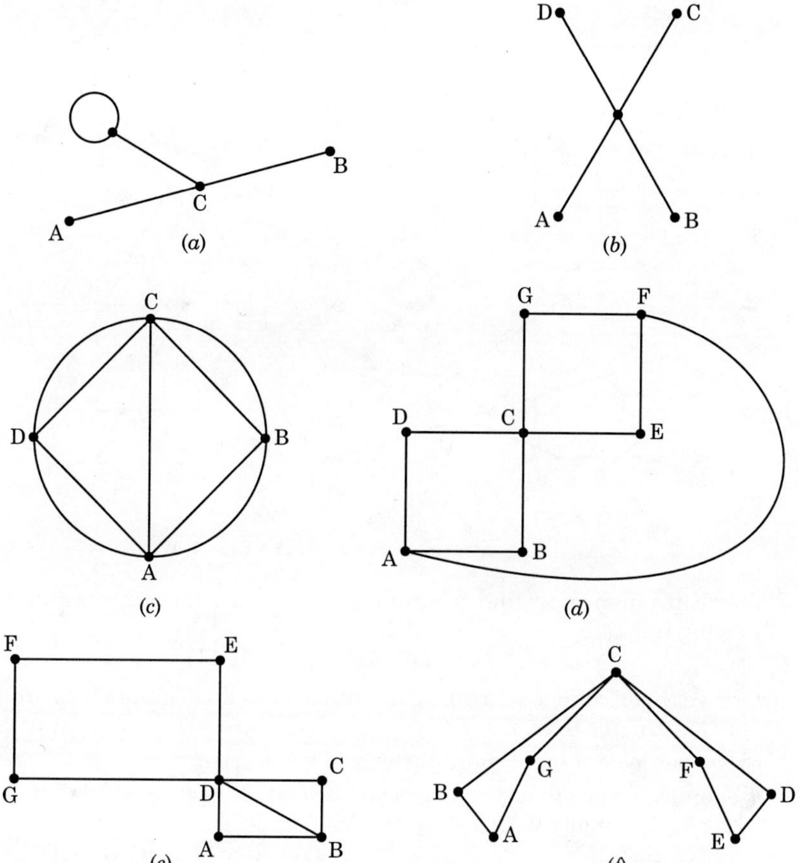

Fig. 9.123.

4. Verify whether the Hamiltonian path is possible in the following fig. 9.124.

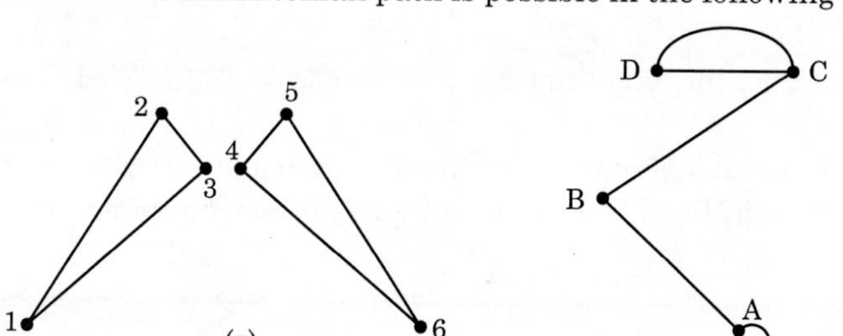

Fig. 9.124.

5. Determine wheterh the graphs in Fig. 9.125. have a Hamiltonian circuit, a Hamiltonian path but not Hamiltonian circuit, or neither. If the graphs have a Hamiltonian circuit, write and sketch the circuits.

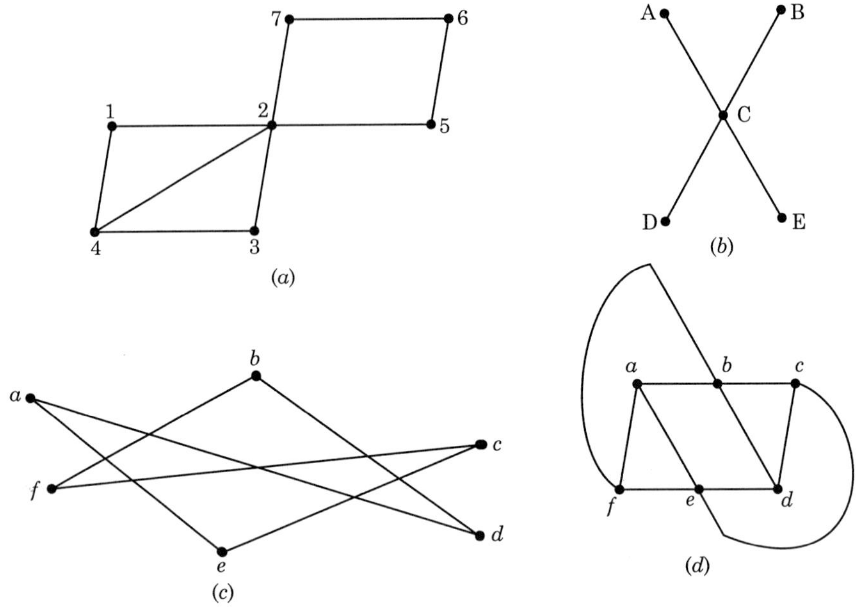

Fig. 9.125.

6. Using Karnaugh map find a minimal sum of product expression of the following function

$$F(A, B, C, D) = \Sigma(0, 1, 4, 5, 9, 11)$$

7. (*a*) Prove that a connected graph *g* has atleast one cycle if the average degree of *G* is greater than 2.
   (*b*) Draw at least 3 regular graphs with 6 vertices.
   (*c*) A connected graph has 12 edges and 6 vertices, each of which has degree 2 or 5. How many vertices are of each degree?

(Terminal Exam., GGSIPU, 2001)

8. (*a*) Define a binary tree, regular binary tree, and path length of a binary tree.
   (*b*) Show that a simple graph can have at most

$$\frac{(n-k)(n-k+1)}{2} \text{ edges}$$

   Interpret the result in the following cases : When $k = 1$, $k = 0$ and $k = n$

OR

   (*a*) Define Euler Circuits, Hamiltonian path and Hamiltonian circuit.
   (*b*) Determine Euler Circuits and Hamiltonian circuits in the following graphs :

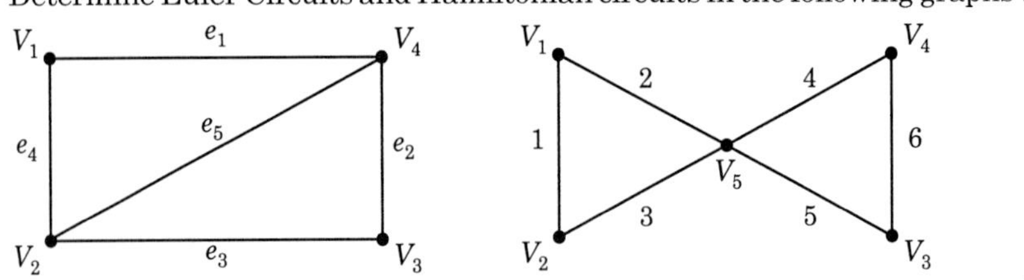

Fig. 9.126. ***(Terminal, Exam. GGSIPU, 2001)***

(*c*) Draw all possible non- isomorphic spanning trees of the labeled graphs *G*.

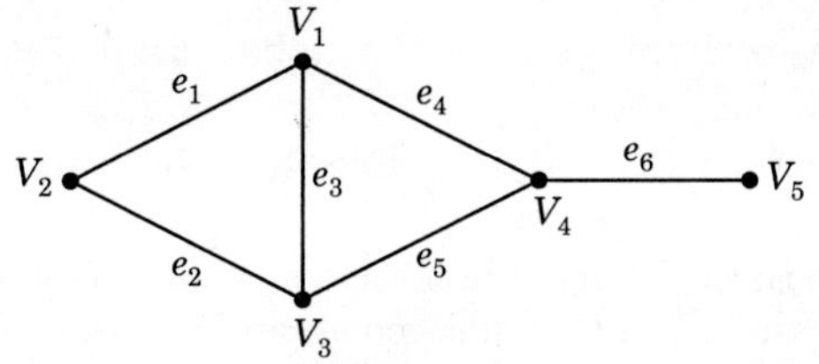

Fig. 9.127. ***(Terminal, Exam. GGSIPU, 2001)***

9. Draw the minimum cost spanning tree for the graph given below according to kruskal's algorithm. Show the acceptance or rejection of edges at each stage of tree formation. What will be the cost of the spanning tree?

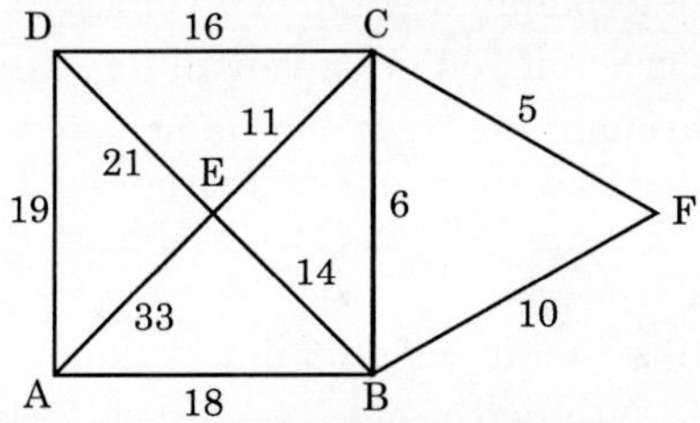

Fig. 9.128. ***(End-term, Exam. GGSIPU, 2001)***

10. (*a*) Define Euler circuit and Hamiltonian circuits in a connected graphs with *n* vertices and *e* edges. Determine these circuit in a complete graph 5 vertices.

    (*b*) If the average of a connected graph *G* exceeds 2, then graph *G* has at least one cycle.

    (*c*) Define sub-graphs, binary tree, spanning tree and chromatic number of a graph. ***(End-term, Exam. GGSIPU, 2002)***

11. Prove that every connected graph n vertices and $(n - 1)$ edges is a tree. ***(End-term, Exam. GGSIPU, 2001)***

DO IT BY YOURSELF

1. Convince yourself that the maximum degree of any vertex in a simple graph with n vertices is n – 1.
2. Show that the maximum number of edges in a simple graph with n vertices is n (n – 1)/2.
3. Define konigsberg Bridge Problem and draw graph of the problem.
4. Prove that the number of vertices of odd degree in a graph is always even.
5. Prove that a graph cannot have single odd vertex. Also prove that if a graph contains exactly two odd vertices, then they are joined by some edge.
6. Prove that if average degree of a connected graph G > 2, then it has at-least one cycle.
7. Prove that a simple graph with n vertices and k components can have at most (n – k) (n – k + 1)/2 edges.
8. Find the number of vertices of degree a and vertices of degree b, if a connected graph has e edges and n vertices each has degree a or b.

9. Prove that a given connected graph G is an Euler graph if and only if all vertices of G are of even degree.
10. Prove that any two simple connected graphs with n vertices, all of degree two, are isomorphic.
11. Prove that a simple graph with n vertices must be connected if it has more that $[(n-1)(n-2)]/2$ edges.
12. Let a, b, and c be three distinct vertices in a graph. There is a path between a and b and also there is a path between b and c. Prove that there is a path between a and c.
13. Draw a graph in which an Euler line is also a Hamiltonian circuit. What can you say about such graph in general?
14. Prove that a three with n vertices has $n - 1$ edges.
15. What is the nullity of a complete graph of n vertices?
16. Can you construct a graph if you are given all its spanning trees? How?
17. Let T1 and T2 be two spanning trees of a connected graph G. If edge e is in T1 but not in T2, prove that there exists another edge f in T2 but not in T1 such that sub graph (T1 – e) U f and (T2 – f) U e are also spanning trees of G.
18. Prove that a graph has dual if and only if it is planar.
19. Prove that every tree with two or more vertices is 2-chromatic.
20. Prove that a graph with at least one edge is 2- chromatic if and only if it has no circuits of odd length.
21. Given a graph, what is the technique to find all maximal independent sets and all minimal dominating sets.
22. Prove that a graph of n vertices is a complete graph if and only if its chromatic polynomial is $P_n(\lambda) = \lambda(\lambda-1)(\lambda-2)....(\lambda-n+1)$.
23. Prove that an n-vertices graph is a tree if and only if its chromatic polynomial $P_n(\lambda) = \lambda(\lambda-1)^{n-1}$.
24. Prove that the vertices of every planer graph can be properly coloured with five colors.
25. Draw all possible simple non-isomorphic graphs with four vertices. Describe the characteristics of each graph.
26. Draw atleast 3 regular graphs 6 vertices.
27. A connected graph has 12 edges and 6 vertices, each of which has degree 2 or 5. How many vertices are of each degree?
28. Define a binary tree, regular binary tree, and path length of a binary tree.
29. Define compliment of a graph, Euler circuits, Hamiltonian path and Hamiltonian circuits.
30. State Dijkstra's algorithm and illustrate with an example.
31. Convince yourself that the maximum degree of any vertex in a simple graph with n vertices is $n - 1$.
32. Show that the maximum number of edges in a simple graph with n vertices is $n(n-1)/2$.
33. Define Konigsberg Bridge Problem and draw graph of the problem.
34. Prove that the number of vertices of odd degree in a graph is always even.

35. Prove that a graph cannot have single odd vertex. Also prove that if a graph contains exactly two odd vertices, then they are joined by some edge.
36. Prove that if average degree of a connected graph $G > 2$, then it has atleast one cycle.
37. Prove that a simple graph with n vertices and k components can have a most $(n - k)(n - k + 1)/2$ edges.
38. Find the number of vertices of degree a and vertices of degree b, if a connected graph has a edges and n vertices, each has degree a or b.
39. Prove that a given connected graph G is an Euler graph if and only if all vertices of G are of even degree.
40. Prove that any two simple connected graphs with n vertices, all of degree two, are isomorphic.
41. Prove that a simple graph with n vertices must be connected if it has more that $[n - 1](n - 2)/2$ edges.
42. Let a, b, and c be three distinct vertices in a graph. There is a path between a and b and also there is a path between b and c. Prove that there is a path between a and c.
43. Draw a graph in which an Euler line is also a Hamiltonian circuit. What can you say about such graphs in general?
44. Prove that a tree with n vertices has $n - 1$ edges.
45. What is the nullity of complete graph of n vertices?
46. Can you construct a graph if you are given all its spanning trees? How?
47. Let T_1 and T_2 be two spanning trees of a connected graph G. If edge e is in T_1 but not in T_2, prove that there exists another edge f in T_2 but not in T_1 such that sub graphs $(T_1 - e) \cup f$ and $(T_2 - f) \cup e$ are also spanning trees of G.
48. Prove that a graph has dual if and only if it is planer.
49. Prove that every tree with two or more vertices is 2- chromatic.
50. Prove that a graph with atleast one edge is 2-chromatic if and only if it has no circuits of odd length.
51. Given a graph, what is the technique to find all maximal independent sets and all minimal dominating sets.
52. Prove that a graph of n vertices is a complete graph if and only if its chromatic polynomial is $\text{Pn}(\lambda) = \lambda(\lambda - 1)(\lambda - 2)....(\lambda - n + 1)$.
53. Prove that an n-vertex graph is a tree if and only if its chromatic polynomial $Pn(\lambda) = \lambda(\lambda - 1)^{n-1}$.
54. Prove that the vertices of every planar graph can be properly coloured with five colours.
55. Draw all possible simple non-isomorphic graphs with four vertices. Describe the characteristics of each graph.
56. Draw atleast 3 regular graphs 6 vertices.
57. A connected graph has 12 edges and 6 vertices, each of which has degree 2 or 5. How many vertices are of each degree?
58. Define a binary tree, regular binary tree, and path length of a binary tree.
59. Define compliment of a graph, Euler circuits, Hamiltonian path and Hamiltonian circuits.
60. State Dijkstra's algorithm and illustrate with an example.

MULTIPLE CHOICE QUESTIONS

1. The number of vertex in a loop is
   (*a*) 0 (*b*) 1 (*c*) 2 (*d*) 3
2. Initial and final vertias of an open walk are
   (*a*) different (*b*) coincident
   (*c*) same (*d*) all of these
3. A cycle is a closed walk in which
   (*a*) neither any edge nor any vertex escept the initial one, is repeated
   (*b*) no edge is repeated, but vertices can be repeated
   (*c*) edges can be repeated but no vertex is repeated
   (*d*) None of these
4. A simple grapyh has
   (*a*) a self leep as well as parallel edges
   (*b*) a self loop but no parallel edges
   (*c*) no self loop but has parallel edges
   (*d*) neither a self loop nor parallel edges
5. A Trivial graph is a finite graph with one vertex and
   (*a*) no edge (*b*) one edge
   (*c*) two edges (*d*) none
6. Which of the following statements are true?
   (*a*) Every cycle is a Hamiltonian graph.
   (*b*) Any graph obtained by adding edges (and not the vertices) to a Hamiltonian graph is also Hamiltonian
   (*c*) Trees are always Hamiltonian
   (*d*) A Hamiltonian graph always has a pendant vertex
7. The number of vertices of odd degree in a graph
   (*a*) is always even (*b*) is always odd
   (*c*) cn be even or odd (*d*) None
8. The sum of degree of all the vertices in a graph with e edges is equal to
   (*a*) $2e$ (*b*) e (*c*) e^2 (*d*) $e + 1$
9. Deletion of a vertex implies
   (*a*) removal of all vertices that has odd degree
   (*b*) removal of all vertices adjacent to this vertex
   (*c*) removal of all edges incident on that vertex
   (*d*) all of these
10. If $G_1 = (V_1, E_1)$ and $G_2 = (V_2, E_2)$ are two graphs, then union of G_1 and G_2 is given by $G = G_1 \cup G_2$ where $G = (V, E)$ in which
   (*a*) $V = V_1 \cup V_2$ and $E = E_1 \cup E_2$ (*b*) $V = V_1 \cup V_2$ and $E = E_1 \cap E_2$
   (*c*) $V = V_1 \cap V_2$ and $E = E_1 \cap E_2$ (*d*) $V = V_1 \cap V_2$ and $E = E_1 \cup E_2$
11. Number of vertices in isomorphic graphs
   (*a*) is always the same (*b*) is always different
   (*c*) may or may not be the same (*d*) may or may not be different
12. The number of vertices with a given degree in two isomorphic graphs
   (*a*) is always the same (*b*) is always different
   (*c*) may or may not be the same (*d*) may or may not be different

13. The eccentricity of centre in a tree is called its
   (*a*) parameter (*b*) radius (*c*) diameter (*d*) height
14. If every vertex of a binary tree has eighter two children or no children, itis called
   (*a*) Full or complete binary tree (*b*) Forest
   (*c*) Null graph (*d*) Binary search tree
15. The number of pendant in a binary trees with n vertices is
   (*a*) 0 (*b*) 1 (*c*) $\frac{n+1}{2}$ (*d*) $\frac{n(n-1)}{2}$
16. A tree with n vertices has
   (*a*) n edges (*b*) $(n + 1)$ edges
   (*c*) $(n - 1)$ edges (*d*) None of these
17. Number of paths between any two vertices of a tree
   (*a*) is exactly 1 (*b*) are 2 (*c*) are more than 2 (*d*) zero
18. The number of labelled trees with 4 vertices is
   (*a*) 2 (*b*) 4 (*c*) 8 (*d*) 16

Answers

1. (*b*)	2. (*a*)	3. (*a*)	4. (*d*)	5. (*a*)	6. (*a*)
7. (*a*)	8. (*a*)	9. (*c*)	10. (*a*)	11. (*a*)	12. (*c*)
13. (*b*)	14. (*a*)	15. (*c*)	16. (*c*)	17. (*a*)	18. (*d*)

❐❐❐

CHAPTER 10

ALGEBRAIC STRUCTURES

This Chapter Covers :

10.1. INTRODUCTION

In the previous chapters we studied the concepts of set theory, relation, function, properties of a relation. Now in this chapter we will explain an algebraic system in which first we will explain the system at axiomatic level, **the group** which occupy a very important role in the study of **abstract algebra** or **algebraic sytem**. After that we will give the concept of **ring** and **field**.

10.2. GROUP

Definition : A **group** G is a non-empty set with a **binary operation** 'o' satisfying the following properties :

(*i*) $\forall\, a, b, c \in G$

$$ao\,(bo\,c) = (a \text{ o } b) \text{ o } c \qquad \textbf{Associative law}$$

(*ii*) $\exists$ an element e in G such that:

$$a\text{o } e = e \text{ o } a = a \ \forall \ a \in G$$

The element e is called **identity element** of G.

(*iii*) For each a in $G \ \exists$ an element a^{-1} in G such that

$$a \text{ o } a^{-1} = a^{-1} \text{ o } a = e$$

a^{-1} is called inverse of a.

For the sake of convenience we, write $a \text{ o } b = ab$

EXAMPLES

1. The set of all integers, Z form a **group** under ***addition***. Here addition '+' is the binary operation.

   '0' is the identity element.

   '$-a$' is the inverse of 'a' in Z.

2. The set of all integers, Z does not form a group under ***multiplication***, as inverse property does not hold here.

   For example, $2 \in Z$ but $\dfrac{1}{2} \notin Z$.

3. The set of all integers, Z does not form a group under ***negative operation***, as associativity does not hold.

   For example, $2, 5, 7, \in Z$

   $$2 - (5 - 7) = 2 - (-2) = 4$$
   $$(2 - 5) - 7 = -3 - 7 = -10$$
   $$\therefore \quad 2 - (5 - 7) \neq (2 - 5) - 7$$

4. The set of all rational numbers Q excluding 0 form a group under multiplication. Such that

   '1' is the identity element and

   $\dfrac{1}{a}$ is the inverse of $a \ \forall \ a \in Q - \{0\}$

5. $G = \{1, -1\}$ is a group under multiplication.

   (*a*) Associativity holds.

   (*b*) 1 is the identity element as:

   $$1 \cdot 1 = 1$$
   $$1 \cdot (-1) = -1 = (-1) \cdot 1$$

   (*c*) each element in G is inverse of itself.

6. The set R of all reals does not form a group under the multiplication. as $0 \in R$

   $$\frac{1}{0} = \infty \text{ does not exists.}$$

   That is inverse of 0, zero does not exists.

EXAMPLE 10.1. Q^+ = The set of positive rational numbers under the binary operation '$*$' defined by

$$a * b = \frac{ab}{2}$$

is a group as

(*i*) **Associativity** $a, b, c \in Q^+$

$$a * (b * c) = a * \left(\frac{bc}{2}\right) = \frac{abc}{4}$$

$$(a * b) * c = \left(\frac{ab}{2}\right) * c$$

$$= \frac{abc}{4}$$

$$\Rightarrow a * (b * c) = (a * b) * c, \qquad \forall\, a, b, c \in Q^+$$

as a,b,c are arbitrary elements in Q^+

(*ii*) $\forall\, a \in Q^+$, an element $x \in Q^+$ is the idenity element if

$$a * x = x * a = a$$

$$\frac{ax}{2} = a$$

$$\Rightarrow x = 2$$

$\Rightarrow$ 2 is the identity element.

(*iii*) **Inverse element** $\forall\, a \in Q^+$, an element $y \in Q^+$ is the inverse of a if

$$a * y = y * a = \text{Idenity.}$$

$$= 2$$

$$\Rightarrow a * y = 2$$

$$\Rightarrow \frac{ay}{2} = 2$$

$$\Rightarrow y = \frac{4}{a}$$

$\Rightarrow \left(\frac{4}{a}\right)$ is the inverse element of a in Q^+ as a is an arbitrary element therefore the result is true for all the element in Q^+.

10.2.1. Abelian Group

Let G be a group and if $\forall\, a, b \in G$

$a \text{ o } b = b \text{ o } a$

then the group G is said to be **commutative or abelian.**

EXAMPLE 10.2. Let M be the set of all square matrices of order two over the set of reals. In M define addition by:

$$\begin{bmatrix} a & b \\ c & d \end{bmatrix} + \begin{bmatrix} p & q \\ r & s \end{bmatrix} = \begin{bmatrix} a+b & b+q \\ c+r & d+s \end{bmatrix}$$

then M is an ***abelian group*** with $\begin{bmatrix} 0 & 0 \\ 0 & 0 \end{bmatrix}$ as identity.

and $\begin{bmatrix} -x & -y \\ -z & -t \end{bmatrix}$ is the inverse of $\begin{bmatrix} x & y \\ z & t \end{bmatrix}$ in M.

EXAMPLE 10.3. Let X be the set of all square matrices of order two over the set of reals of the form $\begin{bmatrix} a & b \\ c & d \end{bmatrix}$, $a, b, c, d \in R$ with $ad - bc \neq 0$. Define multiplication in X by

$$\begin{bmatrix} a & b \\ c & d \end{bmatrix} * \begin{bmatrix} e & f \\ g & h \end{bmatrix} = \begin{bmatrix} ae + bg & af + bh \\ ce + dg & cf + dh \end{bmatrix}$$

Under this multiplication X is a group with identity $\begin{bmatrix} 1 & 0 \\ 0 & 1 \end{bmatrix}$.

as $$\begin{bmatrix} a & b \\ c & d \end{bmatrix} * \begin{bmatrix} 1 & 0 \\ 0 & 1 \end{bmatrix} = \begin{bmatrix} a & b \\ c & d \end{bmatrix}$$

$$\begin{bmatrix} 1 & 0 \\ 0 & 1 \end{bmatrix} * \begin{bmatrix} a & b \\ c & d \end{bmatrix} = \begin{bmatrix} a & b \\ c & d \end{bmatrix}$$

The inverse element of $\begin{bmatrix} a & b \\ c & d \end{bmatrix}$ in M is:

$$\begin{bmatrix} d/p & -b/p \\ -c/p & a/p \end{bmatrix}$$

Where $$p = ad - bc.$$

As $$\begin{bmatrix} a & b \\ c & d \end{bmatrix} * \begin{bmatrix} d/p & -b/p \\ -c/p & +a/p \end{bmatrix} = \begin{bmatrix} \dfrac{ad - bc}{p} & 0 \\ 0 & \dfrac{ad - bc}{p} \end{bmatrix} = \begin{bmatrix} 1 & 0 \\ 0 & 1 \end{bmatrix}$$

$$\begin{bmatrix} d/p & -b/p \\ -c/p & a/b \end{bmatrix} * \begin{bmatrix} a & b \\ c & d \end{bmatrix} = \begin{bmatrix} 1 & 0 \\ 0 & 1 \end{bmatrix}$$

But X is not an abelian group.

For example, let $A, B \in X$

Let $$A = \begin{bmatrix} 1 & 1 \\ 1 & 0 \end{bmatrix} \quad \text{and} \quad B = \begin{bmatrix} 1 & 1 \\ 0 & 1 \end{bmatrix}$$

$$AB = \begin{bmatrix} 1 & 1 \\ 1 & 0 \end{bmatrix}\begin{bmatrix} 1 & 1 \\ 0 & 1 \end{bmatrix} = \begin{bmatrix} 1 & 2 \\ 1 & 1 \end{bmatrix}$$

$$BA = \begin{bmatrix} 1 & 1 \\ 0 & 1 \end{bmatrix}\begin{bmatrix} 1 & 1 \\ 1 & 0 \end{bmatrix} = \begin{bmatrix} 2 & 1 \\ 1 & 0 \end{bmatrix}$$

$$AB \neq BA$$

10.3. PROPERTIES OF GROUPS

Groups can be classified in many ways. In the last section, we have mentioned a classification of groups into **abelian & non-abelian groups**. In this section classification comes from the consideration of the number of elements in a group and also we will define some properties of the group.

***Definition :* Finite Group :**

A finite group is a group containing finite number of elements. The **order of a finite group** G is nothing but the number of elements in G. For example, $\{1, -1\}$ under multiplication is a finite group.

***Definition :* Infinite Group :**

An infinite group is a group containing infinitely many elements. An infinite group is said to have an infinite order. For example Z, the set of integers under the operation addition '+' is an infinite group.

Some Elementray Properties of Groups :

Theorem 10.1: In a group G there is one & only one identity e such that $xe = ex = x \; \forall \; x \in G$. In other words the identity element is unique in a group.

Proof : Let us suppose that f is another identity element in the group G. then

$$ef = e = fe \text{ as } e \in G \quad \text{...}(i)$$

Since e is also an identity element & $f \in G$

$$\therefore \quad ef = f = fe \quad \text{...}(ii)$$

(i) & (ii) $\Rightarrow$

$$e = f$$

Thus, the identity element is unique in a group.

Theorem 10.2. In a group G, the inverse of an element is unique, that is, for each $x \in G$ there is only one element $x^{-1} \in G$ such that $xx^{-1} = x^{-1}x = e$

Proof : Suppose y is also an inverse of x then

$$xy = yx = e$$

Also x^{-1} is the inverse of x then

$$x\,x^{-1} = x^{-1}\,x = e.$$

Now consider

$$\begin{aligned} x^{-1} &= x^{-1}\,e \\ &= x^{-1}(x\,y) \\ &= (x^{-1}\,x)y \\ &= ey = y \\ \Rightarrow x^{-1} &= y \qquad \text{(As associativity holds in } G) \end{aligned}$$

$\Rightarrow$ inverse element is unique in G as x is any arbitrary element in G.

Theorem 10.3. If a and b are elements of a group G then

(i) $\quad (a^{-1})^{-1} = a$

(ii) $\quad (ab)^{-1} = b^{-1}\,a^{-1}$

Proof: (i) Since $aa^{-1} = aa^{-1} = e$, it follows that a is the inverse of a^{-1}.

That is $(a^{-1})^{-1} = a$.

(*ii*) In this part we need to prove that inverse of ab is $b^{-1} a^{-1}$.

Now consider

$$\begin{aligned}(ab)(b^{-1} a^{-1}) &= a(bb^{-1}) a^{-1} \\ &= a\, e\, a^{-1} \qquad \text{(By associative law)} \\ &= a\, a^{-1} \\ &= e \\ (b^{-1} a^{-1})(ab) &= b^{-1} (a^{-1} a)\, b \\ &= b^{-1}\, e\, b \\ &= b^{-1} b \\ &= e\end{aligned}$$

Thus the result is true.

Theorem (10.4) : If a, b, c are elements of a group G then

(*i*) $ab = ac$ implies $b = c$

(*ii*) $ba = ca$ implies $b = c$

Proof : (*i*) $ab = ac$

Now as $a \in G$ therefore $\exists\, a^{-1} \in G : a^{-1} a = a\, a^{-1} = e$

$$\begin{aligned}\Rightarrow a^{-1}(ab) &= a^{-1}(ac) \\ \Rightarrow (a^{-1}a)\, b &= (a^{-1} a)\, c \\ \Rightarrow b &= c\end{aligned}$$

(*ii*) Similarly the result can be proved by post multiplying the equation $ba = ca$ by a^{-1}.

EXAMPLE. 10.4. If for each a and b is group G, $(ab)^2 = a^2 b^2$, Show that G is abelian.

Proof : $\forall\, a, b \in G$, it is given that

$$\begin{aligned}(ab)^2 &= a^2 b^2 \\ \Rightarrow (ab)(ab) &= a^2 b^2 \\ \Rightarrow ab\, ab &= a\, a\, b\, b \qquad \text{[As associativity holds]}\end{aligned}$$

Premultiply by a^{-1} (inverse of a)

$$\begin{aligned}\Rightarrow a^{-1}\, a\, b\, a\, b &= a^{-1}\, a\, a\, b\, b \\ \Rightarrow (a^{-1}\, a)\, b\, a\, b &= (a^{-1}\, a)\, a\, b\, b \\ \Rightarrow e\, b\, a\, b &= e\, a\, b\, b \\ \Rightarrow b\, a\, b &= a\, b\, b\end{aligned}$$

Post multiply by b^{-1} (inverse of b)

$$\Rightarrow b\, a = ab$$

$\Rightarrow G$ is abelian

EXAMPLE 10.5. Prove that if $x^2 = e$ for all elements of a group G, then G is commutative.

Solution : Now let $a, b \in G$

As closure holds in G

$$\Rightarrow ab \in G$$

By the given hypothesis

$$\begin{aligned}(ab)^2 &= e \\ \Rightarrow (ab)(ab) &= e\end{aligned}$$

$$\Rightarrow \quad ab\,ab = e \qquad \text{(By associativity)}$$

$\Rightarrow$ Premultiply by a

$$\Rightarrow a\,a\,ba\,b = ae$$
$$\Rightarrow a^2\,b\,a\,b = ae$$
$$\Rightarrow e\,b\,a\,b = \text{a} \qquad [\text{as } a^2 = e]$$
$$\Rightarrow b\,a\,b = a$$

Post multiply by b.

$$\Rightarrow b\,a\,b\,b = ab$$
$$\Rightarrow b\,a\,b^2 = ab \qquad [\text{as } b^2 = e]$$
$$\Rightarrow b\,a\,e = ab$$
$$\Rightarrow b\,a = ab$$

$\Rightarrow G$ is commutative.

Theorem 10.5. If a and b are elements of a group G then the equations $ax = b$ and $ya = b$ have unique solutions in G.

Proof : Now $ax = b \; \forall \; a, b \in G$.

Pre-multiply by a^{-1} on both the sides, we will get

$$a^{-1}(ax) = a^{-1}\,b$$
$$\Rightarrow (a^{-1}\,a)\,x = a^{-1}\,b$$
$$\Rightarrow ex = a^{-1}\,b$$
$$\Rightarrow x = a^{-1}\,b$$

So $x = a^{-1}\,b$ is a solution of $ax = b$.

Similarly $y = ba^{-1}$ is a solution of $y\,a = b$.

Now we need to show that x is unique.

Suppose $ax = b$ and $az = b$.

that is, z is another solution of $ax = b$.

$$\Rightarrow \quad ax = az$$

Pre-multiply both the sides by a^{-1*}

$$\Rightarrow \quad x = Z$$

$\Rightarrow x$ is unique.

Similary, the uniqueness of y follows in the equation $ya = b$.

Theorem 10.6. Suppose G is a set with a binary operation satisfying the following conditions:

(*i*) If a, b and c are elements of G then

$$a(bc) = (ab)c$$

(*ii*) These exists an element e in G such that

$$ae = a \quad \forall \; a \in G$$

that is , G has a right identity.

(*iii*) For each a in G, $\exists$ an element a^{-1} in G such that, $a\,a^{-1} = e$. That is, there exists a right inverse for each a in G.

Then G is a group.

Proof : Compairing the given conditions with the conditions earlier in the

* **Note :** It is to be noted that inverse of a exists as $a \in G$.

definition of the group, we find that we have to show $a^{-1} a = e$ and $ea = a$, that is, left inverse & left identity.

Let $a \in G$ be any element

By (*iii*) $\exists\, a^{-1} \in G$ such that $a\, a^{-1} = e$

$\therefore$ for $a^{-1} \in G\, \exists (a^{-1})^{-1}$ such that $(a^{-1})\, (a^{-1})^{-1} = e$

Now $\quad a^{-1} a = a^{-1}(a\, e) = (a^{-1}\, a)\, e = (a^{-1}\, a)e = (a^{-1}\, a)\, (a^{-1}(a^{-1})^{-1}$

$= a^{-1}\, (a\, a^{-1})\, (a^{-1})^{-1}$ (By associativity)

$= a^{-1}e(a^{-1})^{-1}$

$= a^{-1}\, (a^{-1})^{-1} = e$

Thus for any $a \in G\, \exists\, a^{-1} \in G$ such that

$$a^{-1}\, a = a\, a^{-1} = e$$

Thus, a^{-1} is the inverse of a.

Again $\quad e\, a = (a\, a^{-1})\, a = a\, (a^{-1}\, a) = a\, e = a$

$\therefore \quad a\, e = e\, a = a \quad \forall\, a \in G$

That is ***e* is identity of *G***

Hence ***G* is group.**

EXAMPLE 10.6. If G is a group in which $(ab)^{i} = a^{i}\, b^{i}$ for three consecutive integers i and any a, b in G, then G is abelian.

Solution: Let n, $n + 1$, $n + 2$ be three consecutive integers for which the given condition holds. Then for any a, $b \in G$

$$(a\, b)^{n} = a^{n}\, b^{n}$$
$$(ab)^{n+1} = a^{n+1}\, b^{n+1}$$
$$(ab)^{n+2} = a^{n+2}\, b^{n+2}$$

Now $\quad (ab)^{n+2} = a^{n+2}\, b^{n+2}$

$= (ab)\, (ab)^{n+1}$

$= ab\, a^{n+1}\, b^{n+1}$

$\Rightarrow a^{n+2}\, b^{n+2} = ab\, a^{n+1}\, b^{n+1}$

$\Rightarrow a\, a^{n+1}\, b\, b^{n+1} = ab\, a^{n+1}\, b^{n+1}$

$\Rightarrow a^{n+1}\, b = b\, a^{n+1}$ ...(1) [By cancellation law]

Similarly

$$(ab)^{n+1} = a^{n+1}\, b^{n+1}$$
$$(ab)\, (ab)^{n} = a^{n+1}\, b^{n+1}$$
$$ab\, a^{n}\, b^{n} = a\, a^{n}\, b\, b^{n}$$

$\Rightarrow \quad ba^{n} = a^{n}\, b$

Post multiply by a on both the sides we will get

$$b\, a^{n+1} = a^{n}\, b\, a$$

(1) $\Rightarrow \quad a^{n+1}\, b = a^{n}\, b\, a$

$\Rightarrow \quad a^{n}\, ab = a^{n}\, ba$

By left cancellation we will get

$$ab = ba$$

$\Rightarrow$ ***G* is abelian.**

---

* ***Note :*** It is to be noted that inverse of a exists as $a \in G$.

EXAMPLE 10.7. For any a, x in a group, show that $(x^{-1}\,ax)^n = x^{-1}\,a^n\,x$ where n is a positive integer.

Solution : We will prove the result by principle of mathematical Induction.

Basic Step :

For $n = 1$ the result is obvious.

For $n = 2$

$$\begin{aligned}(x^{-1}\,ax)^2 &= (x^{-1}\,ax)\,(x^{-1}\,ax)\\ &= x^{-1}\,a(x\,x^{-1})\,ax\\ &= x^{-1}\,a\,e\,a\,x\\ &= x^{-1}\,a^2\,x\end{aligned}$$

That is, for $n = 2$, the result is true.

Inductive Step :

Let us assume that the result is true for k, $k \le n$. Now we will prove the result for $k + 1$.

$$\begin{aligned}(x^{-1}\,ax)^{k+1} &= (x^{-1}\,a\,x)^k\,(x^{-1}\,ax)\\ &= (x^{-1}\,a^k\,x)\,(x^{-1}\,ax) && \text{[as the result is true for } k]\\ &= x^{-1}\,a^k\,(x\,x^{-1})\,a\,x\\ &= x^{-1}\,a^k\,e\,a\,x\\ &= x^{-1}\,a^{k+1}\,x\end{aligned}$$

$\therefore$ the result is true for $k + 1$.

Thus by principle of mathematical induction, the result is true for $\forall n$.

EXAMPLE 10.8. Quaternion Group

Let $G = \{\pm 1, \pm \text{i}, \pm \text{j}, \pm \text{k}\}$.

Define product on G by usual multiplication together

$$\begin{aligned}i^2 &= j^2 = k^2 = -1\\ ij &= k = -ji\\ jk &= i = -kj\\ kj &= j = -ik\end{aligned}$$

Then G forms a group and is called ***Quaternion Group.***

Associativity

$$\begin{aligned}i(jk) &= (i\,j)\,k\\ \Rightarrow ii &= k\,k\\ \Rightarrow\quad -1 &= -1\end{aligned}$$

Identity '1' is identity element

As

$$\begin{aligned}a \cdot 1 &= 1 \cdot a\\ &= a \qquad \forall\, a \in G\end{aligned}$$

Inverse

$$\begin{aligned}(-1\,(-1) &= 1\\ (i)\,(-i) &= -i^2 = 1\\ (j)\,(-j) &= -j^2 = 1\\ (k)\,(-k) &= -k^2 = 1\end{aligned}$$

That is $-i$ is the inverse of i & *vice-verse.*

$-j$ is the inverse of j & *vice-verse.*

$-k$ is the inverse of k & *vice-verse.*

– 1 is the inverse of itself.

Therefore G is a group.

But G is not an abelian group as

$$i\ j = -j\ i$$
$$j\ k = -k\ j$$
$$k\ i = -i\ k$$

EXAMPLE 10.9. Group of Residues

Let $G = \{0, 1, 2, \ldots n-1\}$ under addition modulo n, where n is any positive integer. Then G is a group.

for $a, b \in G$

$a \oplus_n b = c$, where c is the least non-negative integer obtained as remainder when $(a + b)$ is divided by n.

In particular for $n = 5$.

$$G = \{0, 1, 2, 3, 4\} \text{ under addition modulo } 5.$$

The composition table is

$\oplus_5$	0	1	2	3	4
0	0	1	2	3	4
1	1	2	3	4	0
2	2	3	4	0	1
3	3	4	0	1	2
4	4	0	1	2	3

Before doing the following example the students are required to recall permutation functions.

EXAMPLE 10.10. *Klein* 4 group

Consider $K_4 = \{I, (12), (34), (13)\ (24), (14)\ (23)\}$

Let us denote the elements by e, x, y, z then the following table gives us the respective products

	e	x	y	z
e	e	x	y	z
x	x	e	z	y
y	y	z	e	x
z	z	y	x	e

EXAMPLE 10.11. Modular Groups $(Z_n, +, [0]_n)$

$x \equiv y \pmod{m}$ if and only if n divides $(x - y)$. That is if we can find an integer q such that $x - y = q\ m$.

We denote the equivalence class of x modulo m by $[x]_m$.

For example,

$$Z_7 = \{[0]_7, [1]_7, [2]_7, [3]_7, [4]_7, [5]_7, [6]_7,\}$$

We define the operations '+' and '·' on Z_n.

Let $[x]_n$ and $[y]_n$ be equivalence classes in Z_n and let x and y be elements of $[x]_n$ and $[y]_n$ respectively.

We define $[x]_n + [y]_n = [x + y]_n$

$$[x]_n \cdot [y]_n = [xy]_n$$

For example

$$[2]_7 + [6]_7 = [2 + 6]_7 = [1]_7$$

$$[2]_7 \cdot [6]_7 = [12]_7 = [5]_7$$

Order of ***the elements of Z_n***

As we are using additive notation we need to find the smallest value m such that

$$\underbrace{[x]_n + [x]_n + \ldots\ldots + [x]_n}_{m \text{ times}} = [m\ x]_n = [0]_n$$

EXAMPLE 10.12. The order of $[3]_{5 \text{ times}}$ in Z_5 is 5, as:

$$[3] + [3] + \text{- -} + [3]_{5 \text{ times}} = [3 + 3 + 3 + 3 + 3]_5$$
$$= [15]_5$$
$$= [0]_5$$

and no smaller multiple of 3 is congruent to 0 mod 5.

EXAMPLE 10.13. The order of $[5_6]$ is 6

$\because$ $[5]_6 + [5]_6 + [5]_6 + [5]_6 + [5]_6 + [5]_6 = [30]_6 = [0]_6$

$\because$ when we divide 30 by 6, the residue is 6.

EXAMPLE 10.14. In the group Z_{11}, Give the values of following:

(*a*) 3(6)

(*b*) 36(4)

Solution: (*a*)

$$3(6) \equiv 18 \pmod{11}$$
$$\equiv 7 \pmod{11}$$

$\therefore$ $3(6) = 7$

(*b*)

$$(36)\,(4) = 144 \pmod{11}$$
$$= 1 \pmod{11}$$

$\therefore$ $36(4) = 1$

EXAMPLE 10.15. List the additive inverse of the following elements

(*a*) 5, 7, 9 in Z_{10} (*b*) 18, 20, 8 in Z_{40}

Solution :

(*a*) Inverse of 5 is 5 in Z_{10}
Inverse of 7 is 3 in Z_{10}
Inverse of 9 is 1 in Z_{10}.

(*b*) Inverse of 18 is 22 in Z_{40}
Inverse of 20 is 20 in Z_{40}
Inverse of 8 is 32 in Z_{40}.

10.4. SUB-GROUPS

In the section (10.2) and* (10.3) we have defined the group and its properties.

We have seen that Q, the set of rational numbers form a group under addition and R, the set of reals also form a group under addition.

Q is a subset of R. There are lot of similar situations.

That is the same property in the subset of group.

Definition : Sub-group

Let G be a group and S be a non empty subset of G. Then S is said to be sub-group of G if and only if S is a group under the binary operation of G that is if S satisfies all the properties of group under the same operation of G.

If G is a group with the identity element e then the subset $\{e\}$ and G are trivially sub-groups of G and these subgroups are called as trivial sub-groups of G and all other sub-groups will be called non-trivial or proper sub-groups.

We use the notation $H \leq G$ to represent that H is a sub-group of G and $H < G$ is the notation for trivial sub-group of G.

EXAMPLE 10.16. The set of even integers is a subgroup of the group of integers under the binary operation addition . '+'. In Notation $< Z_{2n}, + > < < Z, + >$

EXAMPLE 10.17. The set of integers is a subgroup of set of rationals under the binary operation addition, '+'.

In Notation $<Z, + > < < Q, + >$

EXAMPLE 10.18. The set of rationals Q is a subgroup of the group of reals under addition, '+'.

In notation $< Q, + > < < R, + >$

EXAMPLE 10.19. (Not a subgroup)

The set of $\{1, -1\}$ form a group under multiplication. This is not a subgroup of the additive group of integers.

By using the properties of a group, now we define a theorem which give us a very simple test to determine whether or not a given subset S of a group G is a sub-group of G.

Theorem 10.7. Let G be a group and S be a non-empty subset of G. Then S is a sub-group of G if and only if

(*i*) for all a, b in S , $ab \in S$

That is closure holds in S

(*ii*) For all a in S the inverse a^{-1} of a is in S. That is the property of inverse exists in group.

Proof : Let the conditions (1) and (2) be satisfied.

Since associative law is true in G therefore it is true in S also.

Let $a, b, c \in S \subseteq G$

$$\Rightarrow a(b\ c) = (a\ b)\ c$$

Let $a \in S$ then $a^{-1} \in S$ by (*ii*)

$(1) \Rightarrow a\, a^{-1} \in S$

$\Rightarrow e \in S$

$\Rightarrow$ Identity element belongs to S

Thus all the properties of G are satisfies in S therefore S is a group. Combining the above two properties we can now define the corollary.

10.5. NORMAL SUBGROUPS SECTION

If H is a subgroup of a group G scuh that:

$$Ha = aH \qquad \forall\, a \in G$$

then we say that H is a normal sub-group of a group G. In other words a subgroup H of group G is normal sub-group of G if every left coset of H in G is a right coset of H in G. It is to be noted that if G is an abelian group obviously then all sub-groups of G will be normal.

Theorem 10.8. A subgroup H of a group G is normal sub-group of a group G if and only if

$$g^{-1} Hg = H \qquad \forall\, g \in G$$

or

$$g H g^{-1} = H \qquad \forall\, g \in G.$$

Proof : Let H be a normal subgroup of a group G

$$\Rightarrow \quad Hg = gH \qquad \forall\, g \in G$$

$$\Rightarrow \quad g^{-1} Hg = g^{-1} gH$$

$$\Rightarrow \quad g^{-1} Hg = H$$

Conversely,

$$\text{Let} \quad g^{-1} Hg = H \qquad \forall\, g \in G$$

$$\Rightarrow \quad g(g^{-1} Hg) = gH \qquad \forall\, g \in G$$

$$\Rightarrow \quad (g g^{-1}) Hg = gH \qquad \forall\, g \in G \qquad \text{[By associativity]}$$

$$\Rightarrow \quad eHg = gH$$

$$\Rightarrow \quad Hg = gH$$

As $g \in G$ is arbitrary thus, every left coset of H in G is a right coset of H in G.

EXAMPLE 10.20. Show that $g^{-1} Hg$ is a subgroup of a group G, for a sub-group H of G.

Solution : Let $a, b \in g^{-1} Hg$ be any two elements.

$$a = g^{-1} h_1 g \text{ for some } h_1 \in H$$

$$b = g^{-1} h_2 g \text{ for some } h_2 \in H$$

$$\Rightarrow \quad ab^{-1} = (g^{-1} h_1 g)(g^{-1} h_2 g)^{-1}$$

$$= (g^{-1} h_1 g)(g^{-1} h_2^{-1} g)$$

$$= g^{-1} h_1 (g g^{-1}) h_2^{-1} g \qquad \text{[By associativity]}$$

$$= g^{-1} h_1 (g g^{-1}) h_2^{-1} g$$

$$= g^{-1} h_1 h_2^{-1} g$$

$$= g^{-1} h g ,\ h = h_1 h_2^{-1} \in H \quad \text{[As } H \text{ is subgroup of } G\text{]}$$

$\Rightarrow a\, b^{-1} \in g^{-1} Hg$

$\Rightarrow g^{-1} Hg$ is a sub-group of G.

Theorem 10.9. A sub-group H of a group G is normal if and only if $g^{-1} hg \in H \;\forall\; h \in H, g \in G$.

Proof : Let H be normal in G

then $\quad Hg = gH \quad \forall\, g \in G$

Let $g \in G$ be any element then

$$gh \in gH \qquad \forall\, h \in H$$

$$\Rightarrow \quad gh \in Hg \qquad \text{[As } H \text{ is normal thus } Hg = gH]$$

$$\Rightarrow \quad g\,h\,g^{-1} \in H\,g\,g^{-1} \qquad \forall\, h \in H$$

$$\Rightarrow \quad g\,h\,g^{-1} \in H \qquad \forall\, h \in \text{H}$$

Conversely, Let $\quad g^{-1}\,hg \in H \qquad \forall\, h \in H, g \in G$

As $g \in G$

$$\Rightarrow \quad g(g^{-1}\,hg) \in gH \qquad \forall\, h \in H$$

$$\Rightarrow \quad (g\,g^{-1})\,hg \in gH$$

$$\Rightarrow \quad hg \in gH$$

$$\Rightarrow \quad Hg \subseteq g\,H \qquad \text{...}(i)$$

Again as $g \in G, g^{-1} \in G$

By given hypothesis

$$(g^{-1})^{-1}\,h\,g^{-1} \in H \qquad \forall\, h \in H$$

$$\Rightarrow \quad g\,h\,g^{-1} \in H \qquad \forall\, h \in H$$

$$\Rightarrow \quad (g\,h\,g^{-1})\,g \in Hg \qquad \forall\, h \in H$$

$$\Rightarrow \quad (g\,h)\,(g^{-1}\,g) \in Hg \qquad \forall\, h \in H$$

$$\Rightarrow \quad gh \in Hg \qquad \forall\, h \in H$$

$$\Rightarrow \quad gH \subseteq Hg \qquad \text{...}(ii)$$

Therefore (*i*) and (*ii*) implies that

$$gH = Hg$$

that is all right cosets of H in G is same as left cosets of H in G

$\Rightarrow$ H is a normal subgroup of a group G.

Theorem 10.10. If H and K are subgroups of a group G such that one of them is normal then HK is a subgroup of G.

Proof : Let H is normal subgroup of G.

$$\Rightarrow Hg = gH \qquad \forall\; g \in G$$

$$\Rightarrow Hk = kH \qquad \forall\; k \in K \quad (k \in K \leq G \Rightarrow k \in G)$$

$$\Rightarrow HK = KH$$

$$\Rightarrow HK \text{ is a sub-group}$$

Theorem 10.11. If H and K are two normal subgroups of G, then HK is a normal sub-group of G.

Proof : HK is a sub-group of G by previous theorem. Let $g \in G$, $a \in HK$ be any element

$$a \in HK \;\Rightarrow\; a = h\,k, \qquad h \in H, \quad k \in k$$

$$g^{-1}\,a\,g = g^{-1}\,h\,k\,g$$

$$= g^{-1}\,h\,k\,g$$

$$= g^{-1}\,h\,e\,k\,g$$

$$= g^{-1} h (g\ g^{-1})kg$$
$$= (g^{-1} h\ g)\ (g^{-1} kg)$$
$$\in HK \qquad \text{[As } H \text{ and } K \text{ are normal sub-groups]}.$$

$\Rightarrow HK$ is a normal sub-group of a group G.

Theorem 10.12. *A sub-group H of a group G is normal if and only if product of any two right cosets of H in G is a right coset of H in G.*

Proof : Let H be normal in G

Let Ha and Hb be any two right cosets of H in G.

$$\Rightarrow \quad (Ha)\ (Hb) = H(aH)\ b \text{ [By associativity]}$$
$$= H\ (Ha)\ b \text{ [As } H \text{ is normal in } G]$$
$$= HH\ ab$$
$$= Hab$$
$$= Hc,\ c = ab \in G$$

which is a right coset of H in G.

Conversely :

Let $g \in G$ be any element, then Hg and Hg^{-1} are right cosets of H in G then by given conditions

$Hg^{-1}\ Hg$ is also a right coset of H in G.

In particular

$e\,g^{-1}\,e\,g \in Hg^{-1}\,Hg \qquad [\because e \in H]$

$\Rightarrow \quad e \in Hg^{-1}\,Hg$

Also $\quad e \in H$

$\Rightarrow \quad Hg^{-1}\,Hg = H$

Now for any $h \in H$

$e\,g^{-1}\,hg \in Hg^{-1}\,Hg = H$

$\Rightarrow \quad g^{-1}\,hg \in H \qquad \forall\, h \in H$

$\Rightarrow H$ is normal sub-group of G.

10.6. SEMI-GROUP

There are some sets with binary operation which have great importance in the field of formal language theory, automata, etc. a **semi-group,** a non-empty set S equipped with an associative binary operation 'o'.

Definition :

Let S be a non-empty set with operation '*'. Then S is said to be a semigroup if

1. S satisfies the closure property under '*',
2. S satisfies associativity property also.

It is clear that a semi-group will be a group if it has both identity as well as inverse element.

A semi-group possesing identity element is called a monoid.

Definition :

Let A be a set with an associative binary operation '*' and an identity element 'e'. We call A, a **monoid** and denote it by the $(A, *, e)$.

It is clear that a monoid group possesing inverse for all the elements in it is a group.

EXAMPLE 10.21. Let S be a set and let $P(S)$ be a power set of S. Consider $P(S)$ under the operation of intersection $\cap$. $(P(S), \cap)$ is a semigroup as '$\cap$' is an associative property. For any subset K of S, we have $K \cap S = S \cap K = K$. Thus the set S is the identity element for $P(S)$ under '$\cap$'. Thus $(P(S), \cap, S)$ is a monoid.

EXAMPLE 10.22. Consider N, the set of natural numbers. Then $< N, + >$ and $< N, * >$ are semigroups as both of them satisfies the associative property. $< N, + >$ is not a monoid as $0 \notin < N, + >$ *i.e.*, identity element does not belong to $< N, + >$.

$< N, * >$ is a monoid as $1 \in <N, * >$.

10.7. GROUP HOMOMORPHISM

As we have discussed homomorphism in lattices, similarly we can define above said in groups to get a relationship between two groups.

Definition: Let $(G_1, *)$ & $(G_2, °)$ be groups.

A map : $G_1 \longrightarrow G_2$ is a homomorphism if

$$\theta\,(x * y) = \theta(x)°\theta(y) \qquad \text{for all} \quad x, y \in G_1$$

and the groups G_1 & G_2 are called homomorphic groups. If G_1 & G_2 are having the same binary operation say '*' then

$$\theta(x * y) = \theta(x) * \theta(y)\ \forall\ x, y \in G_1$$

In addition if f is one-to-one then f is called monomorphism.

If f is onto then f is called epimorphism. f is called an isomorphism if f is both one-to-one & onto and the corresponding groups are called isomorphic groups.

EXAMPLE 10.23. Let $< Z, +>$ set of integers under addition is a group and Z_1 be the set of even integers under addition. Now, define

$f : Z \longrightarrow Z_1$ such that

$$f(x) = 2x \qquad \forall\ x \in Z$$

then

$$\begin{aligned} f(x + y) &= 2\,(x + y) \\ &= 2x + 2y \qquad \forall\ x, y \in Z \\ &= f(x) + f(y) \end{aligned}$$

Thus f is a homomorphism and Z & Z_1 are homomorphic groups.

f is also monomorphism as:

$$f(x) = f(y)$$

$$\Rightarrow \qquad 2x = 2y$$

$$\Rightarrow \qquad x = y$$

f is also epimorphic

Let $\qquad y \in Z_1$

$\Rightarrow y$ is an even integer.

Thus, $\exists\, x \in Z$

$$y = 2x$$

$$\Rightarrow \qquad x = \frac{y}{2}$$

such that $f(x) = f\left(\frac{y}{2}\right) = 2 \cdot \frac{y}{2}$

$$= y.$$

Thus f is an isomorphism.

That is Z_1 & Z_2 are isomorphic to each other.

Thus <Z, + > is isomorphic to < 2Z, + >.

Definition :

Homomorphism from $G \to G$, is called endomorphism

An **Iso-morphism** from $G \to G$ is called automorphism

EXAMPLE 10.24. (1) Let G be any group then identity mapping $i : G \to G$ is a homomorphism

$$i(a\,b) = ab = i(a)\,i(b)$$

(2) Let $f : Z \to Z$ *s.t.* where Z is integer.

$$f(x) = 2x$$

$$f(xy) = 2(x + y) = 2x + 2y = f(x) + f(y)$$

(3) Let $G = < Z, +>$, group of integers under addition.

$G' = \{1, -1\}$ under multiplication.

Define a mapping $f : G \to G'$, *s.t.*

$$f(x) = 1 \text{ if } x \text{ is even}$$

$$= -1 \text{ if } x \text{ is odd}$$

Let $x, y \in G$ be any elements

Case I. x, y are both even, then $x + y$ is even

$$f(x + y) = 1 = (-1) \cdot (-1) = f(x)\,f(y)$$

Case II. x, y are both odd, then $x + y$ is even

$$f(x + y) = 1 = (-1)\cdot(-1) = f(x)\,f(x)$$

Case III. x is even & y is odd, then $x + y$ is odd

$$f(x + y) = -1 = (+1)\,(-1) = f(x)\,f(y),$$

Theorem 10.13. If $f : G \to G'$ be a homomorphism then

(*i*) $f(e) = e'$ where e, e' are identities of G & G'

(*ii*) $f(a^{-1}) = (f(a))^{-1}$

Proof : (*i*) $\qquad e\,e = e$

$$f(e\,e) = f(e)$$

Since f is a homomorphism then

$$f(e)\,f(e) = f(e)$$

$$\Rightarrow \quad f(e)\, f(e) = f(e)\, e'$$

$$f(e) = e' \qquad \text{[By left cancellation in } G'\text{]}$$

(*ii*) $a\, a^{-1} = e$

$$f(a\, a^{-1}) = f(e)$$

$$\Rightarrow \quad f(a)\, f(a^{-1}) = f(e) \qquad \text{(Since } f \text{ is a homomorphism)}$$

$$\Rightarrow \quad f(a)\, f(a^{-1}) = e'$$

$$\Rightarrow \quad f(a^{-1}) = [f(a)]^{-1}$$

[Since product of two elements is identity so each one is inverse of another one]

Theorem 10.14. ***Kerf*** is a normal subgroup of a group G.

Proof : Let $f : G \to G'$ be a homomorphism.

Kerf $\neq \phi$

let $x, y \in Kerf$ be any two element

$$\Rightarrow \quad f(x) = e',\ f(y) = e'$$

$$\begin{aligned} f(x\, y^{-1}) &= f(x)\, f(y^{-1}) \\ &= f(x)\, (f(y))^{-1} \\ &= e'(e')^{-1} \\ &= e' \end{aligned}$$

$$\Rightarrow \quad x\, g^{-1} \in Kerf$$

Thus *Kerf* is a sub-group of G.

Now, we show that *Kerf* is Normal in G

let $g \in G$ & $x \in Kerf$ be any element

$$\Rightarrow \quad f(x) = e'$$

To show $g^{-1}\, xg \in Kerf$

$$\begin{aligned} f(g^{-1}\, xg) &= f(g^{-1})\, f(x)\, f(g) \\ &= f(g^{-1})\, e'\, f(g) \\ &= f(g^{-1})\, f(g) \\ &= (f(g))^{-1}\, f(g) \\ &= e' \end{aligned}$$

$$\Rightarrow \quad g^{-1}\, xy \in Kerf$$

$\Rightarrow$ *Kerf* is a normal subgroup of group G.

Note. Corresponding to a homomorphism we have a quotient group, as *Kerf* is a normal subgroup of G, then Quotient group is $G/Kerf$.

Definition : If $f : G \to G'$ is an onto homomorphism then G', is called homomorphic image of G.

EXAMPLE 10.25. Homomorphic image of an abelian group is abelian.

Solution : let $a', b' \in G,'$ be any element. Since f is onto $\exists\ a, b \in G$ *s.t.*

$$f(a) = a' \quad \& \quad f(b) = a'$$

$$a'\, b' = f(a)\, f(b) = f(a, b) = f(b\, a) \qquad \text{[Since } G \text{ is abelian]}$$

$$a' b' = b' a'$$

$\Rightarrow G'$ is abelian.

Theorem 10.15. Homomorphic image of a cyclic group is cyclic.

Proof : let $f : G \to G'$ be onto homomophism.

Suppose G is cyclic, generated by a.

Let $x' \in G'$ be any element

Since f is onto $\exists$ some $x \in G$ $s.t.$

$$f(x) = x'$$

Now $\quad x \in G = \langle a \rangle$

$\Rightarrow \quad x = a^n$ for some n

$\Rightarrow \quad f(a^n) = x'$

$\Rightarrow \quad f(a.a..... a) = x'$

As f is homomorphism.

$\Rightarrow \quad \underbrace{f(a).(f)a. \;....... f(a)}_{n \text{ times}} = x'$

$\Rightarrow \quad (f(a))^n = x'$

Since x' is any arbitrary element of G' we can conclude that every element of G' can be expressed as some power of $f(a)$. *i.e.*, G' is generated by $f(a)$. *i.e.*, G' is cyclic.

Fundamental Theorem of Group Homomorphism

Theorem 10.16. If $f : G \to G'$ is onto homomorphism

then $\quad G' \cong \dfrac{G}{Kerf}$

Proof : Denote $Kerf = K$

Define a mapping $\phi : G/K \to G'$,

$st.\ \phi(Ka) = f(a) \quad$ [Since $Ka \in G/K$ & $a \in G$.

ϕ is **well define.**

$$Ka = Kb$$

$\Rightarrow \quad a\, b^{-1} \in K = Kerf \quad [\because Ha = Hb \Leftrightarrow a\, b^{-1} \in H]$

$\Rightarrow \quad f(ab^{-1}) = e'$

$\Rightarrow \quad f(a)\, f(b^{-1}) = e'$

$\Rightarrow \quad f(a)\, (f(b))^{-1} = e'$

$$f(a) = f(b)$$

$\Rightarrow \quad \phi(Ka) = \phi(Kb)$

$\Rightarrow \phi$ is well defined

ϕ is one-one

$$\phi(Ka) = \phi(Kb)$$

$\Rightarrow \quad f(a) = f(b)$

$\Rightarrow \quad f(a)\, (f(b))^{-1} = e'$

$\Rightarrow \quad f(ab^{-1}) = e'$

$$\Rightarrow \quad ab^{-1} \in Ker f = K$$

$$\Rightarrow \quad Ka = Kb$$

ϕ is a homomorphism

$$\phi\,(Ka \cdot Kb) = \phi(Kab) \qquad [\because K \text{ is normal}]$$

$$= f(ab)$$

$$= f(a)\,f(b)$$

$$\phi(Ka \cdot Kb) = \phi(Ka) \cdot \phi(Kb)$$

$\Rightarrow \phi$ is a homomorphom

ϕ is onto

Let $g' \in G'$ be any element

$f : G \to G'$ is onto therefore

Since $\exists\, g \in G$ *s.t.*

$$f(g) = g'$$

$$f(Kg) = f(g) = g'$$

Where Kg is required preimage of g'.

Hence ϕ is an isomorphism

In notation, $\dfrac{G}{K} \cong$ G'

EXAMPLE 10.26. Show that the relation of isomorphism is an equivalence relation.

Solution : For the equavalence property

We need to show three properties

Relexivity

Symmetry

Transitivity [Discussed in chapter 5]

Refiexivity $G \cong G$

Consider identity mapping

$$i : G \to G. \qquad \text{[Identity mapping]}$$

$$i(x) = x$$

Similarly $G_1 \cong G_2 \Rightarrow G_2 \cong G_1$

$\exists\, f : G_1 \to G_2$ which is one–one, onto homomorphism, then

$f^{-1} : G_2 \to G_1$ which is also isomorphism.

Transitinty. If $G_1 \cong G_2$

& $G_2 \cong G_3$

$\Rightarrow G_1 \cong G_3$

$f : G_1 \to G_2$

$g : G_2 \to G_2$

$gof : G_1 \to G_3$ is also an isomorphism.

Theorem 10.16. Homomorphic image of a finite group is finite.

Proof : Let $f : G \to G'$ be onto homomophism & let G be finite. By fundamental theorem

$$G' \cong \frac{G}{Kerf}$$

As G finite $\Rightarrow \dfrac{G}{\text{Kerf}}$ is finite

$\Rightarrow G'$ is also finite.

EXAMPLE 10.27. Let $f : G \to G'$ be a homomorphism. For any $a \in G$

$$0(f(a)) \mid 0(a) \ \& \ 0(f(a)) = 0(a) \Leftrightarrow f \text{ is } 1-1$$

Solution : Let $\quad 0(a) = n \ \& \ 0(f(a)) = m$

$$\Rightarrow \quad a^n = e$$

$$\Rightarrow \quad f(a^n) = f(e)$$

$$\Rightarrow \quad f(\underbrace{a.\ a.\ \ldots\ldots.a}_{n \text{ times}}) = f(e)$$

As f is homomorphic

$$\underbrace{f(a)\ f(a)\ \ldots\ldots.\ f(a)}_{n \text{ times}} = f(e)$$

$$\Rightarrow \quad (f(a))^n = e'$$

$$\Rightarrow \quad 0(f(a)) \mid n$$

$$\Rightarrow \quad 0(f(a) \mid 0(a) \qquad \ldots(i)$$

Now, let $\quad 0(f(a)) = 0(a)$

We will show that f is one – one

Let $\quad f(x) = f(y)$

$$\Rightarrow \quad f(x)\ (f(y)^{-1} = e'$$

$$\Rightarrow \quad f(xy^{-1}) = e'$$

$$\Rightarrow \quad 0(f(xy^{-1}) = 1$$

$$\Rightarrow \quad 0(x\,y^{-1}) = 1$$

$$\Rightarrow \quad xy^{-1} = e \qquad [\text{As } 0(f(a)) = 0(a)]$$

$$x = y$$

$\Rightarrow$ ***f* is one – one** map.

Conversely. Let f is one – one, we need to show that

$$0(f(a)) = 0(a)$$

$$f(a^m) = f(\underbrace{a\ a\ a\ \ldots\ldots\ldots\ a}_{\text{mtimes}}) = \underbrace{f(a) \cdot f(a)\ \ldots\ldots.\ f(a)}_{\text{mtimes}}$$

$$= (f(a))^m = e' = f(e) \qquad [\text{As } 0(f(a)) = m]$$

Since $\quad f$ is $1 - 1 \Rightarrow a^m = e$

$$\Rightarrow \quad 0(a) \mid m \quad \Rightarrow \quad n \mid m$$

$$\Rightarrow \quad 0(a) \mid\ 0(f(a) \qquad \ldots(ii)$$

from (i) & (ii)

$$\Rightarrow \quad 0(f(a)) = 0(a)$$

Theorem 10.17. If $f : G \to G'$ is a homomorphism then f is $1 - 1$ if and only if $Kerf = \{e\}$

Proof : Let f is 1 – 1

Let $x \in$ Kerf be any element

$$\Rightarrow \quad f(x) = e'$$
$$\Rightarrow \quad f(x) = f(e)$$
$$\Rightarrow \quad x = e \qquad \text{[as } f \text{ is one – one]}$$

As $x \in$ Kerf is arbitrary.

Therefore $\quad \text{Kerf} = \{e\}$

Conversely. Let Kerf = $\{e\}$

let $\quad f(x) = f(y)$

$$\Rightarrow \quad f(x)\,(f(y))^{-1} = e'$$
$$\Rightarrow \quad f(xy^{-1}) = e'$$
$$\Rightarrow \quad xy^{-1} \in \text{Kerf} = \{e\}$$
$$\Rightarrow \quad x\,y^{-1} = e$$
$$\Rightarrow \quad x = y$$
$$\Rightarrow \quad f \text{ is } 1 - 1$$

Theorem 10.17. Let H & K be the two subgroup of G. s.t H is normal in G then.

$$\frac{K}{H \cap K} \cong \frac{HK}{H}$$

Proof : Define a mapping

$$f : K \to \frac{HK}{H}, \text{ such that}$$
$$f(k) = Hk$$

***f* is well defined and one – one**

$$k_1 = k_2$$
$$\Leftrightarrow \quad Hk_1 = Hk_2$$
$$\Leftrightarrow \quad f(k_1) = f(k_2)$$

***f* is homomorphism**

$$\begin{aligned} f(k_1\,k_2) &= Hk_1\,k_2 \\ &= Hk_1\,Hk_2 \\ &= f(k_1)\,f(k_2) \end{aligned}$$

***f* is onto**

Let $Hx \in \dfrac{HK}{H}$ be any member

$$\Rightarrow \quad x \in H\,K \quad \Rightarrow \quad x = hk$$
$$Hx = H\,h\,k$$

So k is required preimage under Hx.

Thus f is an onto homomorphism.

Therefore by fundamental theorem

$$\frac{HK}{H} \cong \frac{K}{\text{Kerf}}$$

We claim $Kerf = H \cap K$

$$\begin{aligned}
\text{Let } k \in Kerf \Leftrightarrow \quad f(k) &= H \\
\Leftrightarrow \quad Hk &= H \\
\Leftrightarrow \quad k &\in H \qquad [\because Ha = H \Leftrightarrow a \in H] \\
\Leftrightarrow \quad k &\in H \cap K \qquad [\because k \in \text{Kerf} \subseteq K] \\
\Rightarrow \quad \text{Kerf} &= H \cap K
\end{aligned}$$

So
$$\frac{K}{H \cap K} \cong \frac{HK}{H}$$

Freshman's Theorem : Let H; K be two normal subgroups of G. *s.t.* $H \subseteq K$ then ${}^{K}/_{H}$ is a normal subgroup of G/H and

$$\frac{G}{K} \cong \frac{G/H}{K/H}$$

Proof : Define a mapping $f : \frac{G}{H} \to \frac{G}{K}$, *s.t.*

$$f(Ha) = Ka$$

f is **well defined and one one**

$$\begin{aligned}
Ha &= Hb \\
\Leftrightarrow \quad a\,b^{-1} &\in H \subseteq K \\
\Leftrightarrow \quad ab^{-1} &\in K \quad \Leftrightarrow \quad Ka = Kb \\
\Leftrightarrow \quad f(Ha) &= f(Hb)
\end{aligned}$$

F is a **homomorphism**

$$\begin{aligned}
f(Ha\,Hb) &= F(Ha\,b) \\
&= Kab \\
&= Ka \cdot Kb \\
&= f(Ha)\, f(Hb)
\end{aligned}$$

***F* is onto** is obvious as for any ka, the element $ha \in G/H$, as $H \subseteq K$ is the required preimage.

By fundamental theorem

$$\frac{G}{K} \cong \frac{G/H}{\text{Kerf}}$$

Now we claim that

$$\text{Kerf} = K/H$$

$$\begin{aligned}
\text{Let} \quad Ha \in \text{Kerf} &\Leftrightarrow \quad f(Ha) = K \\
\Leftrightarrow \quad Ka &= K \\
\Leftrightarrow \quad a &\in K \\
\Leftrightarrow \quad Ha &\in K/H \\
\text{Hence} \quad K/H &= \text{Kerf}.
\end{aligned}$$

EXAMPLE 10.28. Let G be a group of non-zero complex numbers under multiplication. Let N be one set of complex numbers of absolute value. Show that the quotient group $\frac{G}{N}$ is isomorphic to the group of all positive real numbers under multiplication.

Solution: Let R^{+} the group of positive real numbers under multiplication.

Define a mapping $f : G \to R^+$, s.t.,

$$f(a + ib) = a^2 + b^2$$

$$\begin{aligned} f(a + ib)(c + id) &= f((ac - bd) + i(bc + ad)) \\ &= (ac - bd)^2 + (bc + ad)^2 \\ &= a^2 c^2 + b^2 d^2 + b^2 c^2 + a^2 d^2 \\ &= (a^2 + b^2)(c^2 + d^2) \\ &= f(a + ib) f(c + id) \end{aligned}$$

Thus ***f* is a homomorphism.**

Let $x \in R^+$ be any element

then $\qquad f(\sqrt{x} + io) = x$

$\Rightarrow \sqrt{x} + i0$ is required preimage of x

Hence *f* is onto.

By fundamental theorem

$$R^+ \cong \frac{G}{\text{Kerf}} \qquad \text{...}(i)$$

We Claim $\qquad \text{Kerf} = N$

$(a + ib) \in \text{Kerf} \Leftrightarrow f(a + ib) = 1$

$$\begin{aligned} &\Leftrightarrow & a^2 + b^2 &= 1 \\ &\Leftrightarrow & a + ib &\in N \\ & & Kerf &= N \end{aligned}$$

Thus (1) $\qquad \dfrac{G}{N} \cong R^+$

EXAMPLE 10.29. Any infinite cyclic group is isomorphic to the group $< Z, + >$ integers.

Solution : Let $G = < a >$ be any infinite cyclic group

Define $f : G \to Z$ s.t.

$$f(a^m) = m$$

f is **well defined**

$$a^m = a^n$$

as G is infinite

$$\Rightarrow \qquad m = \text{n}$$

$$f(a^m) = f(a^n)$$

***F* is 1 –1**

$$f(a^m) = f(a^n)$$

$$\Rightarrow \qquad m = n$$

$$\Rightarrow \qquad a^m = a^n$$

f **is Homomorphism**

$$\begin{aligned} f(a^m \cdot a^n) &= f(a^{m+n}) \\ &= m + n \\ &= f(a^m) + f(a^n) \end{aligned}$$

$\Rightarrow f$ is homomorphism

f is **onto obviously**, as for any $m \in Z$, a^m is the required preimage.

Hence G is isomorphic to $< Z, + >$.

Kernel of f

If $f : G_1 \longrightarrow G_2$ is a homomorphism, then the Kernel of f, written kerf, is the set of elements whose image is the identity elemet e' of G_2 : that is:

$$\text{Kerf} = \{a \in G, : f(a) = e'\}$$

Theorem 10.18. Let $f : G \to G'$ be a homomorphism. Then f is a monomorphism if and only if Kerf $= \{e\}$.

Proof : Let f be a monomorphism.

To **show that** $Kerf = \{e\}$

Let $x \in Kerf$.

$$\Rightarrow \quad f(x) = e' = f(e)$$

As f is monomorphism *i.e.*, one – one

$$\Rightarrow \quad x = e$$

As x is arbitray

Hence $Kerf = \{e\}$

Conversely,

Let $\quad Kerf = \{e\}$

To show that f is monomorphism

Let $\quad f(x) = f(y)$

$$\Rightarrow \quad f(x)\,(f(y))^{-1} = e'$$

$$\Rightarrow \quad f(x)\,f(y^{-1}) = e'$$

$$\Rightarrow \quad f(x\,y^{-1}) = e'$$

$$\Rightarrow \quad x\,y^{-1} \in Kerf = \{e\}$$

$$\Rightarrow \quad x\,y^{-1} = e$$

$$\Rightarrow \quad x = y \qquad \text{[Post multiplying by } y\text{]}$$

$\Rightarrow f$ is monomorphism.

Theorem 10.19. Let $f : G \longrightarrow G_1$ be an isomorphism.

If G is abelian then G' is also abelian.

Proof : Let $\quad a_1, b_1 \in G_1$

then there exists $a, b \in G$ such that

$$f(a) = a_1{}^*$$

$$\& \quad f(b) = b_1{}^*.$$

Now $\quad a_1\,b_1 = f(a\,f(b)$

$$= f(a\,b) \qquad [f \text{ is an isomorphism}]$$

$$= f(b\,a) \qquad [\text{As } G \text{ is abelian}]$$

$$= f(b)\,f(a)$$

$$= b_1\,a_1$$

$\Rightarrow G_1$ is abelian.

Theorem 10.20. Any finite group is isomorphic to a group of permulations.

10.7.1. Corollary

A non-empty subset S of a group G is a subgroup of G if and only if for each a, b in S $ab^{-1} \in S$.

Proof : Let $a\, b^{-1} \in S \;\forall\; a, b \in s$

Then as $a \in S$

$\Rightarrow \quad a\,a^{-1} \in S$

$\Rightarrow \quad e \in S$

Now as $e \in S \Rightarrow e\,a^{-1} \in S$

$\Rightarrow a^{-1} \in S$

Thus we conclude that if $a \in S$ then $a^{-1} \in S$ that in the second property of the above theorem is satisfied.

So now we need to prove that if $a,b \in S$ then $ab \in S$

Since $b \in S$, $b^{-1} \in S$

So $a(b^{-1})^{-1} \in S$ by given condition

$\Rightarrow ab \in S$

Therefore it follows from the theorem (10.7) that S is a group.

Conversely

Let S is a sub-group of a group G

then (*i*) satisfies as closure holds in S and

(*ii*) satisfies by the inverse property in S.

Theorem 10.21. Intersection of two sub-groups is a sub-group.

Proof : Let G be a group

Let S & T be two sub-groups of G

then we will show that $S \cap T$ is a sub-group of G.

Let $x, y \in S \cap T$

$\Rightarrow x, y \in S$ and $x, y \in T$

$\Rightarrow x\,y^{-1} \in S$ as S is a sub-group of G

and $x\,y^{-1} \in T$ as T is a sub-group of G.

$\Rightarrow x\,y^{-1} \in S \cap T$

By the corollary given above $S \cap T$ is a sub-group of G.

Theorem 10.22. The Intersection of an arbitrary family of a sub-group of a group G is a sub-group of G.

Proof: $p(n)$ = for any inteqer n if $S_1, S_2, \ldots .\, S_n$ are sub-groups of a group G then $S_1 \cap S_2 \cap \ldots \cap S_n$ is also a sub-group of G.

We will prove the result by principle of mathematical induction.

Basic Step :

***p*(2):** If S_1 & S_2 are subgroups of a group G then $S_1 \cap S_2$ is also a sub-group of G. The result is true by last theorem.

Inductive step:

Let $p(k)$ for $k \leq n$, $k \in Z^+$, the set of positive integers is true. We will now prove that $p(k + 1)$ is true.

$p(k + 1)$ = If $S_1, S_2, \ldots S_k, S_{k+1}$ are sub-groups of a group G then $S_1 \cap S_2 \cap \ldots . S_k, \cap S_{k+1}$ is a sub-group of G.

Let $a, b \in S_1 \cap S_2 \cap \ldots \cap S_{k+1}$

$\Rightarrow \quad a, b \in S_1, S_2, \ldots, S_k, S_{k+1}$

$\Rightarrow \quad a, b \in S_1 \cap S_2 \cap \ldots S_k$ & $a, b \in S_{k+1}$

$\Rightarrow \quad ab^{-1} \in S_1 \cap S_2 \cap \ldots \cap S_k$. (By our hypothesis)

& $ab^{-1} \in S_{k+1}$ (as $S_{k+1} \leq G$)

$\Rightarrow \quad a\,b^{-1} \in S_1 \cap S_2 \cap \ldots \cap S_{k+1}$

$\Rightarrow \quad p(k + 1)$ is true.

Therefore by principle of mathematical induction the result is true.

EXAMPLE 10.30. Union of two subgroups may or may not be a sub-group of a group G.

Solution : We will prove the result by taking an example.

Let Z be the group of integers under the binary operation addition, '+'.

Let S_1 = The set of even integers.

$$= \{2n \mid n \in Z\}$$

Let S_2 = Set of integral multiples of 5.

$$= \{5n \mid n \in Z\}$$

S_1 and S_2 are subgroups of the group Z.

Now $2 \in S_1$ and $5 \in S_2$

$$2 + 7 = 7 \notin S_1 \cup S_2$$

$\therefore$ Closure does not hold in $S_1 \cup S_2$

$\Rightarrow S_1 \cup S_2$ is not a sub-group.

EXAMPLE 10.31. The union of two sub-groups of a group G is a sub-group if and only if one is contained in the other.

Solution : Let H & K be two sub-groups of a group G

Let $\quad H \subset K$ or $K \subset H$

If $\quad H \subset K$ then $H \cup K = K$

As K is a subgroup, therefore $H \cup K$ is also a sub-group of G.

Similarly if $K \subset H$ then

$$H \cup K = H \leq G$$

Therefore result holds.

Conversely

Suppose that $H \cup K$ is a sub-group, then we have to show that either $H \subset$

K or $K \subset H$.

Let $H \not\subset K$

then $\exists$ an element, $x \in H$ but $x \notin K$

Let $y \in K$ therefore

$$x \in H \cup K \;\&\; y \in H \cup K$$

As $H \cup K$ is a sub-group of G

$$\Rightarrow xy \in H \cup K$$

$$\Rightarrow x\,y \in H \text{ or } xy \in K$$

Now $xy \in K$ and $y \in K$

$$\Rightarrow (x\,y)\,y^{-1} \in K$$

$$\Rightarrow x\,(y\,y^{-1}) \in K$$

$$\Rightarrow x\,e \in K$$

$$\Rightarrow x \in K$$

But $x \notin K$

Hence $xy \notin K$

Now $x\,y \in H \;\&\; x \in H$ (By our hypothesis)

$\Rightarrow x^{-1}\,(x\,y) \in H$ [As H is a sub-group]

$\Rightarrow (x^{-1}\,x)\,y \in H$ [By associativity]

$$\Rightarrow e\,y \in H$$

$$\Rightarrow y \in H$$

$\Rightarrow K \subseteq H$ (as $y \in k$)

Hence union of two sub-groups is a sub-group if and only if one is contained in the other.

EXAMPLE 10.32. If H is a sub-group of G, show that

$$g^{-1}\,Hg = \{g^{-1}\,hg \mid h \in H\} \text{ is a sub-group of } G$$

Solution : Let $x, y \in g^{-1}\,Hg$

$$\Rightarrow x = g^{-1}\,h_1\,g$$
$$y = g^{-1}\,h_2\,g$$

for $h_1, h_2 \in H$

Now

$$\begin{aligned} xy^{-1} &= (g^{-1}\,h_1\,g)\,(g^{-1}\,h_2\,g) \\ &= g^{-1}\,h_1\,(g\,g^{-1})\,h_2\,g \\ &= g^{-1}\,h_1\,h_2\,g \in g^{-1}\,Hg \quad \text{as} \quad h_1\,h_2 \in H \end{aligned}$$

$\therefore g^{-1}\,Hg$ is a sub-group of a group G.

10.8. ORDER OF AN ELEMENT

Let G be a groups and $x \in G$ be any element. We say order of x is n if n is the least positive integer such that:

$$x^n = e$$

If it is not possible to find such n, we say x has infinite order. Order of x will be denoted by $o(x)$.

Cyclic group : A group G is called a cyclic group if every element in G is generated by an element of G and that element is called generator of G.
In other words,

A group G is called a cyclic group if there exists an element $x \in G$, such that every element of G can be expressed as power of x. In that case x is called **generator of G** and G is denoted as $\mathbf{G = <x>}$.

If x is the **generator** of a group G then so is x^{-1}, therefore it is clear that generator is not unique.

EXAMPLE 10.33. $G = \{1, -1, i, -i\}$ is a group under multiplication is a cyclic group and $G = <i>$ as

$$1 = j^4$$
$$-1 = i^2$$
$$i = i^1$$
$$-i = i^3$$

Thus, We can say

$$G = <i>$$
$$\text{or} \quad G = <-i>.$$

EXAMPLE 10.34. If in a group G, $x^5 = e$, $x\,y\,x^{-1} = y^2$ for $x, y \in G$ then show that $o(y) = 31$.

Solution :

$$y^2 = x\,y\,x^{-1}$$
$$y^4 = (x\,y\,x^{-1})\,(x\,y\,x^{-1})$$
$$= x\,y\,(x^{-1}\,x)\,y\,x^{-1} \qquad \text{(By associativity)}$$
$$= x\,y^2\,x^{-1}$$
$$= x\,(x\,y\,x^{-1})\,x^{-1}$$
$$= x^2\,y\,x^{-2} \qquad \text{(By associativity)}$$
$$y^8 = (x^2\,y\,x^{-2})\,(x^2\,y\,x^{-2})$$
$$= x^2\,y\,(x^{-2}\,x^2)\,y\,x^{-2}$$
$$= x^2\,y^2\,x^{-2}$$
$$= x^2\,(x\,y\,x^{-1})\,x^{-2}$$
$$= x^3\,y\,x^{-3}$$
$$y^{16} = (x^3\,y\,x^{-3})\,x^3\,y\,x^{-3})$$
$$= x^3\,y^2\,x^{-3}$$
$$= (x^3)\,(x\,y\,x^{-1})\,x^{-3})$$
$$= x^4\,y\,x^{-4}$$
$$y^{32} = (x^4\,y\,x^{-4})\,(x^4\,y\,x^{-4})$$
$$= x^4\,y\,x^{-4}$$
$$= x^5\,y\,x^{-5}$$

$$
\begin{aligned}
&= e\, y\, (e)^{-1} \\
y^{32} &= y \\
y^{\,32-1} &= e \\
y^{31} &= e \\
\Rightarrow 0(y) &= 31
\end{aligned}
$$

Theorem 10.23. A subgroup of a cyclic group is cyclic

Proof : Let G be a cyclic group generated by a.

That is $G = \langle a \rangle$

Let H be a subgroup of G.

If $H = \{e\}$

Nothing to prove

Let $H \neq \{e\}$.

Elements of H will be power of a as H is a sub-group of G. Let m be the least positive integer such that $a^m \in H$.

we claim $H = \langle a^m \rangle$

Let $x \in H$ be any element.

Then $x = a^k$ for some k.

then by division algorithm

$$
\begin{aligned}
k &= mq + r \quad [\text{as } m \text{ is the least positive integer such that } a^m \in H] \\
&\quad 0 \le r < m \\
\Rightarrow r &= k - mq \\
\Rightarrow a^r &= a^{k-mq} \\
&= a^k \cdot a^{-mq} \\
&= x \cdot (a^m)^{-q} \in H
\end{aligned}
$$

But m is the least positive integer such that $a^m \in H$,

$$
\begin{aligned}
\Rightarrow r &= 0 \\
\Rightarrow k &= mq \\
\Rightarrow x &= a^k = (a^m)^q
\end{aligned}
$$

as x is arbitrary in H

$$\Rightarrow H = \langle a^m \rangle$$

Theorem 10.24. A cyclic group is abelian.

Proof : Let G be a cyclic group and

$$G = \langle x \rangle$$

If $a, b \in G$

$$
\begin{aligned}
a &= x^n \\
b &= x^m \qquad \text{for some integers } n, m.
\end{aligned}
$$

Now

$$
\begin{aligned}
xy &= x^n\, x^m \\
&= x^{n+m} \\
&= x^m\, x^n \\
&= y\, x
\end{aligned}
$$

Therefore a cyclic group is abelian.

Theorem 10.25. A group of G prime order can have only trivial subgroups.

Proof : Let G be group of order p.

that is

$$0\,(G) = p$$

By Lagrange's theorem.

Let H be a subgroup of G

$\Rightarrow$ $0(H) \mid 0(G)$

$\Rightarrow$ $0(H) \mid p$

$\Rightarrow$ 0(H) divides a prime p.

$\Rightarrow$ $0(H) = 1$ or $0(H) = p$

$\Downarrow \qquad \Downarrow$

$H = \{e\}$ or $H = \{G\}$

$\therefore$ A group of prime order can have only trivial sub-groups.

10.9. COSETS AND LAGRANGE'S THEOREM

Any sub-group of a group gives rise to a partition of the group into disjoint subsets known as cosets.

If H is a subgroup of a group G and a is any arbitrary element of G, then the set $Ha = \{h\,a \mid h \in H\}$ is called ***right coset*** of H in G and $a\,H = \{a\,h \mid h \in H\}$ is called ***left coset*** of H in G.

Theorem 10.26. There is always a $1-1$, onto mapping between any two right cosets of H in G.

Proof : Let G be a group and let H be a subgroup of G.

Let $a, b \in G$.

Let Ha and Hb be any two right cosets of H in G.

Define a mapping

$$f : Ha \to H\,b$$

such that $f(h\,a) = hb$

then $h_1\,a = h_2\,a$

$\Rightarrow$ $h_1\,a\,a^{-1} = h_2\,a\,a^{-1}$

$\Rightarrow$ $h_1 = h_2$

$\Rightarrow$ $h_1 b = h_2 b$

$\Rightarrow$ $f(h_1\,a) = f(h_2 a)$

i.e., f is well defined

$f(h_1\,a) = f(h_2\,a)$

$\Rightarrow$ $h_1\,b = h_2\,b$

$\Rightarrow$ $h_1 = h_2$ [By right cancellation]

$\Rightarrow$ $h_1\,a = h_2\,a$

i.e., f is one-one.

f is onto, as for any $hb \in Hb$, ha would be its pre image.

It is followed from the above theorem that any two right cosets of H in G

have the same number of elements.

Since $H = He$ is also a right coset of H in G. Thus all right cosets of H in G have the same number of elements as are in H.

Lagrange's Theorem

Theorem 10.27. If G is a finite group and H is a subgroup of G then $0(H)$ divides $0(G)$.

Proof : Let $\quad 0(G) = n$

The number of distinct right cosets of H in G must be less than or equal to n.

$$G = Ha_1 \cup Ha_2 \cup \ldots . \cup Ha_k$$

where k = number of distinct right cosets of H in G.

$$\Rightarrow 0(G) = 0(Ha_1) + 0(Ha_2) + \ldots . + 0\,(Ha_k)$$

As all the right cosets of H in G have the same number of elements as are in H, therefore

$$0(G) = 0(H) + 0(H) + \ldots . + 0(H)_{(t \text{ times})}$$

$$\Rightarrow 0(G) = t\, 0(H)$$

$$\Rightarrow 0(H) \mid 0(G)$$

Theorem 10.28. If H and K are finite subgroups of a group G then

$$0(HK) = \frac{0(H) \cdot 0(K)}{0(H \cap K)}$$

EXAMPLE 10.35. If G is a group of order 91, show that it cannot have two subgroups of order 13.

Proof :

$$0(HK) = \frac{0(H) \circ 0(K)}{0(H \cap K)}$$

$$0(G) \geq \frac{0(H)\, 0(K)}{0(H \cap K)}$$

$$91 \geq \frac{(13)\,(13)}{0(H \cap K)}$$

$$0(H \cap K) \geq \frac{169}{91}$$

$$\Rightarrow 0(H \cap K) = 1$$

$$\Rightarrow \quad 91 \geq 169 \qquad \text{NOT POSSIBLE.}$$

Thus a group of order 91 can not have two subgroups of order 13.

10.10. RINGS

Introduction

In the previous sections we have discussed the concept GROUP, an algebraic structure with only one binary composition. Now we will talk about a non-empty set with two binary compositions. Consider, the set of integers, under usual addition forms a group but under the binary composition "multiplication" Z fails to form a group, but Z does have certain specific properties satisfied with respect to multiplication as well. We will consider some of these properties and introduce a new algebraic structure with two binary compositions called **Ring**. In general there can be any two binary compositions says * & o. But for the sake of convenience we will use the set with two binary compositions addition '+' and multiplication '·'.

Ring. A non-empty set R together with the two binary compositions '+' and '·' is said to form a ring if.

(1) $< R, + >$ is an abelian group.

(2) $a \cdot (b \cdot c) = (a \cdot b) \cdot c$ for all $a, b, c \in R$

That is associativity holds in R. In other words $< R, \cdot >$ is a semigroup.

(3) $a \cdot (b + c) = a \cdot b + a \cdot c$

$(a + b) \cdot c = a \cdot c + b \cdot c$

That is multiplication is distributive with respect to addition.

EXAMPLE 10.36. (1) $< Z, +, \cdot >$ is a ring.

(2) $< Q, +, \cdot >$ is a ring.

In Notation the set $<R, +, \cdot>$ is a ring.

It is to be noted that the first binary composition '+' in the notation of **Ring** insists that R is an **abelian group under addition.** Therefore, it is to be noted that

$< R, +, \cdot >$ is not same as $<R, \cdot, +>$

Commutative Ring

A ring $< R, +, \cdot>$ is called a commutative ring if

$a \cdot b = b \cdot a$ for all $a, b \in R$

Ring with Unity

A ring $< R, +, \cdot>$ is said to be **ring with unity** if R has the identity element for its multiplicative composition, '·' that is, there exists $e \in R$ such that

$$a \cdot e = e \cdot a = a \ \forall \ a \in R.$$

Remark: As '+' and '·' are binary compositions on R, it is understood that closure properties holds in R with respect to these composition. In other words, for all $a, b \in R$, $a + b$ and $a \cdot b$ are unique in R.

EXAMPLE 10.37. Set of real numbers, rational numbers, integers form rings with unity, with respect to, usual addition and multiplication are denoted by $< R, + \cdot >$, $< Q, +, \cdot >$, $<Z, +, \cdot >$.

EXAMPLE 10.38. Set of even integer under usual addition and multiplication forms a ***ring without unity***.

EXAMPLE 10.39. The set M of all 2×2 matrices over integers forms a ring under matrix addition and multiplication. As $< M, + >$ is an abelian group and 2nd and 3rd properties of ring are satisfied by the properties of matrix multiplication.

But $< M, +, \cdot >$ is non-commutative ring with unity as $A \cdot B \neq B \cdot A$ for $A, B \in M$.

The matrix $\begin{bmatrix} 1 & 0 \\ 0 & 1 \end{bmatrix}$ is the unit element of $< M, +, \cdot >$

EXAMPLE 10.40. The set $Z_7 = \{0, 1, 2, 3, 4, 5, 6\}$ forms a ring under addition and multiplication modulo 7. In notation

$<Z, \oplus_7, \otimes_7>$ is a ring.

Following tables give results for these two compositions:

Table 1

$\oplus_7$	0	1	2	3	4	5	6
0	0	1	2	3	4	5	6
1	1	2	3	4	5	6	0
2	2	3	4	5	6	0	1
3	3	4	5	6	0	1	2
4	4	5	6	0	1	2	3
5	5	6	0	1	2	3	4
6	6	0	1	2	3	4	5

Table 2

$\oplus 7$	0	1	2	3	4	5	6
0	0	0	0	0	0	0	0
1	0	1	2	3	4	5	6
2	0	2	4	6	1	3	5
3	0	3	6	2	5	1	4
4	0	4	1	5	2	6	3
5	0	5	3	1	6	4	2
6	0	6	5	4	3	2	1

a $\oplus_7$ b is remainder obtained by dividing $(a + b)$ by 7.

a $\otimes_7$ b is remainder obtained by dividing $(a \times b)$ by 7.

From table 1, it is easily clear that $<R, \oplus_7>$ is an abelian group. By the table 2, $<R, \oplus_7>$ is a semi-group as for any three elements $1, 2, 3 \in R$

$(2 \otimes_7 3) \otimes_7 1 = 6 \otimes_7 1 = 6$

$2 \otimes_7 (3 \otimes_7 1) = 2 \otimes_7 3 = 6$

EXAMPLE 10.41. Let R be the set of all real numbers and let F be the set of all continuous functions $f : R \to R$. Then $<F, +, \cdot>$ forms a ring defined as

for any $f, g \in F$

$$(f + g)(x) = f(x) + g(x) \qquad \text{for all } x \in R$$

$$(f \cdot g)(x) = f(x) \cdot g(x) \qquad \text{for all } x \in R$$

Then $<F, +, \cdot>$ is a commutative ring with unity.

Zero of this ring is the mapping

$$0 : R \to R, \text{ such that,}$$

$$0(x) = 0 \text{ for all } x \in R$$

Also additive inverse for any $f \in F$ is the function $(-f) : R \to R$ such that

$$(-f)(x) = -f(x) \quad \text{for all } x \in R$$

The function $I : R \to R$ is the unity in F is given as

$$I(x) = x \qquad \text{for all } x \in R$$

EXAMPLE 10.42. Let X be any non-empty set. Then, $P(X)$, the power set of X (*i.e.*, set of all subsets of x) forms a ring under + and · defined by:

$$A + B = (A \cup B) - (A \cap B) \qquad A, B \in P(X)$$

$$A \cdot B = A \cap B$$

$<P(X), +, \cdot>$ is a commutative ring with unity.

$< P(X), + >$ is an abelian group as:

1. ϕ is the identity element in $<P(X), +>$

$$A + \phi = (A \cup \phi) - (A \cap \phi)$$

$$= A - \phi$$
$$= A = \phi + A$$

2. $(A + B) + C = A + (B + C) \quad \forall A, B, C \subset P(X)$

3. Every set is the inverse of itself as

$$A + A = (A \cup A) - (A \cap A)$$
$$= A - A$$
$$= \phi$$

4. $A + B = B + A$

$\therefore$ $P(X)$ is an abelian group under addition.

for all $A, B, C \in P(X)$

$$A \cdot (B + C) = A \cdot B + A \cdot C$$

As

$$A \cdot (B + C) = A \cdot [(B \cup C) - (B \cap C)]$$
$$= A \cap (B \cup C) - A \cap B \cap C$$
$$= (A \cap B) \cup (A \cap C) - A \cap B \cap C$$
$$A \cdot B + A \cdot C = A \cap B + A \cap C$$
$$= (A \cap B) \cup (A \cap C) - (A \cap B) \cap (A \cap C)$$
$$= (A \cap B) \cup (A \cap C) - A \cap B \cap C$$
$$\Rightarrow A \cdot (B + C) = A \cdot B + A \cdot C$$

Similary $(A + B) \cdot C = A \cdot C + B \cdot C$

Theorem 10.29. In a ring R, the following results hold

(*i*) $a \cdot 0 = 0 \cdot a = 0$ for all $a \in R$

(*ii*) $a(-b) = (-a)\, b = -ab$ for all $a, b \in R$

(*iii*) $(-a)(-b) = ab$

(*iv*) $a(b - c) = ab - ac$

Proof :

(*i*)

$$a \cdot 0 = a \cdot (0 + 0)$$
$$\Rightarrow a \cdot 0 = a \cdot 0 + a \cdot 0$$
$$\Rightarrow a \cdot 0 + 0 = a \cdot 0 + a \cdot 0$$
$$\Rightarrow 0 = a \cdot 0$$

again consider

$$0 \cdot a = (0 + 0) \cdot a$$
$$\Rightarrow 0 \cdot a = 0 \cdot a + 0 \cdot a$$
$$\Rightarrow 0 \cdot a + 0 = 0 \cdot a + 0 \cdot a$$
$$\Rightarrow 0 = 0 \cdot a$$

(*ii*) $a \cdot 0 = 0$ By part (*i*)

$\Rightarrow$ $a \cdot (-b + b) = 0$

$\Rightarrow$ $a \cdot (-b) + a \cdot b = 0$

$\Rightarrow$ $a \cdot (-b) = -(a \cdot b)$

Similarly $(-a) \cdot b = -(a \cdot b)$

(*iii*) $(-a) \cdot (-b) = -[a \cdot (-b)) = -(-ab) = ab$

(iv) $$a \cdot (b - c) = a \cdot (b + (-c)) = a \cdot b + a \cdot (-c) = a \cdot b - a \cdot c \quad \textbf{proved.}$$

We are so much used to the property that whenever $ab = 0$ then either $a = 0$ or $b = 0$. For example $<Z, +, \cdot>$, the ring of integers, this holds.

But now consider $< M, +, \cdot>$ where M is the set of 2×2 matrices, this property may or may not hold that is we can have two non-zero (2×2) matrices A & B such that AB = 0

Let $$A = \begin{bmatrix} 1 & 0 \\ 0 & 0 \end{bmatrix} \qquad B = \begin{bmatrix} 0 & 0 \\ 0 & 2 \end{bmatrix}$$

$$AB = \begin{bmatrix} 0 & 0 \\ 0 & 0 \end{bmatrix} = \text{Zero Matrix}$$

By discussing this now we define Zero-Divisor

Zero-Divisor

Let R be a commutative ring. A non-zero element a in R is called a zero-divisor, if $\exists$ another non-zero element b in R such that $ab = 0$.

Integral-Domain

A commutative ring R is called an integral-domain if $a \cdot b = 0$ in R implies either $a = 0$ or $b = 0$. In other words, a commutative ring R is called an integral-domain if R has no zero-divisors.

EXAMPLE 10.43. In the ring $(Z_7, \oplus_7, \otimes_7)$, the numbers 1 and 7 are divisors of zero as

$1 \oplus_8 7 = 0$. Thus $(Z_7, \oplus_7, \otimes_7)$ is not an Integral Domain.

In general, the ring $(Z_n, \oplus_n, \otimes_n)$, the numbers 1 & n are divisors of zero as $1 \otimes_n n = 0$.

Theorem 10.30. A commutative ring R is an integral domain iff for all $a, b, c \in R$ $(a \neq 0)$

$$ab = ac$$
$$\Rightarrow b = c$$

OR

The (multiplicative) cancellation law holds in a ring $<R, +, \cdot>$ if and only if R has no divisors of zero.

Proof. Let R is an integral domain.

Let, $ab = ac$ $(a \neq 0)$

$$\Rightarrow ab - ac = 0$$
$$\Rightarrow a(b - c) = 0$$

As R is an integral domain

$$\Rightarrow a = 0 \text{ or } b - c = 0$$

as $a \neq 0$ (given)

$$\Rightarrow b - c = 0$$
$$\Rightarrow b = c$$

Conversely Let the given condition hold.

Let $a, b \in R$ be any element with $a \neq 0$

suppose $$ab = 0$$
$$\Rightarrow ab = a \cdot 0$$
$$\Rightarrow b = 0 \quad \text{using given condition}$$

Thus R is an Integral Domain.

Note: Similarly we have right cancellation law. It may be observed that cancellation is of only non-zero elements.

Definition

An element a in a ring R with unity, is called invertible (or a unit) with respect to multiplication if $\exists$ some $b \in R$ such that

$$ab = 1 = b\,a.$$

Division Ring

A ring R with unity is called a ***Division Ring*** or a ***skew field*** if non-zero elements of R form a group with respect to multiplication. That is, a ring R with unity is a Division ring if non-zero elements of R have multiplicative inverse.

10.11. IDEALS

A non-empty subset I of a ring R is called a left ideal of ring R if

(*i*) $a, b \in I \Rightarrow a - b \in I$

(*ii*) $a \in I, r \in R \Rightarrow r\,a \in I$

I is called a right ideal if

(*i*) $a, b \in I \Rightarrow a - b \in I$

(*ii*) $a \in I, r \in R \Rightarrow a\,r \in I$

If I is both left as well as right ideal of the ring R then we call I both sided ideal or two sided ideal or an ideal of R.

EXAMPLE 10.44. For any ring $< R, + ; >$, R and $\{0\}$ are both sided ideals called the trivial ideals or proper ideals

EXAMPLE 10.45. Let M be set of all 2×2 matrices overs integers

$$A = \left\{ \begin{pmatrix} a & 0 \\ b & 0 \end{pmatrix} \middle| \, a, b \in Z \right\} \text{ be subset of } M$$

Then,

$$\begin{pmatrix} a & 0 \\ b & 0 \end{pmatrix} - \begin{pmatrix} a' & 0 \\ b' & 0 \end{pmatrix} = \begin{pmatrix} a - a' & 0 \\ b - b' & 0 \end{pmatrix} \in A$$

for any $\begin{pmatrix} x & y \\ z & u \end{pmatrix} \in M$, $x, y, z, u \in Z$; set of integers

$$\begin{pmatrix} x & y \\ z & u \end{pmatrix} - \begin{pmatrix} a & 0 \\ b & 0 \end{pmatrix} = \begin{pmatrix} K & 0 \\ M & 0 \end{pmatrix} \in A$$

This shows that A is a left ideal of M. It is to be noted that A is not a right ideal of M.

as for example

$$\begin{pmatrix} 1 & 0 \\ 1 & 0 \end{pmatrix}\begin{pmatrix} 2 & 2 \\ 2 & 2 \end{pmatrix} = \begin{pmatrix} 2 & 2 \\ 2 & 2 \end{pmatrix} \notin A$$

EXAMPLE 10.46. Let

$$B = \left\{\begin{pmatrix} a & b \\ 0 & 0 \end{pmatrix} \middle| a, b \in Z\right\}$$

a subset of M.

Now, we show that B is a right ideal of M but not a left ideal

Solution : Let

$$B = \left\{\begin{pmatrix} a & b \\ 0 & 0 \end{pmatrix} \middle| a_1 b \in z\right\}$$

$$\begin{pmatrix} a & b \\ 0 & 0 \end{pmatrix} - \begin{pmatrix} a' & b' \\ 0 & 0 \end{pmatrix} = \begin{pmatrix} a-a' & b-b' \\ 0 & 0 \end{pmatrix} \in B$$

$$\begin{pmatrix} a & b \\ 0 & 0 \end{pmatrix}\begin{pmatrix} x & y \\ z & u \end{pmatrix} = \begin{pmatrix} ax+bz & ay+bu \\ 0 & 0 \end{pmatrix} \in B$$

$\Rightarrow B$ is a right ideal

Now consider.

$$\begin{pmatrix} x & y \\ z & u \end{pmatrix}\begin{pmatrix} a & b \\ 0 & 0 \end{pmatrix} = \begin{pmatrix} xa & xb \\ za & zb \end{pmatrix} \notin B$$

$\Rightarrow B$ is not a left ideal.

Theorem 10.31. An ideal is always a sub-ring.

Let I be an ideal

$$I \neq \phi,$$

$$a, b \in I \Rightarrow a - b \in I$$

Let $\quad a, b \in I \Rightarrow a \in I, b \in I \subseteq R$

$\Rightarrow a\,b \in I$

$\Rightarrow I$ is a sub-ring.

In the above proof we have used the fact that I is a right ideal. Thus a right ideal is a sub-ring. Similarly one can prove a that a left ideal is also a subring.

EXAMPLE 10.47. Example of a subring which is not an ideal.

Let $< Q, +; \cdot >$ be ring of rationals.

$< Z, + , \cdot >$ be the sub-ring of integers.

$1 \in Z$, $1/2 \in R$.

$$(1)\left(\frac{1}{2}\right) = \frac{1}{2} \notin Z.$$

Thus a sub-ring may not assencially an ideal.

Theorem 10.32. Intersection of two right (left) ideals is a right (left) ideals.

Proof : Let I and I' be the two right ideals.

Let $a, b \in I \cap I'$

$\Rightarrow a, b \in I$ and $a, b \in I'$

$$\left.\begin{array}{l} a - b \in I \\ \text{for any } r \in R \\ (a-b)\, r \in I \end{array}\right] \quad ...(i) \qquad \left.\begin{array}{l} \Rightarrow a - b \in I' \\ \text{for and any } r \in R \\ (a-b) r \in I' \end{array}\right] \quad ...(ii)$$

From (i) and (ii)

$a - b \in I \cap I'$

and for any $r \in R$

$(a - b)\, r \in I \cap I'$

$\Rightarrow$ $I \cap I'$ is the **right ideal**

Simillarly for left ideal

Let I and I' be the two left ideals

Let $a, b \in I \cap I'$

$\Rightarrow a, b \in I$ and $a, b \in I'$

$$\left.\begin{array}{l} a - b \in I \\ \text{for any } r \in R \\ r(a-b)\ \&\in I \end{array}\right] \quad ...(1i) \qquad \left.\begin{array}{l} a - b \in I' \\ \text{for any } r \in R \\ r(a-b) \in I' \end{array}\right] \quad ...(ii)$$

from (i) and (ii)

$$a - b \in I \cap I'$$

and for any $r \in R$

$$r(a - b) \in I \cap I'$$

$\Rightarrow I \cap I'$ is the left ideal.

let $ar_1, ar_2 \in a\,R$ be any elements

then $ar_1 - ar_2 = a(r_1 - r_2) \in a\,R$

For any $a\,r \in a\,R$ and $s \in R$

$$(ar)\, s = a(rs) \in aR$$

Hence aR is a right ideal.

Since R is commutative, aR is also its left ideal.

EXAMPLE 10.48. If A is an ideal of R, let $r(A) = \{x \in R \mid xa = 0\ \forall\ a \in A\}$. Show that $r(A)$ is an ideal of R.

Solution : $r(A) \neq Q$ as $0.a = 0\ \forall\ a \in R \Rightarrow 0 \in r(A)$

let $x, y \in r(A)$ be any elements.

By definition of $r(A)$

$$xa = 0 \quad \forall\ a \in A$$

$$ya = 0$$

$$\Rightarrow \qquad xa - ya = 0$$

$\Rightarrow \qquad (x - y)\, a = 0 \Rightarrow (x - y) \in r(A)$

let $x \in r(A)$ and $r \in R$ be any element

$\Rightarrow \qquad xr = 0 \; \forall \; r \in R$

For any $a \in A$, $r \in R$, A being an ideal of R

We find $\qquad ra \in A \subseteq R$

$$(xr)\, a = x(ra) \text{ for any } a \in A$$

$$= 0 \text{ as } ra \in A$$

$$(x\, r)a = 0 \; \forall \; a \in A \Rightarrow x\, r \in r\,(A)$$

for left side $(r\, x)\, a = r\,(x\, a) = (r)\,(0) = 0$

$\Rightarrow rx \in r\,(A)$

Theorem 10.33. If A, B, C are the three ideals of in ring R then

$$A.(B + C) = AB + AC$$

Proof : Let $x \in A.(B + C)$ be any element

let $\qquad B + C = D$

$\Rightarrow \qquad x \in A\, D$

$\Rightarrow \qquad x = a_1 d_1 + a_2\, d_2 + a_3\, d_3 + ... + a_n\, d_n\,,\; a_i \in A$ and $d_i \in D$

$\qquad di \in D = B + C$

$\Rightarrow \qquad di = b_i + c_i\,;\; b_i \in B,\; c_i \in C$

$\Rightarrow \qquad x = a_1\,(b_1 + c_1) + a_2\,(b_2 + c_2) + ... + a_n(b_n + c_n)$

$\Rightarrow \qquad x = (a_1\, b_1 + a_2\, b_2 + ... + a_n\, b_n) + (a_1\, c_1 + a_2\, c_2 + ...+ a_n\, c_n)$

$\qquad \in A\, B + BC$

$A{\cdot}(B + C) \subseteq AB + BC \qquad ...(i)$

let $y \in AB + AC$ be any element

$y = u + v \qquad$ where $u \in AB$ & $v = AC$

$= (\alpha_1\, \beta_1 + \alpha_2\, \beta_2 + ... + \alpha_n\, \beta_n) + (a_1\, c_1 + a_2\, c_2 + ... + a_n\, c_n)$

$y = \alpha_1\,(\beta_1 + 0) + \alpha_2(\beta_2 + 0) + ... + a_n\,(\beta_n + 0) + a_1\,(0 + c_1) + a_2\,(0 + c_2) + ... + a_n\,(0 + c_n) \in A(B + C)$

$\Rightarrow \qquad AB + AC \subseteq A \cdot (B + C) \qquad ...(ii)$

By (i) and (ii)

$$A \cdot (B + C) = AB + AC$$

EXAMPLE 10.49. If R is a ring with unity. A is any ideal of R such that $1 \in A$ then $A = R$.

(If an ideal contains unity then it is ring).

Solution : By difinition $A \subseteq R$

let $r \in R$ be any element.

Since $1 \in A$ & $r \in A$ (then by definition of ideals

$$1.r \in A$$

$\Rightarrow \qquad r \in A$

$$\Rightarrow \qquad R \subseteq A$$
$$\Rightarrow \qquad R = A$$

EXAMPLE 10.50. If F is a field then show that it has only two ideals F and 0.

Solution : Let $A \neq \{0\}$ be any ideal of F

then $\exists$ at least one $0 \neq a \in A$

Since F is a field a^{-1} exists.

Now $a \in A$, $a^{-1} \in F$, and A is an ideal of F

$a\,a^{-1} \in A \;\Rightarrow\; 1 \in A$

$$\Rightarrow \qquad A = F$$

Thus, every non zero ideal of F is F itself.

EXAMPLE 10.51. If R is a commutative ring and $a \in R$, then the set $aR = \{ar \mid r \in R\}$ is both sided ideal.

Theorem 10.34. Union of two ideals may not be an ideal.

Proof : let Z be the ring of integers

$$H_2 = \{2n \mid n \in Z\}$$
$$H_s = \{3n \mid n \in Z\}$$

are ideals of the ring of integers

$$3n - 3m = 3(n - m) \in H_3$$
$$(3n)\,K = 3\,(nK) \in H_3 \;\forall\; k \in Z$$

$\Rightarrow H_3$ is an ideal.

Similarly H_2 is also an ideal.

But $H_2 \cup H_3$ is not an ideal.

As $2, 3 \in H_2 \cup H_3$

$3 - 2 = 1 \notin H_2 \cup H_3$

Theorem 10.35. Union of two ideals is an ideal if and only if one of them is contained in the other.

Proof : Let A, B be two ideals of a ring R.

Suppose $A \subseteq B$

$A \cup B = B$ which is an ideal

Conversely.

Let A, B be two ideals of a ring R, such that

$A \cup B$ is an ideal

Suppose $A \not\subset B$ & $B \not\subset A$

$\Rightarrow \exists\, x \in A$ st $x \notin B$

$\exists\, y \in B$ st $y \notin A$

$x, y \in A \cup B$

$\Rightarrow x - y \in A \cup B$ [as $A \cup B$ is an ideal]

$\Rightarrow x - y \in A$ or $x - y \in B$

Case I : If $x - y \in A$ and also $x \in A$

$x - (x - y) \in A$

$\Rightarrow y \in A$ **a contradiction**

Case II : If $x - y \in B$ then as $y \in B$ [Since ideal is a sub-ring]

$(x - y) + y \in B \Rightarrow x \in B$ (Contradiction)

Hence, either $A \subseteq B$ or $B \subseteq A$

Product of Two Ideals

Let A and B the two ideals of a ring. R we define

$$AB = \left\{ \sum_{i=1}^{h} a_i b_i \,\middle|\, a_i \in A, b_i \in B \right\}$$

called product of two ideals

Theorem 10.36. Product of two ideals is an ideal.

Proof : Let A, B be any two ideals of a ring R.

Then $AB \neq \phi$ as $0 = 0.0 \in AB$

Let $x, y \in A\,B$ be any elements.

$$\Rightarrow \quad x = a_1\, b_1 + a_2\, b_2 + ... + a_n\, b_n,\ a_i \in A,\ b_i \in B$$

$$y = a_1'\, b_1' + a_2'\, b_2' + ... + a_n'\, b_n'$$

$$x - y = (a_1\, b_1 + ... + a_n\, b_n) - (a_1'\, b_1' + ... + a_n'\, b_n'$$

$$\forall\ a'_i \in A,\ b'_i \in B$$

$$\Rightarrow \quad x - y = \alpha_1\beta_1 + \alpha_2\beta_2 + ... \alpha_t\, \beta_t \in AB,\ \alpha_i \in A,\ \beta_i \in B$$

$$\in AB\ a_i \in A_1\ B_i \in B$$

Let $x \in AB$ and $r \in R$ be any element

Suppose $x \in a_1\, b_1 + a_2\, b_2\ + ... + a_n\, b_n$

$$xr = (a_1 b_1 + a_2 b_2 + ... + (a_n b_n)r$$

$$= (a_1 b_1)r + (a_2 b_2)r + ... + (a_n b_n)r \quad \text{(by distributivity)}$$

$$= a_1(br) + (a_2 b_2 r) + ... + a_n(b_n r) \in AB \quad [\text{As } b_i r \in B]$$

Theorem 10.37. Intersection of right and left ideal may not be an ideal.

e.g. Consider the ideals A and B as done in the previous example.

$$A = \begin{pmatrix} a & 0 \\ b & 0 \end{pmatrix},$$

$$B = \begin{pmatrix} a & b \\ 0 & 0 \end{pmatrix}$$

Here A is the left ideal and B is the right ideal of the ring M of 2×2 matrices.

$$A \cap B = \left\{ \begin{pmatrix} x & 0 \\ 0 & 0 \end{pmatrix} \middle| \, x \in Z \right\}$$

$$\begin{pmatrix} 1 & 0 \\ 0 & 0 \end{pmatrix} \begin{pmatrix} 2 & 2 \\ 2 & 2 \end{pmatrix} = \begin{pmatrix} 2 & 2 \\ 0 & 0 \end{pmatrix} \notin A \cap B$$

Implies $A \cap B$ is not a left ideal.

$$\begin{pmatrix} a & b \\ c & d \end{pmatrix}\begin{pmatrix} x & 0 \\ 0 & 0 \end{pmatrix} = \begin{pmatrix} a_x & 0 \\ c_x & 0 \end{pmatrix}$$

$$\begin{pmatrix} 1 & 1 \\ 1 & 1 \end{pmatrix}\begin{pmatrix} 1 & 0 \\ 0 & 0 \end{pmatrix} = \begin{pmatrix} 1 & 0 \\ 1 & 0 \end{pmatrix} \notin A \cap B$$

Implies $A \cap B$ is not a right ideal.

Thus intersection of a right and a left ideal may not be an ideal.

Sum of two Ideals

Let A, B be any ideals then we difine

$$A + B = \{a + b \mid a \in A,\ b \in B\}$$

then $A + B$ is an ideal of R.

$$A + B \neq \phi \text{ as } 0 = 0 + 0 \in A + B$$

$$x, y \in A + B$$

$$\Rightarrow \quad x = a_1 + b_1 \qquad a_1, a_2 \in A$$

$$y = a_2 + b_2 \qquad b_1, b_2 \in B$$

$$x - y = (a_1 + b_1) - (a_2 + b_2)$$

$$= (a_1 - a_2) + (b_1 - b_2) \in A + B$$

let $\quad x = a + b \in A + B \;\&\; r \in R$

$$xr = (a + b)\, r = ar + br \in A + B$$

[As A & B both are ideals]

$$rx = r(a + b) = ra + rb \in A + B$$

Thus it is clear that ***sum of two left (right) ideals is a left (right) ideals. Sum of a left and a right ideal may not be an ideal.***

For example :

Consider same A and B as in prerious example.

$$A + B = \left\{\begin{pmatrix} a & 0 \\ b & 0 \end{pmatrix} + \begin{pmatrix} c & d \\ 0 & 0 \end{pmatrix}\right\}$$

$$A + B = \left\{\begin{pmatrix} x & y \\ z & 0 \end{pmatrix}\right\}$$

$$\begin{pmatrix} a & b \\ c & d \end{pmatrix}\begin{pmatrix} x & y \\ z & 0 \end{pmatrix} = \begin{pmatrix} ax + bz & ay \\ cx + dz & cy \end{pmatrix}$$

$$= \begin{pmatrix} v & v \\ w & k \end{pmatrix} \notin A + B$$

$\Rightarrow$ Sum of a left and a right ideal is not an ideal.

10.12. BOOLEAN RING

A ring R is called a Boolean ring if $x^2 = x \ \forall\ x \in R$.

EXAMPLE 10.52. The ring $<\{0, 1\}, \oplus_2, \otimes_2\}$ forms a Boolean Ring. That is the ring {0 , 1} under addition and multiplication modulo 2 is a division ring.

Theorem 10.38. A Boolean ring is commutative,

Solution: Let a, $b \in R$ be any elements

$\Rightarrow a + b \in R$ (by closure)

By the definition of Boolean Ring

$$(a + b)^2 = a + \text{b}$$
$$\Rightarrow \quad (a + b)(a + b) = a + b$$
$$\Rightarrow \quad a^2 + ab + ba + b^2 = a + b$$
$$\Rightarrow \quad a + ab + ba + b = a + b$$
$$\Rightarrow \quad ab + ba = 0$$
$$\Rightarrow \quad ab = -ba \qquad ...(i)$$

Premultiplying by a

$$\Rightarrow \quad a(ab) = a(-ba)$$
$$\Rightarrow \quad a^2 b = -a\,b\,a$$
$$\Rightarrow \quad ab = -ab\,a \qquad ...(ii)$$
$$(1) \Rightarrow \quad ab = -ba$$

Post multiplying by a

$$(ab)\,a = (-ba)a$$
$$\Rightarrow a\,b\,a = -ba^2$$
$$\Rightarrow a\,b\,a = -ba \qquad ...(iii)$$

(ii) & $(iii) \Rightarrow$

$$ab = ba$$

Thus R is commutative.

Same as in the case of group, we can define the concept of sub-ring.

Subring

A non-empty subset X of a ring R is said to be a sub-ring of R if X forms a ring under the binary composition of R.

For example : The ring $<Z, +, \cdot>$ of integers is a sub-ring of the ring $<R, +, \cdot>$ of real numbers.

For a ring R, {0} & R itself are always subrings of R, called ***trivial subrings*** of R.

A sub-ring of an Integral Domain will be an Integral Domain.

Theorem 10.39. A non-empty subset X of a ring R is a subring of R if and only if a, $b \in S \Rightarrow ab$, $a - b \in S$.

Proof : Let X be a subring of R

then a, $b \in x \Rightarrow ab \in X$ [By closure]

a, $b \in X \Rightarrow a - b \in X$ [as $<X, +>$ is a sub-group of $<R, +>$]

Conversely Since a, $b \in X$

$$\Rightarrow \quad a\,b \in X \;\&\; a - b \in X$$
$$\Rightarrow \quad <X, +> \text{ forms a sub-group of } <R, +>.$$

Again for any $\quad a, b \in X \subseteq R$

$\Rightarrow \quad a, b \in R$

$\Rightarrow \quad a + b = b + a$

$\Rightarrow \quad \langle X, + \rangle$ is an abelian group.

Let $a, b, c \in X \subseteq R$

As R is ring

$\Rightarrow \quad a \cdot (b \cdot c) = (a \cdot b) \cdot c$ [Multiplicative associativity holds in R]

Again $\quad a. (b + c) = a \cdot b + a \cdot c$ [Distributivity holds in R]

$(a + b) \cdot c = a \cdot c + b \cdot c$

Thus X satisfies all the axioms in the definition of a ring.

Hence X is a subring of R.

10.13. FIELD

A commutative division-ring is called a ***Field*** for example, the set of real numbers under usual addition and multiplication form a field. Thus a field contains atleast two elements namely 0 and 1.

Whereas the set of integers does not form a field.

EXAMPLE 10.53. A division ring which is not a field. Let M be the set of all 2 $\times$ 2 matrices of the type $\begin{pmatrix} x & y \\ -\bar{y} & \bar{x} \end{pmatrix}$ where x and y are complex numbers and $\bar{x}$, $\bar{y}$ are complex conjugates of x, y respectively.

Then M is a ring with unity $\begin{pmatrix} 1 & 0 \\ 0 & 1 \end{pmatrix}$ under matrix addition and matrix multiplication.

If $A \in M$ be any nonzero element then

$$A = \begin{bmatrix} a+ib & c+id \\ -(a-ib) & c-id \end{bmatrix}$$

where a, b, c, d are all non-zero.

$$A^{-1} = \begin{bmatrix} \dfrac{a-ib}{k} & -\dfrac{c+id}{k} \\ \dfrac{c-id}{k} & \dfrac{a+ib}{k} \end{bmatrix}$$

where $k = a^2 + b^2 + c^2 + d^2$, will be multiplicative inverse of A.

Thus M is a Division ring.

But M will not be a field as it is not commutative as:

$$\begin{pmatrix} 0 & 1 \\ -i & 0 \end{pmatrix}\begin{pmatrix} i & 0 \\ 0 & -i \end{pmatrix} = \begin{pmatrix} 0 & -i \\ 1 & 0 \end{pmatrix}$$

$$\begin{pmatrix} i & 0 \\ 0 & -i \end{pmatrix}\begin{pmatrix} 0 & 1 \\ -i & 0 \end{pmatrix} = \begin{pmatrix} 0 & i \\ 1 & 0 \end{pmatrix}$$

$$\Rightarrow \quad \begin{pmatrix} 0 & 1 \\ -i & 0 \end{pmatrix}\begin{pmatrix} i & 0 \\ 0 & -i \end{pmatrix} \neq \begin{pmatrix} i & 0 \\ 0 & -i \end{pmatrix}\begin{pmatrix} 0 & 1 \\ -i & 0 \end{pmatrix}$$

Theorem 10.40. A field is an Integral Domain.

Proof : Let $<R, +, \cdot>$ be a field, then R is a commutative ring.

Let for $a, b \in R$

$ab = 0$ in R

We need to show that either $a = 0$ or $b = 0$.

Suppose a $\neq 0$ then a^{-1} exists (As A is a field)

$$\begin{aligned} \text{Thus} \quad & ab = 0 \\ \Rightarrow \quad & a^{-1}(a\,b) = a^{-1}\,0 \\ \Rightarrow \quad & a^{-1}\,a\,b = 0 \\ \Rightarrow \quad & 1 \cdot b = 0 \\ \Rightarrow \quad & b = 0 \end{aligned}$$

$\Rightarrow R$ is an Integral Domain.

Converse of the above theorem may not hold but it holds partially.

Theorem 10.41. A finite Integral Domain is a field.

Proof : Let R be a finite Integral Domain.

Let R' be the subset of R containing non-zero elements of R.

Since associativity holds in R, it will hold in R'. Thus R' is a finite semi group.

Again cancellation laws hold in R and therefore, in R'. Hence R' is a finite semigroup w.r.t multiplication in which cancellation laws hold.

$\therefore <R', +, \cdot>$ forms a gruop.

$\Rightarrow <R, +, \cdot>$ is a field.

EXAMPLE 10.54. An infinite Integral Domain which is not a field is the ring of integers.

EXAMPLE 10.55. Show that $Z_p = [0, 1, 2, \ldots, p-1]$ modular p is a field if p is prime.

Solution: Let Z_p is a field.

Suppose p is not a prime, then $\exists\ a, b$ such that $p = ab$,

$1 < a, b < p$

$\Rightarrow \quad a \otimes_p b = 0$, where a & b are non zero

$\Rightarrow Z_p$ has zero divisor *i.e.*, Z_p is not an integral domain, a contradiction as Z_p being a field must be an integral domain

$\Rightarrow$ must be a prime.

Conversely

Let p be a prime.

We will show that Z_p is an integral domain

Let $a \otimes_p b = 0, a, b \in Z_p$

$\Rightarrow$ ab is a multiple of p.

$\Rightarrow p \mid ab$

$\Rightarrow p \mid a$ or $p(b)$

$\Rightarrow a = 0$ or $b = 0$ [As p is prime and $a, b \in Z_p \Rightarrow a, b < p$]

Same as in the case of ring, group we can difine a new concept called *sub-field*.

Subfield

A non-empty set X of a field F is called subfield if X forms a field under the operations in F.

As every non-zero element in F has a multiplicative inverse. So we can define the following theorem.

Theorem 10.42. A non-empty subset X of a field F is called subfield if and only if for all $a, b \in X, b \neq 0$

$$\Rightarrow a - b \in X \text{ and } a\, b^{-1} \in X.$$

Proof : Let X be a subfield of F

then $a, b \in X$

$\Rightarrow$ $a - b \in X$ [as $< X, + >$ is a group]

$b \neq 0, a, b \in X$

$\Rightarrow a, b^{-1} \in X$

$\Rightarrow a\, b^{-1} \in X$ [By closure]

Conversely

Let $a, b \in X$

$\Rightarrow a\, b^{-1} \in x, a - b \in X$

$\Rightarrow < X, + >$ forms a group of $<F, + >$

Again for any $a, b \in X \subseteq R$

$a + b = b + a$

$\Rightarrow < X, + >$ is an abelian group.

By a similar argument, we can find that multiplicative associativity and distributvity holds in X.

$\Rightarrow < X, +, \cdot >$ is a subring.

$a, b \in X$

$\Rightarrow a\, b^{-1} \in X$ (given)

In particular, for $a \in X \Rightarrow a\, a^{-1} \in X$

$\Rightarrow 1 \in X$

$\Rightarrow < X, + , \cdot >$ is a ring with unity.

and by the given condition $< X, +, \cdot >$ is a field

$\Rightarrow <X, +, \cdot >$ is a subfield.

EXERCISE 1

1. The set of matrices

$$M = \begin{bmatrix} \cos\theta & -\sin\theta \\ \sin\theta & \cos\theta \end{bmatrix}$$

where θ is a real number, forms a group under matrix multiplication.
2. Show that the set {1, 3, 4, 5, 9} is an abelian group under ***multiplication modulo 11*** as composition.
3. Show that the set of six transformations $f_1, f_2, f_3, f_4, f_5, f_6$ on the set of complex numbers defined by

$$f_1(z) = z$$
$$f_2(z) = 1/z$$
$$f_3(z) = 1 - z$$
$$f_4(z) = \frac{z}{z-1}$$
$$f_5(z) = \frac{1}{1-z}$$
$$f_6(z) = \frac{z-1}{z}$$

4. Show that the set $\{1, \omega, \omega^2\}$ where $\omega^3 = 1$ is a group for multiplication composition.
5. Prove that a group G is abelian. If $b^{-1} a^{-1} ba = e \;\forall\; a, b \in G$
6. If G is a semi-group and for all $x, y \in G$,

$$x^2 y = y = yx^2.$$

Prove that G is an abelian group.
7. If H & K are two subgroups of a group G, then HK is a sub-group of G if and only if $HK = KH$.
8. Prove that $G = (\{0, 1, 2, 3, 4\}, \oplus_5)$ is a cyclic group with generators 1 and 4.
9. If H&K are two sub-groups of a finite group G, such that, $H \subset K$, then show that

$$[G : H] = [G : K]\,[K : H]$$

EXERCISE 2

1. For elements a, x in a group G. Show that:
   (*i*) $0(a) = 0(a^{-1})$
   (*ii*) $0(a) = 0\,(x^{-1}\,ax)$
2. Show that a group cannot be written as union of two (proper) sub-groups, although it is possible to express it as union of three sub-groups.
3. If H and K are sub-groups of a group G such that $(0(H), 0(K)) = 1$, then show that $H \cap K = \{e\}$,
4. If H is a sub-group of a group G then show that

$$g^{-1} Hg = \{g^{-1} hg \mid h \in H, g \in G\}$$

is a sub-group of G.

5. Order of a cyclic group is equal to the order of its generator.
6. If G is a finite abelian group then show that $0(ab)$ is a divisor of LCM of $0(a)$ & $0(b)$.
7. Find order of each element in the group $G = \{\pm 1, \pm i\}$ under multiplication.
8. Show that the set $G = Q - \{1\}$ of rational numbers other than 1 is an abelian group under the composition '*' defined as $\forall\ x, y \in G$

$$x * y = x + y - xy$$

9. Show that the set Q^+ of positive rational numbers does not form a group for the composition '*' defined by $a * b = a/b\ \forall\ a, b \in Q^+$
10. Show that the set $C - \{0\}$ of all non-zero complex numbers is a group for multiplication of complex numbers.

EXERCISE 3

1. Show that intersection of two sub-rings (sub-fields) is a sub-ring (sub-field).
2. If R is a division ring and X be a sub-ring of R. Show that $xy = 0$ in $X \Rightarrow$ either $x = 0$ or $y = 0$.
3. The equation $2x = 3$ has a solution in the ring $< R, +, \cdot>$ but does not have a solution in $<Z, +, \cdot>$.
4. Direct product of two sub-rings $< R_1, +, \cdot >$ and $(R_2, +, \cdot >$ is also a subring.
5. Prove that the set C of all complex numbers is a field under usual addition and multiplication. *[K.U.K. BCA I, 2000, PBI, U, BCA]*
6. Give the addition, multiplication and 'inverse' tables for each of the following fields.
   (*a*) $(Z_2, \oplus_2, \otimes_2)$
   (*b*) $(Z_3, \oplus_3, \otimes_3)$
7. Determine all values of x from the given field which satisfies the given equation.
   (*a*) $x + 1 = -1$ over Z_2, Z_3, Z_5 and Z_7
   (*b*) $2x + 1 = 2$ over Z_3 and Z_5.
   (*c*) $5x + 1 = 2$ over Z_5
8. Prove that the set of real numbers of the form $x + \sqrt{2}\, y$ where x & y are integers with ordinary addition and multiplication forms a ring.

EXERCISE 4

1. Define a group and show that $(G, *)$ is a group where $G = \{(a, b : a, b \in R, a \neq 0\{$ and '*' is a binary operation defined on G such that:

$$(a, b) * (c, d) = (ac, bc + d)\ \forall\ (a, b), (c, d) \in G$$

2. (*a*) Prove that the necessary and sufficient condition for a non-empty sub-set H of a group $(G, *)$ to be a sub-group is

$$a \in H, \quad \Rightarrow \quad a * b^{-1} \in H.$$

   (*b*) Show that two left cosets of H in G are either disjoint or identical where $(H, *)$ is a subgroup of group $(G, *)$.

(c) Define a cyclic group and hence show that $(I, +)$ is a cyclic group where I is the set of integers.

(d) Show that every homomorphic image of group $(G, *)$ is isomorphic to some quotient group of $(G, *)$

3. G is a group and there exist two relative prime positive integers m and n such that $a^m b^m = b^m a^n$ and $a^n b^n = b^n a^n$ for all $a, b, \in G$. Prove that G is Abelain.
4. How many generators are there of the cyclic group G of order 8 ?
5. Prove that $\{1, 2, 3, 4\}$ is a group under the operation x_5.
6. Write out the tables for $G = [\{1, -1, i, -i\}]$ where i is the complex number for which $c^2 = -1$. Show that G is isomorphic to $[Z_4 : t_4]$.
7. (a) What do you mean by homomorphism of group ?
   (b) Write down different properties of Boolean lattice.
8. Write short note on the following :
   (a) monoid
   (b) semi-group
   (c) ring
   (d) cyclic group
   (e) abelian group.

MULTIPLE CHOICE QUESTIONS

1. Let G be a group, $a \in G$. Let

$$(a) = \{a^i \mid i = 0, \pm 1, \pm 2, \ldots\ldots\}. \text{ Then}$$

   (a) (a) is an abelian group
   (b) (a) is non-cyclic subgroup
   (c) (a) is a cyclic subgroup of G generated by a.
   (d) None of these.
2. Two groups G, G^* are said to be isomorphic if:
   (a) There is a homomorphism of G into G^*.
   (b) There is an isomorphism of G^* into G.
   (c) There is an isomorphism of G into G^*.
   (d) None of these.
3. A homomorphism ϕ of G into $\overline{G}$ with kernal k_ϕ is an isomorphism of G into $\overline{G}$
   (a) If $k_\phi = (e)$ (b) iff $k_\phi = (e)$
   (c) if $k_\phi \neq (e)$ (d) None of these.
4. Suppose G is a finite abelian group an $p \mid 0(G)$ where p is prime. Then
   (a) There is an element $a \neq e \in G$ such that $a^p \neq e$
   (b) There is no element $a \neq e \in G$ such that $a^p = e$
   (c) There is no element $a \neq e \in G$ such that $a^p \neq e$
   (d) None of these.
5. If G is an abelian group of order $0(G)$, and if p is a prime number, such that:

$$p^\alpha \mid 0(G),\ p^{\alpha+1} \nmid 0(G), \text{ then}$$

(*a*) G has a subgroup of order p
(*b*) G has a subgroup of order p^{α}
(*c*) G has a subgroup of order $p^{\alpha+1}$
(*d*) None of these.

6. If G is an abelian group of order $0(G)$ and $p^{\alpha} \mid 0(G)$, $p^{\alpha+1} \nmid 0(G)$, there is:
(*a*) a subgroup of G of order $p^{\alpha-1}$
(*b*) a unique sub-group of G of order $p^{\alpha-1}$
(*c*) a unique sub-group of G of order p^{α}
(*d*) none of these.

7. The centre of a group
(*a*) is sometimes a normal sub-group
(*b*) is always a normal sub-group
(*c*) is not a sub-group
(*d*) none of these.

8. If $a, b \in G$, then b is said to be a conjugate of a in G if:
(*a*) there exists an element $c \in G$, such that $b = c\,a\,c^{-1}$
(*b*) there exists an element $c \in G$, such that $b = c^{-1}\,a\,c$
(*c*) there exists an element $c \in G$, such that $b = ac$
(*d*) none of these.

9. If p is a prime number and $p^{\alpha} \mid 0(G)$ then
(*a*) G has a subgroup of order p
(*b*) G has a subgroup of order p^{α}
(*c*) G has a subgroup of order $p^{\alpha+1}$
(*d*) none of these.

10. If a ring contains an identity element for the second operation, it is called
(*a*) ring with unity (*b*) field (*c*) null ring (*d*) group

11. A ring $(R, +)$ with at least two elements, which is (*i*) commutative (*ii*) possesses a unit element and (*iii*) is such that each non-zero element possesses an inverse with respect to the second operation is called
(*a*) field (*b*) integral domain
(*c*) group (*d*) none of then

12. If concellation laws hold in a Ring R then it is
(*a*) a ring without zero divisors (*b*) null ring
(*c*) field (*d*) none of these

13. If $\{G, *\}$ is an abelian group, for all $a, b \in G$ when n is a positive integer then what is the value of $(a * b)^n$
(*a*) $(a \oplus b)^n$ (*b*) $(a^n * b^n)$ (*c*) $a^n \oplus b^n$ (*d*) $(a^n * b^n)^n$

14. If $\{U_n\}$ is set of nth roots of unity, $\{U_n, x\}$ is a cyclic group then itis
(*a*) Ring (*b*) Abelian (*c*) Group (*d*) Field

15. A group Homomorphism f is called group isomorphism, if f is
(*a*) one to many onto (*b*) on to
(*c*) one to one onto (*d*) none of these

16. A commutative ring with unity (containing at least 2 elements) and without zero diviros is called

(a) an algebric system (b) ring homomorphism
(c) field (d) integral domain

17. If $S = N \times N$, the set of ordered pairs of positive integers with operation * defined by $(a, b) * (c, d) = (ad + bc, bd)$ and if

$f:(S, *) \rightarrow (Q, T)$ is defined by $f(a, b) = \frac{a}{b}$, then f is a

(a) Ring homomorphism (b) Semigroup homomorphism
(c) A belian homomorphism (d) None of these

18. If $f : x \rightarrow x$, where $x = \{1, 2, 3, 4\}$ is defined by $f = \{(1, 2)\}, (2, 3), (3, 4), (4, 1)$ a n d $F = \{f^0, f^1, f^2, f^3\}$ then $\{F, .\}$ is a
(a) semi group under the operation associative
(b) abelian group under function composition
(c) monoid under the operation (.) of function composition
(d) none of these

19. If H is a subgroup of a such that $x^2 \varepsilon H$ for every $x \varepsilon G$, then what is the relation between G and H.
(a) H is a normal subgroup of G (b) G is normal subgroup of H
(c) H is disjoint subgroup of G (d) G is disjoint subgroup of H

Answers

1. (c)	2. (c), (b)	3. (b)	4. (d)	5. (b)
6. (c)	7. (b)	8. (b)	9. (b)	10. (a)
11. (a)	12. (a)	13. (b)	14. (b)	15. (c)
16. (d)	17. (b)	18. (c)	19. (a)	

Fill in the blanks :

1. If H is a sub-group of G, $a \in G$,
$Ha = \{ha \mid h \in H\}$, then Ha is called a .... of H in G.
2. The sub-group N of G is a normal sub-group of G iff every .... of N in G is a right coset of N in G.
3. If G is a group, N is a normal sub-group of G, then G/N is called .... of G by N.
4. Automorphism of a group G is an .... of G onto itself.
5. Every permutation is the product of its ....
6. If $a \in G$, then $N(a)$, the .... of a in G, is the set $N(a) = \{x \in G \mid xa = ax\}$
7. If $p^m \mid 0(G)$, $p^{m+1} \nmid 0(G)$, then G has a subgroup of order ....

Answers

1. right coset	2. left coset	3. quotient or factor gorup	4. isomorphism
5. cycles	6. normaliser	7. p^{∞}	

CHAPTER 11

MATRICES

This Chapter Covers :

11.1. INTRODUCTION

Definition : A **matrix** is defined as the rectangular array of numbers or functions that obey certain laws. The individual numbers (or functions) of the array are called the **elements of the matrix.**

for example :

$$\begin{pmatrix} 2 & 4 & 5 \\ 3+4i & 3 & 9 \\ 7 & 8 & 9 \end{pmatrix}, \begin{pmatrix} a & b & c \\ d & e & f \\ g & h & i \end{pmatrix}, \begin{pmatrix} f_1(x) & f_2(x) & f_3(x) \\ g_1(x) & g_2(x) & g_3(x) \\ h_1(x) & h_2(x) & h_3(x) \end{pmatrix}$$

The elements of a matrix belongs to certain rows and certain columns of the matrix. Consider the following matrix

$$\begin{pmatrix} a_{11} & a_{12} & \cdots\cdots & a_{1n} \\ a_{21} & a_{22} & \cdots\cdots & a_{2n} \\ a_{31} & a_{32} & \cdots\cdots & a_{3n} \\ \vdots & \vdots & & \vdots \\ a_{m1} & a_{m2} & \cdots\cdots & a_{mn} \end{pmatrix}$$

This matrix can be denoted by (a_{ij}), $1 \leq i \leq m$, $1 \leq j \leq n$ as it has m rows and n columns. Thus every elements of the matrix can be uniquely characterised by a row index and a column index. For example, the element a_{ij} belong to the ith row and the jth column of the matrix. The 1st symbol (i) denotes row and the second symbol (j) denotes column to which the element a_{ij} belongs.

11.2. TYPE OF MATRICES

Comparable Matrices

Two matrices $A = [a_{ij}]$ & $B = [b_{ij}]$ are said to be comparable if they have same order, that is, if the number of rows and columns in A and B are same.

For exmaple,

The matrices $A = \begin{pmatrix} 1 & 2 & 3 \\ a & b & c \end{pmatrix}$ and $B = \begin{pmatrix} f(x) & g(x) & h(x) \\ f'(x) & g'(x) & h'(x) \end{pmatrix}$ are comparable matrices whereas

$$C = \begin{pmatrix} 1 & 2 & 3 \\ 4 & 5 & 6 \\ 7 & 8 & 9 \end{pmatrix}$$

and
$$D = \begin{pmatrix} \sin\theta & \cos\theta & \sin\theta \\ \tan\theta & \cot\theta & \sin\theta \end{pmatrix}$$

are not comparable as the number of rows are not same.

Equal Matrices: The two matrices $A = [a_{ij}]$ and $B = [b_{ij}]$ are said to be equal if they are :

1. Comparable
2. Corresponding elements in the matrices are same.

that is $a_{ij} = b_{ij}$ for all values of i and j.

For instance in the above example the first two matrices (that is A and B) are comparable but not equal.

Addition and Subtraction

One can do the addition (substraction) of two matrices if they are comparable. Addition of the matrices $A = [a_{ij}]$ and $B = [b_{ij}]$ is defined as the matrix $C = [c_{ij}]$ where $c_{ij} = a_{ij} + b_{ij}\ \forall\ i\ \&\ j$, that is, each element of which is the sum of the corresponding elements of A and B.

Similarly, the difference of two matrices can be calculated.

Note: Two matrices of different order can not be added or substracted.

The matrix addition obeys following properties :

(*i*) The **commutative law** holds for matrices *i.e.*,

$$A + B = B + A$$

or
$$a_{ij} + b_{ij} = b_{ij} + a_{ij}\ \forall\ i\ \&\ j$$

(*ii*) The **associative law** holds for matrix addition *i.e.*,

$$A + (B + C) = (A + B) + C$$

or
$$a_{ij} + (b_{ij} + c_{ij}) = (a_{ij} + b_{ij}) + c_{ij}$$

(*iii*) **Scalar multiplication** is also hold in matrices

$$\lambda(A + B) = \lambda A + \lambda B$$

or
$$\lambda (a_{ij} + b_{ij}) = \lambda a_{ij} + \lambda b_{ij}$$

where λ is any number.

Multiplication of Two Matrices

Two matrices $A = [a_{ij}]$ and $B = [b_{ij}]$ may be multiplied together in the order AB only when the number of columns in A is equal to the number of rows in B. If matrix A is of order $(m \times p)$ and matrix B is of order $(p \times k)$ then the product $C = AB$ is of the order $m \times k$.

Its elements are given by

$$C_{ij} = \sum_{r=1}^{p} a_{ir} b_{rj}$$

The C_{ij}, *i.e.*, $(ij)^{th}$ element of the matrix $C = AB$ is formed as a scalar product of the ith row of A with the jth column of B.

The dummy index r takes on all the values 1, 2, ... p in succession *i.e.*, for $r = 4$

$$C_{ij} = a_{i1}\, b_{1j} + a_{i2}\, b_{2j} + a_{i3}\, b_{3j} + a_{i4}\, b_{4j}$$

Properties of Matrix Multiplication

1. Matrix Multiplication is not Commutative

$AB \neq BA$

for example, if $A = \begin{bmatrix} 1 & 5 & 0 \\ 2 & 1 & 5 \\ 3 & 1 & 1 \end{bmatrix}$ and $B = \begin{bmatrix} 2 & 5 & 2 \\ 1 & 0 & 1 \\ 2 & 0 & 0 \end{bmatrix}$

then $$AB = \begin{bmatrix} 6 & 5 & 7 \\ 15 & 10 & 5 \\ 9 & 15 & 7 \end{bmatrix}$$

and $$BA = \begin{bmatrix} 18 & 17 & 27 \\ 4 & 6 & 1 \\ 2 & 10 & 0 \end{bmatrix}$$

Thus $AB \neq BA$,

In fact, for a given pair of matrices.

In fact, for a given pair of matrices A and B, the products AB and BA may not be even comparable. For example, if A be a 3×4 matrix and B be a 4×3 matrix then AB would be 3×3 matrix while BA would be a 4×4 matrix.

It may also happen that for a given pair of matrices A and B, the product AB may be defined, but the product BA may not be defined. For example, if A is 3×4 matrix and B is 4×2 matrix then AB would be 3×2 matrix but BA is not defined.

Associativity in Multiplication

The multiplication of matrices is associative, that is, if the matrices A, B, C be of suitable order for the product $A(BC)$ and $(AB)C$ to be defined, then

$$A(BC) = (AB)C$$

Distributive law of Multiplication

The multiplication of the matrices is distributive with regard to the addition, that is,

$$A(B + C) = AB + AC$$

where A, B and C are the matrices of suitable order of the above equation to be meaningful.

Sub-matrix

The matrix of elements which remain after deleting any number of row and columns of a matrix A is called the sub-matrix of A.

For example, the matrix $\begin{bmatrix} a & b \\ c & d \end{bmatrix}$ is a sub-matrix of

$$\begin{bmatrix} f & a & b & c & d \\ e & c & d & h & i \\ t & r & s & t & u \end{bmatrix}$$

Because it can be obtained from the above matrix by deleting column 1, 4, 5 and third row.

Partitioning Matrices

In some cases it is convenient to partition matrices into rectangular blocks. These blocks containing the elements of the given matrix are defined as the sub-matrices of the given matrix. The dotted lines indicate the partition of the given matrix. Thus the partitioned matrix A is given as:

$$\begin{bmatrix} x_1 & y_1 & z_1 & \vdots & a_1 & b_1 \\ x_2 & y_2 & z_2 & \vdots & a_2 & b_2 \\ \cdots & \cdots & \cdots & \cdots & \cdots & \cdots \\ r_1 & s_1 & t_1 & \vdots & u_1 & v_1 \\ r_2 & s_2 & t_2 & \vdots & u_2 & v_2 \end{bmatrix}$$

and its partition can be given as

$$A_{11} = \begin{bmatrix} x_1 & y_1 & z_1 \\ x_2 & y_2 & z_2 \end{bmatrix}$$

$$A_{12} = \begin{bmatrix} a_1 & b_1 \\ a_2 & b_2 \end{bmatrix}$$

$$A_{21} = \begin{bmatrix} r_1 & s_1 & t_1 \\ r_2 & s_2 & t_2 \end{bmatrix}$$

$$A_{22} = \begin{bmatrix} u_1 & v_1 \\ u_2 & v_2 \end{bmatrix}$$

The matrix can be partition in a number of ways, for example,

$$\begin{bmatrix} x_1 & y_1 & \vdots & z_1 & a_1 & b_1 \\ x_2 & y_2 & \vdots & z_2 & a_2 & b_2 \\ \cdots & \cdots & \cdots & \cdots & \cdots & \cdots \\ r_1 & s_1 & \vdots & t_1 & u_1 & v_1 \\ r_2 & s_2 & \vdots & t_2 & u_2 & v_2 \end{bmatrix} \text{ and so on.}$$

Let $A = [a_{ij}]$ and $B = [b_{ij}]$ be the matrices of order $m \times p$ and $p \times n$ respectively. The matrix A is partitioned in any arbitrary manner, then the matrix

B is partitioned in such a way that the multiplication of submatrices of both A and B exist.

For example, the matrix A may be partitioned into m submatrices of $1 \times p$ and B into n submatrices of order $p \times 1$. As another example the matrices A and B may be partitioned into submatrices of indicated orders by drawing in the dotted lines as

$$A = \begin{bmatrix} m_1 \times p_1 & \vdots & m_1 \times p_2 & \vdots & m_1 \times p_3 \\ \cdots & \cdots & \cdots & \cdots & \cdots \\ m_2 \times p_1 & \vdots & m_2 \times p_2 & \vdots & m_2 \times p_3 \end{bmatrix}$$

$$B = \begin{bmatrix} p_1 \times n_1 & \vdots & p_1 \times n_2 \\ \cdots & \cdots & \cdots \\ p_2 \times n_1 & \vdots & p_2 \times n_2 \\ \cdots & \cdots & \cdots \\ p_3 \times n_1 & \vdots & p_3 \times n_2 \end{bmatrix}$$

or

$$A = \begin{bmatrix} A_{11} & \vdots & A_{12} & \vdots & A_{13} \\ \cdots & \cdots & \cdots & \cdots & \cdots \\ A_{21} & \vdots & A_{22} & \vdots & A_{23} \end{bmatrix},$$

$$B = \begin{bmatrix} B_{11} & \vdots & B_{12} \\ \cdots & \cdots & \cdots \\ B_{21} & \vdots & B_{22} \\ \cdots & \cdots & \cdots \\ B_{31} & \vdots & B_{32} \end{bmatrix}$$

In any such partitioning it is necessary that the column of A and rows of B be partitioned in exactly the same way; however m_1, m_2, n_1, n_2 may be any non-negative integers such that

$$m_1 + m_2 = m \text{ and } n_1 + n_2 = n$$

Then the product

$$AB = \begin{bmatrix} A_{11}B_{11} + A_{12}B_{21} + A_{13}B_{31} & A_{11}B_{12} + A_{12}B_{22} + A_{13}B_{32} \\ A_{21}B_{11} + A_{22}B_{21} + A_{23}B_{31} & A_{21}B_{12} + A_{22}B_{22} + A_{23}B_{32} \end{bmatrix}$$

Special Types of Matrices

Square Matrix: A matrix in which number of rows are same as number of columns is called a square matrix. For example :

$$\begin{bmatrix} a_{11} & a_{12} & \cdots\cdots & a_{1n} \\ a_{21} & a_{22} & \cdots\cdots & a_{2n} \\ \vdots & \vdots & & \vdots \\ a_{n1} & a_{n2} & \cdots\cdots & a_{nn} \end{bmatrix}$$

is a square matrix of order $n \times n$ or of order n.

Diagonal Matrix

In the square matrix the elements a_{11}, a_{22}, a_{33}, - - - - a_{nn} are called **di-**

agonal elements. The sum of the diagonal elements of a square matrix is called the **trace** of that matrix. For example,

$$\begin{bmatrix} a_{11} & 0 & 0 & \cdots\cdots & 0 \\ 0 & a_{22} & 0 & \cdots\cdots & 0 \\ 0 & 0 & a_{33} & \cdots\cdots & 0 \\ \vdots & & & \ddots & \\ 0 & 0 & 0 & \cdots\cdots & a_{nn} \end{bmatrix}$$

is a square diagonal matrix of order n.

Scalar Matrix : A diagonal matrix in which all the diagonal elements are equal, say α, is called a scalar matrix. For example,

$$\begin{bmatrix} \alpha & 0 & 0 & \cdots & 0 \\ 0 & \alpha & 0 & \cdots & 0 \\ 0 & 0 & \alpha & \cdots & 0 \\ : & : & : & & \\ : & 0 & 0 & & \alpha \end{bmatrix}$$

Thus a n-square matrix $A = [a_{ij}]_{n \times n}$ is a scalar matrix if and only if

$$a_{ij} = \begin{cases} \alpha & for \quad i = j \\ 0 & for \quad i \neq j \end{cases}$$

Identity or Unit Matrix

A scalar matrix in which each diagonal element is unity, is called an **identity or unit matrix**. Thus a n-square matrix $[a_{ij}]$ is a unit matrix if and only if

$$a_{ij} = \begin{cases} 1 & for \quad i = j \\ 0 & for \quad i \neq j \end{cases}$$

It is usually denoted by I_n or simply I.

Row matrix (column matrix)

A matrix which contains only one row (column) is called a row (column) matrix or a row (column) vector.

For example,

$$[a_{11}\ a_{12} \ldots\ldots a_n]_{1 \times n} \quad \text{is a row matrix}$$

and

$$\begin{bmatrix} a_{11} \\ a_{12} \\ : \\ a_{1n} \end{bmatrix}_{n \times 1} \quad \text{is a column matrix,}$$

Null Matrix or Zero Matrix

If each elements of a matrix is zero, then the matrix is called **null matrix** or **Zero Matrix**.

Upper Triangular Matrix

A square matrix, whose every element a_{ij} for $i > j$ is zero, is called an **upper triangular matrix**. Thus

$$\begin{bmatrix} a_{11} & a_{12} & a_{13} & \cdots\cdots & a_{1n} \\ 0 & a_{22} & a_{23} & \cdots\cdots & a_{2n} \\ 0 & 0 & a_{33} & \cdots\cdots & a_{3n} \\ \vdots & \vdots & & & \vdots \\ 0 & 0 & 0 & -- & a_{nn} \end{bmatrix}$$

is an upper triangular matrix.

Lower Triangular Matrix

A square matrix whose every element a_{ij} for $i < j$ is zero, is called the **lower triangular matrix**. For example,

$$\begin{bmatrix} a_{11} & 0 & 0 & -- & 0 \\ a_{21} & a_{22} & 0 & -- & 0 \\ a_{31} & a_{23} & a_{33} & -- & 0 \\ \vdots & \vdots & \vdots & & \\ a_{n1} & a_{n2} & a_{n3} & -- & a_{nn} \end{bmatrix}$$

is lower triangular matrix.

Periodic Matrix

A square matrix A for which $A^{k+1} = A$, where k is a positive integer, is called a **periodic matrix**. If k is the least positive integer for which $A^{k+1} = A$, then k is said to be **period of matrix *A***.

Idempotent Matrix

A square matrix for which $A^2 = A$ is called **idempotent matrix**. It is clear that idempotent-matrix is periodic which period 1.

For example, every identity matrix is an idempotent matrix.

Nilpotent Matrix

A square matrix A for which $A^k = 0$, where k is a positive integer is called the **nilpotent matrix**.

If k is the least positive integer for which $A^k = 0$, then k is said to be the **index of the matrix *A***.

For example :

$A = \begin{bmatrix} 0 & 1 \\ 0 & 0 \end{bmatrix}$ is a nilpotent matrix of index 2.As 2 is the least positive integer for which $A^2 = 0$.

Result : If A and B are idempotent matrices, then $A + B$ will be idempotent if and only if

$$AB = BA = 0$$

Proof : As A and B both are idempotent matrices therefore $A^2 = A$ and $B^2 = B$,

If $(A + B)$ is idempotent, then

$$(A+B)^2 = A+B \quad \text{...}(i)$$

Now

$$\begin{aligned}(A+B)^2 &= (A+B)(A+B)\\ &= A^2+AB+BA+B^2\\ &= A+AB+BA+B \quad \text{...}(ii)\end{aligned}$$

[As A and B both are idempotent]

Comparing (i) and (ii)

$$AB+BA = 0 \quad \Rightarrow \quad AB = BA = 0$$

Thus if $(A + B)$ will be idenpotent, then

$$AB = BA = 0$$

Conversely Let $AB = BA = 0$ and A and B are idempotent matrices, consider

$$\begin{aligned}(A+B)^2 &= (A+B)(A+B)\\ &= A^2+AB+BA+B^2 \qquad [\because AB = BA = 0]\\ &= A^2+B^2 = A+B \qquad [\because A \,\&\, B \text{ are Idempotent}]\end{aligned}$$

Transpose of a matrix

A transpose of a matrix A is obtained by interchanging the rows and columns of the matrix A and is denoted by A' or A^T or $\tilde{A}$. Thus if A is a matrix of order $m \times n$ then its transpose will be of the order $n \times m$.

For example :

$$A = \begin{bmatrix} a_{11} & a_{12} & \cdots\cdots & a_{1n} \\ a_{21} & a_{22} & \cdots\cdots & a_{2n} \\ \vdots & \vdots & & \vdots \\ a_{m1} & a_{m2} & \cdots\cdots & a_{mn} \end{bmatrix}_{m \times n}$$

$$A^T = \begin{bmatrix} a_{11} & a_{21} & \cdots\cdots & a_{m1} \\ a_{12} & a_{22} & \cdots\cdots & a_{m2} \\ \vdots & \vdots & & \vdots \\ a_{1n} & a_{2n} & \cdots\cdots & a_{mn} \end{bmatrix}_{n \times m}$$

It is clear that $(i, j)^{th}$ element of the matrix A is the $(j, i)^{th}$ element of the matrix A^T.

Properties of Transpose of a Matrix

If A^T & B^T be the transpose of two given matrices A & B respectively, then

(1) $(A^T)^T = A$

(2) $(\alpha A)^T = \alpha A^T$, α being any scalar.

(3) $(A+B)^T = A^T + B^T$, provided A and B must be comparable.

(4) $(AB)^T = B^T A^T$, provided product AB exist.

Conjugate of a Matrix

If A is any matrix in the complex field. The matrix obtained from A by replacing its each element by its conjugate complex number, is called the conjugate of the matrix A and is denoted by $\overline{A}$

For example

If $$A = \begin{bmatrix} \cos\theta + i\sin\theta & \tan\theta + i\cot\theta \\ x + iy & \cos\theta + i\sec\theta \end{bmatrix}$$

then

$$\overline{A} = \begin{bmatrix} \cos\theta - i\sin\theta & \tan\theta - i\cot\theta \\ x - iy & \cos\theta - i\sec\theta \end{bmatrix}$$

Properties of the Conjugate of a Matrix

If $\overline{A}$ & $\overline{B}$ are conjugates of matrices A and B respectively, then

(1) $\overline{(\overline{A})} = \text{A}$

(2) $\overline{(A+B)} = \overline{A} + \overline{B}$

(3) $\overline{(\alpha A)} = \overline{\alpha}\,\overline{A}$, α being any complex number.

(4) $\overline{(AB)} = \overline{B}\,\overline{A}$

Tranjugate of a Matrix

The conjugate of the transpose of a matrix A is called the **tranjugate** or **conjugate transpose** of A and is denoted by A^*.

Thus $$A^* = (\overline{A})^T = (\overline{A})^T$$

Properties of tranjugate of a matrix are same as conjugate of the matrix A.

For example : If

$$A = \begin{bmatrix} 2+5j & 5i & 3+4i \\ 2-9i & 0 & 2-3i \\ 7+8i & 6i & 7 \end{bmatrix}$$

Its tranjugate is given as

$$A^* = \begin{bmatrix} 2-5i & 2+9i & 7-8i \\ -5i & 0 & -6i \\ 3-4i & 2+3i & 7 \end{bmatrix}$$

Symmetric and Skew Symmetric Matrix

A square matrix A is said to be **symmetric** if its transpose is same as that of itself that is, if $A^T = A$.

$$a_{ij} = a_{ji} \qquad \text{for all values of } i \text{ and } j.$$

For example :

$$\begin{pmatrix} x & y & z \\ y & z & t \\ z & t & w \end{pmatrix}$$

is a symmetric matrix.

A square matrix $A = [a_{ij}]$ is said to be **skew symmetric** provided $a_{ij} = -a_{ji}$ $\forall$ i and j

Thus $A^T = -A$

In particular for $i = j$

$$a_{ii} = -a_{ij}$$

$$\Rightarrow \quad 2a_{ii} = 0$$

Thus diagonal elemets of the skew matrix are all zero.

For example :

$$\begin{bmatrix} 0 & x & y \\ -x & 0 & z \\ -y & -z & 0 \end{bmatrix}$$

EXAMPLE 11.1. Show that every square matrix can be uniquely expressed as the sum of a symmetric and a skew-symmetric matrix.

Proof : Let $A = [a_{ij}]$ be any square matrix.

Then matrix A can be written as

$$A = \frac{1}{2}(A + A) = \frac{1}{2}(A + A + A^T - A^T)$$

$$= \frac{1}{2}(A + A^T) + \frac{1}{2}(A - A^T) \qquad ...(i)$$

where A^T is the transpose of A.

Substituting $X = \frac{1}{2}(A + A^T)$, $Y = \frac{1}{2}(A - A^T)$ in (1)

$$\Rightarrow \quad A = X + Y$$

Now we will prove that X is symmetric and y is skew symmetric.

$$X^T = \left(\frac{1}{2}(A + A^T)\right)^T$$

$$= \frac{1}{2}(A^T + (A^T)^T)$$

$$= \frac{1}{2}(A^T + A)$$

$$= X$$

$\Rightarrow X$ is a symmetric matrix.

Now consider

$$Y^T = \left(\frac{1}{2}(A - A^T)\right)^T = \frac{1}{2}(A^T - (A^T)^T)$$

$$= \frac{1}{2}(A^T - A) = -\frac{1}{2}(A - A^T) = -Y$$

Thus Y is a skew-symmetric matrix.

Thus matrix A can be expressed as sum of symmetric and skew symmetric matrix.

Now, we will prove the uniqueness.

Let $A = P + Q$ be another possible representation of A where P is symmetric and Q is skew symmetric

$$\Rightarrow P^T = P \quad \text{and} \quad Q^T = -Q$$

Then the representation $A = P + Q$ gives

$$A^T = (P + Q)^T$$
$$= P^T + Q^T$$
$$= P - Q$$

Therefore, $$A + A^T = (P + Q) + (P - Q)$$
$$= 2P$$

$$\Rightarrow P = \frac{1}{2}(A + A^T)$$

Similarly, $$Q = \frac{1}{2}(A - A^T)$$

This proves the result.

EXAMPLE 11.2. If A and B are symmetric matrices then AB is symmetric if and only if A and B commute.

Hermitian Matrix

A square matrix $A = (a_{ij})$ is said to be **Hermitian matrix** if

$$A = A^*$$

That is if A is same as its tranjugate.

For example, the following matrices are Hermitian.

$$\begin{bmatrix} l & i & 0 \\ -i & 2 & -5j \\ 0 & 5i & 3 \end{bmatrix}, \begin{bmatrix} 0 & -4i & 3i \\ 4i & 5 & 2 \\ -3i & 2 & 9 \end{bmatrix}$$

It is clear from the above example that **diagonal element of a Hermitian Matrix will always be real numbers.**

As by the definition

$$a_{ij} = a_{ij}^* \qquad \forall\, i \,\&\, j$$

In particular $$a_{ii} = a_{ii}^* \qquad \forall i$$

Thus every diagonal element of a Hermitian matrix is equal to its complex conjugate and it is only possible if all the diagonal elements of a Hermitian matrix are real numbers.

Skew-Hermitian Matrix

A square matrix $A = [a_{ij}]$ is said to be **Skew-Hermitian** (or **Anti-Hermitian**) matrix if

$$A = -A^*$$

That is $a_{ij} = -a^*_{ij}$ $\forall$ i and j

Thus skew Hermitian matrix is one whose tranjugate coincides with negative of the matrix.

For example, the following matrices are Skew-Hermitian.

$$\begin{bmatrix} i & -5 & 2 \\ 5 & i & 3 \\ -2 & -3 & i \end{bmatrix}, \begin{bmatrix} 0 & 1+i \\ -1+i & 0 \end{bmatrix}$$

It is clear from the above examples that **the diagonal elements of a Skew-Hermitian Matrix is either zero or a purely imaginary number.**

As by the definition

$$a_{ij} = -\overline{a_{ij}} \quad \forall\ i \ \&\ j$$

$$\Rightarrow a_{ij} + \overline{a_{ij}} = 0 \qquad \ldots(i)$$

In particular for $i = j$

$$a_{ii} + \overline{a_{ii}} = 0$$

$$2Re\,[a_{ii}] = 0$$

$$\Rightarrow \quad Re\,[a_{ii}] = 0$$

$\Rightarrow a_{ii}$ is purely imaginary.

This implies that every diagonal element of a Skew-Hermitian Matrix is either zero or is purely imaginary.

EXAMPLE 11.3. Every square matrix can be uniquely expressed as the sum of a Hermitian and Skew-Hermition Matrix.

Hint : Same as in the case of symmetric & skew symmetric matrices.

EXAMPLE 11.4. If H is a Hermitian matrix then iH is a Skew Hermition.

11.3. DETERMINANT OF A MATRIX

Every matrix cannot have a determinant, only a square matrix can have a determinant and is denoted by $|A|$.

Some Properties of Determinants :

1. If every element of a row (or column) of a square matrix A is zero, then $|A| = 0$.
2. An interchange of any two rows (or columns) in a determinant gives the same determinant with the negative sign.
3. Thus if matrix P is obtained from matrix Q by interchanging two rows (or columns) then $|P| = -|Q|$
4. If every element of a row (or a column) of a determinant $|A|$ is multiplied by a scalar k, then the determinant is multiplied by k.

For example :

$$\lambda|A| = \lambda\begin{vmatrix} a_{11} & a_{12} & a_{13} \\ a_{21} & a_{22} & a_{23} \\ a_{31} & a_{32} & a_{33} \end{vmatrix} = \begin{vmatrix} a_{11} & \lambda a_{12} & a_{13} \\ a_{21} & \lambda a_{22} & a_{23} \\ a_{31} & \lambda a_{32} & a_{33} \end{vmatrix}$$

$$= \begin{vmatrix} a_{11} & a_{12} & a_{13} \\ a_{21} & a_{22} & a_{23} \\ ka_{31} & ka_{32} & ka_{33} \end{vmatrix}$$

5. If the matrix B is obtained from matrix A by adding to the elements of its ith row (or column), a scalar multiple of the corresponding elements of another row (or clumn), then $|B| = |A|$

EXAMPLE 11.5.

$$\begin{vmatrix} a_{11} & a_{12} & a_{13} \\ a_{21} & a_{22} & a_{23} \\ a_{31} & a_{32} & a_{33} \end{vmatrix} = \begin{vmatrix} a_{11}+ka_{13} & a_{12} & a_{13} \\ a_{21}+ka_{23} & a_{22} & a_{23} \\ a_{31}+ka_{33} & a_{22} & a_{33} \end{vmatrix}$$

$$= \begin{vmatrix} a_{11} & a_{12} & a_{13} \\ a_{21} & a_{22} & a_{23} \\ a_{31}+ka_{11} & a_{32}+ka_{12} & a_{33}+ka_{13} \end{vmatrix}$$

Singular and Non-Singular Matrices :

A square matrix A is said to be **singular** if its determinant is zero otherwise the matrix is said to be **non-singular**.

For example, the identity matrix will always be a non-singular matrix.

Invertible Matrix

Let A be a n-square matrix. If there exists a n-square matrix B such that

$$AB = BA = I_n$$

where I_n is n-square unit matrix.

then the matrix B is called the **inverse of *A*** and A is called invertible matrix.

Inverse of a matrix is denoted by $\boldsymbol{A^{-1}}$.

For example :

$$\begin{bmatrix} 1 & -1 & 3 \\ -1 & 1 & 2 \\ 3 & 2 & -1 \end{bmatrix} \begin{bmatrix} 1/5 & -1/5 & 1/5 \\ -1/5 & 2/5 & 1/5 \\ 1/5 & 1/5 & 0 \end{bmatrix} = I_3$$

Thus both the matrices on the LHS. are the inverse of each other.

Result: If A and B are two n-square, non-singular matrices then

(*i*) $(AB)^{-1} = B^{-1}A^{-1}$

(*ii*) $(A^{-1})^T = (A^T)^{-1}$

(*iii*) $(A^*)^{-1} = (A^{-1})^*$

Orthogonal Matrices

A square finite matrix A is said to be **orthogonal** if

$$A^TA = I = AA^T$$

where A^T is the transpose of A & I is unit matrix.

We know, $\quad |A^T| = |A|$

and $\quad |A^TA| = |A^T|\ |A|$

Hence if A is orthogonal then $A^T A = I$

$$\Rightarrow |A^T|\,|A| = |I| = 1$$

$$\Rightarrow |A|^2 = 1$$

$$\Rightarrow |A| = \pm 1$$

This shows that the determinant of an orthogonal matrix can only have values + 1 or –1.

Also it is clear that A is non singular so A^{-1} exists.

Now, $$A^TA = I$$

Post multiplying by A^{-1}, we will get

$$A^T A A^{-1} = \text{I } A^{-1}$$

$$\Rightarrow A^T = A^{-1}$$

Thus the inverse of an orthogonal matrix is its transpose.

Result : The product of orthogonal matrices are also orthogonal, *i.e.*, if A and B are orthogonal matrices, then AB and BA are also orthogonal matrices.

Proof : By the given condition

$$A^T A = I \;\&\; B^T B = I$$

such that $$|A^T| = |A| \neq 0$$

and $$|B^T| = |B| \neq 0$$

Now we need to show that AB is also orthogonal, consider

$$\begin{aligned}(AB)^{\text{T}} (AB) &= (B^TA^T)(AB) && \text{(By transpose property)}\\ &= B^T(A^TA)\text{ B} && \text{(By multiplication associativity)}\\ &= B^T I B && \text{(As } A \text{ is orthogonal)}\\ &= B^T B\\ &= I\end{aligned}$$

$\Rightarrow AB$ is an orthogonal matrix.

Similarly, we may show that BA is also orthogonal.

Unitary Matrix

A square matrix A is said to be unitary if

$$A^*A = I = A A^*$$

where A^* is conjugate transpose of A and I is a unit matrix.

Same as in the case of orthogonal matrix, the determinant of unitary matrix is always of unit modulus and hence a unitary matrix is **non-singular**, and product of two unitary matrices are also unitary.

Rank of a matrix

The number of linearly independent rows in a matrix is called Rank of a matrix.

In other words, the number of non-zero rows in the echelon form of a matrix after applying elementary row operations is called Rank of a matrix.

Elementary Operations

Following three operations upon rows or columns of a matrix are defined as elementary operations or elementary transformations.

(*i*) The interchange of two rows (or columns) $R_i \leftrightarrow R_j$

(*ii*) The multiplication of any row (or column) by a non-zero number.

$$R_i \to \alpha R_i \qquad \alpha \neq 0$$

(*iii*) The addition of one row (or column) to another row (or column) multiplied by a non-zero number.

$$R_i \to R_i + R_j \qquad i \neq j$$

Echelon form of a Matrix

A matrix is said to be in echelon form, if the following two conditions hold:

(*i*) All zero rows, if any, are on the bottom of the matrix.

(*ii*) Each leading non-zero entry is to the right of the leading non-zero entry in the preceding row.

In other words the 1st non-zero element in a row must be in the zig-zag form as given below:

For example

$$A = \begin{bmatrix} ① & 2 & 3 & 5 & 7 & 8 \\ 0 & ② & 4 & 0 & 9 & 7 \\ 0 & 0 & ③ & 4 & 5 & 0 \\ 0 & 0 & 0 & ⑦ & 8 & 9 \\ 0 & 0 & 0 & 0 & 0 & 0 \end{bmatrix} \qquad B = \begin{bmatrix} ④ & 5 & 1 & 3 \\ 0 & ⑤ & 7 & 9 \\ 0 & 0 & 0 & ⑧ \\ 0 & 0 & 0 & 0 \end{bmatrix}$$

Here the rank of A, $r(A) = 4$, the number of non-zero rows in the Echelon (Zig Zag) form.

Similarly, $r(B) = 3$

EXAMPLE 11.6. Solve the following determinant.

$$\begin{vmatrix} 1 & bc & a(b+c) \\ 1 & ac & b(a+c) \\ 1 & ab & c(a+b) \end{vmatrix}$$

$C_1 \to C_1 + C_2 + C_3$

$$\Rightarrow \quad \begin{vmatrix} 1+ab+bc+ca & bc & a(b+c) \\ 1+ab+bc+ca & ac & b(a+c) \\ 1+ab+bc+ca & ab & c(a+b) \end{vmatrix}$$

$$\Rightarrow \quad (1+ab+bc+ca)\begin{vmatrix} 1 & bc & a(b+c) \\ 1 & ac & b(a+c) \\ 1 & ab & c(a+b) \end{vmatrix}$$

$R_2 \to R_2 - R_1, \ R_3 \to R_3 - R_1$

$$\Rightarrow \quad (1+ab+bc+ca)\begin{vmatrix} 1 & bc & a(b+c) \\ 0 & c(a-b) & c(b-a) \\ 0 & b(a-c) & b(c-a) \end{vmatrix}$$

$\Rightarrow (1 + ab + bc + ca)\ [bc\ (a - b)(c - a) - bc(b - a)(a - c)]$

$\Rightarrow \quad (1 + ab + bc + ca)(a - b)(c - a)(bc - bc) = 0$

EXAMPLE 11.7. Prove that

$$\begin{vmatrix} x & y & z \\ y & z & x \\ z & x & y \end{vmatrix}^2 = (x^3 + y^3 + z^3 - 3xyz)^2$$

Solution : Consider

$$\begin{vmatrix} x & y & z \\ y & z & x \\ z & x & y \end{vmatrix}$$

$R_1 \to R_1 + R_2 + R_3$

$$\Rightarrow \quad \begin{vmatrix} x+y+z & x+y+z & x+y+z \\ y & z & x \\ z & x & y \end{vmatrix}$$

$$\Rightarrow \quad (x+y+z)\begin{vmatrix} 1 & 1 & 1 \\ y & z & x \\ z & x & y \end{vmatrix}$$

$C_2 \to C_2 - C_1, \quad C_3 \to C_3 - C_1$

$$\Rightarrow \quad (x+y+z)\begin{vmatrix} 1 & 0 & 0 \\ y & z-y & x-y \\ z & x-z & y-z \end{vmatrix}$$

$$\Rightarrow \quad (x + y + z)[-(z - y)^2 - (x - y)\,(x - z)]$$

$$\Rightarrow \quad (x + y + z)(-z^2 - y^2 + 2zy - x^2 + xz + xy - yz)$$

$$\Rightarrow \quad (x + y + z)\,(-x^2 - y^2 - z^2 + xy + yz + xz)$$

$$\Rightarrow \quad -(x^3 + y^3 + z^3 + 3xyz)$$

$$\therefore \quad \begin{vmatrix} x & y & z \\ y & z & x \\ z & x & y \end{vmatrix}^2 = (x^3 + y^3 + z^3 - 3xyz)^2$$

EXAMPLE 11.8. Show that:

$$\begin{vmatrix} a & b & c \\ a^2 & b^2 & c^2 \\ b+c & c+a & a+b \end{vmatrix} = (a - b)(b - c)(c - a)(a + b + c)$$

Solution : Applying elementary operations

$R_3 \to R_3 + R_1$

$$\begin{vmatrix} a & b & c \\ a^2 & b^2 & c^2 \\ a+b+c & a+b+c & a+b+c \end{vmatrix}$$

$$\Rightarrow \quad (a+b+c)\begin{vmatrix} a & b & c \\ a^2 & b^2 & c^2 \\ 1 & 1 & 1 \end{vmatrix}$$

$C_2 \to C_2 - C_1, C_3 \to C_3 - C_1$

$$\Rightarrow \quad (a+b+c)\begin{vmatrix} a & b-a & c-a \\ a^2 & b^2-a^2 & c^2-a^2 \\ 1 & 0 & 0 \end{vmatrix}$$

$$\Rightarrow \quad (a+b+c)\,[(b-a)(c^2-a^2)-(c-a)(b^2-a^2)]$$

$$\Rightarrow \quad (a+b+c)(b-a)(c-a)[(c+a)-(b+a)]$$

$$\Rightarrow \quad (b-a)(c-a)(c-b)(a+b+c)$$

$$\Rightarrow \quad (a-b)(b-c)(c-a)(a+b+c)$$

EXAMPLE 11.9. Solve the following Polynomial.

$$\begin{vmatrix} t-2 & 4 & -3 \\ 1 & t+1 & -2 \\ 0 & 0 & t-4 \end{vmatrix}$$

Solution : Expanding the given determinant

$$(t-4)\big[(t-2)(t+1)-4\big]$$
$$= (t-4)(t^2-t-2-4)$$
$$= (t-4)(t^2-t-6)$$
$$= (t-4)(t^2-3t+2t-6)$$
$$= (t-4)(t-3)(t+2)$$

EXAMPLE 11.10. Show that

$$\begin{vmatrix} a+b & b+c & c+a \\ p+q & q+r & r+p \\ u+v & v+w & w+u \end{vmatrix} \div \begin{vmatrix} a & b & c \\ p & q & r \\ u & v & w \end{vmatrix} = 2$$

Solution : $$\begin{vmatrix} a+b & b+c & c+a \\ p+q & q+r & r+p \\ u+v & v+w & w+u \end{vmatrix}$$

$C_1 \to C_1 + C_2 + C_3$

$$= \begin{vmatrix} 2(a+b+c) & b+c & c+a \\ 2(p+q+r) & q+r & r+p \\ 2(u+v+w) & v+w & w+u \end{vmatrix} = 2\begin{vmatrix} a+b+c & b+c & c+a \\ p+q+r & q+r & r+p \\ u+v+w & v+w & w+u \end{vmatrix}$$

$C_2 \to C_2 - C_1, C_3 \to C_3 - C_1$

$$= 2\begin{vmatrix} a+b+c & -a & -b \\ p+q+r & -p & -q \\ u+v+w & -u & -v \end{vmatrix}$$

$C_1 \to C_1 + C_2 + C_3$

$$= 2\begin{vmatrix} c & -a & -b \\ r & -p & -q \\ w & -u & -v \end{vmatrix} = 2\begin{vmatrix} a & b & c \\ p & q & r \\ u & v & w \end{vmatrix}$$

Therefore, the result is true.

EXAMPLE 11.11. If $a \neq b \neq c$, evaluate

$$\Delta = \begin{vmatrix} 1 & a & a^3 \\ 1 & b & b^3 \\ 1 & c & c^3 \end{vmatrix}$$

Solution : Applying elementary row transformation

$$R_2 \to R_2 - R_1, R_3 \to R_3 - R_1$$

$$\Delta = \begin{vmatrix} 1 & a & a^3 \\ 0 & b-a & b^3 - a^3 \\ 0 & c-a & c^3 - a^3 \end{vmatrix}$$

$$= (b-a)(c^3 - a^3) - (b^3 - a^3)(c-a)$$

$$= (b-a)(c-a)(c^2 + ac + a^2 - b^2 - ab - a^2)$$

$$= (b-a)(c-a)\left[(c^2 - b^2) + a(c-b)\right]$$

$$= (b - a)(c - a)(c - b)(a + b + c)$$

EXAMPLE 11.12. Solve the following equation if $a \neq b \neq c$

$$\begin{vmatrix} a+x & b+x & c+x \\ b+x & c+x & a+x \\ c+x & a+x & b+x \end{vmatrix} = 0$$

Solution : Applying elementary column transformation $C_1 \to C_1 + C_2 + C_3$

$$\begin{vmatrix} a+b+c+3x & b+x & c+x \\ a+b+c+3x & c+x & a+x \\ a+b+c+3x & a+x & b+x \end{vmatrix} = 0$$

$$\Rightarrow \quad (a+b+c+3x)\begin{vmatrix} 1 & b+x & c+x \\ 1 & c+x & a+x \\ 1 & a+x & b+x \end{vmatrix} = 0$$

$R_2 \to R_2 - R_1, \ R_3 \to R_3 - R_1$

$$\Rightarrow \quad (a+b+c+3x)\begin{vmatrix} 1 & b+x & c+x \\ 0 & c-b & a-c \\ 0 & a-b & b-c \end{vmatrix} = 0$$

$$\Rightarrow \quad (a+b+c+3x)\left[-(b-c)^2 - (a-c)(a-b)\right] = 0$$

$$\Rightarrow \quad (a + b + c + 3x) \left[-b^2 - c^2 + 2bc - a^2 + ab + ac - bc\right] = 0$$

$$\Rightarrow \quad (a + b + c + 3x \left(a^2 + b^2 + c^2 - ab - ac - bc\right) = 0$$

As $a \neq b \neq c$

Therefore, $(a^2 + b^2 + c^2 - ab - ac - bc) = 0$

$$\Rightarrow \quad a + b + c + 3x = 0$$

$$\Rightarrow \quad x = \frac{-1}{3}(a+b+c)$$

EXAMPLE 11.13. Without evaluating the determinant, show that:

$$D = \begin{vmatrix} \cos(A-P) & \cos(A-Q) & \cos(A-R) \\ \cos(B-P) & \cos(B-Q) & \cos(B-R) \\ \cos(C-P) & \cos(C-Q) & \cos(C-R) \end{vmatrix} = 0$$

Solution: Expanding all the terms

$$D = \begin{vmatrix} \cos A\cos P+\sin A\sin P & \cos A\cos Q+\sin A\sin Q & \cos A\cos R+\sin A\sin R \\ \cos B\cos P+\sin B\sin P & \cos B\cos Q+\sin B\sin Q & \cos B\cos R+\sin B\sin R \\ \cos C\cos P+\sin C\sec P & \cos C\cos Q+\sin C\sin Q & \cos C\cos R+\sin C\sin R \end{vmatrix}$$

$$= \begin{vmatrix} \cos A & \sin A & 0 \\ \cos B & \sin B & 0 \\ \cos C & \sin C & 0 \end{vmatrix} \begin{vmatrix} \cos P & \sin P & 0 \\ \cos Q & \sin Q & 0 \\ \cos R & \sin R & 0 \end{vmatrix}$$

$$= 0$$

EXAMPLE 11.14. Find the inverse of the following matrix A. Where

$$A = \begin{pmatrix} 1 & 0 & 2 \\ 2 & -1 & 3 \\ 4 & 1 & -3 \end{pmatrix}$$

Solution : $A\,A^{-1} = \quad I,\ IA^{-1} = A^{-1}$

$$\begin{pmatrix} 1 & 0 & 2 \\ 2 & -1 & 3 \\ 4 & 1 & -3 \end{pmatrix} A^{-1} = \begin{pmatrix} 1 & 0 & 0 \\ 0 & 1 & 0 \\ 0 & 0 & 1 \end{pmatrix}$$

$R_2 \to R_2 - 2R_1,\ R_3 \to R_3 - 4R_1$

$$\begin{pmatrix} 1 & 0 & 2 \\ 0 & -1 & -1 \\ 0 & 1 & -11 \end{pmatrix} A^{-1} = \begin{pmatrix} 1 & 0 & 0 \\ -2 & 1 & 0 \\ -4 & 0 & 1 \end{pmatrix}$$

$R_2 \to -R_2$

$$\begin{pmatrix} 1 & 0 & 2 \\ 0 & 1 & 1 \\ 0 & 1 & -11 \end{pmatrix} A^{-1} = \begin{pmatrix} 1 & 0 & 0 \\ 2 & -1 & 0 \\ -6 & 1 & 1 \end{pmatrix}$$

$R_3 \to R_3 - R_2$

$$\begin{pmatrix} 1 & 0 & 2 \\ 0 & 1 & 1 \\ 0 & 0 & -12 \end{pmatrix} A^{-1} = \begin{pmatrix} 1 & 0 & 0 \\ 2 & -1 & 0 \\ -4 & 0 & 1 \end{pmatrix}$$

$R_3 \to \frac{-1}{12} R_3$

$$\begin{pmatrix} 1 & 0 & 2 \\ 0 & 1 & 1 \\ 0 & 0 & 1 \end{pmatrix} A^{-1} = \begin{pmatrix} 1 & 0 & 0 \\ 2 & -1 & 0 \\ 1/2 & -1/12 & -1/12 \end{pmatrix}$$

$R_1 \to R_1 - 2R_3,\ R_2 \to R_2 - R_3$

$$\begin{pmatrix} 1 & 0 & 0 \\ 0 & 1 & 0 \\ 0 & 0 & 1 \end{pmatrix} A^{-1} = \begin{pmatrix} 0 & 1/6 & 1/6 \\ 3/2 & -11/12 & 1/12 \\ 1/2 & -1/12 & -1/12 \end{pmatrix}$$

$$\therefore \quad A^{-1} = \begin{pmatrix} 0 & 1/6 & 1/6 \\ 3/2 & -11/12 & 1/12 \\ 1/2 & -1/12 & -1/12 \end{pmatrix}$$

EXAMPLE 11.15. Show that the matrix $A = \begin{pmatrix} 1 & 1 & 1 \\ 1 & 2 & -3 \\ 2 & -1 & 3 \end{pmatrix}$ satisfies the matrix equation $A^3 - 6A^2 + 5A + 11I = 0$. Hence, find A^{-1}.

Solution : Now $A^2 = A.A$

$$= \begin{pmatrix} 1 & 1 & 1 \\ 1 & 2 & -3 \\ 2 & -1 & 3 \end{pmatrix} \begin{pmatrix} 1 & 1 & 1 \\ 1 & 2 & -3 \\ 2 & -1 & 3 \end{pmatrix} = \begin{pmatrix} 4 & 2 & 1 \\ -3 & 8 & -14 \\ 7 & -3 & 14 \end{pmatrix}$$

$$A^3 = A^2 \cdot A$$

$$= \begin{pmatrix} 4 & 2 & 1 \\ -3 & 8 & -14 \\ 7 & -3 & 14 \end{pmatrix} \begin{pmatrix} 1 & 1 & 1 \\ 1 & 2 & -3 \\ 2 & -1 & 3 \end{pmatrix} = \begin{pmatrix} 8 & 7 & 1 \\ -23 & 27 & -69 \\ 32 & -13 & 58 \end{pmatrix}$$

Substituting these values in the given equation, we will get

$A^3 - 6A^2 + 5A + 11\,I$

$$\begin{pmatrix} 8 & 7 & 1 \\ -23 & 27 & -69 \\ 32 & -13 & 58 \end{pmatrix} - 6 \begin{pmatrix} 4 & 2 & 1 \\ -3 & 8 & -14 \\ 7 & -3 & 14 \end{pmatrix} + 5 \begin{pmatrix} 1 & 1 & 1 \\ 1 & 2 & -3 \\ 2 & -1 & 3 \end{pmatrix} + 11 \begin{pmatrix} 1 & 0 & 0 \\ 0 & 1 & 0 \\ 0 & 0 & 1 \end{pmatrix}$$

$$= \begin{pmatrix} 8-24+5+11 & 7-12+5+0 & 1-6+5+0 \\ -23+18+5+0 & 27-48+10+11 & -69+84-15+0 \\ 32-42+10+0 & -13+18-5+0 & 58-84+15+11 \end{pmatrix}$$

$$= \begin{pmatrix} 0 & 0 & 0 \\ 0 & 0 & 0 \\ 0 & 0 & 0 \end{pmatrix}$$

Thus the matrix A satisfies the given equation.

Now to find A^{-1}, consider

$$A^3 - 6A^2 + 5A + 11I = 0$$

Multiply both sides by A^{-1}

$$A^2 - 6A + 5I + 11A^{-1} = 0$$

$$\Rightarrow \qquad A^{-1} = \frac{-1}{11}(A^2 - 6A + 5I)$$

$$= \frac{-1}{11}\begin{pmatrix} 4 & 2 & 1 \\ -3 & 8 & -14 \\ 7 & -3 & 14 \end{pmatrix} - 6\begin{pmatrix} 1 & 1 & 1 \\ 1 & 2 & -3 \\ 2 & -1 & 3 \end{pmatrix} + 5\begin{pmatrix} 1 & 0 & 0 \\ 0 & 1 & 0 \\ 0 & 0 & 1 \end{pmatrix}$$

$$= \frac{-1}{11}\begin{pmatrix} 4-6+5 & 2-6 & 1-6 \\ -3-6 & 8-12+5 & -14+18 \\ 7-12 & -3+6 & 14-18+5 \end{pmatrix}$$

$$\Rightarrow \qquad A^{-1} = \frac{-1}{11}\begin{pmatrix} 3 & -4 & -5 \\ -9 & 1 & 4 \\ -5 & 3 & 1 \end{pmatrix}$$

$$\Rightarrow \qquad A^{-1} = \begin{pmatrix} -3/11 & 4/11 & 5/11 \\ 9/11 & -1/11 & -4/11 \\ 5/11 & -3/11 & -1/11 \end{pmatrix}$$

❐❐❐

CHAPTER 12

FINITE STATE MACHINE

This Chapter Covers :

12.1. FINITE STATE MACHINE

Finite state machines are often used to recognise pattern in an input sequence, a complex counter, a traffic light controller and a digital combination lock.

In this chapter, the main emphasis is only at the idea of designing finite state machine given a description of the strings that we want to accept.

Suppose we want to solve a problem like the following :

Given a string of zeroes and ones, we want to recognize when the input contains the substring 001 but does not start with 11.

We are going to design a simple theoretical "machine" that will be able to solve pattern recognition problems like this one.

A finite state machine (or finite automation) M is an abstract model of a very simple computer (with limited memory). It has the following components :

- a finite set Q of states.
- a finite set Σ of input symbols.
- an initial state $q_0 \in Q$

- a transition function $\delta : Q \times \Sigma \rightarrow Q$
- a set $A \subseteq Q$ of accepting states.

The machine starts in the initial state. It then receives input, one symbol at a time. Upon receiving each input symbol, the transition function determines the next state of the machine. If it was in state q and it receives input symbol a, then $\delta(q, a)$ tells us the next state of the machine. When the end of an input string is reached, the machine will be in some state; Let's call it q_f. If q_f is a member of the set A of accepting states, then we say that the input string is accepted by the machine. Otherwise, the input string is not accepted.

EXAMPLE 12.1. The finite state machine described below will accept all string of zeroes and ones that contain at least two ones. Instead of describing Q, Σ, q_0, δ and A, we almost always represent a finite state machine with a diagram like the following :

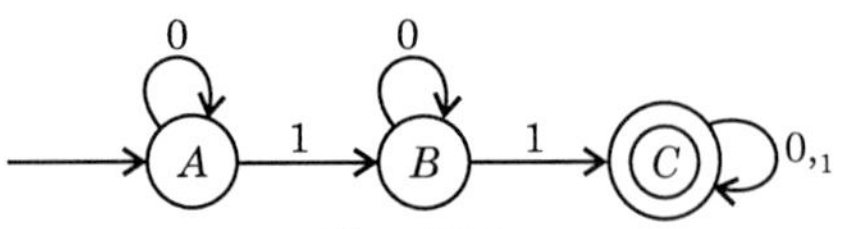

Fig. 12.1.

The diagram above is nice, graphical way of representing all the information that would be in the official, formal description of a finite state machine. The components of the finite state machine above are the following :

- The set of states is $\{A, B, C\}$
- The set of input symbols is $\{0, 1\}$.
- The initial state is A. (This is indicated by the unlabelled arrow going into A).
- The transition function δ can be described by the following transition table:

	0	1
A	A	B
B	B	C
C	C	C

  The top row tells us that, if we are in state A, an input of 0 causes us to stay in state A, while an input of 1 causes us to move to state B.
- The set of accepting states (indicated by a double circle) is $\{C\}$.
  If the sequence of symbols in an input string will take us to state C, then that string is accepted by the machine; otherwise, it is not accepted.
  Let us consider the example input string 1011000 and trace how the finite state machine would process it.
- The machine starts in state A.
- The first input symbol is 1. When we are in state A and we see an input of 1, the diagram (or the transition table) tells us to move to state B.
- The second input symbol is 0. When we are in state B and we see an input of 0, we stay in state B.
- The third input symbol is 1. This causes us to move to state C. Note that, in this example, once we are in state C, there is no way to leave. We will always stay in state C, no matter how much additional input we see. Therefore, in this case, the rest of the input (1000) does not really matter. We know that will end up staying in state C and that the string will be accepted.
- This is the right decision. The string does not contain two or more ones and so we should accept it.

Let us trace a second example through our machine : 0010.

- The machine starts in state A.
- The first input symbol is 0. When we are in state A and we see an input of 0, the diagram (or the transition table) tells us to state in state A.
- This happens again upon seeing the second 0 in the input.
- The third input symbol is 1. This causes us to move to state B.
- Finally the fourth input symbol is 0 and we stay in state B. Because the input is now finished and because we did not end up in an accepting state, this string is not accepted.
- Again, this is the right decision. The string does not have two or more ones and so it should not be accepted.

It is worthwhile at this point to think about why our machine had three states. How would we have designed the machine from scratch, given the problem we were trying to solve ? Given the description of the problem, we know that the only thing we need to remember about a string is the number of ones we have seen so far. This is what the three different state represent. Any time we are in state A, it means that we have not seen any ones yet. In state B, we have seen 1 one so far. In state C, we have seen 2 or more ones.

As we look at more complicated problems, the role of the states will always be the same. As we process an input string, we will want to remember one or more places of information about the part of the string we have seen so far. Each state will represent a different possible case that could arise.

For example, if we wanted to recognize string of letters $a - z$ that contained atleast two $e's$, at least three $s's$, and no $n's$, we would need to read input symbol from the string and keep track of :

1. How many $e's$ have we seen so far (0, 1, at least 2) ?
2. How many $s's$ have we seen so far (0, 1, 2 at least 3) ?
3. How many $n's$ have we seen so far (0, at least 1) ?

We would need 24 different states to keep track of all possible combinations of answers to questions 1, 2, 3 :

State name	*e's seen so far*	*s's seen so far*	*n's seen so far*
S_0	0	0	0
S_1	0	0	≥ 1
S_2	0	1	0
S_3	0	1	≥ 1
S_4	0	2	0
S_5	0	2	≥ 1
S_6	0	≥ 3	0
S_7	0	≥ 3	≥ 1
S_8	1	0	0
S_9	1	0	≥ 1
S_{10}	1	1	0
S_{11}	1	1	≥ 1
S_{12}	1	2	0

S_{13}	1	2	≥ 1
S_{14}	1	≥ 3	0
S_{15}	1	≥ 3	≥ 1
S_{16}	≥ 2	0	0
S_{17}	≥ 2	0	≥ 1
S_{18}	≥ 2	1	0
S_{19}	≥ 2	1	≥ 1
S_{20}	≥ 2	2	0
S_{21}	≥ 2	2	≥ 1
S_{22}	≥ 2	≥ 3	0
S_{23}	≥ 2	≥ 3	≥ 1

So when we asked to design a finite state machine to accept a particular class of strings, ask yourself the questions : what information do we have to keep track of ? What possible cases could arise ? This will tell you what state you must have in your machine.

EXAMPLE 12.2. Design a finite state machine that will accept all strings of zeroes and ones that contain the sub-string 101.

In this case, what do we want to remember ? If we have already seen the substring 101, then we should be in an accepting state that we will never leave. In particular, if we see 101 at the very begining of a string, we should move from the initial state (A) to an accepting state (D) by following three transitions : A to B when we see the first one, B to C when we see the 0 and then C to D when we see the second one.

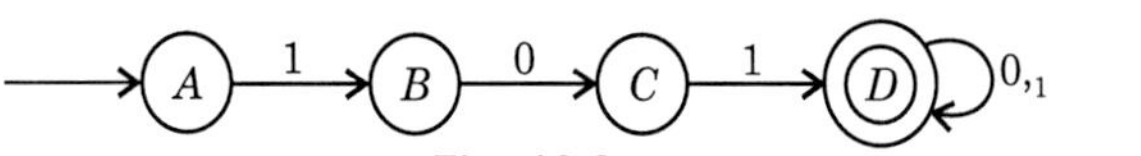

Fig. 12.2.

However, this is not a complete finite state machine. For a finite state machine to be complete (and for it to work correctly on all possible input S), we must define a transition from every state on every possible input symbol.

In our example, what should happen in state C if we see a zero ? What should happen in state B if we see a one ? What should happen in state A if we see a zero ?

Basically, if we are in state C, we are remembering that we have just seen the sequence 10. In case we happen to see a 1 as the next input symbol, we want to be in a position where we can move directly to our accepting state upon seeing a 1.

However, if we see a zero instead, what should we do ? By reading a zero as the next input symbol, we are undoing all the "good" that the sub-string of 10 did in moving us towards the accepting state. We must go all the way back to the begining and in order to make any progress towards the accepting state we must start over with a new instance of the sub-string 101.

EXAMPLE 12.3. Design a finite state machine that will accept all strings of zeroes and ones that contain exactly two ones.

In this case, our machine should be exactly the same as Example 1 (at least two ones), with one exception. In example 1, as soon as we saw a second one, we knew that we would want to accept the string. Therefore, it was OK to move to state C as soon as we saw a second one and then stay there, regardless of any additional input symbols. In this example, we want to accept a string if it has exactly two ones, but not if it has three or more ones, so we should enter state C when we see a second one, but then leave state C (and make sure we never return) as soon as we see a third one.

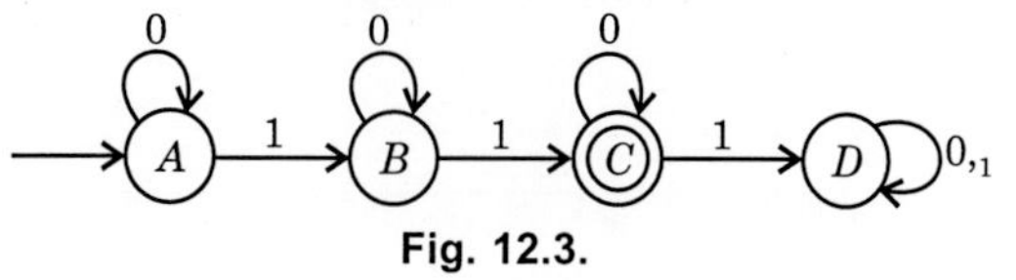

Fig. 12.3.

A state such as D in the example above is often referred to as a dead state. As soon as we enter a dead state, there is no way to exit it and we are guaranteed that we will not accept the string, regardless of any additional input.

In addition to examples like the one above, dead states are appropriate states to add to finite state machines in examples like "Desgin a finite state machine that will accept all strings of zeros and ones that do not contain the substring 101. You can set up a machine to do this by ensuring that, as soon as you see the substring 101, you enter into a dead state and are unable to leave.

Exercise: Design a finite state machine that will accept all strings of zeroes and ones that do not contain the substring 101.

EXAMPLE 12.4. Draw a finite state machine that will accept all strings of zeroes and ones that end in 111.

In this case, we want to make sure that we enter an accepting state whenever we have just seen the substring 111. In case that happens to be the end of the string. We want to make sure that we are in a position to accept it. However, while we are in the accepting state, if we see an input symbol of zero, we should leave the accepting state and go back to the very begining. In order to get back to the accepting state, we should have to see another instance of the substring 111.

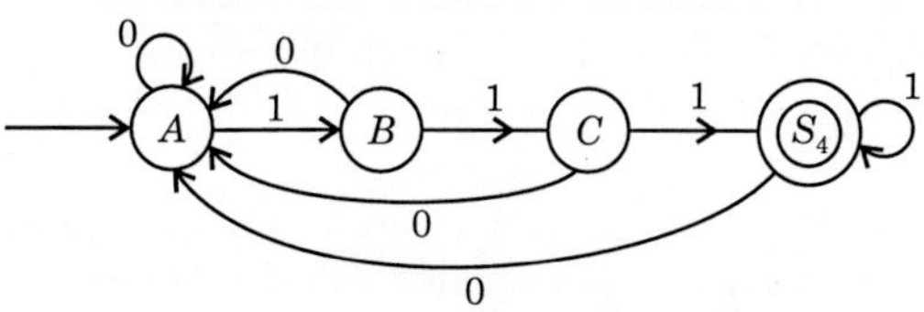

Fig. 12.4.

On the other hand, seeing another one while we are in state D is perfectly fine. We now know that the string ends in 1111, but this is fine; it still ends in 111.

Exercise: Draw a finite state machine that will accept all strings of zeroes and ones that starts with 111.

EXAMPLE 12.5. Draw a finite state machine that will accept all strings of zeroes and ones that contain the substring 00 and contain at least two ones.

In this case, we have to remember two different facts about the string we have seen so far. Have we seen the substring 00 yet (or have we just seen the first 0 in what might end up being a substring of 00) ? How many ones have we seen so far ?

The states in our finite state machine must capture all the possible combinations of answers to those questions.

State Name	*Have we seen 00 yet ?*	*Number of ones Seen so Far*
S_0	No	0
S_1	No	1
S_2	No	≥ 2
S_3	No, but one zero was just seen	0
S_4	No, but one zero was just seen	1
S_5	No, but one zero was just seen	≥ 2
S_6	Yes	0
S_7	Yes	1
S_8	Yes	≥ 2

Of these states, only S_8 should be an accepting state. What should be the entire FSM look like ?

EXAMPLE 12.6. This example is provided just to show that we can construct finite state machine when we have more than two possible input symbols.

Draw a finite state machine that accepts strings of *a's, b's* and *c's* that contain the sub-string *abc*.

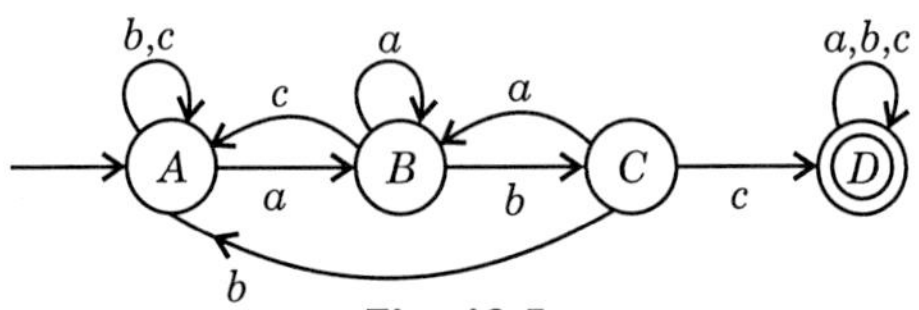

Fig. 12.5.

EXAMPLE 12.7. Something we have not seen yet is an example of a finite state machine that contains more than one accepting state. There is nothing in the definition of a finite state machine that says that we cannot have several accepting states.

Draw a finite state machine that will accept strings of zeroes and ones that contain exactly two ones or exactly one zero.

Again, we can create states that capture all the possible cases that might arise, if we were to count the number of ones and the number of zeroes seen so far.

State Name	*Number of Zeroes Seen*	*Number of Ones Seen*
A	0	0
B	0	1
C	0	2
D	0	≥ 3
E	1	0
F	1	1
G	1	2
H	1	≥ 3
I	≥ 2	0
J	≥ 2	1
K	≥ 2	2
L	≥ 2	≥ 3

One of the big differences between this example and one we have seen before is that the description in this case contains the word *OR*. As long as we have either exactly two ones or exactly one zero, we want to accept the string. So, in our list of states above, we should accept a string if we are in states E, F, G or H (because those states represent situations in which the number of zeroes is exactly one) or if we are in states C, G or *K* (because those states represent situations in which the number of ones is exactly two). Our accepting states should be C, E, F, G, H and K.

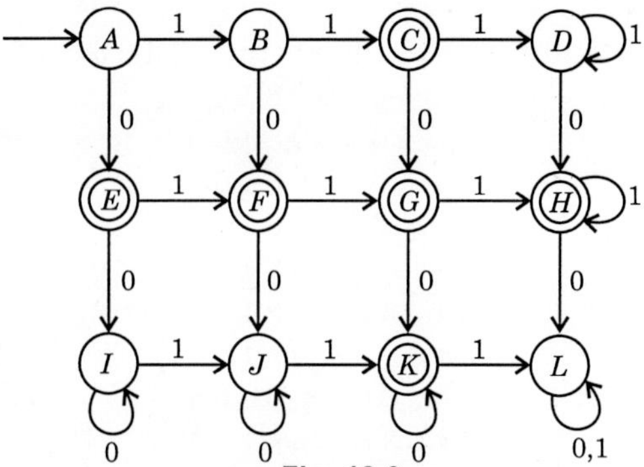

Fig. 12.6.

The finite state machine would look like this Fig. 12.6.

EXAMPLE 12.8. Design a machine which checks whereas a given decimal number is "EVEN".

Solution: There may be two states one for "EVEN" and second one for "ODD". The symbols 0, 2, 4, 8..... indicate that they are "EVEN" numbers because by "2" we get remainder value "0" otherwise it can be considered as "ODD" number. State table for this can be given by:

	State \ Input	0, 2, 4, 5, 8	1, 3, 5, 7, 9
Starting state	$\rightarrow q_0$	q_0	q_1
ODD	(q_1)	q_2	q_1
EVEN	q_2	q_2	q_1

To represent starting state we use '→', symbol, i.e. $\rightarrow q_0$ or q_0 and to represent final state we use '◎' symbol.

Now, we can take $\Sigma = \{0, 1, 2, 3, 4, 5, 6, 7, 8, 9\}$

$Q = \{q_0, q_1, q_2\}$

δ = The transition of states

1_0 = start state

$F = \{q_2]$

M.A.F. (Making function) is given as

Q \ **Σ**	**0, 2, 4, 6, 8**	**1, 3, 5, 7, 9**
q_0	q	0
q_1	q	0
q_2	0	0

1 → even and 0 → odd

Transition diagram will be

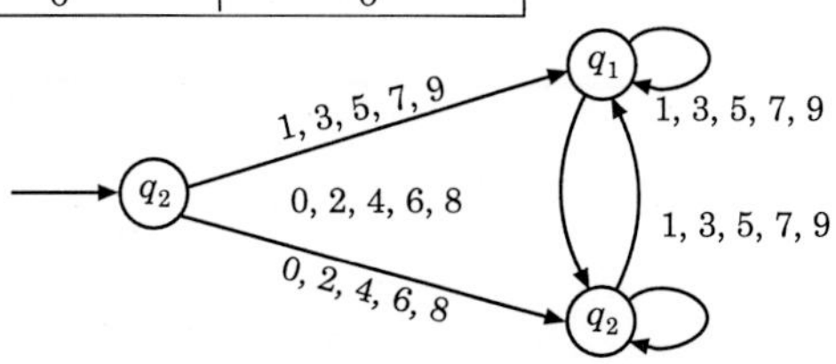

Fig. 12.7.

Suppose the number is 1 2 4 6 3 6 9 8 1 0 5

$\rightarrow$ q_0 1 2 4 6 3 6 9 8 1 0 5
1 q_1 2 4 6 3 6 9 8 1 0 5
1 2 q_2 4 6 3 6 9 8 1 0 5
1 2 4 q_2 6 3 6 9 8 1 0 5
1 2 4 6 q_2 3 6 9 8 1 0 5
1 2 4 6 3 q_1 6 9 8 1 0 5
1 2 4 6 3 6 q_2 9 8 1 0 5
1 2 4 6 3 6 9 q_1 8 1 0 5
1 2 4 6 3 6 9 8 q_2 1 0 5
1 2 4 6 3 6 9 8 1 q_1 0 5
1 2 4 6 3 6 9 8 1 0 q_2 5
1 2 4 6 3 6 9 8 1 0 5 $\boxed{q_1}$

As per end up in the state q_1 which is a non-final state therefore the given number is ODD.

EXAMPLE 12.9. Design a machine which can check that given string is divisible by '4'.

Solution: To chek divisibility by '4' there will be four states with remainder '0', remainder '*q*' remains '2' and remainder '3' and one more state for starting state.

M.A.F. (Machine function) is given as:

Q \ Σ	0, 4, 8	1, 5, 9	2, 6	3, 7
S	q_0	q_1	q_1	q_3
Rem '0' q_0	q_0	q_1	q_2	q_3
Rem '*q*' q_1	q_2	q_3	q_0	q_1
Rem '0' q_2	q_0	q_1	q_2	q_3
Rem '0' q_3	q_2	q_3	q_0	q_1

> ✍ **Note:** We can merge two states by one, if both states have some transition and both of them are final or non-final states.

In above table row-q_1 and row-q_3 both are identical and both of them are non final states. So we can replace state q_3 by q_1.

Now, new table can be represented by

Q \ Σ	0, 4, 8	1, 5, 9	2, 6	3, 7
Start (s)	q_0	$q_{1.3}$	q_2	$q_{1,3}$
q_0	q_0	$q_{1,3}$	q_2	$q_{1,3}$
$q_{1,3}$	q_2	$q_{1,3}$	q_0	$q_{1,3}$
q_2	q_0	$q_{1,3}$	q_2	$q_{1,3}$

We can observe that two columns in a above table are some, i.e., for 1, 5, 9 and 3.7. Hence we can group them into one.

	Q \ E	0, 4, 8	1, 3, 5, 7, 9	2, 6
Start	(S)	q_0	$q_{1,3}$	q_2
	q_0	q_0	$q_{1,3}$	q_2
	$q_{1,3}$	q_2	$q_{1,3}$	q_0
	q_2	q_0	$q_{1,3}$	q_2

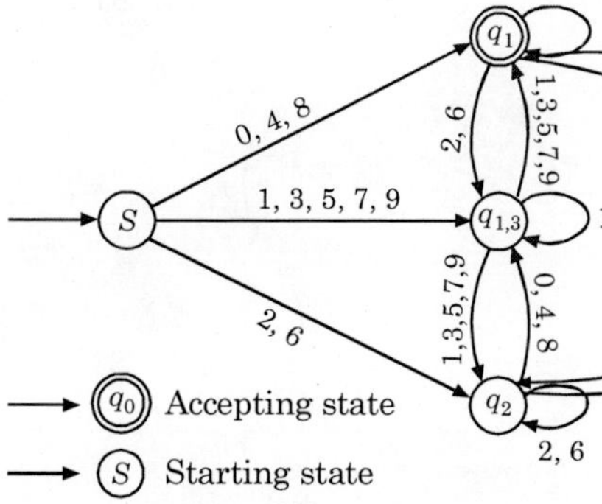

Fig. 12.8.

EXAMPLE 12.10. Design a machine which can check that given binary input string is divisible by '3'.

Solution: In binary number when we want to find the corresponding value receiving a new bit then remembers the original value is doubled and new bit is added. In start state if we receive '0' then we go to state where reaminder is '0', if we get '1' then the next state is remainder '*q*'.

Formula: $2 \times$ odd remainders + next bit

Q \ Σ	0	1
Start → (S)	q_0	q_1
(rem '0') q_0	q_0	q_1
(rem '1') q_1	q_0	q_1
(rem '2') q_2	q_0	q_1

In state q_0: 2×0 (rem '0') + 0 (next bit)

$= 0 + 0 = q_0$

$= 2 \times 0$ (rem '0') + 1 (next bit)

$= 0 + 1 = 1$

In state q_1: 2×1 (rem '1') + 0 (next bit)

$= 2 + 0 = 2$

$= 2 \times 1$ (rem '1') + 1 (next bit)

$= 2 + 1 = 3$

3 is divisible by '3' and remainders is '0', i.e.m state 'q_0'.

In state q_2: 2×2 (old remainder) + 0 (next bit)

= 4 + 0
= 4 (remainder 'q', i.e., i_1)
= 2 × 2 (old remainder '2') + 1 next bit
= 4 + 1
= 5 (remainder '2', i.e., state q_2)

Transition diagram is given as

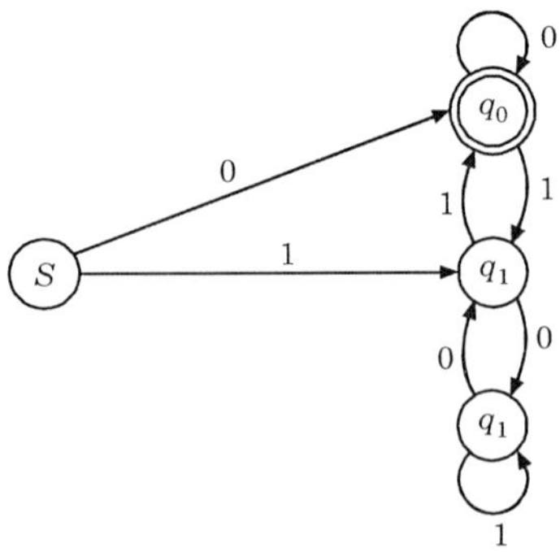

Fig. 12.9.

EXAMPLE 12.11. F.S.M. which can accept strings those can have odd number of 0's and any number of 1's.

Solution: We require only two states

$q_0 \rightarrow$ For even no. of zero's
$q_1 \rightarrow$ For odd no. of zero's

	Q \ Σ	0	1
Start	(s)	q_1	q_0
even '0'	q_0	q_1	q_0
add '0'	q_1	q_0	q_1

Transition diagram is

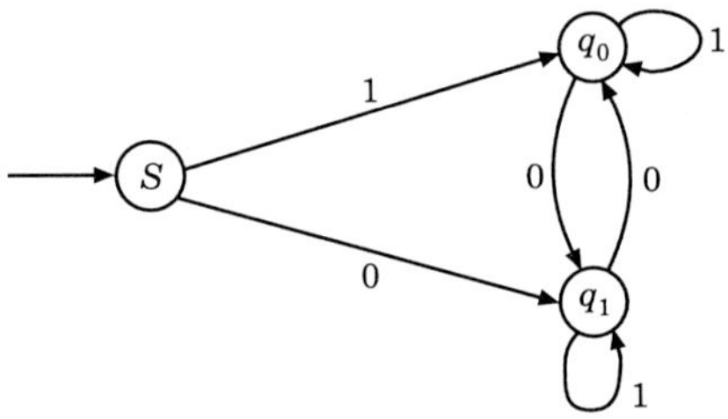

Fig. 12.10.

EXAMPLE 12.12. Design F.S.M. which can accept string contains odd number of 0's and even number of 1's.

Solution: We are consider four states.

q_0 even '0' even '1'

q_1 even '0' odd '1'
q_2 odd '0' even '1'
q_3 odd '0' odd '1'
Therefore, we accepting state is q_2.

	Q \ Σ	0	1
Start	→ (S)	q_2	q_1
Even '0' Even '1'	q_0	q_2	q_1
Even '0' odd '1'	q_1	q_3	q_0
odd '0' even '1'	q_2	q_0	q_3
odd '0' odd '1'	q_1	q_1	q_2

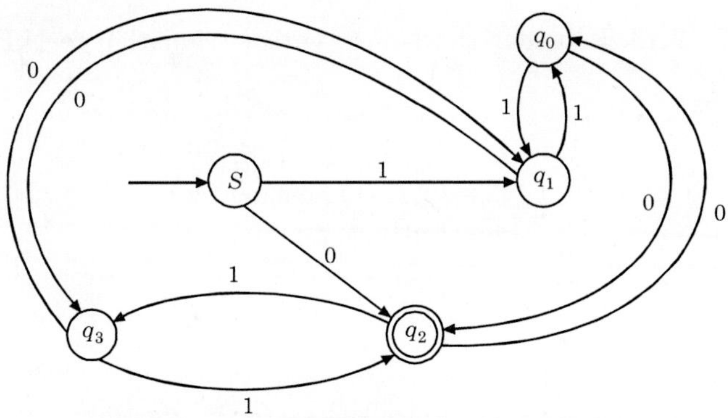

Fig. 12.11.

12.2. N.F.A. WITH ε-TRANSITIONS

If a finite automata is transitives without inputs symbols, then we get a NFA with NULL transitions, because the transition without symbols are called as ε-transitions. Consider the NFA given by following diagram.

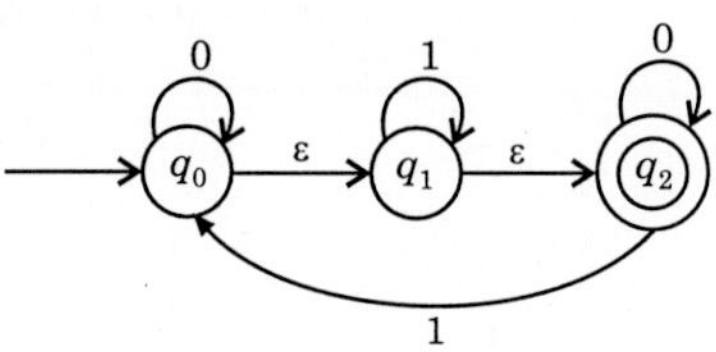

Fig. 12.12.

This is a NFA with ε-transitions because it is possible to make transition from state q_0 to q_1 without consuming any of the input symbol. A finite Automata, NFA with ε-transition will also be denoted as five tuple.

$$(Q, \Sigma, \delta, q_0, F)$$

where, Q = Set of finite states

Σ = Finite input symbol

q_0 = starting state
δ = transition function as mapping
F = Set of final state

and the transition δ can be denoted as

$$\boxed{Q \times (S \cup \varepsilon) \text{ to } 2^Q}$$

12.3. ACCEPTANCE OF A STRING BY THE N.F.A. WITH ε-TRANSITIONS

A string w in $\sum^*$ will be accepted by N.F.A., with ε-transition, if there exist at least on path corresponding to w, which starts in an initial state, and ends in one of the final states. Since this path may formed by ε-transition as well non-ε-transition to find out whether 'w' is accepted or not by the N.F.A. with ε moves, we require to define a function ε-closure (q), where q is a state of the automata. The function ε-closure (q) is defined as follows :

ε-Closure (q) = Set of all these state of the automata (N.F.A. with ε-transition), which can be reached from 'q' on a path labeled by 'ε', *i.e.*, without consuming any input symbol.

EXAMPLE 12.13. Conisder the N.F.A. with ε-transition to N.F.A. without ε-transitions.

$(Q, \Sigma, \delta, q_0, F)$
$Q = \{q_0, q_1, q_2\}$
$\Sigma = \{a, b, c\}$ and ε moves
q_0 = Initial state
$F = \{q_2\}$

Transition diagram is given by,

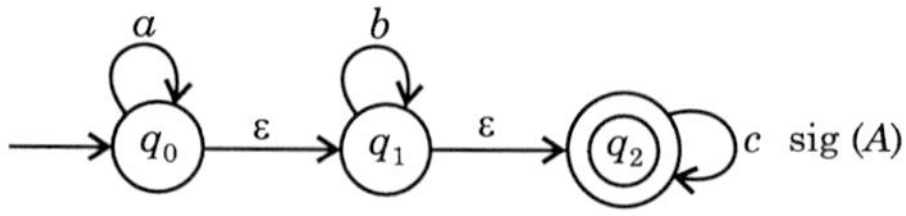

Fig. 12.13.

and Transition Table is given as,

Q \ Σ	a	b	c	ε
q_0	q_0	ϕ	ϕ	q_1
q_0	ϕ	q_1	ϕ	q_2
q_0	ϕ	ϕ	q_2	ϕ

Design of a N.F.A. without ε-move.

Solution: First find the states of N.F.A. without ε-transition including initial state and final states as follows :

Rule 1:

(*a*) The language accepted by the above N.F.A. with ε-transition is the set of strings over $\{a, b, c\}$ including by any member of b's and followed by any number of c's.

(*b*) There will be the same number of states but the names can be constructed by writing the state name as the set of states in the ε-closure.

(*c*) Initial state will be ε-closure of initial state of N.F.A. with ε-transition in figure as ε-closure $(q_0) = \{q_0, q_1, q_2\}$
Now initial state of N.F.A. without ε-transition

(*d*) Rest of the states are

$$\varepsilon\text{-closure } (q_1) = \{q_1, q_2\} \text{ New state}$$
$$\varepsilon\text{-closure } (q_2) = \{q_2\} \text{ New state}$$

(*e*) The final states of N.F.A. without ε-transition are all these new states which contain final states of N.F.A. with ε-transition as number.

So $\{q_0, q_1, q_2\}, \{q_1, q_2\}, \{q_2\}$ are all final states. Therefore, if N.F.A. without ε-transition is,

$$M' = (Q', \Sigma, \delta', q_0, F'\}$$
$$Q' = (\{q_0, q_1, q_2\}, \{q_1, q_2\}, \{q_2\})$$
$$0' = \{q_0, q_1, q_2\}$$
$$F' = (\{q_0, q_1, q_2\}, \{q_1, q_2\}, \{q_2\}$$

Rule 2:

Now we will write the function of δ′. Now, consider each old machine transition in each row, we can ignore 'ϕ' entires and ε-transition column. In the old machine $\delta(q_0, a) = q_0$, thus in new machine, *i.e.*, in

$$M', \delta'(\{q_0, q_1, q_2\}, a) = \{q_0, q_1, q_2\}$$

This is just because the new machine accepts the same language as the old machine and must at least have the same transition for the new state names. Therefore,

$$\begin{aligned}
\delta'(\{q_0, q_1, q_2\}, a) &= \varepsilon\text{-closure } (\delta\{q_0, q_1, q_2\}, a) \\
&= \varepsilon\text{-closure } (\delta(q_0, a) \cup \delta(q_1, a) \cup \delta(q_2, a)) \\
&= \varepsilon\text{-closure } (q_0 \cup \phi \cup \phi) \\
&= \varepsilon\text{-closure } (q_0) \\
&= \{q_0, q_1, q_2\} \\
\delta'(\{q_0, q_1, q_2\}, b) &= \varepsilon\text{-closure } (\delta\{q_0, q_1, q_2\}, b) \\
&= \varepsilon\text{-closure } (\delta(q_0, b) \cup \delta(q_1, b) \cup \delta(q_2, b)) \\
&= \varepsilon\text{-closure } (\phi \cup q_1 \cup \phi) \\
&= \varepsilon\text{-closure } (q_1) \\
&= \{q_1, q_2\} \\
\delta'(\{q_0, q_1, q_2\}, c) &= \varepsilon\text{-closure } (\delta\{q_0, q_1, q_2\}, c) \\
&= \varepsilon\text{-closure } (\delta(q_0, c) \cup \delta(q_1, c) \cup \delta(q_2, c)) \\
&= \varepsilon\text{-closure } (\phi \cup \phi \cup q_2) \\
&= \varepsilon\text{-closure } (q_2) = \{q_2\}
\end{aligned}$$

Similarly, we can write

$$\begin{aligned}
\delta'(\{q_1, q_2\}, a) &= \varepsilon\text{-closure } (\delta\{q_1, q_2\}, a) \\
&= \varepsilon\text{-closure } (\delta(q_1, a) \cup \delta(q_2, a)) \\
&= \varepsilon\text{-closure } (\phi \cup \phi) \\
&= \varepsilon\text{-closure } (\phi) = \phi
\end{aligned}$$

$$\begin{aligned}
\delta'(\{q_1, q_2\}, b) &= \varepsilon\text{-closure } (\delta \{q_1, q_2\}, b) \\
&= \varepsilon\text{-closure } (\delta(q_1, b) \cup \delta(q_2, b)) \\
&= \varepsilon\text{-closure } (q_1 \cup \phi) \\
&= \varepsilon\text{-closure } (q_1) = \{q_1, q_2\} \\
\delta'(\{q_1, q_2\}, c) &= \varepsilon\text{-closure } (\delta \{q_1, q_2\}, c) \\
&= \varepsilon\text{-closure } (\delta(q_1, c) \cup \delta(q_2, c)) \\
&= \varepsilon\text{-closure } (\phi \cup q_2) \\
&= \varepsilon\text{-closure } (q_2) \\
&= \{q_2\} \\
\delta'(q_2, a) &= \varepsilon\text{-closure } (\delta(q_2, a)) \\
&= \varepsilon\text{-closure } (\phi) = \phi \\
\delta'(q_2, b) &= \varepsilon\text{-closure } (\delta(q_2, b)) \\
&= \varepsilon\text{-closure } (\phi) = \phi \\
\delta'(q_2, c) &= \varepsilon\text{-closure } (\delta(q_2, c)) \\
&= \varepsilon\text{-closure } (q_2) = \{q_2\}
\end{aligned}$$

12.4. TYPE OF STATES IN FINITE STATE AUTOMATA

(*a*) **Dead State:** All those non-final states which transit to itself for all input symbols in Σ, are called **Dead State.**

Let $\Sigma = \{0, 1\}$

then q is said to be dead state iff q is not the number of 'F'.

i.e., $\delta(q, 0) = q$

$\delta(q, 1) = q$

(*b*) **Inaccessible States:** All those states which can never be reached from initial state are called **Inaccessible States.**

For example,

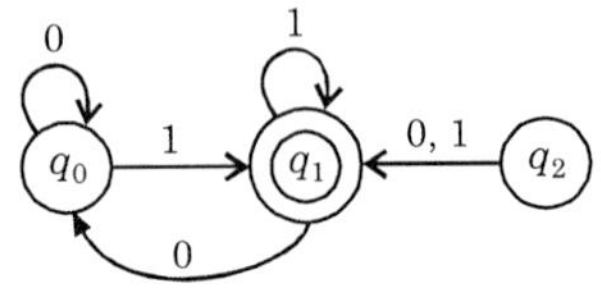

Fig. 12.14.

In the given diagram 'q_2' state is an inaccessible state because we can not reach at state q_2 from initial q_0.

(c) **Indistinguishable States:** Two states q_1, q_2 of D.F.A. are called **Indistinguishable** if

$$\hat{\delta}(q_1, w) \in F \Rightarrow \hat{\delta}(q_2, w) \in F$$

$$\hat{\delta}(q_1, w) \notin F \Rightarrow \hat{\delta}(q_2, w) \notin F$$

and all $\omega \notin \Sigma^*$.

So the transition table for N.F.A. without ε-transition will be as follow:

Q \ Σ	*a*	*b*	*c*
$\{q_0, q_1, q_2\}$ $\{q_1, q_2\}$ $\{q_2\}$	$\{q_0, q_1, q_2\}$ ϕ ϕ	$\{q_1, q_2\}$ $\{q_1, q_2\}$ ϕ	$\{q_2\}$ $\{q_2\}$ $\{q_2\}$

Let $\{q_0, q_1, q_2\}$ as q_A

$\{q_1, q_2\}$ as q_B

$\{q_2\}$ as q_C

So the transition diagram can be represented as,

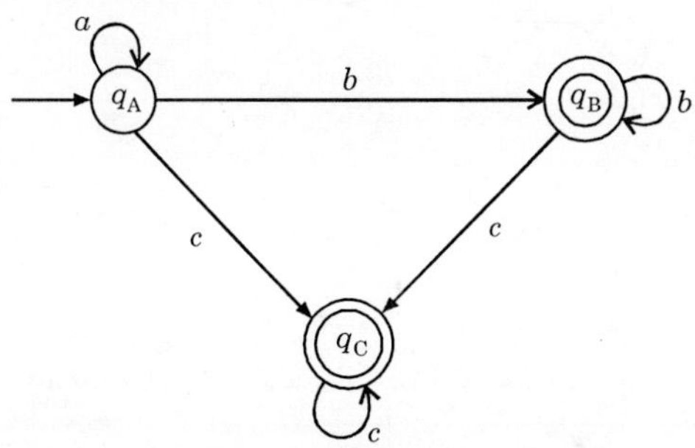

Fig. 12.15.

Distinguishable States: Two states q_1 and q_2 of D.F.A. are called distinguishable iff

$$\hat{\delta}(q_1, w) \in F \quad \text{and} \quad \hat{\delta}(q_1, w) \notin F$$

or Vice-versa for all $w \notin \Sigma^*$.

12.5. MINIMIZATION OF FINITE-AUTOMATA

To minimize a finite automata we have to following steps:

1. First we create two groups of states in which one group is called non-final state and other one is called final states.
2. Now we will draw transition table if transition diagram is given.
3. Try to create now groups as per step '1'.
4. Step '3' is totally based on equivalent phenomenon, i.e., for a particulars states 'q_1' and 'q_2' can be 0's equivalent if both have some transition state belong to one group for input symbol '0' and 1's equivalent if both have transition state belong to any one group. If state q_1, have zero's transition belong to one group and q_2 states zero's transition belong to another group, on that time path q_1, q_2 can not be grouped together same we will check for its equivalent.
5. We now create a new partition.
6. This process terminate when current partition is equivalent to previous partition.

EXAMPLE 12.14. Construct a minimum state automata equivalent to the finite automation given in the figure.

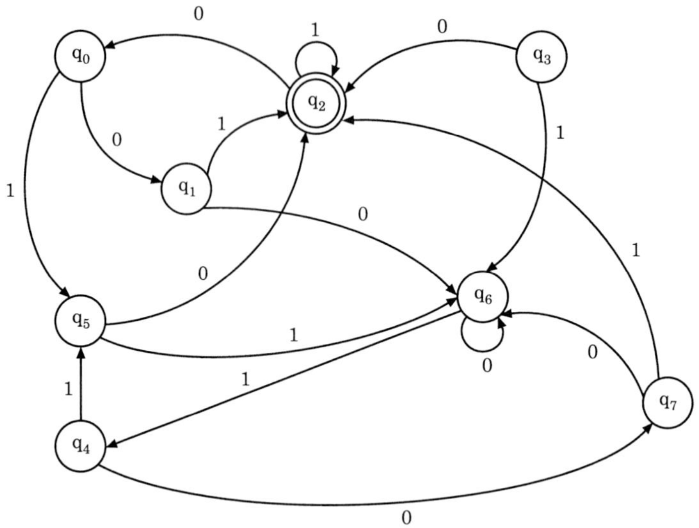

Fig. 12.16.

Solution: First we design transition table for given transition diagram:

Q \ Σ	**0**	**1**
q_0	q_1	q_5
q_1	q_6	q_2
q_2	q_0	q_1
q_3	q_2	q_6
q_4	q_7	q_5
q_5	q_2	q_6
q_6	q_6	q_4
q_7	q_6	q_2

Now we will create two groups one for all final states and another one for all non-final states, *i.e.*,

$$Q_1^0 = F = \{q_2\}$$

$$Q_2^0 = Q - Q_1^0$$

$$= q_0, q_1, q_3, q_4, q_5, q_6, q_7\}$$

So first partition, $$\pi_0 = \left\{\frac{\{q_2\}}{A}, \frac{\{q_0, q_1, 1_3, q_4, q_5, q_6, q_7\}}{B}\right\}$$

Now we check group '*A*' can be further partitioned for input '0' and '1's. We see that for 0's column q_2 belong to group 'B' and for 1's column q_2 belong to same group, therefore we can not further partitioned, because there is only one element in group '*A*'. Now we try to partition for group '*B*'.

If we get two states q_0, q_1. from group '*B*' and we check that transtition of booth, have same group for input 0's i.e., group 'B' it means till now both belong to same group but if we check for input q's, q_0 belong to group 'B' but q_1

belongs to group 'A'. Therefore both states q_0, q_1, cannot be belong to same group. Therefore, here parition is possible.

Now we check q_0 with q_3, q_4, q_5, q_6, q_7 that is they have same group as partition is possible. If we compare 0's column for q_0 with q_3, q_4, q_5, q_6, q_7, we see that q_3, q_5 belong to group '*A*' and q_4, q_6, q_7 belong to group '*B*'. We again compare 1's column for q_0 with q_3, q_4, q_5, q_6, q_7, we see that q_3, q_4, q_5, q_6 belong to one group '*B*' and state q_7 belongs to another group say group 'A'. This is clear that there is a partition for group '*B*' like $\{q_0, q_4, q_6\}$ $\{q_3, q_5\}$ and $\{q_1, q_7\}$.

Therefore, $\pi_1 = \{q_2\}, \{q_0, q_4, q_6\}, \{q_1, q_7\}, \{q_3, q_5\}$

We can say that,

$$Q_1^1 = \{q_2\}$$

$$Q_2^1 = \{q_0, q_4, q_6\}$$

$$Q_3^1 = \{q_1, q_7\}$$

$$Q_4^1 = \{q_3, q_5\}$$

Now, we can check that there is another partition among group *A*, *B*, *C*, D.

$$\pi_1 = \left\{\frac{\{q_2\}}{A}, \frac{\{q_0, q_4, q_6\}}{B}, \frac{\{q_1, q_7\}}{C}, \frac{\{q_3, q_5\}}{D}\right\}$$

As we know that group 'A' has only one element 'q_2' therefore it's partition is not possible. We will try to partition for group '*B*' in which there are three elements $\{q_0, q_4, q_6\}$.

Now we check q_0 with q_4, q_6 with 0's column, that all elements have same group when input symbol is '0'. q_0 goes to group '*C*' and q_4 goes to group '*C*' but q_6 goes to group '*B*' and same for input symbol '1', q_0 and q_4 belong to same group '*D*' but q_6 goes to group '*B*'. Therefore q_0, q_4 will be in different group and 'q_6' will be in different group. Same we will check for $\{q_1, q_7\}$ and $\{q_3, q_5\}$ there is no partition.

$$Q_1^2 = \{q_2\}$$

$$Q_2^2 = \{q_0, q_4\}$$

$$Q_3^2 = \{q_6\}$$

$$Q_4^2 = \{q_1, q_7\}$$

$$Q_5^2 = \{q_3, q_5\}$$

Therefore,

$$\pi_2 = \left\{\frac{\{q_2\}}{A}, \frac{\{q_0, q_4\}}{B}, \frac{\{q_6\}}{C}, \frac{\{q_1, q_7\}}{D}, \frac{\{q_3, q_5\}}{E}\right\}$$

Now, we will again try to partition for group '*A*', '*B*', '*C*', '*D*' and '*E*'. There are no possibilities for again partition for group '*A*', '*B*', '*C*', *D*' and '*E*'.

Therefore $\pi_3 = \{\{q_2\}, \{q_0, q_4\}, \{q_6\}, \{q_1, q_7\}, \{q_3, q_5\}\}$

as $\pi_3\ \pi_2 = \pi_2$ gives the equivalences classes, the minimum state automation is

$$M' = (Q', \Sigma, \delta', q_0', F')$$

$$Q' = \{\{q_2\}, \{q_0, q_4\}, \{q_6\}, \{q_1, q_7\}, \{q_3, q_5\}\}$$

$$q_0' = \{q_0, q_4\}$$

$$F' = \{q_2\}$$

and δ′ is defined as,

Q \ Σ	0	1
$\{q_0, q_4\}$	$\{q_1, q_7\}$	$\{q_3, q_5\}$
$\{q_1, q_7\}$	$\{q_6\}$	$\{q_2\}$
$\{q_3, q_5\}$	$\{q_2\}$	$\{q_6\}$
$\{q_6\}$	$\{q_6\}$	$\{q_0, q_4\}$
$\{q_2\}$	$\{q_0, q_4\}$	$\{q_2\}$

The transition diagram for minimum state automation is,

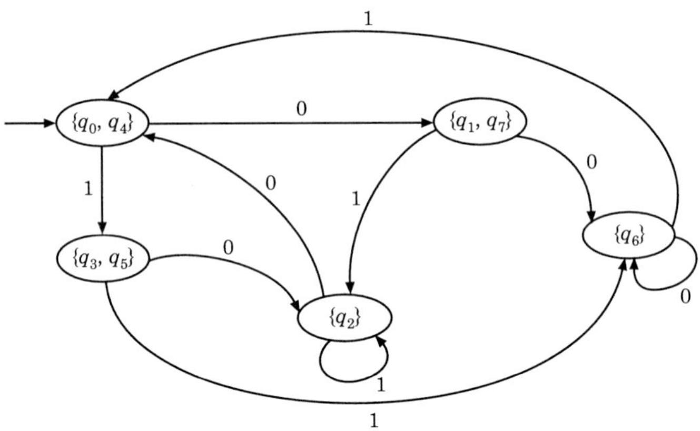

Fig. 12.17.

12.6. FINITE-AUTOMATA WITH OUTPUT

When it is required to give output which is not just "accept" or "reject". We will be required to associate the required output along with each input dependent on the state of mind. Such an output can be given during the transition, *i.e.*, in a particular state q_i on receiving input '1' we may switch to stte q_j and output is '0'. Only the transition is carried out and whenever a state is reached that state will result in same output. There are two such machines which result in output string and two different ways as stated earlier.

(a) Moore Machine : A Moore machine is a collection of five things, which are

1. A finite set of states q_0, q_1, q_2 ... when q_0 is considered as a start state.
2. An alphabet of letters for following the input string, *i.e.*,
$$\Sigma = \{abc \ ...\} \text{ or } \Sigma = 0110 \ .... \}.$$
3. An alphabet of possible output characters
$$\tau = \{x, y, z \ ....\}$$
4. A transition table that shows for each state and each input letter what state is reached next
5. An output table that shows what character τ is printed by each state as it is entered.

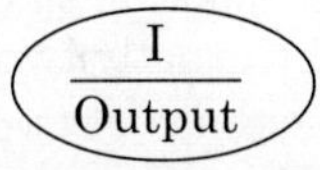

(b) Mealy Machine: A mealy machine is a collection of four things.

1. A finite set of states q_0, q_1, q_2 ... where q_0 is considered as state state.
2. An alphabet of letters $\Sigma = \{a, b \ ....\}$ for forming input string.
3. An alphabet of output character $\tau = \{x, y, z \ ....\}$.
4. Each edge is labelled with a compound symbol of the form Input/Output where q is an input letter and '0' is an output character. Every state must have exactly one outgoing edge for each possible input letter.

EXAMPLE 12.15. Design a Moore machine to get q's complement of given binary number.

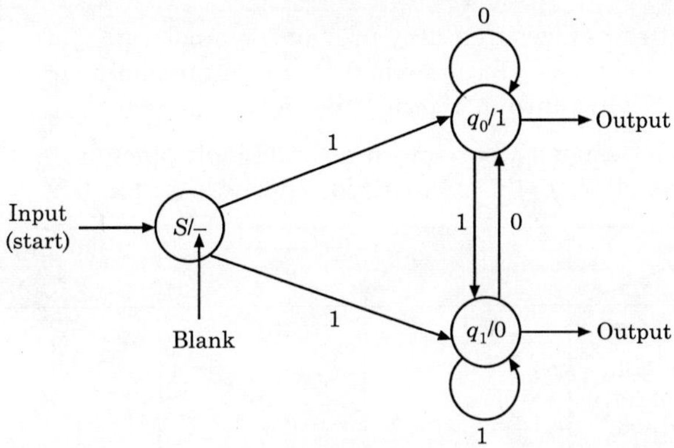

Fig. 12.18.

$q_0 : 0 \rightarrow 1$
$1_1 : 1 \rightarrow 0$

Σ / Q	0	1	Output
S	q_0	q_1	—
q_0	q_0	q_1	1
q_1	q_0	q_1	0

As every state is associated with it's output.

> ✍ Note Down that
> (*i*) States give output is Moore machine.
> (*ii*) Input is associated with output in Mealy machine.

EXAMPLE 12.16. Design a mealy machine to get q's complement of the given number.

Solution:

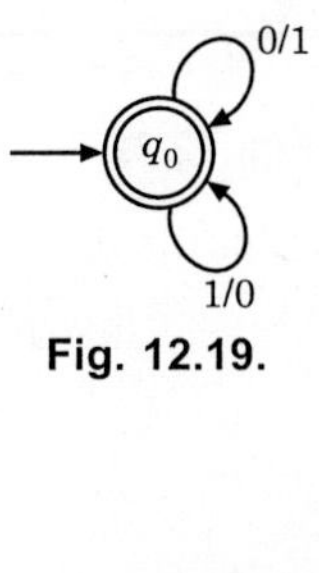

Fig. 12.19.

EXAMPLE 12.17. Design a Moore machine that gives an ouput '1' if input of binary sequence a '1' is preceded by exactly two zero's.

Solution: We should check the sequence of two zero's followed by '1', *i.e.*, for 001 output is '1' but for 0001 output is '0'.

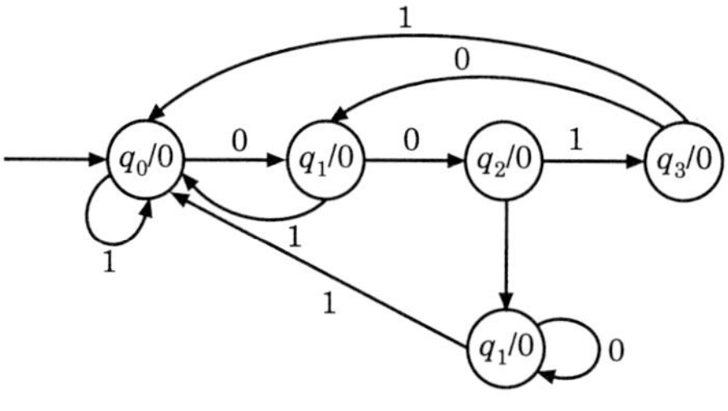

Fig. 12.20.

The state $q1$ recognizes that only previous symbol is '0', q_2 recognizes only previous two symbols are '0' thus when in state q_2 machine encounters '1' it gives to state q_3 giving output '1' otherwise output is zero.

EXAMPLE 12.18. Design a Moore machine and Mealy machine for binary input sequence, output 'A' if '101' is recognized otheriwse output 'B'.

Σ / Q	**0**	**1**	**Output**
end '0' q_0	q_0	q_1	B
end '1' q_1	q_2	q_0	B
end '1_0' q_2	q_0	q_3	B
end '101' q_3	q_2	q_1	A

Moore machine:

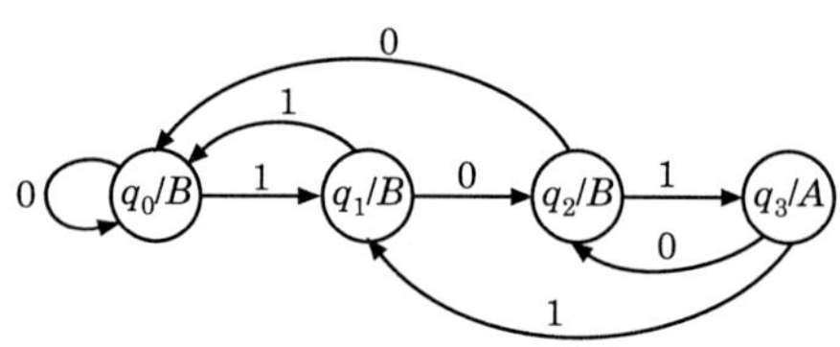

Fig. 12.21.

Mealy machine:

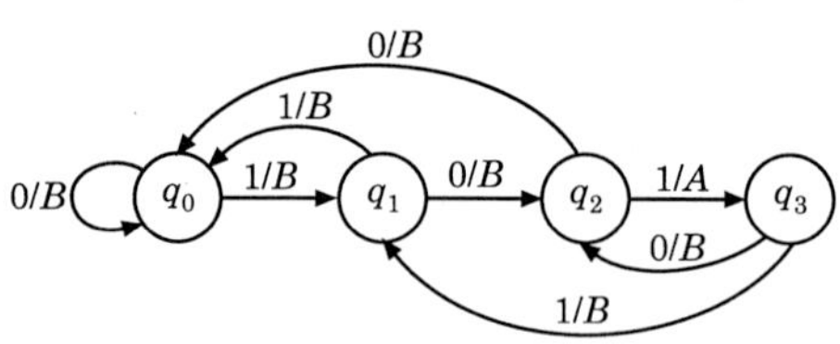

Fig. 12.22.

EXAMPLE 12.19. Design a Mealy machine for the following table and also find the output for the string "*abbabaaa*".

Solution:

Q \ Σ	*a*	*b*	Output
q_0	q_3	q_2	0
q_1	q_1	q_0	0
q_2	q_2	q_3	1
q_3	q_3	q_1	0

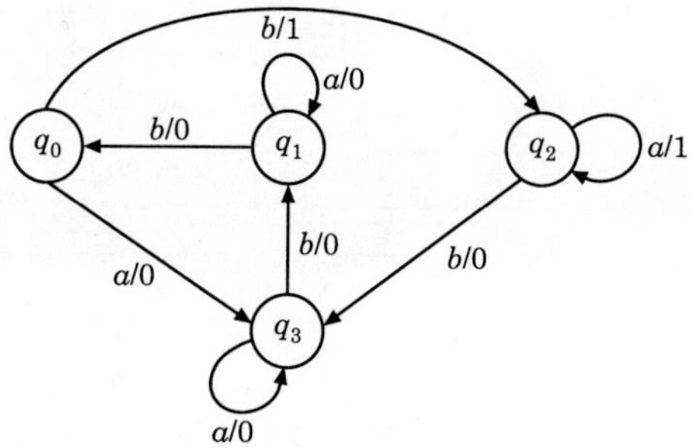

Fig. 12.23.

Input string is "*abbabaaa*"

$\boldsymbol{q_0}$ abba baaa

a $\boldsymbol{q_3}$ bbabaaa output → 0

ab $\boldsymbol{q_1}$ bba baaa output → 0

abb $\boldsymbol{q_0}$ abaaa output → 0

abba $\boldsymbol{q_3}$ baaa output → 0

abbab $\boldsymbol{q_0}$ aaa output → 0

abbaba $\boldsymbol{q_1}$ aa output → 0

abbabaa $\boldsymbol{q_1}$ aa output → 0

abbabaaa $\boldsymbol{q_1}$ → 0 output → 0

abbabaaa $\boldsymbol{q_1}$ → 0

The output sequence is

0 0 0 0 0 0 0 0

12.7. CONVERSION

(*a*) Moore machine to Mealy Machine

EXAMPLE 12.20. Design a mealy machine which is equivalent to Moore machine given in the table:

Present state	Next state		Output
	a = 0	*a* = 1	
→ q_0	q_3	q_1	0
q_1	q_1	q_2	1
q_2	q_2	q_3	0
q_3	q_3	q_0	0

Solution: In Moore machine every input symbol (a) we form a pair consisting of the next state table and the corresponding output and reconstruct the table for Mealy machine for example the states q_0 and q_1 in the next state column should be associated with output 0 and 1.

The transition table for Mealy machine will be

Present state	Next State			
	$a = 0$		$a = 1$	
	Next state	Output	Next state	Output
$\rightarrow q_0$	q_3	0	q_1	1
q_1	q_1	1	q_2	0
q_2	q_2	0	q_3	0
q_3	q_3	0	q_3	0

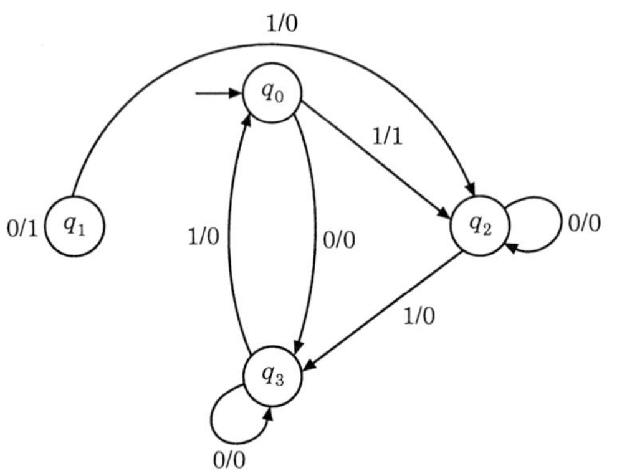

Fig. 12.24. Mealy Machine

EXAMPLE 12.21. Design Mealy machine for correspondence Moore machine which is given in tabular form.

Present state	Next state		Output
	$a = 0$	$a = 1$	
$\rightarrow q_1$	q_1	q_2	0
q_2	q_1	q_3	0
q_3	q_1	q_3	1

Moore Machine

Solution: Transition table for Mealy machine.

Present state	Next State			
	$a = 0$		$a = 1$	
	Next state	Output	Next state	Output
$\rightarrow q_1$	q_1	0	q_2	0
q_2	q_1	0	q_3	1
q_3	q_1	0	q_3	1

Since states q_2 and q_3 are identical, so we can replace one row by another. Here we are replacing row q_3 by q_2. So the transition table for Mealy machine is given by:

Present state	Next State			
	$a = 0$		$a = 1$	
	Next state	Output	Next state	Output
$\rightarrow q_1$	q_1	0	q_2	0
q_2	q_1	0	q_2	1

Transition diagram,

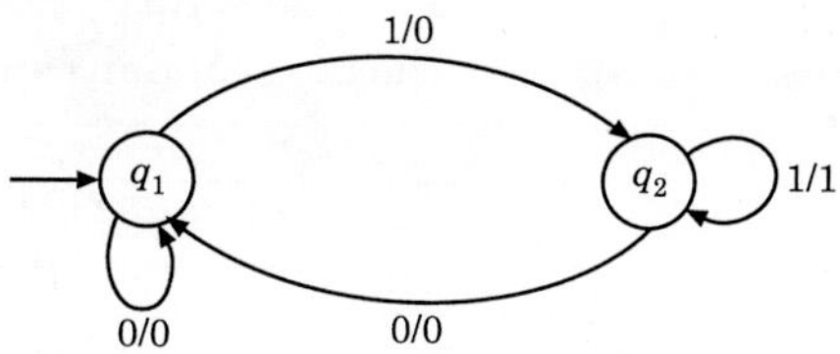

Fig. 12.25.

(*b*) Mealy Machine to Moore Machine

EXAMPLE 12.22. Consider a Mealy machine described by the transition table, construct a Moore machine which is equivalent to the Mealy machine :

Present state	Next State			
	Input $a = 0$		Input $a = 1$	
	Next state	Output	Next state	Output
q_1	q_3	0	q_2	0
q_2	q_1	1	q_4	0
q_3	q_2	1	q_1	1
q_4	q_4	1	q_3	0

Mealy Machine

Solution: We look into the next state column any state, say q_i and determine the number of different outputs associated with q_i in that column we break q_i into several different states, the number of such states being equal to the number of different outputs associated with q_i, for example q_2 is associated with two outputs 0 and 1 and state q_3 is associated only one output 1. So, we will break state q_2 into two parts, first are for output 0 and second one for output 1, but we cannot split state q_3 because q_3 have only one output. Similarly we can break state q_4 put state q_1 cannot be break down because q_1 have only one output. So, new transition table is–

Present state	Next State			
	Input $a = 0$		Input $a = 1$	
	Next state	Output	Next state	Output
q_1	q_3	0	q_{20}	0
q_{20}	q_1	1	q_{40}	0
q_{21}	q_1	1	q_{40}	0
q_3	q_{21}	1	q_1	1
q_{40}	q_{41}	1	q_3	0
q_{41}	q_{41}	1	q_3	0

State q_{20} means state q_2 with output 0 and state q_{21} means q_2 with output 21. The pair of states and outputs in the next column can be arranged as in the new table–

Present state	Next state		
	$a = 0$	$a = 1$	Output
$\rightarrow q_1$	q_3	q_{20}	1
q_{20}	q_1	q_{40}	0
q_{21}	q_1	q_{40}	1
q_3	q_{21}	q_1	0
q_{40}	q_{41}	q_3	0
q_{41}	q_{41}	q_3	1

> ✍ **Note :** If we have m-outputs, n-state, Mealy machine the corresponding m-output. Moore machine has no more than $mn + 1$ states.

Wer observe that the initial state q_1 i s assocated with output '1'. This means that with input '$\wedge$' (Null). We get an output 'q'. If the machine starts at state q_1. Thus the Moore machine accepts a zero length sequence (null sequence) which is not accepted by the Mealy machine. To overcome this situation, either we must neglect the response of a Moore machine to input '$\wedge$', or we must add a new starting state q_0, whose transition are some as state q_1 but output will be '0'.

So final transition table is–

Present state	Next state		Output
	$a = 0$	$a = 1$	
$\rightarrow q_0$	q_3	q_{20}	0
q_1	q_3	q_{20}	1
q_{20}	q_1	q_{40}	0
q_{21}	q_1	q_{40}	1
q_3	q_{21}	q_1	0
q_{40}	q_{41}	q_3	0
q_{41}	q_{41}	q_3	1

Transition Diagram:

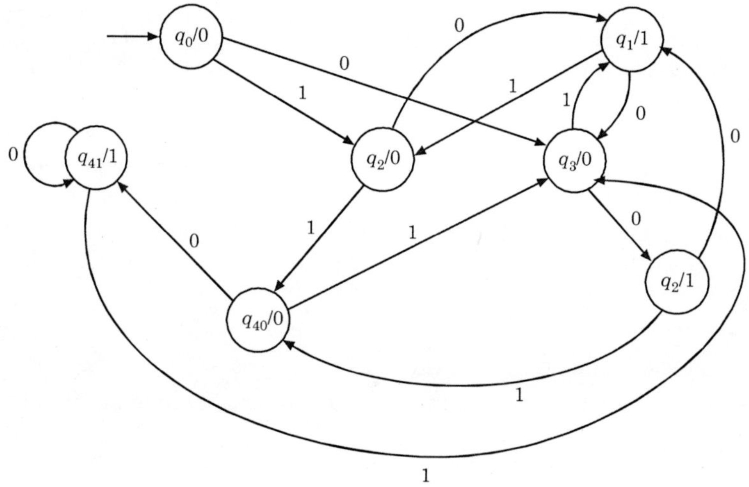

Fig. 12.26.

12.8. PROPERTIES OF FINITE STATE MACHINES

1. **Periodicity :** The fundamental limitation of an F.S.M. is that it does not have the capacity to remember arbitrarily large amount of information. This is so because it has only a fixed number of states and this sets a limit to the length of the sequence it can remember. This means that F.S.M. eventually will always repeat a state or produce a periodic sequence of states.
2. **State Determination :** When the initial state and input sequence is known we can find the current state of F.S.M.
3. **Equivalent Classes of Sequence :** One to periodicity F.S.M. partitions set of input sequence into equivalence classes. The equivalence class or two sequences are said to be equivalent if both of them when receivied in state q_0 take us to the state q_{ff}. Suppose w_1 and w_2 are two strings and we find that,

$$\delta(q_0, w_1) = q_f$$
$$\delta(q_0, w_2) = q_f$$

Then we can say w_1 and w_2 are equivalent.
4. **Impossibililty of Palindrome Recognisation :** We know that finite state machines does not have memory, therefore it can not be recognised the centre, Hence it is impossible to recognise a palindrome.
5. **Impossibility of Multiplication :** Multiplication of operation of long number cannot be carried out in F.S.M. Since full length sequences are to be remembered.
6. **Parenthesis, Tree Representation and Well-formedness :** F.S.M. can not check well formedness because we will have to remember the previous opening brackets.

12.9. GRAMMAR

Grammar is a set of formal rules which check the correctness of the sentences or we can construct the correct sentence using these rules.

These sentences are known as gammatical sentences and the rules as "Syntax" we are also aware that the machine (mic) is formal system which recognizes as well as generates the class of language which depends on its interim states.

Thus are formal languages and we want 'ti' have a relation between the language and the machine. The pioneer work was carried by "CHOMSKY"" in 1950's. The two aspects of the languages are to be considered.

1. Generative capacity
2. Grammatical capacity.

Generative capacity indicates that the grammar should generate all and only the sentences belonging to the languages where as Grammatical capacity gives the 'stress' on primitive structures which compose the sentence. The first will consider the consistency and completeness of the formal system. Whereas second deals with synthesis and analysis.

> ✍ **Note:** The Grammar will generate the strtings, starting for set of non-terminals.

12.10. FORMALIZATION

A grammar which deals with as "Phrase structure grammar".

It is a rule of 'G'.

$$G = \{N, T, P, S\}$$

where, $N \to A$ finite set of non-terminals

$T \to$ First set of terminals

$P \to$ Set of production rules

$S \to$ Start symbol.

> ✍ **Note :** It is context free grammar if the production of the form
> One non-terminal = Finite string of terminals and non-terminals.

The language generated by context free grammar is known as "context free language". The sequence of production used to derive the sentence is known as derivation.

EXAMPLE 12.23. Give the context free grammar which generates a string cotaining only 'a's.

Solution:

$$N = \{S\}$$

$$T = \{a\}$$

$$P = \{S \to a\ S,\ S \to \wedge\}$$

or

$$S \to aS \qquad \text{...(1)}$$

$$S \to \wedge \qquad \text{...(2)}$$

$$S = S$$

Answer : $S \to aS$

$S \to aa\ aS$

$S \to aaa\ aS$

$S \to aaaa\ \wedge$

$\boxed{S \to aaaa}$

EXAMPLE 12.24. Give the context free Grammar to generate a string containing 'a', 'b', in any sequence.

Solution: Let, $N = \{S\}$

$T = \{a, b\}$

$P = \{S \to aS, S \to bS, S \to a, S \to b\}$

$S = S$

or $S \to aS$ ...(1)

$S \to bS$ ...(2)

$S \to a$ ...(3)

$S \to b$ ...(4)

For string : *baabb*

$S \to b\ S$

$S \to b\ aS$

$S \to b\ aaS$

$S \to baa\ bS$

$S \to baa\ bb$

$\boxed{S \to baabb}$

EXAMPLE 12.25. Give the C.F.G. for generating an alternating sequence of 0'*s* and *q*'*s*.

SOLUTION: Production Rule

$S \to 0T$

$S \to qU$

$T \to q\text{U}$

$U \to 0T$

$T \to 1$

$U \to 0$

C.F.G. = $\{\{S, T, U\}, \{0, 1\}, \{S \to 0T, S \to 1U, T \to qU, U \to 0T, T \to 1, U \to 0\}. S\}$

For string 010101

$S \to 0T$ (1) $(S \to 0T)$

$\to 01U$ (3) $(T \to 1U)$

$\to 010T$ (4) $(U \to 0T)$

$\to 0101U$ (3) $(T \to 1U)$

$\to 01010T$ (4) $(U \to 0T)$

$\to 010101$ (5) $(T \to 1)$

For string 1010

$S \to 1\ U$ (2) $S \to 1U$

$S \to 10\ T$ (4) $U \to 0T$

$S \rightarrow 011U$ (3) $T \rightarrow 1U$

$S \rightarrow 1001$ (6) $U \rightarrow 0$

EXAMPLE 12.26. To generate the string in which non-consecutive 'b's can occur but only 'a's be consecutive.

Solution: C.F.G. = $\{\{S, T, U\}, \{a, b\}, \{S \rightarrow aS, S \rightarrow bT, T \rightarrow aS, S \rightarrow a, S \rightarrow b, T \rightarrow a\}, S = S\}\}$

$$\begin{aligned} S &\rightarrow a\,S && (S \rightarrow aS) \\ &\rightarrow a\,bT && (S \rightarrow bT) \\ &\rightarrow ab\,aS && (T \rightarrow aS) \\ &\rightarrow a\,b\,\boxed{a\,a} && (S \rightarrow a) \end{aligned}$$

EXAMPLE 12.27. Write C.F.G. for the string which donot contain 3-consecutive 'b's.

Solution: String donot contain 3-consecutive 'b's means that we can have maximum two consecutive 'b's.

$$N = (S, T, U)$$
$$T = (a, b)$$
$$S = S$$

Production Rules:

$$\begin{aligned} P: \{&S \rightarrow aS \\ &S \rightarrow a \\ &S \rightarrow b \\ &S \rightarrow bT \\ &T \rightarrow bU \\ &T \rightarrow a \\ &T \rightarrow b \\ &U \rightarrow aS \\ &U \rightarrow a\} \end{aligned}$$

Now

$$\begin{aligned} &S \rightarrow aS \\ &S \rightarrow ab\,T \\ &S \rightarrow abbU \\ &S \rightarrow abbaS \end{aligned}$$

$$\boxed{S \rightarrow a\ \boxed{b\,b}\ aa}$$

EXAMPLE 12.28. Write a C.F.G. which generates string having equal number of a's and b's.

Solution: C.F.G. = (V, Σ, R, S)

$$V = \{S\}$$
$$\Sigma = \{a, b\}$$
$$R = S \rightarrow aS\,bS \mid bS\,aS \mid \wedge$$

Derivation of strings $baa\ baa\ bb$

$$\begin{aligned} &S \rightarrow b\,S\,aS && (2) \\ &S \rightarrow b \wedge aS && (3) \\ &S \rightarrow b\,a\,S \end{aligned}$$

$$S \to b\ a\ aS\ bS \quad (1)$$
$$S \to b\ aa\ \wedge\ bS \quad (3)$$
$$S \to b\ aa\ bS$$
$$S \to baa\ b\ aS\ bS \quad (1)$$
$$S \to b\ aa\ ba\ (aS\ bS)\ bS \quad (2)$$
$$S \to b\ aa\ baa\ SbS\ bS$$
$$S \to b\ aa\ baa\ \wedge\ bS\ bS$$
$$S \to b\ aa\ baa\ b\ \wedge\ bS$$
$$S \to b\ aa\ baa\ bb\ \wedge$$

$$\boxed{S \to b\ aa\ b\ aa\ b\ b}$$

12.11. CONTEXT FREE GRAMMAR

A context-free Grammar '*G*' is a quadruple

$$(V, \Sigma, R, S)$$

where $V \to$ is an alphabet

$\Sigma \to$ The set of terminals is a subset of '*V*'

$R \to$ The set of rules is a finite subset of $(V - E) \times V^*$

and $S \to$ The start symbol is an element of '$V - \Sigma$'

Thd numbers of '$V - \Sigma$' are called "Non-terminals"

for any $A \in V - \Sigma$ and $V \in V^*$, we write $A \xrightarrow[a]{} $ whenever $(A, V) \in R$.

The radiation $\Rightarrow_{G^*}$ is the reflexive, transitive closure of $\Rightarrow_G$ finally, $L\,(G)$, the language generated by '*G*', is $\{w \in \sum^*,\ S \Rightarrow_{G^*} w\}$ we also say that '*G*' generates each string in $L\,(G)$. A language '*L*' is said to be a context free language if $L = L\,(G)$ for some context-free Grammar '*G*'.

EXAMPLE 12.29. Let $G = (V, \Sigma, R, E)$ where $V\,\Sigma$ S and R are as follows :

$V = \{+, *, (\ ,), id, T, F, E\}$ [*id* means identifier]

$\Sigma = \{+, *, (\ ,), id\}$

$R = \{E \to E + T, E \to T, T \to T * F, T \to F, F \to (E), F \to id\}$

We can give number to them as,

$$E \to E + T \quad (1)$$
$$E \to T \quad (2)$$
$$T \to T * F \quad (3)$$
$$T \to F \quad (4)$$
$$F \to (E) \quad (5)$$
$$F \to id \quad (6)$$

The *E*, *T* and *F* are expression, term and factor to generate string $\to (id \times id + id) \times (id + id)$.

Solution:

$$E \to T \quad (2)$$
$$\to T * F \quad (3)$$
$$\to T * (E) \quad (5)$$
$$\to T * (E + T) \quad (1)$$

$$\begin{array}{ll}
\rightarrow T * (T + T) & (2)\\
\rightarrow T * (F + T) & (4)\\
\rightarrow T * (id + T) & (6)\\
\rightarrow T * (id + F) & (4)\\
\rightarrow T * (id + id) & (6)\\
\rightarrow F * (id + id) & (4)\\
\rightarrow (E) * (id + id) & (5)\\
\rightarrow (E + T) * (id + id) & (1)\\
\rightarrow (E + F) * (id + id) & (4)\\
\rightarrow (E + id) * (id + id) & (6)\\
\rightarrow (T + id) * (id + id) & (2)\\
\rightarrow (T * F + id) * (id + id) & (3)\\
\rightarrow (T * id + id) * (id + id) & (6)\\
\rightarrow (F * id + id) * (id + id) & (4)\\
\rightarrow (id * id + id) * (id + id) &
\end{array}$$

EXAMPLE 12.30. To design a context free Grammar which can produce balanced parenthesis.

Let,

$$\begin{array}{ll}
G = (V, \Sigma, R, S) & \\
V = \{S, (\,,)\} & \\
\Sigma = \{(\,,)\} & \\
R = \{S \rightarrow e \text{ (Final State)} & (1)\\
S \rightarrow (S) & (2)\\
S \rightarrow SS & (3)
\end{array}$$

or $R = \{S \rightarrow e, S \rightarrow (S), S \rightarrow SS\}$

Solution: To generate (( ) ( ))

$$\begin{array}{ll}
S \rightarrow (S) & (2)\\
S \rightarrow (S\,S) & (3)\\
S \rightarrow ((S)\,S) & (2)\\
S \rightarrow ((e)\,S) & (1)\\
S \rightarrow ((e)\,(S)) & (2)\\
S \rightarrow ((e)\,(e)) & (1)\\
S \rightarrow ((\,)\,(\,)) &
\end{array}$$

EXAMPLE 12.31. To construct context free grammar for.

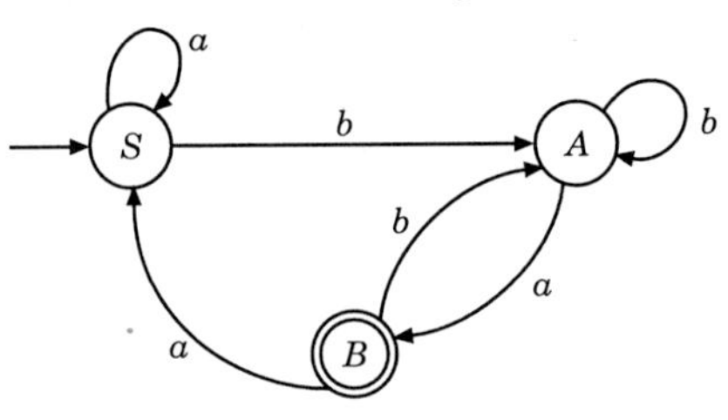

Fig. 12.27.

Solution:

$$\begin{array}{l}
S \rightarrow aS\\
S \rightarrow bA\\
A \rightarrow bA
\end{array}$$

$$A \to aB$$
$$B \to bA$$
$$B \to aS$$
$$B \to e \text{ (Final State)}$$

To terminate for final state.

> ✍ **Note :** The non-terminals are the states of the automation as for rules, for each transaction from 'q' to 'p' on input 'a' we have in 'R' the rule.
>
> $q \to ap$'

EXAMPLE 12.32. Construct a context free grammar that generates the language.

Solution: $L = \{w\, c\, w^R / w \in \{a, b\}^*\}$

$$S \to AcA \quad (1)$$
$$S \to BcB \quad (2)$$
$$A \to aB \mid Ba \quad (3)$$
$$B \to bA \mid Ab \quad (4)$$
$$B \to \wedge \quad (5)$$
$$A \to \wedge \quad (6)$$

For string "$aabcbaa$"

$$S \to A\ c\ A \quad (1)$$
$$\to aB\ c\ Ba \quad (3)$$
$$\to aAb\ c\ bA\ a \quad (4)$$
$$\to a\ aB\ bcb\ Ba\ a \quad (5)$$
$$\to aa \wedge bcb\ Baa \quad (5)$$
$$\to aa\ b\ c\ b \wedge aa \quad (5)$$
$$\to \boxed{aa\ bc\ baa}$$

12.12. NORMAL FORMS OF CONTEXT FREE GRAMMAR

(*a*) Chomsky Normal Form (C.N.F.) : A context for grammar 'C' is in Chomsky Normal Form if every production is of the form $A \to 0$ or $A \to BC$ and $S \to$ '$\wedge$' is in 'G' if $\wedge \in \lfloor (G)$.

When '$\wedge$' is in $\lfloor (G)$, we assume that 'S' does not appear on R.H.S. of any production

e.g.,

$$S \to AB / \wedge$$
$$A \to a$$
$$B \to b$$

Than 'G' is in Chomsky Normal Form.

EXAMPLE 12.33. Convert the grammar with production $S \to ABa$, $A \to aab$, $B \to Ac$ to Chomsky Normal Form.

Solution:

$$S \to ABa$$
$$A \to aab$$
$$B \to Ac$$

Take the first production $S \to ABa$

$$S \to ABB_1 \qquad [\text{where } B_1 \to a]$$

Now $A \to B_1B_1\ b$ [as $B_1 \to a$]

$B \to AC$ [where $C \to c$]

Now $S \to ABB_1$

$S \to DB_1$ [where $D \to AB$]

$A \to B_1\ B_1\ B_2$ [where $B_2 \to b$]

$B \to AC$

$C \to C$

$B_1 \to a$

Now $S \to DB_1$

$D \to AB$

$A \to B_3\ B_2$

$B_2 \to b$ (This is in C.N.F.)

$B \to AC$

$C \to c$

$B_1 \to a$

EXAMPLE 12.34. Convert the given C.F.G. into C.N.F.

$$S \to aSa\ /\ SSa\ /\ a$$

Solution: $S \to aSa$

$S \to SSa$

$S \to a$

Considers $A \to a$

$S \to ASA\ /\ SSA\ /\ A$

Now consider $A_1 \to SA$

$S_1 \to SS$

$S \to AA_1\ /\ S_1A\ /\ A$

$A_1 \to SA$

$S_1 \to SS$

$A \to a$

Above production are in C.N.F.

***(b)* Greibach Normal Form (G.N.F.) :** A context for grammar is said to be in Greibach Normal Form (G.N.F.) if all production have the form :

$$\boxed{A \to aX}$$

where, $a \in T$, $x \in N^*$ means " X can be Null".

EXAMPLE 12.35. Convert the following grammar in "Griebach Normal Form".

$S \to AB$

$A \to aA\ /\ bB\ /\ b$

$B \to b$

Solution: Given grammar is not G.N.F. So by using the substitution.

$S \to AB$

$A \to aA\ /\ bB\ /\ b$

Put all values of 'A'

$S \to aAB\ /\ bBB\ /\ bB$

$$A \to aA \mid bB \mid b$$
$$B \to b$$

$\because$ X may be Null. So above all productions are in Griebach Normal Form.

EXAMPLE 12.36. Consider the grammar

$$S \to abSb \mid aa \text{ in G.N.F.}$$

Solution:

$$S \to aBSB \mid aA$$
$$A \to a$$
$$B \to b$$

which is in G.N.F.

EXAMPLE 12.37. Convert the grammar

$$S \to ABb \mid a$$
$$A \to aaA \mid B$$
$$B \to bAb \quad \text{into G.N.F.}$$

Solution:

$$S \to ABb \mid a$$

We can put the value of A

$$A \to aaA$$

then $$S \to aaABb \mid a$$
$$P \to b$$
$$Q \to a$$

then $$S \to aQABP \mid a$$
$$P \to b$$
$$Q \to a$$
$$A \to aaA \mid B$$
$$B \to bAb$$

Therefore $$A \to aaA \mid bAb$$
$$A \to aQA \mid bAP$$
$$(\because P \to b,\ Q \to a)$$
$$B \to bAP$$

So final production of G.N.F. are

$$S \to aQABP \mid a$$
$$P \to b$$
$$Q \to a$$
$$A \to aQA \mid bAP$$
$$B \to baP$$

EXERCISE (A)

1. A finite state automation is defined by a transition diagram shown in Fig. 12.28 below. Find,

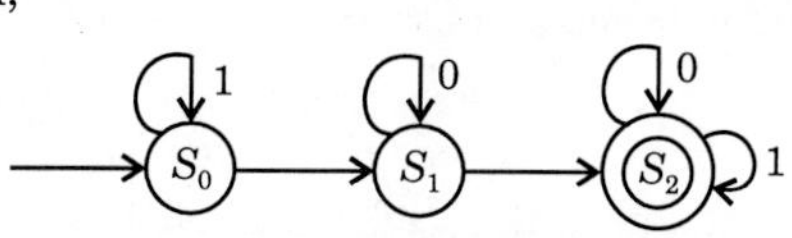

Fig. 12.28.

(*i*) Its states
(*ii*) Its input symbols
(*iii*) Its initial state
(*iv*) Its accepting states
(*v*) $F(S_1, 1)$
(*vi*) Write its next-stable table.

2. Prove that the lattice given by following diagram is modular.

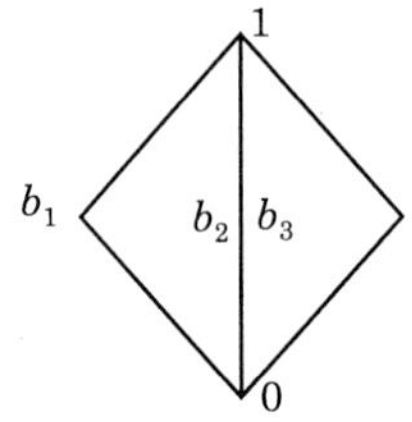

Fig. 12.29.

3. Show that in a complemented, distributed lattice, the following are equivalent
   (*i*) $a \le b$ (*ii*) $a \wedge b' = 0$
   (*iii*) $a' \vee b = 1$
4. (*a*) If *s* be the set of all positive divisors of 30, prove that $(s, \le)$ is a poset where $a \le b$ means '*a* is a divisor of *b*' for *a*, *b*, *f*, *s*. Also draw the Hasse diagram of the poset $(s, \le)$.
   (*b*) Define lattices. Give two examples of lattices also give an example of poset which is not a lattice.
   (*c*) Show that the set *N* of natural numbers under divisibility forms a poset.
5. (*a*) Let *G* be a group of integers under addition and let *N* be the set of all integral multiples of 3. Proves that *n* is a subgroup of a *G* and determine all the cosets of *N* in *G*.
   (*b*) Draw the Hasse diagram for the partial ordering :
   $\{(A, B) \mid A \le B\}$ on the power set $P(s)$ where $s = \{a, b, c\}$.
   (*c*) Find the least and greatest element in the poset $(Z^+, 1)$, if they exist.
6. (*a*) $A = \{2, 4, 6, 8, 12, 18, 24, 36, 72\}$ with the partial order of divisibility.
   (*b*) $A = [2, 3, 4, 6, 12, 18, 24, 36]$ with the partial order of divisibility.
7. Let *P* be relation on the Power Set $P(S)$ of a finite Set *S* of cardinality *n* define *P* by $(A, B) \in P$ iff $A \cap B = \phi$.
   (*a*) Consider the specific case $n = 3$ and determine the cardinality of *P*.
   (*b*) What is the cardinality of *P* for an arbitrary *n*7. Express answer in terms *n*.
8. What do you mean by
   (*a*) Posets
   (*b*) Hasse Diagram
   (*c*) Bounded and Distributed lattice.
9. (*a*) Prove that undirected possesses an eulerian path if it is connected and has either zero or two vertices of old degree.
   (*b*) Let $G = (V, E)$ be an undirected graph with *k* components and $|V| = n$, $|E| = m$. Prove that $m \ge n - k$.

10. (*a*) Find the language $L(m)$ accepted by the automation m in Fig. 12.30.

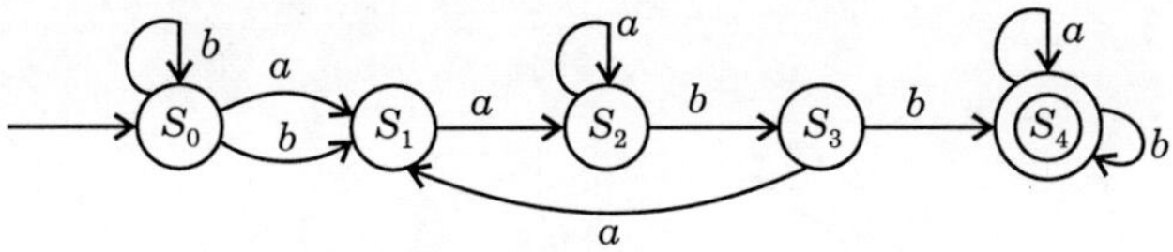

Fig. 12.30.

(*b*) Consider the finite state automation B defined by the following next state table.

	$S \backslash A$	f: a	b	c
$\longrightarrow$	s_0	s_1	s_0	s_2
	s_1	s_0	s_3	s_0
	s_2	s_3	s_2	s_0
O	s_3	s_1	s_0	s_1

(*i*) What are the states of B ?

(*ii*) Draw the transition diagram of B.

EXERCISE (B)

1. Design a finite state machine to check whether a given binary number is divisible by '3'.
2. Design a F.S.M. to set all strings which starts with '0' and end with 1's.
3. Design a F.S.M. to set all strings ending with '101'
4. Design a F.S.M. which accepts strings those have exactly two zero's anywhere.
5. Design a F.S.M. for strings having '101' or '110' as a substring.
6. Consider the N.F.A. given by the following diagram and find the equivalent N.F.A. without ε-transition.

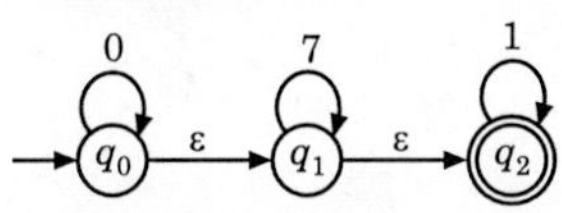

Fig. 12.31.

7. Construct the minimum state automation equivalent to the transition diagram given below ?

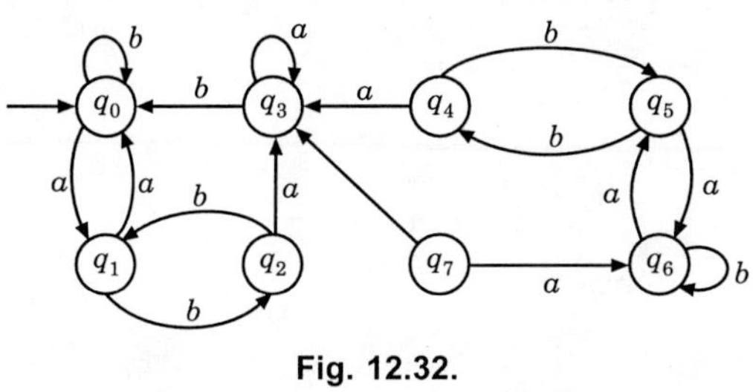

Fig. 12.32.

8. Construct a minimum state automation equivalence to the following diagram.

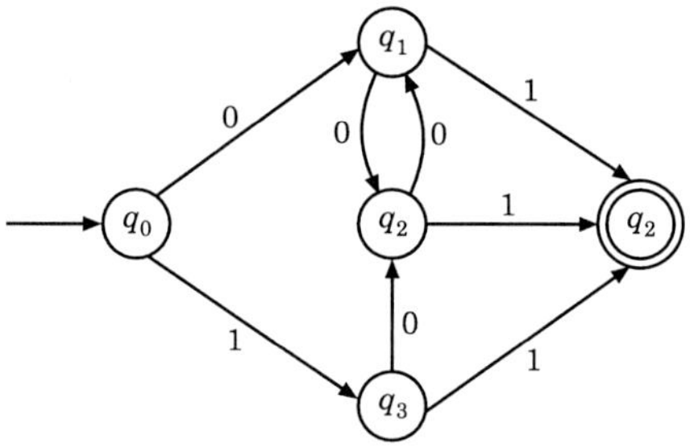

Fig. 12.33.

9. Design a Moore and Mealy machine to convert each occurrence of substring '100' by '101'

10. Design a Mealy machine for the following table and also find the output for the string "*abba baaa*".

Q \ Σ	a	b	**Output**
q_0	q_1	q_2	1
q_1	q_1	q_1	0
q_2	q_1	q_0	1

11. Convert the given Mealy machine into Moore machine.

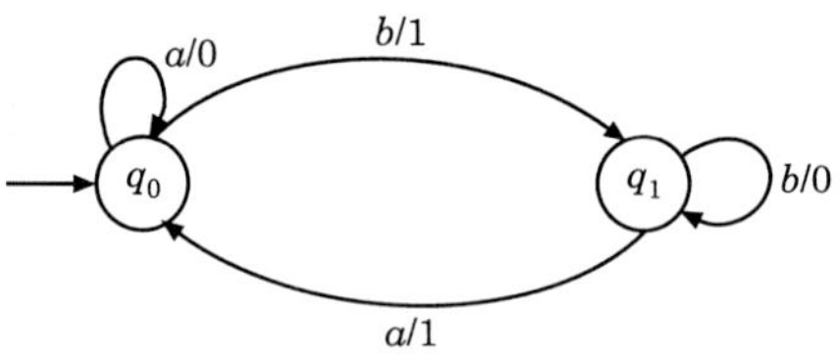

Fig. 12.34.

12. Consider a Mealy machine, construct a Moore machine equivalent to this Mealy machine.

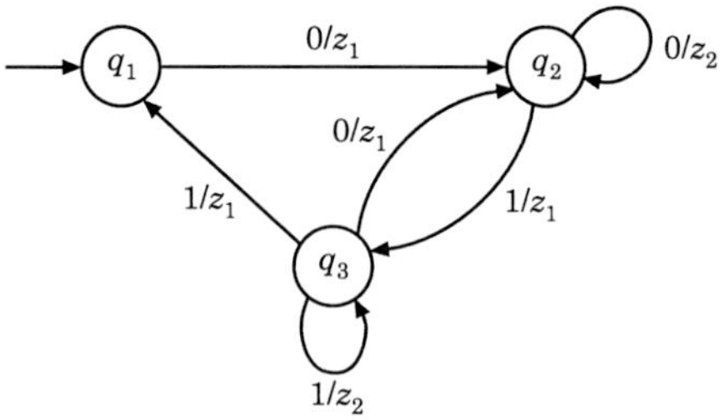

Fig. 12.35.

13. Convert the given Mealy machine.

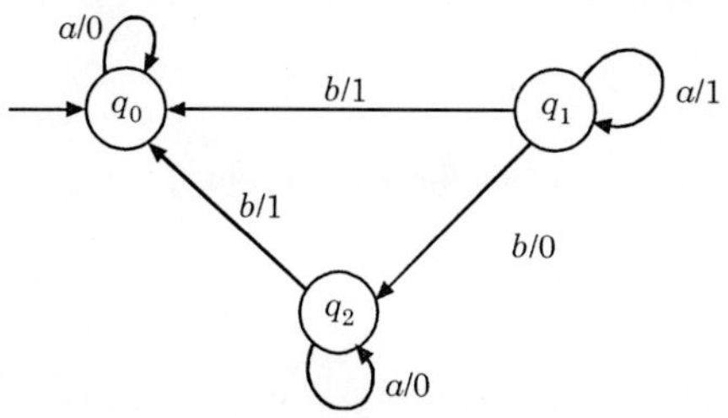

Fig. 12.36.

14. Design a Mealy machine which can give output Even, odd according to the total number of q's encountered is even or odd. The input symbols are 0 and 1.
15. For the input from $(0 + 1 + 2)^*$, construct Mealy machine which will give output 'A' whenever the sum of 3 consecutive symbols is 0, q or 4, will give output 'B' when the sum is 2 or 3 and will given output 'C' when sum is 5 or 6.

MULTIPLE CHOICE QUESTIONS

1. Describe the language accepted by the N.F.A. whose transition diagram is given by

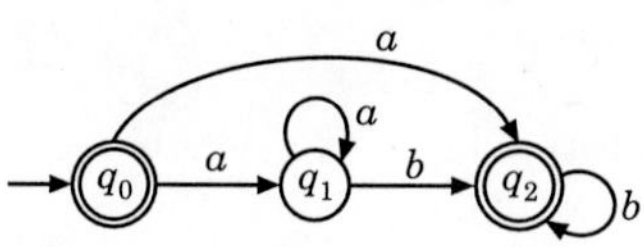

Fig. 1.

(*a*) $L(M) = \{\varepsilon, a\} \cap \{a^m b^n : m, n \leq 1\}$
(*b*) $L(M) = \{\varepsilon, a\} \cup \{a^m b^n : m, n \geq 1\}$
(*c*) $L(M) = \{\varepsilon, a\} \cup \{a^{m-1} b^{n-1} : m, n \leq 1\}$
(*d*) All of these

2. Find the equivalent of following regular expression
$(b + aa^*b) + (b + aa^*b)(a + ba^*b)^*(a + ba^*b)$
(a) $a^*b(a + ba^*b)^*$ (b) $ab^*(a + bab^*)^*$
(*c*) $a^*b^*(a^* + ba^*b)^*$ (*d*) None of these

3. The language $L = \{a^i b^j c^k : 0 \leq i < j < k\}$ is a
(*a*) Regular (*b*) Context free
(*c*) Context sensitive (*d*) None of these.

4. Find the regular expression for the given N.F.A.

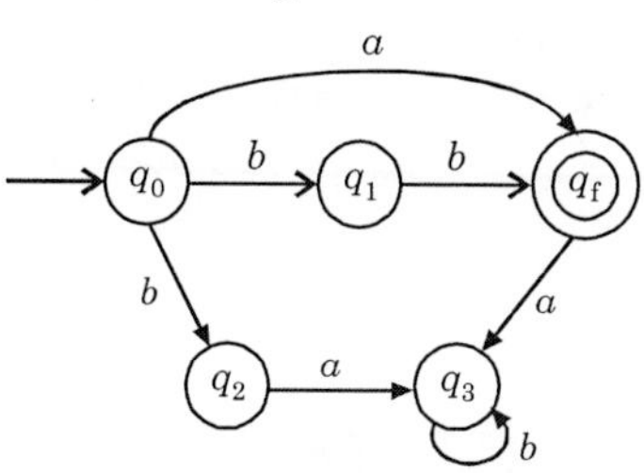

Fig. 2.

(*a*) $a + b^*b + baba^*$ (*b*) $a + bb + ba^*ba$
(*c*) $a + bb + bab^*a$ (*d*) None of these

5. Represent the following set by regular expression
$\{a^n : n \text{ is divisible by 2 or 3 or } n = S\}$
(*a*) $(a^*a) + (aa^*a)^* + (aaaaa)^*$ (*b*) $(aa)^* + (aaa)^* + (aaaaa)^*$
(*c*) $(aa)^* + (aa^*a)^* + (aaa^*aa)$ (*d*) None of these

6. If $\Sigma = \{x\}$ to find the value of Σ^* and Σ^+
(*a*) $\Sigma^* = \{\varepsilon, x, xx, xx, xx...\}$ and $\Sigma^+ = \{x, xx, xxx ...\}$
(*b*) $\Sigma^* = \{x, xx, xxx ...\}$ and $\Sigma^+ = (\varepsilon, x, xx, xxx ...\}$
(*c*) Both (*a*) and (*b*) (*d*) None of these

7. If $M_1 = (Q, \Sigma, O, \delta, f, q_0)$ is Moore machine, then there is a Mealy machine M_2 which is
(*a*) equivalent to M_1 (*b*) equivalent to $M_1 \cap M_2$
(*c*) equivalent to $M_1 \cup M_2$ (*d*) None of these

8. Consider the finite automation whose diagram is given by

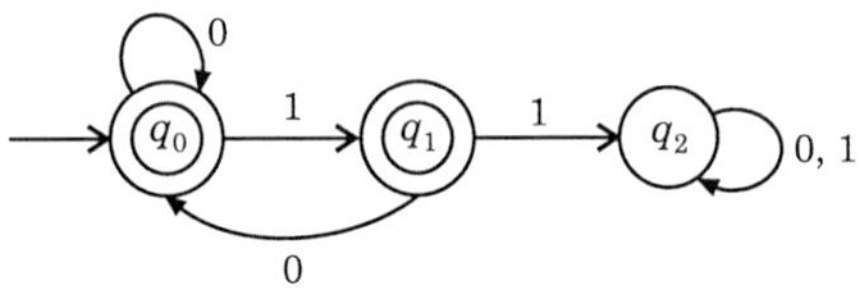

Fig. 3.

What kind of strings are accepted by this machine ?
(*a*) all string which start with 0 and end with 1
(*b*) any string of the form $y \parallel x$, when y ends in *a* 0 or y is ε and x is a string ending 0 or 1.
(*c*) Both (*a*) and (*b*) (*d*) None of these

9. Construct context free grammars to generate the following
$\{0^m 1^n \,/\, m \neq n, n \geq 1\}$
(*a*) $S \to 0A \,/\, A \to 0A \,/\, 0$ (*b*) $S \to 1B \,/\, 1B\,B \,/1$
(*c*) both (*a*) and (*b*) (*a*) None of these

10. A variable A in a context for grammar is nullable if
(*a*) $A \overset{*}{\Rightarrow} \lambda$ (*b*) $A \overset{*}{\Rightarrow} \lambda^*$
(*c*) $A \overset{*}{\Rightarrow} \lambda \cup \{\varepsilon\}$ (*a*) None of these

Answers

1. (*b*)	2. (*a*)	3. (*a*)	4. (*c*)	5. (*b*)	6. (*a*)
7. (*a*)	8. (*b*)	9. (*c*)	10. (*a*)		

❒❒❒

CHAPTER 13

LOGIC PROGRAMMING

This Chapter Covers :

13.1. INTRODUCTION

We are familier with lots of construction for "automatic theorem provers" in proportional and predicate lgoic. Although in the general case such an ambitious goal can not be achieved, as will follow from considerations discussed in earlier sections, nevertheless the hope is that partial theorem provers will play an important role and be useful in certain applications. This is one of the aims of studying artificial intelligence to try to capture formal reasoning efficiently with in a machine and 'logic programming' represents one of the important attacks on this goal.

Superficially, the concern with automatic theorem provers may seen rather restrictive : surely we should be studying systematic methods of getting machines to perform desired '**computations**'. In fact the two problems are very closely related. For example, we may ask whether such and such a conclusion is a consequence of certain hypothesis, and this can be viewed as a desire to **compute** an output ('yes' or 'no') from given inputs. We have already hinted at a not totally superficial resemblance between proofs and programs, and logic progamming seeks to make this more eoxplicit, and actually treat propositional or predicate logic as a 'programming language'.

13.2. WHAT IS LOGIC

Logic means drawing conclusions on the basis of conditions. Logic can be defined as scientific study of the process of reaponing of a system of rules and procedures that help in the reasonance process. The logic process takes input as some information (called premises) and produces some outputs called conclusions.

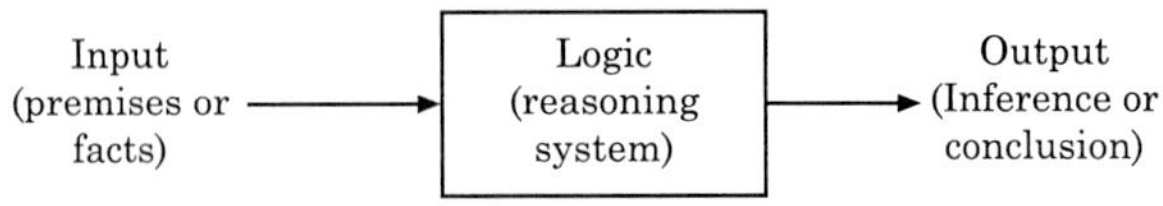

Fig. 13.1.

A logic can be categorized into two types :

(*i*) Propositional logic

(*ii*) Predicate logic

13.3. PROPOSITIONAL LOGIC

Simplest form of logic where all statements are made of propositions. In propositional logic, every proposition connected with some logic connective, can take only two values either True or false. Further propositions are two kinds:

(*a*) Atomic propositions (Simple proposition)

(*b*) Molecular propositions (Compound proposition or propositional calculus)

Molecular propositions are formed by combining one or more atomic propositions using a set of logical connection (*i.e.*, AND, OR, NOT).

Premise: A promise is a claim that is a reason (or element of a set of reasons) for, or objection against some other claim.

Compound propositions symbols used for logical connectives:

$\sim$ Negations
$\wedge$ Conjunction (AND)
$\vee$ Disjunction (OR)
$\rightarrow$ Implication
$\leftrightarrow$ Double implication (iff)
( ) { } Delimiter for functuation

Some commonly used logical equivalence. Normal forms in propositional logic are

CNF(Conjunctive Normal Form)

DNF(Disjunctive Normal Form)

CNF: A formula A is said to be CNF if it has the form

$$A = A_1 \vee A_2 \vee A_3 \ldots A_n \; n \geq 1$$

where each $A_1, A_2, A_3 \ldots A_n$ is a disjunction of an atom or negation of an atom.

DNF: A formula A is said to be in DNF if it has the form

$$A = A_1 \vee A_2 \vee A_3 \ldots A_n \; n \geq 1$$

where $A_1, A_2, A_3 \ldots A_n$ is a conjunction of an atom or negation of an atom.

First Order Predicate Logic (FOPL)

Predicates: A predicate is defined as a relation that binds two atoms together, for example:

"Kabir likes Discrete Mathematics"

This can be represented as

likes (Kabir, Discrete-mathematics)

here likes is a predicate that links two atoms together (Kabir and Discrete-mathematics). This predicate can be generated as like (*X*, *Y*) where X and Y are variables meaning *X* likes *Y*.

It is also possible to have a function as an argument. For example, "Ram's father is Manish's father. This can be represented by

Father ((father Ram), Manish)

here Father is a predicate and father (Ram) is a function to indicate Ram's father.

Quantifiers: A quantifier is a symbol that permits are to declare or identify the range and scope of the variables in the logical expressions. There are two logic quantifiers used in Logic:

(*i*) Universal Quantifier (∀)

(*ii*) Existential Quantifier (∃)

If '*x*' is a variable than $\forall x$ is read as

1. for all *x*
2. for each *x*
3. for every *x*

Similarly if '*y*' is a variable than $\exists y$ is read as

1. There exists a '*y*'.
2. For some '*y*'
3. For at least one '*y*'.

Terms: Terms are arguments in predicate

Father (Father, (Ram)), Manish

(Father, (Ram)) → It is also a term

Terms can be defines as follows:

1. A constant is a term
2. Variable is a term
3. If '*f*' is a function and X_1, X_2, X_3, X_n are terms of $f(X_1, X_2, X_3,$of $\ldots X_n)$ is a term.

Some predicates calculus expressions and their sematics:

Calculus Expression	Meaning
1. $\forall x\ [A(x)]$	– A is true for all x
2. $\exists x\ [A(x)]$	– A is true for some x
3. $\forall x\ [\sim A(x)]$	– A is false for all x
4. $\exists x\ [\sim A(x)]$	– A is false for some x
5. $\sim [\forall x\ [\sim A(x)]]$	– A is low for some x
6. $\sim [\exists x\ [\sim A(x)]]$	– A is low for all x

Normal Form in Predicate Logic: CNF and DNF are two major normal form *S* in propositional logic. In predicate logic there are also two another Normal forms called. **Prenex Normal Form** and **Clausal Normal Form**. A formula A in predicate logic is said to be in **Pronex National Form** if it has the form –

$$(Q_1, x_1), (Q_2, x_2), (Q_3, x_3) \dots (Q_n, x_n)\, B$$

where (Q_i, x_i) is either $\forall$, $\exists$ and B is formula without any Quantifier.

The following formula given as example are in Prenex Normal Form (PNF).

1. $\forall x\ \forall y\ \exists z\ ((A\ (x, y, z)\ \vee B\ (y, z) \rightarrow C\ (x, z))$
2. $\forall x\ \exists z\ ((A\ (x, z) \rightarrow B\ (x))$

To understood the concept of Clausal Normal Form and Prenex Normal Form, you must know the basic concept of skolomization, because when we will talk about the conversion procedure of a Formula to Clausal Normal Form, then we add a new step in conversion, we need to skolomize, so first we will study skolomization then we will study Clausal Normal Form and Prenex Normal Form (PNF).

Clause, Ground Clause and Horm Clause: A clause can be represented as the disjunction of number of literals. A ground clause is one which no variable occurs in negation form in the expression. Horn clause is a clause with almost one positive literal.

Clause Conversion Procedure: There are six steps for converting into clausal normal form.

Step 1: Eliminate all implications of equivalency connective use

(*i*) Replace $P \rightarrow Q$ by $\sim P \cup Q$

(*ii*) Replace $P \rightarrow Q$ by $(\sim P \cup Q) \cap (\sim Q \cup P)$

Step 2: Move all negation into immediately precede an atom use

(*i*) Replace $\sim (\sim P)$ by $\sim$ (double negation)

(*ii*) $(\exists x) \sim F\ (x)$ in place of $\sim (\forall x)\ F\ (x)$

(*iii*) $(\forall x) \sim F\ (x)$ in place of $\sim (\exists x) \sim F\ (x)$.

Step 3: Rename variable, if necessary, so that all quantifiers have diffrerent variables assignments that is rename variables bound by one quantifier are not the same as variables bound by one quantifier are not the same as variables bound by a different quantifier.

Step 4: Skolomize by replacing all existentially quantified variable with skolen functions and delete the corresponding existential quantifiers.

Step 5: Move all universal quantifiers to the left of the expression and put the expressions on the right into CNF.

Step 6: Eliminate all universal quantifiers and conjunction since they are retained implicity, the resulting expressions are clauses and the set of such expressions is said to be in clausal form.

EXAMPLE 13.1. Convert the given clause into the clausal form.

$$\exists x\ \exists y\ (\forall\ x\ P\ (f\ (x), y, z)) \rightarrow (\exists u\ Q\ (\mathrm{x, u}) \wedge \exists v\ R\ (y, v))$$

Solution: Step 1: (Application)

$$\sim \exists x\ \exists y \sim (\forall z\ P\ (f\ (x), y, z)) \vee (\exists u\ Q\ (x, u) \wedge \exists v\ R\ (y, u))$$

Step 2: (Application)

$$\forall x\ \forall y\ (\exists z \sim P\ (f\ (x), y, z)) \vee (\exists u\ Q\ (x, u) \wedge \exists v\ R\ (y, v))$$

Step 3: (Not required)

Step 4: $\forall y\,(\sim P\,(f\,(a), y, g\,(y))) \vee Q\,(a, h\,(y))) \wedge R\,(y, i\,(y)))$

Step 5: (Application)

$\forall y\,(\sim P\,(f\,(a), y, g\,(y))) \vee Q\,(a, h\,(y))) \wedge (\sim P\,(f\,(a), y, g\,(Q) \wedge R\,(y, i\,(y))$

Step 6: $P + QR = (P + Q)\,(P + \text{R})$ distributive law

$[\sim P\,(f\,(a), y, g\,(y) \vee Q\,(a, h\,(y))] \wedge$

$[\sim P\,(f\,(a), y, g\,(y) \vee R\,(y, i\,(y))]$ **Ans.**

Resolution Principle: Resolution is a proof or procedure that carries out a single operation, the variety of processes involved increasing with statements in predicate logic. Resolution operates on statements not have been converted into a very convenient form. Resolution procedure is proved by reputation, in other words to prove a statement (that means) to show that it is valid), in order to prove a statement, resolution attempts to show that negation of statement produces a contraediction with the known facts, *i.e.,* it is unsatifiable.

Basis of Resolution:

1. A simple iterative process, in each step at which two clauses called parent clauses are compared, attempts a new clause that has been infered from than, the new clause represent ways that the two parent clauses interact with each other. Resolution operates by taking two clauses that each contain the same literal.
2. The literal must occur in positive form in one clause and in negative form in other clause.
3. The resolution is obtained by combining all of the literals of the two parent clauses, except the one that cancel each other.
4. If the clause that is produced is the empty clause, then a contradiction has been found.

EXAMPLE 13.2. Consider the following clauses:

$$A : P \vee Q \vee R$$
$$B : \sim P \vee Q \vee R$$
$$C : \sim Q \vee R$$

and find the resolvent.

Solution:

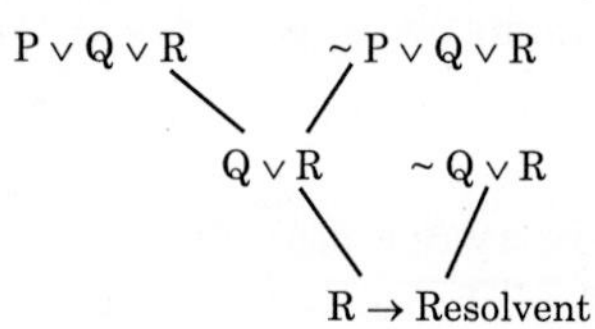

Resolution Principle (Formal Definition)L The resolution principle companies the two ideas:

1. Syllogism principle of the propositional clause
2. Instantiation principle of predicate calculus (Substitution)

Resolution Principle to Propositional Calculus: The procedure (s) with respect to set of axioms the following are the points to be considered.

1. Connect all the prepositions of F to clause form.
2. Negate *S* and Convert the result to clause form. Add it to the set of clauses obtained in Step 1.
3. Repeat while a contradiction is found or no progress can be made.

Substitution (Unification): Resolution works on the principle of identifying complementary literals in two clauses and deleting then hereby forming a new literal (resolved) the process is simple and straight forward when one has identical literals. In other words for clauses containing no variables, resolution is easy when there are variables, the problem becomes complicated and it is necessary to make proper substitution.

> ✍ **Note:** Substitution variables are called Most General Unifer or MGU.

A substitution that makes two clauses, resolvable is called a unifier and process of identifying such unifiers is carried out by the unification algorithm for example $P(x, y)$ for resolving clauses $P(t, k(n))$. Replace $\left(\frac{x}{t}, \frac{y}{k(n)}\right)$ and the process is called unification.

Theorem Proving Using Resolution: Basically these are two methods for theorem proving:

1. Using Rule of Inference
2. Using Reflection.

In first method we start with the given axioms and then use the rules of inference to prove the theorem and in second method we prove the negation of the result cannot be tone (prove by reflection).

The second method that is commonly known as theorem proving using reflection and methodology for that is:

1. Find the negation of the result to be proved.
2. Add it as a valid statement to the given set of statements.
3. Perform resolution in these statements untill contradiction is encountered.
4. Conclude that the contradiction is due to the assumed negation of the result.
5. So the negated assumption that is made in false or the result to be proved is True.

EXAMPLE 13.3. Given axioms are

(*i*) $\forall x$ [Doctor $(x) \rightarrow$ knows – surgery (x)]
(*ii*) Doctor (Kavita)

Proved: Knows – surgery (Kavita).

Solution: Modous Ponen states that if there is an axiom of the term $P \rightarrow Q$ and another of the form P, then Q basically follows :

Assuming Doctor (Kavita) as P and

[Doctor $(x) \rightarrow$ knows $\rightarrow$ surgery (x)] as $(P \rightarrow Q)$

We can prove with the help of modous ponens the Q is true, *i.e.*, knows – surgery (Kavita).

Representation of Facts in Logic (Predicates) Facts: Let us take some facts shich are given as,

(*a*) Marcus was a man.
(*b*) Marcus was pompcian.
(*c*) All pompcian were romans.

(*d*) Caesar was a ruler.
(*e*) All romans were either loyal to caesar or hated him.
(*f*) Everyone is loyal to someone.
(*g*) People only by to assassinate ruler they are loyal to :
(*h*) Marcus fed to assassitional caesar.
(*i*) All men are people.

In predicate calculus (Predicate facts) are

(*a*) Man (Marcus)
(*b*) Pompcian (Marcus)
(*c*) $(\forall x)$ [Pompceian $(x) \rightarrow$ Roman (x)]
(*d*) Ruler (caesar)
(*e*) $(\forall x)$ [Roman $(x) \rightarrow$ [loyal–to (x), caesar) $\vee$ hate $(x$, caesar)]]
(*f*) $(\forall x)$ $(\exists y)$ loyal–to (x, y)
(*g*) $(\forall x)$ $(\forall y)$ person $(x) \wedge$ Rular $(y) \wedge$ try assassinate (x, y)
(*h*) tryassassinate (Marcus, Caesar)
(*i*) $\forall x$ [Man $(x) \rightarrow$ Person (x)]

EXAMPLE 13.4. Consider the following facts,

(*a*) The members of the bridge dub are Joe, Bill, Sally and allen.
(*b*) Joe is married to sally.
(*c*) The spouse of every married person in the dub is also in the dub.
(*d*) The last meeting of the dub was in Joe's house. Represent above facts in predicate logic. Also prove that Allen is not married.

Solution: Facts in predicate logic or well formed form are given as

(*a*) member (Joe)
member (Bill)
member (Sally)
member (Allen)
(*b*) married (Joe, Sally)
(*c*) $\forall x$, $\forall y$ [(married, $(x, y) \wedge$ member (x)) $\rightarrow$ member (y)]
(*d*) meeting (lost, Joe)

To prove allen is not married.

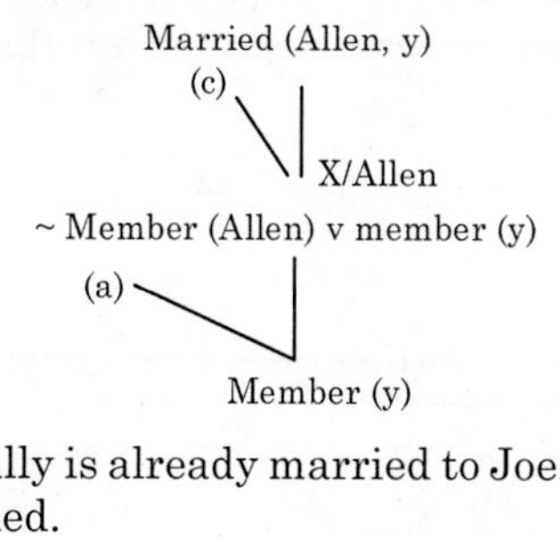

y can be Sally, but Sally is already married to Joe. Therefore contradiction. Hence Allen is not married.

Introduction to Programming: PROLOG: First order predicate logic implies the existence of a set of predicate symbols along with a set of connectives. First order predicate logic implies that there is no means provided for "talking about" the predicate themselves. Prolog is passed on FOPL (first order predicate logic) but uses a restricted version of the clausal form. Clausal form is a particular way of writing the propositions clause form. Prolog is a socalled logic programming languages. Strictly, it is not the only one but most such languages are its descendents. We will spend a little time outlining the basic ideas undertyping both propositional and predicate logic.

Prolog provides for the representation of a subset of first order predicate calculus. The restrictions on what can be done will become clearer later. We will now explain how we can write logical statements in Prolog.

If "the capital of M.P. is Bhopal" then we can represent this is predicate calculus form as:

M.P. has – Capital Delhi

We can also have binary relationship, i.e., two axioms are related with a relationship) written in infix form. That is, the relationship is written between the two things related. The relationship (or predicate) has been given the name "has—capital". Hence we can say that the predicate name is "has–capital" and in Prolog form will be as:

has - Capital (M.P. Bhopal)

When we write a prefix form and say that the relationship takes two arguments. Prefix because the relationship is indicated before two related things.

The exact rule for the termination of a clause is that a clause must end with a "•" followed by white space white space can be any of space, tab, newline, end-of-file. It is safe to simply put "•" followed bgy newline. Also note that relations donot need to hold between exactly two objects. For example,

meets (Ram, Ravi, bridge)

might be read as,

Ram meets Ravi by the bridge

Here three objects are related so it makes little set to think of the relation meets as binary, it is ternary.

If we can relate two objects or there then it is reasonable to relate n where $n \leq 2$. Is there any significance to a relationship that relates one or even zero objects ? A statement likes

Ankur is clever

might be represented either as

clever (Ankur)	or	Ankur (clever)

It is, perhaps, preferable to see cleverness as being property which is possessed by Ankur.

A 'relation' that has no arguments can be seen as a single proposition. Thus the binary relation "M.P. has-capital Bhopal" above might be rewritten as the statement "M.P. has Capital Bhopal" which is a relation with no arguments.

Basic Terminology of Prolog Programming Language

Term: Prolog's single data type is the *term*. Terms are either atoms, number, variables or compound terms.

Atoms: Atoms can be regarded as a special case of compound terms (or arity zero). *e.g.*, x, blue or 'some atoms'.

Numbers: In addition to floats and integers which are prescribed by the prolog. ISO standard, many prolog implementations also provide relational numbers and unbounded integers.

Variables: Variables are denoted by a string consisting of letters, numbers and underscore characters, and beginning with an upper-case letter or underscore. Variables closely resemble variables in logic in that they are place holders for arbitrary terms. A variable can become instantiated via unification. For example,

$$? - \Gamma = f\,(X, g\,(X)),\, x = \text{a}$$
$$T = f\,(a, g(a))$$
$$X = a$$

Compound Terms: A compound terms has a functor (function operator) and a number of arguments, which are again terms. The number of arguments is called the term's arity. Examples for terms are $f\,(a,\, b)$ and g $(x,\, y,\, z)$. Some operators can be used in infix motation. For example, the terms + (a, b) and = (x, y) can also be written as $a + b$ and $x = y$ respectively. Users can also declare arbitray sequences of symbols as new infix or postfix operators to allow for domain-specific motations. Special cases of compound terms are lists, strings and constants.

(*a*) **Lists:** Lists are defined inductively. The atom 'nil' (denoted as [ ]) is a list. A compound term with functor . (dot) and arity 2, whose secondargument is a list, which is itself a list. There exists special syntax for donating lists. For example, • (*A*, *B*) it equivalent to [*A*, *B*] and list . (1, • (2, • (3, [ ])) can also be written as [1, 2, 3].

(*b*) Strings: A sequence of characters surrounded by quotes is equivalent to a list of numeric character codes, generally in a local character encoding or unicode if the system supports unicode. For example,

? – *Xs* = "? ? ? ? ? ?"

Xs = [928, 961, 972, 955, 959, 947]

(c) **Consultants:** If, we have

Father (Yashraj, Aditya).

Then Yashraj and Aditya refer to specific objects. Both Yashraj and Aditya are constants. In particular, in PROLOG Terminology, both are atoms. Also "Father" happens to be an atom too because it refers to a specific relationship. Geneerally speaking if a string of characters starts with a lower-case letter, the DEC-10 family of prolog assume that the entity is an atom.

Goals and Clauses : We distinguishable between a Prolog goal and Prolog clause. A clause is syntactic entity expressing a relationship as required by Prolog.

> ✍ **Note :** We will regard the '•' as terminating clause (this is not strictly correct)

Father (Yashraj, Aditya) is a goal
Father (Yashraj, Aditya), is a unit clause

The objectives unit and non-unit distinguish two kinds of clause – Initiatively, facts and rules respectively.

Multiple Clauses: A predicate may be defined by a set of clauses with the same predicate name and the same number of arguments. We will therefore informally describe the way in which this is handled through an example. The logical statement (in first order form).

Square (2, 4) $\wedge$ Square (3, 6), Square (4, 16)

is to be represented as three distinct Prolog clauses.

Square (2, 4) •
Square (3, 9) •
Square (4, 16) •

This way of turning a conjunctive statement into Prolog is one of the fundamental restrictions previously mentioned. There are more to follow. All the above clauses are unit clauses. We may now introduce a graphical representation which will be used in a number of different ways. The idea we use here is to represent a program (In this case, consisting of a set of unit clauses) as a tree.

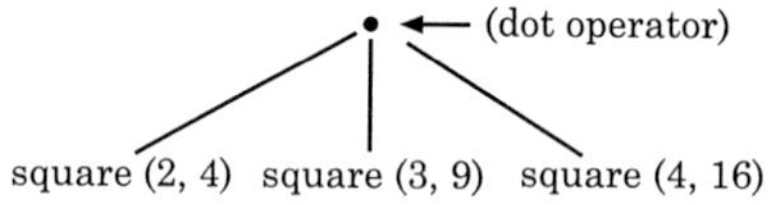

The tree give above is an example of OR true. It might have been expected that we would call this an AND true but, when we are trying to determine whether a statement such as square (2, 4) is true then we might use either the first clause or the second or the third and so on.

Rules: Let us take an example,

Divisible - by - two : even

This is a non-unit clause. In general a clause consists of two parts : The *head* and the *body*. The head is "divisible - by - too" and the body is even. Even is sometimes referred ": –" is read as 'If'. An informal reading of the clause is "divisible-by-two" is true if even is true" which is equivalent to "even divisible-by-two". Any number of symbols may be in the body of the Rule.

This is another way in which Prolog is a restriction of full first order predicate calculus. For example, we cannot translate.

Rich (Ramesh) $\rightarrow$ happy (Ramesh) $\wedge$ powerful (Ramesh)

directly in to Prolog version.

happy (Ramesh), powerful (Ramesh) : rich (Ramesh)

For an example of a clause with more than one subgoal in the body. A fact is effectivley a rule with no subgoals. You may have noticed that, even if it is held that "even" is a relation, it does not seen to relate anything to anything else. The rule is not as much use as it might be because it does not reveal the more interesting relationship that a number is divisible by two if it is even. We can express this with the help of the logical variable. Here is the improved rule:

divisible - by - two (x) : even (x)

This is also a non-unit clause. The named logical variable is X. This Prolog clause is equivalent to the predicate calculus statement.

$(\forall x)$ [even $(X) \rightarrow$ divisible - by - two (X)]

The Logical Variable: In prolog, if object is referred to by a name starting with a capital letter then the object has the status of a logical variable. In the above rule there are two reference to X. All this means is that the two reference are to the same object whatever the object is. The scope rule for prolog is that two uses of an identical name for a logical variable only refer to the same object if the uses of an identical name for a logical variable only refer to the same

happy (X) : healthy (X) wife (X) : – old (X)

object if the uses are with in a single clause. Therefore in the two references to X in the first clause do not refer to the same object as the references to X in the second clause. Do not assume that the word logical is redundant. It is used to distinguish between the nature of the variable as used in predicate calculus and the variable used in imperative languages like ALOL, BASIC, FORTRAN etc. In those languages, a variable name indicates a strong location which may 'Contain' different values of different moments in the execution of the program.

Although this needs some further comments, it is probably better to start with this statement and qualify it later.

Rules and Conjunctions: A man is happy if he is rich and famous might translate to

happy (person) : man (person), rich (person), famous (person).

The ',' indicates the conjunction and is roughly equivalent to the predicate calculus. Therefore, read ',' as 'and'. The whole of the above is one (non-unit) single clause. It has three subgoals in its body—these subgoals are 'conjoined'. In this single clause, the logical variable Person refers to the same object throughout.

By the way, we might have chosen any name for the logical variable other than person. It is common practice to name a logical variable in some way that reminds you of what kind of entity is being handled. We can now describe this clause graphically. In this case we can represent conjuctions using AND tree.

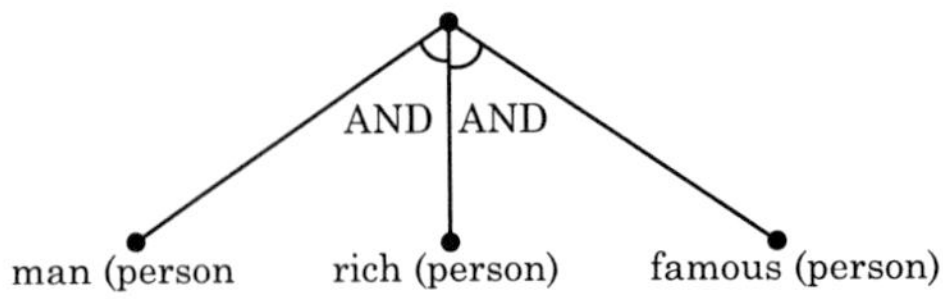

The way in which we discriminate between an OR tree and on AND tree is the use of a conviction between two branches of the subgoals. We need this distinction because we are going to represent the structure of a program using a combined AND/OR tree.

Rules and Disjunctions: Let us take an example : "Someone is happy if they are healthy, wealthy or wise". translates to:

happy (person) : healthy (Person).
happy (person) : wealthy (Person)
happy (person) : wise (person)

Note how we can have had to rewrite the original informal statement into something like:

Someone is happy if they are healthy OR
Someone is happy if they are wealthy OR
Someone is happy if they are wise

We have also assumed that each clause in implicitly universally quantified, i.e. the first are above represents.

$$(\forall x)\ [\text{healthy}\ (x) \rightarrow \text{happy}\ (x)]$$

The predicate name "happy" is known as functor. Any predicate name is a functor. The functor has one argument.

We describe a predicate with name "predname" with arity "n" as predname/n. It has one argument then we say its arity is 1. The predicate happy/1 is defined by three clauses.

Both Disjunctions and Conjunctions: If we combine both conjunctions and disjunctions together, we can find the following informal stlatements and their translation as follows:

happy (person) : healthy (person), woman (person)
happy (person) : wealthy (person), woman (person)
happy (person) : wise (person), woman (person)

It can be informally interpreted as meaning that a woman is happy if she is healthy, wealthy or wise.

We now combine the OR three representation together with the AND tree representation to form an AND/OR three that shows the structure of the definition of happy/1.

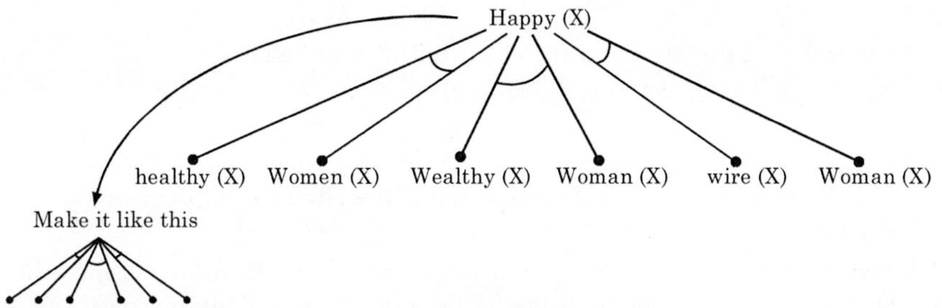

Queries and Disjunctions: Informally, a querry is a goal which is submitted to Prolog in order to determine whether this goal is true of false. As, at top level, Prolog normally expects queries it prints the prompt:

? –

and expects you to type in one or more goals. We tell the Prolog system that we have finished a query or any clause by typing "•" followed by typing the key normally labelled "RETURN". A very common syntax error ofor beginners is to pass "RETURN" (↵) before "•". This is not a problem but just a typing mistake. We look at the case where we only want to solve one goal. Perhaps we would like to determine whether or not.

Woman (Marry).

In this case we would type this in and see (What is actually tyuped is embddened):

? – Woman (Marry).

Now ? – Woman (Marry) . is also a clause. Essentially, a clause with or empty head.

Unification: In proportional logic, it is easy to determine that two literals can not both be true at the same time. For example, at any time L and L will not be true. In predicate logic, this matching process is more complicated since the arguments of predicates must be considered. For example, man (Ravi) and man (Ravi) is a contradiction, while man (Ravi) and man (Ramesh) is not. This in order to determine contradiction we need matching procedure that compares two literals and discovers whether there exists a set of substitution that make them identical. There is a strategic forward recursive procedure, called the **unification algorithm**.

The basic idea of unification is very simple. To attempt to unify two literals, we check if their initial predicate symbols are the some.

If so we can proceed, otherwise there is no way. For example, the two literals can not be unified. If the predicate symbols match, then we must check the

tryassassinate (Ravi, Ramesh) hate (Ravi, Ramesh)

arguments one pair at a time. If the first matches, we can continue with the second and soon. To test each arguments pair, we can simply call the unification procedure recursively.

The matching rules are simply. Different constants or predicates can not match.

For example, Suppose we want to unify the expressions

$$P\,(x, x)$$
$$P\,(y, z)$$

The two instances of P match find. Next we compare x and y decide that if we substitute y for x, they could match. We will write that substution as

$$y/x$$

We could of course, have decided instead to substitute x for y. Since they are both just dummy variable names. The algorithm will simply pick one or these two subsitutions.

But now, if we simply continue and math x and y we produce te substition z/x, but we cannot substitute both y and z for x. So we have not produced a consistent substitution. Finding the first substitution y/x is to make that sbustitution through the literals giving.

$$P\,(x, y)$$
$$P\,(y, z)$$

Now we can attempt to unify arguments y and z, which succeeds with the substitution z/y. We write the composition as

$$(z/y) \quad (y/x)$$

Recursion: We have already met a recursive definition. Here are some more. One of my ancestors is one of my parents or one of their ancestors. A string of characters is a single characters or a single character folowed by a string of characters. A paragraphs is a sentence of a sentence appended to a paragraph. To decouple a train, uncouple the first carriage and then decouple the rest of the train. An example of recursive program is

```
talks–about (A, B) : knows (A, B).
talks–about (P, R) : knows (P, Q), talks–about (Q, R).
```

Roughly translated: "You talk about someone either if you know then or you know someone who talks about them".

If you took at the AND/OR true of the search space you can see that there is a subset which is the same shape as the whole true rejecting the single recursive call to talk–about/2.

The solution of a given problem depends on being able to stop recursing at some point. Because the leftmost path down the true is not infinite in length it is reasonable to hope for a solution.

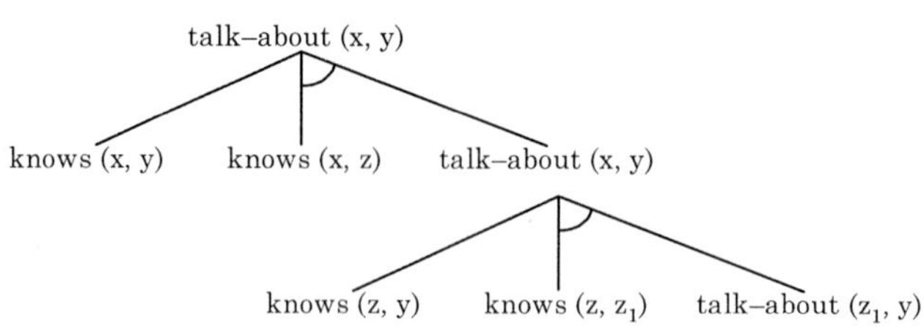

In searching the tree with a number of facts along with the clauses for took–about/2.

Table

talk–about (*A*, B) : knows (*A*, *B*)
talk–about (*P*, R) : knows (*P*, *Q*),
talk–about (*Q*, *R*)

↑ Recursive function

knows (Ravi, Ramesh).
knows (Ramesh, Rohan).
knows (Rohan, Raghav).
knows (Raghav, Ritu).
knows (Ritu, Vikas).
knows (Bhaskar, Kavita).
knows (Ankur, Shardha).

} Data base

Using the goal

talk–about (*x*, *y*)

If we ask for repeated solutions to this goal, we get, in the order shown

X = Ravi	*Y* = Ramesh
X = Ramesh	*Y* = Rohan
X = Rohan	*Y* = Raghav
X = Raghav	*Y* = Ritu
X = Rohan	*Y* = Ritu

and so on.

The search strategy implies that Prolog keep on trying to satisfy the symbol knows (*X*, *Y*) until there are no more solutions to this. Prolog then finds thant, in the second clause for talks–about/2, it can satisfy the talks-about (*X*, *Y*) goal by first–findinig a third party who *X* knows. It satisfies knows (*X*, *Z*) with *X* = Ravi, *Z* = Ram and then recurses looking for a solution to the goal talk–about (Ram, *Z*). It finds the solution by matching against the second knows/2 clouse. The above AND/OR tree was formed by talking the top level goal and, for each clause with the same predicate name and arity, crating an OR choice leading to subgoals constructed from the bodies of the matched clauses. For each subgoal in a conjunction of subgoals. We create an AND choice.

Lists: Lists, for now, can be regarded as space Prolog structures that can be used to represent an ordered sequence of Prolog terms. For examples here are some legal lists :

[*a*, *b*, *c*, *d*, *e*,] a list with six elements (all atoms)

[ ] a list with no elements in it.

[Tea, coffee, milk shake] a list with three atoms.

[happy (Ramesh), [Tea, coffee], [1, (2), 3]] a list with three elements

The list example is a little difficult to dicipher the first element is happy (Ramesh), the second is [Tea, coffee] it is a list with two elements and an atom for original list and the third is [1, [2], 3] is another list. Note that the "," used in the construction of a list is just an argument separator as in the term father (*X*, *Y*) also note that, because order is preserved, the list [*a*, *b*, *c*] is not same as list [*a*, *c*, *b*].

How to Construct/Deconstruct a List: There are so many operation that can be performed on a given list in Prolog for a given arbitrary list, we need ways of adding to it and taking it apart. The basic approach provides a simple way of splitting a list into two parts, the first element (if there is one) and the rest is another list. The corresponding way of joining two atoms to form a list requires taking an element and a list and inserting the element of the front of the list.

List Distruction: In list distruction, first we learn how to remove the first element from the list

$$[X/Y] = [f, g, h, e]$$

will result in, $X = f$

- The first element of the list is known as the HEAD of the list.
- The list formed by deleting the HEAD is called TALK of the list and it is again a list and its first element is also HEAD of the list. This list has been reduced in length and can be further distructed or constructed.

List Construction: The construction of a list is the reverse : take a variable bound to any old list, say $X = [a, b, c]$ and add the element, say m at the front with

```
result–wanted = [m/X]
```

when m is HEAD of list and X is TAIL of the list having $X = [a, b, c]$.

Empty List: In empty list simple can be represented by [ ]. This list ([ ]) has no elements in it. It cannot therefore be destructed. An attempt to do this will fail. The empty list ([ ]) is an atom.

A Recursive Program Using List: We can make us of a built-in predicate called write/1 to write out all the elements of a list in order. Note that the argument of write/1 must be a legal Prolog term. Write/1 is a side-effecting predicate. It captures the logical relation of always being true but it also produces output which has no opart to play in the logical interpretation. It is therefore hard to produce a declarative reading for this predicate despite. Its utility from the procedural point of view. There are a fair number of other predicates which suffer from this problem including consult/1 and reconsult/1. To write out a list of terms, write out the first element and then write out the remainder (the tail).

```
show-a-list ([ ])
show-a-list ( [H, T] ) : Write (H), show-a-list (T).
```

Note that this can be improved by printing a space between elements of the list. This requires you to add the subgoal write (' ') into the body of the second clause and before the recursive call to show-a-list/1. This will write the elements out on a single line. If you wanted to write each element on a different line then you would need the built-in predicate *nl/o*.

The second clause of show-a-list/1 roughly captures the meaning above. Then what does the first clause achieve ? Without the first clause, show-a-list/1 would produce the required output and then fail because it would have to handle the empty list ([ ]) which cannot be deconstructed. Although show-a-list/1 is a side-effecting predicate, the natural procedural reading is that it succeeds once it has printed the list of terms. The first clause handles the case of the empty list so that the predicate will always succeed if it is given a list of terms to print. Quite reasonably, it will fail if given a non-list.

SUMMARY

- Prolog allows a program to be read either declaratively or procedurally. Thus dual semanticxs is attractive. Procedural programming requires that the programmes tell the computer what to do.
- The propositional calculus is based on statements which have truth valves (true or false). The calculus provides a means of determining the truth values associated with statements formed from "atomic" statements.
- The predicate calculus includes a wides range of entities. It permits the description of relations and the use of variables. It also requires an understanding of quantification.
- Prolog's single data type is term. Terms are either atoms, numbers, variables or compound terms.
- A compound term has a functor and a number of arguments, which are again terms.
- Lists are defined inductively. The atom 'nil' denoted as [ ] is a list. A compound term with functor . (dot) and arity 2, whose second argument is a list itself.

EXERCISE

1. Consider the following axioms. Convert to predicate form and solve using resolution:
   (*a*) Anyone whom mary loves is a football star
   (*b*) Any student who does not pass does not play
   (*c*) John is a student
   (*d*) Any student who does not study does not pass
   (*e*) Anyone who does not play is not a football star
   (*f*) (Conclusion) If John does not study, then Mary doesnot love John
2. Consider the following axioms. Convert to predicate form and solve using resolution
   (*a*) Every child Loves santa
   (*b*) Everyone who loves santa loves any reindeer
   (*c*) Rudolph is a reindeer and Rudolph has a red nose
   (*d*) Anything which has a red nose is weird or is a clown
   (*e*) No reindeer is a clmm
   (*f*) Scrooge does not love anything which is weird
   (*g*) (Conclusion) scrooge is not a child

   Represent these axioms in predicate calculus, skolemize as necessary and convert each formula to clause form. Prove the unsatisfiability of the set of clauses by resolution.
3. Explain the following terms:
   (*a*) Term (*b*) Atoms
   (*c*) Numbers (*d*) Variable
   (*e*) Data type (*f*) Lists
4. What are the rules for programming in Prolog?
5. What do you mean by loops and recursion in Prolog.
6. What are implementation technique in Prolog?
7. Discuss comparison of implementations of Prolog.

8. What do you mean by multiple clause?
9. How do you construct/deconstruct a list?
10. Write a program (recursive) using list.

MULTIPLE CHOICE QUESTIONS

1. What is notation of using case to extract the meaning of sentence
   (*a*) Robert's (*b*) Shunk's
   (*c*) Fill more's (*d*) None of these
2. CD provides:
   I. a structure into which nodes representing information can be placed
   II. a specific set of primitives
   III. at a given level of granularity
   (*a*) Only I (*b*) Only I and II
   (*c*) ALL (*d*) None of these
3. Which is represented as a series of diagrams depicting actions using both abstract and real physical situations?
   (*a*) Sentences (*b*) Paragraph
   (*c*) Grammer (*d*) None of these
4. Which language allows a program to be read either declaratively or procedurally?
   (*a*) C/C++ (*b*) LISP
   (*c*) Prolog (*d*) None of these
5. Prolog is based on
   (*a*) Second order predicate logic (*b*) First order predicate logic
   (*c*) Object oriented logic (*d*) None of these
6. Propositional calculus is based on statements which have
   (*a*) True values (*b*) False values
   (*c*) Truth values (*d*) All values
7. A formula with no free variable is
   (*a*) Formula (*b*) Clause
   (*c*) Sentence (*d*) Paragraph
8. Prolog's single data type is the
   (*a*) Term (*b*) Data type
   (*c*) Number (*d*) None of these
9. When the Prolog program/system is entered we are at top level. The system is waiting for us to initiate a
   (*a*) Conversation (*b*) Translation
   (*c*) Procedure (*a*) None of these
10. What name given to the way Prolog does its matching?
   (*a*) Modification (*b*) Translation
   (*c*) Conversation (*a*) Unification

Answers

1. (*c*)	2. (*c*)	3. (*a*)	4. (*c*)	5. (*b*)	6. (*c*)
7. (*c*)	8. (*a*)	9. (*a*)	10. (*d*)		

❐❐❐

APPENDICES

APPENDIX A

Computational Complexity

The concept of complexity analysis was introduced in the 1970s, to evaluate the performance of an algorithm. The worst - case analysis measures the degrees of difficulty in problem solving under the worst scenario. The computational complexity provides us an index of assessing the growth in computational effort of an algorithm as a function of the size of the problem in the worst-case analysis. The complexity of an algorithm is usually measured in this context by the number of elementary operations such as additions, multiplications and comparisons, which depends on the algorithm and the total size of the input data in binary representation.

The complexity theory makes a fundamental distinction between two kinds of algorithms. One kind-polynomial algorithms - is guaranteed to solve each problem instance in an amount of time that is polynomial in some measure of the problem size. In complexity theory, the term efficient often is used to denote polynomial algorithms. The other kind - exponential algorithms - may require an amount of time, that is, exponential in the problem size. Exponential bounds grow at a spectacularly faster rate than polynomial bounds. For example, the number of elementary particles in the visible universe is estimated at between e^{78} and e^{80}, yet a problem of size 80 would be regarded as small by almost any measure.

Through much of its history, complexity theory has been concerned with the worst - case behaviour of algorithms, that is with finding an upper bound on the time required to solve any problem instance from the chosen class. Because some algorithms show a discrepancy between the worst - case bound and performance on typical problem instances, a fair amount of recent research has focused on average - case performance. The general technique of average - case analysis is to impose a certain probability distribution on the class of problem instances and then to identify the "expected" performance of the algorithm under this distribution.

For a general iterative scheme complexity is determined by the product of the total number of iterations and the number of operations at each iteration. The total number of iterations certainly depends on the accuracy level required, while the number of elementary operations depends upon the binary representation of the input size.

In the binary system it takes $(r + 1)$ bits to represent a positive integer $\zeta \in [2^r, 2^{r+1}]$ for a non-negative integer r. Therefore for a positive integer ζ, we require $\lceil \log (1 + \zeta) \rceil$ binary bits to represent it, where $\lceil \cdot \rceil$ denotes the round - up integer value. Adding one more bit for signs, a total of $1 + \lceil \log (1 + |\zeta| \rceil$ binary bits are needed for encoding an arbitrary integer ζ.

The turing machine is a conceptual device that is used in complexity theory as a model of the computing process. The power of turing machine lies in its simplicity and generality but for the complexity analysis of numerical algorithms it is easier to work with higher - level models of computation. Rational arithmetic can be carried out by performing integer arithmetic on the numerator and denominators of the operands. For instance the operation

$$\frac{151}{210} \div \frac{79}{23} = \frac{3473}{16590}$$

is carried out *via.,* a two integer multiplications $151 \times 23 = 3473$ and $210 \times 79 = 16590$. It is not too hard to imagine how integer arithmetic like this could be performed on a turing machine. The machine could simply be programmed to go through the same pencil - and - paper procedures for integer multiplication addition and subtraction. In fact the entire rational number model of computation can be emulated on a turing machine not just the arithmetic itself but also the process of retrieving the operands from storage and storing the result. The number of steps needed by the turning machine to perform each unit of rational computation is a polynomial function of the number of bits required to store the operands, so the emulation is efficient in the sense of complexity theory.

It is obvious that the simple arithmetic operations of $+$, $-$, $\times$ and $\div$ performed in two rational arguments produce a rational result, whereas the square root operation does not preserve rationally. The process of moving from one iterate to the next involve large scale computations such as matrix multiplications and the solution of linear system. Since almost all these computations are composed from the basis $+$, $-$, $\times$ and $\div$ scalar operations, they preserve rationally.

One operation in the algorithms that does not preserve rationality is the determination of the step length α. Often, to find α, we must solve quadratic or quartic equations a process that involves taking square roots. In all cases, however, the algorithms can be modified to allow for an approximate value of α to be used instead - a rational value that is found by using only rational arithmetic yet is accurate enough to produce the same theoretical behaviour of the algorithm is the exact α.

GROWTH OF FUNCTION

Different running time can be obtained for the algorithm. The running time of an algorithm depends upon various characteristics and slight variation in the characteristics varies the running time. The algorithm efficiency and performance in comparison to alternate algorithm is best described by the order of growth of the running time of an algorithm.

Suppose one algorithm for a problem has time complexity as c^3n^2 and another algorithm has $c_1n^3 + c_2n^2$. Then, it can be easily observed that the algorithm with complexity c_3n^2 will be faster than the one with compelxity $c_1x^3 + c_2n^2$ for sufficiently large values of n. While considering the small value of n, either of the algorithm could be faster depending upon the values of c_1, c_2 and c_3. If we consider $c_3 = 100$, $c_1 = 1$ and $c_2 = 2$ then then $c_1n^3 + c_2n^2 \leq c_3\, n^2$ for $n \leq 98$ and $c_3n^2 \geq c_1n^3 + c_2\, n^2$ for $n > 98$.

From the above discussion, it can be easily observed that whatever be the value of c_1, c_2 and c_3. There will be an 'n' beyond which the algorithm having complexity c_3n^2 will be faster then the algorithm having complexity $c_1n^3 + c_2n^2$.

We will refer '*n*' as 'Break-even point for zero break-even point the algorithm with complexity c_3n^2 is faster (or at least as fast) then the algorithm with complexity $c_1n^3 + c_2n^2$. It is difficult to measure the correct breakeven point and analytically. Now let us introduce certain notations which describes the algorithm efficiency and performance in a meaningful way. These notations are referred to as "Asymptotic Notations". It describes the behaviour of time or space complexity for large instance characteristics.

Arymptotic Notation: We should know that the behaviour of the function being compared is ignored for small values of *n*. The asymptotic running time of an algorithm is defined in terms of functions. The domains of these functions are set of natural numbers and real numbers. The asymptotic notations for natural numbers and real numbers.

- The natural number set is denoted as $N = \{0, 1, 2 ...\}$
- The positive integer set is denoted as $I^+ = \{1, 2, 3 ...\}$
- The red number set is denoted as R
- The positive real number set is denoted as R^+.
- The non-negative real set is denoted as R^*.

Here we will discuss various arymptotic notations like big oh, big theta and other "order set *s*". Now consider, two functions '*f*' and '*g*'. These functions are from *N* to R^*, then figure *A* informally illustrates the relationships between the orders of functions.

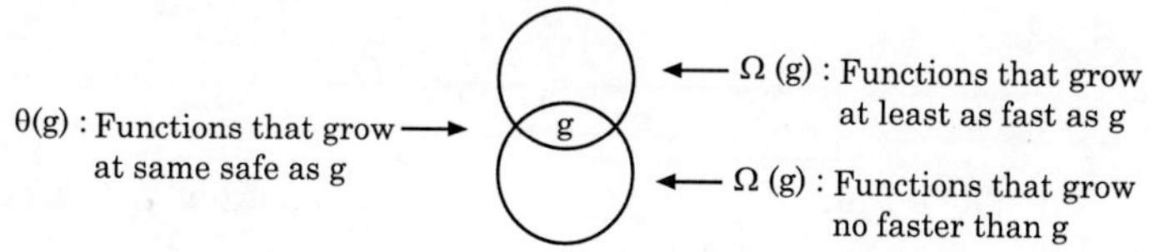

Fig. A. Asymptotic notations: Big omega (W), Big theta (θ), Big oh (O)

The asymptotic notations that we will study will permit us to constitute upper and / or lower bounds on the value of '*f*' for suitably large values of '*n*'.

Big oh Notation (*O*): The upper bound for the function '*f*' is provided by the big oh notation(*O*).

Definition [Big oh (*O*)]: Considering '**g**' to be a function from the non-negative integers into the positive real numbers. Then *O*(*g*) isthe set of function *f*, also from the non-negative integers to the positive real numbers, such that for some real constant $c > 0$ and some non-negative integers constant number $f(n) \leq c\, g(n)$ for all $n \geq n_0$.

For all values of $n > n_0$ function '*g*' is at most *c* times the function '*g*'. It can be noticed that for all '*n*' a function may be in *O*(*g*) even if $f(n) > g(n)$. Thus, '*g*' provides an upper bound by some constant multiple, on the value of '*f*' for all suitably large '*n*' (*i.e.*, $n \geq$ ho). The set *O*(*g*) is usually called as 'oh of *g*' or big '*oh* of *g*'. As *O*(*g*) is described as a set it is a good practice to say '*f* is oh of *g*' or '*f* is big oh of *g*'. Some common asymptotic functions are as follows:

Constant: 1, ...

algorithmic : log *n* ...

linear : n ...
quadratic : n^2 ...
exponential : 2^n ...
factorial : $n!$ ...
cubic : n^3 ...

In general, $O(g(n)) = \{f(n)\}$: there exists positive constants such that $O \le f \le c\, g(n)$ for all n, $n \ge n_o$.

Constant function: Suppose given constant functions:

- $f(n) = 16$
- $f(n) = 25$
- $f(n) = 1526$

Then for satisfying the big *oh* condition, the above function can be expressed as follow:

$f(n) \le 16 * 1$ where $c = 16$ and $n_0 = 0$
$f(n) \le 25 * 1$ where $c = 25$ and $n_0 = 0$
$f(n) \le 1526 * 1$ where $c = 1526$ and $n_0 = 0$

Thus, for above function we can specify the big oh notation as $O(1)$. So, $f(n) = 0(1)$

Linear function consider following linear functions

- $f(n) = 3n + 6$
- $f(n) = 2n + 5$
- $f(n) = 7n + 9$

For $f(n) = 3n + 6$, when 'n' is at least 6, $n \ge 6$

$3n + 6 \le 3x + n \le 4n$

So, $\boxed{f(n) = O(n)}$

Thus, the above function is bound by a linear function. We can also prove the above function is another way as:

$3n + 6 \le 3n + 6n \le 8n$, for $n \ge 1$

where $c = 8$, $n_o = 1$ which satisfies the big oh condition. Similarly

$\Rightarrow$ $f(n) = 2n + 5 \le 2n + 2n \le 3n$ $(c = 3, n_o = 3)$

So $\boxed{f(n) = O(n)}$

$f(n) = 7n + 9$ for $n \ge 9$

$\Rightarrow$ $7n + 9 \le 7n + n \le 8(c = 8, n = 9)$

So $\boxed{f(n) = O(n)}$

Quadratic function: Consider the following quadratic functions as:

- $f(n) = 27n^2 + 16n$
- $f(n) = 27n^2 + 16$
- $f(n) = 10n^2 + 7$
- $f(n) = 27n^2 + 16n + 25$

Taking $f(n) = 27n^2 + 16n$, for $n^2 \ge 16n$ {or $16\,n \le n^2$}

$27n^2 + 16n \le 27n^2 + n^2 \le 28n^2$, $(c = 28, n_o = 16)$

so $\boxed{f(n) = O(n^2)}$

if we consider $n \leq n^2$ then
$27n^2 + 16n \leq 27n^2 + 16n^2 \leq 43n^2$ ($c = 43$, $n_o = 1$)

So, $\boxed{f(n) = O(n^2)}$

$\Rightarrow$ $f(n) = 27\, n^2 + 16$ for $n \geq 16$
$27n^2 + 16 \leq 27n^2 + n$
Now, for $n \leq n^2$
$27n^2 + n \leq 27\, n^2 + n^2 \leq 28n^2$, {(c = 28, $n_o = 1$)}

So, $\boxed{f(n) = O(n^2)}$

$f(n) = 10\, n^2 + 7$ for $n \geq 7$
$10\, n^2 + 7 \leq 10^2 + n$
Now for $n \leq n^2$
$10n^2 + n \leq 10n^2 + n^2 \leq 11n^2$ ($c = 11$, $n_o = 1$)

So, $\boxed{f(n) = O(n^2)}$

$\Rightarrow$ $f(n) = 27n^2 + 16n + 25$, for $n \geq 25$
$27\, n^2 + 16n + 25 \leq 27n^2 + 16n + n \leq 27n^2 + 17n$
Now, for $n \geq 17$, $n^2 \geq 17n$, thus
$27n^2 + 17n \leq 27n^2 + n^2 \leq 28n^2$ ($c = 28$, $n_o = 17$)

So, $\boxed{f(n) = O(n^2)}$

Cubic Functions: Consider the following cubic functions.

- $f(n) = 2n^3 + n^2 + 2n$
- $f(n) = 4n^3 + 2n + 3$
- $f(n) = 3n^3 + 4n$
- $f(n) = 5n^3 + n^2 + 3n + 2$

$\Rightarrow$ $f(n) = 2n^3 + n^2 + 2n$
For $n^2 \geq 2n$
$2n^3 + n^2 + 2n \leq 2n^3 + n^2 + n^2 \leq 2n^3 + 2n^2$
Now, for $n^3 \geq 2n^2$
$2n^3 + 2n^2 \leq 2n^3 + n^3 \leq 3n^3$ ($c = 3$, $n_o = 2$)

So, $\boxed{f(n) = O(n^3)}$

$\Rightarrow$ $f(n) = 4n^3 + 2n + 3$ for $n \geq 3$
$4n^3 + 2n + 3 \leq 4\, n^3 + 2n + n \leq 4n^3 + 3n$
Now, for $n^3 \leq 3n$
$4n^3 + 3n \leq 4n^3 + n^3 \leq 4n^3$ ($c = 5$, $n_o = 3$)

So, $\boxed{f(n) = O(n^3)}$

$\Rightarrow$ for $f(n) = 3n^3 + 4n$ for $n^3 \geq 4n$
$3n^3 + 4n \leq 3n^3 + n^3 \leq 4n$ ($c = 4$, $n_o = 4$)

So, $\boxed{f(n) = O(n^3)}$

$\Rightarrow$ $f(n) = 5n^3 + n^2 + 3n + 2 \quad \text{for } n \geq 2$

$5n^3 + n^2 + 3n + 2 \leq 5n^2 + n^2 + 3n + n \leq 5n^3 + n^2 + 4n$ for $n^2 \geq 4n$

$5n^3 + n^2 + 4n \leq 5n^3 + n^2 + n^2 \leq 5n^3 + 2n^2$ for $n^3 \geq 2n^2$

$5n^3 + n^2 \leq 5n^3 + n^3 \leq 6n^3$ $(c = 6, n_o = 2)$

So, $\boxed{f(n) = O(n^3)}$

Exponential Functions: Consider the following exponential functions:

- $f(n) = 2^n + 6n^2 + 3n$
- $f(n) = 4 * 2^n + 3n$
- $f(n) = 5 * 2^n + 3n + 5$
- $f(n) = 6 * 2^n + 6$
- $f(n) = 3 * 2^n + 4n^2 + 5n + 3$

$\Rightarrow$ $f(n) = 2^n + 6n^2 + 3n$ for $n^2 \geq 3n$

$2^n + 6n^2 + 3n \leq 2n + 6n^2 + n^2 \leq 2^n + 7n^2$

for $2n \geq n^2$ $\{n \geq 4\}$

$2^n + 7n^2 \leq 2^n + 7 * 2^n \leq 8 * 2^n$ $\{c = 8, n_o = 4\}$

So, $\boxed{f(n) = O(2^n)}$

$\Rightarrow$ $f(n) = 4 * 2^n + 3n$ for $2^n > n$ $\{n \geq 1\}$

$4 * 2^n + 3n \leq 4 * 2^n + 3 * 2^n = 7 * 2^n$ $\{c = 7, n_o = 1\}$

So, $\boxed{f(n) = O(2^n)}$

$\Rightarrow$ $f(n) = 5 * 2^n + 3n + 5$ for $n \geq 5$

$5 * 2^n + 3n + 5 \leq 5 * 2^n + 3^n + n \leq 5 * 2^n + 4n$

for $2n > n$ $\{n \geq 1\}$

$5 * 2^n + 4n \leq 5 * 2n + 4 * 2n \leq 9 * 2n$ $\{c = 9, n_o = 1\}$

So, $\boxed{f(n) = O(2^n)}$

$\Rightarrow$ $f(n) = 6 * 2^n + 6$

for $2n \leq 6$ $\quad \{n \geq n_o = 0\}$

$6 * 2n + 6 \leq 6 * 2^n + 2n \leq 7 * 2n$ $\{c = 7, n_o = 0\}$

So, $\boxed{f(n) = O(2^n)}$

$\Rightarrow$ $f(n) = 3 * 2^n + 4n^2 + 5n + 3$ for $n \geq 3$

$3 * 2^n + 4n^2 + 5n + 3 \leq 3 * 2^n + 4n^2 + 5n + n \leq 3 * 2^n + 4n^2 + 6n$

for $n \geq 6$, $n^2 \geq 6n$

$3 * 2^n + 4n^2 + 6n \leq 3 * 2^n + 4n^2 + n^2 \leq 3 * 2^n + 5n^2$

for $n \geq 5$, $n^2 \leq 2n$

$3 * 2^n + 5n^2 \leq 3 * 2^n + 5 * 2n \leq 8 * 2^n$ $\{c = 8, n_o = 5\}$

So, $\boxed{f(n) = O(2^n)}$

Incorrect Bounds:

- $7n + 5 \neq O(1)$
- $2n + 3 \neq O(1)$
- $10n^2 + 7 \neq O(n)$
- $27n^2 + 16n + 25 \neq O(n)$
- $3n^3 + 4n \neq O(n^2)$
- $5n^3 + n^2 + 3n + 2 \neq O(n^2)$

$\Rightarrow \quad 7n + 5 \neq O(1)$

Let us assume that $7n + 5 = O(1)$ then, $7n + 5 < 0$ for all n, $n \geq n_o$, thus $n < c\ (c - 5) / 7$ for all $n \geq n_o$.

The above condition is false for

$$\boxed{n > \max \{n_o, (c - 5) / 7\}}$$

because the left side increase as 'n' increases and there is no effect on the right hand side.

Thus, $7n + 5 \neq O(1)$ is **True**

$\Rightarrow \quad 2n + 3 \neq O(1)$

Consider $2n + 3 = O(1)$ then,

$2n + 3 < c$ for all n, $n \geq n_o$

Thus $\boxed{n < (c - 3)/2}$ for all n, $n \geq n_o$

The above condition is **false** for,

$$\boxed{n > \max \{n_o, (c - 3) / 2\}}$$

Thus, $2n + 3 \neq O(1)$ is **True**

$\Rightarrow \quad 10n^2 + 7 \neq O(n)$

Consider $10n^2 + 7 = O(n)$ then

$10n^2 + 7 < cn$ for all n, $n \geq n_o$

now divide both the sides by 'n'

$\boxed{10\ n + 7/n < c}$ for all n, $n \geq n_o$.

The above relation is **false** as left side increases is 'n' increases and there will be no change on the right hand side.

Thus, $10\ n^2 + 7 \neq O(n)$ is **true**

$\Rightarrow 27n^2 + 16n + 25 \neq O(n)$

Consider $27n^2 + 16n + 25 = O(n)$ then,

$27n^2 + 16n + 25 < cn$ for all n, $n \geq n_o$

Divide both the side of the relation by 'n'

$\boxed{27n + 16 + 25/n < c}$ for all n, $n \geq n_o$.

But the above relation is **incorrect** as left hand side of the relation increases as 'n' increases and there will be no change on the right hand side of the relation.

Thus, $27n^2 + 16\ n + 25 \neq O(n)$ is **True.**

$\Rightarrow \quad f(n) = 3n^3 + 4n \neq O(n^2)$

Consider, $3n^3 + 4n = O(n^2)$, then
$3n^3 + 4n < c\, n^2$ for all n, $n \leq n_o$
Divide both sides by n^2

$3n + 4/n < c$	for all n, $n \geq n_o$

The above relation is incorrect as 'n' increases, left hand side of the relation is also increases and there will be no change on the right hand side of the relation.

Thus, $3n^3 + 4n \neq O(n)$ is True

$\Rightarrow$ $$f(n) = 5n^3 + n^2 + 3n + 2 \neq O(n^2)$$

Consider $5n^3 + n^2 + 3n + 2 = O(n^2)$ then,
$5n^3 + n^2 + 3n + 2 < Cn^2$ for all n, $n \leq n_o$
divide both the sides of the relation by 'n^2'.

$5n + 1 + 3/n + 2/n^2 < c$

But the above relation is **incorrect** as 'n' increases the value of the left side of the relation increases where as the right side does not change.

Thus, $5n^3 + n^2 + 3n + 2 \neq O(n)$ is **True.**

Loose Bounds:

- $2n + 3 = O(n^2)$
- $4n^2 + 5n + 6 = O(n^4)$
- $3n^3 + 4n^2 + n = O(n^6)$
- $7n * 2^n + 22 = O(n^2.2^n)$

$\Rightarrow f(n) = 2n + 3$, since $2n + 3 \leq n^2$ for all n, $n \geq n_o = 3$.

But the above bound is not the tighter bound, as we have a smaller function (in this case linear) that also satisfies for the big oh relation. Hence, **True**.

$\Rightarrow$ $$f(n) = 4n^2 + 5n + 6 = O(n^4)$$

Since, $4n^2 + 5n + 6 < 4n^4$ for all n, $n \geq n_o = 2$

In this case also, n^4 does not provide the tight upper bound, as we have 'n^2' that also satisfies the big-oh relation.

$\Rightarrow$ $$f(n) = 3n^3 + 4n^2 + n = O(n^6)$$

Since, $3n^3 + 4n^2 + n < 3n^6$ for all n, $n \geq n_o = 2$ again, 'n^6' does not provide the tight upper bound, as we have 'n^3' (cubic function) a smaller function that also satisfies the big oh relation

$\Rightarrow$ $$7n * 2n + 22 = O(n^2\, 2^n)$$

Since, $7n * 2^n + 22 < 7n^2\, 2^n$ for all n, $n \geq n_o = 2$ again, '$7^n\, 2^n$' does not provide the tight upper bound as we have 'n^{2n}' that also satisfies the big-oh relation.

From the above examples we can depict that the statement $f(n) = O(g(n))$ describes that $c\, g(n)$ is an upper bound on the value of $f(n)$ for all n, $n \geq n_o$. It does not say anything about how good or tight this bound is.

> ✍ **Note:** When $f(n)$ is polynomial of n, then order of $f(n)$, i.e., the $g(n)$ is $f(n) = O(g(n))$.

Theorem 1: If $f(n) = a_m n^m + a_{m-1} n^{n-1} + \ldots$
$\ldots a_1 n + a_0$ and $a_m > 0$ then $f(n) = O(n^m)$

Proof: $$f(n) \le \sum_{k=0}^{m} |a_k| n^k$$

$$\le n^m \sum_{0}^{m} |a_k| n^{k-m}$$

$$\le n^m \sum_{0}^{m} |a_k| \text{ for } n \ge 1$$

So $\boxed{f(n) = O(n^m)}$

Applying Theorem 1

- $f(n) = 2n + 3$ $\quad m = 1$, thus $f(n) = O(n^m) = O(n)$
- $f(n) = 7n + 5$ $\quad m = 1$, thus $f(n) = O(n^m) = O(n)$
- $f(n) = 27n^2 + 16n + 25$ $\quad m = 2$, thus $f(n) = O(n^m) = O(n^2)$
- $f(n) = 10n^2 + 7$ $\quad m = 2$, thus $f(n) = O(n^m) = O(n^2)$
- $f(n) = 3n^3 + 4\,n$ $\quad m = 2$, thus $f(n) = O(n^m) = O(n^3)$
- $f(n) = 5n^3 + n^2 + 3n + 2$ $\quad m = 3$, thus $f(n) = O(n^m) = O(n^3)$

Theorem 2:

Let $f(n)$ and $g(n)$ be such that $\lim_{n\to\infty} \frac{f(n)}{g(n)}$ exists, then a function of $\varepsilon\ O(g)$ if

$\lim_{n\to\infty} \frac{f(n)}{g(n)} = c < \infty$, also including the case in which limit is O.

Proof: If $f(n) = O(g(n))$ then, for all n, $n \ge n_o$ positive 'c' and 'n_o' exist such that $\frac{f(n)}{g(n)} \le c$. Hence, $\lim_{n\to\infty} \frac{f(n)}{g(n)}$.

Now suppose $\lim_{n\to\infty} \frac{f(n)}{g(n)} \le c$, then that means an '$n_0$' exists such that

$$f(n) \le \max\{1, c\} \times g(n) \text{ for all } n,\ n \ge n_o.$$

Explanation: Growth rates,

$$\lim_{n\to\infty} \frac{f(n)}{g(n)} \begin{cases} O\ f(n) \text{ grows slower than } g(n) \\ \infty\ f(n) \text{ grows faster than } g(n) \\ \text{otherwise } f(n) \text{ and } g(n) \text{ have the same} \\ \text{growth rates} \end{cases}$$

The growth rates of some functions are shown in Fig. B.

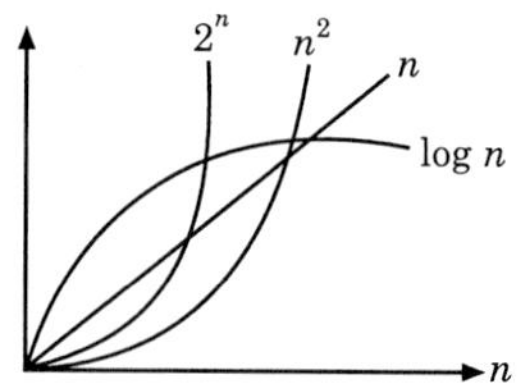

Fig. B. Growth rates of functions

Applying Theorem 2:

- $f(n) = 2n + 3 = O(n)$
- $f(n) = 7n + 5 = O(n)$
- $f(n) = 27n^2 + 16n + 25 = O(n^2)$
- $f(n) = 10n^2 + 7 = O(n^2)$
- $f(n) = 3n^3 + 4n = O(n^3)$
- $f(n) = 5n^3 + n^2 + 3n + 2 = O(n^3)$
- $f(n) = 2^n + 6n^2 + 3n = O(2^n)$
- $f(n) = 5 \times 2^n + 3n + 5 = O(2^n)$

$\Rightarrow$ $f(n) = 2n + 3 = O(n)$

$$\lim_{n \to \infty} \frac{2n + 3}{n} = 2$$

Thus, $\boxed{f(n) = O(n)}$

$\Rightarrow$ $f(n) = 7n + 5 = O(n)$

$$\lim_{n \to \infty} \frac{7n + 5}{n} = 7$$

Thus, $\boxed{f(n) = O(n)}$

$\Rightarrow$ $f(n) = 27n^2 + 16n + 25 = O(n^2)$

$$\lim_{n \to \infty} \frac{27n^2 + 16n + 25}{n^2} = 27$$

Thus, $\boxed{f(n) = O(n)}$

$\Rightarrow$ $f(n) = 10n^2 + 7 = O(n^2)$

$$\lim_{n \to \infty} \frac{10n^2 + 7}{n^2} = 10$$

Thus, $\boxed{f(n) = O(n^2)}$

$\Rightarrow$ $f(n) = 3n^3 + 4n = O(n^3)$

$$\lim_{n \to \infty} \frac{3n^3 + 4n}{n^3} = 3$$

Thus, $\boxed{f(n) = O(n^3)}$

$\Rightarrow \qquad f(n) = 5n^3 + n^2 + 3n + 2$

$$\lim_{n \to \infty} \frac{5n^3 + n^2 + 3n + 2}{n^3} = 5$$

Thus, $\boxed{f(n) = O(n^3)}$

$\Rightarrow \qquad f(n) = 2^n + 6n^2 + 3n$

$$\lim_{n \to \infty} \frac{2n + 6n^2 + 3n}{2n} = 1$$

Thus, $\boxed{f(n) = O(2^n)}$

$\Rightarrow \qquad f(n) = 5 * 2^n + 3n + 5,$

$$\lim_{n \to \infty} \frac{5 \times 2^n + 3n + 5}{2^n} = 5$$

Thus, $\boxed{f(n) = O(2^n)}$

SUMMARY

- For an algorithm different running time can be obtained.
- The running time of an algorithm depends upon various characteristics and slight variation in the characteristics varies the running time.
- Order of growth of the algorithm running time can describe the algorithm efficiency and performance in comparison to alternate algorithm.
- Asymptotic notations are used to describe the algorithm efficiency and performance in a meaningful way.
- Asymptotic notations describe the behaviour of time or space complexity for large instance characteristics.
- Some asymptotic notations are Big oh(O), Big theta(θ), Big omega (Ω), little oh(o) and little omega (ω).
- Asymptotic notation properties are:
  - Reflexivity
  - Symmetry
  - Transpose symmetry
  - Transitivity

APPENDIX B

Dijkastra's Algorithm for Shortest Path in Weighted Graphs

Step 1: set $\lambda(s) = 0$ and for all vertices $V \neq s$, $\lambda(v) = \infty$ set $T = V$
where T = set of "uncoloured" vertices.
V = vertex set of G.

Step 2 : Let u be vertex in T for which $\lambda(u)$ is minimum.

Step 3 : If $u = t$, stop.

Step 4 : For every edge $e = uv$, incident with u, if $v \in T$ *i.e.* v is from uncoloured set of vertices, change $\lambda(v)$ to min $(\lambda(v), \lambda(u) + w/e))$.

Step 5 : Change T to T-{4} [colour u] and go to step 2 to find an "uncoloured" vertex with minimum label.

Problem : Apply Dijkastra's algorithm to the graph given below and find the shortest path from a to f.

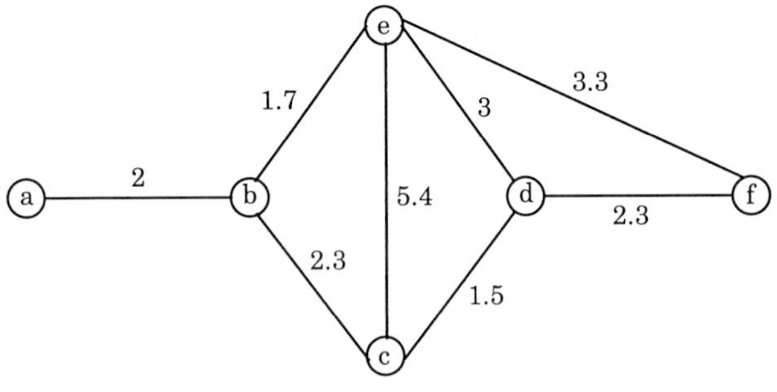

Fig. 1.

Solution : The initial labelling is given b

Vertex V	a	b	c	d	e	f
$\lambda(v)$	0	∞	∞	∞	∞	∞
T	$\{a$	b	c	d	e	$f\}$

Iteration 1 : $u = a$ has $\lambda(4) = 0$ (min),

There are two edges incident with a *i.e.*, ab and ac. Both b, c are in T (uncoloured), $\lambda(b) = \min(\infty, 0 + 2.0)$

$\Rightarrow \lambda(b) = 2.0$

Similarly, $\lambda(c) = 3.0$

Now T becomes T-$\{a\}$ *i.e.*, we colour a,

Vertex V	a	b	c	d	e	f
$\lambda(v)$	0	2.0	3.0	∞	∞	∞
T		{b	c	d	e	f}

Iteration 2. Again $u = b$ has $\lambda(4)$ a min and $u \in T$. There are two edges be and bc incident on b. where e and $c \in T$.

Now $\quad \lambda(e) = \min \{\infty, 2.0 + 1.7 = 3.7\} = 3.7$

$\lambda(e) = \min \{3, 2 + 2.3 = 4.3\} = 3$

Thus T become T-{b} *i.e.*, we colour b.

Vertex V	a	b	c	d	e	f
$\lambda(v)$	0	2.0	3.0	∞	3.7	∞
T			{c	d	e	f}

Iteration 3. Again $u = c$ has $\lambda(u)$ a min for u in T.

There are two edges incident at c *i.e.*, cd and ce

$\lambda(d) = \min \{\infty, 3 + 1.5 = 4.5\}$ 4.5

$\lambda(e) = \min \{3.7, 3 + 5.4 = 8.4\} = 3.7$

and T become T-{c} *i.e.*, we colour c

Vertex V	a	b	c	d	e	f
$\lambda(v)$	0.	2.0	3.0	4.5	3.7	∞
T				{d	e	f}

Iteration 4. Again $u = e$ has $d(u)$ minimum for u in T. There are two edges incident on e from T *i.e.*, ed and ef.

Since $\quad \lambda(d) = \min\{4.5, \lambda(d) + 3.0)$

$= \min \{4.5, 3.7 + 3 = 6.7\} = 4.5$

$\lambda(f) = \min \{\infty, 3.7 + 3.7 = 7\} = 7$

T become T-{e} *i.e.* we colour e.

Vertex V	a	b	c	d	e	f
$\lambda(v)$	0	2.0	3.0	4.5	3.7	7
T				{d		f}

Iteration 5. $u = d$ as $\lambda(4)$ is min for u in T.

The edge incident on d *i.e.*, df.

$\lambda(f) = \min\{7, 4.5 + 2.3 = 6.8\}$

$= 6.8$

and T become T-{d}, *i.e.*, colour d.

Vertex V	a	b	c	d	e	f
$\lambda(v)$	0	2.0	3.0	4.5	3.7	6.8
T						{f}

$u = f$, the only choice.

Hence stop

Thus the length of the shortest path from a to b, c, d, e, f are 2.0, 3.0 4.5, 3,7 and 6.8 rest.

Problem. Find a minimal spanning tree for the connected weighted graph given below using both kruskal's and Prim's algorithm.

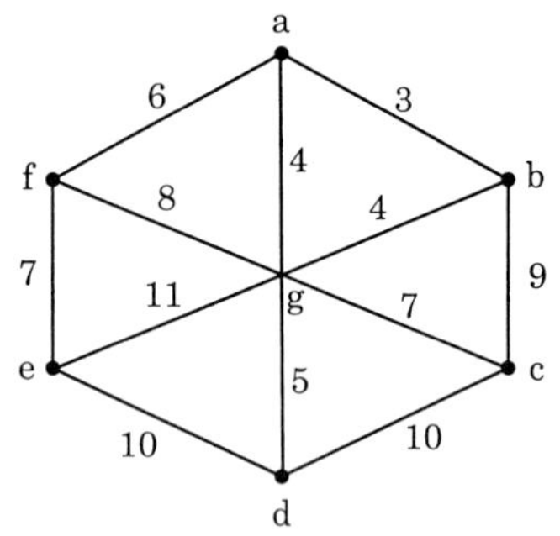

Fig. 2.

Problem. Apply Dijkastra's algorithm to find the shortest from s to t in the graph given below.

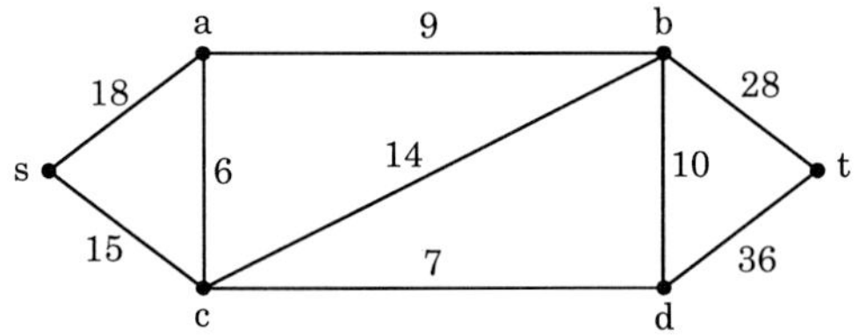

Fig. 3.

Divide and conquer ;

Algorithm D And C(P)

If small (P) then return S(P) ;

else

Divide P into smaller instance P1, P2 .... PK K [3]

Apply D And C to each of them subproblems return combine (D And C (P1), D And C(P2) ... D And C)

❐❐❐

APPENDIX C

Divide and Conquenr Technique

Divide and Conquer's

```
{
Algorithm  D and C(P)
If small (P) then return S(P) ;
else → boolean valued function
{
Divide P into smaller instance P1, P2, .... PK. K ≥ 1.
```

Apply D and C to each of these sub-problems return combine D and C (P_1), D and C $P_2)$ .... and D and $C(P_K)$.

Given a function to compute an n inputs the divide and conquer strategy suggests splitting the input into K distinct subsets. $1 \leq K \leq n$ yielding K subproblems. These subproblems must be solved, and then method must be found to combine sub-solution into a solution of the whole. If the subproblems are still relatively large then the divide and conquer strategy can possibly by reapplied often the subproblems resulting from a divided and conquer design are of the same type as the original problem. Now smaller and smaller subproblems of the same land are generated until eventually subproblems that are small enough to be solved without splitting are produced.

$$\begin{aligned} T(n) &= g(n) \\ &= T(n_1) + T(n_2) + \ldots\ldots + T(n_K) + f(n) \quad \text{otherwise} \end{aligned}$$

T(n) is the time for D and C or any input of sin n and g(n) is the time to compute the answer directly for small inputs. The function f(n) is the time for dividing p and compining the solution to sub-problems.

The complexity of many divide and conquer algorithm is given by recurrence of the form

$$\begin{aligned} T(n) &= T(1) \quad n = 1 \\ &= aT(n/b) + f(n) \quad n > 1 \end{aligned}$$

Binsearch :

```
Algorithm  Binsearch (a, i, l, n)
{
If (l = i) then
{ If (x = a(i)), then return i ;
else return 0 '
}
```

```
else
{ mid = [(i + 1)/2]
IF (x = a[mid]) then return mide ;
else if (n < a[mid]) then
return Binsearch (a, i, mid-1, x) ;
else return Binsearch (a, mid + 1 l, n) ]
}
}
```

Recursive Binary Search.

Searching problem

Let a_1 $1 \le i \le n$ be a list of elements that are worked in nondecreasing order, consider the problem of determining whether a given element x is present in the list. If n is present, we are to determine a value j such that $a_j = x$. If n is siot in the list then j is to be set to 0. Divide and conquer can be used to solve this problem. Let small (P) be true if $n = 1$ where $P = (n_i\ a_i\ \ldots.., a_l, x)$ denote an arbitrary instance of this search problem.

In this case $S(P)$ will take the value i if $n = a_i$ else it will than the value 0 than $g(1) = 0(1)$. If P has more than one element, it can be divided into a new subproblem as follows :

(1) $n = a_q$ problem p immediately solved.

(2) $x < a_q$ in this case x has to be searched for only in the subset a_i, a_{i+1} $a_q -$ 1 any in the subset a_i, a_{i+1}, a_{q-1}.

(3) $x > a_q$ x has to be searched for only in the sublist a_{q+1} ..... a_l.

Space Complexity

Storage is required for storing n element variables low, high, mid, and x —4 so total space location required $n + 4$.

Time Complexity

–15, –6, 0, 7, 9, 23, 54, 82, 101, 112, 125, 131, 142, 151

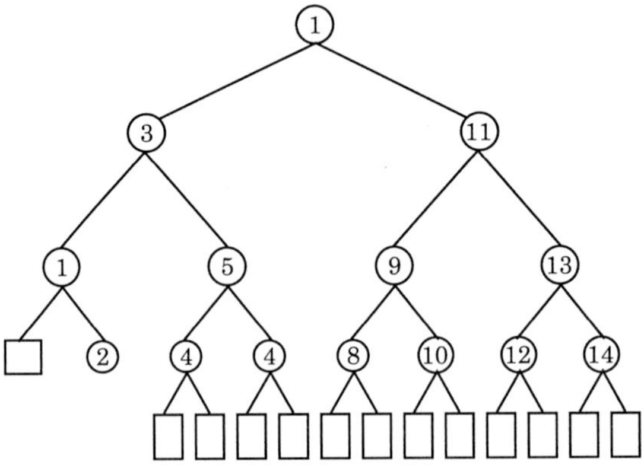

Successful search will stop at circular node

best $Q(1)$

averge $Q(\log n)$

worst $Q(\log n)$

unsuccessful search will stop at square node

$$\begin{array}{ccc} Q & (\log & n) \\ \downarrow & \downarrow & \downarrow \\ \text{best} & \text{average} & \text{worst} \end{array}$$

Best case

If the element to be found at this root or at the first position than the time complexity is $Q(1)$ because we are finding the time complexity in a form of no. of comparision because $Q(1)$ are $\Omega(1)$ so (1).

Worst Case

The complexity is measured by the number $f(n)$ of comparisons to locate item in Data set where data set contains n element. Observe that each elemetn comparison reduces the sample six in half. Hence we require atmost $f(n)$ comparison to locate item where

$$2^{f(n)} > n$$

$$f(n) = [\log = n] + 1$$

running time for the worst case is approximately equal to $\log_2(n)$. One can easily show that the running time for the average case is approximately equal to running time for the worst case.

❐❐❐

MISCELLANEOUS EXERCISES

MISCELLANEOUS EXERCISE–I

1. Answer the following:
   (*a*) What is equivalence relation ? Explain with the help of suitable example.
   (*b*) Differentiate between Partial Ordered Sets and Partial Ordered Relation.
   (*c*) Elaborate different mapping techniques in functions.
2. Attempt the following :
   (*a*) Prove that Power set of a given set $S = \{a, b, c\}$ is a Lattice.
   (*b*) Show that any subgroup of a cyclic group is cyclic.
   (*c*) $F = XY' + YZ'$
   Implement given Boolean expression F using single NOT gate and n number of AND/OR gates.
3. Attempt the following :
   (*a*) Prove that both Join and Meet operations are associative.
   (*b*) Let A be a set of books and R is a binary relation on A such that (a, b) is in R if book a costs more and contains fewer pages than b. In general R is Reflexive ? Symmetric ? Antisymmetric ? Transitive ? Equivalence ? Partially Ordered ?
   (*c*) If R^{-1} and S^{-1} be the inverse of the relation R and S respectively then $(SOR)^{-1}\ R^{-1}\ OS^{-1}$.
   (*d*) Convert following expression into another canonical form.
   $XY'Z + X'Y' + XY$
4. Attempt the following :
   (*a*) Let $(L, \leq)$ be a lattice in which * and $\oplus$ denote the operations Meet & Join respectively. For any a, b belongs to L then prove that
   $a \leq b \Leftrightarrow a * b = a \Leftrightarrow a \oplus b = b$
   (*b*) Let $(L, \leq)$ be a lattice. For any a, b, c belongs to L, the following inequalities, called the distributed inequalities prove that
   $a \oplus (b * c) \leq (a \oplus b) * (a \oplus c)$
   $a * (b \oplus c) \geq (a * b) \oplus (a * c)$
   (*c*) Check the validity of following arguments :
   (*i*) If Mary runs for office, she will be elected.
   (*ii*) If Mary attends the meeting, she will run for office.
5. Attempt the following :
   (*a*) If he is wealthy then he is happy or he is unhappy. Prove that this statement is tautology and also write contradiction of the given statement.
   (*b*) Construct truth table for $(P \vee Q) \vee {\sim}P$
   (*c*) Prove that $(p \rightarrow Q) \Leftrightarrow ({\sim}P \vee Q)$

(*d*) Explain recursively defined function with suitable examples.

(*e*) Let (*A*, .) be a group show that $(ab)^{-1} = b^{-1} a^{-1}$.

6. (*a*) Show that $5n^3 - 3n + 4$ is $O(n^3)$.

(*b*) Prove the following identities for regular expressions

(*i*) $(a + b)^* = a^* + b^*$ (*b*) $b(ab + b)^* a = aa^*b(aa^*b)^*$

(*c*) Describe the sets represented by following expressions

(*i*) $ab(bc)^*$ (*ii*) $a(b + c)$

7. Write short note of the following :

(*a*) Distributed Law in Boolean algebra

(*b*) Complemented Lattice

(*c*) Greatest Lower bound

(*d*) Universal upper bound

(*e*) Monoid

(*f*) Combinations with repetitions.

❒❒❒

MISCELLANEOUS EXERCISE–II

1. (*a*) Thirteen persons have first names Dennis, Evita and Ferdinand and last names oh, Pietro, qunie and Rostukowski. Show that at last two persons have the seine first and last names.

   (*b*) Prove that the necessary and sufficient condition that a non-empty subset H of a group G by a subgroup is $a \in H,\ b \in H \Rightarrow a * b^{-1} \in G$.

   (*c*) Prove that any two right (or left) cosets of H in G are either disjoint or identical.

   (*d*) If for any element a in a ring $a^2 = a$. Show that it is always commutative.

   (*e*) Find the Boolean expression for:
   $x_1 = 1,\ x_2 = 1,\ x_3 = 0,\ x_4 = 1$
   $(x_1 \wedge \bar{x}_2) \vee (x_1 \vee \bar{x}_3)$

   (*f*) Expand the following functions into their canonical sum-of-products form
   $f_1\ (w, x, y, z) = (xy + y\overline{w}z)$.

2. (*a*) Show that every nonempty subset of a lattice has a least upper bound and a greatest lower bound.

   (*b*) State and prove pumping lemma.

   (*c*) Prove that every distribution lattice is a chain.

   (*d*) Let $X = \{1, 2, 3\}$ and the set P is the set of all subsets of power set. Draw the Hasse diagram for the elements of P under the partial order of relation.

   (*e*) Show that in a bounded lattice, the elements which have compliments form a sublattice.

   (*f*) Show that the direct product of any two distributive lattices in a distributive lattice.

3. (*a*) Find the solution of the recurrence relation
   $a_r - 6a_{r-1} + 8a_{r-2} = 4^r$
   $a_0 = 8$ and $a_1 = 22$.

   (*b*) Solve the following recurence relation using generating functions
   $a_n - 9a_{n-1} + 20a_{n-2} = 0$ for $n \geq 2$, $a_0 = 3$ and $a_1 = -10$.

   (*c*) A committee of 5 is to be formed out of 6 gentlement and 4 ladies. In how many ways can this be done if.
   (*i*) At least two ladies are to be included ?
   (*ii*) Atmost two ladies are to be included ?

   (*d*) Find the number of natural number less than 1000 and divisible by 5 which can be formed with the digits 0, 1, 2, 3....9 each digit not occuring more than over in each number.

   (*e*) In how many ways can 7 boys and 5 girls be seated together in a row, so that no two girls are together ?

(*f*) Find the closed form impression for $a_n = \frac{5}{2} . 2^n + \frac{3}{2} n . 2^n$.

4. (*a*) Prove that the relation R of similarity in the set of all triangles in a plane is an equivalence relation.
   (*b*) Prove that $(fog)^{-1} = g^{-1} of^{-1}$ where f and g are functions (one-one onto).
   (*c*) Show that closure of a relation R with respect to a property if it exist is the intersection of all the relation with the property P that contain R.
5. (*a*) If $A_1, A_2 ......, A_h, P, Q$ are propositions and $(A_1 \wedge A_2 \wedge A_3 ..... A_n \wedge P) \Rightarrow Q$ then prove that $(A_1 \wedge A_2 \wedge A_3 ..... A_n) \rightarrow (P \rightarrow Q)$
   (*b*) Show that following implicatons
   (*i*) $Q \Rightarrow (P \rightarrow Q)$
   (*ii*) $P \wedge (P \rightarrow Q) \Rightarrow Q$
   (*c*) Form a truth table for
   (*i*) $((P \vee Q) \rightarrow R) \equiv (P \rightarrow R) \wedge (Q \wedge R)$
   (*ii*) $\sim(P \leftrightarrow Q) \equiv (P \wedge \sim Q) \vee (\sim P \wedge Q)$

❐❐❐

MISCELLANEOUS EXERCISE–III

1. (*a*) A servey of 500 television Watches produced the following Inforamtion. 285 watch football Game, 195 Watch Hockey Games, 115 Watch Basketball Games, 145 Watch football and basket ball games, 70 wtach football and hockey games, 50 watch hockey and basketball games and 50 do not watch any of the three kinds of games.

   (*i*) How many people in the survey watch all three kinds of games ?

   (*ii*) How many people watch exactly one of the sports?

   (*b*) Let $A = \{1, 2, 3, 4, 5, 6\}$ Construct Pictorial descriptions of the relation R on A for the following cases.

   (*i*) $R = \{(J, k) \mid \text{divides } k\}$

   (*ii*) $R = \{(J, k) \mid J \text{ is multiple of } k\}$

   (*iii*) $R = \{(J, k) \mid (J - k) \wedge 2 \text{ belongs to } A\}$

   (*iv*) $R = \{(J, k) \mid J/k \text{ is a Prime}\}$

2. (*a*) Prove by Induction that the Sum of the Cubes of Three Consecutive Integers is Divisible by 9.

   (*b*) Show that by Mathematical induction

   $1^2 - 2^2 + 3^2 - 4^2 + ... (-1)^{n-1} n^2 = (-1)(n - 1) * n(n + 1)/2$

3. (*a*) An island has two tribes of natives. Any native from the first tribe always Tells the truth. While any Native from the other Tribe Always Lies. You arrive at the island and ask a native if there is Gold on the island. He Answers, "There is gold on the island if and only if I always tell the Truth". Which tribe is he from ? Is there Gold on the island ?

   (*b*) Let f, g, h be functions from N to N, where N is the set of natural numbers so that

   $f(n) = n + 1, g(n) = 2n$

   $h(n) = 0, n$ is even

   $\quad\;\; = 1, n$ is odd

   Determine *fof, fog, gof, goh, hog, (fog) oh.*

4. (*a*) If 13 people are a Member in a room, Show that at least 2 of them must have three birthday in the Same Month.

   (*b*) Let G be a group and let a and b be elements of G then

   (*i*) The equation $ax = b$ has a unique Solution in G.

   (*ii*) The equation $ya = b$ has a unique solution in G.

5. (*a*) Solve each of the recurrence Relations

   (*i*) $a_n = 4a_{n-1} + 5a_{n-2}a_1 = 2, a_2 = 6$

   (*ii*) $c_n = -6c_{n-1} - 9c_{n-2}, c_1 = 2.5, c_2 = 4.7$

   (*b*) How many seven-letter words can be formed using the letters of the word "BENZENE"?

6. (*a*) State whether the argument given is valid or not.

   If I try hard and I have talent, then I will become a musician.

   If I become a musician, then I will be happy.

   $\rightarrow$ If I will not be happy, then I did not try hard or I do not have talent.

   (*b*) Find a minimal sum of products from for the Boolean expression *e* with the following truth table.

   (*i*) $T(00001111, 00110011, 01010101) = 10100110$

   (*ii*) $T(00001111, 00110011, 01010101) = 00101111$.

7. (*a*) Consider the set Q of rational numbers and let * be the operation Q defined by

$$a * b = a + b - ab$$

   (*i*) Find 3 * 4, 2 * (–5) and 7 * 1/2

   (*ii*) Is (Q, *) a semi group ? Is it commutative ?

   (*iii*) Find the identity element for *

   (*iv*) Do any of the elements in Q have an inverse ? What is it ?

   (*b*) Show that the following argument is a fallacy $p \rightarrow q, \sim p \vdash \sim q$.

8. (*a*) Let $A = \{1, 2, 3, 4\}$ for the relation R whose matrix is given. Find the matrix of the transitive closure by using Warshall's algorithm.

$$M_r = \begin{pmatrix} 1 & 0 & 0 & 1 \\ 0 & 1 & 1 & 0 \\ 0 & 1 & 1 & 0 \\ 1 & 0 & 0 & 1 \end{pmatrix}$$

   (*b*) Let $f : S \rightarrow S'$ be a semigroup homomorphism $a \sim b$ if $f(a) = f(b)$ then (*i*) $\sim$ is a congruence relation (*ii*) $S/\sim$ is isomorphic to $f(s)$.

9. (*a*) Write short notes on the following :

   (*i*) Homomorphism and isomorphism.

   (*ii*) Group and subgroup.

   (*iii*) Pigeonhole principle.

   (*b*) Define the following :

   (*i*) Bounded lattices and distributive lattices.

   (*ii*) Hasse diagram of partially ordered set.

   (*iii*) Isomorphic ordered set.

10. (*a*) Use mathematical induction to show that

$$n < ((n + 1)/2) \wedge n \text{ where } n > 1 \text{ and } n \in N$$

   (*b*) Use mathematical induction to show that

$$2.7^n + 3.5^n - 5 \text{ is divisible by 24 for all } n > 0.$$

❐❐❐

MISCELLANEOUS EXERCISE–IV

1. (*a*) Define the terms reflexive, symmetric, irreflexive and anti-symmetric relations with examples.
   (*b*) Determine whether the relation R on A = the set of real no. s: defined by a Rb iff $a^2 + b^2 = 4$ is reflexive, Symmetric, irreflexive and transitive.
2. (*a*) Let $S = \{1, 2, 3, 4, 5\}$ and let $A = S \times S$. Define the following relation R on $A : (a, b)\ R\ (a' . b')$ iff $ab' = a'b$, a) show that R is an equivalence relation.
   (*b*) Compute the set A/R of equivalence classes generated by the elements of A.
3. Given $S = \{1, 2, 3, 4\}$ and a relation R on S defined by $R = \{(1, 2), (4, 3), (2, 2), (3, 1)\}$. Show that R is not transitive. Find a relation $R_1 \supseteq R$ such that R_1 is transitive. Can you find another relation $R_2 \supseteq R$ that is also transitive ? Draw that graphs of R and R_1.
4. (*a*) Do the following sets define functions ? If so give their domains and range in each case
   (*i*) $\{(1, (2, 3)), (2, (3, 4)), (3, (1, 4)), (4, (1, 4))\}$
   (*ii*) $\{(1, (2, 3)), (2, (3, 4)), (3, (3, 2))\}$
   (*b*) List all possible functions from $X = \{a, b, c\}$ to $Y = \{0, 1\}$ and indicate in each case whether the function is one-one, is onto & is one-one and onto.
5. Define the term primitive recursive function. Show that the function $f(x, y) = x + y$ is primitive recursive.
6. State Peano's Axioms. Show that $n^3 + 2n$ is divisible by 3 using Mathematical induction.
7. Let R^+ be the set of all positive real no.s. Show that the function $f : R^+ \to R$ defined by $f(x) = 1n\ x$ is an isomorphism of the Semigroup (R^+, x) to the Semigroup $(R, +)$ where x and $+$ are ordinary multiplication and additioin respectively.
8. Show that a group G is abelian iff $(ab)^2 = a^2b^2$ for all elements a and b in G.
9. (*a*) Show that every cyclic group is an abelian group.
   (*b*) Express the following permutations as the product of disjoint cycles
   (*i*) $a = (1\ 2\ 3)\ (4\ 5)\ (1\ 6\ 7\ 8\ 9)\ (1\ 5)$
   (*ii*) $c = (1\ 3\ 2\ 5)\ (1\ 4\ 3)\ (2\ 5\ 1)$
10. Find Karnaugh maps and simplify the expressions
   (*a*) $AB' + A'B'$
   (*b*) $AB' + A'B$
   (*c*) $AB' + A'B + A'B'$
11. (*a*) Determine the Hasse diagram of the partial order having the given Digraph.

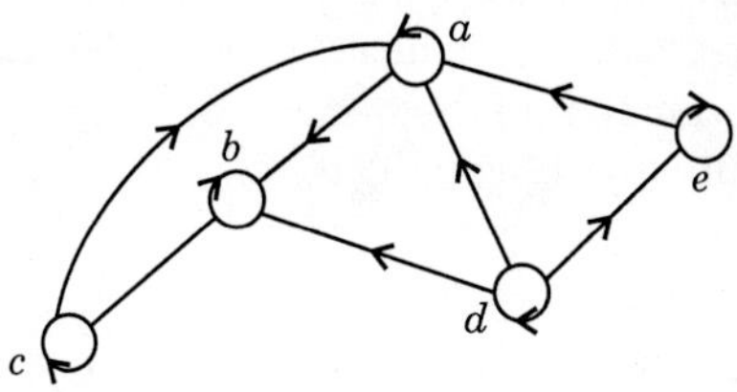

Fig.1

(*b*) Determine the greatest and least elements, if they exist of the Poset. $A = \{2, 4, 6, 8, 12, 18, 24, 36, 72\}$ with the partial order of divisibility.

12. Check which of the following are (*a*) Lattices (*b*) Distributive Lattices.

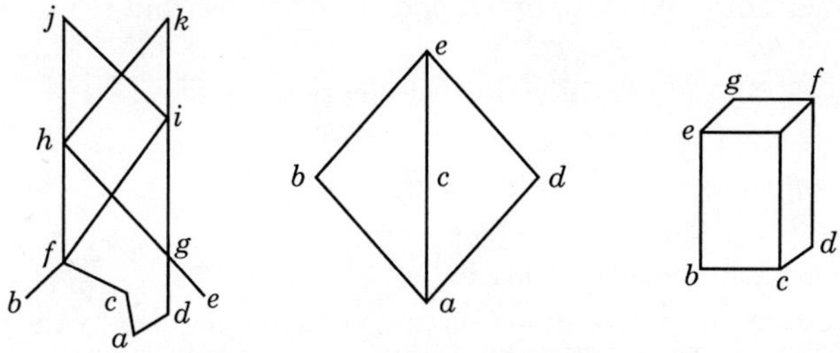

Fig.2

13. Show that in a Lattice $a \vee (a \wedge b) = a$ and $a \wedge (a \vee b) = a$.
14. Define the term Compliment of an element. Show that in a distributive Complimented Lattice if an element has a complement then that is unique.
15. Let *M* be the finite state Machine with state table as given below,
    (*a*) Find the *i/p* set *A*, the state set *S*, the *o/p* set *O* and initial state of *M*.
    (*b*) Draw the state diagram of *M*.
    (*c*) Find the *o/p* string for the *i/p* string aabbcc.

	f			*g*		
S \ *A*	*a*	*b*	*c*	*a*	*b*	*c*
S_0	S_0	S_1	S_2	0	1	0
S_1	S_1	S_1	S_0	1	1	1
S_2	S_2	S_1	S_0	1	0	0

16. Show that if $f(n) = 3n^4 - 5n^2$ and $g(n) = n^4$ for all positive integer n, then f & g have the same order.
17. Solve the recurrence relation $a_n - 3\,a_{n-1} + 2\,a_{n-2} = n^2 + 1$
18. (*a*) State Polya's counting Theorem.
    (*b*) Determine the numeric function corresponding to the generating function $G(z) = (1 + z)^2/(1 - z)^4$.
19. Suppose that a circular disc is divided into *n* sectors, take on would cut a pie with the boundaries of all sectors meeting at a point in the center of the disc. Suppose, further, that we have 10 different colored paints with which we are required to paint the sectors on their disc in such a way that no

adjacent sectors have the same color paint on them. Find the number of ways to paint the n sectors of the disc with the 10 different paints.

20. Use generating function to find the number of ways to select 10 balls from a large pile of red, white and blue balls if
    (*a*) the section has at least two balls of each color
    (*b*) the section has at most two red balls.
21. Write the negation, converse, inverse and contrapositive of the following statement in good English "If you drive then I will walk."
22. Construct truth table to determine whether the following is a tautology, contingency or an absurdity or not $(P \wedge \neg Q) \rightarrow (P \wedge R) \vee (Q \wedge R)$ where P, Q, R are propositions.
23. Define logical equivalence of two propositions. Show that $(P \rightarrow Q) \wedge (R \rightarrow Q) \Leftrightarrow (P \vee R) \rightarrow Q$.
24. Check the validity of the following implication.
    I have headache if I study logic.
    I take an aspirin when ever I have headache.
    If I take aspirin then I drink tea.
    $\therefore$ I drink tea when ever I study logic.
25. When do you say the proposition Q is tautologically implied by the proposition P? Show $(P \rightarrow Q) \rightarrow Q \Rightarrow (P \vee Q)$ without constructing the truth table.

❐❐❐

Previous Years Solved Question Papers

END TERM EXAMINATION

Third Semester (B.Tech) Decemeber, 2007

Time: 3 Hours *Maximum Marks : 75*

Note : Q. 1 is compulsory and attempt one question form each unit.

1. (*a*) If $X = \phi$ and $Y = \{\{\{a\}\}, \phi\}$ then Find $P(P(P(X)))$ and $P(Y)$. (*P* denotes Power Set).
   (*b*) Write down the Inverse, Converse and contrapositive of the following Statement "The crop will be destroyed if there is a flood"
   (*c*) Show the relation, "Divides in positive integers is a POSET"
   (*d*) Show that ~ (p ^ q) follows from ~ P ^ ~ q.
   (*e*) How many ways can the letters in word MISSISSIPI can be arranged?
   (*f*) Test the validity of the following argument:
   If I study, then I will not fail in Mathematics
   If I do not play basket ball, then I will study
   But I failed Mathematics
   Therefore, I must have played basketball.
   (*g*) Show that a cycle of even length is an odd permulation.
   (*h*) Prove that Join and Meet Operations in lattice algebra are associative

UNIT – I

2. (*a*) Find the Principlal conjunctive Normal form of the formula (~ p → R) ^ ($Q \leftrightarrow P$).
   (*b*) Find the general solution of the Fibonacci Relation, Fn which satisfies the Fibonacci relation $F_n = F_{n-1} + F_{n-1}$ for $n > 2$, where $F_0 = F_1 = 1$
   (*c*) Prove by Induction, $3^n > n^3$. for $n > 4$.

3. (*a*) Find the Disjunctive Normal form of the formula ((P V Q) ^ R) v ((PVQ) ^
   (~ P))
   (*b*) Symbolise the following statements / expression.
   (*i*) "All the world loves a lover"
   (*ii*) "For every positive integer there exists a greater positive integer"
   (*c*) Find the approximate cost of sorting an input list by Mergesort by solving the recurrence relation which represents the mergesort.

UNIT– II

4. (*a*) In a room containing 28 females, there are 18 females who speak English, 15 females speak french and 22 speak German.9 females speak both English and French, 11 females speak both French and German whereas 13 speak both German and English. How many females speak all the three languages.

(*b*) Find the various closures of relations

(*c*) Draw the Hase diagram of the POSET (P (A), C) Where A = {a, b, c} and P (A) denote the poweer set and the relation, C is subset

5. (*a*) Define Compatibility Relation.

(*b*) Show that $A \cap (B \cup C) = (A \cap B) \cup (A \cap C)$ by using characteristic function.

(*c*) Define the followings with reference to POSET

(*i*) Maximal Member II) Upper Bound III) Least Member.

UNIT – III

6. (*a*) Prove that in a lattice (L, Relation $<$), $a, b, c, d \varepsilon L$, If $a < b$ and $c < d$, then $a \cup b < c \cup d$ and $a \cap b < d$, where $\cup$ and $\cap$ Join and Meet operations respectively.

(*b*) Prove that in a distributive latticve, if an element has a complement then this complement is Unique.

(*c*) Define direct product of Lattices.

7. (*a*) Find a minimal sum for the Boolean expression:

$E = x + x\,yz + xy\,z$

(*b*) Prove that for any lattice $a \cap (a \cup b) = a$ and $a \cup (a \cap b) = a$, Where $\cup$ and $\cap$ Join and Meet operations respectively

(*c*) Define Lattice Homorphism.

❐❐❐

END TERM EXAMINATION

Third Semester (B.Tech) - Dec. 2008

Time : 3 Hours *Maximum Marks : 75*

Note: Question number 1 is compulsory. Attempt any one question from each unit.

Q.1. (*a*)Verify that the proposition $p \vee \sim (p \wedge q)$ is a tutology.

Ans.

p	q	$p \wedge q$	$\sim (p \wedge q)$	$p \vee \sim (p \wedge q)$
T	T	T	F	T
T	F	F	T	T
F	T	F	T	T
F	F	F	T	T

As we can see that last column has all value True, so given expression is a Tutology.

Q.1. (*b*) Let A = {1, 2, 3, 4, 5}. Find the truth value of each of the following statements.

(*i*) $(\exists x \in A)(x + 3 = 10)$

(*ii*) $(\forall x \in A)(x + 3 < 10)$

(*iii*) $(\exists x \in A)(x + 1 < 5)$

(*iv*) $(\forall x \in A)(x + 3 \geq 10)$

(*v*) $(\exists x \in A)(2x = x + 6)$

Ans. (*i*) **False**; as for no number in A is a solution to $x + 3 = 10$

(*ii*) **True**; Every number in A satisfies $x + 3 < 10$

(*iii*) **True**; for $x = 1, 2, 3$, then it is a solution

(*iv*) **False**; for no number in A satisfies $x + 3 \geq 10$

(*v*) **False**; No number in A is a solution to $2x = x + 6$

Q.1. (*c*) Symbolize the following argument:

S_1 : If a man is a bachelor, he is happy

S_2 : If a man is unhappy, he dies young

S_3 : Bachelors die young

Ans. Let p : He is a bachelor

q : He is happy

r : He dies young

Now we can say the argument S_1, S_2 and S_3 are valid argument and we can symbolize them as

(*i*) $P \to q$
(*ii*) $\sim q \to r$
(*iii*) $p \wedge r$ conclusion
(*iv*) $q \to \sim r$ using transposition of (*ii*)
(*v*) $p \to \sim r$ hypothetical syllogism of (*i*) an (*iv*)
(*vi*) $p \wedge r$ hence given arguments are valid

Q.2. (*a*) Prove by induction : $n^3 < 3^n$ for $n \geq 4$

Ans. Inductive step $n = 1$

$p(n) : n^3 < 3^n$

$p(1) \equiv 1^3 < 3^1$

$1 < 3$ True

Let $n = R$ (k is a positive integer) and we assume that it is true.

$p(k) \equiv k^3 < 3k$ True

Now $n = k + 1$

$p(k + 1) \equiv (k + 1)^3 < 3^{k+1}$

$(k + 1)^3 < 3^k . 3$

$k^3 + 1 + 3k^2 + 3k < 3^k . 3$

$k^3 + 3k^2 + 3k + 1 < 3^k . 3$

$k^3 + 3m < 3^k . 3$

$1 + k^3 + 3m < 3k . m$ where $m = k^2 + k$

$k^3 + 3\text{m} < 3.3^k$

As 3m is added into k^3 and if it is complatiely very less. In that case this condition is also true.

Hence, proved by mathematical induction.

Q.2. (*b*) Deivse an algorithm to calculate the lowest common multiple cm (*a*, *b*) of two positive integers *a* and *b*.

Ans. Let a and b be two positive integers. Then gcd (a, b) lcm $(a, b) = ab$.

Let us assume $d = gcd\ (a, b)$ and $m = \text{lcm}\ (a, b)$ when gcd is greatest common division and lcm least common multiple.

Now $d = gcd\ (a, b)\ (a, b) \to$ then exists r, u and v such that

$$a = dr,\ b = dr,\ gcd\ (r, l) = 1 \quad ...(1)$$

and $$d = au + bv$$

$m = lcd\ (a, b) \to$ There exists t and w such that

$$m = at \text{ and } m = bw \quad ...(2)$$

Now
$$\begin{aligned} d &= au + bv \\ md &= m\ (au + ab) \\ &= mau + mbv \\ &= bwau + atbv \\ &= \text{ab}\ (wu + tv) \text{ by (2)} \end{aligned}$$

Which implies ab divides md, i.e., $\dfrac{ab}{md}$ ...(3)

also $$\frac{ab}{d} = \frac{drb}{d} = rb \text{ and}$$

$$\frac{ab}{d} = \frac{ads}{d} = as \quad \text{by (1)}$$

Hence, $\frac{a}{ab/d}, \frac{b}{ab/d}$ that is $\frac{ab}{d}$ is a common multiple of a and b

This implies $\frac{m}{ab/d}$ or $\frac{md}{ab}$ ...(4)

Thus, $md = ab$ by (3) and (4).

Q.2. (c) Let *a* and *b* be integers and suppose that θ(*a*, *b*) is defined by

$$\theta(a, b) = \begin{cases} 5 & \text{if } a < b \\ \theta(a-b, b+2) + a & \text{if } a \geq b \end{cases}$$

Find θ(2, 7), θ(5, 3), θ(15, 2).

Ans. Given the function θ on a and b as

$$\theta(a, b) = \begin{cases} 5 & \text{if } a < b \\ \theta(a-b, b+2) + a & \text{if } a \geq b \end{cases}$$

(*i*) $\theta(2, 7) = 5$ from first condition $a < b$

(*ii*) $\theta(5, 3)$ first part is not valid as $a \nless b$. Now from part second $a = 5$, $b = 3$.

$$\begin{aligned} \theta(5, 3) &\equiv \theta(5-3, 3+2) + 5 \\ &\equiv \theta(2, 5) + 5 \\ &\equiv 5 + 5 = 10 \quad \textbf{Ans.} \end{aligned}$$

(*iii*)
$$\begin{aligned} \theta(15, 2) &\equiv \theta(15-2, 2+2) + 15 \\ &\equiv \theta(13, 4) + 15 \quad \ldots(1) \end{aligned}$$

Now,
$$\begin{aligned} \theta(13, 4) &\equiv \theta(13-4, 4+2) + 13 \\ &\equiv \theta(9, 6) + 13 \quad \ldots(2) \end{aligned}$$

Now,
$$\begin{aligned} \theta(9, 6) &\equiv \theta(9-6, 6+2) + 9 \\ &\equiv \theta(3, 8) + 9 \\ &\equiv 5 + 9 = 14 \quad \ldots(3) \end{aligned}$$

From (3) and (2)

$$\theta(13, 4) = 14 + 13 = 27 \quad \ldots(4)$$

From (4) and (1)

$$\theta(15, 2) = 27 + 15 = 42 \quad \textbf{Ans.}$$

Q.3. (*a*) What are the properties of numbers in a Pascal triangle? Form the pascal's triangle for n = 7.

Ans. Pascal's formula can be given by,

$$\boxed{C(n+1, r) = C(n, r) + C(n, r-1)}$$

where n and r are positive integers wit $r \leq n$. It gives a recurrence relation for the computation of Binomial coefficient given the initial data $C(n, o) = C(n, n) = 1$ for all n. Note that no multiplication is needed for this computation. One can obtain the numbers by constructing a triangular array using very siple arithmetic. The triangular array is usually called *Pascal's triangle*. The nth row of the triangle, which contains $n + 1$ numbers, can be formed from the preceding row by the following rules.

(*a*) The first ($k = 0$) and the last ($k = n$) entries are both equal to 1.

(*b*) For $1 \le k \le n - 1$, the kth entry in the nth row is the sum of the $(k - 1)^{th}$ and k^{th} entries in the $(n - 1)$ rows.

The first equal rows ($n = 7$) of Pascal triangle are showm in the following diagram.

$$\begin{array}{c} 1 \\ 1 \quad 1 \\ 1 \quad 2 \quad 1 \\ 1 \quad 3 \quad 3 \quad 1 \\ 1 \quad 4 \quad 6 \quad 4 \quad 1 \\ 1 \quad 5 \quad 10 \quad 10 \quad 5 \quad 1 \\ 1 \quad 6 \quad 15 \quad 20 \quad 15 \quad 6 \quad 1 \\ 1 \quad 7 \quad 21 \quad 35 \quad 35 \quad 21 \quad 7 \quad 1 \end{array}$$

Q.3. (*b*) Write down all possible partitions of the set $S = \{1, 2, 3, 4\}$

Ans. Each partition of S contains eight, 1, 2, 3 or 4 distinct set 1 as shown below.

1. [{1, 2, 3, 4}]
2. (*i*) [{1}, {2, 3, 4}], [{2}, {1, 3, 4}], [{3}], [{1, 2, 4}]
   (*ii*) [{4}, {1, 2, 3}], [{1, 2}, {3, 4}], [{2, 3}, {1, 4}]
   (*iii*) [{1, 3}, {2, 4}]
3. [{1}, {2}, {3, 4}], [{1}, {3}, {2, 4}], [{1}, {4}, {2, 3}]
   [{b}, {3}, {9, 4}], [{2}, {4}, {1, 3}], [{3}, {4}, {1, 2}]
4. [{1}, {2}, {3}, {4}]

Q.3. (*c*) Let A and B any set 8. Prove the following results.

(*i*) A is the disjoint union of A/B and $A \cap B$

(*ii*) $A \cup B$ is the disjoint union of A/B and B/A and $A \cap B$ where $A / B \equiv A \sim B \equiv A$ difference B

Ans. (*i*) A is the disjoint union of A/B and $A \cap B$ can also be written as

$$\boxed{A - (A \cap B) = A - B}$$

To prove this,

Solution (*i*), Let $(x \,\varepsilon\, A$ and $(x \not\varepsilon\, A \cap B)$ ...(1)

$x \,\varepsilon\, A$ and $x \not\varepsilon\, A$ or $x \not\varepsilon\, B$

$[(x \,\varepsilon\, A)$ and $(x \not\varepsilon\, A)]$ or $[(x \,\varepsilon\, A)$ and $(x \not\varepsilon\, A)]$

$$\equiv (x \not\varepsilon\, \phi) \text{ or } (x \not\varepsilon\, \phi) \rightarrow x \in \phi \cup (A - B)$$

$$\equiv x \in (A - B) \quad ...(2)$$

$$A - (A \cap B) \subseteq A - B \quad ...(3)$$

Again let,

$x \in A - B$ ...(4)

$\Rightarrow x \in A$ and $x \notin B$

$\Rightarrow x \in A$ and $x \notin A \cap B$ (as $x \notin B$)

$\Rightarrow \qquad x \in A - (A \cap B)$ ...(5)

From (4) and (5), we have

$(A - B) \subseteq A - (A \cap B)$

From (3) and (6), we have

$$\boxed{(A - B) = A - (A \cap B)} \quad \textbf{Hence proved.}$$

Solution (*ii*), We can re write the equation as

$(A - B) \cup (B - A) = (A \cup B) - (A \cap B)$

or $(A / B) \cup (B / A) = (A \cup B) / (A \cap B)$

We have $x \in (A - B) \cup (B - A)$

$\Rightarrow x \in A$ and $(A - B)$ or $x \varepsilon (B - A)$

$\Rightarrow (x \in A$ and $x \notin B)$ or $(\Rightarrow x \in B$ and $x \notin A)$

$\Rightarrow [x \in A$ or $(x \in A$ and $x \notin B)]$ and $[x \notin A$ or $(x \in A$ and $x \notin B)]$

$\Rightarrow [x \in (A \cup B)$ and $x \notin A \cap B]$

$\Rightarrow x \in (A \cup B) - (A \cap B)$

Hence, $(A - B) \cup (B - A) = (A \cup B) - (A \cap B)$

or $(A / B) \cup (B / A) \equiv (A \cup B) / (A \cap B)$ **Hence proved.**

Q.4. (*a*) Let *S* be a set and *P*(*s*) be the set of all subsets of *S*. Show that (*P*(*s*) ∩, ∪) is a lattice. Draw Hasse diagram of the Lattice *P*(*s*) of all subsets of *S* = {*a*, *b*, *c*}.

Ans. In the lattice, $a \leq b$ whenever $a \wedge b = a$. Thus in this case $a \leq b$ whenever $a \wedge b = a$ so we can make Hasse diagram by the lattice $(P(s), \cap, \cup)$ that will be same under the partial ordering of set inclusion.

The partial order $\leq$ on s may be described by the diagraph that will be converted into Hasse diagram by following these steps.

(*a*) In first step we remove all cycles of length 1 at all vertices.

(*b*) In next step we delete all edges corresponding to the transitivity property of partial order.

(*c*) Arrange each edge so that its starting vertex is below its terminal vertex.

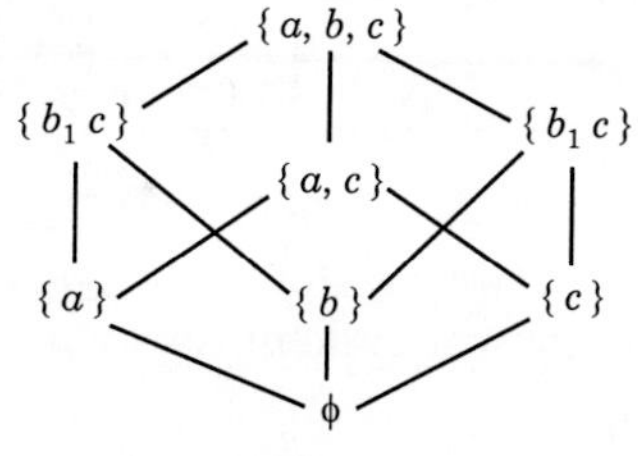

Fig. 1.

(*d*) Remove all direction arrows on the directed edges.

Q.4. (*b*) Express $E(x, y, z) = ((x, y)'\, z)'\, ((x' + z)\, (y' + z)')$ in its complete sum of products form.

Ans. Given expression is,

$$= ((x\,y)'\,z)'\,((x' + z)\,(y' + z)')$$
$$\equiv [((x\,y)' + z\,[(x' + z)'\,(y' + z)']$$

Using DeMorgon's law

$$\equiv [(x\,y + z')\,[x\,.\,z' + y\,z']$$

Using DeMorgon's law

$$\equiv (x.y)\,(x.\,z') + z'.\,(x\,.\,z') + (x\,.\,y)\,(yz') + z'\,(yz')$$
$$\equiv x.x\,.y.\,z' + x\,.\,z'.\,z' + x\,.\,y.\,y.\,z' + y.z'.\,z'$$
$$\equiv xy\,z' + x\,.\,z' + x\,y\,z' + y\,z'$$
$$\equiv x\,yz' + x\,z' + y\,z'$$
$$\equiv x.y.\,z' + x\,.\,z'.\,z'\,.1 + y.\,z'\,.1 \qquad \text{as } (a + a' = 1)$$
$$\equiv x.y.\,z' + x\,.\,z'.(y + y') + y.z'.(x + x')$$
$$\equiv x.y.\,z' + x\,.\,z'.y + x.z'.y' + x.y.z' + x\,.\,y\,.\,z'$$
$$\equiv x.y.\,z' + x\,.y.z' + x.y'.z' + x.y.z' + x'.\,y\,.\,z'$$
$$\equiv x.y.\,z' + x\,.y'.z' + x'.y.z'$$

So complete sum of product (disjunctive normal form is)

$$\boxed{x.y.z' + x.y'.z' + x.y.z'}$$

Q.4. (*c*) Find the fundamental product *P* represented by the rectangle in Karnaugh map.

$x \backslash yz$	yz	yz'	$y'z'$	$y'z$
x	1			1
x'	1			1

Fig. 2.

Ans. In this case find those literals which appear in all the squares of the basic rectangle then *P* is the product of such literals. Now taking all 1's to form a square

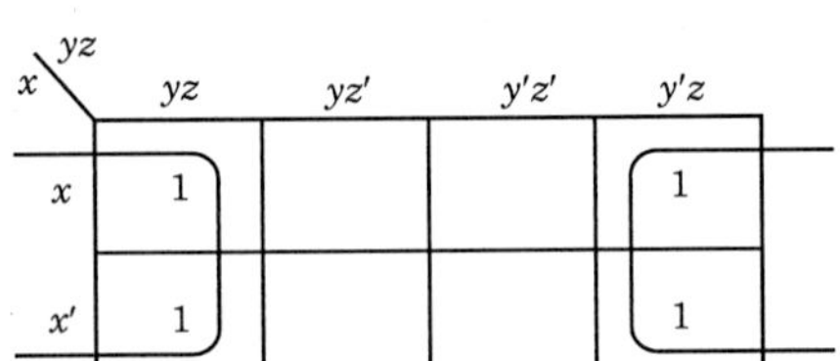

Fig. 3.

Only z is appear in all the four squares so we can say that fundamental product *P* represented by the rectangle in Karanaugh map is

$$\boxed{P = z.}$$

Q.5. (*a*) For the graph

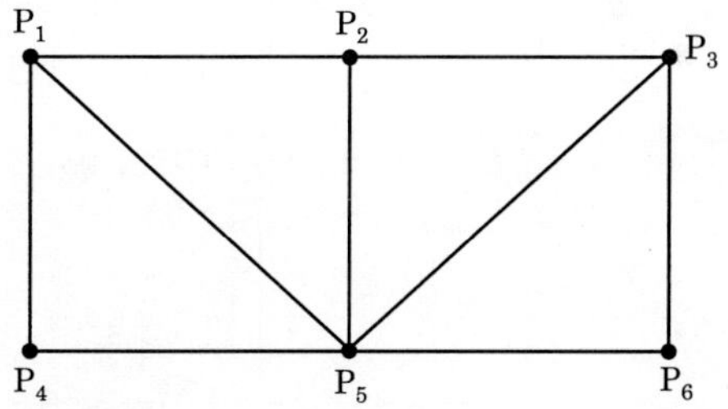

Fig. 4.

Classify the following sequences as simple path, path as Trail:

(*i*) $\{P_4, P_1, P_2, P_5, P_1, P_2, P_3, P_6\}$

(*ii*) $\{P_4, P_1, P_5, P_2, P_6\}$

(*iii*) $\{P_4, P_1, P_5, P_1, P_2, P_3, P_6\}$

(*iv*) $\{P_4, P_1, P_5, P_3, P_6\}$

(*v*) $\{P_1, P_4, P_5, P_3, P_6\}$

Ans.

(*i*) In sequence (*i*) $\{P_1, P_2\}$ edge is appear twice so it can not be a trail. So it is a path from P_4 to P_6.

(*ii*) There is not edge from P_2 to P_4, so its not a path.

(*iii*) This sequence is a trail since no edge is twice but it is not a simple path as the vertex P_5 is used twice.

(*iv*) It is a trail from P_4 to P_6 but it is not having repeated edge.

(*v*) It is also a trail from P_1 to P_6 because it does not have any repeated edge.

Q.5. (*b*) Using shortest path algorithm, find the shortest path matrix for the weighted graph *G*.

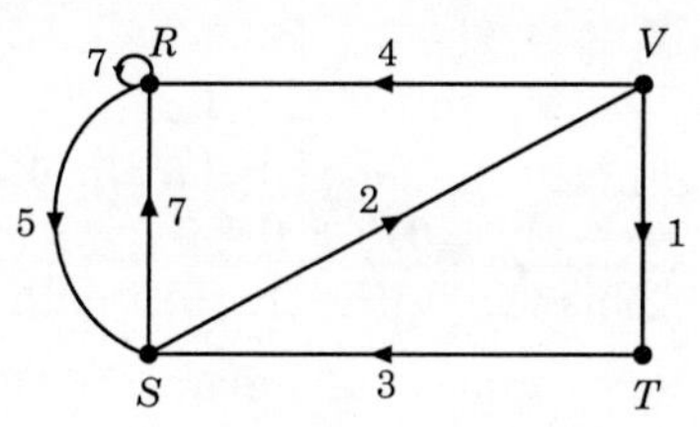

Fig. 5.

Ans. Weighted graph matrix can be given by,

$$W = \begin{array}{c} \\ R \\ S \\ T \\ U \end{array}\begin{array}{c} \begin{array}{cccc} R & S & T & U \end{array} \\ \begin{bmatrix} 7 & 5 & 0 & 0 \\ 7 & 0 & 0 & 2 \\ 0 & 3 & 0 & 0 \\ 4 & 0 & 1 & 0 \end{bmatrix} \end{array}$$

On apply Warshall's algorithm for shortest path we obtain

$$W_0 = \begin{bmatrix} 7 & 5 & \infty & \infty \\ 7 & \infty & \infty & 2 \\ \infty & 3 & \infty & \infty \\ 4 & \infty & 1 & \infty \end{bmatrix} \begin{bmatrix} RR & RS & - & - \\ SR & - & - & SU \\ - & TS & - & - \\ UR & - & UT & - \end{bmatrix}$$

$$W_1 = \begin{bmatrix} 7 & 5 & \infty & \infty \\ 7 & 12 & \infty & 2 \\ \infty & 3 & \infty & \infty \\ 4 & 9 & 1 & \infty \end{bmatrix} \begin{bmatrix} RR & RS & - & - \\ SR & SRS & - & SU \\ - & TS & - & - \\ UR & URS & UT & - \end{bmatrix}$$

$$W_2 = \begin{bmatrix} 7 & 5 & \infty & 7 \\ 7 & 12 & \infty & 2 \\ 10 & 3 & \infty & 5 \\ 4 & 9 & 1 & 11 \end{bmatrix} \begin{bmatrix} RR & RS & - & RSU \\ SR & SRS & - & SU \\ TSR & TS & - & TSU \\ UR & URS & UT & URS \end{bmatrix}$$

$$W_3 = \begin{bmatrix} 7 & 5 & \infty & 7 \\ 7 & 12 & \infty & 2 \\ 10 & 3 & \infty & 5 \\ 4 & 9 & 1 & 16 \end{bmatrix} \begin{bmatrix} RR & RS & - & RSU \\ SR & SRS & - & SU \\ TSR & TS & - & TSU \\ UR & UTS & UT & UTSU \end{bmatrix}$$

$$W_4 = \begin{bmatrix} 7 & 5 & 8 & 7 \\ 7 & 11 & 3 & 2 \\ 9 & 3 & 6 & 5 \\ 4 & 4 & 1 & 6 \end{bmatrix} \begin{bmatrix} RR & RS & RSUT & RSU \\ SR & SURS & SUT & SU \\ TSUR & TS & TSUT & TSU \\ UR & UTS & UT & UTSU \end{bmatrix}$$

$$\begin{aligned} W_1 &= \{4, 2\} = \mathrm{M_{in}}\,[W_1\,\{4, 2\}],\ W_0\,\{4, 1\} + W_0\,\{1, 2\} \\ &= \mathrm{M_{in}}\,[\infty, 4 + 5] = 9 \\ W_2 &= \{1, 3\} = \mathrm{M_{in}}\,[W_1\,\{1, 3\}],\ W_1\,\{1, 2\} + W_1\,\{2, 3\} \\ &= \mathrm{M_{in}}\,[\infty, 5 + \infty] = \infty \\ W_3 &= \{4, 2\} = \mathrm{M_{in}}\,[W_2\,\{4, 2\}],\ W_2\,\{4, 3\} + W_2\,\{3, 2\} \\ &= \mathrm{M_{in}}\,[9, 3 + 1] = 4 \\ W_4 &= \{3, 1\} = \mathrm{M_{in}}\,[W_3\,\{3, 1\}],\ W_3\,\{3, 4\} + W_3\,\{4, 1\} \\ &= \mathrm{M_{in}}\,[10, 5 + 4] = 9 \end{aligned}$$

As W and W_4 are same, this will be the shortest path matrix.

Q.6. (*a*) Define the Adjacency matrix of a graph *G*. Obtain the adjacency matrix of the graph *G*. Given as:

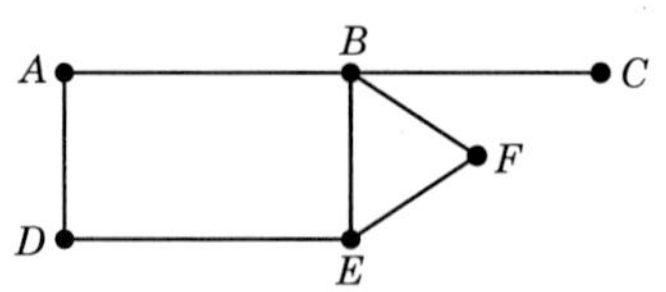

Fig. 6.

Ans. The adjacency matrix $A = \{a_{ij}\}$ of any graph G can be defined as a $m \times m$ matrix which is defined by

$$a_{ij} = \begin{bmatrix} 1 & \text{if } V_i \text{ is adjacents } V_j \\ 0 & \text{otherwise} \end{bmatrix}$$

The adjacency matrix of given graph G can be given by

	A	B	C	D	E	F
A	0	1	0	1	0	0
B	1	0	1	0	1	1
C	0	1	0	0	0	0
D	1	0	0	0	1	0
E	0	1	0	1	0	1
F	0	1	0	0	1	0

Q.6. (*b*) Find the minimal spanning tree for the weighted graph given by

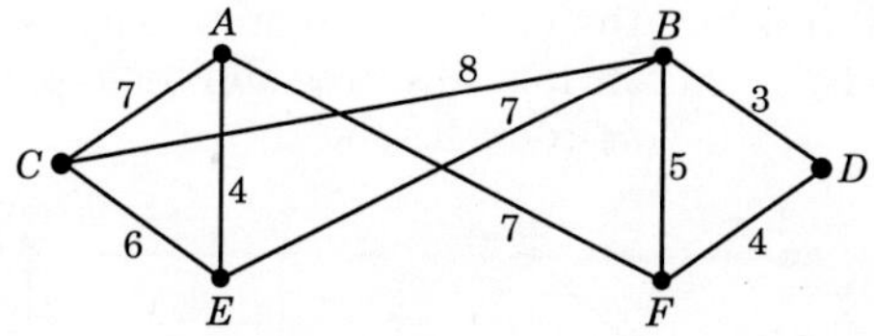

Fig. 7.

Ans. First we order the edges by increasing weights and then we remove the edges (high weight) without disconnecting G.

Edges	BD	DF	AE	BF	CE	BE	AC	AF	BC
Weight	3	4	4	5	6	7	7	7	8
Delete	No	No	No	Yes	No	No	Yes	Yes	Yes

The minimal spanning Tree of G which is obtained by removing edges, is given as

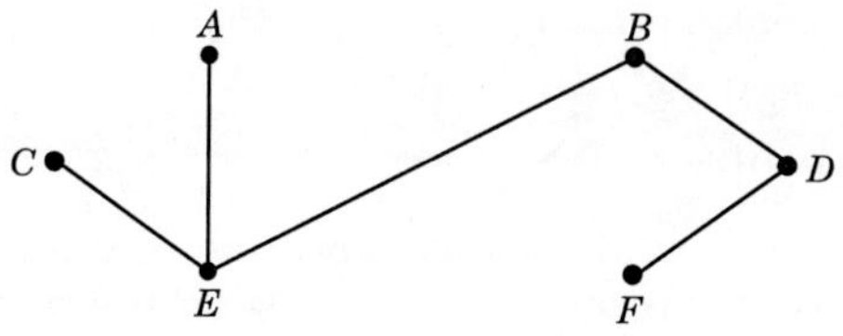

Fig. 8.

Q.6. (*c*) Find the graph given below, find the order or the vertices of the graph:

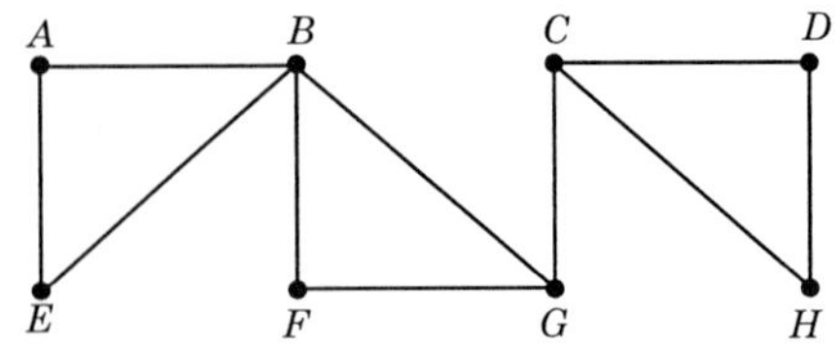

Fig. 9.

In which these are processes using a DFS (depth first search) algorithm beginning at vertex *B*.

Ans. In depth first search (DFS) we choose arbitrarily a vertex in graph (*G*) [in above figure it is *B*] as the root. A path may be formed by adding edges successfully such that each edge incorporated in the path is incident with the last vertex which is not traversed earlie. If the path is such that all vertices of graph *G* are being traversed then it also forms a spanning tree. If all vertices of *G* are not there in the path so constructed, then move back to the vertex next to the last vertex reached in the path and form a new path starting from it, if it is possible to do so. If it is not, then once again move back to vertex next to it and try to form a path from this vertex. The procedure is repeated till all the edges so formed includes all the vertex of Graph *G*.

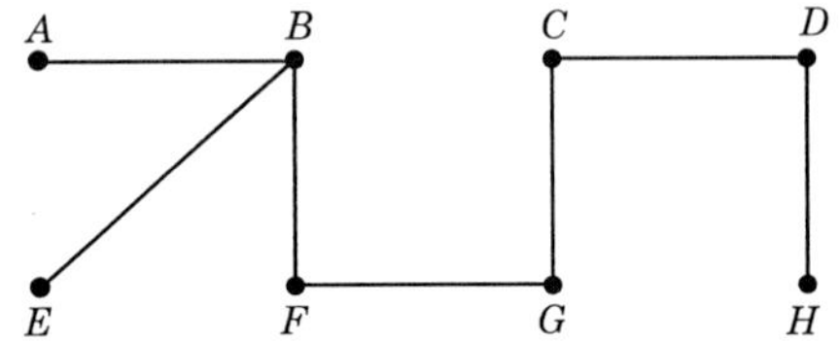

Fig. 10.

And the order of vertices is = B, A, E, F, G, C, D, H. **Ans.**

Q.7. (*a*) Express the sentences in predicate logic:

(*i*) Everyone in the room spokes French or German

(*ii*) Someone in the room either spoken French or German

Ans. $p(x)$: x can speak french

$g(x)$: x can speak German

(*i*) $(\forall x)\ [p(x) \vee g(x)]$

(*ii*) $(\exists x)\ [p(x) \vee g(x)]$

Q.7. (*b*) Let *z* be the set of all integers. For *a*, *b* ε, *z*, defined *a R b* by $b = a^r$ for some positive integer *r*. Show that *R* is a partial order relation on *z*.

Ans.

(*i*) as $a = a'$ for $r = 1$, so we can say that R is reflexive.

(*ii*) Suppose $a\ R\ b$ and $b\ R\ a$ say $b = a^r$ and $a = b^s$ then $a = (a^r)^s = a^{rs}$.

Now there are three possibilities to be true

(*a*) $rs = 1$ then $r = 1$ and $s = 1$ and so $a = b$ if $a = 1$ then $b = 1^r = 1 = a$ and similarly if $b = 1$ then $a = 1$.

And lastly if $a = -1$ then $b = -1$ (since $b \neq 1$) and $a = b$ if all three cases then $a = b$. Thus, R is anti-symmetric.

(*iii*) Suppose aRb and bRc and $b = a^s$ and $c = b^s$ then $c = (a^r)^s = a^{rs}$ therefore aRc, then a R is transitive.

As R is reflexive, anti-symmetric and transitive too. So we can say that relation R is a partial order on Z.

Q.7. (*c*) Let $A = \{a, b, c\}$ and $B = \{1, 2, 3\}$.

Determine the nuber of:

(*i*) Relations from *A* to *B*

(*ii*) Functions from *A* to *B*

Ans. (*i*) Cardinality of $|A| = 3 = \text{m}$

Cardinality of $|B| = 2 = \text{n}$

So total number of relations from A to B

$$= \text{power set } A \times \text{power sets of } B$$
$$= 2^m \times 2^n = 2^{m.n} \quad (m = 3, n = 2)$$
$$= 2^{3.2} = 2^6 \quad \textbf{Ans.}$$

(*ii*) Let m and n are cardinality of set A and B respectively.

So, total number of functions from A to B can be given by $= n^m$

$$\boxed{\text{Total functions} = 2^3}$$

$$m = 3, n = 2 \quad \textbf{Ans.}$$

Q.8. (*a*) Find the Boolean matrices *AB* and *B* + *A* where

$$A = \begin{bmatrix} 0 & 1 & 0 \\ 1 & 0 & 1 \\ 1 & 0 & 0 \end{bmatrix} \text{ and } B = \begin{bmatrix} 1 & 0 & 0 \\ 1 & 0 & 0 \\ 0 & 1 & 1 \end{bmatrix} \text{ are Boolean Matrices.}$$

Ans.

$$A = \begin{bmatrix} 0 & 1 & 0 \\ 1 & 0 & 1 \\ 1 & 0 & 0 \end{bmatrix} \text{ and } B = \begin{bmatrix} 1 & 0 & 0 \\ 1 & 0 & 0 \\ 0 & 1 & 1 \end{bmatrix}$$

$$A \,.\, B = \begin{bmatrix} 1 & 0 & 0 \\ 1 & 1 & 1 \\ 1 & 0 & 0 \end{bmatrix} \quad \textbf{Ans.}$$

and

$$A + B = \begin{bmatrix} 0 & 1 & 0 \\ 1 & 0 & 1 \\ 1 & 0 & 0 \end{bmatrix} + \begin{bmatrix} 1 & 0 & 0 \\ 1 & 0 & 0 \\ 0 & 1 & 1 \end{bmatrix}$$

$$A + B = \begin{bmatrix} 1 & 1 & 0 \\ 1 & 0 & 1 \\ 1 & 1 & 1 \end{bmatrix} \quad \textbf{Ans.}$$

Q.8. (*b*) State and prove Euler's formula regarding a finite connected planner graph.

Ans. Suppose the connected Map M consists of a simple vertex P as in Fig. 11. (*a*). Then $V = 1$, $E = 0$ and $R = 1$. Hence, $V - E + R = Z$. Otherwise M can be built up from a single vertex by the following two constructions.

(*i*) Add a new vertex Q_2 and connect it to an existing vertex Q_1 by an edge which does not across any existing edge as in Fig. 11. (*b*).

(*ii*) Connect two existing vertices Q_1 and Q_2 by an edge e which does not cross any existing edge as in Fig. 11. (*c*).

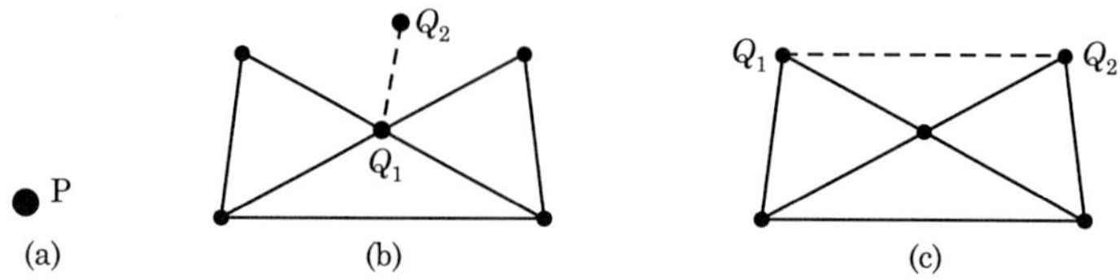

Fig. 11.

Neither operation changes the value of $V - E + R$. Hence, M has the same value of $V - E + R$ as the map consisting of a single vertex, that is $V - E + R = 2$, thus the theorem is proved.

Q.8. (*c*) Let $A = \{1, 2, 3, 4\}$ and Relation given by

$R = \{(1, 1), (2, 2), (2, 3), (3, 2), (4, 2), (4, 4)\}$ which R is a relation on R. Examine whether R is an equivalence relation. Also draw its directed graph and find $R^2 = R_0R$.

Ans. For equivalence relation, it should be reflexive, symmetric and transitive.

(*i*) R is not reflexive below $3 \varepsilon A$ but $3 \not R\, 3$, *i.e.*, $(3, 3) \notin R$.

(*ii*) R is not symmetric because $4\, R\, 2$ but $2 \not R\, 4$ *i.e.*, $(4, 2)\ \varepsilon R$ but $(2, 4) \notin R$.

(*iii*) R is not transitive because $4\, R\, 2$ and $2\, R\, 3$ but $4 \not R\, 3$, *i.e.*, $(4, 2)\ \varepsilon R$ and, εR but $(4, 3) \notin R$.

So, R is not symmetric reflexive and transitive, hence, R is not an equivalence relation.

(*a*) Diagraph of R can be given by

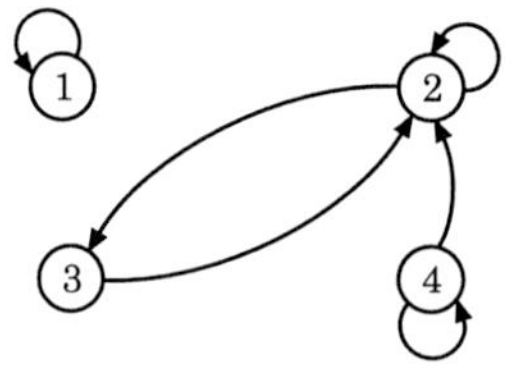

Fig. 12.

(*b*) For each pair $(a, b)\ \varepsilon\ R$. Find all $(b, c)\ \varepsilon\ R$ since $(a, c)\ \varepsilon R^2$.

$R_oR = R^2 = \{(1, 1), (2, 2), (2, 3), (3, 2), (3, 3), (4, 2), (4, 3), (4, 4)\}$ **Ans.**

❒❒❒

END TERM EXAMINATION

Third Semester (B.Tech) - Dec. 2009

Time : 3 Hours *Maximum Marks : 75*

Note: Question number 1 is compulsory. Attempt any one question from each unit.

Q.1. (*a*) A relation *R* is defined on the set *Z* of integers as follows:xRy if and only if $x^2 + y^2 = 25$. Express *R* and R^{-1} as the sets of ordered pairs and hene find their respective domains.

Ans. 1(*a*) Given relation xRy, where x related with y if an only if $x^2 + y^2 = 25$ such that $x, y \varepsilon z$ where Z is set of integers.

Now, $x^2 + y^2 = 25$ (given)

on solving the given equation, we have two set

$$x = 0; y = 5$$

and $y = 5 ; x = 0$

So the relation $\boxed{R = \{(0, 5), (5, 0)\}}$

To find the reverse of relation R, interchange the value of x with y and y with x, we get

$$\boxed{R^{-1} = \{5, 0\}, (0, 5)}$$

Domain of relaiton R is 0 and 5 *i.e.*, {0, 5} and

Domain of Reverse relaiton R^{-1} is 5 an 0 *i.e.*, {0, 5} **Ans.**

Q.1. (*b*) Show that $\sim (p \leftrightarrow q) \equiv (p \wedge \sim q) \vee (p \sim \wedge q)$.

Ans. To prove the given logical equivalence, we first take left hand side logical expression.

Method I: Using truth table

LHS = $\sim (p \leftrightarrow q)$

Truth table of given logical expression is:

p	g	p ↔ g	~ (p ↔ g)
F	F	T	T
F	T	F	F
T	F	F	F
T	T	T	T

RHS: $(p \wedge \sim q) \vee (p \sim \wedge q)$

Truth table of given logical expression is:

p	g	~ p	~ g	p ∧ ~ g	~ p ∧ g	(p ∧ ~ g) ∨ (~ p ∧ g)
F	F	T	T	F	F	F
F	T	T	F	F	T	T
T	F	F	T	T	F	T
T	T	F	F	F	F	F

as L.H.S. = R.H.S. Hence proved

Method II (without using Truth table)

$\sim (p \leftrightarrow q) \equiv (p \wedge \sim q) \vee (p \sim \wedge q)$

Taking left hand side:

$\sim (p \leftrightarrow q) \equiv \sim [(p \rightarrow q) \wedge (q \rightarrow p)]$

$\because \quad (p \leftrightarrow q) \equiv (p \rightarrow q) \wedge (q \rightarrow p)$

$\equiv \sim [\sim p \vee q \wedge (q \vee p)]$

$\because \quad (p \leftrightarrow q \equiv \wedge q \rightarrow p)$ and $(g \rightarrow p = \sim g \vee p)$

$\equiv \sim (\sim p \vee q) \vee \sim (q \vee \sim p)$ [using de morgans law]

$\equiv \sim (\sim p \vee q) \vee \sim (\sim p \vee q) \quad \because (p \vee q \equiv q \vee p)$

$\equiv$ R.H.S. **Hence proved.**

Q.1. (*c*) Construct that a tree with *n* vertices has exactly n – 1 edges.

Ans. Construct an input and output table for the expression $x_1 . (x_2 + x_3) + x_2$.

Q.1. (*d*) Show that a tree with *n* vertices has exactly *n* – 1 edges.

Ans. Proof (by induction) – we can see that the theorem is true for

n = 1 [1 — node, No edge]

n = 2  [2 — node, 1 — edge]

n = 2  [3 — node, 2 — edge]

We assume that the theorem is true for all trees wiht less than *n*(up to *n* – 1) vertices. Now we shall show that if it is true upto (*n* – 1) vertices, it is also true for *n* vertices as follows:

Let v_i and v_j be the end vertices of an edge e_k in a tree *T* by definition, there is no other path between v_i and v_j except e_k.

Deletion of e_k from *T* shall make T-ek as disconnected graph; considering of two components say T_1 and T_2. As there is no cycle in *T*, there must not be any cycle in T_1 and T_2 also. So T_1 as well as T_2 is an acyclic connected graph and hence each is a Tree.

Let number of vertices in T_1 and T_2 be *p* and *q* respectively. As deletion of e_k from *T* does not affect the number of vertices in *T*, the number of vertices in T_1 and T_2 taken together shall be the same as in *T*. *i.e.*, $p + q = n$.

In T_1, number of vertices = $p(< n)$, so number of edges in $T_1 = p - 1$ (by our assumption that the theorem holds for less then n vertices).

Similarly in T_2, number of vertices = $q(< n)$, so the number of edges in $T_2 = q - 1$.

Therefore, the total number of edges in T_1 and T_2

$$= (p - 1) + (q - 1)$$
$$= p + q - 2 \qquad (\text{as } p + q = n)$$
$$= n - 2$$

As the number of edges in T

= (number of edges in T_1) + (number of edges in T_2) + 1

(as one edge e_k is deleted from T to convert it into T_1 and T_2)

$$= (n - 2) + 1$$
$$= (n - 1) \text{ by equation (2)}$$ **Hence proved.**

Q.1. (*e*) *G* is a non-empty connected graph whose vertices are all of even degree. Show that *G* is an Eulerian graph.

Ans. Suppose the G is a connect graph and every vertex is even degree. We can struct an Eulerian trail. We begin a trail T_1 at any edge e. We extend T_1 by adding an edge after the other. If T_1 is not closed at any steps, say T_1 begins at u but ends at $v \neq u$, then only an odd number of the edges incident on v appear in T_1; hence we can extend T_1 by another edge incident on v. Thus, we can continue to extend T_1 until T_1 returns to its initial vertex u *i.e.*, until T_1 is closed. If T_1 includes all the edges of G, then T_1 is an Eulerian Trail.

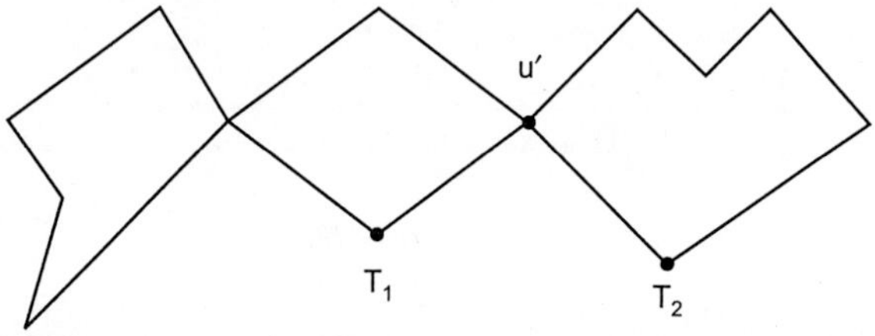

Suppose T_1 does not include all edges of G, consider the graph H obtained by deleting all edges of T_1 from G. H may not be connected, but each vertex of H has even degree since T_1 contains an even number of the edge, incident on any vertex. Since, G is connected,there is an edge e' H which has an end point u' in T_1.

We construct a trail T_2 in H beginning at u' and using e'. Since all vertices in H have even degree. We can continue to extent T_2 untill T_2 returns to u' as shwn in figure. We can clearly put T_1 and T_2 together to form a larges closed trail in G. We continue this process until all edges of G are used. We finally obtain an Euler i an trail and so G is Eulerian.

> **Note: Euler path:** A path in a connected graph G is called Euler path if it includes every edge excostly once. Since, the path contains every edge exactly once, it is also called Euler trail.
>
> **Eules Circuit:** An Euler path that is circuit is called Euler circuit. A connected graph which has a Eulerian circuit is called an Eullerian graph.

Q.1. (*f*) For the statement "If *N* is an odd number then N^2 is also an odd number". What are the inverse, converse and contrapositive statements of above statement?

Ans. Given statement is in the form $p \leftrightarrow q$ "If *N* is an odd number then N^2 is also an odd number".

Let p : *N* is an odd number

q : N^2 is also an odd number

(*a*) Inverse will be in the form of

so, Inverse of the given statement will be:

(*b*) Converse will be in the form of

so, converse of the given statement will be:

(*c*) Contrapositive will be in the form of

so, contrapositive of the given statement will be

Q.1. (*g*) Solve the recurrence relation $U_{n+1} - aUn = 0$ where a ¹ 1 is a constant and initially $u_0 = 3$.

Ans. Give recurrence relation is

$U_{n+1} - a\ U_{n=0}$ when $a \neq 1$ and $u_0 = 3$ using substitution method

$$U_{n+1} = a\ U_n$$

put $n = 0$

$$U_1 = \text{a}\ U_0 \text{ (as } U_0 = 3)$$

using substitution method

$$U_{n+1} = a\ U_n$$

put $n = 0$

$$U_1 = a\ U_0 \text{ (as } U_0 = 3)$$

$$U_1 = a\,.3 \qquad \text{...}(i)$$

put $n = 1$

$U_2 = a.U_1$ [but $U_1 = a.3$ given equation (*i*)]

so $U_2 = a.[a\ .\ 3]$

$$U_2 = a^2.\ 3$$

Similarly for *n* terms

$$U_n = a^n\ .\ 3$$

Solution is $[U_n = 3.a^n]$ **Ans.**

Q.1. (*h*) Prove that if $m + n \geq 125$, then $m \geq 63$, or $n \geq 63$, m, n being positive integers.

Ans. There seems to be no way to prove the given fact directly. Instead, one can prove by taking the contra positive:

not "$m \geq$ 63 or $n \geq 63$" implies not "$m + n \geq 125$". By de mogons law, the negation of

"$m \geq 63$ or $n \geq 63$" is "not $m \geq 63$ and $n \geq 63$" *i.e.* "$m \leq 63$ and $n \leq 63$" so the contrapositive proposition is if $m \leq 63$ and $n \leq 63$ then $m + n \leq 125$.

This follows immediately from property of in equalities : $a \leq c$ and $b \leq d$ imply that $a + b \leq c + d$ for all real numbers a, b, c, d.

The second type of indirect proof is known as "proof by contradiction (reduction od absurdum)". In this type of proof, we assume the opposite of what we trying to prove assume the opposite of what we trying to prove and get a logical contradiction. Hence, our assumption must have been false, and therefore what we originally required to prove must be true. We give a proof of the statement of the example by contradiction.

UNIT-I

Q.2. (*a*) Prove by mathematical induction that $T_n = n^3 + 3n^2 = 5n + 3$ is divisible by 3 for all n e N.

Ans. (*i*) Basis of Induction for $n = 1$

$T(1) = (1)3 + 3(1)^2 + 5(1) + 3 = 12$ which is divisible by 3.

(*ii*) Induction step for $n = k$

$$T(k) = k^3 + 3\,k^2 + 5\,k + 3$$

Let us assume that it is divisible by 3

So $\quad T(k) = k^3 + 3\,k^2 + 5\,k + 3 = 3.r \quad ...(i)$

Now, for $\quad n = k + 1$

$$\begin{aligned} T(k+1) &= (k+1)3 + 3\,(k+1)2 + 5\,(k+1) + 3 \\ &= (k^3 + 3\,k^2 + 3k + 1) + 3\,(k^2 + 1 + 2k) + 5k + 5E \\ &= k^3 + 3k^2 + 3k + 1 + 3\,k^2 + 3 + 6k + 5k + 5 + 3 \\ &= k^3 + 6\,k^2 + 14\,k + 12 \\ &= (k^3 + 3\,k^2 + 5\,k + 3) + 3k^2 + 9k + 9 \\ &= 3.r + 3(k^2 + 3k + 3\,) \\ &= 3\,(r + k^2 + 3k + 3) \\ &= \text{multiple of } 3 \end{aligned}$$

So we can say tht is divisible by 3 **Hence proved.**

Q.2. (*b*) (*i*) Obtian the Disjunctive Normal Form of the Boolean function $F = x.y'.z + x'.z$.

Ans. Given $F = x.y'.z + x'.\,z$ (find DNF)

as f is a function on xy an z and in part two third variable y is missing so we can modify $x'z \equiv (x'.\,z).1 \quad$ [as ... $x_i.1 \equiv x_i$]

$$\begin{aligned} &\equiv (x'.\,2)\,.\,(y + y') && [\text{as } y + y' \equiv 1] \\ &= (x'\,z\,y) + (x'\,z\,y') && [\text{by distributive between}] \end{aligned}$$

Now F will become $x.y'.z + x'\,z\,y + x'\,z\,y'$

Which is required Disjunctive Normal Form

Or

$x\,y'\,z + x'\,y\,z + x'\,y'\,z \quad$ [By commutative law] **Ans.**

Q.2. (*b*) (*ii*) Obtain the Conjunctive Normal Form in three variables of the Boolean function $F_1 = (x + y).(x + z')$.

Ans. Obtain the Conjunctive Normal Form (CNF) in three variables of the Boolean function.

$$= (x + y)\,.\,(x + 2')$$

Given $\quad F = (x + y)\,.\,(x + z')$

$$\equiv (x + y + 0)\,.\,(x + z' + 0)$$

$$= (x + y + z.z')\,.\,(x + z' + yy') \qquad \begin{bmatrix} \text{as } x_i + 0 = x_i\ \forall_i \\ \text{and } x_i.x_i' = 0 \end{bmatrix}$$

$$= (x + y + z)(x + y + z')(x + z' + y)(x + z' + y')$$

[By distributive law]

$$\equiv (x + y + z)(x + y + z')(x + y + z')(x + y' + z') \quad \textbf{Ans.}$$

Q.2. (c) Give example of relations R on $A = \{a, b, c, d\}$ having the stated property.

(i) R is both symmetric and antisymmetric.

(ii) R is neither symmetric nor antisymmetric.

(iii) R is transitive but RUR^{-1} is not transitive.

Ans. $A = \{a, b, c, d\}$

(i) R is both symmetric and anti-symmetric

$R = \{(a, a), (b, b), (c, c), (d, d)\}$

(ii) R is neither symmetric nor antisymmetric

$R = \{(a, a), (a, b), (b, a), (a, c)\}$

(iii) R is transitive but $R \vee R^{-1}$ is not transitive

$R = \{(a, c), (d, d)\}$

Q.3 (a) Assume the statements:

$A(x) \equiv \{x - 2 = 0\}$

$B(x) \equiv \{x - 2 \geq 0\}$

Where x ε R. Determine the underlying set of real numbers on which the following statements are true:

(i) $A(x) \cup B(x)$

(ii) $A(x) \cap B(x)$

(iii) $\overline{B(x)} \Rightarrow \overline{A(x)}$

(iv) $B(x) \Rightarrow A(x)$.

Ans. From above definition of set $A(x)$ and $B(x)$, we can have

$A(x) = \{0, 1, 2\}$

$B(x) = \{-2, -1, 0\}$

(i) $A(x) \cup B(x) = \{1, 2, -2, -1\}$

(ii) $A(x) \cap B(x) = \{\phi, 0\}$

(iii) $\overline{B(x)} \Rightarrow \overline{A(x)} =$

(iv) $B(x) \Rightarrow A(x) =$

Q.3 (b) Let Z be the set of integers. Show that the relations xRy if and only if $x + y$ is even, is an equivalence relation Z. Determine the equivalence class of 1 ε Z.

$A(x) \equiv \{x - 2 = 0\}$

$B(x) \equiv \{x - 2 \geq 0\}$

Where x ε R. Determine the underlying set of real numbers on which the following statements are true: (i) $A(x) \cup B(x)$ (ii) $A(x) \cap B(x)$ (iii) $B(x) \Rightarrow A(x)$ (iv) $B(x) \Rightarrow A(x)$.

Ans. Let $A = \{0, 1, 2, 3, 4...\}$ set of integers and R is a relation on defined on A such that xRy if and only if $x + y$ is even then

$R = \{0, 2\}, (0, 4), (0, 5), (1, 1) (1, 3), (1, 5), (1,7),$
$\{2, 0\}, (2, 2), (2, 4)... \}$

Or

$R = \{(x, y) : x + y = 2m,\ m \text{ is any integer}\}$

$[0] = \{x : x \,\varepsilon\, A$ and Or y *i.e.,* $0 + x = 2m\}$ where m is integer

$[0] = \{0, 2, 4, 6, 8 ...\}$

Now,

$[1] = \{x : x \,\varepsilon\, A$ and $f\, R\, y$ *i.e.,* $1 + x = 2m\}$ where m is integer

$[1] = \{x : x \,\varepsilon\, A$ and $f\, R\, y$ *i.e.,* $x = 2m - 1\}$ $m = 0, 1, 2$

$[1] = \{-1\}\ 1, 3, 5, 7, ...\}$

Now

$[2] = \{x : x \,\varepsilon\, A$ and $2\, R\, y$ *i.e.,* $2 + x = 2m$ Or $x = 2m - 2\}$

$[2] = \{2, 0, 2, 4, 6, 8...\}$

Now $[3] = \{$x $: x \,\varepsilon\, A$ and $3\, R\, y$ *i.e.,* $3 = x = 2$m Or $x = 2m - 3$ Or $x = 2m - 3\}$

$[3] = \{-1, 1, 3, 5, 7, ...\}$

UNIT-II

Q.4. (*a*) *A*, *B*, *C* are three study programmes in a university each of 3 years duration. The programme *A* has 26 papers, programme *B* has 32s papers and programme *C* has 40 papers. The programmes *A* and *C* have 6 papers in common and *B* and *C* have 3 papers in common. Further *A*, *B* and *C* have 2 papers in common.

(*i*) How many different papers in the three programmes are being taught?

(*ii*) How many papers are in the programme *B* but not in *A* and *C*?

Ans. (*i*) A has 26 papers *i.e.,* $|A| = 26$

B has 32 papers *i.e.,* $|B| = 32$

C has 40 papers *i.e.,* $|C| = 40$

A and *C* have 6 papers in common *i,.e,* $A \cap B = 6$

B and *C* have 3 papers in common *i,.e,* $B \cap C = 3$

and *A, B* and *C* have 2 papers in common *i,.e,* $A \cap B \cap C = 2$

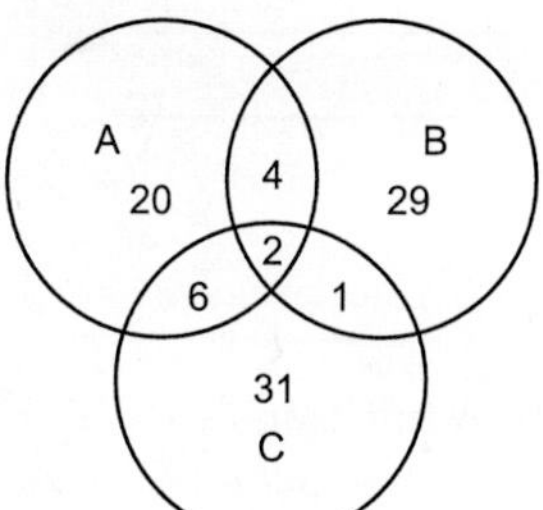

Q.4. (*b*) Let *X* be a binary operation in *R*, the set of all real numbers, defined by $a \times b = a + b + ab$ for all *a*, *b* ε *R*. Find the inverse of a ε $R(a \neq -1)$.

Ans. $a \times b = a + b + ab$ (given)

as $a \neq -1$

let if $a = 1$ then $b = -1$

$a = 1$

$$1 \times b = 1 + b + b$$
$$b = 1 + b + b$$
$$b = 1 + 2b = 1$$

Now $a = 2$

$$2 \times b = 2 + b + 2 \times b \quad \Rightarrow b = -2$$

Now take $a = 3 \quad \Rightarrow b = -3$

So $R = \{(0, 0)\ (1, -1), (2, -2), (3, -3)...\}$

Now, $R^{-1} = \{(0, 0), (-1, 1), (-2, 2), (-3, 3)...\}$

Q.5. (*a*) Find the total number of ways in which 10 students can be seated in a (*i*) row (*ii*) circle.

Ans. (*i*) 10 students can be seated in a row as 10.9.8.7.6.5.4.3.2.1 = 10! **Ans.**

(*ii*) 10 students can be seated in a circle is (10 – 1)! = 9! Ans.

(circular perpulation).

Q.5. (*b*) Let $S = \{1, 2, 3\}$. Find the number of different functions that can be defined from *S* to *S*. How many of these functions are one one, onto.

Ans. $S = \{1, 2, 3\}$

(*i*) Injective function (one-to-one):

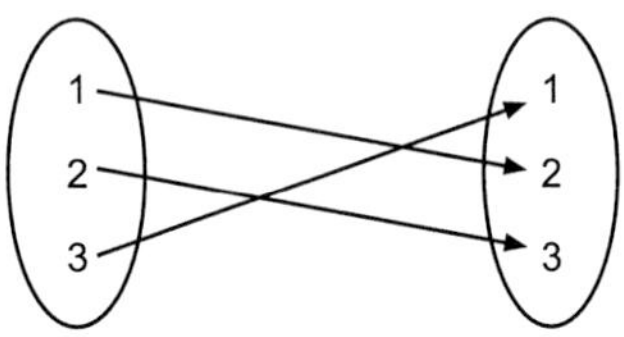

(Injective)

Note: A function from *A* into *B* is one-to-one or injective, if for all elements in $A = \{x_1, x_2\}$ such that $f(x_1) = f(x_2)$ implies $x_1 = x_2$

(*ii*) Bijective function:

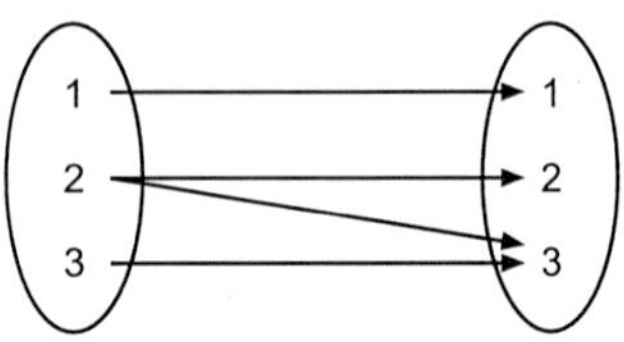

(Surjective)

Note: A function *f* from *A* to *B* said to be bijective or one-to-one correspondence, if *f* is both injective and surjective *i.e.*, both one-to-one and onto.

(*iii*) On to function: A function *f* from *A* to *B* is onto or surjective if every element of *B* is the image of some element in *A*, that is if *B* = range of *f*.

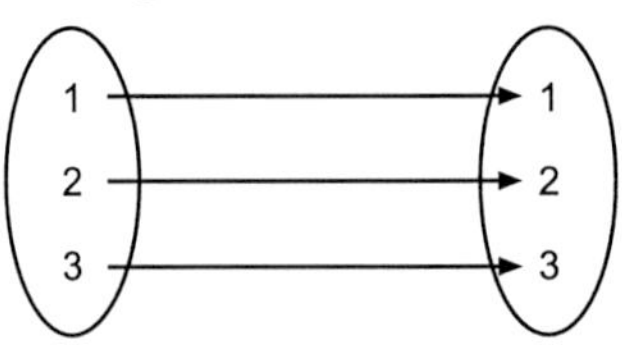

(Surjective as well as bijective)

Q.5. (*c*) Show that in the set of integers *I*, the relation R given by aRb if a divides *b* is reflexive, transitive and antisymmetric.

Ans. (*i*) Reflexive: As every number is divisible by itself

$a/a,\ c/c \Rightarrow aRa,\ bRb,\ cRc$

$\therefore$ R is reflexive

(*ii*) Anti symmetric: A relation R is symmetric if $(a, b) \varepsilon R$ implies $(b, a) \varepsilon R$. So, if x is divisible by y then obviously y is not divisible by x unless $x = y$, So symmetric

(*iii*) Transitive: If for $x, y, z \varepsilon N$, x is divisible by y and y is divisible by z, then automatically x divisible by z. Hence, R is transitive.

UNIT-III

Q.6. (*a*) Define a Boolean expression and minimize the following Boolean expression: $F = \overline{AB} + \bar{A} + AB$.

Ans. Boolean expression: Consider a Boolean algebra $(B, \vee, \wedge, 0, 1)$. A Boolean expression over Boolean algebra B is defined as –

(*i*) Every element of B is a Boolean expression

(*ii*) Every variable name is a Boolean expression

(*iii*) If a_1 and a_2 are Boolean expression, then a′1, $a_1 \vee a_2$ and $a_1 \vee a_2$ are Boolean expression.

Give Boolean expression is

$$F = \overline{AB} + \bar{A} + AB$$

$$\equiv (\bar{A} + \bar{B}) + \bar{A} + AB \quad \text{Using Demogans law}$$

$$\equiv \bar{A} + \bar{B} + \bar{A} + AB \equiv (\bar{A} + \bar{A}) + \bar{B} + AB \equiv \bar{A} + \bar{B} + AB$$

Q.6. (*b*) Simplify the Boolean function $F(A, B, C, D) = \Sigma\,(0, 1, 2, 3, 4, 5, 6, 7, 8, 9, 11)$.

Ans. Given $F = (A, B, C, D) = \Sigma\,(0, 1, 2, 3, 4, 5, 6, 7, 8, 9, 11)$

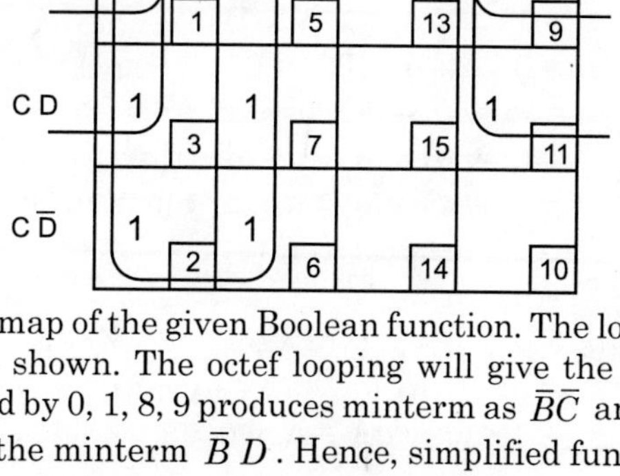

The Karnaugh map of the given Boolean function. The looping of one octant and two guads are shown. The octef looping will give the min term $\bar{A}$. The looping quad formed by 0, 1, 8, 9 produces minterm as $\bar{B}\bar{C}$ and the quad formed by 1, 3, 9, 11 gives the minterm $\bar{B}\,D$. Hence, simplified function

$$F = \bar{A} + \bar{B}\bar{C} + \bar{B}D \quad \textbf{Ans.}$$

Q.7. (*a*) Let *R* be a relation defined in the set Z^+ of all positive integers such that *aRb*, for any *a*, *b*, ε Z^+ if a divides *b*. Let D_n be the set of all divisions of *n*, *n* is a positive integer. Show that (*Dn*, *R*) is a sublattice of (Z^+, *R*).

Ans. Let (L, ∧ ,∨) be a lattice and *S* be a non-empty sub set of *L*. Then *S* is said to be a sub-lattice of *L* if *S* is a lattice with respect to the operation of *L* *i.e.*, (L,∧ ,∨) is a lattice. It means tht *S* is a sub lattice of *L* if *S* is closed under the operations of ∧ and ∨ of *L*.

We know that L = (*N*, gcd, l.c.m) is a lattice. Let D_n be the set of all positive interval divisors of an integer n ε *N*.

Now L.C.M as well as G.C.D of any pair of elements in D_n say (d_1, d_2) divides *n*. Therefore, (Dn, g.c.d, l.c.m) shall be a sub-lattice of l let *n* be a positive integer and S_n be the set of all divisiors of *n*.

For $n = 6$, $S_6 = \{1, 2, 3, 6\}$ and for $n = 24$,

$S_{24} = \{1, 2, 3, 4, 6, 12, 24\}$. Let *D* denote the relation of *n* division in Z^+ such that for any *a*, *b*, *e* Z^+, *a* R_b iff a divides *b* when Z^+ be the set of all positive integers. The lattice < S_6, *D* >, < S_{24}, *D* > , < S_2, *D* > and < S_3, *D* > are shown as

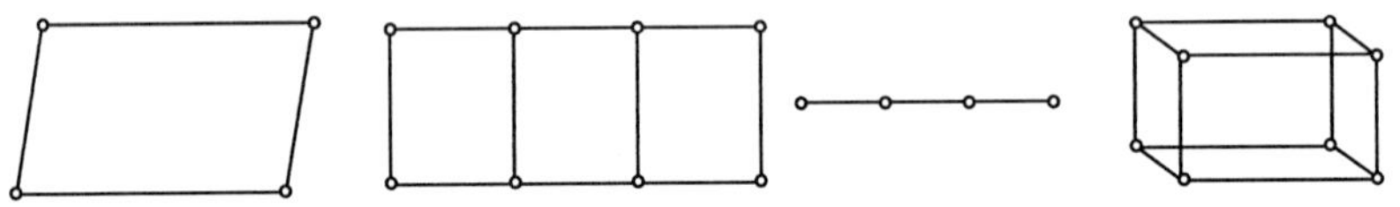

Q.7. (*b*) Prove that in a distributive lattice, if an element has a complement then this complement is unique.

Ans. Suppose that an element a has two complements *b* and *c* then

$$a \vee b = 1 \quad a \wedge b = 0$$
$$a \vee c = 1 \quad a \wedge c = 0$$

we have
$$\begin{aligned} b &= b \wedge 1 \\ &= b \wedge (a \vee c) \quad (\because a \vee c = 1) \\ &= (b \wedge a) \vee (b \wedge c) \text{ by distributivity} \\ &= 0 \vee (b \wedge c) \qquad (\because a \vee b = 0) \\ &= (a \wedge c) \vee (b \wedge c) \\ &= (a \wedge b) \vee c \quad (\because a \vee b = 1) \\ &= 1 \wedge c = c \end{aligned}$$

Alternative method: Let a_1 and a_2 be the complement of a ε *L*, then

$$a \vee a_1 = 1 \quad a \wedge a_2 = 1 \quad \ldots(1)$$
$$a \vee a_1 = 0 \quad a \wedge a_2 = 0 \quad \ldots(2)$$

Now we have
$$\begin{aligned} a_1 &= (a_1 \vee 0) = a_1 \vee (a \wedge a_2) \text{ from (2)} \\ &= (a_1 \vee a) \wedge (a_1 \vee a_2) \text{ (distributive law)} \\ &= (a \vee a_1) \wedge (a_1 \wedge a_2) \text{ (by commutative)} \\ &= 1 \times (a_1 \vee a_2) \text{ by (1)} \\ &= a_1 \vee a_2 \end{aligned}$$

Similarly
$$\begin{aligned} a_2 &= a_2 \vee 0 = a_2 \vee (a \wedge a_1) \\ &= (a_2 \vee a) \wedge (a_2 \vee a_1) = (a \vee a_1) = (a \vee a_2) \wedge (a \vee a_2) \\ &= 1 \wedge (a_1 \vee a_2) \;= a_1 \vee a_2 \end{aligned}$$

UNIT-IV

Q.8. (*a*) Show that a simple non-directed graph G is a tree if an only if it is connected and has no cycles.

Ans. A simple non directed/undirected graph *G* is a tree if *G* satisfies any of the following conditions.

I : *G* has No cycle then it is a tree

Proof = As *G* has no cycle *i.e.*, then is a path between every pair of vertices of *G*. It is a connected graph. Since, *G* has only one path between every pair of its vertices, it cannot have any cycle (as existence of cycle requires two or more path between a pair of vertices). Hence, *G* is a connected acyclic graph and is therefore a tree.

II: *G* is a connected graph then it is a tree proof: Minimum number of edges to make a graph of n vertices as a connected graph is n^{-1} we see that removal of a single edge from the graph *G* shall make it disconnected. So $(n - 1)$ is the minimum no of edges required to make a graph of *n* vertices as connected and acydic.

Therefore if *G* is a minimally connected graph (a connected graph with n vertices and n^{-1} edges). Thus, the theorem can also be stated as "A minimally connected graph is a tree". Moreover if a graph is not minimally connected then deletion of an edge from *G* shall also peep G – e as connected. Therefore, this *e* shall be an edge in some cycle and *G* will not be a tree.

Q.8. (*b*) Find the minimal spanning tree of the following:

Ans. $AB = 12 AD = 1$

$BC = 3 \quad AE = 5$

$CD = 8 \quad AC = 4$

$DE = 9 \quad BE = 1$

$EF = 2 \quad FC = 11$

$AF = G \quad EC = 7$

$BF = 12$

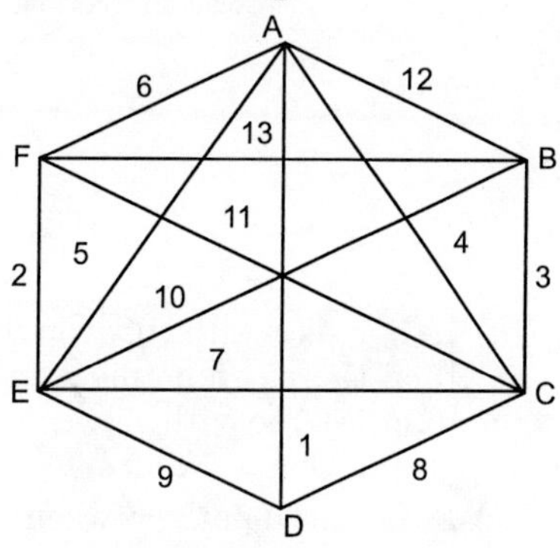

(1)

(2)
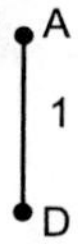

(3)
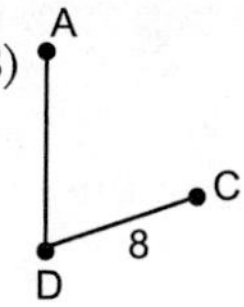

(4)
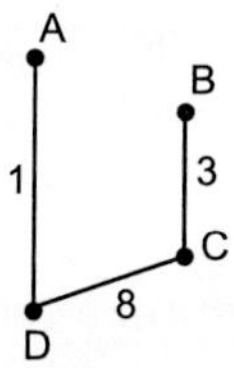

(5)
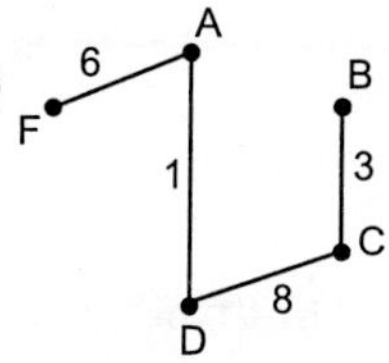

(6)
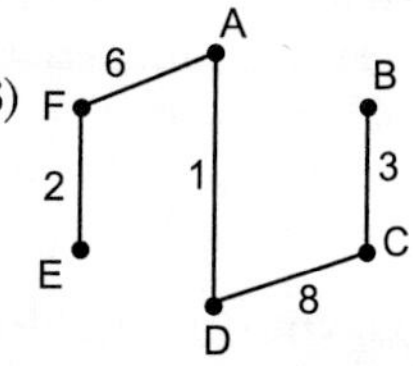

This is minimum spanning tree. **Ans.**

Q.9. (*a*) State and prove five colour theorem.

Ans. Five colour theorem: It states that "The vertices of every planar graph can be properly coloured with five colours" proof (By induction): we know that the vertices of a simple planar graph *G* having vertices up to 5 can be properly coloured with five colours *i.e.*, we know that the theorem is true for number of vertices $n = 1, 2, 3, 4, 5$. Let us assume that the theorem is true for the number of vertices ofG equal to $n - 1$ and shall prove on this basis, the theorem to be true for the graph with *n* vertices also.

As *G* is a planar graph it must have at last one vertex with degree ≤ 5. We call this vertex as *V*. If we delete the vertex *υ* along with all the edges incident on it, from the graph *G*, then the remaining graph $G_1 = (G - \vee)$ shall contain $n - 1$ vertices and therefore can be properly coloured wth five colours (on the basis of the assumption that the theorem is true for a planar graph with $(n - 1)$ vertices). Now, we have to show that this vertex can *υ* can also be properly coloured wiht any one of the same 5 colours.

If the degree of *υ* is 1, 2, 3 or 4 then it is quite evident that *υ* can be properly coloured with the remaining colour (out of 5) not assigned to its 4 adjacent vertices.

But if the degree of *υ* is 5 then all the 5 vertices adjacent to *υ* (*i.e.*, υ_1, υ_2, υ_3, υ_4, υ_5) would have been assigned all the five colours as shown in figure

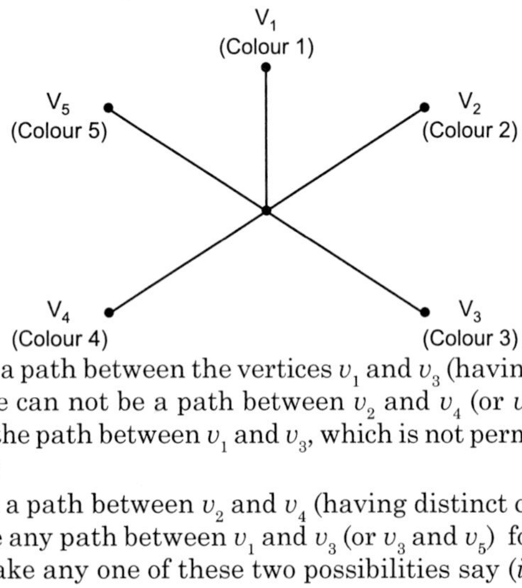

(*i*) If there is a path between the vertices υ_1 and υ_3 (having distinct colours), then there can not be a path between υ_2 and υ_4 (or υ_2 and υ_5) as it will intersect the path between υ_1 and υ_3, which is not permissible in a planar graph. Or

(*ii*) If there is a path between υ_2 and υ_4 (having distinct colours) then these can not be any path between υ_1 and υ_3 (or υ_3 and υ_5) for the same reason we may take any one of these two possibilities say (*i*) *i.e.*, υ_1 and υ_3 are connected and then υ_2 and υ_4 can be assigned the same colour say copur 2 and then the remaining colour 4 can be assigned for proper colouring of *υ*.

If we assume the possibility (*ii*) as true *i.e.*, υ_2 and υ_4 are connected and υ_1 and υ_3 are not connected then the two vertices υ_1 and υ_3 can be assigned the same colour say colour 1 and the left our colour-3 can be used for properly colouring of *υ*.

Hence, theorem is proved.

Q.9. (*b*) Explain an algorithm that finds the shortest distances between all pairs of vertices in a graph.

Ans. Let us take Dijkstras Algorithm for finding the shortest path in a weighted graph between two vertex a and f in the figure given.

Dijkstras Algorithm:

Input: A connected weighted graph

Output: $L(2)$, the length of shortest distance from a to Z.

Step 1: Set $L(a) = 0$ and for all vertices $v \neq a$ and $L(v) = \infty$

Set $T = v$ where T = set of vertices having temporary labels and V = vertex set of G.

Step 2: Let u be a vertex in T for which $L(u)$ is minimum and hence the permanent label of u.

Step 3: if $u = z$ the stop.

Step 4: For every edge $e = (u, v)$, incident with u, if $v \varepsilon T$, change $L(v)$ to min $(L(v) + W(e))$

Step 5: Change T to $T - \{u\}$ and go to step 2.

Let us now illustrate this algorithm with the weighted graph in the following example.

For a given find shortest path between a to f

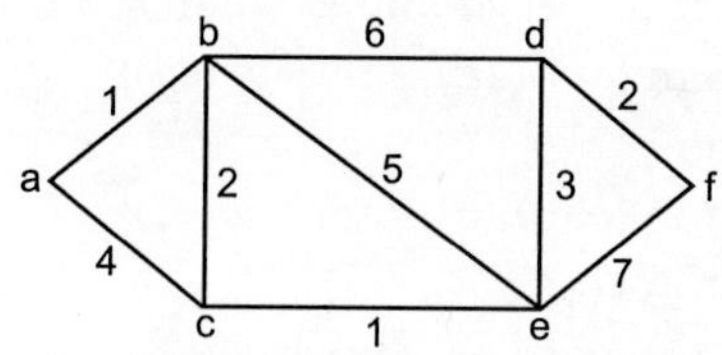

The initial labelling is given by

vertex v	a	b	c	d	e	f
$L(v)$	0	∞	∞	∞	∞	∞
T	a	b	c	d	e	f

Iteration 1 : $u = a$ has $L(u) = 0$, T becomes $T - \{a\}$. There are two edges incident with a *i.e.*, ab and ac where b and $c \varepsilon T$.

$$L(b) = \min \{\text{old } L(b), L(a) + w\,(ab)\}$$
$$= \min \{\infty, 0 + 1\} = 1$$
$$L(c) = \min \{\text{old } L(c), L(a) + w\,(ac)\}$$
$$= \min \{\infty, 0 + 4\} = 4$$

Hence, minimum label is $L(b) = 1$

vertex v	a	b	c	d	e	f
$L(v)$	0	1	4	∞	∞	∞
T	a	b	c	d	e	f

Iteration 2: $u = b$, the permanent label of b is 1. T becomes $T - \{b\}$ there are three edges incident with b *i.e.*, bc, bd and be when c, d, e, ε s

$$L(c) = \min \{\text{old } L(c), L(b) + w\,(bc)\}$$
$$= \min \{4, 1 + 2\} = 3$$
$$L(d) = \min \{\text{old } L(d), L(b) + w\,(bd)\}$$
$$= \min \{\infty, 1 + c\} = 7$$

$$L(e) = \min \{\text{old } L(e), L(b) + w\ (be)\}$$
$$= \min \{ \infty, 1 + 5\} = 6$$

vertex v	a	b	c	d	e	f
$L(v)$	0	1	3	7	6	∞
T	{		c	d	e	f

Thus minimum label if $L(c) = 3$

Iteration 3: $u = c$ the permanent label of e is 3, T becomes $T - \varepsilon c$. There is one edge with c *i.e.*, ce where $e \; \varepsilon \; T$

$$L(c) = \min \{\text{old } L(e), L(c) + w(ce)\}$$
$$= \min \{6, 3 + 1\} = 4$$

Thus minimum label is $L(c) = 4$

vertex v	a	b	c	d	e	f
$L(v)$	0	1	3	7	4	∞
T	{			d	e	f

Iteration 4: $u = e$ the permanent label of e is 3, is 4. T becomes $T - \varepsilon 3$. There is two edge incident with e *i.e*, ed and ef when $d, f \; \varepsilon \; T$

$$L(d) = \min \{\text{old } L(d), L(e) + w(ed)\}$$
$$= \min \{7, 4 + 3\} = 7$$
$$L(d) = \min \{\text{old } L(d), L(e) + w(ed)\}$$
$$= \min \{7, 4 + 3\} = 7$$
$$L(d) = \min \{\text{old } L(f), L(e) + w(ef)\}$$
$$= \min \{\infty, 4 + 7\} = 11$$

vertex v	a	b	c	d	e	f
$L(v)$	0	1	3	7	4	11
T	{			d		f

Thus, minimum label $L(d) = 7$

Iteration 5: $u = d$, the permanent label of d is 7. T becomes $T - \{d\}$. There is one edge incident with d *i.e*, df and when $f \; \varepsilon \; T$

$$L(f) = \min \{\text{old } L(f), L(d) + w(d)\}$$
$$= \min \{11, 7 + 2\} = 9$$

vertex v	a	b	c	d	e	f
$L(v)$	0	1	3	7	4	9
T	{					f}

The minimum label if $L(f) = 9$

Since, $u = f$, the only choice. Iteration stops. Thus, the shortest path/distance between a and f is a, and moreover, the shortest path is (a, b, c, d, e, f).

❐❐❐

END TERM EXAMINATION

THIRD SEMESTER (B.TECH) - Dec. 2010

Time : 3 Hours *Maximum Marks : 75*

Note: Question number 1 is compulsory. Attempt any one question from each unit.

Q.1. (*a*) Show that $p \Rightarrow q \equiv \sim p \cup q$

Ans. Show that $p \Rightarrow q \equiv \sim p \cup q$

p	g	p→g	~p	~p∨g
F	T	T	F	T
F	F	F	F	F
T	T	T	T	T
T	F	T	T	T

From truth tables column 2 and column 5, we can say that $p \Rightarrow q \equiv \sim p \cup q$.

Q.1. (*b*) Define a Poset by giving an example.

Ans. A non-empty set A together with the binary relation R is said to be poset or partially ordered set if R is reflexive, anti symmetric and trantitive.

Example: The relation $\geq$ is a partial ordering on the set of Integers z.

(*i*) Since $a \geq a$ for every a, $\geq$ is reflexive.

(*ii*) $a \geq b$ and $b \geq a$ imply $a = b$, so we can conclude that $\geq$ is anti-symmetric too.

(*iii*) $a \geq b$ and $b \geq c$ imply $a \geq b \geq c$ or $a \geq c$, so we can say that, $\geq$ is transitive. Hence, relation $\geq$ is a POSET.

Q.1. (*c*) Discuss the commutativity and associativity of the binary operation defined on the set of integers as

$a \times b = a - b + ab$

for all $a, b, \varepsilon I$.

Ans. Note that:

(*i*) Commutative Property: A binary operation on a set A is said to be commutative if $a \times b = b \otimes a, \ \forall \ a, b, \varepsilon A$.

(*ii*) Associative Property: A binary operation $\otimes$ on a set A is said to be associated if

$$(a \otimes b) \otimes c = a \otimes (b \otimes c), \ \forall \ a, b, c, \varepsilon A.$$

Now given binary operation is

$$a \otimes b = a - b + ab \ \forall \ a, b \ \varepsilon \ I \quad \text{(set of Integers)}$$

and $b \otimes a \equiv b - a + ba$

$\equiv b - a + ab$ (as multiplication is commulative)

$\equiv -(a - b) + ab$ (as commulative is not defined on subtractic as $a - b \neq b -$ a a, b $\forall$ $a, b, \varepsilon I$) operation defined on a and b are not commulative operation.

Now the operation $\otimes$ defined as:

$$a \otimes b = a - b + ab \text{ on the set of Integers}$$

To check the associativity, let

$$(a \otimes b) \otimes c \equiv (a - b + ab) \otimes c$$
$$\equiv (a - b + ab) - c + (a - b + ab) . c$$
$$= a - b + ab - c + ac - bc + abc \quad ...(i)$$

and

$$a \otimes (b \otimes c) = a \otimes (b - c + bc)$$
$$= a - (b - c + bc) + a . (b - c + bc)$$
$$= \text{a} - \text{b} + c - bc + ab - ac + abc \quad ...(ii)$$

From equation (i) and (ii). It is clear that

(1) ≠ (2) so the given relation not associative.

Q.1.(*d*) Let *S* = {*a*, *b*, *c*, *d*}. How many functions can be defined from the set *S* to itself? How many of these are one-one, onto?

Ans. $S = \{a, b, c, d\}$ and function is defined from S to it say so, $f : S \to S$ and we can write all possible function from $S \to S = 4^4 = 256$.

So all possible functions from $S \to S$ are

$$F_1 = \{(a, a), (b, a), (c, a), (d, a)\}$$
$$F_2 = \{(a, a), (b, a), (c, a), (d, a)\}$$
$$F_3 = \{(a, c), (b, c), (c, a), (d, a)\}$$
$$F_4 = \{(a, d), (b, d), (c, d), (d, d)\}$$
$$F_5 = \{(a, a), (b, b), (c, c), (d, d)\}$$
$$F_6 = \{(a, b), (b, c), (c, d), (d, a)\}$$
$$F_7 = \{(a, c), (b, d), (c, a), (d, b)\}$$
$$F_8 = \{(a, d), (b, a), (c, b), (d, c)\}$$

There is a function

$F = \{a, a\}, (b, d), (c, b), (d, c)\}$ is one-to-one and on to.

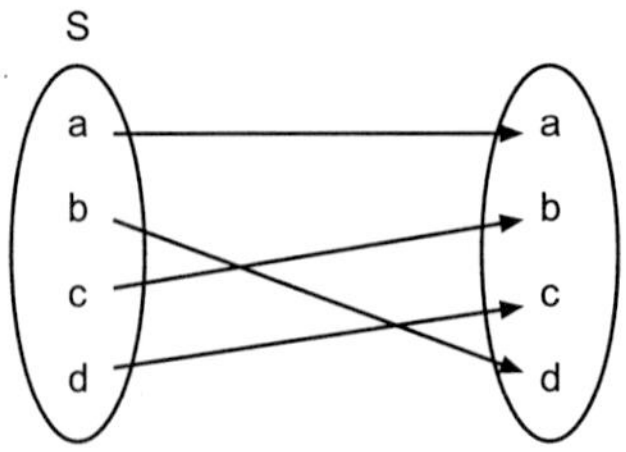

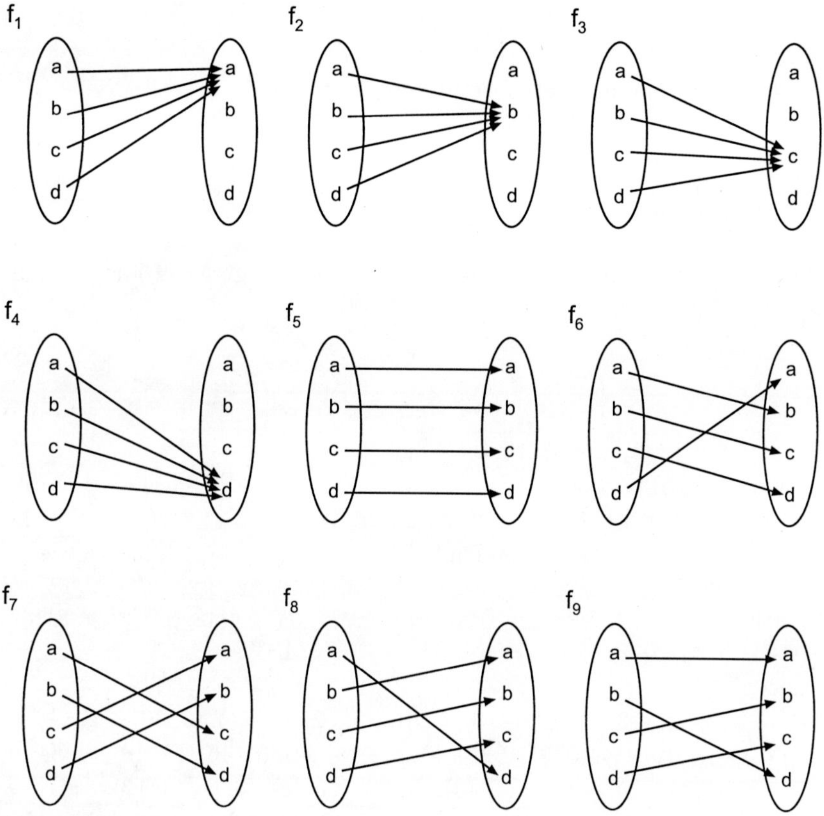

Q.1.(*e*) For a planar graph given below, verify the Eulers formula

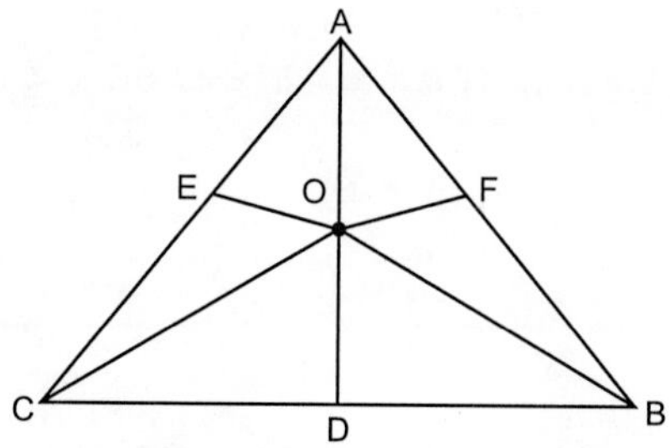

Ans. Let a be a connected planar graph with V vertices and E edges. Where $V \geq 3$. Then we can prove that $E \geq 3V - 6$. Not that it is not true for k1 where $V = 1$ and $e = 0$ and it is also not true for k_2 where $V = 2$ and $e = 1$. To prove or verify the Eulers formula in it, let R be the number of regions in a planar representation of G. So by Euler's formula $V - E + R = 2$

Now the sum of the degrees of the regions canals to $2E$ (by theorem a. The sum of the degrees of the regions of a map is equal to twice the number of edges). But each region has degree 3 or more, hence, $2E \geq 3R$. Thus

$$R \geq \frac{2E}{3}$$

Substituting this in Eulers formula gives

$$2 = V - E + R \leq V - E + \frac{2E}{3}$$

or $\quad 2 \leq V - \frac{E}{3}$

Multiply the inequality by 3 gives $6 \leq 3V - E$ which gives us our result. So for the given figure/graph.

$$V = 7,\ E = 12,\ R = 7$$

So $\quad = V - E + R$

$= 7 - 12 + 7 = 2$ **Hence proved.**

Q.1. (*f*) Let $S = \{2, 3, 4, 6, 8, 12, 24, 36, 48\}$.

Define the relation

$\leq$ in S by $X \leq Y$ if Xdivides Y.

Ans. Find those numbers in S divisible by 2, 3, 4, 6, 8, 12, 24, 36, 48 are 2/2, 2/4, 2/6, 2/8, 2/12, 2/24, 2/36, 2/48, 3/3, 3/6, 3/12, 3/24, 3/36, 3/48, 4/14, 4/18, 4/12, 4/24, 4/36, 4/48, 6/6, 6/12, 6/24, 6/36, 6/48, 8/8, 8/24, 8/48, 12/12, 12/24, 12/36, 12/48, 24/24, 24/48, 36/36, 48/48

Hence,

$R = \{(2, 2), (2, 4), (2, 6), (2, 8), (2, 12), (2, 24), (2, 36),$
$(2, 48), (3, 3), (3, 6), (3, 12), (3, 24), (3, 36), (3, 48), (4, 4)$
$(4, 8), (4, 12), (4, 24), (4, 36), (4, 48), (6, 6), (6, 12),$
$(6, 24), (6, 16), (6, 48), (8, 8), (8, 24), (8, 48), (12, 12),$
$(12, 24), (12, 36), (12, 48), (24, 24), (24, 48), (36, 36),$
$(48, 48)$ **Ans.**

Q.1. (*g*) In a Boolean Algebra if $a + x = b + x$ and $x' = b \times x'$ then prove that $a = b$.

Ans. Dual if $ax = bx$ and $ax' = bx'$ then $a = b$

assume $\quad ax = bx$ and $ax' = bx'$

$a = a.1 = a.(x + x')$
$= ax + ax'$
$= bx + bx' \quad$ [as $ax = bx$ and $ax' = bx'$]
$= b(x + x')$
$= b\ .1\ = b$

assume $\quad a = a + 0$
$= a + x.\ x'$
$= (a + x)\ .\ (a + x')$
$= (b + x)\ (b + x') \quad$ [as $a + z = b + x$ and $a + x' = b + x'$]
$= b + xx'$
$= \text{b} + 0 = \text{b}$

$a = b$ **Hence proved.**

Q.1. (*h*) Find a minimal spanning tree of the following connected graph:

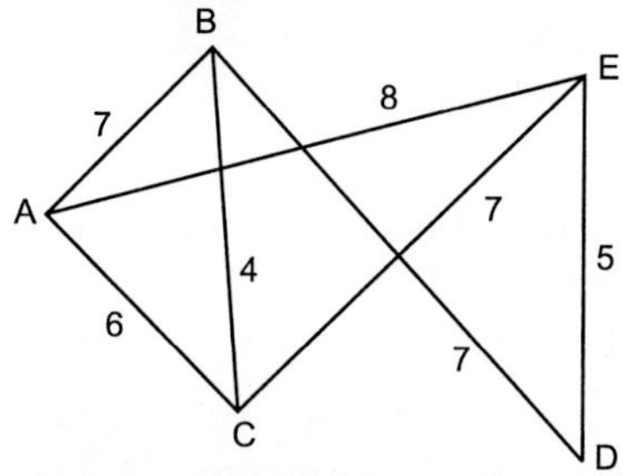

Ans. $AB = 7$

$BC = 4$

$AC = 6$

$BD = 7$

$AE = 8$

$DE = 5$

$CE = 7$

For minimum spanning tree

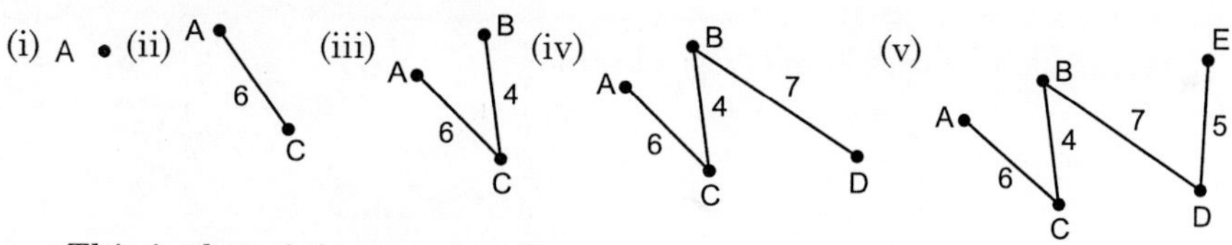

This is the minimum spanning.

Q.2. (*a*) Define a relation *R* on the set of all natural numbers *N* given by:

$R = \{(a, b) / a - b\}$ is an odd positive integer}

Which of the following properties are satisfied by *R*:

(*i*) Reflexive (*ii*) Symmetric (*iii*) Antisymmetric

(*iv*) Transitive (*v*) Equivalance relation

(*vi*) Partial ordering relation

Ans. (*i*) Let $a \varepsilon I$, then $a - a = 0$, which is even integer, *i.e.*, $a \not R a$, so R is not reflexive

(*ii*) Let $a, b \varepsilon (2N + 1)$ (odd positive integers) then if $a - b$ is an odd integer, then $(b - a) = -(a - b)$ is also an odd integer *i.e.*,

$a R b \rightarrow bRa$, So R is symmetric

(*iii*) Let $a, b \varepsilon I$, then $a - b = (2N + 1)I$

and $(b - a) = -(a - b)$ which is a negative odd integer. So $(a - b)$ both are odd positive integer if $a = b$, so R is antisymmetric.

(*iv*) Let $a, b \varepsilon I$ and $(b, c) \varepsilon I$, then

$$(a - b) = (2N + 1) \quad ...(i)$$

and $(b - c) = (2M + 1)$ ...(*ii*)

adding (1) and (2)

$$(a - c) = 2N + 2M + 2$$
$$(a - c) = 2(M + N + 1)$$

which is even so R is not transitive.

(*v*) As R is not reflexive, is symmetric is antisymmetric and not transitive, so R is not an equivalent relation.

(*vi*) As R is not reflexive, is antisymmetric and not transitive, so R is not a POSET.

Q.2. *(b)* Solve the linear difference equation:

$u_1 = 2; u_n = u_{n-1} + 6; n \geq 2$

Ans. We backtrack the value of u_n by substituting the expression of u_{n-1}, u_{n-2} and so on, until a pattern is clear.

given, $u_n = u_{n-1} + 6$ ...(*i*)

replacing n by $n - 1$ in we obtain

$$u_{n-1} = u_{n-2} + 6$$

from (1) $u_n = (u_{n-2} + 6) + 6$

$u_n = u_{n-2} + 2.6$ ...(*ii*)

replacing n by $(n - 2)$ in (1) we obtain

$$u_{n-2} = u_{n-3} + 6$$

so from (2) $u_n = u_{n-2} + 2.6$

$$u_n = (u_{n-3} + 6) + 2.6$$

$u_n = u_{n-3} + 3.6$ ...(*iii*)

In general $u_n = u_{n-k} = k.6$

for $k = n - 1$

so $u_n = u_{n-(n-1)} + (n - 1).6$

$$u_n = u_1 + (n - 1) . 6 \text{ (as } u_1 = 2 \text{ given)}$$
$$u_n = 2 + (n - 1) . 6$$

which is an explicit formula.

Q.3.(*a*) Using the Principle of Mathematic Induction, prove that $A(n) = 5 \times 2^{3n-2} + 3^{3n-1}$ is divisible by 19.

Ans. Step I (Basic step): For $n = 1$

$A(1) = 5 \times 2^{3.1-2} + 3^{3.1-1}$ ...(*i*)

$$A(1) = 5 \times 2 + 3^2$$
$$A(1) = 10 \times 9 + 19$$

Step II (Inductive Hypothesis): We have to prove that whenever $A(m)$ is true, $A(m + 1)$ is also true

or $5 \times 2^{3(m+1)-2} + 3^{3m-1}$ is divisible by 19

or $5 \times 2^{3m-2-2} + 3^{3m-1}$ is divisible by 19

or $5 \times 2^{3m-2} + 3^{3m-1} = 19k$ where $k \in I$

Step III (Inductive step): We have to prove that whenever $A(m)$ is true, $A(m + 1)$ is also true.

or $5 \times 2^{3(m+1)-2} + 3^{3(m+1)-1}$ is divisible by 19

Now from L.H.S of (3)

$$= 5 \times 2^{3(m+1)-2} + 3^{3(m+1)-1}$$
$$= 5 \times 2^{(3m-2)+3} + 3^{(3m-1)+3}$$
$$= 5 \times 2^{3m-2}.\ 2^3 + 3^{m-1}.3^3$$
$$= 8 \times 5 \times 2^{3m-2} + 27 \times 3^{m-1}$$
$$= 8 \times 5 \times 2^{3m-2} + 8 \times 3^{3m-1} + 19 \times 3^{3m-1}$$
$$= 8\ [5 \times 2^{3m-2} + 3^{3m-1}] + 19 \times 3^{3m-1}$$
$$= 8..[A(m) + 19.3^{m-1}]$$

Now from equation (2)

$$= 8\ [19\ k] + 19.3^{m-1}$$
$$= 19\ [8\ k] + 19.3^{m-1}$$
$$= 19[8\ k + 3^{m-1}]$$

Which is divisible by 19 *i.e.*, $A(m)$ is true

$\Rightarrow A(m + 1)$ is true also $A(1)$ is true. Hence, by the principal of mathemaial induction $A(n)$ = true $\forall\ n \ \varepsilon\ N$.

Q.3. (*b*) Given two sentences *A*, *B* as :

$$A \equiv \{5 = 8\}$$
$$B \equiv \{5 < 8\}$$

What is the meaning of the sentences:

(*i*) $A \cup B$, (*ii*) $A \cap B$ (*iii*) $A \Rightarrow B$ (*iv*) $\sim A$, (*v*) $\sim B$

Which of these sentences are True and which are False?

Ans. (*i*) $A \cup B = \{5 \leq 8\}$; False

(*ii*) $A \cup B = \{\phi\}$; True

(*iii*) $A \Rightarrow B = \{5 < 8\}$; True

(*iv*) $\sim A = \{5 \neq 8\}$; True

(*v*) $\sim B = \{5 \geq 8\}$; False

UNIT -2

Q.4. (*a*) Seventy-five tourists out of a group of 120 speak German, 50 speak French, 40 speak Spanish, 30 speak both German and French, 15 speak both German and Spanish and 12 speak both French and Spanish. How many tourists in the group speak all the three languages?

Ans. $|\cup|$ = 120 total tourists

$|G|$ = 75, speak German

$|F|$ = 50, speak french

$|S|$ = 40, speak spanish

$|G \cup F|$ = 30, speak both German and French

$|G \cup S| = 15$, speak both German and Spanish

$|F \cup S| = 12$, speak both french and spanish

Find $|G \cap F \cap S| = ?$ As $|G \cup F \cup S| = 120$ because 120 of the tourists speak at last one of the language, using Indusion-Exclusion rule.

$|G \cup F \cup S| = |9| + |F| + |S| - |G \cup F| - |G \cup S| - |F \cup S| + |G \cap F \cap S|$

$120 = 75 + 50 + 40 - 30 - 15 - 12 + |G \cap F \cap S|$

$|G \cap F \cap S| = 120 - 108 = 12$

So total 12 tourists can speak all the three languages.

Q.4. (*b*) Let *S* = {1, 2, 3, 4, 5, 6, 7, 8, 9, 10, 11, 12}. Define a relation *R* on *S* by xRy if and ony if $x^2 - y^2$ is a multiple of 3. Show that *R* is an equivalence relation on *S*. Determine the equivalence classes in *S* under the relation *R*.

Ans. $x = 1, y = 1 \Rightarrow x^2 - y^2 = 1^2 - 1^2 = 0 = 3.0 \Rightarrow 1\,R\,1$

$x = 1, y = 2 \Rightarrow 1^2 - 2^2 = 1 - 4 = -3 \Rightarrow 1\,R\,2 \quad 1\,\not R\,3$

$x = 1, y = 4 \Rightarrow 1^2 - 4^2 = 1 - 16 = -15 \Rightarrow 1\,R\,3$

$x = 1, y = 5 \Rightarrow 1^2 - 5^2 = 1 - 25 = -24 \Rightarrow 1\,R\,5$

$x = 1, y = 7 \Rightarrow 1^2 - 7^2 = 1 - 49 = -48 \Rightarrow 1\,R\,7 \quad 1\,\not R\,6$

$x = 1, y = 8 \Rightarrow 1^2 - 8^2 = 1 - 64 = -63 \Rightarrow 1\,R\,8$

$x = 1, y = 10 \Rightarrow 1^2 - 10^2 = 1 - 100 = -99 \Rightarrow 1\,R\,10 \quad 1\,\not R\,9$

$x = 1, y = 11 \Rightarrow 1^2 - 11^2 = 1 - 121 = -120 \Rightarrow 1\,R\,11$

2 *R* 2

$x = 2, y = 4 \Rightarrow 2^2 - 4^2 = 4 - 16 = 12 \Rightarrow 2\,R\,4 \quad 2\,\not R\,3$

$x = 2, y = 5 \Rightarrow 2^2 - 5^2 = 4 - 25 = 21 \Rightarrow 2\,R\,5$

$x = 2, y = 7 \Rightarrow 2^2 - 7^2 = 4 - 49 = 45 \Rightarrow 2\,R\,7 \quad 2\,\not R\,6$

$x = 2, y = 18 \Rightarrow 2^2 - 10^2 = 4 - 100 = 96 \Rightarrow 2\,R\,10 \quad 2\,\not R\,8$

$x = 2, y = 11 \Rightarrow 2^2 - 11^2 = 4 - 121 = 117 \Rightarrow 2\,R\,11 \quad 2\,\not R\,12$

3 *R* 3

$x = 3, y = 6 \Rightarrow 3^2 - 6^2 = 9 - 36 = 27 \Rightarrow 3\,R\,6 \quad 3\,\not R\,4$

$x = 3, y = 9 \Rightarrow 3^2 - 9^2 = 9 - 81 = 72 \Rightarrow 3\,R\,9 \quad 3\,\not R\,5$

$x = 3, y = 12 \Rightarrow 3^2 - 12^2 = 9 - 144 = 135 \Rightarrow 3\,R\,12 \quad 3\,\not R\,12$

4 *R* 4

$x = 4, y = 5 \Rightarrow 4^2 - 5^2 = 16 - 25 = -9 \Rightarrow 4\,R\,5$

$x = 4, y = 7 \Rightarrow 4^2 - 7^2 = 16 - 49 = -33 \Rightarrow 4\,R\,7 \quad 4\,\not R\,6$

$x = 4, y = 8 \Rightarrow 4^2 - 8^2 = 16 - 64 = -48 \Rightarrow 4\,R\,8 \quad 4 \not R\, 9$

$x = 4, y = 10 \Rightarrow 4^2 - 10^2 = 16 - 100 = -84 \Rightarrow 4\,R\,10$

$x = 4, y = 11 \Rightarrow 4^2 - 11^2 = 16 - 121 = -105 \Rightarrow 4\,R\,11 \quad 4 \not R\, 12$

5 *R* 5

$x = 5, y = 7 \Rightarrow 5^2 - 7^2 = 25 - 49 = -24 \Rightarrow 5\,R\,7 \quad 5 \not R\, 6$

$x = 5, y = 8 \Rightarrow 5^2 - 8^2 = 25 - 64 = -39 \Rightarrow 5\,R\,8$

$x = 5, y = 10 \Rightarrow 5^2 - 10^2 = 25 - 100 = -75 \Rightarrow 5\,R\,10 \quad 5 \not R\, 9$

$x = 5, y = 11 \Rightarrow 5^2 - 11^2 = 25 - 121 = -96 \Rightarrow 5\,R\,11 \quad 5 \not R\, 12$

6 *R* 6

$x = 6, y = 7 \Rightarrow 6^2 - 9^2 = 36 - 81 = -45 \Rightarrow 6\,R\,9 \quad 6 \not R\, 7$

$x = 6, y = 12 \Rightarrow 6^2 - 12^2 = 36 - 144 = -108 \Rightarrow 6\,R\,12 \quad 6 \not R\, 8$

7 *R* 7

$x = 7, y = 8 \Rightarrow 7^2 - 8^2 = 49 - 64 = -15 \Rightarrow 7\,R\,8$

$x = 7, y = 10 \Rightarrow 7^2 - 10^2 = 49 - 100 = -51 \Rightarrow 7\,R\,10 \quad 7 \not R\, 9$

$x = 7, y = 11 \Rightarrow 7^2 - 11^2 = 49 - 121 = -72 \Rightarrow 7\,R\,11 \quad 7 \not R\, 12$

8 *R* 8

$x = 8, y = 10 \Rightarrow 8^2 - 10^2 = 64 - 100 = -36 \Rightarrow 8\,R\,10 \quad 8 \not R\, 9$

$x = 8, y = 11 \Rightarrow 8^2 - 11^2 = 64 - 121 = -57 \Rightarrow 8\,R\,12 \quad 8 \not R\, 12$

9 *R* 9

$x = 9, y = 12 \Rightarrow 9^2 - 12^2 = 81 - 144 = 63 \Rightarrow 9\,R\,12 \quad 9 \not R\, 10$

$9 \not R\, 11$

10 *R* 10

$x = 10, y = 11 \Rightarrow 10^2 - 11^2 = 100 - 121 = -21 \Rightarrow 10\,R\,11 \quad 9 \not R\, 11$

$9 \not R\, 12$

11 *R* 11

$x = 11, y = 11 \not R\, 12$

12 *R* 12

So relation R can be defined as:

R = { (1, 1), (1, 2), (1, 3), (1, 5), (1, 7), (1, 8), (1, 10), (1, 11),(2, 2), (2, 4), (2, 5), (2, 7), (2, 10), (2, 11), (3, 3), (3, 6), (3, 9), (3, 12), (4, 4), (4, 5), (4, 7), (4, 8), (4, 10), (4, 11), (5, 7), (5, 8), (5, 10), (5, 11), (5, 5), (6, 6), (6, 9), (6, 12), (7, 7), (7, 8), (7, 10), (7, 11), (8, 8), (8, 10), (8, 11), (9, 9), (9, 12), (10, 10), (10, 11), (11, 11), (12, 12) }

Reflexive: $\forall\ x, y, \varepsilon\ R$, $x^2 - x^2$ is multiple of 3

also $(x, x)\ \varepsilon\ R\ \forall\ x, y, \varepsilon\ R$, R is reflexive.

Symmetric: Let $x, y \varepsilon R$ and $(x, y) \varepsilon R$

$\Rightarrow x^2 - y^2$ is multiple of 3

$\Rightarrow x^2 - y^2 = 3\,m_1 = (m_1$ is being a Natural number)

$\Rightarrow y^2 - x^2 = -3\,m_1 = 3\,m_2$ (m_2 is a Natural number)

So $y^2 - x^2$ multiple of 3 or R is symmetric

Transitive: Let $x, y, z \varepsilon R$

$$\left.\begin{array}{l} x^2 - y^2 = 3\,m_1 \\ y^2 - z^2 = 3\,m_2 \end{array}\right\} m_1 \text{ and } m_2 \text{ are natural numbers}$$

adding (1) and (2) we have $x^2 - z^2 = 3\;3(\boldsymbol{m_1} + m_2)$

so $x^2 = z^2$ is also multiple of 3 or $(x, z) \varepsilon R$

so R is transitive

Hence, R is an equivalence relation

we have equivalence classes

$[0] = [0, 3, 6, 9, 12]$ and $[3] = [3, 6, 9, 12]$ **Ans.**

$[1] = [1, 4, 7, 10]$ as $[0] = [3]$

$[2] = [2, 5, 8, 11]$ **Ans.**

Q.5. (*a*) Let f: $R \rightarrow R$ be given by $f(x) = (x + 1)^2 - 1$. Find f^{-1} and its domain. Also find the set $S = \{x : f(x) = f^{-1}(x)\}$.

Ans. $x = 0\; f(0) = (0 + 1)^2 - 1 = 1 - 1 = 0$

$x = 1\; f(1) = (1 + 1)^2 - 1 = 2^2 - 1 = 3$

$x = 2\; f(2) = (2 + 1)^2 - 1 = 3^2 - 1 = 8$

$x = 3\; f(3) = (3 + 1)^2 - 1 = 4^2 - 1 = 15$

$x = 4\; f(4) = (4 + 1)^2 - 1 = 5^2 - 1 = 24$

and so on so function of can be written as

$f = \{(0, 0), (1, 3), (2, 8), (3, 15), (4, 24)\;,,,\}$

and $f^{-1} = \{(0, 0), (3, 1), (8, 2), (15, 3), (24, 4)...\}$

domain of f is $\{0, 3, 8, 15, 24 ...\}$

now the set $S = \{x : f(x) = f^{-1}(x)\}$ can

$S = \{0\}$ and for this element the condition $f(x) = f^{-1}(x)$ true.

Q.5. (*b*) Let N be the set of natural numbers and $X = N \times N$. Define a binary operation x on X by $(a, b)\; x\; (c, d) = (ac, bd)$; $a, b, c, d\; \varepsilon\; N$. Show that x B. O. on X is both commutative and associative.

Ans. Suppose $x = \{a, b\}\; y = (c, d)\; z = (e, f)$

we have

$(xy)\, z = (ac, bd) \otimes (e, f)$

$(xy)z = [(ac)\, e, (bd)\, f]$ ...(*i*)

and

$x(yz) = (a, b) \otimes (ce, df)$

$x(yz) = [a(c, e), b(df)]$ ...(*ii*)

Since, a, b, c, d, e, f all Natural numbers,

$(ac)e = a(ce)$ and $(bd)f = b(df)$

Thus, $(xy)\,z = x(yz)$ and hence $\otimes$ is associative

(*ii*) For commutative property we have

$$(a, b) \otimes (c, d) = (ac, bd) \quad \text{...}(i)$$

$$(c, d) \otimes (a, b) = (ca, db) \quad \text{...}(ii)$$

Since, a, b, c, $d \in N$ so

$$ac = ca \text{ and } bd = db$$

Thus, from (1) and (2) it is proved that $\otimes$ B.O (Binary operation) is commutative.

Q.5. (*c*) Three films *A*, *B*, *C* were shown in a cinema hall for a week to 40 students (each of whom saw either al the three films or only one film). 13 students saw film *A*, 16 students saw film *B* and 19 students saw film *C*. How many students saw all the three films?

UNIT -3

Q.6. (*a*) For the ordered set A given in the figure

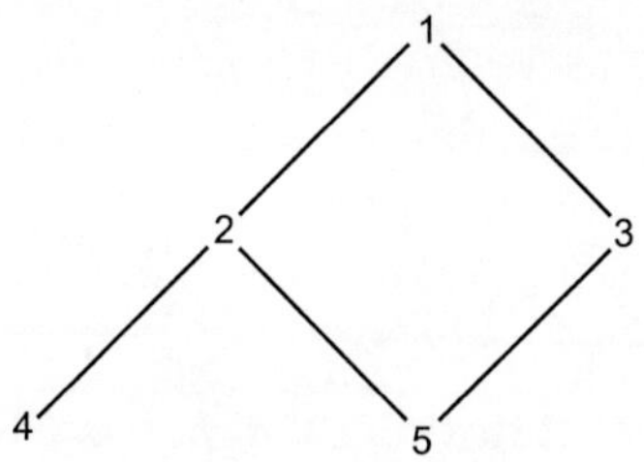

Let *L(A)* be the set collection of all linearly ordered subsets of A wiht 2 or more elements. Let *L(A)* be ordered by inclusion. Draw the Hasse diagram of *L(A)*.

Ans. There the elements of L(A) are as follows:

{1, 2, 4}, {1, 2, 5}, {1, 3, 5}, {1, 2,}, {1, 4}, {1, 3}, {1, 5}, {2, 4}, {2, 5}, {3, 5}.

The Hasse diagram of L(A) is shown below:

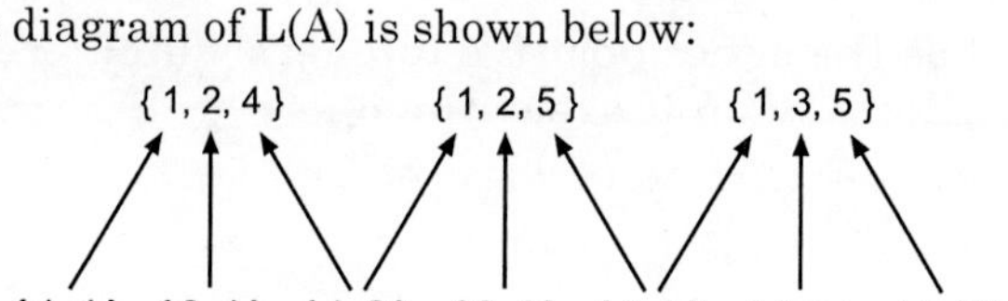

Q.6. (*b*) Convert the function.

$F(x, y, z) = (x + y + z)\,(xy + x'\ z)$ into conjunctive and disjunctive normal forms.

Ans. Given function on x, y, z is

$$F(x, y, z) = (x + y + z)\,(xy + x'z)$$

DNF form:

$$
\begin{aligned}
F(x, y, z) &= (x + y + z)\,(xy + x'z) \\
&= (x + y + z)\,xy + (x + y + z)\,x'z \text{ By distribution lane} \\
&= (xxy + yxy + zxy) + (xx'z + yx'z + zx'z) \\
&= (xy + xy + xyz) + (o.z + x'yz + x'z) \\
&= (xy + xyz) + (x'yz + x'z) \\
&\qquad [\text{as } xx = x,\ yy = y,\ xx' = 0,\ xy + xy = xy] \\
&= xy + xyz + x'yz + x'z \\
&= xy.1 + xyz + x'yz + x'z.1 \\
&= xy(z + z') + xyz + x'yz + x'z(y + y') \\
&= xyz + xyz' + xyz + x'yz + xyz + x'y'z \\
&= xyz + xyz' + x'yz + z'y'\text{z which is required DNF}
\end{aligned}
$$

CNF form:

$$
\begin{aligned}
f(x, y, z) &= (x + y + z)\,(xy + xz') \\
&= (x + y + z)\,(x.(y + z)) \\
&= (x + y + z)\,.\,(x + 0)\,.\,(y + z' + 0) \\
&= (x + y + z)\,.\,(x = yy')\,.\,(y + z' + xx') \\
&= (x + y + z)\,.\,(x + y)\,(x + y')\,(y + z' + x)\,(y + z + x') \\
&= (x + y + z)\,(x + y + 0)\,(x + y' + 0)\,(y + z' + x)\,(y + z' + x') \\
&= (x + y + z)\,(x + y + zz')\,(x + y + zz')\,(y + z' + x)\,(x + y' + z') \\
&= (x + y + z)\,(x + y + z)\,(x + y + z')\,(x + y' + z)\,(x + y' + z') \\
&\qquad (y + z' + x)\,(y + z + x') \\
&= (x + y + z)\,(x + y + z')\,(x + y' + z)\,(x + y + z')\,(x' + y + z')
\end{aligned}
$$

which is required CNF

Q.7. (*a*) In a lattice (*L*, Relation ≤), *a*, *b*, ε *L*. Show tht $a \cap b = b$ if and only if $a \cup b = a$.

Ans. If $a \wedge b$ = a, then by considering $b \leq (a \wedge b)$

we find $b \vee (a \wedge b) = b \vee a = a \vee b$

since $a \wedge b = a$ and $a \vee b = b$ if $a \leq b = b$ we also have

$b \vee (a \wedge b) = a$ then $a \vee b = b$. The converse may be proved similarly.

Q.7. (*b*) Let *B* be the set of positive integers which are divisors of 30 and the operation ∧, ∨ on it are defined as :

$$
\begin{aligned}
a \vee b &= \text{L. C. M. } (a, b) \\
a \wedge b &= \text{H. C. F. } (a, b)
\end{aligned}
$$

$$a' = \frac{30}{a} \text{ for all } a, b, \varepsilon B$$

Prove that (B ∧, ∨ ,′ ,1, 30) is a Boolean Algebra.

Ans. In D_{30} it is easily verified that for 10, 15, 2, ε D_{30}

(*a*) Commutative law:

$$
\begin{aligned}
10 \vee 15 &= \text{LCM } (10, 15) = 30 = 15 \vee 10 \\
10 \wedge 15 &= \text{HCF } (10, 15) = 5 = 5 \wedge 10
\end{aligned}
$$

(b) Complementation operation:

$$10' = \frac{30}{10} = 3; \qquad 15' = \frac{30}{15} = 2;$$

(c) Distributive law:

$$10 \vee (15 \wedge 2) = 10 \vee 1 = 10 \text{ and } (10 \vee 15) \wedge (10 \vee 2)$$
$$= 30 \wedge 10 = 10 \quad \text{and}$$
$$10 \wedge (15 \vee 2) = 10 \wedge 30 = 10 \text{ and } (10 \wedge 15) \vee (10 \wedge 2) = 5 \vee 2 = 10$$

(c) Associative law:

$$(10 \vee 15) \vee 2 = 30 \vee 2 = 30 \text{ and } 10 \vee (15 \wedge 2) = 10 \vee 30 = 30 \quad \text{and}$$
$$(10 \wedge 15) \vee 2 = 5 \wedge 2 = 1 \text{ and } 10 \wedge (15 \wedge 2) = 10 \wedge 1 = 1$$

Similarly it is possible to show that the rules of the boolean algebra are satisfied by any choice of the element a, b, c belonging to D_{30}. Hence, is a Boolean Algebra.

UNIT -4

Q.8. (a) Examine the isomerphism of graphs G and H given below:

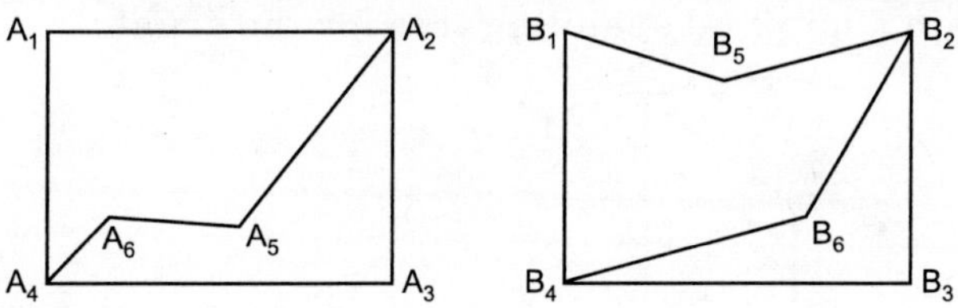

Ans. Two graphs G and H are said to be isomorphic to each other if there is a one to one correspondence between their vertices and between their edges so that the incidence relationshihp is maintained

Some properties of isomorphic graphs:

1. Number of vertices in isomorphic graphs is the same.
2. Number of edges in isomorphic groups is also the same.
3. Each one of the isomorphic graphs has an equal number of vertices with a given degree.

Now take graph G and HG as in graph G. We have six vertices

$V(G) = \{A_1, A_2, A_3, A_4, A_5, A_6\}$ and in graph H we also have six vertices.

$V(H) = \{B_1, B_2, B_3, B_4, B_5, B_6\}$. Let we can give name to the edges in gaph G and H as

$$A_1 - A_2 = e_1 A_4 - A_6 = e_5$$
$$A_2 - A_3 = e_2 A_5 - A_5 = e_6$$
$$A_3 - A_4 = e_3 A_5 - A_2 = e_7$$
$$A_4 - A_1 = e_4$$

So in graph G we have seven edges

$$E(G) = \{e_1, e_2, e_3, e_4, e_5, e_6, e_7,\}$$

Now in graph H we can also give name to all edges as:

$B_1 - B_2 = e'_1 \qquad B_4 - B_1 = e'_4$

$B_5 - B_2 = e'_7 \qquad B_4 - B_6 = e'_5$

$B_2 - B_3 = e'_2 \qquad B_6 - B_2 = e'_6$

$B_3 - B_4 = e'_3$

so in graph H we also have seven edges

$$E(H) = \{e'_1, e'_2, e'_3, e'_4, e'_5, e'_6, e'_7\}$$

So two properties are satisfied. Now for third property for isomorphism each one of the graphs has an equal number of vertices with a given degree.

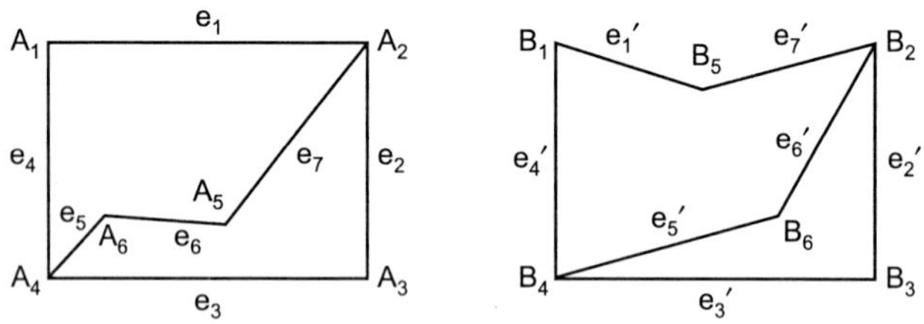

We can see that vertixes A_1, A_2, A_3 and A_4 and in graph H vertices B_1, B_2, B_3 and B_4 have the incidence relationship. But vertices A_5 and A_6 in G and B5 and B_6 in graph H don't preserve the incidence relationship and adjacency so they are not isomorphic.

Q. 8. (*b*) Obtain a minimal spanning tree for the grpah:

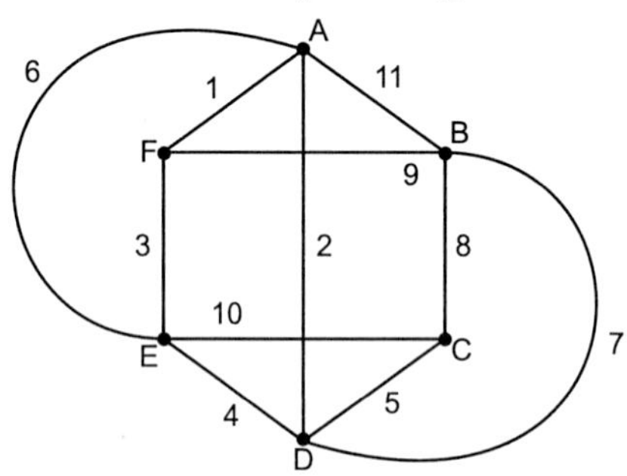

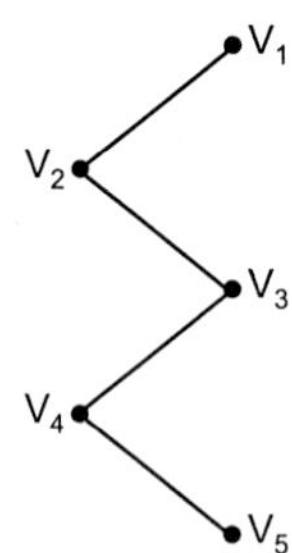

Ans. Given that

$AB = 11$	$BC = 8$	$CD = 5$
$DE = 4$	$EF = 3$	$FA = 1$
$AD = 2$	$BD = 7$	$AE = 6$

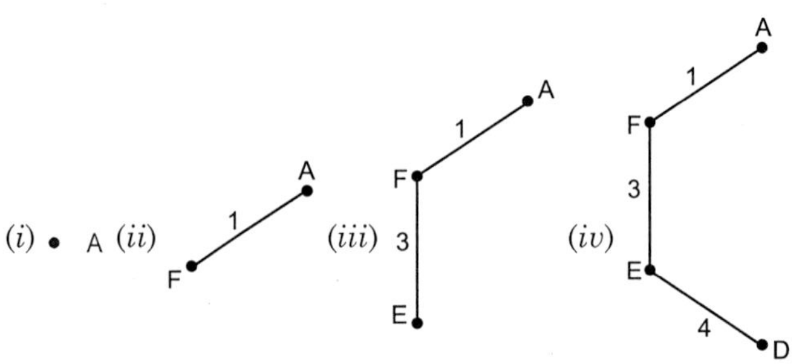

(v)

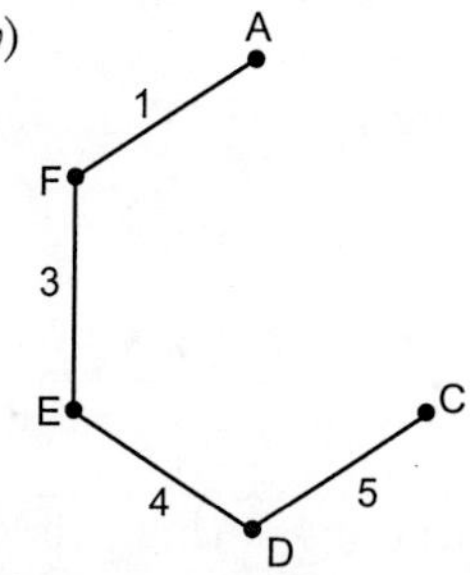

(vi)

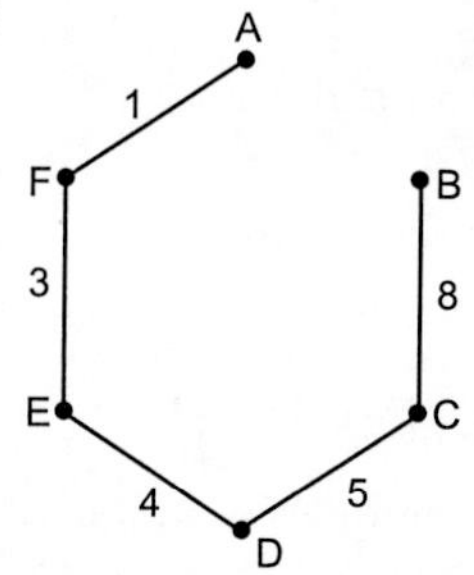

This is one opanning tree. **Ans.**

Q.9. (*a*) Define the following types of graphs with examples:

(*i*) Euler Path

(*ii*) Hamiltonian circuits

(*iii*) Directed graph

Ans. (*i*) Euler path: Euler in a graph *G* is a path that traverse every edge exactly once and only once.

Conversely for a graph *G* having more than two vertices of odd degree there can not be an euler path for example consider the graph given,

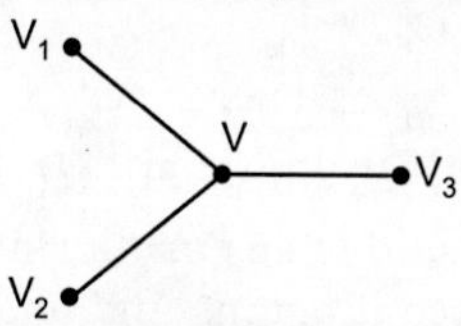

In the graph above, we have three vertices of odd degree say v_1, v_2 and v_3. It is clear from the figure tht we cannot find a path in which each edge should traverse exactly once.

(*ii*) Hamiltonian circuits: is a circuit that contains each vertex exactly once except for the first vertex exactly once except for the first vertex which is also the last vertex from the figure given here, represents the Hamiltonian path $\{v_1, v_2, v_3, v_4, v_5\}$

while the figure shown here,

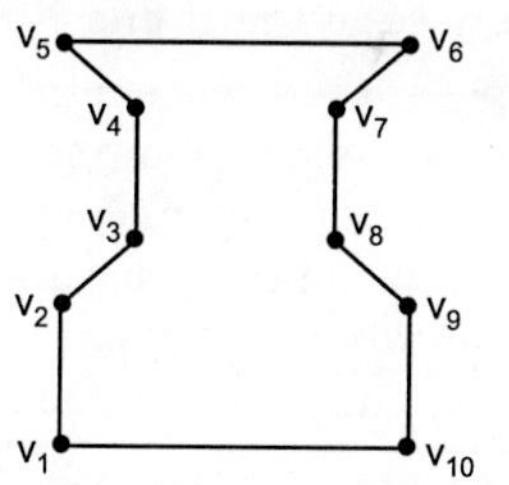

This figure represents the Hamiltonian circuits $\{v_1, v_2, v_3, v_4, v_5, v_6, v_7, v_8, v_9, v_{10}, v_{11},\}$

(*iii*) Directed gaph: A directed gaph *G* or a diagraph (or simply graphy) consists of two things:

(*i*) A set V whose elements are called vertices, nodes, or points.

(*ii*) A set E of ordered pairs (x, v) of vertices called arcs or directed edges or simply edges.

Suppose $e = (x, v)$ is a directed edge in a diagraph G. Then following terminology is used

(*a*) e begins at u and ends at v.

(*b*) u is the origin or initial point of e and v is the distination or terminal point of e.

(*c*) v is the success of u.

(*d*) u is adjacent to v and v is adjacent from u, if $u = v$ then e is called a loop.

Consider the directed graph G given below. It consists of four vertices and seven edges as follow.

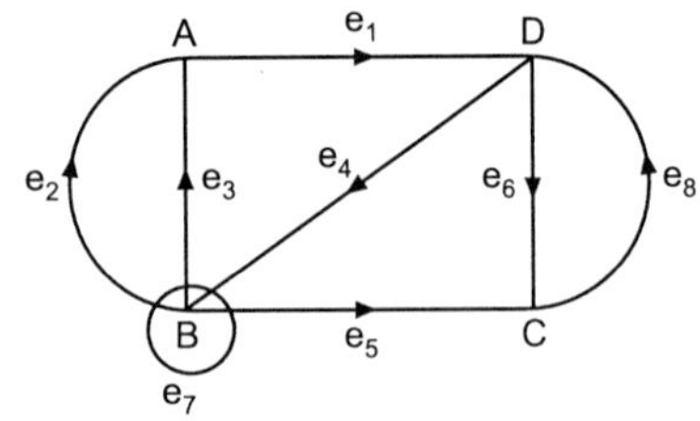

$$V(G) = \{A, B, C, D\}$$

and $$E(G) = \{e_1, e_2 \ldots e_7\} \text{ Or}$$

$$E(G) = \{(A, D), (B, B), (B, A), (D, B), (B, C), (D, C), (B, A)\}$$

The edges e_2 and e_3 are said to be parallel since both begin at B and end at A. The edge e_7 is a loop since it begins and ends at B.

Q.9. (*b*) Define a tree and show that a tree with n vertices has exactly (n – 1) edges.

Ans. Tree: A tree is a connected acyclid graph *i.e.*, a connected graph having no cycle. Its edges is called a trivial tree other wise T is a non trivial tree.

To prove that a tree T with n vertices has edge, we use the mathematical induction on n.

Basis of Induction: When $n = 1$ then T has only one vertex. Since, it has no cycle, T can not have any edge *i.e.*, it has $e = 0 = n - 1$.

Inductive step: Suppose the theorem is true for $n = k \geq 2$ where k is some positive integers. We use this to show that the result is true for $n = k + 1$. Let T be a tree with $k + 1$ vertices and let uv be edge of T. Then if we remove the edge uv from T we obtain the graph T-uv. Then the graph is disconnected since $T - uv$ contain no (u, v) path. If there were a path, say $u, v, u, v_3 \ldots u$, from u to v then when we added back the edge uv there would be a cycle $u_1\ v_1\ v_2 \ldots v, u$ in T.

Thus $T - uv$ is disconnected. The removal of an edge from a graph can disconnect the graph into at most two components. So, T-uv has two components, say T_1 and T_2. Since there were no cycles in T to begin with, both components are connected and are without cycles. Thus, T_1 and T_2 are trees and each has fewer than n vertices. This means that we can apply the induction hypothesis to T_1 and T_2 to give

$$e(T_1) = v(T_1) - 1$$
$$e(T_2) = u(T_2) - 1$$

But the construction of T_1 and T_2 by removal of a single edge from T gives that

$$e(T) = e(T_1) + e(T_2) + 1 \text{ and that } u(T) = v(T_1) +$$

it follows that

$$e(T) = u(T_1) - 1 + v(T_2) - 1 + 1$$
$$= v(T) - 1$$
$$= k + 1 - 1 = k$$

Thus T has k edges, as required. Hence, by the principle of mathematical induction the theorem is proved.

Q. 9. (c) Show that a non-empty connected graph G is Eulerian if an only if its vertices are all of even degree.

Ans. Let G be Eulerian. Then G has an eulerian trailwhich begins and ends at u, say. If we travel along the trail then each time we visit a vertex we use two edges, one in and one out. This is also true for the start vertex because we also ends then. Since an Eulerian trail uses every edge, once, each occurance of u represents a contribution of 2 to its degree. Thus, deg (v) is even.

Conversely, suppose that G is connected and every vertex is even. We construct an Euleria trail. We begin a trail T_1 any edge e,we extend T_1 by adding a edge after the other. If T_1 is not closed at any step, say T_1 begins at u but ends at $v \neq u$, then only an odd number of the edges incident on v, appear in T_1; hence, we can extend T_1 by another edge incident on v. Thus, we can continue to extend T_1 untill T_1 returns to its initial vertex u. *i.e.*, until T_1 is returns to its initial vertex u. *i.e.*, untill T_1 is closed. If T_1 includes all the edges of G, then T_1 is an Eulerian Trail. Suppose T_1 does not include all edges of G. Consider the graph H obtained by deleting all edges of T_1 from G. H may not be connected, but each vertex of H has even degree since T_1 contains an even numbers of th e edges incident on any vertex. Since, G is connected there is an edge e' of H which has an end point u' in T_1. We construct a trail T_2 in H beginning at u' and using e'. Since, all vertices in H have even degree. We can continue to extend T_2 untill T_2 returns to u' as shown in figure.

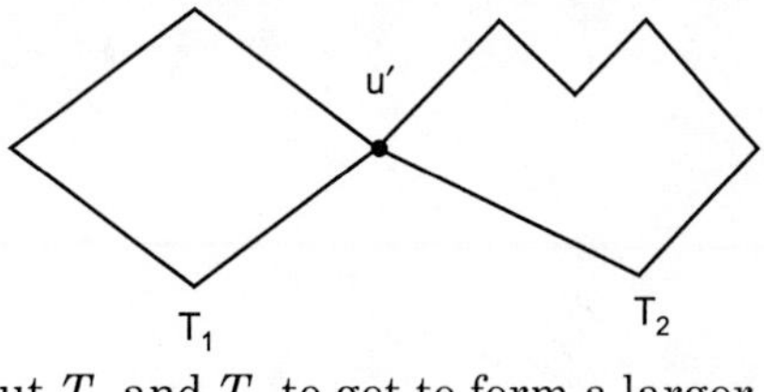

We can clearly put T_1 and T_2 to get to form a larger closed trail in G. We continue this process untill all the edges of G are used. We finally obtain an Eulerian trial, and so G is Eulerian.

In the konigsberg bridge problem, the degree of every vertex (in figure (*b*) below) is odd

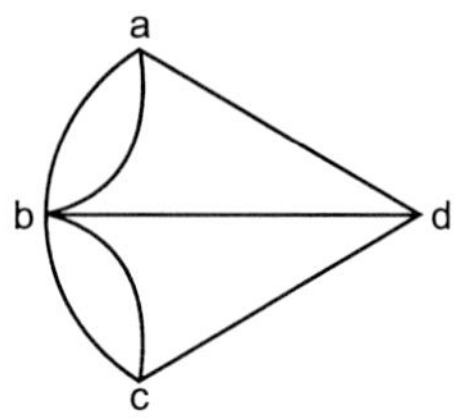

So the graph is not Eulerian, Consequently, it is possible to walk through the city crossing each bridge exactly once and return home.

❐❐❐

INDEX

INDEX

❐❐❐

Win Prizes

ATTENTION READERS

Win Prizes

We request for your frank assessment regarding various aspects of the book as given below :

DISCRETE MATHEMATICS

By: Anjana Gupta and Bhaskar Kapoor

Please Fill up the Sheet in Neat and Clean Writing

(i) What topics of your syllabus, important for your examination point of view, are not covered in this book ?

....

....

....

....

(ii) What are the chapters wherein the treatment of the subject matter is not systematic or organised or updated ?

....

....

....

....

(iii) Have you come across misprints/mistakes/factual inaccuracies in the book? Please specify the chapters and the page numbers.

....

....

....

....

(iv) What other books on the same subject have you found/heard better than the present book? Please specify in terms of price and quality.

....

....

....

....

(v) Further suggestions and comments for the improvement of this book.

....

....

....

....

(PS: If need be please attach a separate sheet to write your views)

OTHER DETAILS

(i) Who recommended you this book ? (Please tick (✓) in the box near the option relevant to you.

☐ Teacher ☐ Friends ☐ Bookseller

....

....

....

....

(ii) Name of the recommending teacher, his designation and address :

....

....

....

....

(iii) Name and address of the bookseller you purchased the book from :

....

....

....

....

(iv) Name and address of your institution (Please mention the University or Board as the case may be)

....

....

....

....

(v) Your name and complete postal address :

....

....

....

....

(vi) Course you are enrolled in :

....

....

....

....

The best assessment will be awarded each month: The award will be in the form of our publications as decided by the Editorial Board.

Please mail the filled up coupon at your earliest to :

S.K. KATARIA & SONS®

4885/109, Prakash Mahal, Dr. Subhash Bhargav Lane,
Opposite Delhi Medical Association, Daryaganj, New Delhi-110002 (India)
Phone: +91-11-43551243 Mobile: +91-9871775858, 7678425209
e-mail: katariabook@gmail.com, katariabooks@yahoo.com
Website: www.skkatariaandsons.com